Seymour/Carraher's

Polymer Chemistry

UNDERGRADUATE CHEMISTRY

A Series of Textbooks

Edited by

J. J. LAGOWSKI
Department of Chemistry
The University of Texas at Austin

Additional Volumes in Preparation

Seymour/Carraher's

Polymer Chemistry

AN INTRODUCTION

Fourth Edition

Revised and Expanded

Charles E. Carraher, Jr.

College of Science
Florida Atlantic University
Boca Raton, Florida
and
Florida Center for Environmental Studies
Palm Beach Gardens, Florida

Marcel Dekker, Inc.   New York • Basel • Hong Kong

Library of Congress Cataloging-in-Publication Data

Seymour / Carraher's polymer chemistry : an introduction. — 4th ed., rev. and
  expanded / Charles E. Carraher, Jr.
    p. cm. — (Undergraduate chemistry ; 13)
  Rev. ed. of : Polymer chemistry / Raymond B. Seymour, Charles E. Carraher, Jr.
3rd ed., rev. and expanded. c1992.
  Includes bibliographical references and index.
  ISBN 0-8247-9752-3 (alk. paper)
  1. Polymers. 2. Polymerization. I. Carraher, Charles E. II. Polymer Chemistry.
III. Title. IV. Series: Undergraduate chemistry; v.13.
  QD381.S483 1996
  547.7—dc20   96-22846
                                                              CIP

The publisher offers discounts on this book when ordered in bulk quantities. For
more information, write to Special Sales/Professional Marketing at the address
below.

This book is printed on acid-free paper.

Marcel Dekker, Inc.
270 Madison Avenue, New York, New York 10016

Current printing (last digit):
10 9 8 7 6 5 4 3 2 1

PRINTED IN THE UNITED STATES OF AMERICA

*To Raymond Seymour—educator, scientist,
pioneer, prophet, historian, family man,
and friend—we will miss you*

Foreword

Polymer science and technology has developed tremendously over the last few decades and the production of polymers and plastics products has increased at a remarkable pace. Nearly 150 million tons per year of plastic materials are produced worldwide (about 2% of the wood used) to fulfill the ever-growing needs of the *plastic age*; in the industrialized world plastic materials are used at a rate of nearly 100 kg per person per year. The total use of plastic materials contributes about 4% to the gross domestic product in the United States. Plastics have no counterpart in other materials in terms of weight, ease of fabrication, high utilization, and economics.

It is no wonder that the demand and the need for teaching in polymer science and technology have increased rapidly. To teach polymer science, a readable and up-to-date introductory textbook is required that covers the entire field of polymer science, engineering, and technology. This goal has been achieved in Carraher's textbook. It is eminently useful for teaching polymer science in departments of chemistry, chemical engineering, and material science, and also for teaching polymer science and technology in polymer science institutes, which concentrate entirely on the science and technologies of polymers.

This fourth edition addresses the important subject of polymer science and technology, with emphasis on making it understandable to students. The book is ideally suited not only for graduate courses but also for an undergraduate curriculum. It has not become more voluminous, simply by adding information—less important subjects have been removed and more important issues have been introduced.

Some applications of polymer science and technology have become of high value. The authors have interwoven discussion of these subjects with the basics in polymer science and technology. Testimony to the high acceptance of this book is that early demand required reprinting and updating of the third edition. We see the result in this new significantly changed and improved edition.

Otto Vogl
Herman F. Mark Professor
of Polymer Science
Polytechnic University
Brooklyn, New York

Preface to the Fourth Edition

The fourth edition continues the emphasis on being "user friendly" and providing a balance between theory and the "real world." It contains the construction of "bridges" and extensions of concepts and principles covered in other introductory chemistry courses. It continues to be in agreement with the material and level recommended by the American Chemical Society Committee on Professional Training.

The following features are continued, enhanced, and updated.

Chapter summaries
Extensive glossary at the end of each chapter
Exercises and questions with answers included in a special appendix
Extensive bibliography
Special section on nomenclature
Extensive listing of trade names
Appendix with structures of common polymers

The book is written to show the intercorrelation between synthetic and natural polymers and organic, organometallic, and inorganic polymers.

The initial edition was written utilizing an extensive survey of about 100 schools; the results are given in the following table. The full results of this survey appear in the Journal of Chemical Education, 57(6), 436 (1980). As shown in this table, *Polymer Chemistry: An Introduction* includes all the major and optional topics recommended in the syllabus adopted by the joint polymer education committee. A fuller discussion of the suggested syllabus is given in Appendix D. We have emphasized the topics given high points in the national survey.

In addition to the general topic areas listed in the survey, this text includes relevant chapters on inorganic polymers (Chapters 11 and 12) and synthesis of

Suggested Syllabus

	Suggested percentage of course time	Corresponding chapter in *Polymer Chemistry*	Survey[a] points
Major topics			
1. Introduction	5	1	—
2. Polymer Structure	10	2	104
(Morphology)			86
3. Molecular Weights	10	3, 4	104
4. Step-Reaction Polymers	10	7	100
5. Chain-Reaction Polymers:	10		
Ionic and Complex Coordinative		8	83
Free Radical		9	100
6. Copolymerization	10	10	86
7. Testing and Characterization	10	5	51
Optional topics			
8. Rheology	5	3	85
9. Solubility	5	3	85
10. Natural and Biomedical Polymers	5	6	11
11. Additives	5	13, 14	4
12. Reactions of Polymers	5	15	61
13. Synthesis of Reactants	5	16	—
14. Polymer Technology	5	17	28

[a]From 127 replies from 97 schools, based on +1, topic to be included in introductory courses; 0, no strong feeling; −1, topic should be omitted from introductory courses.

reactants (Chapter 16). While no specific chapter is devoted to commercial polymer science, these important aspects, as well as emerging and projected future areas of activity, are presented throughout the text.

Information already presented in traditional undergraduate courses of organic chemistry, physical chemistry, and so on, is interrelated with information that focuses on polymer topics. This assists students in integrating their chemical knowledge and illustrates the connection between theoretical and applied chemical knowledge. Further, industrial practices and testing procedures are integrated with the theoretical treatment of the various topics, allowing the reader to bridge the gap between industrial practice and the classroom. Also incorporated into the text are uses and amounts of particular polymers and industrial procedures, enabling the reader to better judge the importance and potential applications of the presented material.

A section interrelating condensation, stepwise kinetics, vinyl polymers, and chain-wise kinetics is included, since these topics are critical to the understanding of polymer science, but often are confusing to students.

The initial chapter is shorter than the others to provide time for student orientation. However, the other chapters should require an average of a week each and hence should fit in with the time allotted for a one-semester course. Some of the topics listed as optional in the suggested syllabus may be omitted when the course is taught on the quarter system or expanded if taught over a two-quarter period.

Each chapter is largely self-contained but is related to preceding chapters. Whenever possible, difficult concepts are distributed over several chapters.

In addition to serving as a one-semester or one- or two-quarter undergraduate textbook or introductory graduate-level text, this book should also be of considerable value to biologists, environmental scientists, engineers, and technologists who are concerned with plastics, fibers, elastomers (rubbers), coatings (paints), adhesives, biopolymers, or other macromolecules.

The fourth edition adds the latest important topics of polymer science including

DNA profiling
Recycling codes
Smart materials
Liquid crystals
Composites
Ionomers
Dendrites and
Soluble stereoregulating catalysts.

The chapter on additives has been rewritten and the chapter dealing with monomer synthesis revised to emphasize the synthesis of basic materials from petroleum and coal.

Sections on kinetics, polyethylene, high-performance materials, and molecular-weight concepts have been updated. Two new appendixes have been added—one dealing with the ever-important topic of health and one with the new management and environmental frameworks designated as ISO 9000 and ISO 14000.

This is the first edition compiled in the absence of the senior coauthor, who is missed by all of us who owe him so much.

Charles E. Carraher, Jr.

Preface to the Third Edition

The third edition continues the presentation of polymer chemistry as a unified and cohesive area of study that includes inorganic, synthetic, and natural macromolecules. Thus, sections dealing with

Superconductors
Enzymes
Comparative natural–synthetic macromolecule structure

have been added. To continue the "real-world" emphasis of polymers, sections on

Geotextiles
Solid waste
Room-temperature vulcanizing agents

have also been included. Sections dealing with polymer characterization have been updated, and descriptions of newer surface determination techniques have been added. As expected, "bridges" and extensions of material already offered in other core courses are emphasized.

Raymond B. Seymour
Charles E. Carraher, Jr.

Preface to the Second Edition

The revised text continues the emphasis on being "user friendly" and providing a balance between theory and "the real world." Updating occurs throughout the text. Additions, reflecting the expanding role of polymers, include sections on genetic engineering, the amorphous bulk state, mechanisms of energy absorption and polymeric mixtures. Sections on polymer testing and technology were completely redone and expanded. The emphasis on polymer structure–property relationships has been increased. A separate chapter on inorganic polymers, including concrete, sand, diamonds, graphite, asbestos, soil, and rocks, has been added to highlight the importance of polymers in the world of inorganics. Additional "helps" have been added including integrated sample (worked) problems and appendixes related to polymer models, polymer core course committees, and structures of common polymers. Again, transitions and extensions from material already covered in traditional core chemistry courses are emphasized.

Raymond B. Seymour
Charles E. Carraher, Jr.

Preface to the First Edition

While the eminent Professor Mark, who wrote the foreword, offered graduate courses in general polymer chemistry at the Polytechnic Institute in Brooklyn in 1940, the senior author of this new book offered his first undergraduate course in polymer chemistry at the University of Chattanooga in 1945. The only undergraduate textbook available at that time was *Synthetic Resins and Rubbers* by Dr. Paul Powers.

This pioneering book was acceptable in the early days of polymer chemistry, but it was no longer acceptable when the senior author joined the Los Angeles Trade Tech College as an Adjunct Professor and Sul Ross State University as a Professor less than 15 years later. Hence, *Introduction to Polymer Chemistry* was written using classroom notes from lectures at Trade Tech and Sul Ross.

Introduction to Polymer Chemistry was published by McGraw-Hill and translated into several foreign languages. Then it was published as an international edition by McGraw-Hill Kogakusha (Tokyo) and reprinted by Krieger Publishing Co. However, that book is no longer acceptable as an undergraduate text and does not meet the newer ACS guidelines nor the syllabus suggested by the polymer education committees of that society. The latter is shown below with appropriate chapters from our new book and points assigned from 127 replies received from 97 schools in answer to a survey by members of the American Chemical Society. (The full results of this survey are given in the Journal of Chemical Education, 57(6), 436 (1980), Core curriculum in introductory courses of polymer chemistry.)

As shown by the chapters cited in the table, *Polymer Chemistry* includes all the major and optional topics recommended in the syllabus adopted by the joint polymer education committee of the American Chemical Society. (A fuller discussion of the suggested syllabus is found in Appendix D: Syllabus). We have em-

Suggested Syllabus

	Suggested percentage of course time	Corresponding chapter in *Polymer Chemistry*	Survey[a] points
Major topics			
1. Introduction	5	1	—
2. Polymer Structure	10	2	104
(Morphology)			86
3. Molecular Weights	10	3, 4	104
4. Step-Reaction Polymers	10	7	100
5. Chain-Reaction Polymers:	10		
Ionic and Complex Coordinative		8	83
Free Radical		9	100
6. Copolymerization	10	10	86
7. Testing and Characterization	10	5	51
Optional topics			
8. Rheology	5	3	85
9. Solubility	5	3	85
10. Natural and Biomedical Polymers	5	6	11
11. Additives	5	13, 14	4
12. Reactions of Polymers	5	15	61
13. Synthesis of Reactants	5	16	—
14. Polymer Technology	5	17	28

[a]From 127 replies from 97 schools, based on +1, topic to be included in introductory courses; 0, no strong feeling; −1, topic should be omitted from introductory courses.

phasized the topics given high points in the national survey. A similar viewpoint was taken in the development of the ACS Standard Examination of Polymer Chemistry. This book also includes relevant chapters on inorganic polymers (11 and 12) and the synthesis of reactants (16) and emphasis on history and additives in an attempt to make it a complete textbook on polymer chemistry.

While no specific chapter is devoted to commercial polymers, these important products, as well as emerging and projected future areas of research activity, are discussed throughout the text.

Information already presented in traditional undergraduate courses of organic chemistry, physical chemistry, etc., is interrelated with information presented that focuses on polymer topics. This assists the student in integrating his or her chemical knowledge and illustrates the interrelationship between theoretical and applied chemical knowledge. Further, industrial practices and testing procedures are integrated with the theoretical treatment of the various topics allowing the reader to bridge the gap between industrial practice and the classroom. Also incorporated into the text are uses and amounts used of particular polymers and procedures allowing the reader to better judge the importance and potential applications of the presented material.

A section relating to the topics of condensation, stepwise kinetics, vinyl polymers, and chainwise kinetics is added, since these topics are critical to an understanding of polymer chemistry but are often sources of confusion to students.

The first chapter is shorter than the others in order to provide time for student orientation. However, the other 16 chapters should not require more than a week's time each and hence should fit in well with the time allotted for a one-semester course. Some of the topics listed as optional in the suggested syllabus may be omitted when the course is taught on the quarter system, or expanded if taught over a two-quarter period.

Each chapter in this book is essentially self-contained, but each relates to preceding chapters. Whenever possible, difficult concepts are distributed over several chapters. A glossary, suggested questions, and a list of references are included at the end of each chapter. Some relevant trade names may be found in the glossaries, and many trade names are listed in a separate section at the end of the book. The toxicity of monomers and reactants is discussed throughout the book.

In addition to serving as a one semester or one- or two-quarter undergraduate textbook, this book should also be of considerable value to biologists, environmental scientists, engineers, and technologists who are concerned with plastics, fibers, elastomers (rubbers), coatings (paints), adhesives, biopolymers, or other macromolecules.

Raymond B. Seymour
Charles E. Carraher, Jr.

Acknowledgments

The authors gratefully acknowledge the contributions of Herman Mark of the Polytechnic Institute of New York; Charles L. McCormick, University of Southern Mississippi; William Feld, Wright State University; Eli Pearce, Polytechnic Institute of New York; Fredinard Rodriguez, Cornell University and Otto Vogl, University of Massachusetts, for their reviewing, advising, and counseling efforts; and to Charles Carraher III and Shawn Carraher for their help in proofing and indexing. A further special thanks to Maurits Dekker, for his encouragement and many contributions to the polymer literature.

The authors also thank the following for their special contributions to the book: Charles Gebelein, Les Sperling, Angelo Volpe, Stam Israel, Carl Wooten, Rita Blumstein, Eckhard Hellmuth, Frank Millich, Norman Miller, Rudy Deanin, Guy Donaruma, Leo Mandelkern, R. V. Subramanian, Charles Pittman, Brian Currell, C. Bamford, Roger Epton, Paul Flory, Charles Overberger, William Bailey, Jim O'Donnell, Rob Burford, Edgar Hardy, John H. Coates, Don Napper, Frank Harris, G. Allan Stahl, John Westerman, William A. Field, Nan-Loh Yang, Sheldon Clare, E. N. Ipiotis, D. H. Richards, G. Kirshenbaum, A. M. Sarquis, Lon Mathias, Sukumar Maiti, S. Temin, Yoshinobu Naoshima, Eberhard Neuse, John Sheats, George Hess, David Emerson, Kenneth Bixgorin, Thomas Miranda, M. B. Hocking, Marsha Colbert, Joseph Lagowski, Dorothy Sterling, Amanda Murphy, John Kloss, Qingmao Zhang, Bhoomin Pandya, Ernest Randolph, Alberto Rivalta, and Fengchen He.

This book could not have been written without the long-time efforts of Professor Herman Mark, who was one of the fathers of polymer science.

For the fourth edition, a special thanks for the assistance of Colleen Carraher.

We acknowledge the kind permission of Gerry Kirshenbaum and *Polymer News* for allowing us to use portions of articles that have appeared in *Polymer News*.

A Note on the Nomenclature

As with most areas of science, names associated with reactions, particular chemical and physical tests, etc., were historically derived with few overall guiding principles. Further, the wide diversity of polymer science permitted a wide diversity in naming polymers. Even though the International Union of Pure and Applied Chemistry, IUPAC, has a long-standing commission associated with the nomenclature of polymers [reports include "Report on nomenclature in the field of macromolecules," Journal of Polymer Science, 8, 257 (1952); "Report on nomenclature dealing with steric regularity in high polymers," Pure and Applied Chemistry, 12, 645 (1966); "Basic definitions of terms relating to polymers," IUPAC Information Bull. App., 13, 1 (1971); and "Nomenclature of regular single-strand organic polymers," Macromolecules, 6(2), 149 (1973)], most of these suggestions for naming of simple polymers have not yet been accepted by many in the polymer science community.

Although there is wide diversity in the practice of naming polymers, we will concentrate on the three most utilized systems.

COMMON NAMES

Little rhyme or reason is associated with the common names of polymers. Some names are derived from the place of origin of the material, such as *Hevea brasiliensis*—literally "rubber from Brazil"—for natural rubber. Other polymers are named after their discoverer, as is Bakelite, the three-dimensional polymer produced by condensation of phenol and formaldehyde, which was commercialized by Leo Baekeland in 1905.

For some important groups of polymers, special names and systems of nomenclature were invented. For example, the nylons were named according to the number of carbons in the diamine and carboxylic acid reactants (monomers) used in their syntheses. The nylon produced by the condensation of 1,6-hexanediamine (6 carbons) and sebacic acid (10 carbons) is called nylon-6,10.

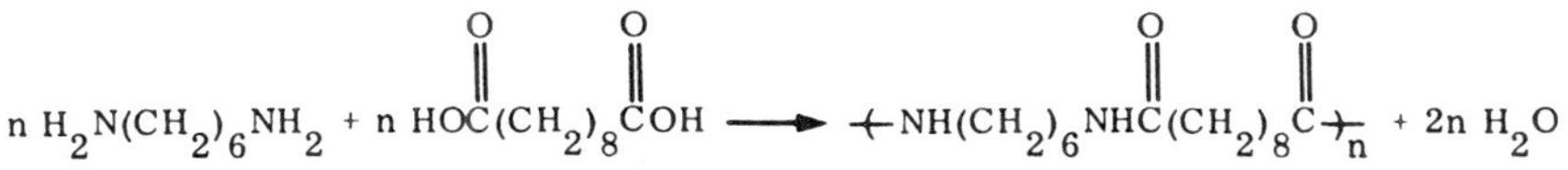

$$n\ H_2N(CH_2)_6NH_2 + n\ HO\overset{O}{\overset{\|}{C}}(CH_2)_8\overset{O}{\overset{\|}{C}}OH \longrightarrow \ (NH(CH_2)_6NH\overset{O}{\overset{\|}{C}}(CH_2)_8\overset{O}{\overset{\|}{C}})_n + 2n\ H_2O$$

1,6-Hexanediamine   Sebacic acid   Nylon-6,10

Similarly, the polymer from 1,6-hexanediamine and adipic acid (each with 6 carbons) is called nylon-6,6 or nylon-66, and the nylon from the single reactant caprolactam (6 carbons) is called nylon-6.

SOURCE-BASED NAMES

Most polymer names used by polymer scientists are source-based, that is, they are based on the common name of the reactant monomer, preceded by the prefix "poly." For example, polystyrene is the most frequently used name for the polymer derived from the monomer 1-phenylethene, which has the common name styrene.

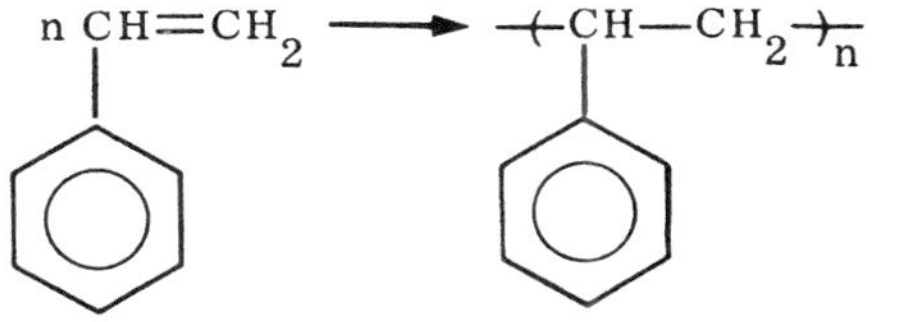

$$n\ CH{=}CH_2 \longrightarrow\ (CH{-}CH_2)_n$$

Styrene   Polystyrene

The vast majority of polymers based on the vinyl group ($CH_2{=}CHX$) or the vinylidene group ($CH_2{=}CX_2$) as the repeat unit are known by their source-based names. For example, polyethylene is derived from the monomer ethylene, poly(vinyl chloride) from the monomer vinyl chloride, and poly(methyl methacrylate) from methyl methacrylate:

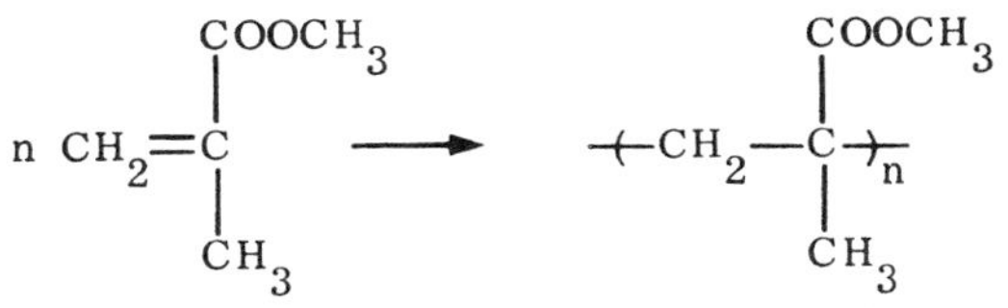

$$n\ CH_2{=}\overset{COOCH_3}{\underset{CH_3}{C}} \longrightarrow\ (CH_2{-}\overset{COOCH_3}{\underset{CH_3}{C}})_n$$

Methyl   Poly(methyl
methacrylate   methacrylate)

Many condensation polymers are also named in this manner. In the case of poly(ethylene terephthalate), the glycol portion of the name of the monomer, ethylene glycol, is used in constructing the polymer name, so that the name is actually a hybrid of a source-based and a structure-based name.

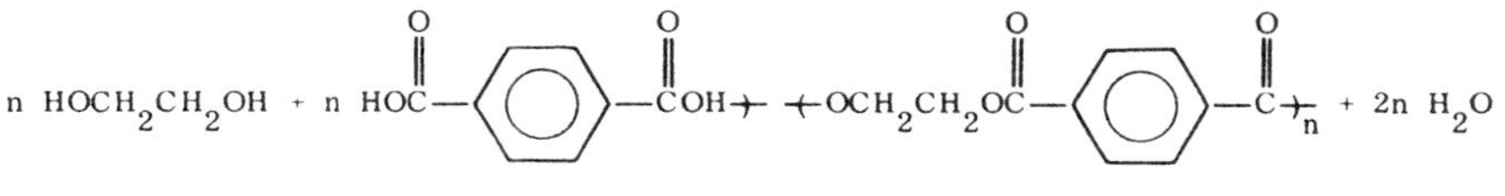

$$n\ HOCH_2CH_2OH + n\ HO\overset{O}{\overset{\|}{C}}{-}\langle\bigcirc\rangle{-}\overset{O}{\overset{\|}{C}}OH + (OCH_2CH_2O\overset{O}{\overset{\|}{C}}{-}\langle\bigcirc\rangle{-}\overset{O}{\overset{\|}{C}})_n + 2n\ H_2O$$

Ethylene   Terephthalic   Poly(ethylene
glycol   acid   terephthalate)

This polymer is well known by trade names, such as Dacron, or its common name, polyester.

Although it is often suggested that parentheses be used in naming polymers of more than one word [like poly(vinylidene chloride)] but not for single-word polymers (like polyethylene), many authors omit entirely the use of parentheses for either case (like polyvinylidene chloride). Thus there exists a variety of practices with respect to even source-based names.

Copolymers are composed of two or more monomer units. Source-based names are conveniently used to describe copolymers by using an appropriate term between the names of the monomers. Any of a half dozen or so connecting terms may be used, depending on what is known about the structure of the copolymer. When no information is specified about the sequence of monomer units in the copolymer, the connective term *co* is used in the general format poly(A-*co*-B), where A and B are the names of the two monomers. An unspecified copolymer of styrene and methyl methacrylate would be called poly[styrene-*co*-(methyl methacrylate)].

Kraton, the yellow rubberlike material on the bottom of many running shoes, is an example of a copolymer about which structural information is available. It is formed from a group of styrene units, i.e., a "block" of polystyrene, attached to a group of butadiene units, or a block of polybutadiene, which is attached to another block of polystyrene forming a triblock copolymer. The general representation of such a block copolymer is –AAAAABBBBBAAAAA–, where each A or B represents an individual monomer unit. The proper source-based name for Kraton is polystyrene-*block*-polybutadiene-*block*-polystyrene, with the prefix "poly" being retained for each block.

STRUCTURE-BASED NAMES

Although source-based names are generally employed for simple polymers, the international body responsible for systematic nomenclature of chemicals, IUPAC, has published a number of reports for the naming of polymers, now being accepted for more complex polymers. The IUPAC system names the components of the repeat unit, arranged in a prescribed order. The rules for selecting the order of the components to be used as the repeat unit are found elsewhere [Macromolecules, 6(2), 149 (1973); Pure & Applied Chemistry, 48, 373 (1976), 57, 149 (1985), and 57, 1427 (1985)]. However, once the order is selected, the naming is straightforward for simple linear molecules, as indicated in the following examples:

Polystyrene

$+CH—CH_2+_n$

Poly(1-phenylethylene)

$+C—CH_2+_n$ with CH_3 and $CH_3OC=O$

Poly[1-(methoxycarbonyl)-1-methylethylene]

A listing of source- and structure-based names for some common polymers follows:

Comparison of Polymer Names

Source-based	Structure-based
Polyacrylonitrile	Poly(1-cyanoethylene)
Poly(ethylene oxide)	Poly(oxyethylene)
Poly(ethylene terephthalate)	Poly(oxyethyleneoxyterephthaloyl)
Polyisobutylene	Poly(1,1-dimethylethylene)
Poly(methyl methacrylate)	Poly[(1-methoxycarbonyl)-1-methylethylene]
Polypropylene	Poly(propylene)
Polystyrene	Poly(1-phenylethylene)
Poly(tetrafluoroethylene)	Poly(difluoromethylene)
Poly(vinyl acetate)	Poly(1-acetoxyethylene)
Poly(vinyl alcohol)	Poly(1-hydroxyethylene)
Poly(vinyl chloride)	Poly(1-chloroethylene)
Poly(vinyl butyral)	Poly[(2-propyl-1,3-dioxane-4,6-diyl)methylene]

Portions adapted from C. Carraher, G. Hess, and L. Sperling, J. Chem. Ed., *64*(1), 36 (1987).
Used with permission of the Journal of Chemical Education and the authors.

Additional information on the naming of polymers can be found in:

Compendium of Macromolecular Nomenclature, CRC Press, Boca Raton, Florida, 1991.
Polymer Nomenclature, Polymer Preprints, *33*(1), 6–11 (1992).
Basic Classification and Definitions of Polymerization Reactions, Pure Appl. Chem., *66*: 2483–2486 (1994).
Graphic Representations (Chemical Formulae) of Macromolecules, Pure Appl. Chem., *66*: 2469–2482 (1994).
Structure-Based Nomenclature for Irregular Single-Strand Organic Polymers, Pure Appl. Chem., *66*: 873–889 (1994).
Nomenclature of Regular Double-Strand (Ladder and Spiro) Organic Polymers, Pure Appl. Chem., *65*: 1561–1580 (1993).
A. D. Jenkins and K. L. Loening, Nomenclature; in *Comprehensive Polymer Science* (G. Allen, J. Bevington, C. Booth, and C. Price, Eds.), Vol. 1, pp. 13–54, Pergamon Press, Oxford, 1989.
N. M. Bikales, Nomenclature; in *Encyclopedia of Polymer Science and Engineering*, 2nd ed. (H. F. Mark, N. M. Bikales, C. G. Overberger, and G. Menges, Eds.), Vol. 10, Wiley, New York, 1987, pp. 191–204.
Definitions of Terms Relating to Crystalline Polymers, Pure Appl. Chem., *61*: 769–785 (1989).
A Classification of Linear Single-Strand Polymers, Pure Appl. Chem., *61*: 243–254 (1989).
Definitions of Terms Relating to Individual Macromolecules, Their Assemblies, and Dilute Polymer Solutions, Pure Appl. Chem., *61*: 211–241 (1989).

Use of Abbreviations for Names of Polymeric Substances, Pure Appl. Chem., *59*: 691–693 (1987).

Source-Based Nomenclature for Copolymers, Pure Appl. Chem., *57*: 1427–1440 (1985).

Nomenclature for Regular Single-Strand and Quasi Single-Strand Inorganic and Coordination Polymers, Pure Appl. Chem., *57*: 149–168 (1985).

Notes on Terminology for Molar Masses in Polymer Science: Makromol. Chem., *185*, Appendix to No. 1 (1984). J. Polymer Sci., Polymer Lett. Ed., *22*, 57 (1984). J. Macromol. Sci. Chem., *A21*, 903 (1984). J. Colloid Interface Science, *101*, 277 (1984). Br. Polym. J., *17*, 92 (1985).

Stereochemical Definitions and Notations Relating to Polymers, Pure Appl. Chem., *53*: 733–752 (1981).

Nomenclature of Regular Single-Strand Organic Polymers, Pure Appl. Chem., *48*: 373–385 (1976).

Basic Definitions of Terms Relating to Polymers, Pure Appl. Chem., *40*: 479–491 (1974).

Thanks to William Work and W. V. (Val) Metanomski for their assistance with polymer nomenclature.

Contents

Contents

Contents

1

Introduction to Polymer Science

1.1 HISTORY OF POLYMERS

Since most chemists and chemical engineers are now involved in some phase of polymer science or technology, some have called this the polymer age. Actually, we have always lived in a polymer age.

The ancient Greeks classified all matter as animal, vegetable, and mineral. Minerals were emphasized by the alchemists, but medieval artisans emphasized animal and vegetable matter. All are largely polymers and are important to life as we know it.

Polymer is derived from the Greek *poly* and *meros*, meaning many and parts, respectively. Some scientists prefer to use the word *macromolecule*, or large molecule, instead of polymer. Others maintain that naturally occurring polymers, or *biopolymers*, and synthetic polymers should be studied in different courses. However, the same principles apply to all polymers. If one discounts the end uses, the differences between all polymers, including plastics, fibers, and elastomers or rubbers, are determined primarily by the intermolecular and intramolecular forces between the molecules and within the individual molecule, respectively, and by the functional groups present.

In addition to being the basis of life itself, protein is used as a source of amino acids and energy. The ancients degraded or depolymerized the protein in tough meat by aging and cooking, and they denatured egg albumin by heating or adding vinegar to the eggs.

Early humans learned how to process, dye, and weave the natural proteinaceous fibers of wool and silk and the carbohydrate fibers of flax and cotton. Early South American civilizations such as the Aztecs used natural rubber (*Hevea brasiliensis*) for making elastic articles and for waterproofing fabrics.

There has always been an abundance of natural fibers and elastomers but few plastics. Of course, early humans employed a crude plastic art in tanning the protein in animal skins to make leather and in heat-formed tortoise shells. They also used naturally occurring tars as caulking materials and extracted shellac from the excrement of small coccid insects (*Coccus lacca*).

Until Wöhler synthesized urea from inorganic compounds in 1828, there had been little progress in organic chemistry since the alchemists emphasized the transmutation of base metals to gold and believed in a vital force theory. Despite this essential breakthrough, little progress was made in understanding organic chemistry until the 1850s, when Kekulé developed the presently accepted technique for writing structural formulas.

However, polymer scientists displayed a talent for making empirical discoveries before the science was developed. Thus, Charles and Nelson Goodyear transformed hevea rubber from a sticky thermoplastic to a useful elastomer (vulcanized rubber) or a hard thermoset plastic (ebonite), respectively, by heating it with small or large amounts of sulfur long before Kekulé developed his formula-writing techniques.

Likewise, Schönbein reacted cellulose with nitric acid, and Menard, in 1846, made collodion by dissolving the cellulose nitrate product of that re-reaction in a mixture of ethanol and ethyl ether. Collodion, which was used as a liquid court plaster, also served in the 1860s as Parkes and Hyatt's reactant for making celluloid, which was the first artificial thermoplastic, and Chardonnet's reactant for making artificial silk.

While most of these early discoveries were empirical, they may be used to explain some terminology and theory in modern polymer science. It is important to note that all these inventors, like ancient humans, converted a naturally occurring polymer to a more useful product. Thus, Charles Goodyear transformed the heat-softenable thermoplastic hevea rubber to a less heat-sensitive product by using sulfur to form a relatively small number of connecting links or cross-links between the long individual polyisoprene chainlike molecules.

Nelson Goodyear used sulfur to produce many cross-links between the polyisoprene chains so that the product was no longer thermoplastic but was a thermoset plastic. *Thermoplastics* are two-dimensional molecules that may be softened by heat and returned to their original state by cooling, while *thermosetting plastics* are three-dimensional network polymers that cannot be reshaped by heating.

Both cellulose and cellulose nitrate are linear, or two-dimensional, polymers, but the former cannot be softened because of the presence of multitudinous hydrogen bonds between the chainlike molecules. When used as an explosive, the cellulose nitrate is completely nitrated and is essentially a trinitrate of cellulose. In contrast, Parkes and Hyatt used a dinitrate, or secondary cellulose nitrate, which contained many residual hydrogen bonds and in addition was highly flammable.

Parkes added castor oil in order to plasticize—to reduce the effect of—the intermolecular hydrogen bonds. Hyatt used camphor for the same purpose. Count Hilaire de Chardonnet forced Menard's collodion through very small holes called spinnerets and obtained filaments by evaporating the mixture of solvents. The flammability of these cellulosic filaments was reduced by denitrification using sodium bisulfite.

It is of interest to note that the Goodyears converted a thermoplastic elastomer to a thermoset elastomer and a hard thermoset plastic by the addition of small and large amounts of sulfur cross-links. Schönbein reduced the number of intermolecular hydrogen bonds present in cellulose by reacting it with nitric acid, and while neither cellulose nor cellulose dinitrate was soluble in ethanol or ethyl ether, Menard dissolved the latter in an equimolar solution of these two solvents.

Hyatt softened the flammable cellulose dinitrate by adding camphor, which reduced the effectiveness of the intermolecular hydrogen bonds. Chardonnet regenerated cellulose in the form of continuous filaments. After denitrification and stretching, these filaments possessed all the chemical and physical properties of the original cellulose.

However, since no one at that time knew what a polymer was, they had no idea of the complex changes that had taken place in the pioneer production of useful rubber, plastics, and fibers. Even today, some organic chemists find it difficult to visualize these large macromolecules. Over a century ago, Graham coined the term colloid for aggregates with dimensions in the range of 10^{-9} to 10^{-7} m. Unfortunately, the size of many polymer molecules is in this range, but it is important to note that, unlike colloids, polymers are individual molecules whose size cannot be reduced without breaking the covalent bonds which hold the atoms together in these chainlike molecules.

An oligomer, a very low molecular weight polymer of ethylene glycol, was prepared and the correct structure of $[HO(OCH_2CH_2)_8OH]$ was assigned in 1860. Nevertheless, Fittig and Engelhorn incorrectly assigned a cyclic structure to poly(methacrylic acid) $[-CH_2C(CH_3)COOH]$, which they prepared in 1880. By use of the Raoult and van't Hoff concepts, several chemists obtained high molecular weight values for these and other linear polymers, but since they were unable to visualize such things as macromolecules, they concluded that the Raoult technique was not applicable to the determination of the molecular weight of these molecules.

It was generally recognized by the leading organic chemists of the nineteenth century that phenol would condense with formaldehyde. Since they did not recognize the concept of functionality, Baeyer, Michael, and Kleeberg produced useless cross-linked goos, gunks, and messes and then returned to their research on reactions of monofunctional reactants. However, by use of a large excess of phenol, Smith, Luft, and Blumer were able to obtain thermoplastic condensation products.

While there is no evidence that Baekeland recognized the existence of macromolecules, he did understand functionality, and by the use of controlled amounts of phenol and formaldehyde, he produced thermoplastic resins which could be converted to thermosetting plastics. He coined the term A-stage resole resin to describe the thermoplastic Bakelite produced by the condensation of an excess of formaldehyde and phenol under alkaline conditions. This A-stage resole resin was converted to a thermoset (infusible) cross-link *C-stage Bakelite* by additional heating or advancement of the resin. Baekeland also prepared thermoplastic resins called *novolacs* by the condensation of phenol with a small amount of formaldehyde in acidic solutions. The novolacs were converted to thermosets by the addition of formaldehyde from hexamethylenetetramine. While other polymers had been synthesized in the laboratory before 1910, Bakelite was the first truly synthetic plastic. The fact that the recipes used today are essentially the same as those revealed in the original patents demonstrates Baekeland's ingenuity and knowledge of the chemistry of the condensation of trifunctional phenol with difunctional formaldehyde.

Prior to World War I, celluloid, shellac, Galalith (casein), Bakelite, and cellulose acetate plastics; hevea rubber, cotton, wool, silk, and rayon fibers; Glyptal polyester coatings, bitumen or asphalt, and coumarone-indene and petroleum resins were all commercially available. However, as evidenced by the chronological data shown in Table 1.1, there was little additional development in polymer technology prior to World War II because of the lack of knowledge of polymer science.

Nobel laureate Hermann Staudinger laid the groundwork for modern polymer science in the 1920s when he demonstrated that natural and synthetic polymers were not aggregates like colloids or cyclic compounds like cyclohexane, but were long chainlike molecules with characteristic end groups. The advice given to Dr. Staudinger by his colleagues was "Dear Colleague, Leave the concept of large molecules well alone. . . . There can be no such thing as a macromolecule." Unfortunately, some nonpolymer scientists who no longer believe in the vital force concept, like Staudinger's colleagues, still question the existence or importance of macromolecules.

In 1928, Meyer and Mark used X-ray techniques to determine the dimensions of crystallites in cellulose and natural rubber. During the following year, Carothers synthesized and characterized linear aliphatic polyesters. Since these were not suitable for fibers, he synthesized polyamides that are known by the generic name of nylon. It is of interest to note that naturally occurring wool and silk protein fibers are also polyamides.

The leading polymer scientists of the 1930s agreed that polymers were chainlike molecules and that the viscosities of solution of these macromolecules were dependent on the size and shape of the molecules in the solution. While it is true that the large-scale production of many polymers was accelerated by World War II, it must be recognized that the production of these essential products was dependent on the concepts developed by Staudinger, Carothers, Mark, and other leading polymer scientists.

The development of polymer technology since the 1940s has been extremely rapid. In some instances, such as polymerization in aqueous emulsion systems, the art has preceded the science, but much theory has been developed, so that polymer science today is relevant and no longer largely empirical.

1.2 TODAY'S MARKETPLACE

As shown in Tables 1.2, 1.3, and 1.4, almost 70 billion pounds (35 million tons) of synthetic polymers are produced annually in the United States, and the growth of the industry is continuing at a faster rate than any other industry. There is every reason to believe that this polymer age will continue as long as petroleum and other feedstocks are available and as long as consumers continue to enjoy the comfort provided by elastomers, fibers, plastics, adhesives, and coatings. The 70 billion pounds translates to over 200 pounds of synthetic polymers for every man, woman, and child in the United States. This does not include paper and wood-related products, natural polymers, such as cotton and wool, or inorganic polymers.

The number of professional chemists directly involved with polymers is estimated to be between 40 and 60% of all of the chemists employed. The number of chemical industrial employees involved with synthetic polymers is reported by the

Table 1.1 Chronological Development of Commercial Polymers (to 1970)

Date	Material
Before 1800	Cotton, flax, wool, and silk fibers; bitumen caulking materials; glass and hydraulic cements; leather and cellulose sheet (paper); natural rubber (Hevea brasiliensis), gutta percha, balata, and shellac
1839	Vulcanization of rubber (Charles Goodyear)
1846	Nitration of cellulose (Schönbein)
1851	Ebonite (hard rubber; Nelson Goodyear)
1860	Molding of shellac and gutta percha
1868	Celluloid (plasticized cellulose nitrate; Hyatt)
1889	Regenerated cellulosic fibers (Chardonnet)
1889	Cellulose nitrate photographic films (Reichenbach)
1890	Cuprammonia rayon fibers (Despeisses)
1892	Viscose rayon fibers (Cross, Bevan, and Beadle)
1907	Phenol-formaldehyde resins (Bakelite; Baekeland)
1907	Cellulose acetate solutions (dope; Doerfinger)
1908	Cellulose acetate photographic fibers
1912	Regenerated cellulose sheet (cellophane)
1923	Cellulose nitrate automobile lacquers
1924	Cellulose acetate fibers
1926	Alkyd polyester (Kienle)
1927	Poly(vinyl chloride) (PVC) wall covering
1927	Cellulose acetate sheet and rods
1929	Polysulfide synthetic elastomer (Thiokol; Patrick)
1929	Urea-formaldehyde resins
1931	Poly(methyl methacrylate) (PMMA) plastics
1931	Polychloroprene elastomer (Neoprene)
1935	Ethylcellulose
1936	Poly(vinyl acetate)
1936	Poly(vinyl butyral) safety glass

Table 1.1 Continued

Date	Material
1937	Polystyrene
1937	Styrene-butadiene (Buna-S) and styrene-acrylonitrile (Buna-N) copolymer elastomers
1938	Nylon-66 fibers (Carothers)
1939	Melamine-formaldehyde resins
1940	Isobutylene-isoprene elastomer (butyl rubber; Sparks and Thomas)
1941	Low-density polyethylene (LDPE)
1942	Unsaturated polyesters (Ellis and Rust)
1943	Fluorocarbon resins (Teflon; Plunkett)
1943	Silicones
1943	Polyurethanes (Baeyer)
1947	Epoxy resins
1948	Copolymers of acrylonitrile, butadiene and styrene (ABS)
1950	Polyester fibers (Whinfield and Dickson)
1950	Polyacrylonitrile fibers
1956	Polyoxymethylene (acetals)
1957	High-density (linear) polyethylene (HDPE)
1957	Polypropylene
1957	Polycarbonate
1959	cis-Polybutadiene and cis-polyisoprene elastomers
1960	Ethylene-propylene copolymer elastomers
1962	Polyimide resins
1964	Poly(phenylene oxide)
1965	Polysulfone
1965	Styrene-butadiene block copolymers
1970	Poly(butylene terephthalate)
1971	Polyphenylene sulfide

Table 1.2 U.S. Production of Plastics (Millions of Pounds)

	1975	1980	1985	1990	1994
Thermosetting resins					
Epoxies	200	320	390	510	600
Polyesters	800	950	1,200	1,300	1,460
Ureas	690	1,200	1,200	1,500	1,890
Melamines	120	170	200	220	290
Phenolics	1,100	1,500	2,600	2,900	3,210
Thermoplastics					
Polyethylene—low density	4,700	7,300	8,900	9,700	12,590
Polyethylene—high density	2,500	4,400	6,700	8,100	10,950
Polypropylene	1,900	3,600	5,100	7,200	9,460
Styrene-acrylonitrile	110	100	70	—	—
Polystyrene	2,700	3,500	4,100	5,000	5,500
Acrylonitrile-butadiene-styrene	1,200	1,700	2,000	—	—
Polyamides	140	270	400	—	—
Poly(vinyl chloride) and copolymers	3,600	5,500	6,800	9,100	10,880

U.S. Department of Labor to be over one million, or almost 60% of the chemical industrial work force (Table 1.5).

Polymeric materials, along with the majority of the chemical industrial products, contribute positively to the balance of trade (Table 1.6). In fact, plastics and resins show the greatest value increase of exports minus imports.

Table 1.7 contains a listing of the major chemical producers. All of these producers are involved directly and/or indirectly with polymers.

Thus, polymers play a critical role in our everyday lives, actually forming the basis for both plant and animal life, and represent an area where chemists continue to make major important contributions.

SUMMARY

After reading this chapter, you should understand the following concepts.

1. Polymers or macromolecules are giant molecules with molecular weights at least 100 times greater than those of smaller molecules like water or methanol.

2. If we disregard metals and inorganic compounds, we observe that practically everything else in this world is polymeric. This includes the protein and nucleic acid in our bodies, the fibers we use for clothing, the protein and starch we eat, the elastomers in our automobile tires, the paint, plastic wall and floor coverings, foam insulation, dishes, furniture, pipes, and so forth, in our homes.

3. In spite of the many varieties of fibers, elastomers, and plastic, they all have a similar structure and are governed by the same theories. Linear polymers, such as high-density polyethylene (HDPE), consist of long chains made up of thousands of covalently bonded carbon atoms. The repeating unit for HDPE is represented as $[CH_2CH_2]_n$, where n is the number of repeating units.

4. Most linear polymers such as HDPE are thermoplastic, that is, they may be softened by heat and hardened by cooling in a reversible physical process. However,

Table 1.3 U.S. Production of Manmade Fibers (Millions of Pounds)

	1975	1980	1984	1990	1994
Noncellulosic					
Acrylics	500	780	670	510	440
Nylons	1800	780	2400	2700	2750
Olefins	500	750	1000	1800	2410
Polyesters	3000	4000	3400	3200	3860
Cellulosic					
Acetate	310	320	200	210	230
Rayon	440	490	420	300	270
Glass	550	870	1400	—	—

linear polymers like cellulose, which have very strong intermolecular forces (hydrogen bonds), cannot be softened by heating below the decomposition temperature.

5. Thermoset polymers are cross-linked and cannot be softened by heating. Thermoplastics such as natural rubber and A-stage resole Bakelite resins can be transformed to thermosetting polymers by the introduction of cross-links between the polymer chains.

6. Early developments in polymer technology were empirical because of a lack of knowledge of polymer science. Advancements in polymer technology were rapid in the 1930s and 1940s because of the theories developed by Staudinger, Carothers, Mark, and other polymer scientists.

7. Thermoplastic resole resins that may be thermoset by heating are obtained by heating phenol and formaldehyde on the alkaline side. Novolacs are obtained when an insufficient amount of formaldehyde is reacted in acid solution. Novolacs are converted to infusible plastics by heating with hexamethylenetetramine.

GLOSSARY

ABS: A polymer produced by the copolymerization of acrylonitrile, butadiene, and styrene.

acetal: A polymer produced by the polymerization of formaldehyde.

Table 1.4 U.S. Production of Synthetic Rubber (Millions of Pounds)

	1975	1980	1985	1990	1994
Synthetic rubber					
Styrene-butadiene	2400	2200	1600	1900	2180
Polybutadiene	580	620	660	890	1090
Nitrile	120	120	120	120	150
Ethylene-propylene	160	280	440	560	660
Other	660	840	1000	1200	1400

Table 1.5 U.S. Chemical Industrial Employment (in Thousands)

	1975	1985	1994
Industrial inorganic	149	143	131
Drugs	167	205	264
Soaps, cleaners, etc.	142	148	152
Industrial organics	150	164	144
Agricultural	65	60	56
Synthetic polymers	888	1026	1150

Source: U.S. Department of Labor.

alkyd: A polyester produced by the condensation of a dihydric alcohol, such as ethylene glycol, and a dicarboxylic acid, such as phthalic acid, in the presence of controlled amounts of an unsaturated monofunctional organic acid, such as oleic acid.

A-stage: A linear resole resin.

Bakelite: A polymer produced by the condensation of phenol and formaldehyde.

biopolymer: A naturally occurring polymer, such as protein.

buna: Copolymer of butadiene and styrene or acrylonitrile.

butyl rubber: A copolymer of isobutylene and isoprene.

cellulose: A naturally occurring carbohydrate polymer consisting of repeating glucose units.

cellulose acetate: The ester of cellulose and acetic acid.

cellulose nitrate: The product obtained by the reaction of nitric acid and sulfuric acid with cellulose; erroneously called nitrocellulose. The product is classified as primary, secondary, or tertiary according to how many groups in each repeating anhydroglucose unit in cellulose are nitrated.

cis: A geometric isomer with both groups on the same side of the ethylenic double bond.

Table 1.6 U.S. Chemical Trade—Imports and Exports, 1994 (Millions of Dollars)

	Exports	Imports
Organic chemicals	12,800	10,800
Inorganic chemicals	4,100	4,100
Oils and perfumes	3,500	2,000
Dyes, colorants	2,300	1,900
Medicinals and pharmaceuticals	6,100	4,700
Fertilizers	2,700	1,300
Plastics and resins	8,500	3,300
Others	7,600	2,700
Total chemicals[a]	51,600	33,400
Total	502,800	669,100

[a]Includes nonlisted chemicals.

Table 1.7 Major Chemical Producers Based on Sales (in Millions of U.S. Dollars)

Producer	1985	1990	1994
United States			
DuPont	11,300	15,600	16,800
Dow Chemical	9,500	14,700	14,100
Exxon	6,700	11,200	11,000
Monsanto	5,200	5,700	5,900
Union Carbide	4,000	7,600	4,900
Shell Oil	3,300	3,700	4,100
Celanese (now Hoechst Celanese)	3,300	5,500	7,000
W. R. Grace	2,900	3,600	3,200
Chevron	2,600	3,300	3,100
BASF	2,600	4,400	4,300
Eastman Kodak (now Eastman Chemical)	2,300	3,600	4,300
General Electric	2,300	5,200	5,700
Phillips Petroleum	2,300	2,100	2,800
Mobil	2,300	4,100	4,200
Allied-Signal	2,100	2,800	3,300
Rohm & Haas	2,000	2,800	3,500
American Cyanamid	1,800	2,800	—
Air Products	1,700	2,600	3,200
Occidental Petroleum	1,600	5,000	4,700
Ciba-Geigy	1,500	2,000	2,800
Ashland Oil	1,500	2,200	2,900
B. F. Goodrich	1,400	1,900	1,100
FMC	1,300	1,600	1,900
Amoco	—	—	4,700
ICI Americas	—	—	4,000
Arco Chemical	—	—	3,400
Huntsman Chemical	—	—	3,400
Akzo-Nobel	—	—	3,000
Europe			
Bayer (West Germany)	15,600		26,800
BASF (West Germany)	15,100		28,700
Hoechst (West Germany)	14,500		30,600
Imperial Chemical Inds. (U.K.)	13,900		14,100
DSM (Netherlands)	7,300		4,900
Rhone-Poulence (France)	6,200		15,600
Akzo (Netherlands)	5,400		12,200
Norsk Hydro (Norway)	4,900		10,100
Solvay (Belgium)	3,800		7,800
Roche (Switzerland)	3,600		10,800
Sandoz (Switzerland)	3,400		11,600
BOC International (U.K.)	2,500		5,300
Ciba (Switzerland)	—		16,100
Degussa (Germany)	—		8,500
EniChem (Italy)	—		7,100
Japan			
Mitsubishi Chemical Industries	3,960		8,500
Asahi Chemical Industries	3,910		9,600
Sumitomo Chemical	3,220		5,500

Table 1.7 Continued

Producer	1985	1990	1994
Toray Industries	3,200		5,100
Mitsui Toatsu Chemicals	2,100		3,700
Teijin	2,100		3,200
Showa Denko	2,100		4,400
Sekisui Chemical	—		6,900
Takeda Chemical Industries	—		5,600
Toray Industries	—		5,100
Canada			
DuPont Canada	870		1,230
Celanese Canada	270		460
Nova Corp.	—		2,700
Methanex	—		1,500

collodion: A solution of cellulose nitrate in an equimolar mixture of ethanol and ethyl ether.

colloid: An aggregate 10^{-9} to 10^{-7} m long; an aggregate of smaller particles.

coumarone-indene resins: Polymers produced by the polymerization of distillation residues.

covalent bond: Chemical bonds formed by the sharing of electrons of the bonded atoms such as the carbon atoms in graphite or diamonds.

cross-links: Covalent bonds between two or more linear polymeric chains.

crystallites: Aggregates of polymers in crystalline form.

C-stage: A cross-linked resole resin.

cuprammonia rayon: Rayon produced from a solution of cellulose in a copper ammonia hydroxide solution.

denaturation: The change in properties of a protein resulting from the application of heat or the addition of a foreign agent, such as ethanol.

ebonite: Hard rubber, highly cross-linked natural rubber (NR).

elastomer: A rubber.

filament: The individual extrudate emerging from the holes in a spinneret.

functionality: The number of reactive groups in a molecule.

Galalith: A plastic produced by molding casein.

glyptal: A polyester produced by the condensation of glycerol and phthalic anhydride.

hydrogen bonds: Very strong forces resulting from the attraction of hydrogen atoms in one molecule with an oxygen or nitrogen atom in another molecule. These forces may also be present as intramolecular forces in macromolecules.

intermolecular forces: Secondary valence forces between molecules.

intramolecular forces: Secondary valence forces within a molecule.

linear: A continuous chain, such as HDPE.

macromolecule: A polymer.

natural rubber, or NR (*Hevea brasiliensis*): Polyisoprene obtained from rubber plants.

novolac: A thermoplastic phenol-formaldehyde resin prepared by the condensation of phenol and a small amount of formaldehyde on the acid side. Novolacs may be advanced to thermoset by heating with formaldehyde, which is usually obtained by the thermal decomposition of hexamethylenetetramine.

NR: Natural rubber.

Nylon-66: A polyamide produced by the condensation of adipic acid [$HOOC(CH_2)_4COOH$] and hexamethylenediamine [$H_2N(CH_2)_6NH_2$].

oligomer: A very low molecular weight polymer in which the number of repeating units (n) equals 2 to 10 (*oligos* is the Greek term for few).

plasticizer: An additive that reduces intermolecular forces in polymers.

polymer: A giant or macromolecule made up on multiple repeating units, such as polyethylene, in which at least 1000 ethylene units (—CH_2CH_2—) are joined together by covalent bonds. The word is derived from the Greek words meaning many parts.

polymer age: An age when the use of polymers is emphasized, as in the twentieth century.

protein: A polymer made up of many amino acid repeating units, i.e., a polyamide.

Raoult's law: A law that states that colligative properties of solutions, such as osmotic pressure and changes in vapor pressure, are related to the number of solute molecules present.

rayon: Regenerated cellulose in the form of filaments.

resole: A condensation product produced by the reaction of phenol and formaldehyde under alkaline conditions.

shellac: A resin secreted by small coccid insects.

spinneret: Small holes through which a solution or molten polymer is extruded in order to form filaments for fibers.

tanning: Cross-linking of proteins by the reaction with cross-linking agents such as tannic acid.

thermoplastic: A linear polymer that may be softened by heat and cooled in a reversible physical process.

thermoset plastic: A network polymer usually obtained by cross-linking a linear polymer. Thiokol: A polyethylene sulfide elastomer.

viscose: A solution of cellulose xanthate produced by the reaction of alkali cellulose and carbon disulfide.

viscosity: The resistance to flow as applied to a solution or a molten solid.

vital force concept: A hypothesis that stated that organic compounds could be produced only by natural processes and not in the laboratory.

vulcanization: The process in which NR is cross-linked by heating with sulfur.

EXERCISES

1. Name six polymers that you encounter daily.

2. Which would be more likely to be softened by heat?
   (a) (1) unvulcanized rubber, or (2) ebonite
   (b) (1) A-stage, or (2) C-stage resole
   (c) (1) cellulose, or (2) cellulose acetate

3. Name a polymer having the following repeating units:
   (a) ethylene (—CH_2CH_2—)
   (b) phenol and formaldehyde residual units
   (c) amino acid residual units

4. In which of the following polymers will hydrogen bonding predominate?
   (a) natural rubber (NR)
   (b) linear polyethylene (HDPE)
   (c) cellulose
   (d) cellulose nitrate

5. Which of the following products are polymeric?
   (a) water
   (b) wood
   (c) meat
   (d) cotton
   (e) rubber tires
   (f) paint

6. Which of the following is a thermoset or cross-linked polymer?
   (a) cellulose
   (b) unvulcanized rubber
   (c) A-stage resole
   (d) cellulose nitrate
   (e) molded Bakelite
   (f) ebonite

7. Which of the items in question 6 are thermoplastic?

8. If you were Staudinger, how would you answer your critical colleagues?

9. Why are so many outstanding polymer scientists alive today?

10. What percentage of polymer science students receive job offers after graduation?

11. What is the molecular weight of $H(CH_2CH_2)_{1000}H$?

12. What is the principal difference between rayon and cellophane?

13. Which is the more heat stable, a resole or a novolac phenolic (Bakelite) resin?

BIBLIOGRAPHY

Adkins, R. T. (1989): *Information Sources in Polymers and Plastics*, Butterworth, Kent, England.
Alcacer, L. (1987): *Conducting Polymers*, Kluwer Academic, Hingham, Massachusetts.
Allcock, H. R. (1981): *Contemporary Polymer Chemistry*, Prentice-Hall, Englewood Cliffs, New Jersey.

Allen, G., Bevington, J. C. (1988): *Comprehensive Polymer Science*, Pergamon, New York.

Allen, P. W. (1972): *Natural Rubber and Synthetics*, Halsted, New York.

———. (1974): Palmerton, New York.

Alper, J., Nelson, G., *Polymeric Materials: Chemistry for the Future*, ACS, Washington, D.C.

Anderson, J. C., Leaver, K. D., Alexander, J. M., Rawlins, R. D. (1975): *Material Science*, Halsted, New York.

Bassett, D. C. (1988): *Developments in Crystalline Polymers*, Elsevier, New York.

Bauh, C. E. H. (1972): *Macromolecular Science*, Wiley, New York.

Billmeyer, F. W. (1984): *Textbook of Polymer Science*, 3rd ed., Wiley Interscience, New York.

Bloor, D., Chance, R. R. (1985): *Polydiacetylenes: Synthesis, Structure and Electronic Properties*, Kluwer Academic, Hingham, Massachusetts.

Bolker, H. I. (1974): *Natural and Synthetic Polymers*, Marcel Dekker, New York.

Bovy, F., Winslow, F. (1977): *Macromolecules*, Academic, New York.

Briston, J. H. (1974): *Plastic Fibers*, Halsted, New York.

Brown, R. P. (1989): *Handbook of Plastics Test Methods*, 2nd ed., Wiley, New York.

Brydson, J. A. (1989): *Plastic Materials*, 5th ed., Newmes-Butterworth, Kent, England.

Carraher, C. E., Gebelein, C. G. (1982): *Biological Activities of Polymers*, ACS, Washington, D.C.

Carraher, C. E., Seymour, R. B. (1985): Fundamentals of polymer science, in *Applied Polymer Science*, ACS Washington, D.C.

Challa, I. (1993): *Polymer Chemistry: An Introduction*, Prentice-Hall, Englewood Cliffs, New Jersey.

Collinan, T. D., Navitz, A. E. (1972): *Fundamentals of Polymer Chemistry*, Hayden, New York.

Cowie, J. M. G. (1991): *Polymers: Chemistry and Physics of Modern Materials*, 2nd ed., Blackie, London.

Cross, J. A. (1974): *Plastics: Resource and Environmental Profile Analysis*, Manufacturing Chemists Assoc., Washington, D.C.

Culbertson, B. M., Pittman, C. U. (1983): *New Monomers and Polymers*, Plenum, New York.

Dubois, J. H. (1972): *Plastics History U.S.A.*, Cahners Books, Boston.

Dubois, J. H., John, F. W. (1974): *Plastics*, Van Nostrand-Reinhold, New York.

Elias, H. G. (1984): *Macromolecules*, Vols. 1 and 2, Plenum, New York.

Flory, P. J. (1953): *Principles of Polymer Chemistry*, Chap. 1, Cornell University Press, Ithaca, New York.

Gebelein, C. G., Carraher, C. E. (1985): *Bioactive Polymeric Systems: An Overview*, Plenum, New York.

———. (1985): *Polymeric Materials in Medication*, Plenum, New York.

Glanville, A. B. (1973): *The Plastics Engineers Data Book.* Industrial Press, New York.

Golding, B. (1959): *Polymers and Resins*, Van Nostrand, Princeton, New Jersey.

Goodman, S. H. (1986): *Handbook of Thermoset Plastics*, Noyes, Park Ridge, New Jersey.

Hall, C. (1981): *Polymer Materials*, Wiley, New York.

Harper, C. A. (1975): *Handbook of Plastics and Elastomers*, McGraw-Hill, New York.

Heinisch, K. F. (1974): *Dictionary of Rubber*, Halsted, New York.

Huggins, M. A. (1976): *Annual Review of Material Science*, Annual Reviews, Inc., Palo Alto, California.

Iwayanngi, T., et al. (1989): *Polymers in Microlithography*, ACS, Washington, D.C.

Jenkins, A. D. (1972): *Polymer Science*, American Elsevier, New York.

Kaufman, H., Falcetta, V. (1977): *Introduction to Polymer Science and Technology*, Wiley, New York.

Kirshenbaum, G. S. (1973): *Polymer Science Study Guide*, Gordon and Breach, New York.

Korschwitz, J. (1990): *Polymer Characterization and Analysis*, Wiley, New York.

Lai, J. H. (1989): *Polymers for Electronic Applications*, CRC, Boca Raton, Florida.

Lee, L. (1990). *Adhesive Bonding*, Plenum, New York.

Lemstra, P. J. (1989): *Integration of Fundamental Polymer Science and Technology*, Elsevier, New York.

Lynch, C. T. (1976): *Handbook of Material Science*, CRC Press, Cleveland, Ohio.

Mandelkern, L. L. (1983): *An Introduction to Macromolecules*, 2nd ed., Springer-Verlag, New York.

Mark, H. F. (1985): *Applied Polymer Science*, Chap. 1 (R. W. Tess and G. Paehlein, Eds.), Polymeric Materials Division of ACS, Washington, D.C.

McArdle, C. B. (1989): *Side Chain Liquid Crystal Polymers*, Routledge Chapman Hall, Boston.

Mebane, R., Rybolt, T. (1995): *Plastics and Polymers*, TFC Pub., New York.

Misral, G. S. (1993): *Introductory Polymer Chemistry*, Halsted, New York.

Moncreif, R. W. (1975): *Manmade Fibers*, Halsted, New York.

Morawetz, H. (1985): *Origins and Growth of a Science, Polymers*, Wiley, New York.

Morton, M. (1987): *Rubber Technology*, 3rd ed., Van Nostrand-Reinhold, New York.

Odian, G. (1991): *Principles of Polymerization*, Wiley, New York.

Parker, D. B. V. (1975): *Polymer Chemistry*, Applied Science, Essex, England.

Paul, D. R., Sperling, L. H. (1986): *Multicomponent Polymer Materials*, ACS, Washington, D.C.

Plate, N. A., Shibaev, V. P. (1987): *Comb-Shaped Polymers and Liquid Crystals*, Plenum, New York.

Powers, P. O. (1943): *Synthetic Resins and Rubbers*, Chap. 1, John Wiley, New York.

Prasad, P., Nigam, J. (1992): *Frontiers of Polymer Research*, Plenum, New York.

Rodriguez, F. (1989): *Principles of Polymer Systems*, 3rd ed., Hemisphere, New York.

Rosen, S. L. (1993): *Fundamental Principles of Polymeric Materials*, 2nd ed. Wiley, New York.

Rudin, A. (1982): *Elements of Polymer Science and Engineering*, Academic Press, New York.

Saunders, K. J. (1988): *Organic Polymer Chemistry*, Routledge Chapman & Hall, Boston.

Seeman, J. (1993): *Herman Mark—From Small Organic Molecules to Large: A Century of Progress*, ACS, Washington, D.C.

Seymour, R. B. (1971): *Introduction to Polymer Chemistry*, Chap. 1, McGraw-Hill, New York.

———. (1975): *Modern Plastics Technology*, Reston Publishing, Reston, Virginia.

———. (1982). *Plastics vs Corrosives*, Wiley, New York.

———. (1990). *Engineering Polymer Sourcebook*, McGraw-Hill, New York.

Seymour, R. B., Carraher, C. E. (1984): *Structure-Property Relationships in Polymers*, Plenum, New York.

Seymour, R. B., Carraher, C. E. (1990): *Giant Molecules*, Wiley, New York.

Seymour, R. B., Kirshenbaum, G. S. (1986): *High Performance Polymers*, Elsevier, New York.

Seymour, R. B., Mark, H. F. (1988): *Applications of Polymers*, Plenum, New York.

Seymour, R. B., Stahl, G. (1982): *Macromolecular Solutions*, Pergamon, Elmsford, New York.

———. (1987): *Pioneers in Polymer Science*, Litarvan Lit., Denver, Colorado.

Sperling, L. H. (1986): *Introduction to Physical Polymer Science*, Wiley, New York.

Staudinger, H. (1932): *Die Hochmolekularen*, Springer-Verlag, Berlin.

Stevens, M. P. (1990): *Polymer Chemistry*, 2nd ed. Oxford University Press, Oxford, England.

Sun, S. (1994): *Chemistry of Macromolecules: Basic Principles and Issues*, Wiley, New York.

Tadokoro, H. (1990): *Structure of Crystalline Polymers*, Krieger.

Treloar, L. R. G. (1975): *Introduction to Polymer Science*, Wykeham, England.

Tschoegl, N. W. (1989): *Phenomenological Theory of Linear Viscoelastic Behavior*, Springer-Verlag, New York.

Ulrich, H. (1988): *Raw Materials for Industrial Polymers*, Oxford University Press, Oxford, England.

Van Krevelen, D. W. (1972): *Properties of Polymers*, American Elsevier, New York.

Vogl, O. (1988): *Polymer Science in the Next Decade*, Wiley, New York.

Yasuda, H. (1985): *Plasma Polymerization*, Academic Press, New York.

Young, R. J. (1981): *Introduction to Polymers*, Chapman and Hill, London.

2

Polymer Structure (Morphology)

2.1   STEREOCHEMISTRY OF POLYMERS

High-density polyethylene (HDPE), formerly called low-pressure polyethylene [$H(CH_2CH_2)_nH$], like other alkanes [$H(CH_2)_nH$], may be used to illustrate polymer structure. As in introductory organic chemistry, we can comprehend almost all the complex organic compounds if we understand the simplest chemistry, i.e., alkane chemistry.

High-density polyethylene, like decane [$H(CH_2)_{10}H$] or paraffin [$H(CH_2)_{\sim 50}H$], is a linear chainlike molecule consisting of catenated carbon atoms bonded by covalent bonds. The carbon atoms in all alkanes, including HDPE, are joined at characteristic tetrahedral bond angles of approximately 109.5°. While decane consists of 10 methylene groups (CH_2), HDPE may contain more than 1000 of these groups. While we use the term normal straight chain or linear for alkanes, we know that, because of the characteristic bond angles, the chains are zigzag-shaped.

The distance between the carbon atoms is 1.54 angstroms (Å) or 0.154 nanometers (nm). The apparent zigzag distance between carbon atoms in a chain of many carbon atoms is 1.26 Å, or 0.126 nm. Therefore, the length of an extended nonane chain would be 8 (1.26 Å), or 10.08 Å, or 1.008 nm. Likewise, the length of an extended chain of HDPE having 1000 repeat ethylene units or structural elements [$H(CH_2CH_2)_{1000}H$ or $H(CH_2)_{2000}H$] would be 2520 Å or 252 nm. However, as shown by the magnified simulated structure in the diagram for HDPE (Fig. 2.1, top) because of rotation of the carbon-carbon bonds, these chains are seldom extended to their full contour length but are present in many different shapes, or *conformations*. Since the size of the hydrogen atoms and bond angles are less significant in giant molecules, they are not usually shown in HDPE and other polymers.

17

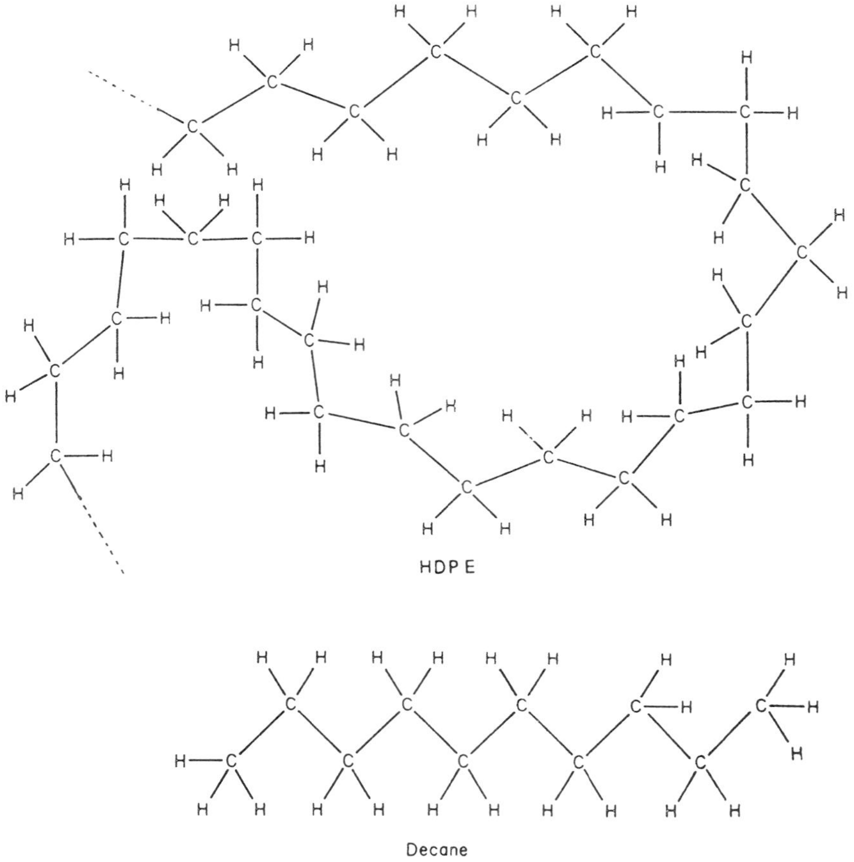

Figure 2.1 Magnified simulated structure of high-density polyethylene (HDPE), contrasted with the structural formula of decane.

Problem

Determine the contour length of a polyethylene chain 1300 ethylene units long given the average zigzag distance between carbon atoms of 0.126 nm. There are two carbon atoms within the polymer chain backbone per repeat unit. Thus, per unit the average contour length is 2×0.126 nm $= 0.252$ nm. The contour length is then 0.252 nm/unit $\times$ 1300 units $\simeq 330$ nm.

Each specific protein molecule has a specific molecular weight, like the classic small molecules, and is said to be monodisperse. However, commercial synthetic polymers, such as HDPE, are made up of molecules of different molecular weights. Thus, the numerical value for n, or the degree of polymerization (DP), should be considered as an average DP and designated with an overbar, i.e., $\overline{DP}$. Accordingly, the average molecular weight ($\overline{M}$) of a polydisperse polymer will equal the product of $\overline{DP}$ and the molecular weight of the repeating unit or mer.

In classic organic chemistry, it is customary to call a nonlinear molecule, like isobutane, a branched chain. However, the polymer chemist uses the term pendant group to label any group present on the repeating units. Thus, polypropylene

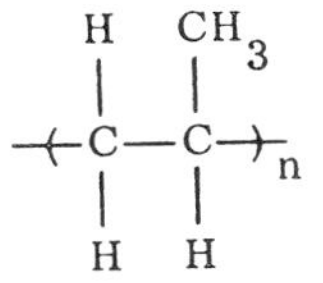

has a pendant methyl group but is designated as a linear polymer. In contrast, low-density polyethylene (LDPE), which was formerly called high-pressure polyethylene, is a branched polymer because chain extensions or branches of polyethylene sequences are present on branch points, irregularly spaced along the polymer chain, as shown in Fig. 2.2. The number of branches in nonlinear polyethylene (LDPE) may vary from 1 per 20 methylene groups to 1 per 100 methylene groups. This branching, like branching in simple alkanes like isobutane, increases the specific volume and thus reduces the density of the polymer.

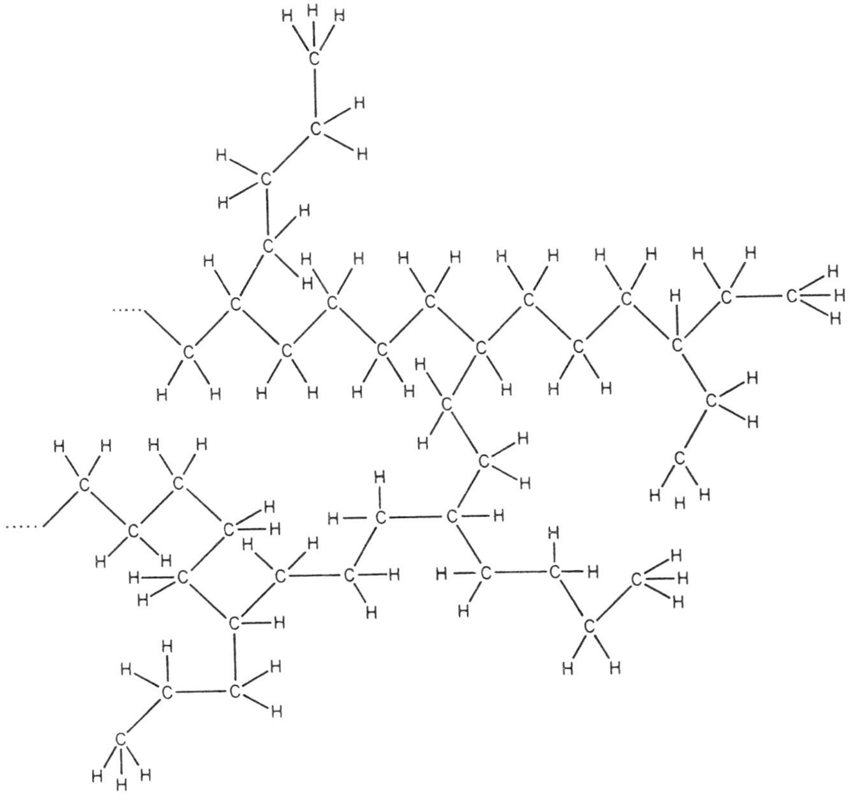

Figure 2.2 Simulated structural formula of branched low-density polyethylene (LDPE; compare to Fig. 2.1, HDPE).

Recently, low-pressure processes have been developed which produce linear low-density polyethylene (LLDPE). LLDPE is largely linear but does have some branching. The linearity provides strength while the branching provides toughness (Table 2.1).

Problem

Determine the approximate number of repeat units (degree of polymerization) for a polypropylene chain with a molecular weight of 5.4×10^4. The formula weight of a polypropylene unit is 42 amu. The number of repeat units = 54,000 amu/42 amu/unit $\simeq$ 1300 units.

Both linear and branched polymers are thermoplastics. However, cross-linked three-dimensional, or network, polymers are thermoset polymers. As shown in Fig. 2.3, the cross-linked density may vary from the low cross-linked density in vulcanized rubber to the high cross-linked density observed in ebonite.

While there is only one possible arrangement for the repeat units in HDPE, these units in polypropylene (PP) and many other polymers may be arranged in a head-to-tail or a head-to-head configuration, as shown in Fig. 2.4. Fortunately, the usual arrangement is head to tail, so that the pendant groups are usually on every other carbon atom in the chain.

Proteins are polyamides in which the building units consist of the L (levo) optical isomers of amino acids. In contrast, the building units in starch and cellulose are D (dextro)-glucose, which are joined by α and β acetal groups. All D and L enantiomers are structurally related to the D and L glyceraldehyde structures shown in Fig. 2.5.

In the early 1950s, Nobel laureate Giulio Natta used stereospecific coordination catalysts to produce stereospecific isomers of polypropylene. Natta used the term tacticity to describe the different possible structures. As shown in Fig. 2.6, the isomer corresponding to the arrangement DDDD or LLLL is called isotactic PP. The isomer corresponding to the arrangement DLDL is called syndiotactic PP, and polypropylene having a random arrangement of building units corresponding to DDLDDLD, etc., is called atactic PP. Isotactic PP, which is available commercially, is a highly crystalline polymer with a melting point of 160°C, while the atactic isomer is an amorphous (noncrystalline) soft polymer with a melting point of 75°C. The term eutactic is used to describe either an isotactic polymer, a syndiotactic polymer, or a mixture of both.

While most polymers contain only one chiral or asymmetric center in the repeating units, it is possible to have diisotacticity when two different substituents

Table 2.1 Types of Commercial Polyethylene

	General structure	Crystallinity (%)	Density (g/cc)
LDPE	Linear with branching	ca. 50	0.92–0.94
LLDPE	Linear with less branching	50	0.92–0.94
HDPE	Linear with little branching	90	0.95

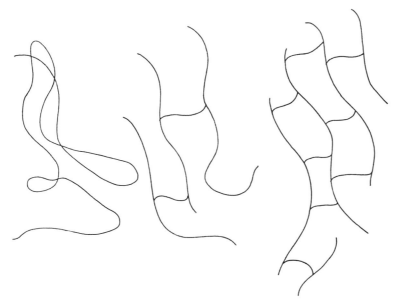

Figure 2.3 Simulated skeletal structural formulas of a linear polymer (left) and network polymers with low (middle) and high (right) cross-linked density.

(R and R′) are present at the chiral centers. These isomers are labeled erythro- and threodiisotactic and erythro- and threosyndiotactic isomers as shown in Fig. 2.7.

The many different conformers resulting from rotation about the carbon-carbon bond in a simple molecule like n-butane [H(CH₂)₄H] may be shown by Newman projections. As shown in Fig. 2.8, the most stable form is the anti or trans (t) conformer in which the two methyl groups (Me) are as far apart as possible. The

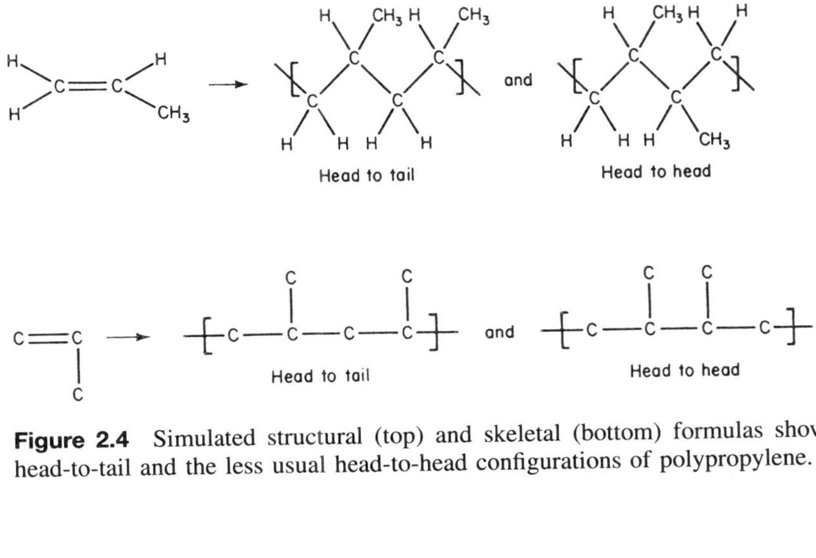

Figure 2.4 Simulated structural (top) and skeletal (bottom) formulas showing the usual head-to-tail and the less usual head-to-head configurations of polypropylene.

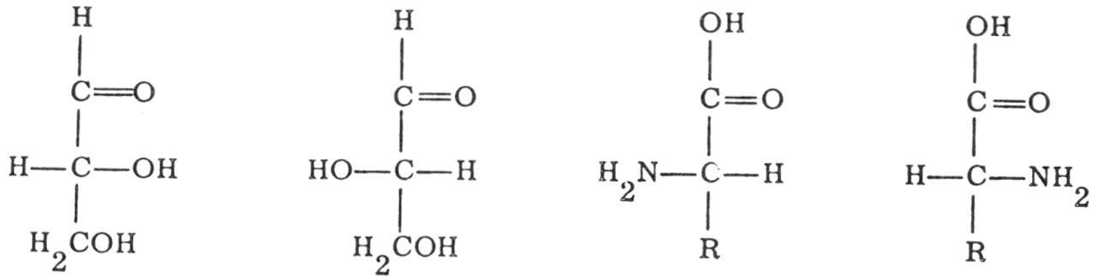

D-Glyceraldehyde L-Glyceraldehyde L-Amino acids D-Amino acids

Figure 2.5 Optical isomers of glyceraldehyde and amino acids.

difference in energy between the anti and eclipsed conformer is at least 3 kcal, and, of course, there are numerous conformations between these two extremes (0 and 180°). Among these are the two mirror-image gauche (g) conformers in which the methyl groups are 60° apart.

In a polymer such as HDPE, the methyl groups shown in Fig. 2.8 would be replaced by methylene groups in the chain. The flexibility in a polymer would be related to the ease of conversion from t to g. This ease is dependent on the lack of hindering groups and increased temperature. Thus, poly(methyl methacrylate) (PMMA) is hard at room temperature because of the polar ester groups that restrict rotation. In contrast, polyisobutylene is flexible at room temperature. The flexibility of both polymers will be increased as the temperature is increased.

2.2 MOLECULAR INTERACTIONS

The forces present in nature are often divided into primary forces (typically greater than 50 kcal/mol of interactions) and secondary forces (typically less than 10 kcal/mol of interactions). Primary bonding forces can be further subdivided into ionic (characterized by a lack of directional bonding; between atoms of largely

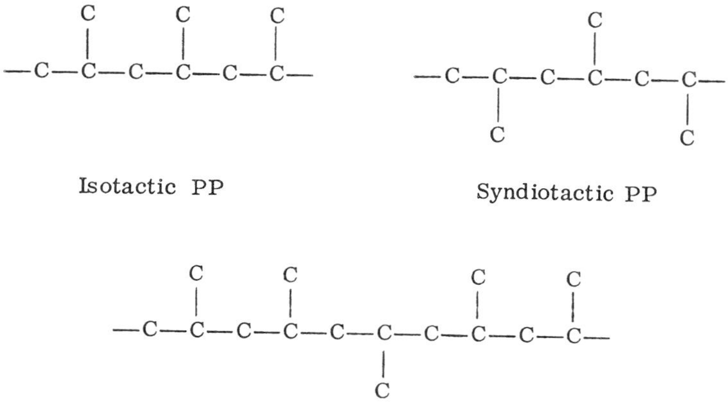

Isotactic PP Syndiotactic PP Atactic PP

Figure 2.6 Skeletal formulas of isotactic, syndiotactic, and atactic polypropylene (PP).

R   R' R   R'
| | | |
—C—C—C—C—

Erythrodiisotactic

```
    R       R
    |       |
—C—C—C—C—
    |       |
    R'      R'
```

Threodiisotactic

```
    R   R''
    |   |
—C—C—C—C—
        |   |
        R   R'
```

Erythrodisyndiotactic

```
    R       R'
    |       |
—C—C—C—C—
        |   |
        R'  R
```

Threodisyndiotactic

Figure 2.7   Skeletal formulas of ditactic isomers.

differing electronegativities; not typically present within polymer backbones), metallic (the number of outer, valence electrons is too small to provide complete outer shells; often considered as charged atoms surrounded by a potentially fluid sea of electrons; lack of bonding direction; not typically found in polymers), and covalent (including coordinate and dative) bonding (which are the major means of bonding within polymers; directional). The bonding lengths of primary bonds are usually about 0.90 to 2.0 Å (0.09–0.2 nm) with the carbon-carbon bond length being about 1.5 to 1.6 Å (0.15–0.16 nm).

Secondary forces, frequently called van der Waals forces, since they are the forces responsible for the van der Waals corrections to the ideal gas relationships, are of longer distance in interaction, generally having significant interaction between 2.5 and 5 Å (0.25–0.5 nm). The force of these interactions is inversely proportional to some power of r, generally 2 or greater [force $\propto 1/(\text{distance})^r$] and thus is quite dependent on the distance between the interacting molecules. Thus, many physical properties of polymers are indeed quite dependent on both the conformation (arrangements related to rotation about single bonds) and configuration (arrangements related to the actual chemical bonding about a given atom), since

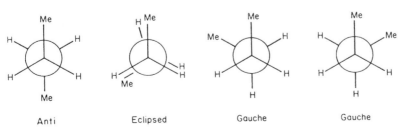

Figure 2.8   Newman projections of designated conformers of n-butane.

both affect the proximity one chain can have relative to another. Thus, amorphous polypropylene is more flexible than crystalline polypropylene (compare linear polymers a and b of Fig. 2.9).

Atoms in individual polymer molecules are thus joined to each other by relatively strong covalent bonds. The bond energies of the carbon-carbon bonds are on the order of 80 to 90 kcal/mol. Further, polymer molecules, like all other molecules, are attracted to each other (and for long-chain polymer chains even between segments of the same chain) by intermolecular, secondary forces.

These intermolecular forces are also responsible for the increase in boiling points within a homologous series such as the alkanes, for the higher than expected boiling points of polar organic molecules such as alkyl chlorides, and for the abnormally high boiling points of alcohols, amines, and amides. While the forces

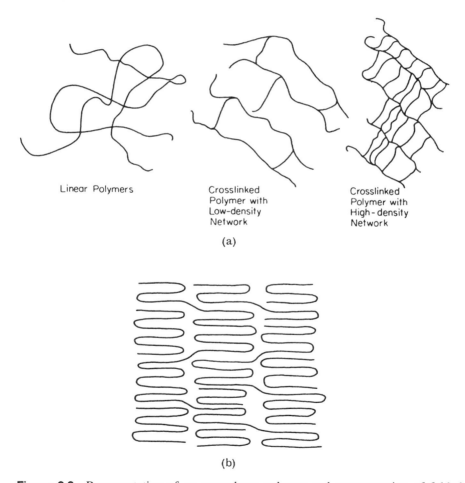

Linear Polymers   Crosslinked   Crosslinked
                         Polymer with   Polymer with
                         Low-density   High-density
                         Network   Network

(a)

(b)

Figure 2.9 Representation of an amorphous polymer and representation of folded polymer chains in polymer crystals. (From *Modern Plastics Technology* by R. Seymour, 1975, Reston Publishing Company, Reston, Virginia. Used with permission of Reston Publishing Company.)

responsible for these increases in boiling points are all called van der Waals forces, these forces are subclassified in accordance with their source and intensity. Secondary, intermolecular forces include London dispersion forces, induced permanent forces, and dipolar forces, including hydrogen bonding.

Nonpolar molecules such as ethane [$H(CH_2)_2H$] and polyethylene are attracted to each other by weak London or dispersion forces resulting from induced dipole-dipole interaction. The temporary or transient dipoles in ethane or along the polyethylene chain are due to instantaneous fluctuations in the density of the electron clouds. The energy range of these forces is about 2 kcal per mole unit in nonpolar and polar polymers alike, and this force is independent of temperature. These London forces are typically the major forces present between chains in largely nonpolar polymers present in elastomers and soft plastics.

It is of interest to note that methane, ethane, and ethylene are all gases; hexane, octane, and nonane are all liquids (at room conditions); while polyethylene is a waxy solid. This trend is primarily due to both an increase in mass per molecule and to an increase in the London forces per molecule as the chain length increases. Assuming that the attraction between methylene or methyl units is 2 kcal/mol, we calculate an interaction between methane to be 2 kcal/mol, hexane to be 12 kcal/mol, and for a mole of polyethylene chains of 1000 units to be 4000 kcal.

Problem

Determine the approximate interaction present for a single polyethylene chain of 1500 repeat units within a liquid hexane solution assuming that the interactions are about 2 kcal/mol repeat methylene unit. There are 6×10^{23} units per mol; thus the interactive energy per methylene moiety is 2000 cal/mol/6×10^{23} units/mol = 3.3×10^{-21} cal/unit. There are two methylene units per repeat unit, so there are 2×1500 or 3000 methylene units. The interaction present in the polyethylene is then about 3000 units $\times$ 3.3×10^{-21} cal/unit $\simeq 1 \times 10^{-17}$ cal.

Polar molecules such as ethyl chloride (H_3C-CH_2Cl) and poly(vinyl chloride) [$-(CH_2-CHCl)_n$, PVC, see Fig. 2.10] are attracted to each other by dipole-dipole interactions resulting from the electrostatic attraction of a chlorine atom in one

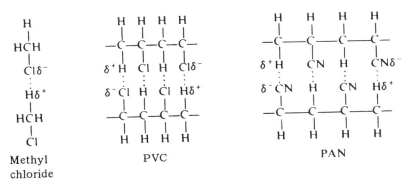

Methyl chloride   PVC   PAN

Figure 2.10 Typical dipole-dipole interaction between molecules of methyl chloride and segments of chains of poly(vinyl chloride) (PVC) and polyacrylonitrile (PAN). (From *Modern Plastics Technology* by R. Seymour, 1975, Reston Publishing Company, Reston, Virginia. Used with permission of Reston Publishing Company.)

molecule to a hydrogen atom in another molecule. Since this dipole-dipole inter-action, which ranges from 2 to 6 kcal per mol repeat unit in the molecule, is temperature dependent, these forces are reduced as the temperature is increased in the processing of polymers. While the London forces are typically weaker than the dipole-dipole forces, the former are also present in polar compounds, such as ethyl chloride and PVC. These dipole-dipole forces are characteristic of many plastics.

Strong polar molecules such as ethanol, poly(vinyl alcohol), and cellulose are attracted to each other by a special type of dipole-dipole interaction called hydrogen bonding, in which the oxygen or nitrogen atoms in one molecule are attracted to the hydrogen atoms in another molecule. These are the strongest of the intermolecular forces and may have energies as high as 10 kcal per mol repeat unit. Intermolecular hydrogen bonds are usually present in fibers, such as cotton, wool, silk, nylon, Acrylan, polyesters, and polyurethanes. Intramolecular hydrogen bonds are responsible for the helices observed in starch and globular proteins.

It is important to note that the high melting point of nylon-66 (265°C, Fig. 2.11) is the result of a combination of London, dipole-dipole, and hydrogen bonding forces between the polyamide chains. The hydrogen bonds are decreased when the hydrogen atoms in the amide groups in nylon are replaced by methyl groups and when the hydroxyl groups in cellulose are esterified.

In addition to the contribution of intermolecular forces, chain entanglement is also an important contributory factor to the physical properties of polymers. While paraffin wax and HDPE are homologs with relatively high molecular weights, the chain length of paraffin is too short to permit entanglement and hence it lacks the strength and other characteristic properties of HDPE.

The critical chain length (z) required for the onset of entanglement is dependent on the polarity and shape of the polymer. Thus, the number of atoms in the critical chain lengths of poly(methyl methacrylate) (PMMA), polystyrene (PS), and polyisobutylene are 208, 730, and 610, respectively. The melt viscosity (η) of a polymer is often found to be proportional to the 3.4 power of the critical chain length as related in Eq. (2.1), regardless of the structure of the polymer. The constant K is temperature dependent.

$$\log \eta = 3.4 \log z + \log K \tag{2.1}$$

Viscosity is a measure of the resistance to flow. The latter, which is the result of cooperative movement of the polymer segments from hole to hole in a melt, is

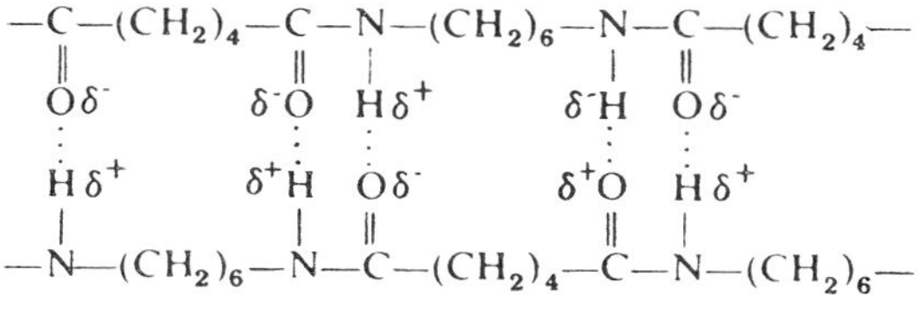

Figure 2.11 Typical hydrogen bonding between hydrogen and oxygen or nitrogen atoms in nylon-66. (From *Modern Plastics Technology*, by R. Seymour, 1975, Reston Publishing Company, Reston, Virginia. Used with permission of Reston Publishing Company.)

impeded by chain entanglement, intermolecular forces, the presence of reinforcing agents, and cross-links.

The *flexibility* of amorphous polymers above the glassy state is governed by the same forces as melt viscosity and is dependent on a wriggling type of segment motion in the polymer chains. This flexibility is increased when many methylene groups (CH_2) or oxygen atoms are present between stiffening groups in the chain. Thus, the flexibility of aliphatic polyesters usually increases as m is increased.

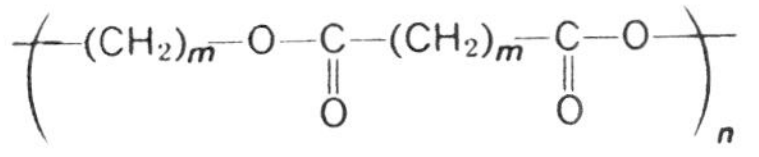

Aliphatic polyester

In contrast, the flexibility of amorphous polymers above the glassy state is decreased when stiffening groups such as

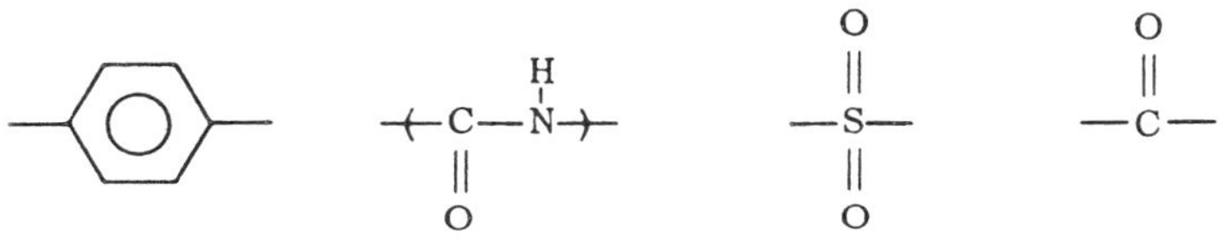

p–phenylene   amide   sulfone   carbonyl

are present in the polymer backbone. Thus, poly(ethylene terephthalate) is stiffer and higher melting than poly(ethylene adipate), and the former is stiffer than poly(butylene terephthalate) because of the presence of fewer methylene groups between the stiffening groups.

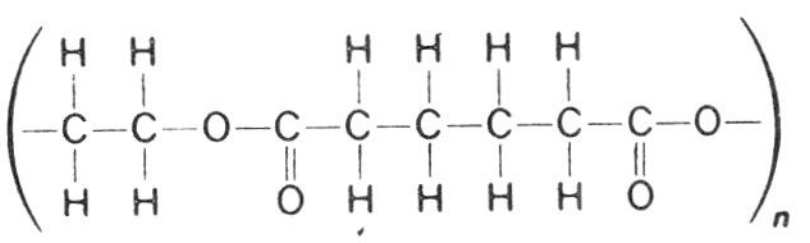

Poly(ethylene adipate) 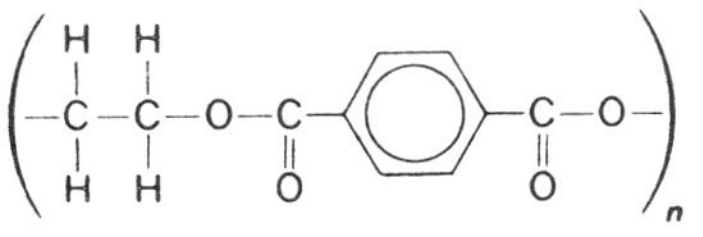   Poly(ethylene terephthalate)

The flexibility of amorphous polymers is reduced drastically when they are cooled below a characteristic transition temperature called the *glass transition temperature* (T_g). At temperatures below T_g, there is no segmental motion and any dimensional changes in the polymer chain are the result of temporary distortions of the primary valence bonds. Amorphous plastics perform best below T_g, but elastomers must be used above the brittle point, or T_g.

The *melting point* (T_m) is called a first-order transition temperature, and T_g is sometimes called a second-order transition temperature. The values for T_m are usually 33 to 100% greater than T_g, and symmetrical polymers like HDPE exhibit the greatest difference between T_m and T_g. As shown by the data in Table 2.2, the T_g values are low for elastomers and flexible polymers and relatively high for hard amorphous plastics.

As shown in Table 2.2, the T_g value of isotactic polypropylene (PP) is 373°K or 100°C, yet because of its high degree of crystallinity it does not flow to any great extent below its melting point of 438°K (165°C). In contrast, the highly amorphous polyisobutylene, which has a T_g value of 203°K (−70°C), flows at room

Table 2.2 Approximate Glass Transition Temperatures (T$_g$) for Selected Polymers

Polymer	T$_g$ (°K)
Cellulose acetate butyrate	323
Cellulose triacetate	430
Polyethylene (LDPE)	148
Polypropylene (atactic)	253
Polypropylene (isotactic)	373
Polytetrafluoroethylene	160, 400[a]
Poly(ethyl acrylate)	249
Poly(methyl acrylate)	279
Poly(butyl methaceylate) (atactic)	339
Poly(methyl methacrylate) (atactic)	378
Polyacrylonitrile	378
Poly(vinyl acetate)	301
Poly(vinyl alcohol)	358
Poly(vinyl chloride)	354
Cis-poly-1,3-butadiene	165
Trans-poly-1,3-butadiene	255
Poly(hexamethylene adipamide) (nylon-66)	330
Poly(ethylene adipate)	223
Poly(ethylene terephthalate) (PET)	342
Polydimethylsiloxane (silicone)	150
Polystyrene	373

[a]Two major transitions observed.

temperature. Also, as shown in Table 2.2, T$_g$ decreases as the size of the ester groups increases in polyacrylates and polymethacrylates. The effect of the phenylene stiffening group is also demonstrated by the T$_g$ of poly(ethylene terephthalate), which is 119°K higher than that of poly(ethylene adipate).

Since the specific volume of polymers increases at T$_g$ in order to accommodate the increased segmental chain motion, T$_g$ values may be estimated from plots of the change in specific volume with temperature. Other properties, such as stiffness (modulus), refractive index, dielectric properties, gas permeability, X-ray adsorption, and heat capacity, all change at T$_g$. Thus, T$_g$ may be estimated by noting the change in any of these values, such as the increase in gas permeability. Since the change in the slope of the specific volume-temperature or index of refraction-temperature curves is not always obvious, it is best to extrapolate the curves linearly and designate the intersection of these curves as T_g, as shown in Fig. 2.12.

As shown in Fig. 2.13, values for both T_g and T_m are observed as endothermic transitions in calorimetric measurements, such as differential thermal analysis (DTA) or differential scanning calorimetry (DSC). It is important to note that since the values observed for T_g are dependent on the test method and on time, the values obtained by different techniques may vary by a few degrees. While the T_g value reported in the literature is related to the onset of segmental motion in the principal chain of polymer backbone, separate values, called α, β, . . . or secondary, tertiary T_g, may be observed for the onset of motion of large pendant groups or branches on the polymer chain.

While no motion exists, except the stretching or distortion of covalent bonds, at temperatures below T_g, the onset of segmental motion leads to many different conformations. Thus, the full contour length (nl) of a polymer chain obtained by

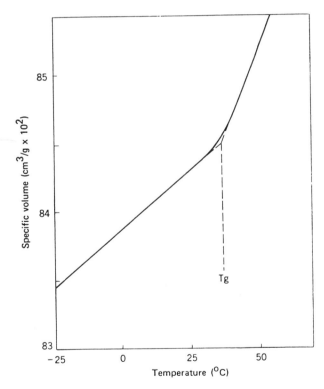

Figure 2.12 Determination of T_g by noting abrupt change in specific volume. Please remember that values such as those appearing here for specific volume are 1/100 of the values shown as designated by the multiplier 10^2. For example, the value where the break in the curve occurs is not 84.5 or 8450, but is 0.845. (From *Introduction to Polymer Chemistry* by R. Seymour, 1971, McGraw-Hill, New York. Used with permission of McGraw-Hill Book Company.)

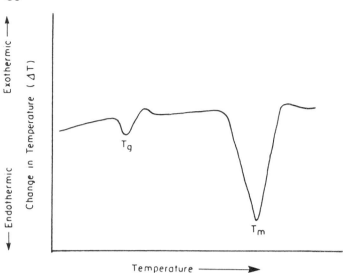

Figure 2.13 Typical DTA thermogram of a polymer.

multiplying the length of each mer, or repeat unit (l), by the number of units in the chain (n) provides a value of the length of only one of the many possible conformers present.

It is not always possible or generally useful to calculate the length of other conformers, but it is important to know the average end-to-end distance of polymer chains. The statistical method for this determination, called the random flight technique, was developed by Lord Raleigh in 1919. The classic statistical approach may be used to show the distance traveled by a blindfolded person taking n number of steps of length l in a random walk or the distance flown by a confused moth or bird.

The distance traveled from start to finish is not the straight-line path measured as nl but the root-mean-square distance $(\sqrt{\bar{r}^2})$, which is equal to l $\sqrt{n}$. Nobel laureate Paul Flory and others have introduced several corrections so that this random flight technique could be applied to polymer chains having a full contour length of ln.

When we calculate the distance values for HDPE [H(CH$_2$CH$_2$)$_n$H], where DP or n equals 1000 using a C—C bond length of 1.26 Å or 0.126 nm, we will find approximate values of ln of 252 nm, i.e., [0.126(2)(1000)], and of l $\sqrt{n}$ of 8.1 nm, i.e., (0.256 $\sqrt{1000}$). Thus, the calculated root-mean-square distance $\sqrt{\bar{r}^2}$, where r is the vector distance from end to end, is only about 3% of the full contour end-to-end distance.

Since there are restrictions in polymer chain motions that do not apply to the blind-folded walker, corrections must be made that increase the value found by the Raleigh technique. Thus, the value of $\sqrt{\bar{r}^2}$ increases from 8.1 to 9.8 nm when one corrects for the fixed tetrahedral angles in the polymer chain.

A still higher value of 12.2 nm is obtained for the root-mean-square end-to-end distance when one corrects for the hindrance to motion caused by the hydrogen atoms. Since the hydrogen atoms of the first and fifth carbon atom overlap when the methylene groups assume a cyclopentanelike shape, another correction must be made for this so-called pentane interference. The corrected value for $\sqrt{\bar{r}^2}$ is 18.0 nm.

Problem

Determine the average (root-mean-square average) distance for polypropylene chains with DP = 1300. The end-to-end distance between carbon atoms is 0.126 nm or (2) (0.126 nm) for each ethylene unit. The relationship between the root-mean-square distance, $\sqrt{\bar{r}^2}$, and number of carbon-carbon distances, n, is $\sqrt{\bar{r}^2} = \ell \sqrt{n}$.

$$\sqrt{\bar{r}^2} = 0.256 \sqrt{1300} \approx 9.2 \text{ nm} \tag{2.2}$$

This is much less than the contour length of 330 nm (see problem, page 18).

While corrections should also be made for the excluded volume, the approximate value of 18.0 nm can be used for the root-mean-square, end-to-end distance. The excluded volume results from the fact that in contrast to the blindfolded walker who may backtrack without interference, only one atom of a three-dimensional carbon-carbon chain may occupy any specific volume at any specified time, and thus the space occupied by all other atoms must be excluded from the walker's available path.

The number of possible conformers increases with chain length and can be shown statistically to equal 2^{2n}. Thus, when n = 1000, the number of possible conformers of HDPE is 2^{2000}, or 10^{600}. As shown in Fig. 2.14, the end-to-end distance (r) of a linear molecule such as HDPE may be readily visualized and must be viewed statistically as an average value.

However, since there are many ends in a branched polymer, it is customary to use the radius of gyration (S) instead of r for such polymers. The radius of gyration

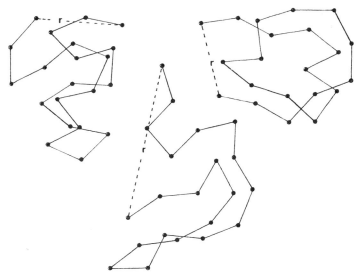

Figure 2.14 End-to-end distances (r) of linear polymer chains containing the same number of units.

is actually the root-mean-square distance of a chain end from the polymer's center of gravity. S is less than the end-to-end distance (r), and for linear polymers, $\bar{r}^2$ is equal to $6\bar{S}^2$.

In addition to the restrictions to free rotation noted for HDPE, free rotation of polymer chains will be hindered when the hydrogen atoms in polyethylene are replaced by bulky groups. Because the energy barrier (E) restricting the rotation from trans to gauche conformers is low (3 kcal per mer) in HDPE, these polymers are flexible, and this flexibility increases with temperature (T) in accordance with the Arrhenius equation shown in Eq. (2.3). The flexibility is related to the orientation time (τ_m), which is a measure of the ease of uncoiling of polymer coils. The constant A is related to the polymer structure, and R is the ideal gas constant.

$$\tau_m = A \, e^{E/RT} \quad \text{or} \quad \log \tau_m = \log A + \frac{E}{2.3RT} \tag{2.3}$$

The bulky phenyl group in polystyrene (PS) restricts rotation, and hence its T_g and τ_m are higher than the values for HDPE. When substituents such as chlorine atoms are present in polystyrene, T_g and τ_m values are even higher. Likewise, aromatic nylons, called aramids, have greater T_g and τ_m's than aliphatic nylons.

The intermolecular bonds in these polyamides and other fibers, including β-keratin, produce strong pleated sheets. Hair, fingernails and toenails, feathers, and horns have a β-keratin structure. Polyurethanes, polyacrylonitrile, and polyesters are characterized by the presence of strong hydrogen bonds. In contrast, isotactic polypropylene, which has no hydrogen bonds, is also a strong fiber as a result of the good fit of the regularly spaced methyl pendant groups on the chain. Since this molecular geometry is not present in atactic polypropylene (A-PP), it is not a fiber.

α-Keratin (composed of parallel polypeptide α-helices) and most globular proteins are characterized by intramolecular bonds. These and many other polymers, including nucleic acids, may form helices. Ribonucleic acid (RNA) exists as a single helix, while deoxyribonucleic acid (DNA) may exist as a double helix.

2.3 POLYMER CRYSTALS

Prior to 1920, leading chemistry researchers not only stated that macromolecules were nonexistent, but they also believed that the products called macromolecules, i.e., proteins, hevea elastomer, and cellulose, could not exist in the crystalline form. However, in the early 1920s, Haworth used X-ray diffraction techniques to show that elongated cellulose was a crystalline polymer consisting of repeat units of cellobiose. In 1925, Katz jokingly placed a stretched natural rubber band in an X-ray spectrometer and to his surprise observed an interference pattern typical of a crystalline substance.

This phenomenon may also be shown qualitatively by the development of opacity when a rubber band is stretched and by the abnormal stiffening and whitening of unvulcanized rubber when it is stored for several days at 0°C. The opacity noted in stretched rubber and cold rubber is the result of the formation of crystallites, or regions of crystallinity. The latter were first explained by a fringed micelle model which is now outmoded because single crystals of stereoregular polymers have been observed.

In contrast to the transparent films of amorphous polymers, relatively thick films of LDPE are translucent because of the presence of crystals. This opacity is readily eliminated when the film is heated above 100°C. It is of interest to note that Sauter produced single crystals of polymers in 1932, and Bunn produced single crystals of LDPE in 1939, but the existence of single crystals was not generally recognized until the 1950s, when three experimenters, namely, Fischer, Keller, and Till, reproduced Bunn's work independently.

It is now recognized that ordered polymers may form lamellar crystals with a thickness of 10 to 20 nm in which the polymer chains are folded back upon themselves to produce parallel chains perpendicular to the face of the crystals, as shown in Fig. 2.15.

Amorphous polymers with irregular bulky groups are seldom crystallizable, and unless special techniques are used, ordered polymers are seldom 100% crystalline. The rate of crystallization may be monitored by X-ray diffraction techniques or by dilatometry (measurement of change in volume).

The crystallites in polymers are small and usually organized into larger, shallow pyramidlike structures called spherulites, which may be seen by the naked eye and viewed as Maltese cross-like structures with polarized light and crossed Nicol prisms in a microscope, as shown in Fig. 2.16.

As in most crystallization processes, in the absence of nucleation, there is an induction period during which disentanglement of chains takes place. This step is

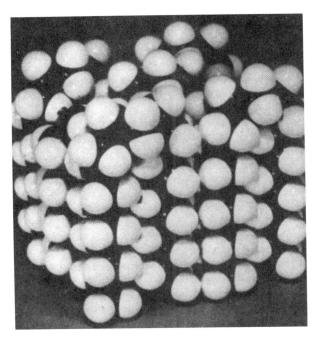

Figure 2.15 Model representation of a folded-chain lamellar crystal for polyethylene at the surface of a single crystal. (From P. Geil and D. Reneker, J. Appl. Physics, *31*:1921 (1960). With permission from the American Institute of Physics.)

Figure 2.16 Maltese cross–like pattern for spherulites viewed under a polarizing micro-scope with crossed Nicol prisms in a siliconelike polymer. The large and small spherulites are the result of crystallization occurring at different temperatures. [From F. Price, in *Growth and Perfection of Crystals* (R. Doremus, B. Roberts, and D. Turnbull, eds.). John Wiley, New York, 1958, p. 466. With permission from John Wiley and Sons, Publishers.]

followed by a slow rate of crystal growth. However, the rate of crystallization increases and then slows down towards the end of the crystallization process. The rate of crystalline growth may be followed by dilatometry using the Avrami equation, which was developed to follow the rate of crystallization of metals.

As shown by Eq. (2.4), the quotient of the difference between the specific volume V_t at time t and the final specific volume V_f divided by the difference between the original specific volume V_o and the final volume is equal to an experimental expression in which K is a kinetic constant related to the rate of nucleation and growth and n is an integer related to nucleation and growth of crystals. The value of n can vary (Table 2.3).

$$\frac{V_t - V_f}{V_o - V_f} = e^{-Kt^n} \tag{2.4}$$

It should be noted that reports of noninteger values for n have been made.

It is now believed that the crystalline and amorphous domains may be represented by the lamellae models shown in Fig. 2.17 or the related fringed-micelle model shown in Fig. 2.18.

The crystalline regions may be disrupted by processing techniques such as thermoforming and extrusion of plastics and biaxial orientation and cold drawing of fibers. In the last process, which is descriptive of the others, the crystallites are ordered in the direction of the stress, the filament shrinks in diameter (necks down),

Table 2.3 Avrami Values for Particular Crystallization Growth for Sporadic Nucleation

Crystallization growth pattern	n-Value
Fibril	2
Disc	3
Spherulite	4
Sheaf	6

and heat is evolved and reabsorbed as a result of additional orientation and crystallization.

In addition to crystallization of the backbone of polymers, crystallization may also occur in regularly spaced bulky groups even when an amorphous structure is maintained in the backbone. In general, the pendant group must contain at least 10 carbon atoms in order for this side-chain crystallization to occur. Rapid crystallization to produce films with good transparency may be brought about by the addition of a crystalline nucleating agent, such as benzoic acid, and by cooling rapidly.

Ordered polymers with small pendant groups crystallize more readily than those with bulky groups, such as poly(vinyl acetate), $-(CH_2CHOOCCH_3$. Thus, the hydrolytic product of the latter [poly(vinyl alcohol), $-CH_2-CHOH]$ crystallizes readily. Crystallization also occurs when different groups with similar size, like CH_2 and CH_3, are present (see Fig. 2.19).

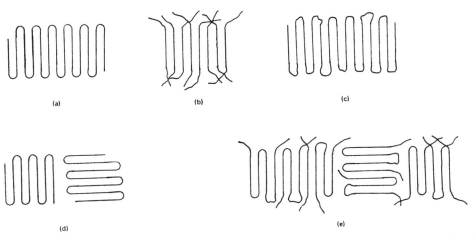

Figure 2.17 Schematic two-dimensional representations of models of the fold surface in polymer lamellae: (a) sharp folds, (b) switchboard model, (c) loops with loose folds, (d) buttressed loops, and (e) a combination of these features.

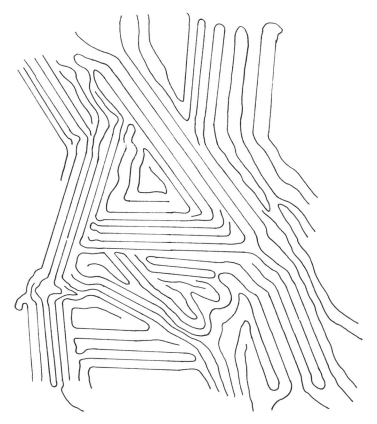

Figure 2.18 Schematic two-dimensional representation of a modified micelle model of the crystalline-amorphous structure of polymers incorporating features from Fig. 2.17.

While polymeric hydrocarbons have been used as illustrations for simplicity, it is important to note that the principles discussed apply to all polymers, organic as well as inorganic and natural as well as synthetic, and to elastomers, plastics, and fibers. The principal differences among the last are related to T_g, which is governed by the groups present in the chain and by pendant groups and the relative strength of the intermolecular bonds.

The precise mechanistic steps involved with polymer crystal formation are gradually evolving. As with much of science, the polymer scientist tackles the seemingly simplest examples first. For polymers, these are the simple vinyl polymers. When such polymers are crystallized from dilute solution, they form lamellar-shaped single crystals (as depicted in Fig. 2.15) with the distances between folds being on the order of 25 to 50 units (100–200 Å). In more concentrated solutions more complex multilayered dendritic structures are formed.

For polymers crystallized from the bulk, the end result is typically formation of spherulites. For polyethylene, the initial structure formed is a single crystal with folded-chain lamellae as shown in Fig. 2.15 and depicted in Fig. 2.20a. These quickly lead to the formation of sheaflike structures (Fig. 2.20d) called axialites or

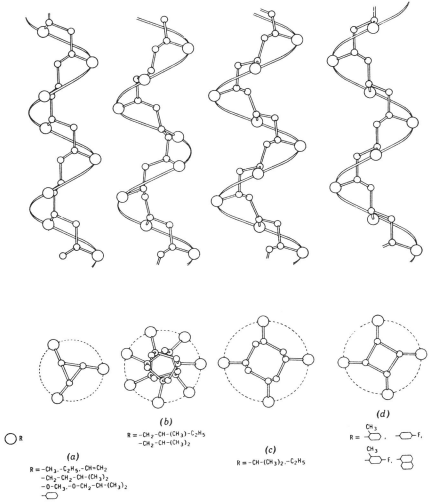

Figure 2.19 Helical conformations of isotactic vinyl polymers. [From N. Gaylord, in *Linear and Stereoregular Addition Polymers* (N. Gaylord and H. Mark, eds.). Wiley Interscience, New York, 1959. With permission from the Interscience Division of John Wiley and Sons, Publishers.]

hedrites. As growth proceeds, the lamellae develop on either side of a central reference plane. The lamellae continue to fan out, occupying increasing volume sections through the formation of additional lamellae at appropriate branch points. The result is the formation of spherulites as pictured in Figs. 2.16 and 2.20.

2.4 AMORPHOUS BULK STATE

An amorphous bulk polymer contains chains that are arranged in something less than a well-ordered, crystalline manner. Physically, amorphous polymers exhibit a glass transition temperature but not a melting temperature and do not give a clear

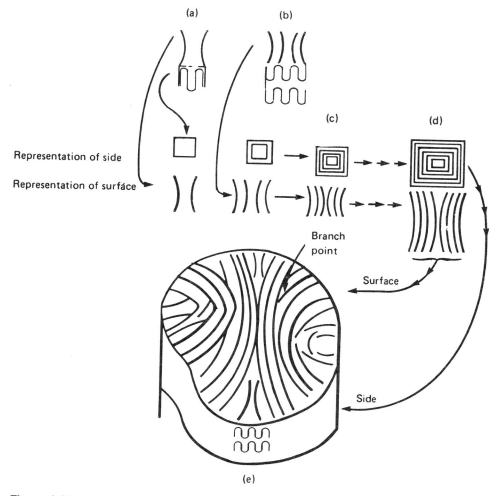

Figure 2.20 Steps in the formation of a spherulite from the bulk.

X-ray diffraction pattern. Amorphous polymer chains have been likened to spaghetti strands within a pot of spaghetti, but in actuality the true extent of disorder that results in an amorphous polymer is still not well understood.

Section 5.14 contains a discussion of a number of techniques employed in the search for the real structure of the amorphous bulk state. Briefly, there is evidence to suggest that little order exists in the amorphous state, the order being similar to that observed with low molecular weight hydrocarbons in the case of vinyl polymers for short-range interactions. For long-range interactions, there is evidence that the chains approximate a random coil with some portions paralleling one another.

In 1953 Flory and Mark suggested a random coil model where the chains had conformations similar to those present if the polymer were in a theta solvent (similar to Fig. 2.9a, left). In 1957 Kargin suggested that amorphous polymer chains existed as aggregates in parallel alignment. Models continue to be developed, but all contain the elements of disorder suggested by Flory and Mark and the elements of order suggested by Kargin.

Products with properties intermediate between elastomers and fibers are grouped together under the heading "plastics."

2.5 POLYMER STRUCTURE–PROPERTY RELATIONSHIPS

Throughout the text we will relate polymer structure to the properties of the polymer. Polymer properties are related not only to the chemical nature of the polymer but also to such factors as extent and distribution of crystallinity, distribution of polymer chain lengths, and nature and amount of additives, such as fillers, reinforcing agents, and plasticizers, to mention only a few. These factors influence essentially all the polymeric properties to some extent, such as hardness, flammability, weatherability, chemical resistance, biologic responses, comfort, appearance, dyeability, softening point, electrical properties, stiffness, flex life, moisture retention, etc. Chapters 1 through 12 concentrate on the chemical nature of the polymer itself, whereas Chapters 13 and 14 deal with the nature and effect on polymer properties by addition of plasticizers, fillers, stabilizers, etc. Chapter 17 deals with the application of both the polymers themselves and suitable additives aimed at producing polymers exhibiting desired properties.

Here we will briefly deal with only the chemical and physical nature of polymeric materials that permit their division into three broad divisions—elastomers or rubbers, fibers, and plastics.

Elastomers are high polymers possessing chemical and/or physical cross-links. For industrial application the "use" temperature must be above the T_g (to allow for "chain" mobility), and its normal state (unextended) must be amorphous. The restoring force, after elongation, is largely due to entropy. As the material is elongated, the random chains are forced to occupy more ordered positions. On release of the applied force the chains tend to return to a more random state. Gross, actual mobility of chains must be low. The cohesive energy forces between chains should be low to permit rapid, easy expansion. In its extended state a chain should exhibit a high tensile strength, whereas at low extensions it should have a low tensile strength. Cross-linked vinyl polymers often meet the desired property requirements. The material, after deformation, should return to its original shape because of the cross-linking. This property is often referred to as a elastic "memory." Figure 2.21 illustrates force versus elongation for a typical elastomer.

Fiber properties include high tensile strength and high modulus (high stress for small strains). These can be obtained from high molecular symmetry and high cohesive energies between chains, both requiring a fairly high degree of polymer crystallinity. Fibers are normally linear and drawn (oriented) in one direction, producing high mechanical properties in that direction. Typical condensation polymers, such as polyester and nylon, often exhibit these properties. If the fiber is to be ironed, its T_g should be above 200°C, and if it is to be drawn from the melt, its T_g should be below 300°C. Branching and cross-linking are undesirable since they disrupt crystalline formation, even though a small amount of cross-linking may increase some physical properties, if effected after the material is drawn and processed.

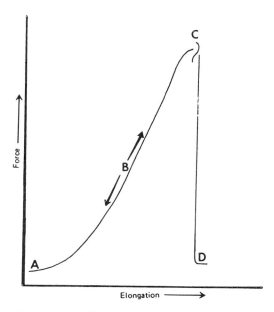

Figure 2.21 Elongation of an elastomer as a function of applied force where A is the original "relaxed" state, B represents movement to full extension, and C is point at which the elastomer "breaks."

Some polymers can be classified within two categories, with properties being greatly varied by varying molecular weight, end groups, processing, cross-linking, plasticizer, and so on. Nylon in its more crystalline form behaves as a fiber, whereas less crystalline forms are generally classified as plastics. Selected property-structure relationships are summarized in Tables 2.4 and 2.5.

Many polymers can be treated to express more than one behavior. Thus, nylon-66 provides a good fibrous material when aligned and behaves as a plastic if it is not subjected to orientation. Polyesters also exhibit the same tendencies. Other materials, such as poly(vinyl chloride) and siloxanes, can be processed to act as plastics or elastomers.

2.6 CRYSTALLINE AND AMORPHOUS COMBINATIONS

Most polymers consist of a combination of crystalline and amorphous regions. Even within polymer crystals such as spherulites (Fig. 2.20), the regions between the ordered folded crystalline lamellae are less ordered, approximating amorphous regions. This combination of crystalline and amorphous regions is important for the formation of materials that have both good strength (contributed to largely by the crystalline portions) and some flexibility or "softness" (derived from the amorphous portions). Figure 2.22 shows the general relationship between material "hardness/softness" and the proportion that is crystalline (for linear polymers).

Through the use of specific treatment(s) the crystalline/amorphous regions can vary from being largely random to being preferentially oriented in one direction

Table 2.4 Selected Property-Structure Relationships

Glass transition temperature

  Increases with presence of bulky pendant groups

    Stiffening groups as 1,4-phenylene

    Chain symmetry

    Polar groups

    Cross-linking

  Decreases with presence of additives like plasticizers

    Flexible pendant groups

    Nonpolar groups

    Dissymmetry

Solubility

  Favored by lower chain lengths

    Increased amorphous content

    Low interchain force

    Disorder and dissymmetry

    Increased temperature

    Compatible solvent

Crystallinity

  Favored by high interchain forces

    Regular structure; high symmetry

    Decrease in volume

    Increased stress

    Slow cooling from melt

    Homogeneous chain lengths

(Fig. 2.23) and in the proportion of crystalline/amorphous regions. Thus, polymers can be oriented through the unidirectional "pulling" of the bulk material either during the initial synthesis (such as the "pulling" of fibers as they exit a spinneret) or during the processing phase where preferential application of stress ("pulling") in one direction results in the preferential orientation of the chains, including both crystalline and amorphous regions. This preferential orientation results in fibers or bulk material with anisotropic properties, with the material generally showing greater strength along the axis of applied stress (in the direction of the "pull"). These crystalline sites may be on a somewhat molecular level involving only a few

Table 2.5 General Property Performance-Structure Relationships[a]

	Increased crystallinity	Increased crosslinking	Increased mol. wt.	Increased mol. wt. distribution	Addition of polar backbone units	Addition of backbone stiffening groups
Abrasion resistance	+	+	+	−	+	−
Brittleness	−	M	+	+	+	+
Chemical resistance	+	V	+	−	−	+
Hardness	+	+	+	+	+	+
T_g	+	+	+	−	+	+
Solubility	−	−	−	0	−	−
Tensile strength	+	M	+	−	+	+
Toughness	−	−	+	−	+	−
Yield	+	+	+	+	+	+

[a] + = increase in property; 0 = little or no effect; − = decrease in property; M = property passes through a maximum; V = variable results dependent on particular sample and temperature.

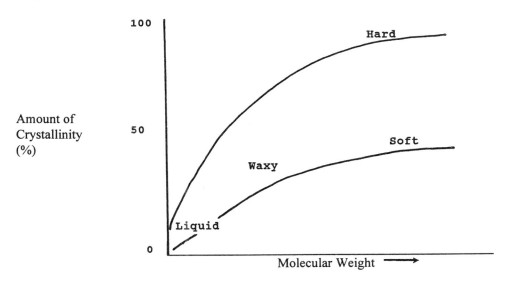

Figure 2.22 General physical nature of materials as a function of the amount of crystallization.

chains (Fig. 2.23), or they may exist as larger units such as spherulites (Figs. 2.16 and 2.20).

The amount of orientation is dependent on a number of factors. Increased mobility of the crystalline and amorphous regions typically results in greater reorientation for a specified applied stress. Thus, materials with little or no cross-linking, materials with lowered inter- and intramolecular attraction, and materials that are near (or above) their glass transition temperature will respond with greater reorientation (per unit of stress) in comparison to materials where mobility is more limited. Application of increased stress will eventually lead to distortion of both the crystalline (including spherulites) and amorphous regions and finally breakage of primary chains.

SUMMARY

1. Polymers, or macromolecules, are high molecular weight compounds with chain lengths greater than the critical length required for the entanglement of these chains. There is an abrupt change in melt viscosity and other physical properties of high molecular weight compounds when the chain length exceeds the critical chain length.

2. While some naturally occurring polymers, such as proteins, are monodisperse, i.e., all have the same molecular weight, other natural and synthetic polymers, such as cellulose and polyethylene, are polydisperse, i.e., they consist of a mixture of polymer chains with different molecular weights. Hence, one uses the term $\overline{DP}$ to indicate an average degree of polymerization, where $\overline{DP}$ is equal to the number of mers (repeating units) in the polymer chain.

3. Many polymers, such as cellulose and HDPE, are linear polymers consisting of long, continuous, covalently bonded atoms. Others such as amylodextrin and

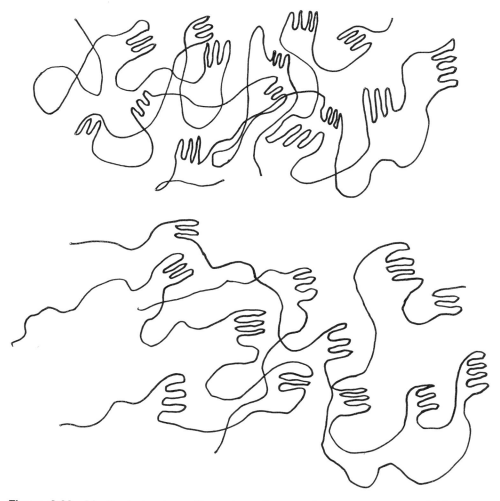

Figure 2.23 Idealized structures illustrating mixtures of crystalline (⁓⁓⁓) and amorphous (⁓⁓⁓) regions unoriented (top) and oriented (bottom).

LDPE have branches or chain extensions from the polymer backbone and hence have greater volume and lower density than linear polymers. Both linear polymers and those with branches are thermoplastics. In contrast, network polymers such as ebonite, in which individual chains are joined to each other by covalently bonded cross-links, are infusible thermoset polymers.

4. Functional groups in the polymer backbone, such as the methyl group in polypropylene and hevea rubber, are called pendant groups.

5. Many rubberlike polymers are flexible because the free rotation of carbon-carbon single bonds allows the formation of many different shapes, or conformations. This segmental motion is restricted by bulky pendant groups, by stiffening groups in the polymer chains, and by strong intermolecular forces. Hydrogen bond-

ing, which is the strongest of these intermolecular forces, is essential for most strong fibers.

6. Free rotation of covalently bonded atoms is also prevented by the presence of double bonds. Thus, stable trans and cis configurations are possible for polymers such as polyisoprene. The cis and trans isomeric forms of polyisoprene are known as flexible hevea rubber and hard plastic gutta percha, respectively.

7. When a chiral center is present in a polymer such as polypropylene, many different configurations or optical isomers are possible. The principal configurations with ordered arrangements of the pendant groups are high-melting, strong molecules known as isotactic and syndiotactic isomers. Lower-melting isomers in which the pendant groups are randomly oriented in space are known as atactic polymers.

8. The temperature at which segmental motion occurs because of free rotation of the covalent bonds is a characteristic temperature called the glass transition temperature. To be useful as plastics and elastomers, the polymers must be at a temperature below and above the glass transition temperature, respectively.

9. Since the specific volume, index of refraction, gas permeability, and heat capacity increase because of the onset of segmental motion at T_g, abrupt changes in these properties may be used to determine T_g.

10. The first-order transition, or melting point (T_m), is 33 to 100% greater than T_g, which is sometimes called the second-order transition. The greatest difference between T_m and T_g is demonstrated by symmetrical polymers like HDPE.

11. A polymer chain stretched out to its full contour length represents only one of the myriad of conformations present in a polymer at temperatures above T_g. Hence, the chain length is expressed statistically as the root-mean-square distance $\sqrt{\overline{r^2}}$, which is about 7% of the full contour length of the polymer chain.

12. Since branched chains like LDPE have many chain ends, it is customary to use the radius of gyration (S), which is the distance of a chain end from the polymer's center of gravity, instead of r.

13. The flexibility, which is related inversely to the orientation time (τ_m) increases as the temperature increases and may be calculated from the Arrhenius equation:

$$\tau_m = A\ e^{E/RT}$$

14. Fibers and stretched elastomers are translucent because of the presence of spherulites consisting of organized crystallites or regions of crystallinity.

15. Since single-lamellar crystals consisting of folded chains of symmetrical polymers can be prepared, it is now assumed that crystalline polymers may be represented by a switchboard model consisting of crystalline and amorphous domains.

16. Additional orientation of crystalline polymers occurs and physical properties are improved when films are biaxially oriented or when fibers are stretched.

17. The principal differences between elastomers, plastics, and fibers are the presence and absence of stiffening groups in the chain, the size of the pendant groups on the chain, and the strength of the intermolecular forces. Elastomers are usually characterized by the absence of stiffening groups in the polymer backbone, the presence of bulky pendant groups, and the absence of strong intermolecular forces. In contrast, fibers are characterized by the presence of stiffening groups in

the polymer backbone and of intermolecular hydrogen bonds and the absence of branching or irregularly spaced pendant groups. The structure and properties of plastics are between these two extremes.

GLOSSARY

amorphous: Noncrystalline polymer or noncrystalline areas in a polymer.

anti form: Trans (t) or low-energy conformer.

aramides: Aromatic nylons.

Arrhenius equation: An equation showing the exponential effect of temperature on a process.

atactic: A polymer in which there is a random arrangement of pendant groups on each side of the chain, as in atactic PP:

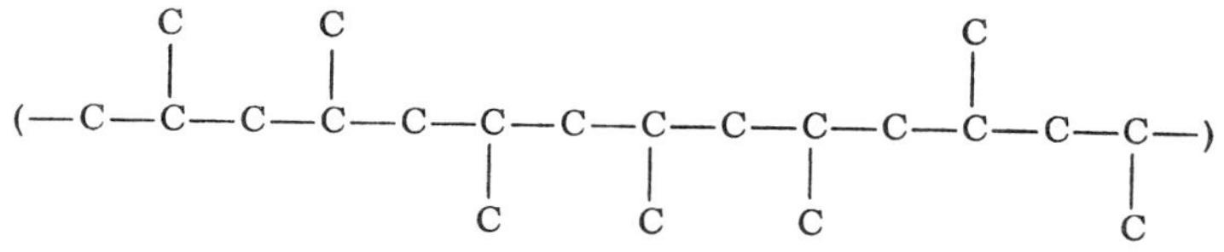

$$(-C-C-C-C-C-C-C-C-C-C-C-C-C-)$$

Avrami equation: An equation used to describe the crystallization rate.

backbone: The principal chain in a polymer molecule.

biaxially oriented film: A strong film prepared by stretching the film in two directions at right angles to each other. This strong film will shrink to its original dimensions when heated.

branched polymer: A polymer having extensions of the polymer chain attached to the polymer backbone, such as LDPE. Polymers having pendant groups, such as the methyl groups in polypropylene, are not considered to be branched polymers.

bulky groups: Large pendant groups on a polymer chain.

cellulose: A polymer in which cellobiose is the repeating unit.

chiral center: An asymmetric center such as a carbon atom with four different groups.

cold drawing: The stretching of a fiber or fibers to obtain products with high tensile strength.

configurations: Related chemical structures produced by the breaking and making of primary valence bonds.

conformations: Various shapes of polymers resulting from the rotation of single bonds in the polymer chain.

conformer: A shape produced by a change in the conformation of a polymer.

contour length: The fully extended length of a polymer chain, equal to the product of the length of each repeating unit (l) times the number of units, or mers (n), i.e., nl is the full contour length.

critical chain length (z): The minimum chain length required for entanglement of the polymer chains.

cross-linked density: A measure of the relative degree of cross-linking in a network polymer.

crystalline polymer: A polymer with ordered structure which has been allowed to disentangle and form crystals such as HDPE. Thus, isotactic polypropylene, cellulose, and stretched rubber are crystalline polymers.

crystallites: Regions of crystallinity.

differential scanning calorimetry (DSC): An instrumental thermal analytical technique in which the difference in the amount of heat absorbed by a polymer sample and a standard is measured by the power consumed as the temperature is increased.

differential thermal analysis (DTA): A thermal instrumental analytical technique in which the rate of absorption of heat by a polymer is compared with that of a standard such as glass or alumina.

dilatometry: A technique in which changes in specific volume are measured.

dipole-dipole interactions: Moderate secondary valence forces between polar groups in different molecules or in different locations in the same molecule.

dispersion forces: Same as London forces.

DNA: Deoxyribonucleic acid.

DP: Degree of polymerization or the number of repeating units (mers) in a polymer chain.

$\overline{\text{DP}}$: Average degree of polymerization in a polydisperse polymer.

η: Viscosity or coefficient of viscosity.

end-to-end distance (r): The shortest distance between chain ends in a polymer.

endothermic: A process in which energy is absorbed.

eutactic: An isotactic or syndiotactic polymer.

excluded volume: The volume that must be disregarded because only one atom of a chain may occupy any specific space at any specified time.

fiber: A polymer with strong intermolecular hydrogen bonding.

flexibilizing groups: Those groups in the polymer backbone that increase the segmental motion of polymers, e.g., oxygen atoms or multiple methylene groups.

fringed micelle model: An outmoded model showing amorphous and crystalline domains in a polymer.

gauche forms (g): Conformers in which the methylene groups in the polymer chain are 60° apart relative to rotation about a C—C bond.

glass transition temperature (T_g): A characteristic temperature at which glassy amorphous polymers become flexible or rubberlike because of the onset of segmental motion.

glassy state: Hard, brittle state.

gutta percha: Naturally occurring trans isomer of polyisoprene.

head-to-tail configuration: The normal sequence of mers in which the pendant groups are regularly spaced like the methyl groups in polypropylene, i.e.,

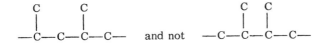

high-density polyethylene (HDPE): Formerly called low-pressure polyethylene, a linear polymer produced by the polymerization of ethylene in the presence of Ziegler-Natta or Phillips catalysts.

hydrogen bonding: Strong secondary valence forces between a hydrogen atom in one molecule and an oxygen, nitrogen, or fluorine atom in another molecule. These forces may also exist between hydrogen atoms in one location and oxygen, nitrogen, or fluorine atoms in another location in the same molecule. Intermolecular hydrogen bonds are responsible for the high strength of fibers. Helices are the result of intramolecular hydrogen bonds.

intermolecular forces: Secondary valence, or van der Waals, forces between different molecules.

intramolecular forces: Secondary valence, or van der Waals, forces within the same molecule.

isotactic: A polymer in which the pendant groups are all on the same side of the polymer backbone, as in isotactic PP:

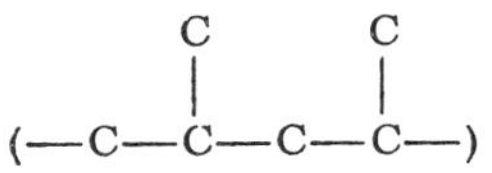

$$
\begin{matrix}
C & & C & \\
| & & | & \\
(-C-C-C-C-) &
\end{matrix}
$$

lamellar: Platelike in shape.

linear polymer: A polymer like HDPE that consists of a linear chain without chain-extending branches.

London forces: Weak transitory dispersion forces resulting from induced dipole-induced dipole interaction.

low-density polyethylene (LDPE): Formerly called high-pressure polyethylene, a branched polymer produced by the free radical-initiated polymerization of ethylene at high pressure.

Maltese cross: A cross with arms like arrowheads pointing inward.

melting point (T_m): The first-order transition when the solid and liquid phases are in equilibrium.

mer: The repeating unit in a polymer chain.

methylene: $-CH_2-$.

modulus: The ratio of stress to strain, as of strength to elongation, which is a measure of stiffness of a polymer.

monodisperse: A polymer made up of molecules of one specific molecular weight, such as a protein.

n: Symbol for the number of mers (repeating units) in a polymer.

nanometer (nm): 10^{-9} m.

nylon: A synthetic polyamide.

pendant groups: Groups attached to the main polymer chain or backbone, like the methyl groups in polypropylene.

pentane interference: The interference with free motion caused by the overlap of the hydrogen atoms on the terminal carbon atoms in pentane.

polydisperse: A polymer consisting of molecules of many different molecular weights, such as commercial HDPE.

r: Symbol for end-to-end distance.

radius of gyration (S): The root-mean-square distance of a chain end to a polymer's center of gravity.

random flight technique: A statistical approach used to measure the shortest distance between the start and finish of a random flight.

RNA: Ribonucleic acid.

root mean square distance: $\sqrt{\overline{r^2}} = 1 \sqrt{n}$, the average end-to-end distance of polymer chains.

S: The radius of gyration.

side-chain crystallization: Crystallization related to that of regularly spaced long pendant groups.

single polymer crystals: A lamellar structure consisting of folded chains of a linear polymer, such as polyethylene.

spherulites: Aggregates of polymer crystallites.

stiffening groups: Those groups in the polymer backbone that decrease the segmental motion of polymers, e.g., phenylene, amide, carbonyl, and sulfonyl groups.

switchboard model: A model resembling a switchboard used to depict crystalline and amorphous domains in a polymer.

syndiotactic: A polymer in which the pendant groups are arranged alternately on each side of the polymer backbone, as in syndiotactic PP:

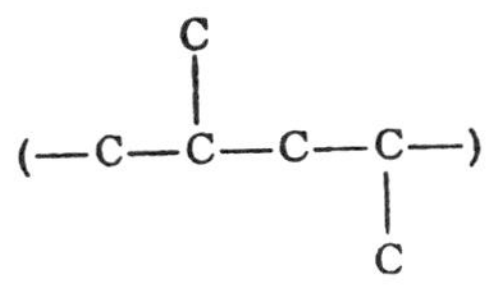

$$
\begin{array}{c}
\text{C} \\
| \\
(-\text{C}-\text{C}-\text{C}-\text{C}-) \\
| \\
\text{C}
\end{array}
$$

τ_m: The orientation time, a measure of the ease of uncoiling.

tacticity: The arrangement of the pendant groups in space. Examples are isotactic or syndiotactic polymers.

van der Waals forces: Forces based on attractions between groups in different molecules or in different locations in the same molecule.

viscosity: A measure of the resistance of a polymer to flow, either as a melt or as a solution.

EXERCISES

1. Make crude sketches or diagrams showing (a) a linear polymer, (b) a polymer with pendant groups, (c) a polymer with short branches, (d) a polymer with long branches, and cross-linked polymers with (e) low, and (f) high cross-linked density.

2. Which has (a) the greater volume, and (b) the lower softening point: HDPE or LDPE?

3. What is the approximate bond angle of the carbon atoms in (a) a linear, and (b) a cross-linked polymer?

4. What is the approximate length of an HDPE chain when n equals 2000? of a PVC chain of the same numer of repeating units?

5. Which of the following is a monodisperse polymer: (a) hevea rubber, (b) corn starch, (c) cellulose from cotton, (d) casein from milk, (e) HDPE, (f) PVC, (g) β-keratin, (h) nylon-66, (i) DNA?

6. What is the degree of polymerization ($\overline{DP}$) of LDPE having an average molecular weight ($\overline{M}$) of 27,974?

7. What is the structure of the repeating unit (mer) in (a) polypropylene, (b) poly(vinyl chloride), (c) hevea rubber?

8. Which of the following is a branched chain polymer: (a) HDPE, (b) isotactic PP, (c) LDPE, (d) amylose starch?

9. Which of the following is a thermoplastic: (a) ebonite, (b) Bakelite, (c) vulcanized rubber, (d) HDPE, (e) celluloid, (f) PVC, (g) LDPE?

10. Which has the higher cross-linked density, (a) ebonite or (b) soft vulcanized rubber?

11. Do HDPE and LDPE differ in (a) configuration, or (b) conformation?

12. Which is a trans isomer: (a) gutta percha, or (b) hevea rubber?

13. Which will have the higher softening point: (a) gutta percha, or (b) hevea rubber?

14. Show (a) a head-to-tail, and (b) a head-to-head configuration for poly(vinyl alcohol).

15. Show the structure of a typical portion of the chain of (a) syndiotactic PVC, (b) isotactic PVC.

16. Show Newman projections of the gauche forms of HDPE.

17. Name polymers whose intermolecular forces are principally (a) London forces, (b) dipole-dipole forces, (c) hydrogen bonding.

18. Which will be more flexible: (a) poly(ethylene terephthalate), or (b) poly-(butylene terephthalate)?

19. Which will have the higher glass transition temperature (T_g): (a) poly(methyl methacrylate), or (b) poly(butyl methacrylate)?

20. Which will have the higher T_g: (a) isotactic polypropylene, or (b) atactic polypropylene?

21. Which will be more permeable to a gas at room temperature: (a) isotactic polypropylene, or (b) atactic polypropylene?

22. Which will have the greater difference between T_m and T_g values: (a) HDPE, or (b) LDPE?

23. What is the full contour length of a molecule of HDPE with a DP of 1500?

24. Which would be more flexible: (a) poly(methyl acrylate), or (b) poly(methyl methacrylate)?

25. Would you expect the orientation time of HDPE to increase by approximately 5 or 50% when it is cooled from 90 to 80°C?

26. Which would have the higher melting point: (a) nylon-66, or (b) an aramide?

27. What type of hydrogen bonds are present in a globular protein?

28. Which would have the greater tendency to cold flow at room temperature: (a) poly(vinyl acetate) ($T_g = 301°K$), or (b) polystyrene ($T_g = 375°K$)?

29. Which would be more transparent: (a) polystyrene, or (b) isotactic polypropylene?

30. Which would be more apt to produce crystallites: (a) HDPE, or (b) poly(butyl methacrylate)?

31. How would you cast a nearly transparent film of LDPE?

32. Which would tend to be more crystalline when stretched: (a) unvulcanized rubber, or (b) ebonite?

33. Which would be more apt to exhibit side-chain crystallization (a) poly(methyl methacrylate), or (b) poly(dodecyl methacrylate).

BIBLIOGRAPHY

Alfrey, T., Gurnee, E. F. (1956): Dynamics of viscoelastic behavior, in *Rheology—Theory and Applications* (F. R. Eirich, ed.), Academic, New York.

Bailey, R. T., North, A., Pethrick, R. (1981): *Molecular Motion in High Polymers*, Oxford University Press, Oxford, England.

Bassett, D. C. (1981): *Principles of Polymer Morphology*, Cambridge University Press, Cambridge, England.

Bicerano, J. (1992): *Computational Modeling of Polymers*, Marcel Dekker, New York.

Bicerano, J. (1993): *Prediction of Polymer Properties*, Marcel Dekker, New York.

Brandrup, J., Immergut, E. H. (1975): *Polymer Handbook*, 2nd ed., Wiley, New York.

Chan, C.-M. (1993): *Polymer Surface Techniques*, Hauser-Gardner, Cincinnati, Ohio.

Deanin, R. D. (1972): *Polymer Structure, Properties and Applications*, Cahners Books, Boston.

Dosiere, M. (1993): *Crystallization of Polymers*, Kluwer, Dordrecht, Netherlands.

Flory, P. J. (1953): *Principles of Polymer Chemistry*, Cornell University Press, Ithaca, New York.

Geil, P. H. (1963): *Polymer Single Crystals*, Wiley-Interscience, New York.

Hall, I. H. (1984): *Structure of Crystalline Polymers*, Applied Science, Essex, England.

Hiltner, A. (1983): *Structure-Property Relationships of Polymeric Solids*, Plenum, New York.

Katz, J. R. (1925): Crystalline structure of rubber, Kolloid, *36*:300.

Koenig, J. L. (1980): *Chemical Microstructure of Polymer Chains*, Wiley-Interscience, New York.

Lenz, R. W. (1967): *Organic Chemistry of High Polymers*, Wiley-Interscience, New York.

Mark, H. F. (1967): Giant molecules, Sci. Am., *197*:80.

Marvel, C. S. (1959): *An Introduction to the Organic Chemistry of High Polymers*, Wiley, New York.

McGrew, F. C. (1938): Structure of synthetic high polymers, J. Chem. Ed., *35*:178.

Natta, G. (1955): Stereospecific macromolecules, J. Polymer Sci., *16*:143.

Pauling, L, Corey, R. B., Branson, H. R. (1951): The structure of proteins, Proc. Natl. Acad. Sci. USA, *37*:205.

Raleigh, Lord. (1929): Random flight problem, Phil. Mag., *37*:321.

Sabbatini, L, Zambonin, P. G. (1993): *Surface Characterization of Advanced Polymers*, VCH, New York.

Sanchez, I. (1992): *Physics of Polymer Interfaces*, Butterworth-Heinemann, London.

Seymour, R. B. ———. (1975): *Modern Plastics Technology*, Chap. 1, Reston, Reston, Virginia.

Seymour, R. B., Carraher, C. E. (1984): *Structure-Property Relationships in Polymers*, Plenum, New York.

Watson, J. D., Crick, F. H. C. (1953): A structure for DNA, Nature, *171*:737.

Woodward, A. E. (1995): *Understanding Polymer Morphology*, Hanser-Gardner, Cincinnati, Ohio.

3

Rheology and Solubility

3.1 RHEOLOGY

The branch of science related to the study of deformation and flow of materials was given the name rheology by Bingham, whom some have called the father of modern rheology. The prefix *rheo* is derived from the Greek term *rheos*, meaning current or flow. The study of rheology includes two vastly different branches of mechanics called fluid and solid mechanics. The polymer chemist is usually concerned with viscoelastic materials that act as both solids and fluids.

The elastic component is dominant in solids, hence their mechanical properties may be described by Hooke's law [Eq. (3.1)], which states that the applied stress (s) is proportional to the resultant strain (γ) but is independent of the rate of this strain (dγ/dt).

$$s = E\gamma \tag{3.1}$$

Stress is equal to force per unit area, and strain or elongation is the extension per unit length. For an isotropic solid, i.e., one having the same properties regardless of direction, the strain is defined by Poisson's ratio, $V = \gamma_1/\gamma_w$, the percentage change in longitudinal strain, γ_1, to the percentage change in lateral strain, γ_w.

When there is no volume change, as when an elastomer is stretched, Poisson's ratio is 0.5. This value decreases as the T_g of the substance increases and approaches 0.3 for rigid PVC and ebonite. For simplicity, the polymers will be considered to be isotropic viscoelastic solids with a Poisson's ratio of 0.5, and only deformations in tension and shear will be considered. Thus, a shear modulus (G) will usually be used in place of Young's modulus of elasticity (E) [refer to Eq. (3.2), Hooke's law for shear], where $E \simeq 2.6G$ at temperatures below T_g. The moduli (G) for steel, HDPE, and hevea rubber (NR) are 86, 0.087, and 0.00067 m^{-2}, respectively.

$$s = G\gamma \text{ and } ds = Gd\gamma \tag{3.2}$$

The viscous component is dominant in liquids, hence their flow properties may be described by Newton's law [Eq. (3.3)] (where η is viscosity), which states that the applied stress s is proportional to the rate of strain $d\gamma/dt$, but is independent of the strain γ or applied velocity gradient. The symbol $\dot{\gamma}$ is sometimes used for strain rate.

$$s = \eta \frac{d\gamma}{dt} \tag{3.3}$$

Both Hooke's and Newton's laws are valid for small changes in strain or rate of strain, and both are useful in studying the effect of stress on viscoelastic materials. The initial elongation of a stressed polymer below T_g is the reversible elongation due to a stretching of covalent bonds and distortion of the bond angles. Some of the very early stages of elongation by disentanglement of chains may also be reversible.

However, the rate of flow, which is related to slower disentanglement and slippage of polymer chains past each other, is irreversible and increases (and η decreases) as the temperature increases in accordance with the following Arrhenius equation (3.4) in which E is the activation energy for viscous flow.

$$\eta = A\, e^{E/RT} \tag{3.4}$$

It is convenient to use a simple weightless Hookean, or ideal, elastic spring with a modulus of G and a simple Newtonian (fluid) dash pot or shock absorber having a liquid with a viscosity of η as models to demonstrate the deformation of an elastic solid and an ideal liquid. The stress-strain curves for these models are shown in Fig. 3.1.

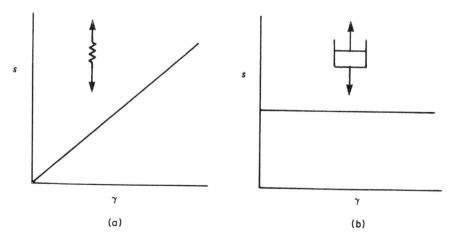

(a)   (b)

Figure 3.1 Stress-strain plots for (a) a Hookean spring where E (Eq. 3.1) is the slope, and (b) a Newtonian dash pot where s is constant.

Since polymers are viscoelastic solids, combinations of these models are used to demonstrate the deformation resulting from the application of stress to an isotropic solid polymer. As shown in Fig. 3.2, Maxwell joined the two models in series to explain the mechanical properties of pitch and tar. He assumed that the contributions of both the spring and dash pot to strain were additive and that the application of stress would cause an instantaneous elongation of the spring, followed by a slow response of the piston in the dash pot. Thus, the relaxation time (τ), when the stress and elongation have reached equilibrium, is equal to η/G.

In the Maxwell model for viscoelastic deformation, it is assumed that $\gamma_{total} = \gamma_{elastic} + \gamma_{viscous}$. This may be expressed in the form of the following differential equation from Eqs. (3.2) and (3.3).

$$\frac{d\gamma}{dt} = \frac{s}{\eta} + \frac{ds}{dt}\frac{1}{G} \tag{3.5}$$

The rate of strain $d\gamma/dt$ is equal to zero under conditions of constant stress (s), i.e.,

$$\frac{s}{\eta} + \frac{ds}{dt}\frac{1}{G} = 0 \tag{3.6}$$

Then, assuming that $s = s_o$ at zero time, gives

$$s = s_o e^{-tG/\eta} \tag{3.7}$$

And, since the relaxation time $\tau = \eta/G$, then

$$s = s_o e^{-t/\tau} \tag{3.8}$$

Thus, according to Eq. (3.8) for the Maxwell model or element, under conditions of constant strain, the stress will decrease exponentially with time and at the relaxation time $t = \tau$, s will be equal to $1/e = 1/2.7$, or 0.37 of its original value (s_o).

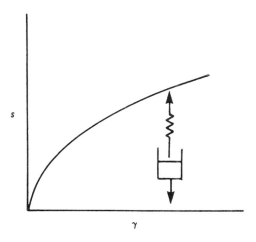

Figure 3.2 Stress-time plot for stress relaxation in the Maxwell model.

As shown in Fig. 3.3, the spring and dash pot are parallel in the Voigt-Kelvin model. In this model or element, the applied stress is shared between the spring and the dash pot, and thus the elastic response is retarded by the viscous resistance of the liquid in the dash pot. In this model, the vertical movement of the spring is essentially equal to that of the piston in the dash pot. Thus, if G is much larger than η, the retardation time (η/G) or τ is small, and τ is large if η is much larger than G.

In the Voigt-Kelvin model for viscoelastic deformation, it is assumed that the total stress is equal to the sum of the viscous and elastic stress $s_v + s_e$, as shown in Eq. (3.9).

$$s = \eta \frac{d\gamma}{dt} + G\gamma \tag{3.9}$$

On integration one obtains

$$\gamma = \frac{s}{G}(1 - e^{-tG/\eta}) = \frac{s}{G}(1 - e^{-t/\tau}) \tag{3.10}$$

The retardation time τ is the time for the strain to decrease to $1 - (1/e)$ or $1 - (1/2.7) = 0.63$ of the original value. The viscoelastic flow of polymers is explained by appropriate combinations of the Maxwell and Voigt-Kelvin models.

While polymer melts and elastomers flow readily when stress is applied, structural plastics must resist irreversible deformation and behave as elastic solids when relatively small stresses are applied. These plastics are called ideal or Bingham plastics as described by

$$s - s_o = \eta \frac{d\gamma}{dt} \tag{3.11}$$

As shown in Fig. 3.4, a Bingham plastic exhibits Newtonian flow above the stress yield or stress value (s_o). The curves for shear thickening (dilatant) and shear thinning (pseudo-plastic) materials are also shown in Fig. 3.4.

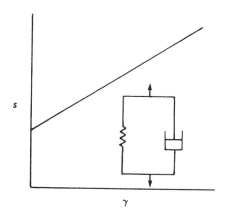

Figure 3.3 Stress-time plot for a Voigt-Kelvin model.

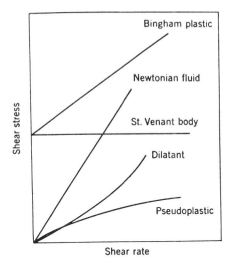

Figure 3.4 Various types of polymer flow. (From *Introduction to Polymer Chemistry* by R. Seymour, McGraw-Hill, New York, 1971. Used with permission of McGraw-Hill Book Company.)

Liquids that undergo a decrease in viscosity with time are called thixotropic, or false-bodied. Those that undergo an increase in viscosity with time are called rheopectic (shear thickening). The term *creep* is used to describe the slow slippage of polymer chains over a long period of time. The Herschel-Buckley equation [Eq. (3.12)] is a general equation which reduces to the Bingham equation when $\eta = 1$ and to the Newtonian equation when $\eta = 1$ and $s_o = 0$:

$$(s - s_o)\,\eta = \phi \frac{d\gamma}{dt} \tag{3.12}$$

where ϕ is related to viscosity.

Eyring has explained liquid flow using a random hole-filling model in which the holes account for about 15% of the volume at room temperature. The size of these holes is similar to that of small molecules, and hole filling is restrained by an energy barrier which is about one-third the value of the heat of evaporation. The number of holes increases as the temperature increases, and thus flow or hole filling is temperature dependent.

Small molecules jump into the holes and leave empty holes when their energy exceeds the activation energy (E_a). This last value is smaller for linear polymers which fill the holes by successive correlated jumps of chain segments along the polymer chain. The jump frequency ϕ is governed by a segmental factor (f_0), and both ϕ and f_0 are related to molecular structure and temperature.

For convenience and simplicity, polymers have been considered to be isotropic in which the principal force is shear stress. While such assumptions are acceptable for polymers at low shear rates, they fail to account for stresses perpendicular to the plane of the shear stress, which are encountered at high shear rates. Thus, an

extrude such as a pipe or filament expands when it emerges from the die in what
is called the Barus or Weissenberg effect, or die swell.

As illustrated later in Figure 5.3, viscoelasticity can be subclassified into five
types: (1) viscous glass, Hookean elastic, or Hookean glass region, where chain
segmental motion is quite restricted and involves mainly only bond bending and
bond angle deformation; the material behaves as a glass like a glass window; while
flow occurs it is detectable only with delicate, exacting instruments; stained-glass
windows in the old churches in Europe are typically thicker at the bottom of each
segment due to the slow flow; (2) glassy transition and (3) the rubbery flow region
are both related to what is often referred to as the viscoelastic region where polymer
deformation is reversible but time dependent and associated with both side-chain
and main-chain rotation; (4) rubbery, highly elastic rubbery, or rubberlike elasticity,
where local segmental mobility occurs but total chain flow is restricted by physical
and/or chemical network matrix structure; and (5) rubbery flow or viscous flow
region where irreversible bulk deformation and slippage of chains past one another
occur. Each of these viscoelastic types is time dependent. Thus, given a short
interaction time, window glass acts as a Hookean glass or like a solid, yet observation of glass over many years would permit the visual observation of flow, with
the window glass giving a viscous flow response and thus acting as a fluid. In fact,
most polymers give a response as noted, for example, in Fig. 3.5 if a rubber ball
were dropped onto the material either as the temperature of the sample increased

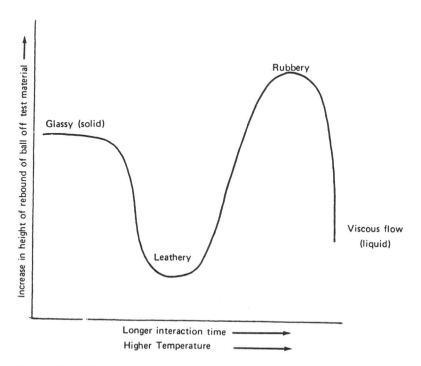

Figure 3.5 Regions of material response as a function of interaction (reaction) time and
temperature.

or the interaction time increased. Commercial Silly Putty or Nutty Putty easily illustrates three of these regions. When struck rapidly it shatters as a solid, when dropped at a moderate rate it acts as a rubber, and when allowed to reside in its container it will flow to occupy the container contour, acting as a liquid. The exact response (illustrated in Fig. 3.5) varies from material to material.

The study of these types of viscoelasticity may be simplified by application of the Boltzman time-temperature superposition principle which provides master curves. This transposition for amorphous polymers is readily accomplished by use of a shift factor a_T calculated relative to a reference temperature T_R, which may be equal to T_g.

The relationship of the shift factor a_T to the reference temperature T_R and some other temperature T, which is between T_g and $T_g + 50°K$, may be approximated by the Arrhenius equation:

$$\log a_T = -\frac{b}{2.3TT_g} (T - T_g) \tag{3.13}$$

According to the more widely used empirical Williams, Landel, and Ferry (WLF) equation, all linear, amorphous polymers have similar viscoelastic properties at T_g and at specified temperatures above T_g such as $T_g + 25°C$; and the constants C_1 and C_2 are related to holes or free volume. When T_g is the reference temperature, $C_1 = 17.44$ and $C_2 = 51.6$.

$$\log a_T = -\frac{C_1(T - T_g)}{C_2 + (T - T_g)} \tag{3.14}$$

Equations such as (3.13) and (3.14) have been used to predict the long-range (long-time) behavior of materials from short-range (short-time) measurements. While this can be done under certain conditions with some success, such approaches have been employed in attempts to predict "long-term" failure of materials. One of the arguments employed to justify this involves the relationship between such equations and properties such as the flow of materials so that it can be argued that there is some relationship between long-range properties dependent on properties such as the flow of the material and its subsequent failure. Such arguments might be valid where failure is approached in some orderly fashion, but most failures are catastrophic dependent on the "weak link within the chain" rather than overall structure. Thus, care should be employed when utilizing such time-temperature superposition approaches to predict failure.

3.2 SOLUBILITY

The physical properties of polymers, including T_g values, are related to the strength of the covalent bonds, the stiffness of the segments in the polymer backbone, and the strength of the intermolecular forces between the polymer molecules. The strength of the intermolecular forces is equal to the cohesive energy density (CED), which is the molar energy of vaporization per unit volume. Since intermolecular attractions of solvent and solute must be overcome when a solute dissolves, CED values may be used to predict solubility.

When a polymer dissolves, the first step is a slow swelling process called solvation in which the polymer molecule swells by a factor δ, which is related to

CED. Linear and branched polymers dissolve in a second step, but network polymers remain in a swollen condition.

In order for solution to take place, it is essential that the free energy G, which is the driving force in the solution process, decrease as shown in the Gibbs free energy equation for constant temperature [Eq. (3.15)]. ΔH and ΔS are equal to the change in enthalpy and the change in entropy in this equation.

$$\Delta G = \Delta H - T \, \Delta S \tag{3.15}$$

By assuming that the sizes of polymer segments were similar to those of solvent molecules, Flory and Huggins obtained an expression for the partial molar Gibbs free energy of dilution, which included the dimensionless Flory-Huggins interaction parameter, $\chi_1 = Z \, \Delta H / RT$ in which $Z =$ a lattice coordination number. It is now recognized that χ_1 is composed of enthalpic and entropic contributions.

While the Flory-Huggins theory has its limitations, it may be used to predict the equilibrium behavior between liquid phases containing an amorphous polymer. The theory may also be used to predict the cloud point, which is just below the critical solution temperature T_c at which the two phases coalesce. The Flory-Huggins interaction parameter may be used as a measure of solvent power. The value of χ_1 for poor solvents is 0.5 and decreases for good solvents.

Some limitations of the Flory-Huggins lattice theory were overcome by Flory and Krigbaum, who assumed the presence of an excluded volume—the volume occupied by a polymer chain that exhibited long-range intramolecular interactions. These interactions were described in terms of free energy by introducing the enthalpy and entropy terms K_1 and ψ_1. These terms are equal when ΔG equals zero. The temperature at which these conditions prevail is the θ temperature at which the effects of the excluded volume are eliminated and the polymer molecule assumes an unperturbed conformation in dilute solutions. The θ temperature is the lowest temperature at which a polymer of infinite molecular weight is completely miscible with a specific solvent. The coil expands above the θ temperature and contracts at lower temperatures.

As early as 1926, Hildebrand showed a relationship between solubility and the internal pressure of the solvent, and in 1931 Scatchard incorporated the CED concept into Hildebrand's equation. This led to the concept of a solubility parameter which is the square root of CED. Thus, as shown below, the solubility parameter δ for nonpolar solvents is equal to the square root of the heat of vaporization per unit volume:

$$\delta = \left(\frac{\Delta E}{V} \right)^{1/2} \tag{3.16}$$

According to Hildebrand, the heat of mixing a solute and a solvent is proportional to the square of the difference in solubility parameters, as shown by the following equation in which ϕ is the partial volume of each component, namely, solvent γ_1 and solute ϕ_2. Since typically the entropy term favors solution and the enthalpy term acts counter to solution, the general objective is to match solvent and solute so that the difference between their δ values is small.

$$\Delta H_m = \phi_1 \phi_2 (\delta_1 - \delta_2)^2 \tag{3.17}$$

The solubility parameter concept predicts the heat of mixing liquids and amorphous polymers. Hence, any nonpolar amorphous polymer will dissolve in a liquid or a mixture of liquids having a solubility parameter that does not differ by more than ± 1.8 (cal cm^{-3})$^{0.5}$. The Hildebrand (H) is preferred over these complex units.

The solubility parameters concept, like Flory's θ temperature, is based on Gibbs free energy. Thus, as the term ΔH in the expression ($\Delta G = \Delta H - T \Delta S$) approaches zero, ΔG will have the negative value required for solution to occur. The entropy (S) increases in the solution process and hence the emphasis is on negative or low values of ΔH_m.

For nonpolar solvents, which have been called regular solvents by Hildebrand, the solubility parameter δ is equal to the square root of the difference between the enthalpy of evaporation (ΔH_v) and the product of the ideal gas-constant (R) and the Kelvin temperature (T) all divided by the molar volume (V), as shown below:

$$\delta = \left(\frac{\Delta E}{V}\right)^{1/2} = \left(\frac{\Delta H_v - RT}{V}\right)^{1/2} \tag{3.18}$$

Since it is difficult to measure the molar volume, its equivalent, namely, the molecular weight M divided by density D, is substituted for V as shown below:

$$\delta = \left[D \frac{(\Delta H_v - RT)}{M}\right]^{1/2} \quad \text{or} \quad \left[\frac{D(\Delta H_v - RT)}{M}\right]^{1/2} \tag{3.19}$$

As shown by the following illustration, this expression may be used to calculate the solubility parameter δ for any nonpolar solvent such as n-heptane at 298°K. n-Heptane has a molar heat of vaporization of 8700 cal, a density of 0.68 g cm^{-3}, and a molecular weight of 100.

$$\delta = \left\{\frac{0.68[8700 - 2(298)]}{100}\right\}^{1/2} = (55.1 \text{ cal cm}^{-3})^{1/2} = 7.4 \text{ H} \tag{3.20}$$

The solubility parameter (CED)$^{1/2}$ is also related to the intrinsic viscosity of solutions ($[\eta]$) as shown by the following expression:

$$[\eta] = \eta_o \, e^{-V(\delta - \delta_o)^2} \tag{3.21}$$

The term intrinsic viscosity or limiting viscosity number is defined later in this chapter.

Since the heat of vaporization of solid polymers is not readily obtained, Small has supplied values for molar attraction constants (G) which are additive and can be used in the following equation for the estimation of the solubility parameter of nonpolar polymers:

$$\delta = \frac{D\Sigma G}{M} \tag{3.22}$$

Typical values for G at 25°C are shown in Table 3.1.

Table 3.1   Small's Molar Attraction Constants (at 25°C)

Group	$G[(cal/cm^3)^{1/2} \, mol^{-1}]$
$-CH_3$	214
$>CH_2$	133
$>CH$	28
$-C-$	-93
$=CH_2$	190
$=CH-$	111
$=C<$	19
$HC\equiv C-$	285
Phenyl	735
Phenylene	658
$-H$	80-100
$-C\equiv N$	410
F or Cl	250-270
Br	340
$>CF_2$	150
$-S-$	225

The use of Small's equation may be illustrated by calculating the solubility parameter of amorphous polypropylene (D = 0.905), which consists of the units CH, CH_2, and CH_3 in each mer. Polypropylene has a mer weight of 42.

$$\delta = \frac{0.905(28 + 133 + 214)}{42} = 8.1 \text{ H} \tag{3.23}$$

Since CED is related to intermolecular attractions and chain stiffness, Hayes has derived an expression relating δ, T_g, and a chain stiffness constant M as shown below:

$$\delta = [M(T_g - 25)]^{1/2} \tag{3.24}$$

Since the polarity of most solvents except the hydrocarbons decreases as the molecular weight increases in a homologous series, δ values also decrease, as shown in Fig. 3.6.

Since "like dissolves like" is not a quantitative expression, paint technologists attempted to develop more quantitative empirical parameters before the Hildebrand solubility parameter had been developed. The Kauri-Butanol and aniline points are still in use and are considered standard tests by the American Society for Testing and Materials (ASTM).

The Kauri-Butanol value is equal to the minimum volume of test solvent that produces turbidity when added to a standard solution of Kauri-Copal resin in 1-butanol. The aniline point is the lowest temperature at which equal volumes of aniline and the test solvent are completely miscible. Both tests are measurements of the relative aromaticity of the test solvent, and their values may be converted to δ values.

Since the law of mixtures applies to the solubility parameter, it is possible to blend nonsolvents to form a mixture which will serve as a good solvent. For example, an equimolar mixture of n-pentane (δ = 7.1 H) and n-octane (δ = 7.6 H) will have a solubility parameter value of 7.35 H.

The solubility parameter of a polymer may be readily determined by noting the extent of swelling or actual solution of small amounts of polymer in a series of solvents having different δ values. Providing the polymer is in solution, its δ

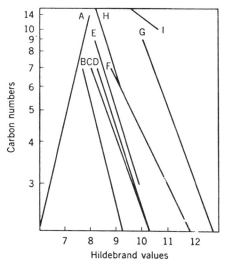

A = Normal alkanes, B = Normal chloroalkanes,
C = Methyl esters, D = Other alkyl formates and
acetates, E = Methyl ketones, F = Alkyl nitriles,
G = Normal alkanols, H = Alkyl benzenes, I =
Dialkyl phthalates.

Figure 3.6 Spectrum of solubility-parameter values for polymers (δ = 6.2 to 15.4). (From *Introduction to Polymer Chemistry* by R. Seymour, McGraw-Hill, New York, 1971. Used with permission of McGraw-Hill Book Company.)

value may be determined by turbidimetric titration using as titrants two different nonsolvents, one that is more polar and one that is less polar than the solvent present in the solution.

Since dipole-dipole forces are present in polar solvents and polar molecules, these must be taken into account when estimating solubilities with such "nonregular solvents." A third factor must be considered for hydrogen-bonded solvents or polymers. Domains of solubility for nonregular solvents or solutes may be shown on three-dimensional plots showing the relationships between the "regular solvents," dipolar and "H"-bonding contributions, and the solubility parameter values.

Plasticizers are typically nonvolatile solvents with $\Delta\delta$ values between the polymer and the plasticizer of less than 1.8 H. Plasticizers reduce the intermolecular attractions (CED and δ) of polymers such as cellulose nitrate (CN) and PVC and make processing less difficult. While camphor and tricresyl phosphate, which are plasticizers for CN and PVC, were discovered empirically, it is now possible to use δ values to screen potential plasticizers.

Complete data for solubility parameters may be found in the *Polymer Handbook* (Burrell, 1974). Typical data are tabulated in Tables 3.2 and 3.3.

3.3 VISCOMETRY

In looking at the relationship between the force, f, necessary to move a plane of area, A, relative to another plane a distance, d, from the initial plane, it is found that the force is proportional to the area of the plane and inversely proportional to the distance, or

$$f \propto A/d$$

In order to make this a direct relationship, a *proportionality factor* is introduced. This factor is called the *coefficient of shear viscosity* or simply *viscosity*:

$$f = \eta \, (A/d)$$

Viscosity can be considered as a measure of the resistance of a material to flow. In fact, the inverse of viscosity is given the name *fluidity*. As a material's resistance to flow increases, its viscosity increases.

Viscosities have been reported using a number of units. The CGS (centigrams, grams, seconds) unit of viscosity is called the Poise, which is dyne seconds per square centimeter. Another widely employed unit is Pascals (or Pas), which is Newton seconds per square centimeter. The relationship is 10 Poise = 1 Pas.

Table 3.4 gives the (general) viscosity for some common materials. It is important to note the wide variety of viscosities of materials from gases such as air to viscoelastic solids such as glass.

In polymer science, we typically do *not* utilize direct measures of viscosity, but rather employ relative measures—measuring the flow rate of one material relative to that of a second material.

Viscometry is one of the most widely utilized methods for the characterization of polymer molecular weight since it provides the easiest and most rapid means of obtaining molecular weight-related data and requires a minimum amount of instrumentation. A most obvious characteristic of polymer solutions is their high viscosity, even when the amount of added polymer is small.

Table 3.2 Solubility Parameters (δ) for Typical Solvents

Poorly hydrogen-bonded solvents (δ_p)		Moderately hydrogen-bonded solvents (δ_m)		Strongly hydrogen-bonded solvents (δ_s)	
Hydrogen	3.0	Diisopropyl ether	6.9	Diethylamine	8.0
Dimethylsiloxane	5.5	Diethyl ether	7.4	n-Amylamine	8.7
Difluorodichloromethane	5.5	Isoamyl acetate	7.8	2-Ethylhexanol	9.5
Ethane	6.0	Diisobutyl ketone	7.8	Isoamyl alcohol	10.0
Neopentane	6.3	Di-n-propyl ether	7.8	Acetic acid	10.1
Amylene	6.9	sec-Butyl acetate	8.2	Meta-cresol	10.2
Nitro-n-octane	7.0	Isopropyl acetate	8.4	Aniline	10.3
n-Pentane	7.0	Methyl amyl ketone	8.5	n-Octyl alcohol	10.3
n-Octane	7.6	Butyraldehyde	9.0	tert-Butyl alcohol	10.6
Turpentine	8.1	Ethyl acetate	9.0	n-Amyl alcohol	10.9
Cyclohexane	8.2	Methyl ethyl ketone	9.3	n-Butyl alcohol	11.4
Cymene	8.2	Butyl cellosolve	9.5	Isopropyl alcohol	11.5
Monofluorodichloromethane	8.3	Methyl acetate	9.6	Diethylene glycol	12.1
Dipentene	8.5	Dichloroethyl ether	9.8	Furfuryl alcohol	12.5
Carbon tetrachloride	8.6	Acetone	9.9	Ethyl alcohol	12.7
n-Propylbenzene	8.6	Dioxane	10.0	N-ethylformamide	13.9
p-Chlorotoluene	8.8	Cyclopentanone	10.4	Methanol	14.5
Decalin	8.8	Cellosolve	10.5	Ethylene glycol	14.6
Xylene	8.8	N,N-dimethylacetamide	10.8	Glycerol	16.5
Benzene	9.2	Furfural	11.2	Water	23.4
Styrene	9.3	N,N-dimethylformamide	12.1		
Tetralin	9.4	1,2-Propylene carbonate	13.3		
Chlorobenzene	9.5	Ethylene carbonate	14.7		
Ethylene dichloride	9.8				
p-Dichlorobenzene	10.0				
Nitroethane	11.1				
Acetonitrile	11.9				
Nitroethane	12.7				

Table 3.3 Approximate Solubility Parameter Values for Polymers

Polymer	δ_p	δ_m	δ_s
Polytetrafluoroethylene	5.8–6.4		
Ester gum	7.0–10.6	7.4–10.8	9.5–10.9
Alkyd 45% soy oil	7.0–11.1	7.4–10.8	9.5–11.8
Silicone DC-1107	7.0–9.5	9.3–10.8	9.5–11.5
Poly(vinyl ethyl ether)	7.0–11.0	7.4–10.8	9.5–14.0
Poly(butyl acrylate)	7.0–12.5	7.4–11.5	
Poly(butyl methacrylate)	7.4–11.0	7.4–10.0	9.5–11.2
Silicone DC-23	7.5–8.5	7.5–8.0	9.5–10.0
Polyisobutylene	7.5–8.0	—	—
Polyethylene	7.7–8.2	—	—
Gilsonite	7.9–9.5	7.8–8.5	—
Poly(vinyl butyl ether)	7.8–10.6	7.5–10.0	9.5–11.2
Natural rubber	8.1–8.5	—	—
Hypalon 20 (sulfochlorinated LDPE)	8.1–9.8	8.4–8.8	—
Ethylcellulose N-22	8.1–11.1	7.4–10.8	9.5–14.5
Chlorinated rubber	8.5–10.6	7.8–10.8	—
Dammar gum	8.5–10.6	7.8–10.0	9.5–10.9
Versamid 100	8.5–10.6	8.5–8.9	9.5–11.4
Polystyrene	8.5–10.6	9.1–9.4	—
Poly(vinyl acetate)	8.5–9.5	—	—
Poly(vinyl chloride)	8.5–11.0	7.8–10.5	—
Phenolic resins	8.5–11.5	7.8–13.2	9.5–13.6
Buna N (butadiene-acrylonitrile copolymer)	8.7–9.3	—	—
Poly(methyl methacrylate)	8.9–12.7	8.5–13.3	—
Carbowax 4000 (poly(ethylene oxide))	8.9–12.7	8.5–14.5	9.5–14.5
Thiokol (poly(ethylene sulfide))	9.0–10.0	—	—
Polycarbonate	9.5–10.6	9.5–10.0	—
Pliolite P-1230 (cyclized rubber)	9.5–10.6	—	—
Mylar (poly(ethylene terephthalate))	9.5–10.8	9.3–9.9	—
Vinyl chloride-acetate copolymer	9.5–11.0	7.8–13.0	—
Polyurethane	9.8–10.3	—	—
Styrene acrylonitrile copolymer	10.6–11.1	9.4–9.8	—
Vinsol (rosin derivative)	10.6–11.8	7.7–13.0	9.5–12.5
Epon 1001 (epoxy)	10.6–11.1	8.5–13.3	—
Shellac	—	10.0–11.0	9.5–14.0
Polymethacrylonitrile	—	10.6–11.0	—
Cellulose acetate	11.1–12.5	10.0–14.5	—
Nitrocellulose	11.1–12.5	8.0–14.5	12.5–14.5
Polyacrylonitrile	—	12.0–14.0	—
Poly(vinyl alcohol)	—	—	12.0–13.0
Nylon-66 (poly(hexamethylene adipamide))	—	—	13.5–15.0
Cellulose	—	—	14.5–16.5

Table 3.4 Viscosities of Selected Common Materials

Substance	General Viscosity (MPas)
Air	0.00001
Water	0.001
Polymer latexes/paints	0.01
PVC plastisols	0.1
Glycerol	10.
Polymer resins and "pancake" syrups	100.
Liquid polyurethanes	1000.
Polymer "melts"	10000.
Pitch	100000000.
Glass	1000000000000000000000.

The ratio of the viscosity of a polymer solution to that of the solvent is called relative viscosity (η_r). This value minus 1 is called the specific viscosity (η_{sp}), and the reduced viscosity (η_{red}), or viscosity number, is obtained by dividing η_{sp} by the concentration of the solution (c). The intrinsic viscosity, or limiting viscosity number, is obtained by extrapolating η_{red} to zero concentration. These relationships are given in Table 3.5, and a typical plot of η_{sp}/c and ln η_r/c as a function of concentration is given in Fig. 3.7.

Staudinger showed that the intrinsic viscosity of a solution ($[\eta]$), like the viscosity of a melt (η), was related to the average molecular weight of the polymer (M). The present form of this relationship is expressed by the Mark-Houwink equation [Eq. (3.25)], in which the proportionality constant K is characteristic of the polymer and solvent and the exponential a is a function of the shape of the polymer coil in a solution. In a θ solvent, the a value for the ideal statistical coil is 0.5. This value, which is actually a measure of the interaction of the solvent and

Table 3.5 Commonly Used Viscometry Terms

Common name	Recommended name (IUPAC)	Definition	Symbol
Relative viscosity	Viscosity ratio	η/η_0	$\eta_{rel} = \eta_r$
Specific viscosity	—	$(\eta/\eta_0) - 1$ or $(\eta - \eta_0)/\eta_0$ or $\eta_r - 1$	η_{sp}
Reduced viscosity	Viscosity number	η_{sp}/c	η_{red} or η_{sp}/c
Inherent viscosity	Logarithmic viscosity number	ln $(\eta_r)/c$	η_{inh} or ln $(\eta_r)/c$
Intrinsic viscosity	Limiting viscosity number	$\lim(\eta_{sp}/c)_{c \to 0}$ or $\lim(\ln \eta_r/c)_{c \to 0}$	$[\eta]$ or LVN

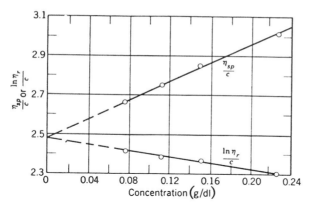

Figure 3.7 Reduced and inherent viscosity-concentration curves for a polystyrene in benzene. (From R. Ewart, in *Advances in Colloid Science*, Vol. II (H. Mark and G. Whitby, eds.). Wiley Interscience, New York, 1946. With permission from the Interscience Division of John Wiley and Sons, Publishers.)

polymer, increases as the coil expands in good solvents, and the value is between 1.8 and 2.0 for a rigid polymer chain extended to its full contour length and 0 for spheres. When a equals 1.0, the Mark-Houwink equation becomes the Staudinger viscosity equation. However, the value of a is usually 0.5 to 0.8 in polymer solutions. K generally has values in the range of 10^{-2} to 10^{-4} ml/g. Sample values are given in Table 3.6. A more complete collection of K and a values can be found in the *Polymer Handbook* (Burrell, 1974).

Since the relative viscosity η_{rel} is a ratio of viscosities, it is dimensionless, as is the specific viscosity. However, since the value for reduced viscosity is obtained by dividing η_{sp} by the concentration, η_{red} is expressed in reciprocal concentration units, such as milliliters per gram. The intrinsic viscosity will have the same units.

$$[\eta] = K\overline{M}^a \tag{3.25}$$

Unlike most of the methods discussed in the following chapter, viscometry does not yield absolute molecular weight values, but rather is only a relative measure of a polymer's molecular weight. The reason may be stated in several ways. First, an exact theory of polymer solution viscosity as related to chain size is still in the formulation stage. Second, an expression, such as [Eq. (3.25)], cannot be directly used to relate (absolute) polymer viscosity and polymer molecular weight using only viscometry measurements since two additional unknowns, K and a, must be determined.

Thus, viscometry measurements must be correlated with an "absolute molecular weight method" such as light scattering. Taking the log of Eq. (3.25) yields Eq. (3.26):

$$\log [\eta] = a \log \overline{M} + \log K \tag{3.26}$$

This predicts a linear relationship between log $[\eta]$ and log $\overline{M}$ with a slope of a and intercept log K. Experimentally the viscosity is determined for several pol-

Table 3.6 Typical K Values for the Mark-Houwink Equation

Polymer	Solvent	Temp. (°K)	$K \times 10^5$ dl g^{-1}
LDPE (low-density polyethylene)	Decalin	343	39
HDPE (high-density polyethylene)	Decalin	408	68
Polypropylene (isotactic)	Decalin	408	11
Polystyrene	Decalin	373	16
Poly(vinyl chloride)	Chlorobenzene	303	71
Poly(vinyl acetate)	Acetone	298	11
Poly(methyl acrylate)	Acetone	298	6
Polyacrylonitrile	Dimethylformamide	298	17
Poly(methyl methacrylate)	Acetone	298	10
Poly(ethylene terephthalate)	m-Cresol	298	1
Nylon-66	90% aqueous formic acid	298	110

ymer samples varying only in molecular weight. Then the molecular weight of each sample is determined using an absolute method. A plot of log $[\eta]$ versus log $\overline{M}$ is constructed enabling the determination of a and K. Often K is determined by inserting a known $[\eta]$ and $\overline{M}$ value along with the calculated a value and solving for K. It is customery to distinguish the type of a and K value determined. For instance, if light scattering were employed to determine molecular weights then the a and K values and subsequent $\overline{M}$ values are designated as weight-average values.

After calculation of a and K values for a given polymer-solvent pair, $\overline{M}$ can be easily calculated using a determined $[\eta]$ and Eq. (3.25).

Flory, Debye, and Kirkwood have shown that $[\eta]$ is directly proportional to the effective hydrodynamic volume of the polymer in solution and inversely proportional to the molecular weight (M). The effective hydrodynamic volume is the cube of the root-mean-square end-to-end distance $(\sqrt{\overline{r^2}})^3$. The proportionality constant (ϕ) in the Flory equation for hydrodynamic volume [Eq. (3.27)] has been considered to be a universal constant independent of solvent, polymer, temperature, and molecular weight.

$$[\eta] = \phi(\bar{r})^{23/2}M^{-1} \tag{3.27}$$

The actual average end-to-end distance, $\bar{r}$, is related to the nonsolvent expanded average end-to-end distance, $\bar{r}_o$, using the Flory expansion factor, α, as follows:

$$\bar{r} = \bar{r}_o\alpha \tag{3.28}$$

Substitution of Eq. (3.28) into Eq. (3.27) and rearrangement gives

$$[\eta]M = \phi(\bar{r}^{-2})^{3/2}\alpha^3 \tag{3.29}$$

Values of α range from 1 for Flory theta-solvents to about 3 for high polymers in good solvents.

In Eq. (3.25) it is found that for random coils, "a" ranges from 0.5 for theta solvents to 0.8 for good solvents, 0 for hard spheres, about 1 for semicoils, and 2 for rigid rods.

The θ temperature corresponds to the Boyle point in an imperfect gas and is the range in which the virial coefficient B in the expanded gas law becomes zero. This same concept applies to the modification of the gas law (PV = nRT) used to determine the osmotic pressure (π) of a polymer solution as shown below:

$$\pi = \frac{RT}{\overline{M}} C + BC^2 + \cdots \tag{3.30}$$

where R is the universal gas constant, T is the temperature in degrees Kelvin, $\overline{M}$ is the number-average molecular weight, and C is the concentration of polymer in solution.

For linear chains at their θ temperature, i.e., the temperature at which the chain attains its unperturbed dimensions, the Flory equation resembles the Mark-Houwink equation in which α is equal to 1.0, as shown below:

$$[\eta] = KM^{1/2}\alpha^3 = KM^{1/2} \tag{3.31}$$

The intrinsic viscosity of a solution, like the melt viscosity, is temperature dependent and will decrease as the temperature increases as shown by the following Arrhenius equation:

$$[\eta] = Ae^{E/RT} \tag{3.32}$$

However, if the original temperature is below the θ temperature, the viscosity will increase when the mixture of polymer and solvent is heated to a temperature slightly above the θ temperature.

Provided the temperature is held constant, the viscosity of a solution may be measured in any simple viscometer such as an Ubbelohde viscometer, a falling ball viscometer, or a rotational viscometer, such as a Brookfield viscometer. It is customary to describe the viscosity of a solid plastic in terms of the melt index, which is the weight in grams of a polymer extruded through an orifice in a specified time.

Viscosity measurements of polymer solutions are carried out using a viscometer, such as any of those pictured in Fig. 3.8, placed in a constant temperature bath with temperature controlled to plus or minus 0.1°C. The most commonly used viscometer is the Ubbelohde viscometer, which, because of the side arm, gives flow times independent of the volume of liquid in the reservoir bulb.

The following relationship exists for a given viscometer:

$$\frac{\eta}{\eta_o} = \frac{\rho t}{\rho_o t_o} = \eta_r \tag{3.33}$$

where t and t_o are the flow times for the polymer solution and solvent, respectively, and ρ is the density of the polymer solution.

Viscometry measurements are generally made on solutions that contain 0.01 to 0.001 g of polymer per milliliter of solution. For such dilute solutions, $\rho \simeq \rho_o$, giving

$$\frac{\eta}{\eta_o} = \frac{t}{t_o} = \eta_r \tag{3.34}$$

Thus, η_r is simply a ratio of flow times for equal volumes of polymer solution and solvent. Reduced viscosity is related to $[\eta]$ by a virial equation as follows:

$$\eta_{sp}/c = [\eta] + k_1 [\eta]^2 c + k'[\eta]^3 c^2 + \cdots \tag{3.35}$$

For most systems, Eq. (3.35) reduces to the Huggins viscosity relationship

$$\eta_{sp}/c = [\eta] + k_1 [\eta]^2 c \tag{3.36}$$

which allows $[\eta]$ to be determined from the intercept of a plot of η_{sp}/c versus c and is the basis for the top plot given in Fig. 3.7.

Another relationship often used in determining $[\eta]$ is called the inherent viscosity equation and is as follows:

$$\frac{\ln \eta_r}{c} = [\eta] - k_2 [\eta]^2 c \tag{3.37}$$

Again, in a plot of $\ln \eta_r/c$ versus c, an extrapolation to c equals zero and allows the calculation of $[\eta]$. Plotting using Eq. (3.36) is more common than plot-

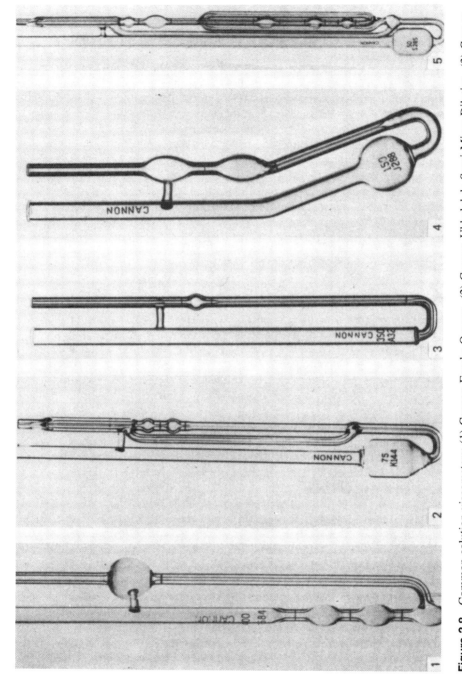

Figure 3.8 Common solution viscometers: (1) Cannon-Fenske Opaque, (2) Cannon-Ubbelohde Semi-Micro Dilution, (3) Cannon-Manning Semi-Micro, (4) Cannon-Fenske Routine, (5) Cannon-Ubbelohde Shear Dilution. (With permission Cannon Instrument Co.)

ting using Eq. (3.37), even though the latter probably yields more precise values of η since k_1 is generally larger than k_2.

While k_1 and k_2 are mathematically related by

$$k_1 + k_2 = 0.5 \tag{3.38}$$

many systems appear not to follow this relationship.

Problem

Calculate the relative viscosity, specific viscosity, reduced viscosity, and inherent viscosity of a 0.5% (made by dissolving 0.25 g of polymer in 50 ml of solvent) solution where the time for solvent flow between the two appropriate marks was 60 s and the time of flow for the solution was 80 s. Using the relationship $t/t_o = \eta/\eta_o = \eta_r$, a value of η is calculated as follows: 80 s/60 s = 1.3 = η_r. The specific viscosity is determined employing any of the following three relationships.

$$\frac{\eta}{\eta_o} - 1 = \frac{\eta - \eta_o}{\eta_o} = \eta_r - 1$$

Selecting the first one yields

$$\frac{\eta}{\eta_o} - 1 = \frac{t}{t_o} - 1 = \frac{80}{60} - 1 = 1.3 - 1 = 0.3$$

To determine the reduced viscosity, an appropriate concentration unit must be selected. The most widely employed concentrations in viscosity determinations are g/ml (g/cc) and g/dl or %. The units g/cc are recommended by IUPAC, while the units of % or g/dl are the most commonly used units. Since the problem gives the units in %, this unit will be employed.

Reduced viscosity is then calculated as follows:

$$\frac{\eta_{sp}}{c} = \frac{0.3}{0.5} = 0.06\%^{-1} \quad \text{or} \quad 0.6 \frac{dl}{g}$$

Problem

Determine the molecular weight of a polystyrene sample which has an a value of 0.60, a K value of 1.6×10^{-4} dl/g, and a limiting viscosity number or intrinsic viscosity of 0.04 dl/g. The molecular weight can be found by the relationship:

$$[\eta] = K\overline{M}^a$$
$$\log [\eta] = a \log M + \log K$$
$$\log M = [\log [\eta] - \log K]/a = [\log (0.04) - \log (1.6 \times 10^{-4}]/0.60 \simeq 4$$
$$\therefore M = 1 \times 10^4$$

SUMMARY

1. Since polymers have properties of both solids and fluids, they are called viscoelastic materials. Both Hooke's law for elastic solids $(S = E\gamma)$ and Newton's law for ideal liquids $[S = \eta(d\gamma/dt)]$ are used with modifications to describe polymers.

2. If there is no change in volume when an isotropic solid is stretched, it has a Poission's ratio of 0.5, but this value decreases as motion is restricted by cross-linking or by increased cohesive energy density (CED). The latter is a measure of intermolecular forces. An isotropic solid is one having similar properties in all directions. The viscosity $[\eta]$ decreases as the temperature is increased.

3. The behavior of viscoelastic materials may be described using combinations of Maxwell and Voigt-Kelvin models or elements. In the Maxwell element, which consists of a Hookean spring and a Newtonian dash pot in series, the application of stress causes an instantaneous elongation of the spring followed by a slow response of the piston in the cylinder or dash pot. The term relaxation time (τ), which is equal to η/G, is used to describe the time when the stress and elongation have reached equilibrium or when the stress is equal to 1/e of its original value.

4. In the Voigt-Kelvin model or element, which consists of a Hookean spring and Newtonian dash pot or shock absorber in parallel, the elastic response to the application of stress is retarded by the resistance of the liquid in the dash pot. The term retardation time (τ), which is equal to η/G, is used to describe the time for the model to decrease to $1 - (1/3)$, or 0.63 of its original value.

5. A Bingham plastic is one which does not flow until the applied stress exceeds a threshold stress value. In non-Newtonian fluids, the viscosity may increase with time (rheopectic) or decrease with time (thixotropic). If the shear rate does not increase as rapidly as the applied stress, the system is dilatant, and if it increases more rapidly, the system is pseudoplastic.

6. Eyring has described polymer flow as a random hole-filling process in which the chain segments overcome an energy barrier and fill the holes by successive correlated jumps of segments along the polymer chain.

7. Since the viscoelastic properties of polymers are similar at their glass transition temperatures, these properties are also similar at temperatures up to 100°K above T_g. The shift factor for a reference temperature such as T_g may be calculated from the WLF equation, which includes constants for the free volume.

8. A polymer dissolves by a swelling process followed by a dispersion process or disintegration of the swollen particles. This process may occur if there is a decrease in free energy. Since the second step in the solution process involves an increase in entropy, it is essential that the change in enthalpy be negligible or negative to assure a negative value for the change in free energy.

9. Flory and Huggins developed an interaction parameter (χ_1) which may be used as a measure of the solvent power of solvents for amorphous polymers. Flory and Krigbaum introduced the term θ temperature at which a polymer of infinite molecular weight exists as a statistical coil in a solvent.

10. Hildebrand used solubility parameters which are the square root of CED to predict the solubility of nonpolar polymers in nonpolar solvents. This concept is also applicable to mixtures of solvents. For polar solvents it is also necessary to consider dipole-dipole interactions and hydrogen bonding in predicting solubility.

11. The molecular weight of a polymer is proportional to the intrinsic viscosity of its solution ($[\eta]$) when the polymer chain is extended to its full contour length. The value $[\eta]$, or limiting viscosity number, is proportional to the square root of the molecular weight M when the polymer is in a θ solvent. In general, $[\eta] = KM^a$ where a, which is a measure of the shape of the polymer chain, is usually 0.5 to 0.8. The constant K is dependent on the polymer and solvent studied.

12. The intrinsic viscosity is the limiting reduced viscosity (η_{red}) at zero concentration. The reduced viscosity of viscosity number is equal to the specific viscosity (η_{sp}) divided by concentration. The value η_{sp} is obtained by subtracting 1 from the relative viscosity (η_r), which is the ratio of the viscosities of the solution and the solvent. Plasticizers are nonvolatile good solvents which reduce the CED and T_g of a polymer.

GLOSSARY

a: Symbol for exponent in Mark-Houwink equation; a measure of solvent polymer interaction.

a_T: Constant in the WLF equation related to holes or free volume of a polymer.

a_T: Shift factor relative to a reference temperature.

alpha (α): The linear expansion ratio of a polymer molecule in a Flory θ solvent.

aniline point: A measure of the aromaticity of a solvent.

Arrhenius equation for viscosity: $\eta = A\ e^{E/RT}$.

biaxially stretching: Stretching of a film in two directions perpendicular to each other.

Bingham, E. C.: The father of rheology.

Bingham equation: $s - s_o = \eta \dfrac{d\gamma}{dt}$.

Bingham plastic: A plastic that does not flow until the external stress exceeds a critical threshold value (s_o).

chi (χ): Flory-Huggins interaction parameter.

cloud point: The temperature at which a polymer starts to precipitate when the temperature is lowered.

CN: Symbol for cellulose nitrate.

cohesive energy density (CED): The heat of vaporization per unit volume $\Delta E(V^{-1})$.

creep: Cold flow of a polymer.

critical solution temperature (T_c): The temperature at which the two liquid phases, containing an amorphous polymer, coalesce.

dash pot: A model for Newtonian fluids consisting of a piston and a cylinder containing a viscous liquid.

delta (δ): Symbol for solubility parameter.

dilatant: Shear thickening agent; describes a system where the shear rate does not increase as rapidly as the applied stress.

e: Symbol for base of Napierian logarithms (e = 2.718).

E: Energy of activation.

E: Symbol for Young's modulus of elasticity.

effective hydrodynamic volume [$(\sqrt{\overline{r^2}})^3$]: The cube of the root-mean-square end-to-end distance of a polymer chain.

eta (η): Symbol for viscosity.

Flory-Huggins theory: A theory used to predict the equilibrium behavior between liquid phases containing an amorphous polymer.

G: Symbol for shear modulus.

G: Symbol for Small's molar attraction constants.

gamma (γ): Symbol for strain.

Gibbs equation: The relationship between free energy (ΔG), enthalpy (ΔH), and entropy (ΔS); $\Delta G = \Delta H - T \Delta S$ at constant T.

H: Symbol for enthalpy or heat content and for Hildebrand units.

Herschel-Buckley equation: $(s - s_o)^n = \phi(d\gamma/dt)$.

Hildebrand (H): Unit used in place of $(cal\ cm^{-3})^{0.5}$ for solubility parameter values.

Hookean: Obeys Hooke's law.

Hooke's law: $s = E\gamma$.

isotropic: Having similar properties in all directions.

K: Symbol for Kelvin or absolute temperature scale.

Kauri-Butanol values: A measure of the aromaticity of a solvent.

log: Common logarithm based on the number 10.

M: Symbol for Hayes chain stiffening constant.

M: Symbol for molecular weight.

$\overline{M}$: Symbol for average molecular weight.

Mark-Houwink equation: $[\eta] = K\overline{M}^a$, where a is 0.5 for a statistical coil in a θ solvent and 2.0 for a rigid rod.

Maxwell element or model: A model in which an ideal spring and dash pot are connected in series, used to study the stress relaxation of polymers.

melt index: A measure of flow related inversely to melt viscosity, the time for 10 g of a polymer, such as a polyolefin, to pass through a standard orifice at a specified time and temperature.

modulus: Stress per unit strain. A high modulus plastic is stiff and has very low elongation.

Newtonian fluid: A fluid whose viscosity (η) is proportional to the applied viscosity gradient $d\gamma/dt$.

Newton's law: Stress is proportional to flow; $s = \eta(d\gamma/dt)$.

osmotic pressure (π): The pressure exerted by a solvent when it is separated from a solution by a membrane.

phi (ϕ): Proportionality constant in Flory equation $= 2.5 \times 10^{23}\ mol^{-1}$.

plasticizer: A nonvolatile solvent which is compatible with a hard plastic and reduces its T_g.

Poisson's ratio: The ratio of the percentage change in length of a specimen under tension to its percentage change in width.

pseudoplastic: Shear thinning agent; system where the shear rate increases faster than the applied stress.

relaxation time (τ): Time for stress of a polymer under constant strain to decrease to $1/e$ or 0.37 of its original value.

retardation time (τ): Time for the stress in a deformed polymer to decrease to 63% of the original value.

rheology: The science of flow.

rheopectic: A rheopectic liquid is one whose viscosity increases with time.

s: Symbol for applied stress.

S: Symbol for entropy or measure of disorder.

shear: Stress caused by planes sliding by each other, as in a pair of shears or the greasing or polishing of a flat surface.

solubility parameter (δ): A numerical value equal to $\sqrt{\text{CED}}$ which can be used to predict solubility.

stress (s): Force per unit area.

stress relaxation: The relaxation of a stressed specimen with time after the load is removed.

theta (θ) solvent: A solvent in which the polymer exists as a statistical coil and where the second virial constant B equals zero at the θ temperature.

theta temperature: The temperature at which a polymer of infinite molecular weight starts to precipitate from a solution.

thixotropic: A thixotropic liquid is one whose viscosity decreases with time.

turbidimetric titration: A technique in which a poor solvent is added slowly to a solution of a polymer and the point of incipient turbidity is observed.

V: Symbol for molar volume.

velocity gradient: $d\gamma/dt$, $\dot{\gamma}$, or flow rate.

viscoelastic: Having the properties of a liquid and a solid.

viscosity: Resistance to flow.

viscosity, intrinsic [η]: The limiting viscosity number obtained by extrapolation of the reduced viscosity to zero concentration.

viscosity, reduced: The specific viscosity divided by the concentration.

viscosity, relative: The ratio of the viscosities of a solution and its solvent.

viscosity, specific: The difference between the relative viscosity and 1.

Voigt-Kelvin element or model: A model consisting of an ideal spring and dash pot in parallel in which the elastic response is retarded by viscous resistance of the fluid in the cylinder.

WLF: Williams, Landel, and Ferry equation for predicting viscoelastic properties at temperatures above T_g when these properties are known for one specific temperature such as T_g.

EXERCISES

1. What is the difference between morphology and rheology?

2. Which of the following is viscoelastic: (a) steel, (b) polystyrene, (c) diamond, or (d) neoprene?

3. Define G in Hooke's law.

4. Which would be isotropic: (a) a nylon filament, (b) an extruded pipe, or (c) ebonite?

5. Would Poisson's ratio increase or decrease when (a) a plasticizer is added to rigid PVC and (b) the amount of sulfur used in the vulcanization of rubber is increased?

6. How would the slopes in Fig. 3.1a differ for polyisobutylene and polystyrene?

7. Which will have the higher relaxation time (τ): (a) unvulcanized rubber, or (b) ebonite?

8. Which would be more readily extruded through a die: (a) a pseudoplastic, or (b) a dilatant substance?

9. Which would increase in volume when stretched: (a) plasticized PVC, or (b) rigid PVC?

10. According to the Eyring theory, which would have the higher percentage of holes: (a) polystyrene at its T_g, or (b) hevea rubber at its T_g?

11. In designing a die for a pipe with an outside diameter of 5 cm, would you choose inside dimensions of (a) less than 5 cm, (b) 5 cm, or (c) greater than 5 cm for the die?

12. At what temperature would the properties of polystyrene resemble those of hevea rubber at 35°K above its T_g?

13. What is the significance of the constants in the WLF equation?

14. Define the proportionality constant in Newton's law.

15. What term is used to describe the decrease of stress at constant length with time?

16. In which element or model for a viscoelastic body, (a) Maxwell, or (b) Voigt-Kelvin, will the elastic response be retarded by viscous resistance?

17. According to Hildebrand, what is a regular solvent?

18. Which of the two steps that occur in the solution process, (a) swelling and (b) dispersion of the polymer particles, can be accelerated by agitation?

19. Define CED.

20. For solution to occur ΔG must be: (a) 0, (b) < 0, or (c) > 0.

21. Will a polymer swollen by a solvent have higher or lower entropy than the solid polymer?

22. Define the change in entropy in the Gibbs free energy equation.

23. Is a liquid that has a value of 0.3 for its interaction parameter (χ_1) a good or a poor solvent?

24. What is the value of ΔG at the θ temperature?

25. What term is used to describe the temperature at which a polymer of infinite molecular weight precipitates from a dilute solution?

26. At which temperature will the polymer coil be larger in a poor solvent: (a) at the θ temperature, (b) above the θ temperature, or (c) below the θ temperature?

27. If δ for water is equal to 23.4 H, what is the CED for water?

28. What is the heat of mixing of two solvents having identical δ values?

29. If the density (D) is 0.85 g/cm³ and the molar volume (V) is 1,176,470 cm³, what is the molecular weight?

30. Use Small's molar attraction constants to calculate δ for polystyrene.

$$\begin{array}{c} \text{H} \quad \text{H} \\ | \quad \ | \\ +\!\!\text{C}\!-\!\text{C}\!\!+ \\ | \quad \ | \\ \text{H} \quad \text{C}_6\text{H}_5 \end{array}$$

31. Calculate M for a polymer having a δ value of 10 H and a T_g value of 325°K (M = chain stiffness).

32. Why do δ values decrease as the molecular weight increases in a homologous series of aliphatic polar solvents?

33. Which would be the better solvent for polystyrene: (a) n-pentane, (b) benzene, or (c) acetonitrile?

34. Which will have the higher slope when its reduced viscosity or viscosity number is plotted against concentration: a solution of polystyrene (a) in benzene, or (b) in n-octane?

35. What is the value of the virial constant B in Eq. (3.28) at the θ temperature?

36. When is the Flory equation [Eq. (3.27)] similar to the Mark-Houwink equation?

37. What is the term used for the cube root of the hydrodynamic volume?

38. Explain why the viscosity of a polymer solution decreases as the temperature increases.

39. Which sample of LDPE has the higher average molecular weight: (a) one with a melt index of 10, or (b) one with a melt index of 8?

BIBLIOGRAPHY

Aklonis, J. J., MacKnight, W. J. (1983): *Introduction to Polymer Viscoelasticity*, Wiley-Interscience, New York.

Alfrey, T. (1948): *Mechanical Behavior of High Polymers*, Interscience, New York.

Allen, P. W. (1959): Solubility and choice of solvents, in *Techniques of Polymer Characterization* (P. W. Allen, ed.), Butterworths, London.

Barton, A. F. M. (1983): *Solubility Parameters and Other Cohesion Parameters*, CRC Press, Boca Raton, Florida.

Bohdanecky, M., Kovar, J. (1982): *Viscosity of Polymer Solutions*, Elsevier, Amsterdam.

Burrell, H. (1974): Solubility parameter values, in *Polymer Handbook* (J. Brandrup and E. H. Immergut, eds.), Wiley, New York.

Carraher, C. E. (1970): Polymer models, J. Chem. Ed., *47*:58.

Christensen, R. M. (1971): *Theory of Viscoelasticity, an Introduction*, Academic, New York.

Dack, M. R. J. (1975): *Solutions and Solubilities Techniques of Chemistry*, Wiley, New York.

Doolittle, A. K. (1954): *The Technology of Solvents and Plasticizers*, Wiley, New York.

Harris, F., Seymour, R. B. (1977): *Solubility Property Relationships in Polymer*, Academic, New York.

Hearle, J. W. S. (1982): *Polymers and Their Properties*, Ellis Horwood, Chichester, England.

Hildebrand, J. H., Prausnetz, J. M., Scott, R. L. (1970): *Regular and Related Solutions*, Van Nostrand-Reinhold, New York.

Huggins, M. L. (1942): Some properties of solutions of long chain compounds, J. Phys. Chem., *46*:151.

Janeschitz-Kriegl, H. (1983): *Polymer Melt Rheology and Flow Birefringence*, Springer-Verlag, New York.

Kennedy, P. K. (1995). *Flow Analysis of Injection Molds*, Hanser-Gardner, Cincinnati, Ohio.

Kinloch, A. J., Young, R. J. (1983): *Fracture Behavior of Polymers*, Applied Science, New York.

Kurñta, M. (1982): *Thermodynamics of Polymer Solutions*, Gordon and Breach, New York.

Leonov, A., Prokunin, A. (1993): *Nonlinear Viscoelastic Effects in Flows of Polymer Melts*, Elsevier, New York.

Mark, J. E., Eisenberg, A., Graessley, W. W., Mandelkern, L., Koenig, J. L. (1984): *Physical Properties of Polymers*, ACS, Washington, D.C.

Matsuoka, S. (1992): *Relaxation Phenomena in Polymers*, Hanser-Gardner, Cincinnati, Ohio.

Morawetz, H. (1983): *Macromolecules in Solution*, Krieger Publishing Co., Malabar, Florida.

Nielsen, L. E. (1974): *Mechanical Properties of Polymers*, Marcel Dekker, New York.

Seymour, R. B. (1975): Solubility parameters of organic compounds, in *Handbook of Chemistry and Physics*, C-720, 56th ed., CRC Press, Cleveland.

Small, P. A. (1953): Some factors affecting the solubility of polymers, J. Appl. Chem., *3*: 71.

Solc, K. (1982): *Polymer Compatibility and Incompatibility*, Gordon and Breach, New York.

Sperling, L. (1981): *Interpenetrating Polymer Networks and Related Materials*, Plenum, New York.

Tobolsky, A. V. (1960): *Properties and Structure of Polymers*, Wiley, New York.

Tschoegl, N. W. (1981): *The Theory of Linear Viscoelastic Behavior*, Academic, New York.

Ward, I. M. (1983): *Mechanical Properties of Solid Polymers*, 2nd ed., Wiley, New York.

Williams, J. G. (1984): *Fracture Mechanics of Polymers*, Horwood, Chichester, England.

4

Molecular Weight of Polymers

4.1 INTRODUCTION

The average molecule weight ($\overline{M}$) of a polymer is the product of the average number of repeat units or mers expressed as $\overline{n}$ or $\overline{DP}$ times the molecular weight of these repeating units. $\overline{M}$ for a group of chains of average formula $(CH_2CH_2)_{1000}$ is $1000(28) = 28,000$.

Polymerization reactions, both synthetic and natural, lead to polymers with heterogeneous molecular weights, i.e., polymer chains with a different number of units. Molecular weight distributions may be relatively broad (Fig. 4.1), as is the case for most synthetic polymers and many naturally occurring polymers. It may be relatively narrow for certain natural polymers (because of the imposed steric and electronic constraints), or may be mono-, bi-, tri-, or polymodal. A bimodal curve is often characteristic of a polymerization occurring under two distinct pathways or environments. Most synthetic polymers and many naturally occurring polymers consist of molecules with different molecular weights and are said to be polydisperse. In contrast, specific proteins and nucleic acids, like typical small molecules, consist of molecules with a specific molecular weight (M) and are said to be monodisperse.

Since typical small molecules and large molecules with molecular weights less than a critical value (Z) required for chain entanglement are weak and are readily attacked by appropriate reactants, it is apparent that the following properties are related to molecular weight. Thus, melt viscosity, tensile strength, modulus, impact strength or toughness, and resistance to heat and corrosives are dependent on the molecular weight of amorphous polymers and their molecular weight distribution (MWD). In contrast, density, specific heat capacity, and refractive index are essen-

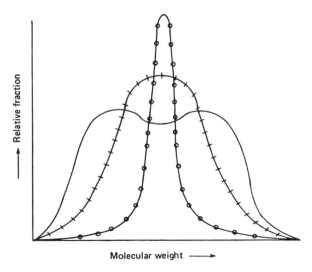

Molecular weight ⟶

Figure 4.1 Representative differential weight distribution curves: (┼┼┼┼┼┼┼) relatively broad distribution curve; (⊖⊖⊖⊖) relatively narrow distribution curve; (————) bimodal distribution curve.

tially independent of the molecular weight at molecular weight values above the critical molecular weight.

The melt viscosity is usually proportional to the 3.4 power of the average molecular weight at values above the critical molecular weight required for chain entanglement, i.e., $\eta \; \alpha \; \overline{M}^{3.4}$. Thus, the melt viscosity increases rapidly as the molecular weight increases and more energy is required for the processing and fabrication of these large molecules. However, as shown in Fig. 4.2, the strength of polymers increases as the molecular weight increases and then tends to level off.

Thus, while a value above the threshold molecular weight value (TMWV; lowest molecular weight where the desired property value is achieved) is essential for most practical applications, the additional cost of energy required for processing extremely high molecular weight polymers is seldom justified. Accordingly, it is customary to establish a commercial polymer range above the TMWV but below the extremely high molecular weight range. However, it should be noted that since toughness increases with molecular weight, extremely high molecular weight polymers, such as ultrahigh molecular weight polyethylene (UHMPE), are used for the production of tough articles such as trash barrels.

Oligomers and other low molecular weight polymers are not useful for applications where high strength is required. The word oligomer is derived from the Greek word *oligos*, mean a few. The value for TMWV will be dependent on T_g, the cohesive energy density (CED) of amorphous polymers, the extent of crystallinity in crystalline polymers, and the effect of reinforcements in polymeric composites. Thus, while a low molecular weight amorphous polymer may be satisfactory for use as a coating or adhesive, a $\overline{DP}$ value of at least 1000 may be required if the polymer is used as an elastomer or plastic. With the exception of polymers with highly regular structures, such as isotactic polypropylene, strong

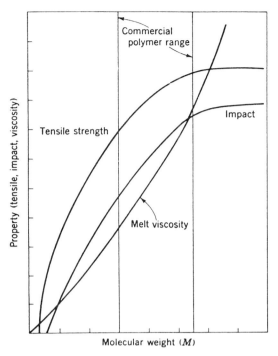

Figure 4.2 Relationship of polymer properties to molecular weight. (From *Introduction to Polymer Chemistry* by R. Seymour, McGraw-Hill, New York, 1971. Used with permission of McGraw-Hill Book Company.)

hydrogen intermolecular bonds are required for fibers. Because of their higher CED values, lower $\overline{DP}$ values are satisfactory for polar polymers used as fibers.

4.2 AVERAGE MOLECULAR WEIGHT VALUES

Small molecules such as benzene, ethylene, and glucose have precise structures such that each molecule of benzene will have 6 atoms of carbon and 6 atoms of hydrogen, each molecule of ethylene will have 2 atoms of carbon and 4 atoms of hydrogen, and each molecule of glucose will have 12 atoms of hydrogen, 6 atoms of carbon and 6 atoms of oxygen. By comparison, each molecule of poly-1,4-phenylene may have a differing number of benzene-derived moieties, while single molecules (single chains) of polyethylene may vary in the number of ethylene units, the extent and frequency of branching, the distribution of branching and the length of the branching. Finally, glucose acts as a basic unit in a whole host of naturally available materials including cellulose, lactose, maltose, starch, and sucrose (some polymeric and others oligomeric). While a few polymers, such as enzymes and nucleic acids, must have very specific structures, most polymeric materials consist of molecules, individual polymer chains, that can vary in a number of features. Here we will concentrate on the variation in the number of units composing the individual polymer chains.

While there are several statistically described averages, we will concentrate on the two that are most germane to polymers: number average and weight average. These are averages based on statistical approaches that can be described mathematically and which correspond to measurements of specific factors.

The number average value, corresponding to a measure of chain length of polymer chains, is called the *number-average molecular weight*. Physically, the number-average molecular weight can be measured by any technique that "counts" the molecules. These techniques include vapor phase and membrane osmometry, freezing point lowering, boiling point elevation, and end-group analysis.

We can describe the number average using a jar filled with plastic capsules such as those that contain tiny prizes (Fig. 4.3). Here, each capsule contains one polymer chain. All of the capsules are the same size, regardless of the size of the polymer chain contained therein. Capsules are then withdrawn, opened, and the individual chain length determined and recorded. The probability of drawing a capsule containing a chain with a specific length is dependent on the fraction of capsules containing such a chain and independent of the length of the chain. (In point of fact, this is an exercise in fantasy since the molecular size of single molecules is not easily measured.) After a sufficient number of capsules have been withdrawn and the chain size recorded, a graph like the one shown in Fig. 4.4 is constructed. The most probable value is the number-average molecular weight or number-average chain length. It should be apparent that the probability of drawing out a chain of a particular length is independent of the length or size of the polymer chain, but the probability is dependent on the number of chains of various lengths.

Figure 4.3 Jar with capsules, each of which contains a single polymer chain where the capsule size is the same and independent of the chain size, illustrating the number-average dependence on molecular weight.

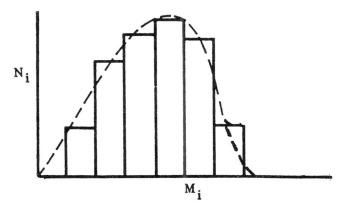

Figure 4.4 Molecular weight distribution for a polydisperse polymeric sample constructed from "capsule-derived" data.

The *weight-average molecular weight* is similarly described, except that the capsules correspond in size to the size of the polymer chain (Fig. 4.5). Thus, a capsule containing a long polymer chain will be larger than one containing a smaller chain, and the probability of drawing a capsule containing a long polymer chain will be greater because of its greater size. Again, a graph can be constructed and the maximum value is the weight-average molecular weight.

Figure 4.5 Jar with capsules, each of which contains a single polymer chain where the capsule size is directly related to the size of the polymer chain contained within the capsule.

Several mathematical moments (about a mean) can be described using the differential or frequency distribution curve, and can be described by equations. The first moment is the number-average molecular weight, $\overline{M}_n$. Any measurement that leads to the number of molecules, functional groups, or particles that are present in a given weight of sample allows the calculation of $\overline{M}_n$. The number-average molecular weight $\overline{M}_n$ is calculated like any other numerical average by dividing the sum of the individual molecular weight values by the number of molecules. Thus, $\overline{M}_n$ for three molecules having molecular weights of 1.00×10^5, 2.0×10^5, and 3.00×10^5 would be $(6.00 \times 10^5)/3 = 2.00 \times 10^5$. This solution is shown mathematically:

$$\overline{M}_n = \frac{\text{total weight of sample}}{\text{no. of molecules of } N_i} = \frac{W}{\sum\limits_{i=1}^{\infty} N_i} = \frac{\sum\limits_{i=1}^{\infty} M_i N_i}{\sum\limits_{i=1}^{\infty} N_i} \tag{4.1}$$

Most thermodynamic properties are related to the number of particles present and thus are dependent on $\overline{M}_n$.

Colligative properties are dependent on the number of particles present and are obviously related to $\overline{M}_n$. $\overline{M}_n$ values are independent of molecular size and are highly sensitive to small molecules present in the mixture. Values for $\overline{M}_n$ are determined by Raoult's techniques that are dependent on colligative properties such as ebulliometry (boiling point elevation), cryometry (freezing point depression), osmometry, and endgroup analysis.

Weight-average molecular weight, $\overline{M}_w$, is determined from experiments in which each molecule or chain makes a contribution to the measured result relative to its size. This average is more dependent on the number of heavier molecules than is the number-average molecule weight, which is dependent simply on the total number of particles.

The weight-average molecular weight $\overline{M}_w$ is the second moment or second power average and is shown mathematically as

$$\overline{M}_w = \frac{\sum\limits_{i=1}^{\infty} M_i^2 N_i}{\sum\limits_{i=1}^{\infty} M_i N_i} \tag{4.2}$$

Thus, the weight-average molecular weight for the example used in calculating $\overline{M}_n$ would be 2.33×10^5:

$$\frac{(1.00 \times 10^{10}) + (4.00 \times 10^{10}) + (9 \times 10^{10})}{6.00 \times 10^5} = 2.33 \times 10^5$$

Bulk properties associated with large deformations, such as viscosity and toughness, are particularly affected by $\overline{M}_w$ values. $\overline{M}_w$ values are determined by light scattering and ultracentrifugation techniques.

However, melt elasticity is more closely dependent on $\overline{M}_z$—the z-average molecular weight which can also be obtained by ultracentrifugation techniques. M_z is the third moment or third power average and is shown mathematically as

$$\overline{M}_z = \frac{\sum_{i=1}^{\infty} M_i^3 N_i}{\sum_{i=1}^{\infty} M_i^2 N_i} \qquad (4.3)$$

Thus, the $\overline{M}_z$ average molecular weight for the example used in calculating $\overline{M}_n$ and $\overline{M}_w$ would be 2.57×10^5:

$$\frac{(1 \times 10^{15}) + (8 \times 10^{15}) + (27 \times 10^{15})}{(1 \times 10^{15}) + (4 \times 10^{10}) + (9 \times 10^{10})} = 2.57 \times 10^5$$

While $z + 1$ and higher average molecular weights may be calculated, the major interests are in $\overline{M}_n$, $\overline{M}_v$, $\overline{M}_w$, and $\overline{M}_z$, which as shown in Fig. 4.6 are listed in order of increasing size. For heterogeneous molecular weight systems, $\overline{M}_z$ is always greater than $\overline{M}_w$ and $\overline{M}_w$ is always greater than $\overline{M}_n$. The ratio $\overline{M}_w/\overline{M}_n$ is a measure of polydispersity and is called the polydispersity index. The most probable distribution for polydisperse polymers produced by condensation techniques is a polydispersity index of 2.0. Thus, for a polymer mixture which is heterogeneous with respect to molecular weight, $\overline{M}_z > \overline{M}_w > \overline{M}_n$. As the heterogeneity decreases, the various molecular weight values converge until for homogeneous mixtures $\overline{M}_z = \overline{M}_w = \overline{M}_n$. The ratios of such molecular weight values are often used to describe the molecular weight heterogeneity of polymer samples.

Typical techniques for molecular weight determination are given in Table 4.1. The most popular techniques will be considered briefly.

All classic molecular weight determination methods require the polymer to be in solution. To minimize polymer-polymer interactions, solutions equal to and less than 1 g of polymer to 100 ml of solution are utilized. To further minimize solute interactions, extrapolation of the measurements to infinite dilution is normally practiced.

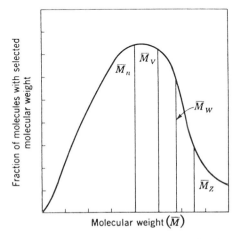

Figure 4.6 Molecular weight distributions. (From *Introduction to Polymer Chemistry* by R. Seymour, McGraw-Hill, New York, 1971. Used with permission of McGraw-Hill Book Company.)

Table 4.1 Typical Molecular Weight Determination Methods[a]

Method	Type of mol. wt. average	Applicable wt. range	Other information
Light scattering	$\overline{M}_w$	to ∞	Can also give shape
Membrane osmometry	$\overline{M}_n$	2×10^4 to 2×10^6	
Vapor phase osmometry	$\overline{M}_n$	to 40,000	
Electron and X-ray microscopy	$\overline{M}_{n,w,z}$	10^2 to ∞	Shape, distribution
Isopiestic method (isothermal distillation)	$\overline{M}_n$	to 20,000	
Ebulliometry (boiling point elevation)	$\overline{M}_n$	to 40,000	
Cryoscopy (melting point depression)	$\overline{M}_n$	to 50,000	
End-group analysis	$\overline{M}_n$	to 20,000	
Osmodialysis	$\overline{M}_n$	500–25,000	
Centrifugation			
Sedimentation equilibrium	$\overline{M}_z$	to ∞	
Archibald modification	$\overline{M}_{z,w}$	to ∞	
Trautman's method	$\overline{M}_w$	to ∞	
Sedimentation velocity	Gives a real M only for monodisperse systems	to ∞	

[a]"to ∞" means that the molecular weight of the largest particles soluble in a suitable solvent can be determined in theory.

When the exponent a in the Mark-Houwink equation is equal to 1, the average molecular weight obtained by viscosity measurements ($\overline{M}_v$) is equal to $\overline{M}_w$. However, since typical values of a are 0.5 to 0.8, the value $\overline{M}_w$ is usually greater than $\overline{M}_v$. Since viscometry does not yield absolute values of $\overline{M}$ as is the case with other techniques, one must plot $[\eta]$ against known values of $\overline{M}$ and determine the constants K and a in the Mark-Houwink equation. Some of these values are available in the *Polymer Handbook* (Burrell, 1974), and simple comparative effluent times or melt indices are often sufficient for comparative purposes and quality control where K and a are known.

For polydisperse polymer samples, molecular weight values determined from colligative properties (4.5–4.7), light scattering photometry (4.9), and the appropriate data treatment of ultracentrifugation (4.10) are referred to as "absolute molecular weights," while those determined from GPC (4.4) and viscometry (3.3) are referred to as relative molecular weights. An absolute molecular weight is one that can be determined experimentally and where the molecular weight can be related, through basic equations, to the parameter(s) measured. GPC and viscometry require calibration employing polymers of known molecular weight determined from an absolute molecular weight technique.

4.3 FRACTIONATION OF POLYDISPERSE SYSTEMS

The data plotted in Fig. 4.6 were obtained by the fractionation of a polydisperse polymer. Prior to the introduction of gel permeation chromatography (GPC), polydisperse polymers were fractionated by the addition of a nonsolvent to a polymer solution, by cooling a solution of polymer, solvent evaporation, zone melting, extraction, diffusion, or centrifugation. The molecular weight of the fractions may be determined by any of the classic techniques previously mentioned and discussed subsequently in this chapter.

The least sophisticated but most convenient technique is fractional precipitation, which is dependent on the slight change in the solubility parameter with molecular weight. Thus, when a small amount of miscible nonsolvent is added to a polymer solution at a constant temperature, the product with the highest molecular weight precipitates. This procedure may be repeated after the precipitate is removed. These fractions may also be redissolved and fractionally precipitated.

For example, isopropyl alcohol or methanol may be added dropwise to a solution of polystyrene in benzene until the solution becomes turbid. It is preferable to heat this solution and allow it to cool before removing the first and subsequent fractions. Extraction of a polymer in a Soxhlet-type apparatus in which fractions are removed at specific time intervals may also be used as a fractionation procedure.

4.4 GEL PERMEATION CHROMATOGRAPHY (GPC)

As will be noted shortly, colligative and scattering methods allow calculation of absolute molecular weight values. These methods are often time-consuming and can be expensive. For routine analysis it is often customary to determine molecular weight employing calibrated viscometry-related values (see Sec. 3.3) and gel permeation chromatography columns.

Gel permeation chromatography, which has been called gel filtration, is a type of liquid-solid elution chromatography that separates polydisperse polymers into fractions by means of the sieving action of a cross-linked polystyrene gel or other sievelike packing. The polystyrene gel, which serves as the stationary phase, is commercially available with a wide distribution of pore sizes (1 to 10^6 nm). Since the smaller molecules permeate the gel particles preferentially, the highest molecular weight fractions are eluted first. Thus, the fractions are separated on the basis of size by GPC.

As shown in Fig. 4.7, a solution using a solvent such as tetrahydrofuran (THF) and the solvent alone (THF) are pumped through separate columns at a rate of about 1 ml/min. The differences in refractive index between the solvent and solution are determined by a differential refractometer and recorded automatically. Each column unit must be calibrated using polymers of known molecular weights. The details of the packed column are shown in Figure 4.8.

The unautomated procedure was used first to separate protein oligomers (polypeptides) by use of Sephadex gels. Silica gels are also used as the GPC sieves. The efficiency of these packed columns may be determined by calculating the height in ft equivalent to a theoretical plate (HETP) which is the reciprocal of the plate count per ft (P). As shown by the expression in Eq. (4.4), P is directly proportional to the square of the elution volume (V_e) and inversely proportional to the height of the column in ft and the square of the baseline (d). As shown in Fig. 4.9, the latter is the width of the baseline of an idealized peak obtained by drawing

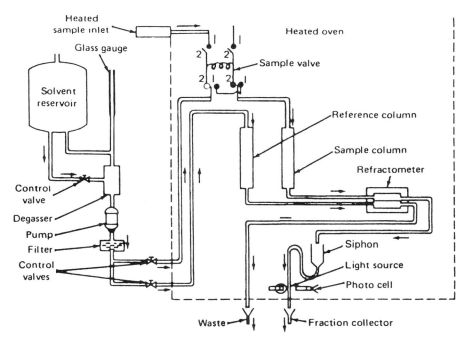

Figure 4.7 Sketch showing flow of solution and solvent in gel permeation chromatograph (GPC). (With permission of Waters Associates.)

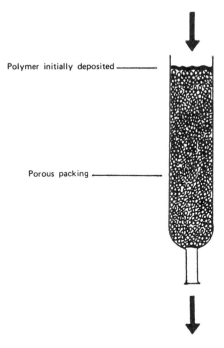

Figure 4.8 Sketch showing solvent-swollen cross-linked polymer packing in GPC column.

lines tangent to the sides of the actual GPC bell-shaped peak of a pure substance, such as THF.

$$p = \frac{16}{f} \left(\frac{V_e}{d}\right)^2 \tag{4.4}$$

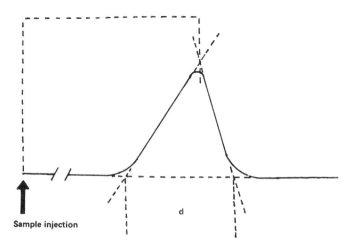

Figure 4.9 Sketch showing the width (d) of a peak in GPC.

The molecular weight of a polymer is related to the effective size occupied by the polymer chain in solution or its hydrodynamic volume (V). For example, V for polystyrene in THF at 25°C is equal to $1.5 \times 10^{-2} \, \overline{M}_w^{1.70}$. In addition to fractionating polydisperse polymers and providing information on polymer size, GPC may be coupled with other instruments, such as an ultraviolet spectrophotometer, for analysis of the eluted fractions.

The motion in and out of the stationary phase is dependent on a number of factors including Brownian motion, chain size, and conformation. The latter two are related to the polymer chain's hydrodynamic volume, i.e., the real, excluded volume occupied by the polymer chain.

The quantities on the right side of Eq. (3.29) are related to the hydrodynamic volume of the polymer. As noted before, the rate of flow of a polymer chain is related to its size and hydrodynamic volume. It has been suggested that it is possible to generate, for a specific column, a simple plot of $[\eta]M$ versus elution volume for polymer samples of known molecular weight and that this "calibration line" can then be employed to obtain $[\eta]M$, and subsequently M when $[\eta]$ is determined; that is, $[\eta]M \propto$ elution volume. Preliminary results have been promising.

4.5 END-GROUP ANALYSIS

While early experimenters were unable to detect the end-groups present in polymers, appropriate techniques are now available for detecting and analyzing quantitatively functional end-groups of linear polymers, such as those in nylon. The amino end-groups of nylon dissolved in m-cresol are readily determined by titration with a methanolic perchloric acid solution. The sensitivity of this method decreases as the molecular weight increases. Thus, this technique is limited to the determination of polymers with a molecular weight of less than about 20,000. Other titratable end-groups are the hydroxyl and carboxyl groups in polyesters and the epoxy end-groups in epoxy resins.

4.6 EBULLIOMETRY AND CRYOMETRY

These techniques, based on Raoult's law, are similar to those used for classic low molecular weight compounds and are dependent on the sensitivity of the thermometry available. The number-average molecular weight $\overline{M}_n$ in both cases is based on the Clausius-Clapeyron equation using boiling point elevation and freezing point depression (ΔT), as shown:

$$\overline{M}_n = \frac{RT^2V}{\Delta H}\left(\frac{C}{\Delta T}\right)_{C\to 0} \tag{4.5}$$

Results obtained using the Clausius-Clapeyron equation, in which T is the Kelvin temperature and ΔH is the heat of transition, must be extrapolated to zero concentration. This technique, like end-group analysis, is limited to low molecular weight polymers. By use of thermistors sensitive to 1×10^{-4} °C, it is possible to measure molecular weight values up to 40,000 to 50,000, although the typical limits are about 5000.

4.7 OSMOMETRY

A measurement of any of the colligative properties of a polymer solution involves a counting of solute (polymer) molecules in a given amount of solvent and yields a number-average. The most common colligative property that is conveniently measured for high polymers is osmotic pressure. This is based on the use of a semipermeable membrane through which solvent molecules pass freely, but through which polymer molecules are unable to pass. Existing membranes only approximate ideal semipermeability, the chief limitation being the passage of low molecular weight polymer chains through the membrane.

There is a thermodynamic drive toward dilution of the polymer-containing solution with a net flow of solvent toward the cell containing the polymer. This results in an increase in liquid in that cell causing a rise in the liquid level in the corresponding measuring tube. The rise in liquid level is opposed and balanced by a hydrostatic pressure resulting in a difference in the liquid levels of the two measuring tubes—the difference is directly related to the osmotic pressure of the polymer-containing solution. Thus, solvent molecules tend to pass through a semipermeable membrane to reach a "static" equilibrium, as illustrated in Fig. 4.10.

Since osmotic pressure is dependent on colligative properties, i.e., the number of particles present, the measurement of this pressure (osmometry) may be applied

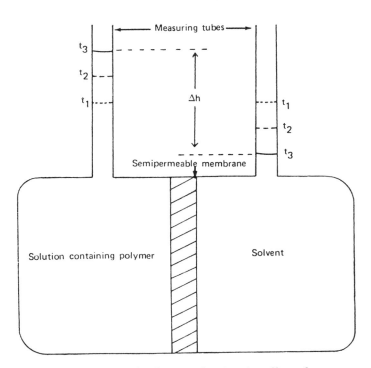

Figure 4.10 Schematic diagram showing the effect of pressure exerted by a solvent separated by a semipermeable membrane from a solution containing a nontransportable material (polymer) as a function of time, where t_1 represents the initial measuring tube levels, t_2 the levels after an elapsed time, and t_3 the levels when the "static" equilibrium occurs.

to the determination of the osmotic pressure of solvents versus polymer solutions. The difference in height (Δh) of the liquids in the columns may be converted to osmotic pressure (π) by multiplying the gravity (g) and the density of the solution (ρ), i.e., $\pi = \Delta h\,\rho g$.

In an automatic membrane osmometer, such as the one shown in Fig. 4.11, the unrestricted capillary rise in a dilute solution is measured in accordance with the modified van't Hoff equation:

$$\pi = \frac{RT}{\overline{M}_n}C + BC^2 \tag{4.6}$$

As shown in Fig. 4.12, the reciprocal of the number average molecular weight ($\overline{M}_n^{-1}$) is the intercept when data for π/RTC versus C are extrapolated to zero concentration.

The slope of the line in Fig. 4.12, i.e., the virial constant B, is related to CED. The value for B would be 0 at the θ temperature. Since this slope increases as the solvency increases, it is advantageous to use a dilute solution consisting of a polymer and a poor solvent. Semipermeable membranes may be constructed from hevea rubber, poly(vinyl alcohol), or cellulose nitrate.

The static head (Δh) developed in the static equilibrium method is eliminated in the dynamic equilibrium method in which a counterpressure is applied to prevent the rise of solvent in the measuring tubes, as shown in Fig. 4.11. Since osmotic pressure is large (1 atm for a 1 M solution), osmometry is useful for the determination of the molecular weight of large molecules.

Static osmotic pressure measurements generally require several days to weeks before a suitable equilibrium is established to permit a meaningful measurement of osmotic pressure. The time required to achieve equilibrium is shortened to several minutes to an hour in most commercial instruments utilizing dynamic techniques.

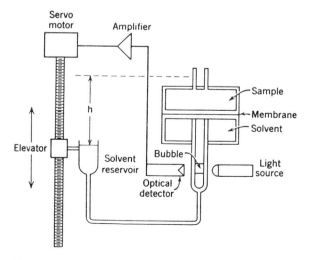

Figure 4.11   Automatic membrane osmometer. (Courtesy of Hewlett-Packard Company.)

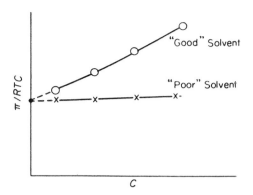

Figure 4.12 Plot of π/RTC versus C used to determine $1/\overline{M}_n$ in osmometry. (From *Modern Plastics Technology* by R. Seymour, Reston Publishing, Reston, Virginia, 1975. Used with permission of Reston Publishing Co.)

Classic osmometry is useful and widely used for the determination of a range of M_n values from 5×10^4 to 2×10^6. New dynamic osmometers expand the lower limit to 2×10^4. The molecular weight of polymers with lower molecular weights which may pass through a membrane may be determined by vapor pressure osmometry (VPO) or isothermal distillation. Both techniques provide absolute values for $\overline{M}_n$.

In the VPO technique, drops of solvent and solution are placed in an insulated chamber in proximity to thermistor probes. Since the solvent molecules evaporate more rapidly from the solvent than from the solution, a difference in temperature (ΔT) is recorded. Thus, the molarity (M) may be determined by use of Eq. (4.7) if the heat of vaporization per gram of solvent (λ) is known.

$$\Delta T = \left(\frac{RT^2}{\lambda 100}\right) \underline{M} \tag{4.7}$$

A sketch of a vapor pressure osmometer is shown in Fig. 4.13.

Problem

Insulin, a hormone that regulates carbohydrate metabolism in the blood, was isolated from a pig. A 0.200-g sample of insulin was dissolved in 25.0 ml of water, and at 30°C the osmotic pressure of the solution was found to be 26.1 torr. What is the molecular weight of the insulin?

Rearrangement of the first terms of Eq. (4.6) gives $M = RTC/\pi$. Appropriate units of concentration, osmotic pressure, and R need to be chosen. The following units of grams per liter, atmospheres, and liter atmosphere per mol°K are employed, giving

$$M = \frac{(0.08206 \text{ 1 atm mol}^{-1}\,°K^{-1})(303°K)(8.0 \text{ gram l}^{-1})}{\left(\dfrac{26.1 \text{ torr}}{760 \text{ torr/atm}}\right)} = 5800$$

This is an "apparent" molecular weight since it is for a single concentration and is not extrapolated to zero concentration.

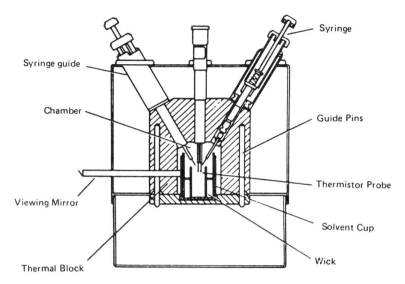

Figure 4.13 A sketch of a vapor pressure osmometer. (Courtesy of Hewlett-Packard Company.)

4.8 REFRACTOMETRY

The index of refraction decreases slightly as the molecular weight increases, and as demonstrated by the techniques used in GPC, this change has been used for the determination of molecular weight after calibration, using samples of known molecular weight distribution.

4.9 LIGHT-SCATTERING MEASUREMENTS

Ever watch a dog or young child chase "moonbeams"? The illumination of dust particles is an illustration of light scattering, not of reflection. Reflection is the deviation of incident light through one particular angle such that the angle of incidence is equal to the angle of reflection. Scattering is the radiation of light in all directions. Thus, in observing the "moon beam," the dust particle directs a beam toward you regardless of the angle you are with relation to the scattering particle. The energy scattered per second (scattered flux) is related to the size and shape of the scattering particle and to the scattering angle.

The measurement of light scattering by polymer molecules in solution is a widely used technique for the determination of absolute values of $\overline{M}_w$. This technique, which is based on the optical heterogeneity of polymer solutions, was developed by Nobel laureate Peter Debye in 1944.

In 1871, Rayleigh showed that induced oscillatory dipoles were developed when light passed through gases and that the amount (intensity) of scattered light (τ) was inversely proportional to the fourth power of the wavelength of light. This investigation was extended to liquids by Einstein and Smoluchowski in 1908. These oscillations reradiate the light energy to produce turbidity, i.e., the Tyndall effect.

Other sources of energy, such as X-rays or laser beams, may be used in place of visible light waves.

For light-scattering measurements, the total amount of the scattered light is deduced from the decrease in intensity of the incident beam, I_o, as it passes through a polymer sample. This can be described in terms of Beer's law for the absorption of light as follows:

$$\frac{I}{I_o} = e^{-\tau l} \tag{4.8}$$

where τ is the measure of the decrease of the incident-beam intensity per unit length 1 of a given solution and is called the turbidity.

The intensity of scattered light or turbidity (τ) is proportional to the square of the difference between the index of refraction (n) of the polymer solution and of the solvent n_o, to the molecular weight of the polymer ($\overline{M}$), and to the inverse fourth power of the wavelength of light used (λ). Thus,

$$\frac{Hc}{\tau} = \frac{1}{\overline{M}_w P_\theta} (1 + 2Bc + Cc^2 + \ldots) \tag{4.9}$$

where the expression for the constant H is as follows:

$$H = \frac{32\pi^2}{3} \frac{n_o{}^2 (dn/dc)^2}{\lambda^4 N} \quad \text{and} \quad \tau = K'n^2 \left(\frac{i_{90}}{i_0}\right) \tag{4.10}$$

where n_o = index of refraction of the solvent, n = index of refraction of the solution, c = concentration, the virial constants B, C, etc., are related to the inter-action of the solvent, P_θ is the particle scattering factor, and N is Avogadro's number. The expression dn/dc is the specific refractive increment and is determined by taking the slope of the refractive index readings as a function of polymer concentration.

In the determination of the weight-average molecular weight of polymer molecules in dust-free solutions, one measures the intensity of scattered light from a mercury arc lamp or laser at different concentrations and at different angles (θ), typically 0, 90, 45, and 135° (Fig. 4.14). The incident light sends out a scattering envelope which has four equivalent quadrants. The ratio of scattering at 45° compared with that for 135° is called the dissymmetry factor or dissymmetry ratio Z.

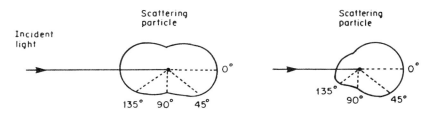

Figure 4.14 Light-scattering envelopes. Distance from the scattering particle to the boundaries of the envelope represents the magnitude of scattered light as a function of angle.

The reduced dissymmetry factor Z_0 is the intercept of the plot of Z as a function of concentration extrapolated to zero concentration. For polymer solutions containing polymers of moderate to low molecular weight, P_θ is 1 and Eq. (4.9) reduces to

$$\frac{Hc}{\tau} = \frac{1}{\overline{M}_w}(1 + 2Bc + Cc^2 + \ldots) \tag{4.11}$$

At low concentrations of polymer in solution, Eq. (4.11) reduces to an equation of a straight line (y = b + mx):

$$\frac{Hc}{\tau} = \frac{1}{\overline{M}_w} + \frac{2Bc}{\overline{M}_w} \tag{4.12}$$

When the ratio of the concentration c to the turbidity τ (related to the intensity of scattering at 0 and 90°) multiplied by the constant H is plotted against concentration, the intercept of the extrapolated curve is the reciprocal of $\overline{M}_w$ and the slope contains the virial constant B, as shown in Fig. 4.15. Z_0 is directly related to P_θ, and both are related to both the size and shape of the scattering particle. As the size of the polymer chain approaches about one-twentieth the wavelength of incident light, scattering interference occurs, giving a scattering envelope which is no longer symmetrical. Here the scattering dependency on molecular weight reverts back to the relationship given in Eq. (4.9), thus, a plot of Hc/τ versus c extrapolated to zero polymer concentration gives as the intercept $1/\overline{M}_w P_\theta$, not $1/\overline{M}_w$. The molecular weight for such situations is typically found using one of two techniques.

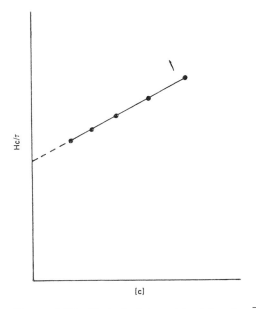

Hc/τ

[c]

Figure 4.15 Typical plot used to determine $\overline{M}_w^{-1}$ from light-scattering data.

Problem

Determine the apparent weight-average molecular weight for a polymer sample where the intensity of scattering at 0° is 1000 and the intensity of scattering at 90° is 10 for a polymer (0.14 g) dissolved in DMSO (100 ml) which had a dn/dc of 1.0.

Most of the terms employed to describe H and τ are equipment constants, and their values are typically supplied with the light-scattering photometer and redetermined periodically. For the sake of calculation we will use the following constant values: $K' = 0.100$ and $\lambda = 546$ nm. DMSO has a measured refractive index of 1.475 at 21°C.

$$\tau = K'n^2 \frac{i_{90}}{i_0} = (0.100)(1.475)^2 \left(\frac{10}{1000}\right) \simeq 2.18 \times 10^{-3}$$

At 546 nm $H = 6.18 \times 10^{-5} n_0^2 (dn/dc)^2 = 6.18 \times 10^{-5}(1.475)^2(1.0)^2 = 1.34 \times 10^{-4}$.

Typically, concentration units of g/ml or g/cc are employed for light-scattering photometry. For the present solution the concentration is 0.14 g/100 ml = 0.0014 g/ml.

$$\frac{Hc}{\tau} = \frac{1.34 \times 10^{-4} \times 1.4 \times 10^{-3}}{2.18 \times 10^{-3}} \simeq 8.6 \times 10^{-5}$$

The apparent molecular weight is then the inverse of 8.6×10^{-5} or 1.2×10^4. This is called "apparent" since it is for a single point and not extrapolated to zero.

The first of the techniques is called the dissymmetric method or approach because it utilizes the determination of Z_0 versus P_θ as a function of polymer shape. $\overline{M}_w$ is determined from the intercept $1/\overline{M}_w P_\theta$ through substitution of the determined P_θ. The weakness in this approach is the necessity of having to assume a shape for the polymer in a particular solution. For small Z_0 values, choosing an incorrect polymer shape results in a small error, but for larger Z_0 values, the error may become significant, i.e., greater than 30%. The positive feature is that the dissymmetric approach is simple compared to the second approach, the Zimm method, where no assumption of polymer shape is necessary.

The Zimm method utilizes the Zimm plot, which is a double extrapolation of the light-scattering data to zero concentration and zero scattering angle. The plot utilizes a redistribution of the Hc/τ factors designating this combination as Kc/R. Figure 4.16 contains a representative Zimm plot.

Low-angle and multiangle light-scattering photometers are available that allow not only the determination of the weight-average molecular weight but also additional values under appropriate conditions. For instance, the new multiangle instrument obtains data using a series of detectors as shown in Fig. 4.16. From data obtained from this instrument a typical Zimm plot is constructed as shown in Fig. 4.17 that gives both the weight-average molecular weight and the mean radius independent of the molecular conformation and branching. The actual shape of the Zimm plot shown in Fig. 4.17 is dependent on various "fit" constants such as the 1000 used as the multiplier for "c" in Fig. 4.17. Obtaining "good' plots can be routine through the use of special computer programs.

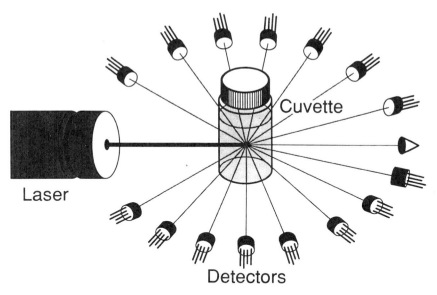

Figure 4.16 Detector arrangement showing a sample surrounded by an array of detectors that collect scattered laser light by the sample. (Used with permission of Wyatt Technology Corporation, P.O. Box 3003, Santa Barbara, CA 93130.)

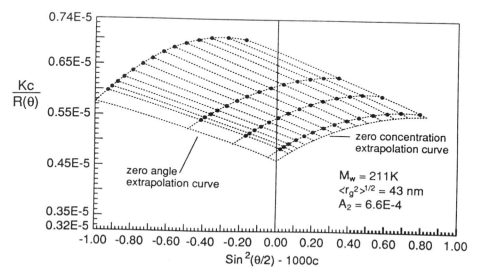

Figure 4.17 Zimm plot for a polymer scaled with a negative-concentration coefficient to improve data aesthetics and accessibility. (Used with permission of Wyatt Technology Corporation, P.O. Box 3003, Santa Barbara, CA 93130.)

A major error involved in the determination of weight-average molecular weight using light-scattering photometry involves determination of dn/dc, since any error in its determination is magnified because it appears as the squared value in the equation relating light scattering and molecular weight, such as Eq. (4.10).

One of the more important advances is the "coupling" of HPLC and light scattering. As noted in Section 4.4, HPLC allows the determination of the molecular weight distribution, but it does not itself allow the calculation of an absolute molecular weight but relies on calibration with polymers of known molecular weight. By coupling HPLC and light scattering, the molecular weight of each fraction is determined, giving a molecular weight distribution where the molecular weight distribution and various molecular weight (weight-average, number-average, Z-average) values can be calculated (since it can be assumed that the fractionated samples approach a single molecular weight so that the weight-average molecular weight is equal to the number-average molecular weight is equal to the Z-average molecular weight, etc.).

4.10 ULTRACENTRIFUGATION

Since the kinetic energy of solvent molecules is much greater than the sedimentation force of gravity, polymer molecules remain suspended in solution. However, this traditional gravitation field, which permits Brownian motion, may be overcome by increasing this force by use of high centrifugal forces, such as the ultracentrifugal forces developed by Nobel laureate The Svedberg in 1925.

Both $\overline{M}_w$ and $\overline{M}_z$ may be determined by subjecting dilute solutions of polymers in appropriate solvents to ultracentrifugal forces at high speeds. Solvents with densities and indices of refraction different from the polymers are chosen to ensure polymer motion and optical detection of this motion.

In the sedimentation velocity experiments, the ultracentrifuge is operated at extremely high speeds up to 70,000 rpm in order to transport the denser polymer molecules through the less dense solvent to the cell bottom or to the top if the density of the solvent is greater than the density of the polymer. The boundary movement during ultracentrifugation can be followed by using optical measurements to monitor the sharp change in index of refraction (n) between solvent and solution.

The rate of sedimentation is defined by the sedimentation constant s, which is directly proportional to the mass m, solution density ρ, and specific volume of the polymer $\overline{V}$, and inversely proportional to the square of the angular velocity of rotation ω, the distance from the center of rotation to the point of observation in the cell r, and the frictional coefficient f, which is inversely related to the diffusion coefficient D extrapolated to infinite dilution. These relationships are shown in the following equations in which $(1 - \overline{V}\rho)$ is called the buoyancy factor since it determines the direction of macromolecular transport in the cell.

$$s = \frac{1}{\omega^2 r}\frac{dr}{dt} = \frac{m(1 - \overline{V}\rho)}{f} \tag{4.13}$$

$$D = \frac{RT}{Nf} \text{ and } mN = M \tag{4.14}$$

$$\frac{D}{s} = \frac{RT}{\overline{M}_\omega\,(1 - \overline{V}\rho)}$$ (4.15)

The sedimentation velocity determination is dynamic and can be completed in a short period of time. It is particularly useful for monodisperse systems and provides qualitative data and some information on molecular weight distribution for polydisperse systems.

The sedimentation equilibrium method yields quantitative results, but long periods of time are required for centrifugation at relatively low velocities to establish equilibrium between sedimentation and diffusion.

As shown in the following equation, the weight-average molecular weight $\overline{M}_w$ is directly proportional to the temperature T and the ln of the ratio of concentration c_2/c_1 at distances r_1 and r_2 from the center of rotation and the point of observation in the cell and inversely proportional to the buoyancy factor, the square of the angular velocity of rotation and the difference between the squares of the distances r_1 and r_2.

$$\overline{M}_w = \frac{2RT \ln c_2/c_1}{(1 - \overline{V}\rho)\omega^2(r_2{}^2 - r_1{}^2)}$$ (4.16)

4.11 SMALL-ANGLE X-RAY SCATTERING

The theoretical basis for light-scattering photometry applies to all radiation. X-ray scattering or diffraction techniques are typically divided into two categories wide-angle X-ray scattering (WAXS) and small-angle X-ray scattering (SAXS). Typically, SAXS gives information on a scale of 1 nm and smaller, while SAXS gives information on a scale of about 1 to 1000 nm. While both employ X-ray scattering, the instrumentation is generally very different. WAXS is employed to measure crystal structure and related parameters, including percentage crystallinity.

Atoms are of the order of 0.1 nm while an extended polyethylene (PE) chain (typically PE exists as a modified random coil in solution) with a molecular weight of 50,000 (and a degree of polymerization of about 1800) would have an end-to-end distance of about 150 nm, well within the range typically employed for SAXS. Thus, SAXS can be utilized to determine weight-average molecular weights because the scattering distance is dependant on the molecular size of the polymer.

SUMMARY

1. Some naturally occurring polymers such as proteins consist of molecules with a specific molecular weight and are called monodisperse. However, cellulose, natural rubber, and most synthetic polymers consist of molecules with different molecular weights and are called polydisperse. Many properties of polymers are dependent on their molecular weight above that required for entanglement. Since the melt viscosity increases exponentially with molecular weight, the high energy costs of processing high molecular weight polymers are not usually justified.

2. The distribution of molecular weights in a polydisperse system may be represented on a typical probability curve. The number-average molecular weight $\overline{M}_n$, which is smallest in magnitude, and a simple arithmetic mean may be deter-

mined by techniques based on colligative properties, such as osmotic pressure, boiling point increase, and freezing point depression. The weight-average molecular weight ($\overline{M}_w$), which is larger than $\overline{M}_n$ for polydisperse systems, is the second-power average. This value may be determined by light scattering which, like techniques based on colligative properties, yields absolute values for molecular weights.

3. Since $\overline{M}_w = \overline{M}_n$ for a monodisperse system, the polydispersity index $\overline{M}_w/\overline{M}_n$ is a measure of polydispersity. The most probable value for this index for polymers synthesized by condensation techniques is 2.0. The viscosity average $\overline{M}_v$, which is not an absolute value, is usually between $\overline{M}_n$ and $\overline{M}_w$, but when the exponent a in the Mark-Houwink equation is equal to 1, $\overline{M}_w = \overline{M}_v$.

4. Various molecular weight distributions may be obtained by special analytical techniques—centrifugation, gel permeation chromatography, and solvent fractionation. In the last technique one adds a small amount of nonsolvent to precipitate the highest molecular weight polymer from solution. In gel permeation chromatography, smaller cross-linked polymers in a column act as a sieve and allow the larger molecules to elute first. After calibration, the molecular weight of the eluted polymer is determined automatically by measuring the difference in index of refraction of the solution and solvent.

5. The use of boiling point increase and freezing point lowering is limited to the determination of the molecular weights of relatively small polymers. However, since the osmotic pressure of a 1 M solution is large (1 atm), osmometry is readily used for characterizing large molecules. The possibility of oligomers and other small polymers passing through the semipermeable membrane is avoided by use of vapor pressure osmometry for characterization of polymers having molecular weights less than 40,000. This technique utilizes the difference in temperatures noted during the evaporation of a solvent from a solution and the pure solvent.

6. Absolute molecular weight values reported as weight-average molecular weights are obtained by light-scattering techniques. $\overline{M}_w$ values for nonpolar polymers with molecular weights less than 100,000 are determined by extrapolating the plot of the product of the ratio of the concentration and the turbidity and a constant H versus concentration, the intercept is $\overline{M}_w^{-1}$.

7. Since the gravitational force present in the ultracentrifuge is sufficient to precipitate polymer molecules in accordance with their size, ultracentrifugation is used as a technique for the determination of molecular weight, especially for monodisperse systems such as proteins.

GLOSSARY

a: Exponent in Mark-Houwink equation, usually 0.6 to 0.8.

B: symbol for the second virial constant.

Brownian motion: Movement of large particles in a liquid as a result of bombardment by smaller molecules. This phenomenon was named after the botanist Robert Brown, who observed it in 1827.

buoyancy factor: $(1 - \overline{V}\rho)$ or 1 − the specific volume × density of a polymer. This term, used in ultracentrifugal experiments, determines the direction of polymer transport under the effect of centrifugal forces in the cell.

CED: Cohesive energy density.

Clausius-Clapeyron equation: For small changes in pressure, the unintegrated expression $dT/dp = RT^2/Lp$ may be used where L is the molar heat of vaporization. The form used for molecular weight determination is as follows:

$$\overline{M}_n = \frac{RT^2V}{\Delta H}\left(\frac{C}{\Delta T}\right)_{C\rightarrow 0}$$

colligative properties: Properties of a solution which are dependent on the number of solute molecules present and are usually related to the effect of these molecules on vapor pressure lowering.

commercial polymer range: A molecular weight range high enough to have good physical properties but not too high for economical processing.

critical molecular weight: The threshold value of molecular weight required for chain entanglement.

cryometry: Measurement of $\overline{M}_n$ from freezing point depression.

d: The width of the baseline of idealized peaks in a gel permeation chromatogram.

ebulliometry: Measurement of $\overline{M}_n$ from boiling point elevation of a solution.

end-group analysis: The determination of molecular weight by counting the number of end groups.

end groups: The functional groups at chain ends such as the carboxyl groups (COOH) in polyesters.

f: Symbol for frictional coefficient in ultracentrifugal experiments.

fractional precipitation: Fractionation of polydisperse systems by adding small amounts of a nonsolvent to a solution of the polymer and separating the precipitate.

fractionation of polymers: Separation of a polydisperse polymer into fractions of similar molecular weights.

gel permeation chromatography: A type of liquid-solid elution chromatography which automatically separates solutions of polydisperse polymers into fractions by means of the sieving action of a swollen cross-linked polymeric gel.

GPC: Gel permeation chromatography.

H: Symbol for the proportionality constant in a light-scattering equation.

HETP: Height equivalent to a theoretical plate (in feet).

isothermal distillation: Same as vapor pressure osmometry.

lambda (λ): Symbol for the wavelength of light.

ln: Symbol for natural or naperian logarithm.

$\overline{M}_n$: Number-average molecular weight.

monodisperse: A system consisting of molecules of one molecular weight only.

$\overline{M}_w$: Weight-average molecular weight.

$\overline{M}_z$: z-Average molecular weight.

MWD: Molecular weight distribution.

n: Index of refraction.

N: Symbol for Avogadro's number, 6.23×10^{23}.

number-average molecular weight: The arithmetical mean value obtained by dividing the sum of the molecular weights by the number of molecules.

oligomer: Very low molecular weight polymer usually with DP less than 10 (the Greek *oligos* means few).

omega (ω): Symbol for the angular velocity of rotation.

osmometry: The determination of molecular weight $\overline{M}_n$ from measurement of osmotic pressure.

osmotic pressure: The pressure a solute would exert in solution if it were an ideal gas at the same volume.

P: Symbol for plate count in GPC.

pi (π): Symbol for osmotic pressure.

plate count (P): The reciprocal of HETP.

polydisperse: A mixture containing polymer molecules of different molecular weights.

polydispersity index: $\overline{M}_w / \overline{M}_n$.

polypeptides: A term used to describe oligomers of proteins, which of course are also polypeptides or polyamino acids.

Raoult's law: The vapor pressure of a solvent in equilibrium with a solution is equal to the product of the mole fraction of the solvent and the vapor pressure of the pure solvent at any specified temperature. Osmotic pressure, boiling point elevation, and freezing point depression are related to this decrease in vapor pressure.

rho (ρ): Symbol for density.

sedimentation equilibrium experiment: An ultracentrifugal technique which provides quantitative information on molecular weights. Long times are required for the attainment of equilibrium in this method.

sedimentation velocity experiment: A dynamic experiment with the ultracentrifuge which provides qualitative information on molecular weight in a short period of time.

semipermeable membranes: Those membranes that will permit the diffusion of solvent molecules but not large molecules such as polymers.

tau (τ): Symbol for turbidity.

theta (θ): Symbol for the angle of the incident beam in light-scattering experiments.

ultracentrifuge: A centrifuge which by increasing the force of gravity by as much as 100,000 times causes solutes to settle from solutions in accordance with their molecular weights.

$\overline{V}$: Symbol for specific volume.

V: Hydrodynamic volume in GPC.

V_e: Elution volume in GPC.

van't Hoff's law: Osmotic pressure

$$\pi = \frac{C}{M}\frac{RT}{V} = \frac{RTC}{M}.$$

vapor pressure osmometry: A technique for determining the molecular weight of relatively small polymeric molecules by measuring the relative heats of evaporation of a solvent from a solution and pure solvent.

weight-average molecular weight: The second-power average of molecular weights in a polydisperse polymer.

z-average molecular weight: The third-power average of molecular weights in a polydisperse polymer.

Zimm plot: A type of double extrapolation used to determine $\overline{M}_w$ in light-scattering experiments. Both the concentration of the solution and the angle of the incident beam of light are extrapolated to zero on one plot.

zone melting: The fractionation of polydisperse systems by heating over long periods of time.

EXERCISES

1. Which of the following is polydisperse: (a) casein, (b) commercial polystyrene, (c) paraffin wax, (d) cellulose, or (e) *Hevea brasiliensis*?

2. If $\overline{M}_n$ for LDPE is 1,400,000, what is the value of $\overline{DP}$?

3. What are the $\overline{M}_n$ and $\overline{M}_w$ values for a mixture of five molecules each having the following molecular weights: 1.25×10^6, 1.35×10^6, 1.50×10^6, 1.75×10^6, 2.00×10^6?

4. What is the most probable value for the polydispersity index for (a) a monodisperse polymer, and (b) a polydisperse polymer synthesized by condensation techniques?

5. List in order of increasing values: $\overline{M}_z$, $\overline{M}_n$, $\overline{M}_w$, and $\overline{M}_v$.

6. Which of the following provides an absolute measure of the molecular weight of polymers: (a) viscometry, (b) cryometry, (c) osmometry, (d) light scattering, (e) GPC?

7. What is the relationship between the intrinsic viscosity or limiting viscosity number $[\eta]$ and average molecular weight $\overline{M}$?

8. What molecular weight determination techniques can be used to fractionate polydisperse polymers?

9. Which of the following techniques yields a number-average molecular weight, $\overline{M}_n$: (a) viscometry, (b) light scattering, (c) ultracentrifugation, (d) osmometry, (e) ebulliometry, (f) cryometry?

10. What is the relationship of HETP and plate count per foot in GPC?

11. What is the value of the exponent a in the Mark-Houwink equation for polymers in θ solvents?

12. How many amino groups are present in each molecule of nylon-66 made with an excess of hexamethylenediamine?

13. What is the value of the exponent a in the Mark-Houwink equation for a rigid rod?

14. If the value of K and a in the Mark-Houwink equation are 1×10^{-2} cm^3 g^{-1} and 0.5, respectively, what is the average molecular weight of a polymer whose solution has an intrinsic viscosity of 150 cm^3 g^{-1}?

15. Which polymer of ethylene will have the highest molecular weight: (a) a trimer, (b) an oligomer, or (c) UHMWPE?

16. What is a Zimm plot?

17. What type of molecular weight average, $\overline{M}_w$ or $\overline{M}_n$, is based on colligative properties?

18. What principle is used in the determination of molecular weight by vapor pressure osmometry?

19. What does the melt viscosity increase faster with molecular weight increase than other properties such as tensile strength?

20. In spite of the high costs of processing, ultrahigh molecular weight polyethylene is used for making trash cans and other durable goods. Why?

21. Under what conditions are $\overline{M}_v$ and $\overline{M}_w$ equal for a polydisperse system?

22. What nonsolvent would you use to fractionate a polydisperse polymer in a solution?

23. Which colligative property technique would you use to determine the molecular weight of a polymer having a molecular weight of (a) 40,000, (b) 80,000?

24. What is the advantage of vapor pressure osmometry when measuring relatively low molecular weight polymers?

25. Which will yield the higher apparent molecular weight values in the light-scattering method: (a) a dust-free system, or (b) one in which dust particles are present?

26. Which is the more rapid ultracentrifugation technique for the determination of molecular weight of polymers: (a) sedimentation velocity method, or (b) sedimentation equilibrium method?

27. Which ultracentrifugation technique is more accurate: (a) sedimentation velocity, or (b) sedimentation equilibrium method?

28. What is the significance of the virial constant B in osmometry and light-scattering equations?

BIBLIOGRAPHY

Allen, P. W. (1959): *Techniques of Polymer Characterization*, Butterworths, London.

Benoit, H. (1968): Use of light scattering and hydrodynamic methods for determining the overall conformation of helical molecules, J. Chem. Phys., *65*:23–30.

Billmeyer, Jr., F. W. (1966): Measuring the weight of giant molecules, Chemistry, *39*:8–14.

Billmeyer, Jr., F. W. Kokle, V. (1964): The molecular structure of polyethylene. XV. Comparison of number-average molecular weights by various methods, J. Am. Chem. Soc., 86:3544–3546.

Braun, D., Cherdon, H., Keru, W. (1972): *Techniques of Polymer Synthesis and Characterization*, Wiley-Interscience, New York.

Burrell, H. (1974): Solubility parameter values, in *Polymer Handbook* (J. Brandrup and E. H. Immergut, eds.), Wiley, New York.

Carr, Jr., C. I. Zimm, B. H. (1950): Absolute intensity of light scattering from pure liquids and solutions, J. Chem. Phys., *18*:1616–1626.

Collins, E. A., Bares, J., Billmeyer, F. W. (1973): *Experiments in Polymer Science*, Wiley-Interscience, New York.

Debye, P. J. (1944): Light scattering analysis, J. Appl. Phys., *15*:338.

Debye, P., Bueche, A. M. (1948): Intrinsic viscosity, diffusion, and sedimentation rates of polymers in solutions, J. Chem. Phys., *16*:573–579.

Einstein, A. (1910): Theory of the opalescence of homogeneous liquids and liquid mixtures in the neighborhood of the critical state, Ann. Physik., *33*:1275–1298.

Francuskiewicz, F. (1994): *Polymer Fractionation*, Springer-Verlag, New York.

Huggins, M. L. (1942): The viscosity of dilute solutions of long-chain molecules. IV. Dependence on concentration, J. Am. Chem. Soc., *64*:2716–2718.

Krigbaum, W. R., Flory, P. J. (1952): Treatment of osmotic pressure data, J. Polymer Sci., *9*:503–588.

Mark, H. F. (1948): *Frontiers in Chemistry*, Vol. 5, Interscience, New York.

Mark, H. F., Whitby, G. S. (eds.) (1940): *Collected Papers of Wallace Hume Carothers on High Polymeric Substances*, Interscience, New York.

McCaffery, E. M. (1970): *Laboratory Preparation for Macromolecular Chemistry*, McGraw-Hill, New York.

Morgan, P. W. (1965): *Condensation Polymers by Interfacial and Solution Methods*, Wiley-Interscience, New York.

Provder, T. (1993): *Chromatography of Polymers*, ACS, Washington, D.C.

Rayleigh, Lord. (1871): On the light from the sky, its polarization and color, Phil. Mag., *41*: 107–120, 274–279.

———. (1914): On the diffraction of light by spheres of small relative index, Proc. Roy. Soc., *A90*:219–225.

Slade, P. E. (1975): *Polymer Molecular Weights,* Marcel Dekker, New York.

Staudinger, H. (1928): Ber. Bunsenges. Phys. Chem., *61*:2427.

———. (1932): *Die Hochmolekularen Organischen Verbindungen*, Springer-Verlag, Berlin.

Staudinger, H., Heuer, W. (1930): Highly polymerized compounds. XXXIII. A relation between the viscosity and the molecular weight of polystyrenes, Ber. Bunsenges. Phys. Chem., *63B*:222–234.

Svedberg, T., Pederson, K. O. (1940): *The Ultracentrifuge*, Clarendon, Oxford.

Williams, J. W., van Holde, Kensal, E., Baldwin, R. L., Fujita, H. (1958): The theory of sedimentation analysis, Chem. Rev., *58*:715–806.

Zimm, B. H. (1948): The scattering of light and the radical distribution function of high polymer solutions, J. Chem. Phys., *16*:1093–1099.

————. (1948): Apparatus and methods for measurement and interpretation of the angular variation of light scattering: Preliminary results on polystyrene solutions, J. Chem. Phys. *16*:1099–1116.

Zimm, B. H., Kelb, R. W. (1959): J. Polymer Sci., *37*:19.

5

Testing and Characterization of Polymers

Public acceptance of polymers is usually associated with an assurance of quality based on a knowledge of successful long-term and reliable tests. In contrast, much of the dissatisfaction with synthetic polymers is related to failures that might have been prevented by proper testing, design, and quality control. The American Society for Testing and Materials (ASTM), through its committees D-1 on paint and D-20 on plastics, has developed many standard tests, which may be referred to by all producers and consumers of polymeric materials. There are also cooperating groups in many other technical societies: the American National Standards Institute (ANSI), an International Standards Organization (ISO), and standards societies such as the British Standards Institution (BSI) in England, the Deutsche Normenausschuss (DNA) in Germany, and comparable groups in every nation with developed polymer technology throughout the entire world.

A number of testing techniques have already been considered in Chaps. 3 and 4. We will concentrate on the areas of physical testing and spectroscopy, including thermal characterization, in this chapter. Most of these techniques can be directly applied to nonpolymeric materials, such as small organic molecules, inorganic salts, and metals.

The testing of materials can be based on whether the tested material is chemically changed or is left unchanged. Nondestructive tests are those that result in no chemical change in the material. Nondestructive tests include many electrical property determinations, infrared and ultraviolet spectroanalysis, simple melting point determinations, density and color determinations, and most mechanical property determinations.

Destructive tests are those where there is a change in the chemical structure of at least a portion of the tested material. Examples include flammability property determination and chemical resistance tests where the test material is not resistant to the tested material.

5.1 TYPICAL STRESS-STRAIN BEHAVIOR

As noted above, most physical tests involve nondestructive evaluations. For our purposes, three types of mechanical stress measurements can be described as pictured in Fig. 5.1. The ratio of stress to strain is called Young's modulus. This ratio is also called the modulus of elasticity and tensile modulus. It is calculated by dividing the stress by the strain:

$$\text{Young's modulus} = \frac{\text{stress (Pa)}}{\text{strain (mm/mm)}} \tag{5.1}$$

Large values for Young's modulus indicate that the material is rigid and resistant to elongation and stretching. Many synthetic organic polymers have a Young's modulus in the general range of about 10^5 psi, fused quartz has a Young's modulus of about 10^6, cast iron, tungsten, and copper have values in the range of 10^7, and diamond a value of about 10^8.

As shown in Fig. 5.2, Carswell and Nason assigned five classifications to polymers. The soft and weak class (a) of polymers, including polyisobutylene, is

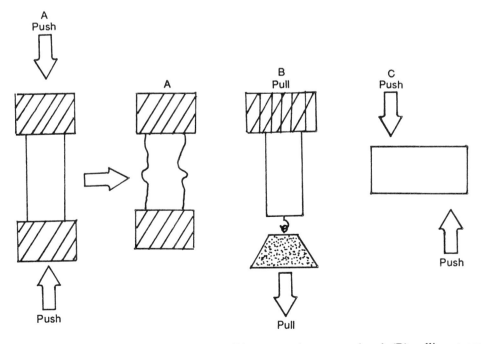

Figure 5.1 Major types of stress tests: (A) compressive start and end, (B) pulling stress or tensile strength, and (C) shear stress.

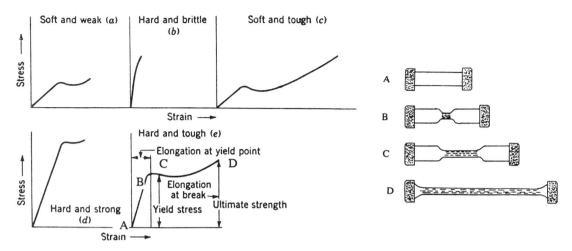

Figure 5.2 Typical stress-strain curves for plastics. (From *Introduction to Polymer Chemistry* by R. Seymour, McGraw-Hill, New York, 1971. Used with permission of McGraw-Hill Book Company.)

characterized by a low modulus of elasticity, low yield (stress) point, and moderate time-dependent elongation. The Poisson ratio, i.e., the ratio of contraction to elongation, for class (a) polymers is 0.5, which is similar to that of liquids.

In contrast, the Poisson ratio of hard and brittle class (b) polymers, such as polystyrene, approaches 0.3. Class (b) polymers are characterized by a high modulus of elasticity, a poorly defined yield point, and little elongation before failure. However, class (c) polymers, such as plasticized PVC, have a low modulus of elasticity, high elongation, a Poisson ratio of about 0.5–0.6, and a well-defined yield point. Since class (c) polymers stretch after the yield point, the area under the entire curve, which represents toughness, is greater than that in class (b).

Rigid PVC is representative of hard and strong class (d) polymers. These polymers have a high modulus of elasticity and high yield strength. The curve for hard and tough class (e) polymers, such as ABS copolymers, shows moderate elongation prior to the yield point followed by nonrecoverable elongation. In general, the behavior of all classes is Hookean prior to the yield point. The reversible recoverable elongation prior to the yield point, called the elastic range, is primarily the result of bending and stretching covalent bonds in the polymer backbone. This useful portion of the stress-strain curve may also include some recoverable uncoiling of polymer chains. Irreversible slippage of polymer chains is the predominant mechanism after the yield point.

Since these properties are time dependent, class (a) polymers may resemble class (d) polymers if the stress is applied rapidly, and vice versa. These properties are also temperature dependent. Hence, the properties of class (c) polymers may resemble class (b) polymers when the temperature is decreased. The effects of temperature and the mechanisms of elongation are summarized in Fig. 5.3.

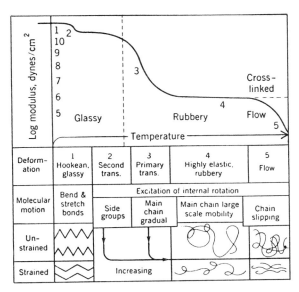

Figure 5.3 Characteristic effects of temperature on the properties of a typical polymer. (From *Introduction to Polymer Chemistry* by R. Seymour, McGraw-Hill, New York, 1971. Used with permission of McGraw-Hill Book Company.)

5.2 STRESS-STRAIN RELATIONSHIPS

The stress-strain behavior of materials is dependent on how the precise test is performed and the physical state of the material, such as whether the material is above or below its T_g.

Elastic solids under a constant rate of strain give responses as pictured in Fig. 5.4A. The overall input is shown on the left and the response on the right. Thus, constant strain begins at a and continues to b, where strain remains constant. The increase in strain produces analogous increases in stress (below the point of fracture) as shown in the corresponding graph on the right. The ordinates of such plots are scaleless, emphasizing the relative nature of the responses. The second set of plots in Fig. 5.4B and C describes what occurs when a stress is applied rapidly from a to b, followed by release of this stress at c and returning to a. For materials below their T_g, the response is shown in B, where the strain response is immediate and the compliance is constant. The response shown in C is for materials between their T_g and T_m. There is some rapid response with strain from a to b (right side, C) at which point creep begins. Over the period of b to c, the creep is time (t) dependent and can be described as follows:

$$\text{Compliance(t)} = \frac{\text{strain(t)}}{\text{stress}} \tag{5.2}$$

Creep behavior is similar to viscous liquid flow, and materials exhibiting creep are described as viscoelastic materials. The linear behavior shown by Eq. (5.2) describes a linear time-dependent response, i.e., compliance and strain are linearly

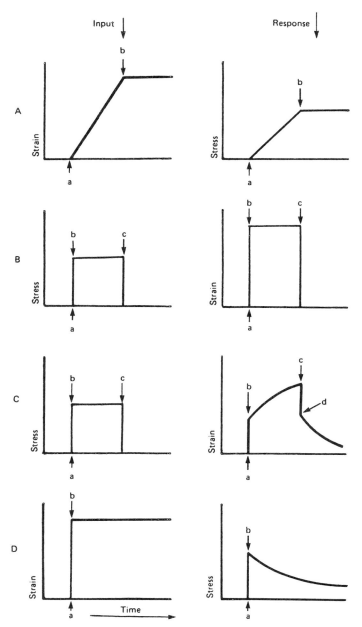

Figure 5.4 Stress-strain responses of elastic and viscoelastic materials and for various models (ϵ = strain).

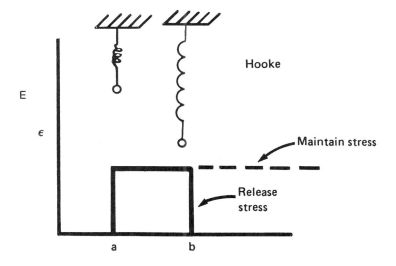

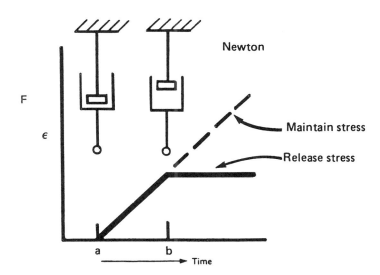

Figure 5.4 Continued.

related to strain through inverse stress. This linear behavior is typical of most amorphous polymers for small strains over short periods of time. Further, the over-all effect of a number of such imposed stresses are additive.

Recovery occurs when the applied stress is relieved. Thus, amorphous polymers and polymers containing amorphous portions act both as elastic solids and viscous liquids above their T_g.

The plots in Fig. 5.4D show the response where fixed strain is rapidly applied to a specimen. This stress is time dependent, and a decrease in stress, known as

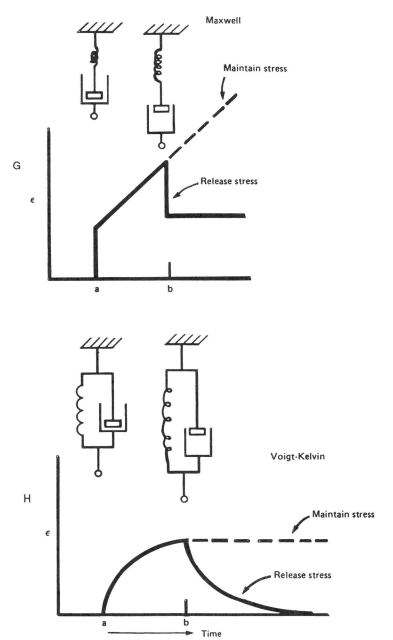

Figure 5.4   Continued.

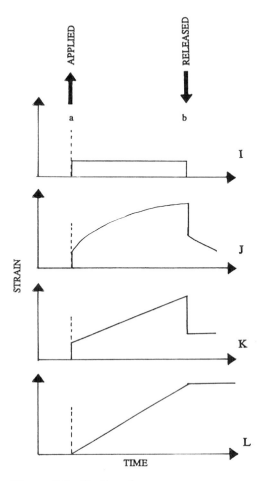

Figure 5.4 Continued.

stress relaxation, occurs. This stress relaxation is also time related and linearly related to stress as follows:

$$\text{Stress relaxation} = \frac{\text{stress(t)}}{\text{strain}} = S(t)/\gamma \qquad (5.3)$$

As noted in Sec. 3.1, models have been developed in an attempt to illustrate stress-strain behavior. Plots for the stress-strain-time behavior of two of these models are shown in Fig. 5.5. The Maxwell model (Fig. 3.2) consists of a combination of a spring and a dash pot in series. The response of a constant stress by the Maxwell model given in Fig. 5.5A is described as follows:

$$\text{Strain} = (\text{compliance} + t/\eta) \text{ stress} \qquad (5.4)$$

Thus the strain increases linearly with time from b to d.

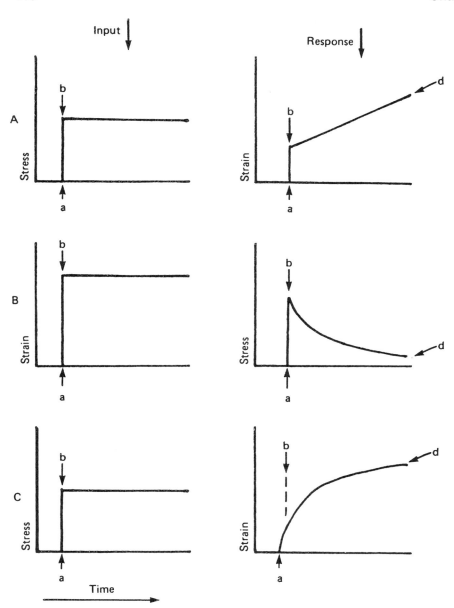

Figure 5.5 Stress-strain relationships for the Maxwell model (A and B) and the Voigt-Kelvin model (C).

The response on a Maxwell model to application of rapid strain is illustrated in Fig. 5.5B and is described by

$$\text{Stress} = \text{strain } e^{-(t/\eta \text{ compliance})/\text{compliance}} = \text{strain } e^{-t/\eta(\text{compliance})^2} \qquad (5.5)$$

A slow relaxation time, t, is a result of a weak spring or high viscosity.

The Voigt-Kelvin model (Fig. 3.3) contains the dash pot and spring in parallel. This combination prevents creep from occurring at a constant rate under a constant load. The response of the Voigt model to rapid stress is shown in Fig. 5.5C and described mathematically as follows:

$$\text{Strain} = \text{stress} \cdot \text{compliance}[1 - e^{-(t/\eta \text{ compliance})}] \qquad (5.6)$$

For long periods of time, strain approaches η. Compliance and the ultimate strain at d approach the strain developed in an elastic solid (Fig. 5.5C) of the same compliance. The strain response is retarded and not instantaneous. The product of η and compliance is characteristic of a retardation time for the system.

Additional models have been developed to attempt to illustrate the real behavior of materials and attempt to determine the particular critical structural parameters which determine physical properties.

Phase transition refers to a change of state. Most low molecular weight materials exist as solids, liquids, and gases. The transitions that separate these states are melting (or freezing) and boiling (or condensing). By contrast, polymers do not vaporize "intact" to a gaseous state, nor do they boil. The state of a polymer, i.e., its physical response character, depends on the time allotted for interaction and the temperature as well as the molecular organization (crystalline, amorphous, mix). The term relaxation refers to the time required for response to a change in pressure, temperature, or other applied parameter. The term dispersion refers to the absorption or emission of energy at a transition. In practice, the terms relaxation and dispersion are often employed interchangeably.

Stress-strain plots can also be given for the various regions shown in Fig. 5.3. It is important to understand that many polymers with amorphous and crystalline regions will give a relationship such as pictured in Fig. 5.3. The region between 1 and 2 (Fig. 5.3) is called the Hookean glassy region or simply the glassy region (see also Fig. 3.5). Here there is a direct and relatively rapid response as stress is applied (a) and as stress is released (b) (Fig. 5.4I). The primary response to an applied stress is the "flexing" of bonds. The region between 2 and 3 (Fig. 5.3) is referred to as the leathery region, where there is both movement of side chains and the beginnings of main chain movement. Here the stress-strain behavior (Fig. 5.4J) can be considered to be sluggish and incomplete with respect to the application and release of the stress. The third and fourth regions (Fig. 5.3, from 3 to 4) are referred to as the rubbery regions and involve main chain movement. In region 3, from the area marked 3 to the plateau, the chain movement is more gradual, while the movement along the plateau is larger in scale. The stress-strain response is given in Fig. 5.4K and shows a rapid initial response followed by a more gradual response due to continued chain slippage. Some of this chain slippage is reversible, while some is not recoverable, resulting in a more or less permanent deformation of the material, often referred to as "creep." The fifth region, between 4 and 5 (Fig. 5.3), is referred to as the viscous or "flow" region where both "local" and "wholesale" chain movement (slippage) occurs in a largely nonreversible fashion (Fig. 5.4L).

These five regions (Fig. 5.3) make up the typical viscoelastic behavior of materials. It must be stated again that for a material to behave as a viscoelastic material it must be *above* the glass transition temperature, but not at the melting point. Thus, viscoelastic behavior occurs between the glass transition temperature (or range) and the melting temperature (or range). As noted in Sec. 3.1 and Fig. 3.5, these responses are time dependent. Thus, an amorphous material or a material with both amorphous and crystalline regions above the glass transition temperature can behave as a glassy solid if the interaction is rapid, such as striking the material with a hammer, or it can act as a liquidlike material if a constant pull is exerted over a long time.

It must be remembered when dealing with a plot such as shown in Fig. 5.3, considering only the temperature aspect, that the glass transition temperature or range occurs around the transition from the "glassy" to the "rubbery" region and that the melting point occurs around the transition to the melt or "flow" region. Further, a material that is largely crystalline will act like a "glass" throughout the temperature region until reaching the melting temperature, at which juncture it will "flow."

Finally, for Figs. 3.5 and 5.3, the thermodynamic driving forces for a deformed material to return to its original shape within the "glassy" region are mainly derived from enthalpy or energy, while the driving forces within the rubbery regions are largely entropy- or "order"-derived forces.

5.3 SPECIFIC PHYSICAL TESTS

Tensile Strength

Tensile strength can be determined by applying force to the test material until it breaks. It is defined by the following relationship:

$$\text{Tensile strength (Pascals)} = \frac{\text{force required to break sample (N)}}{\text{cross-sectional area (m}^2)} \qquad (5.7)$$

Tensile strength, which is a measure of the ability of a polymer to withstand pulling stresses, is usually measured by pulling a dumbbell specimen, such as the one shown in Fig. 5.6, in accordance with ASTM-D638-72. These test specimens, like all others, must be conditioned under standard conditions of humidity (50%) and temperature (23°C) before testing.

The behavior of materials under applied stress and strains is given in Eq. (5.2) and Eq. (5.3). For typical tensile strength measurements, both ends of the specimen are clamped in the jaws of the test apparatus pictured in Fig. 5.6. One jaw is fixed and the other moves at specific increments. The stress and applied force is plotted against strain and elongation.

The elastic modulus (also called tensile modulus or modulus of elasticity) is the ratio of the applied stress to the strain it produces within the region where the relationship between stress and strain is linear. The ultimate tensile strength is equal to the force required to cause failure divided by the minimum cross-sectional area of the test sample.

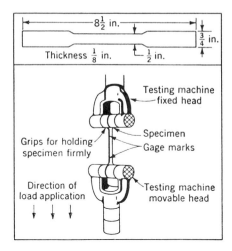

Figure 5.6 Typical tensile test. (From *Introduction to Polymer Chemistry* by R. Seymour, McGraw-Hill, New York, 1971. Used with permission of McGraw-Hill Book Company.)

When rubbery elasticity is required for ample sample performance, a high ultimate elongation is desirable. When rigidity is required, it is desirable that the material exhibit a lower ultimate elongation. Some elongation is desirable since it allows the material to absorb rapid impact and shock. The total area under a stress-strain curve is indicative of overall toughness.

Pulling stress is the deformation of a test sample caused by application of specific loads. It is specifically the change in length of the test sample divided by the original length. Recoverable strain or elongation is called elastic strain. Here, stressed molecules return to their original relative locations after release of the applied force. Elongation may also be a consequence of wholesale movement of chains past one another. The wholesale movement of polymer chains is called creep or plastic strain. Here the strain is nonreversible with an end result of permanent deformation, elongation of the test material. Many samples undergo both reversible and irreversible strain.

As noted above, creep is the irreversible elongation caused by the application of force (such as a weight) over a specified time. If the change in length occurs at room temperature, it is called cold creep.

Elongation is measured by the tensile test shown in Fig. 5.6. The percentage of elongation is equal to the change in length divided by the original length of the specimen multiplied by 100, i.e.,

$$\% \ El = \frac{\Delta l}{l} \cdot 100 \tag{5.8}$$

Problem

Determine the percentage elongation of a 10.0-cm polystyrene sample that increases in length to 10.2 cm when subjected to a tensile stress. The percentage elongation is determined employing % El = $(\Delta l/l) \times 100$ = [(10.2 − 10.0)/10.0] × 100 = 2%.

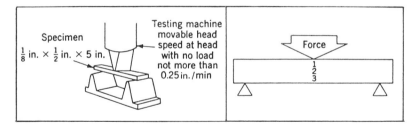

Figure 5.7 Typical flexural strength test. (From *Introduction to Polymer Chemistry* by R. Seymour, McGraw-Hill, New York, 1971. Used with permission of McGraw-Hill Book Company.)

Flexural strength, or cross-breaking strength, is a measure of the bending strength or stiffness of a bar test specimen used as a simple beam in accordance with ASTM-D790.

The specimen is placed on the supports as pictured in Fig. 5.7. A load is applied in the center at a specified rate, and the loading at failure is called the flexural strength. However, since many materials do not break even after being greatly deflected, the actual flexural strength cannot be calculated. Instead, by agreement, the modulus at 5% strain is used for these specimens as a measure of flexural strength, i.e., the ratio of stress to strain when the strain is 5%.

Tensile Strength of Inorganic and Metallic Fibers and Whiskers

The tensile strength of materials is dependent on the treatment and form of the material. Thus, the tensile strength of isotropic bulk nylon-6,6 is less than that of anisotropic oriented nylon-6,6 fiber. Inorganics (Chapter 12) and metals also form fibers and whiskers with varying tensile strengths (Table 5.1). Fibers are generally less crystalline and larger than whiskers.

Many of these inorganic fibers and whiskers are polymeric including many of the oxides (including the so-called ceramic fibers), carbon and graphite materials,

Table 5.1 Tensile Strengths of Inorganic and Metallic Materials as a Function of Form

Material	Form	Tensile strength (MPa)
Graphite	Bulk	1,000
	Fiber	2,800
	Whisker	15,000
Glass	Bulk	1,000
	Fiber	4,000
Steel	Bulk	2,000
	Fiber	4,000
	Whisker	10,000

and silicon carbide. Carbon and graphite materials are similar but differ in the starting materials and the percentage carbon. Carbon fibers, typically derived from polyacrylonitrile (PAN), are about 95% carbon, while graphite fibers are formed at higher temperatures yielding a material with about 99% carbon.

These specialty fibers and whiskers exhibit some of the highest tensile strengths recorded (Table 5.2). They are employed in applications where light weight and high strength are required. Organics offer weight advantages, typically being less dense than most of inorganic and metallic fibers (Table 5.2). Uses include in dental fillings, the aircraft industry, production of lightweight fishing poles, automotive antennas, bicycles, turbine blades, heat-resistant reentry vessels, golf club shafts, etc.

Many of these inorganic fibers and whiskers are utilized as reinforcing agents in composites.

Compressive Strength

Compressive strength is defined as the pressure required to crush a material. Compressive strength is defined by the relationship

$$\text{Compressive strength (Pa)} = \frac{\text{force (Newtons)}}{\text{cross-sectional area (m}^2)} \tag{5.9}$$

Compressive strength can be calculated by dividing the maximum compressive force in Newtons divided by the material's area in meters squared giving the compressive strength in units of Pascals.

Problem

Calculate the compressive strength for a material with a cross-section of 2 millimeters squared where the force required to crush the test sample is 200 kg.

Force = mass times acceleration = 200 kg $\times$ 9.8 m/sec^2

Table 5.2 Ultimate Tensile Strength of Representative Organic, Inorganic, and Metallic Fibers

Material	Tensile strength (MPa)	Tensile strength/density
Aluminum silica	4,100	1,060
Aramid	280	200
Beryllium carbide	1,000	400
Beryllium oxide	500	170
Boron-tungsten boride	3,450	1,500
Carbon	2,800	1,800
Graphite	2,800	1,800
PET	690	500
Quartz	900	400
Steel	4,000	500
Titanium	1,900	400
Tungsten	4,300	220

Since 9.8 m/sec² is the gravity constant,

$$\text{Compressive strength} = \frac{200 \text{ kg} \times 9.8 \text{ m/sec}^2}{2 \text{ mm}^2 \times 10^{-6} \text{ m}^2/\text{mm}^2} = 9.8 \times 10^8 \text{ Pa}$$

Compressive strength, or the ability of a specimen to resist crushing force, is measured by crushing a cylindrical specimen in accordance with ASTM-D695.

The test material is mounted in a compression tool as shown in Fig. 5.8. One of the plungers advances at a constant rate. The ultimate compression strength is

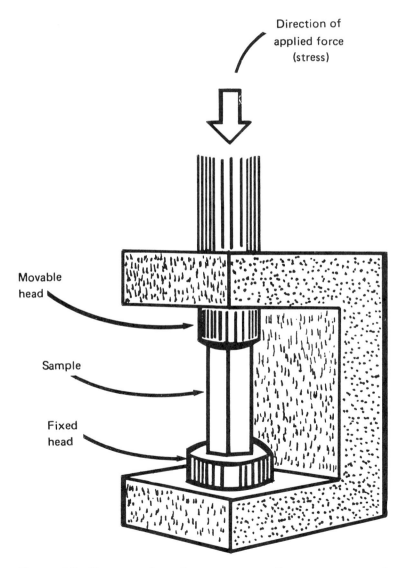

Figure 5.8 Representation of test apparatus for measurement of compression-related properties.

equal to the load that causes failure divided by the minimum cross-sectional area. Since many materials do not fail in compression, strengths reflective of specified deformation are often reported.

Impact Strength

Impact strength is a measure of the energy needed to break a sample—it is not a measure of the stress needed to break the material. The term toughness is typically employed in describing the impact strength of a material. Toughness is a measure of the energy required (or absorbed) to break a sample. Toughness does not have a precise definition but is often described as the area under stress-strain curves such as shown in Fig. 5.2.

Impact strength tests fall within two main categories: (1) falling-mass tests, and (2) pendulum tests. Figure 5.9 illustrates two common falling-mass tests. The falling-mass assembly is typically employed for bulk specimens, whereas the dart test is used for films.

Izod and Charpy Impact Resistance

The Izod impact test (ASTM D256) is a measure of the energy required to break a notched sample under standard conditions. A sample is clamped in the base of a pendulum testing apparatus so that it is cantilevered upward with the notch facing the direction of impact. The pendulum is released from a specified height. The procedure is repeated until the sample is broken. The force needed to break the sample is then calculated from the height and weight of the pendulum required to break the specimen. The Izod value is useful when comparing samples of the same polymer but is not a reliable indicator of toughness, impact strength, or abrasive resistance. The Izod test may indicate the need to avoid sharp corners in some manufactured products. Thus, nylon is notch sensitive, gives relatively low Izod impact values, but is considered a tough material.

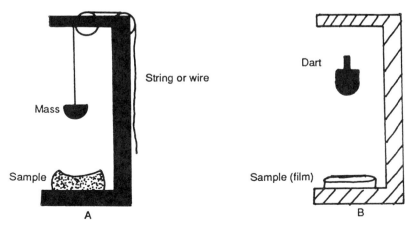

Figure 5.9 Assemblies used to measure the impact strength of (A) solids and (B) films.

An unnotched or oppositely notched specimen (Fig. 5.10) is employed in the Charpy test (ASTM-D256).

Overall, impact strength tests are measures of toughness or the ability of a specimen to withstand a sharp blow, such as being dropped from a specific height or, as already noted, determined by the energy required to break a specimen.

Hardness

The term hardness is a relative term and a general term. Properties such as the ability to resist scratching and marring and abrasion resistance are all related to hardness. There exists a number of instruments that measure a material's ability to resist scratching and marring. There are several relative scales of hardness, such as the Mohs scale used widely by geologists.

For practical purposes, when dealing with solid materials, hardness is the resistance to penetration and indention, or, in other words, it is the ability of a material to resist compression, scratching, and indentation.

Also related to hardness is abrasion resistance—resistance to the process of wearing away the surface of a material. The major test for abrasion resistance involves rubbing an abrader against the surface of the material under specified conditions (ASTM D-1044) and noting the relative amount of weight loss.

Rockwell Hardness

Rockwell hardness tests [ASTM-D785-65 (1970)], depicted in Fig. 5.11, measure hardness in progressive numbers on different scales corresponding to the size of the ball indentor used. The scale symbols correspond to the loads of 60 to 150 kg.

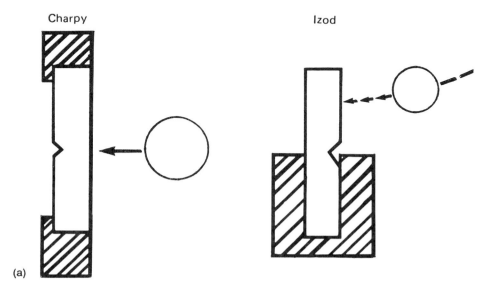

Figure 5.10 Pictorial description of (a) Izod- and Charpy-type pendulum impact tests, and (b) Charpy assembly using an unnotched sample.

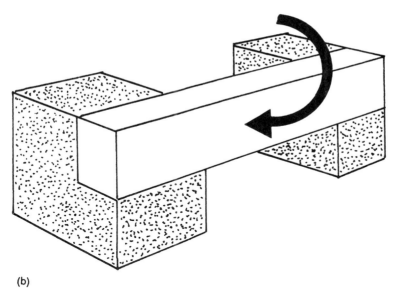

(b)

Figure 5.10 Continued.

As described in Fig. 5.11, the distance RB is used to calculate the Rockwell hardness figure. While the Rockwell hardness allows differentiation between materials, factors such as creep and elastic recovery are involved in determining the Rockwell hardness as seen in the description of the procedure given in Fig. 5.11. Rockwell hardness is not a good measure of wear qualities or abrasion resistance. For instance, polystyrene has a relatively high Rockwell hardness value, yet it is easily scratched.

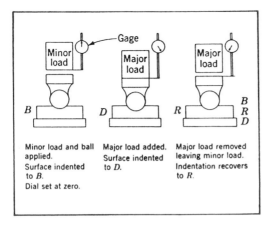

Figure 5.11 Rockwell hardness test. (From *Introduction to Polymer Chemistry* by R. Seymour, McGraw-Hill, New York, 1971. Used with permission of McGraw-Hill Book Company.)

Scratch hardness may be measured on the Mohs scale, which ranges from 1 for talc to 10 for diamond, or by scratching with pencils of specified hardness (ASTM-D-3363). Hardness may also be measured by the number of bounces of a ball or the extent of rocking by a Sward Hardness Rocker. Abrasion resistance may be measured by the loss in weight caused by the rubbing of the wheels of a Tabor-abrader (ASTM-D-1044).

Shear Strength

This test utilizes a punch-type shear fixture for testing flat specimens. The shear strength is equal to the load divided by the area. Thus, the sample is mounted in a punch-type shear fixture and punch-pushed down at a specified rate until shear occurs. This test is important for sheets and films but is not typically employed for extruded or molded products (Figure 5.12).

Deformation Underload

The specimen, usually a 1/2 in. (12.7 mm) cube, is subjected to a specific pressure, typically 1000 psi. The deflection is measured shortly after the load is applied and again 24 hr later. The deformation is calculated as the difference in sample height divided by the original height. This is a good measure of a material's ability to withstand constant, short-term compression but is not a satisfactory measure of long-term creep resistance.

Indentation Hardness

An indenter is pressed into the specimen, typically 1/4 in. (6.35 mm) thick. The indenter is applied to the face of the sample under different loading and the amount

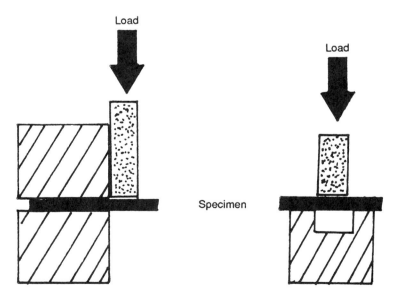

Figure 5.12 Two assemblies employed to measure shear strength.

of indentation is measured. Many of the hardness tests, including the Indentation hardness test (ASTM D2240) and the Rockwell hardness test, are time dependent due to creep. The Indentation hardness test is preferred for softer materials where creep is significant when measurements are taken shortly after the indenter has been applied to the sample.

Hardness is a general term which describes a combination of properties, such as the resistance to penetration, abrasion, and scratching. Indentation hardness of thermosets may be measured by a Barcol Impresser as described in ASTM-D-2583-67. A shore durometer is used to measure the penetration hardness of elastomers and soft thermoplastics (ASTM D2240) (Fig. 5.13).

The tests for change in dimensions of a polymer under long-term stress, called creep or cold flow (ASTM-D-674), are no longer recommended by ASTM. ASTM-D-671 describes suggested tests for fatigue or endurance of plastics under repeated flexure.

Additional Physical Tests

The brittleness temperature of plastics (ASTM-D-746) and plastic film (ASTM-D-1790-62) is the temperature at which 50% of test samples fail an impact test.

The environmental stress cracking (ESR) test pictured in Fig. 5.14 measures the time of failure of scored test samples that are bent 180° and inserted in a solution of standard detergent. Many failures of polymers are the result of molecular changes which can be detected by the use of appropriate instrumentation and testing.

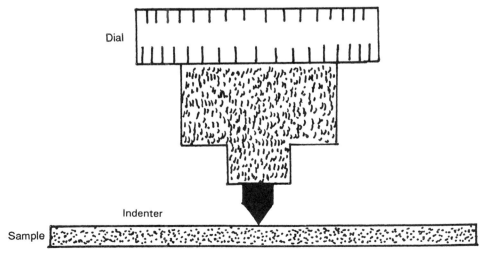

Figure 5.13 Sharp-pointed indenter employed in the measure of indentation hardness [Barcol instrument (ASTM-D-2583-67)].

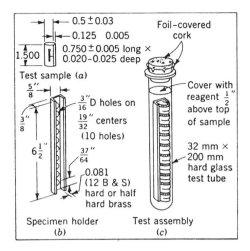

Figure 5.14 Environmental stress cracking test. (From *Introduction to Polymer Chemistry* by R. Seymour, McGraw-Hill, New York, 1971. Used with permission of McGraw-Hill Book Company.)

Failure

The failure of materials can be associated with a number of parameters. Two major causes of failure are creep and fracture. The tensile strength is the nominal stress at the failure of the material. Toughness is related to ductility and, for stress-strain curves, toughness is measured by the area under a stress-strain curve taken to failure. For a material to be tough, it typically has a combination of stiffness and give. The shape of stress-strain curves and, subsequently, the toughness that will control the usefulness of the material where toughness is the important feature is due to many factors, including the interaction time or speed of the test. The Charpy and Izod impact tests interact with the test material rapidly, so that features related to brittleness are important relative to slower interactive tests, such as load creep tests, where factors associated with deformation play a more important role.

Calculations have been made to determine the theoretical upper limits with respect to the strength of polymeric materials. Real materials show behaviors near to those predicted by the theoretical calculation during the initial stress-strain determinations but vary greatly near the failure of the material. It is believed that the major reasons for the actual tensile strengths at failure being smaller than calculated are related to imperfections, the nonhomogeneity of the polymeric structure. These flaws—molecular irregularities—act as the "weak link" in the polymer's behavior. These irregularities can be, for example, dislocations, voids, physical cracks, and energy concentrations.

Even with polymeric materials performing, strengthwise, below the theoretical limits, they have high strength-to-mass ratios. For instance, steel has a tensile strength of about 2000 MPa and a density of 7800 kg/m^3, giving a strength-to-mass ratio of 0.26. Ultrahigh molecular weight polyethylene has a tensile strength of about 380 MPa and a density of 940 kg/m^3, giving a strength-to-mass ratio of 0.40—about double that of steel.

5.4 ELECTRICAL PROPERTIES: THEORY

Some important dielectric properties are dielectric loss, loss factor, dielectric constant (or specific inductive capacity), direct current (DC) conductivity, alternating current (AC) conductivity, and electric breakdown strength. The term dielectric behavior usually refers to the variation of these properties within materials as a function of frequency, composition, voltage, pressure, and temperature.

The dielectric behavior of materials is often studied by employing charging or polarization currents. Since polarization currents depend on the applied voltage and the dimensions of the condenser, it is customary to eliminate this dependence by dividing the charge Q by the voltage V to obtain a parameter C, called the capacitance (capacity),

$$C = \frac{Q}{V} \qquad (5.10)$$

and then using the dielectric constant ϵ, which is defined as

$$\epsilon = \frac{C}{C_0} \qquad (5.11)$$

where C is the capacity of the condenser when the dielectric material is placed between its plates in a vacuum and C_0 is the empty condenser capacity.

Dielectric polarization is the polarized condition in a dielectric resulting from an applied AC or DC field. The polarizability is the electric dipole moment per unit volume induced by an applied field or unit effective intensity. The molar polarizability is a measure of the polarizability per molar volume, thus it is related to the polarizability of the individual molecules or polymer repeat unit.

Conductivity is a measure of the number of ions per unit volume and their average velocity in the direction of a unit applied field. Polarizability is a measure of the number of bound charged particles per cubic unit and their average displacement in the direction of the applied field.

There are two types of charging currents and condenser charges, which may be described as rapidly forming or instantaneous polarizations, and slowly forming or absorptive polarizations. The total polarizability of the dielectric is the sum of contributions due to several types of displacement of charge produced in the material by the applied field. The relaxation time is the time required for polarization to form or disappear. The magnitude of the polarizability, k, of a dielectric is related to the dielectric constant ϵ as follows:

$$k = \frac{3 (\epsilon - 1)}{4 \pi (\epsilon + 2)} \qquad (5.12)$$

The terms "polarizability constant" and "dielectric constant" have been used interchangeably in the qualitative discussion of the magnitude of the dielectric constant. The k values obtained utilizing DC and low-frequency measurements are a summation of electronic, E, atomic, A, dipole, P, and interfacial, I, polarizations as shown in Fig. 5.15. Only the contribution by electronic polarizations is evident at high frequencies. The contributions to dielectric constant at low frequencies for

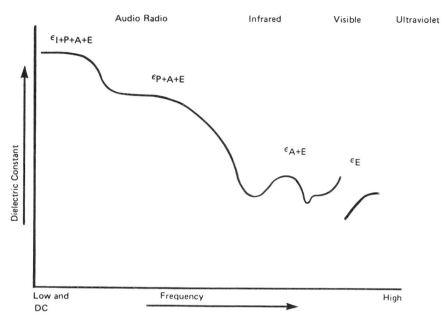

Audio Radio   Infrared   Visible   Ultraviolet

$\epsilon_{I+P+A+E}$

ϵ_{P+A+E}

ϵ_{A+E}

ϵ_E

Dielectric Constant

Low and
DC   Frequency   High

Figure 5.15 Relationship of dielectric constant with frequency emphasizing interfacial, I, dipole, P, atomic, A, and electronic, E, polarization contributions.

a material having interfacial, dipole, atomic, and electronic polarization contributions are additive.

Instantaneous polarization occurs when rapid (less than 10^{-10} sec) transitions take place, i.e., at frequencies greater than 10^{10} Hz or at wavelengths less than 1 cm. Electronic polarization falls within this category and is due to the displacement of charges within the atoms. Electronic polarization is directly proportional to the number of bound electrons in a unit volume and inversely proportional to the forces binding these electrons to the nuclei of the atoms.

Electronic polarization occurs so rapidly that there is no observable effect of time or frequency on the dielectric constant until frequencies are reached which correspond to the visible and ultraviolet spectra. For convenience, the frequency range of the infrared through the ultraviolet region is called the optical frequency range, and the radio and the audio range is called the electric frequency range.

Electronic polarization is an additive property dependent on the atomic bonds. Thus the electronic polarizations and related properties per unit volume are similar for both small molecules and polymers.

Accordingly, values obtained for small molecules can be appropriately applied to analogous polymeric materials. This does not apply in cases where the polymeric nature of the material plays an additional role in the conductance of electric charges, as is the case for whole chain resonance electric conductance.

Atomic polarization is attributed to the relative motion of atoms in the molecule effected by perturbation by the applied field of the vibrations of atoms and ions having a characteristic resonance frequency in the infrared region. The atomic polarization is large in inorganic materials which contain low-energy-conductive bonds and approaches zero for nonpolar organic polymers.

The atomic polarization is rapid, and this as well as the electronic polarizations constitute the instantaneous polarization components. The remaining types of polarization are absorptive types with characteristic relaxation times corresponding to relaxation frequencies.

In 1912, Debye suggested that the high dielectric constants of water, ethanol, and other highly polar molecules are due to the presence of permanent dipoles within each individual molecule. There is a tendency for the molecules to align themselves with their dipole axes in the direction of the applied electric field.

The major contributions to dipole polarizations are additive and are similar whether the moiety is within a small or a large (polymeric) molecule. Even so, the secondary contributions to the overall dipole polarization of a sample are dependent on both the chemical and the physical environment of the specific dipole unit and on the size and the mobility of that unit. Thus dipole contributions can be utilized to measure glass transition temperature T_g and melting point T_m.

These polarizations are the major types found in homogeneous materials. Other types of polarization, called interfacial polarizations, are the result of heterogeneity. Ceramics, polymers containing additives, and paper are considered to be electrically heterogeneous.

5.5 ELECTRICAL MEASUREMENTS

Material response is typically studied by utilizing either direct (constant) applied voltage (DC) or alternating applied voltage (AC). The AC response as a function of frequency is characteristic of the material. In the future these "electrical spectra" may be utilized as a product identification tool, much like infrared spectroscopy. Factors such as current strength, duration of measurements, specimen shape, temperature, and applied pressure are typical factors that affect the electrical responses of materials.

The electric properties of a material vary with the frequency of the applied current. The response of a polymer to an applied current may be delayed because of a number of factors including the interaction between polymer chains, the presence within the chain of specific molecular groupings, and effects related to interactions within the specific atoms themselves. A number of parameters, such as relaxation time, power loss, dissipation factor, and power factor, are employed as measures of this lag.

The movement of dipoles (related to the dipole polarization, P) within a polymer can be divided into two types, an orientation polarization, P', and a dislocating or induced polarization.

The relaxation time required for the charge movement of electronic polarization E to reach equilibrium is extremely short ($\simeq 10^{-15}$ sec), and this type of polarization is related to the square of the index of refraction, n^2. The relaxation time for atomic polarization A is about 10^{-3} sec. The relaxation time for induced orientation polarization P' is dependent on molecular structure and is temperature dependent.

The electric properties of polymers are related to their mechanical behavior. The dielectric constant and dielectric loss factor are analogous to the elastic compliance and mechanical loss factor. Electric resistivity is analogous to viscosity.

Polar polymers, such as ionomers, possess permanent dipole moments. Polar molecules are capable of storing more electric energy than nonpolar polymers. Nonpolar polymers are dependent almost entirely on induced dipoles for electric energy storage. Thus, orientation polarization is produced in addition to the induced polarization, when the polar polymers are placed in an electric field.

The induced dipole moment of a polymer in an electric field is proportional to the strength of the field, and the proportionality constant is related to the polarizability of the atoms in the polymer. The dielectric properties of polymers are affected adversely by the presence of moisture, and this effect is greater in hydrophilic than in hydrophobic polymers.

As shown by the Clausius-Mossotti equation,

$$P = \left(\frac{\epsilon - 1}{\epsilon + 2}\right)\frac{M}{\rho} \tag{5.13}$$

the polarization P of a polymer in an electric field is related to the dielectric constant ϵ, the molecular weight M, and the density ρ.

At low frequencies, the dipole moments of polymers are able to keep in phase with changes in a strong electric field, and the power losses are low. However, as the frequency is increased, the dipole moment orientation may not occur rapidly enough to maintain the dipole in phase with the electric field.

Dielectric Constant

As noted before, the dielectric constant (ASTM-D150-74) is the ratio of the capacity of a condenser made with or containing the test material to the capacity of the same condenser with air as the dielectric. Materials to be employed as insulators in electrical applications should have low dielectric constants, whereas those to be employed as conductors or semiconductors should exhibit high dielectric constants.

The dielectric constant is independent of the frequency at low to moderate frequencies but is dependent on the frequency at high frequencies. The dielectric constant is approximately equal to the square of the index of refraction and to one-third the solubility parameter.

Electrical Resistance

There are a number of electrical properties related to electrical resistance (ASTM-D257). These include insulation resistance, volume resistivity, surface resistivity, volume resistance, and surface resistance.

The bulk (or volume) specific resistance is one of the most useful electrical properties. Specific resistance is a physical quantity that may differ by more than 10^{23} in readily available materials. This unusually wide range of conductivity is basic to the wide use of electricity and many electrical devices. Conductive materials, such as copper, have specific resistance values of about 10^{-6} ohm cm, whereas good insulators, such as polytetrafluoroethylene (PTFE) and low-density polyethylene (LDPE), have values of about 10^{17} ohm cm.

Specific resistance is calculated from Eq. (5.14), in which R is the resistance in ohms, a is the pellet area in cm^2, t is the pellet thickness in cm, and p is the specific resistance in ohm cm.

$$p = R(a/t) \qquad (5.14)$$

Dissipation Factor and Power Factor

The dissipation factor (ASTM-D150) can be defined in several ways, including the following:

1. Ratio of the real (in phase) power to the reactive (90° out of phase) power
2. Measure of the conversion of the reactive power to real power or heat
3. Tangent of the loss angle and the cotangent of the phase angle
4. Ratio of the conductance of a capacitor (in which the material is the dielectric) to its susceptibility

Both the dielectric constant and dissipation factor are measured by comparison or substitution in a classical electric bridge.

The power factor is the energy required for the rotation of the dipoles of a polymer in an applied electrostatic field of increasing frequency. These values, which typically range from 1.5×10^4 for polystyrene to 5×10^{-2} for plasticized cellulose acetate, increase at T_g because of increased chain mobility. The loss factor is the product of the power factor and the dielectric constant and is a measure of the total electric loss in a dielectric material.

Arc Resistance

The arc resistance is a measure of the resistance of a material to the action of an arc of high voltage and low current close to the surface of the sample to form a conducting path on this surface. Arc resistance values are of use in surface quality control since small surface changes will affect the arc resistance value.

Dielectric Strength

No steady current flows in a perfect insulator in a static electric field, but energy is "stored" in the sample as a result of dielectric polarization. Thus the insulator acts as a device to store energy. In actuality, some leakage of current does occur even for the best insulators.

The insulating property of materials breaks down in strong fields. This breakdown strength, called the electric or dielectric strength (DS), i.e., the voltage required for failure, is inversely related to the thickness l of the material as shown in Eq. (5.15).

$$DS \, \alpha \, l^{-0.4} \qquad (5.15)$$

Breakdown may occur below the measured DS as a result of an accumulation of energy through inexact dissipation of the current; this leads to an increase in temperature and thermal breakdown. Breakdown means sudden passage of excessive current through the material, which often results in visible damage to the specimen.

The DS is high for many insulating polymers and may be as high as 10^3 MV /m. The upper limit of the DS of a material is dependent on the ionization energy present in the material. Electric or intrinsic decomposition (breakdown) occurs when electrons are removed from their associated nuclei; this causes secondary ionization and accelerated breakdown. The DS is reduced by mechanical loading of the specimen and by increasing the temperature.

Typically dielectric strength (ASTM-D149) is an indication of the electrical strength of an insulating material. DS is dependent on test conditions.

5.6 WEATHERABILITY

Outdoor exposure of polymer samples, mounted at a 45° angle and facing south (in the northern hemisphere), has been used to measure the resistance of polymers to outdoor weathering (ASTM-D1345). Since these tests are expensive and time-consuming, tests such as ASTM-GS23 have been developed in an attempt to gain "accelerated" test results. Tests related to the accelerated exposure to light are described in ASTM 625 and 645.

There are several accelerated tests which differ in the selection of light source and cyclic exposure to varying degrees of humidity. Some accelerated tests include salt spray, heat, cold, and other weather factors. The ESCR test described in Fig. 5.14 can also give a measure of weatherability.

5.7 OPTICAL PROPERTY TESTS

Since polymers are often used as clear plastics or coatings and have many applications in which transparency is an important property, a knowledge of the optical properties of specific polymers is essential. The radiation scale, of course, includes microwave, infrared, ultraviolet, and visible regions.

It is important to recognize the difference between refraction (associated with properties such as refractive index) and reflection (associated with properties such as haze). This difference is illustrated in Fig. 5.16.

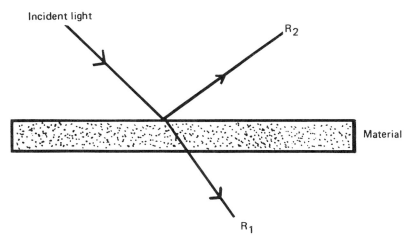

Figure 5.16 Refraction and reflection of incident light at the surface of a solid. The path of refracted light is indicated by R_1 and that of reflected light by R_2.

Index of Refraction

Optical properties are related to both the degree of crystallinity and the actual polymer structure. Most polymers are transparent and colorless, but some, such as phenolic resins and polyacetylenes, are colored, translucent, or opaque. Polymers that are transparent to visible light may be colored by the addition of colorants, and some become opaque as the result of the presence of additives such as fillers, stabilizers, flame retardants, moisture, and gases.

Many of the optical properties of a polymer are related to the refractive index n, which is a measure of the ability of the polymer to refract or bend light as it passes through the polymer. The refractive index n is equal to the ratio of the sine of the angles of incidence, i, and refraction, r, of light passing through the polymer:

$$n = \frac{\sin i}{\sin r} \tag{5.16}$$

The magnitude of n is related to the density of the substance and varies from 1.000 and 1.333 for air and water, respectively, to about 1.5 for many polymers and 2.5 for the white pigment titanium dioxide. The value of n is often high for crystals and is dependent on the wavelength of the incident light and on the temperature; it is usually reported for the wavelength of the transparent sodium D line at 298°K. Typical refractive indices for polymers range from 1.35 for polytetrafluoroethylene to 1.67 for poly(aryl sulfone).

The velocity of light passing through a polymer is affected by the polarity of the bonds in the molecule. Polarizability P is related to the molecular weight per unit volume, M, and density ρ as follows (the Lorenz-Lorenz relationship):

$$P = \left(\frac{n^2 - 1}{n^2 + 2}\right) \frac{M}{\rho} \tag{5.17}$$

The polarizability P of a polymer is related to the number of molecules present per unit volume, and the polarizability of each molecule is related to the number and mobility of the electrons present in the molecule. The P value of carbon is much greater than that of hydrogen, and the value of the latter is usually ignored in calculating the polarizability of organic polymers.

Heterogeneity of n values is related to a number of factors, including end groups, differences in density between amorphous and crystalline regions, anisotropic behavior of crystalline portions, incorporation of additives, and the presence of voids.

The index of refraction of transparent plastics may be determined by placing a drop of a specified liquid on the surface of the polymer before measuring the index with a refractometer. An optical microscope is used to measure the index of refraction in an alternative method (ASTM-D542).

Optical Clarity

Optical clarity or the fraction of illumination transmitted through a material is related by the Beer-Lambert relationship

$$\log \frac{I}{I_o} = -AL \qquad \text{and} \qquad \frac{I}{I_o} = e^{-AL} \tag{5.18}$$

where the fraction of illumination transmitted through a polymer, I/I_o, is dependent on the path length of the light, L, and the absorptivity of the polymer at that wavelength of light, A [see Eq. (4.8)].

Clarity is typical for light passing through a homogeneous material, such as a crystalline, ordered polymer or a completely amorphous polymer. Interference occurs when the light beam passes through a heterogeneous material in which the polarizability of the individual units varies, such as a polymer containing both crystalline and amorphous regions.

Absorption and Reflectance

Colorless materials range from being almost totally transparent to being opaque. The opacity is related to the light-scattering process occurring within the material.

Incident radiation passes through nonabsorbing, isotropic, and optically homogeneous samples with essentially little loss of radiation intensity. Actually, all materials scatter some light. The angular distribution of the scattered light is complex because of the scattering due to micromolecular differences in values.

Transparency is defined as the state permitting perception of objects through a sample. Transmission is the light transmitted. In more specific terms transparency is the amount of undeviated light, i.e., the original intensity minus all light absorbed, scattered, or lost through any other means.

The ratio of reflected light to the incidental light is called the reflectance coefficient, and the ratio of the scattered light to the incident light is called the absorption coefficient. The attenuation coefficient is the term associated with the amount of light lost. The transmission factor is the ratio of the amount of transmitted light to the amount of incident light. The transmission factor for weakly scattering colorless materials decreases exponentially with the sample thickness l and the scattering coefficient S as follows:

$$\ln (\text{transmission factor}) = -Sl \tag{5.19}$$

The light scattering reduces the contrast between light, dark, and other colored parts of objects viewed through the material and produces a milkiness or haze in the transmitted image. Haze is a measure of the amount of light deviating from the direction of transmittancy of the light by at least 2.5°.

The visual appearance and optical properties of a material depend on its color and additives, as well as on the nature of its surface. Gloss is a term employed to describe the surface character of a material responsible for luster or shine, i.e., surface reflection.

A perfect mirrorlike surface reflects all incident light, and this represents one extreme. At the other extreme is a highly scattering surface which reflects light equally in all directions at all angles of incidence. The direct reflection factor is the ratio of the light reflected at the specular angle to the incident light for angles of incidence from 0 to 90°.

Total reflectance is observed at angle theta (Brewster's angle), which is related to the ratio of the n values of the polymer, n_1, and air, n_2.

$$\tan B = \frac{n_1}{n_2} \tag{5.20}$$

Absorption of light or loss of intensity I of light passing through a path of distance l may be calculated from the Lambert relationship

$$I = I_o e^{-4\pi Nkl/\lambda} \tag{5.21}$$

where k is the absorption index, λ is the wavelength of the light, I_o is the original intensity, and N is Avogadro's number.

Luminous reflectance, transmittance, and color can be obtained experimentally as follows. Samples are generally prepared as films, sheets, or molded specimens that have parallel plane surfaces. The sample is mounted as noted in Fig. 5.17 along with a comparison surface, typically white chalk. The sample and comparison surface are exposed to light of varying wavelength. Reflected or transmitted light is then measured. The precise requirements are described in ASTM-E308.

A Hardy-type spectrophotometer may also be used for determining luminous reflectance, transmittance, and color of polymers (ASTM-791). The transmittance of plastic films is measured by ASTM-D1746.

5.8 CHEMICAL RESISTANCE

The classic test for chemical resistance (ASTM-D543) measures the percentage weight change (PWC) of test samples after immersion in many different liquid systems. Tests for chemical resistance have been extended to include changes in mechanical properties of the polymer test sample after immersion. Although there is no standard test of changes in mechanical properties of the samples, changes in the following have been investigated: hardness, tensile strength, stress relaxation, stress rupture, impact strength, compressive strength, flexural strength, and flexural modulus. Since chemical attack involves changes in chemical structure, it can be readily observed by many instrumental methods that measure chemical structure.

Other related ASTM tests include those under accelerated service conditions [ASTM-D756-78 (1971)], water absorption [ASTM-D570-63 (1972)], and ESCR of ethylene plastics (ASTM-D1693-70).

5.9 SPECTRONIC CHARACTERIZATION OF POLYMERS

The index of refraction (n), which is the ratio of the velocity of light in a vacuum to the velocity of light in a transparent medium, e.g., polymer, is characteristic for each polymer. This value, which is also a function of molecular weight, may be determined by use of an Abbe refractometer [ASTM-D542-50 (1970)].

Differences in indices of refraction may be measured by phase-contrast microscopy, and the structure of spherulites may be studied using crossed polarizers in a polarized light microscope. Melting points may be determined when the latter is equipped with a hot stage. Thickness may be measured in nanometers using interference microscopy.

The morphology of polymers may be investigated by electron microscopy and by scanning electron microscopy (SEM). While SEM is limited to images in the 5 to 10 nm range, magnifications of over 200,000 are possible with electron microscopy.

Sample position for
transmittance
measurement

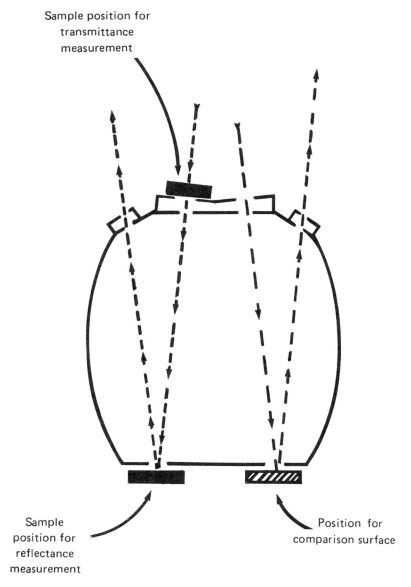

Sample
position for
reflectance
measurement

Position for
comparison surface

Figure 5.17   Instrument for measuring reflectance and transmittance.

Most monomers and polymers may be identified by infrared (IR) spectroscopy in which the energy, in the wavelength range of 1 to 50 nm, is associated with molecular vibration and vibration-rotation spectra of polymer molecules. These motions are comparable to those of small molecules of similar structure (model compounds).

For example, as shown in Fig. 5.18, the IR spectrum of polystyrene is sufficiently characteristic that it is used as a standard for checking instrumental oper-

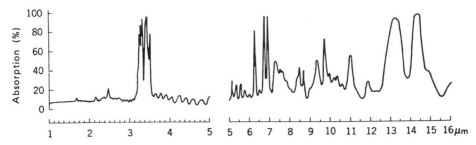

Figure 5.18 Infrared spectrum of a polystyrene film. (From *Introduction to Polymer Chemistry* by R. Seymour, McGraw-Hill, New York, 1971. Used with permission of McGraw-Hill Book Company.)

ation. The repeating unit in polystyrene (C_8H_8) has 16 atoms, and since it has no symmetry, all vibrations are active, i.e., 3° each of rotational and translational freedom and 42° of vibrational freedom (3n − 6).

The bands in the range of 8.7 to 9.7 μm are characteristic wavelengths, but not identified as to origin, and are said to be in the "fingerprint" region. Typical C—H stretching vibrations are at 3.8, 3.4, and 3.g μm, and out-of-plane bending of aromatic C—H bonds occurs at 11.0 and 14.3 μm. Characteristic C—C stretching vibrations are at 6.2 and 6.7 μm. Characteristic bands for other typical groups in polymers are shown in Table 5.3.

The relative amounts of styrene and acrylonitrile in the copolymer in the ultraviolet (UV) spectrogram in Fig. 5.19 may be determined from the relative areas of absorption bands for styrene and acrylonitrile at wave numbers 1600 and 2240 cm^{-1}, respectively.

Ultraviolet spectroscopy has less applicability to the characterization of polymers than IR, but it is useful in detecting aromatics, such as polystyrene, and appropriate additives, such as antioxidants, which exhibit characteristic absorption in the UV region. The absorption of the charge transfer constant of styrene-acrylonitrile-$ZnCl_2$ at different temperatures is shown in Fig. 5.19.

While IR spectroscopy is most useful for the identification of polymers, proton magnetic resonance (pmr) spectroscopy is more useful for elucidating polymer structure. Nuclear protons in the hydrogen atoms in polymers have random orientation, but these protons tend to be oriented, i.e., aligned with or against a strong applied magnetic field.

Absorption of energy by these protons under proper conditions of field strength and frequency, called resonance, causes a spin flip which is displayed on a recorder. The absorption of energy at different frequencies is influenced by neighboring electrons. Thus, as shown in Fig. 5.20, maleic anhydride, vinyl acetate, and the charge transfer complex of these two monomers have different characteristic spectra relative to the internal reference standard, tetramethylsilane, which is given a value of 0 and 10 ppm on the δ scale.

The characteristic spectra and structures of polyolefins are shown in Fig. 5.21 and Table 5.4. As shown in Fig. 5.21, the ratio of the area of the pmr peaks may be used to determine molecular structure based on the ratios of methyl to methylene groups present.

Table 5.3   Absorption Bands for Typical Groups in Polymers

Group	Type of vibration	Wavelength (λ, μm)	Wave number (ν, cm^{-1})
CH$_2$	Stretch	3.38–3.51	2,850–2,960
	Bend	6.82	1,465
	Rock	13.00–13.80	725–890
CH$_3$	Stretch	3.38–3.48	2,860–2,870
	Bend	6.9	1,450
H R $\mid$ $\mid$ C=C $\mid$ $\mid$ H H	C—H stretch	3.25–3.30	3,030–3,085
	C—H bend in plane	7.10–7.68	1,300–1,410
	C—H bend out of plane	10.10–11.00	910–990
	C—C stretch	6.08	1,643
H R $\mid$ $\mid$ C=C $\mid$ $\mid$ H R	C—H stretch	3.24	3,080
	C—H bend in plane	7.10	1,410
	C—H bend out of plane	11.27	888
	C—C stretch	6.06	1,650
Benzene	C—H bend out of plane	14.50	690
OH	Stretch	2.7–3.2	3,150–3,700
SH	Stretch	3.9	2,550
Aliphatic acid	C=O stretch	5.85	1,710
Aromatic acid	C=O stretch	5.92	1,690
CCl	Stretch	12–16	620–830
CN	Stretch	4.8	2,200

Because of the small but consistent concentrations of carbon-13 present in all organic compounds, it is necessary to use more sophisticated nmr spectroscopy (^{13}C-nmr) for determining the effect of neighboring electrons on these nuclei. However, ^{13}C-nmr spectroscopy is an extremely valuable tool for the investigation of polymer structure. A representative ^{13}C-nmr spectrum of the alternating copolymer of styrene and acrylonitrile is shown in Fig. 5.22.

Electron paramagnetic resonance (epr), or electron spin resonance (esr), spectroscopy is a valuable tool for measuring the relative abundance of unpaired electrons present in macroradicals. For example, as shown in Fig. 5.23, macroradicals are formed by the homogeneous cleavage of nylon chains when these filaments are broken, and the concentration of macroradicals increases as the stress is increased.

X-ray diffraction has been used to determine crystalline structure and conformations in polymers. Raman spectroscopy, including laser-Raman spectroscopy, has

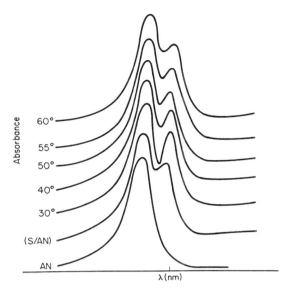

Figure 5.19 Ultraviolet spectra of styrene-acrylonitrile (SAN ZnCl$_2$ in t-butanol at 25, 30, 40, 50, 55, and 60°C, and styrene at 25°C. [From R. Seymour, G. Stahl, D. Garner, and R. Knapp, Polymer Preprints, *17*(1):219 (1976). With permission from the Division of Polymer Chemistry, ACS.]

been used to study the microstructure of polymers such as carbon-carbon double bond stretching vibrations.

5.10   THERMAL ANALYSIS

Major instrumentation involved with the generation of thermal property behavior of materials includes thermal gravimetric analysis (TGA), differential scanning calorimetry (DSC), differential thermal analysis (DTA), torsional braid analysis (TBA), thermal mechanical analysis (TMA), and pyrolysis gas chromatography (PGC).

One of the simplest techniques is PGC, in which the gases, resulting from the pyrolysis of a polymer, are analyzed by gas chromatography. This technique may be used for qualitative and quantitative analysis. The latter requires calibration with known amounts of a standard polymer pyrolyzed under the same conditions as the unknown. Representative PGC pyrograms of poly(vinyl acetate), alternating and random copolymers of vinyl acetate and maleic anhydride, and the copolymer of styrene and a acrylonitrile are shown in Figs. 5.24 and 5.25.

There are several different modes of thermal analysis described as DSC. DSC is a technique of nonequilibrium calorimetry in which the heat flow into or away from the polymer is measured as a function of temperature or time. This is different from DTA where the temperature difference between a reference and a sample is measured as a function of temperature or time. Presently available DSC equipment measures the heat flow by maintaining a thermal balance between the reference and sample by changing a current passing through the heaters under the two cham-

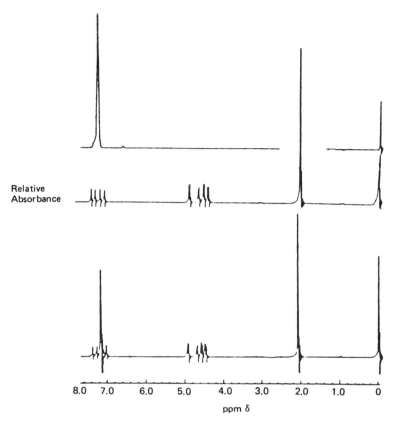

Figure 5.20 PMR spectra of maleic anhydride and vinyl acetate and the charge transfer complex of these two monomers at 25°C. [From R. Seymour, D. Garner, G. Stahl, and L. Sanders, Polymer Preprints, *17*(2):663 (1976). With permission from the Division of Polymer Chemistry, ACS.]

bers. For instance, the heating of a sample and reference proceeds at a predetermined rate until heat is emitted or consumed by the sample. If an endothermic occurrence takes place, the temperature of the sample will be less than that of the reference. The circuitry is programmed to give a constant temperature for both the reference and the sample compartments. Excess current is fed into the sample compartment to raise the temperature to that of the reference. The current necessary to maintain a constant temperature between the sample and reference is recorded. The area under the resulting curve is a direct measure of the heat of transition.

The advantages of DSC and DTA over a good adiabatic calorimeter include speed, low cost, and the ability to use small samples. Sample size can range from 0.5 mg to 10 g. A resultant plot of ΔT as a function of time or temperature is known as a thermogram. Since the temperature difference is indirectly proportional to the heat capacity, the curves resemble inverted specific heat curves. A typical DSC thermogram of a block copolymer of vinyl acetate and acrylic acid is shown

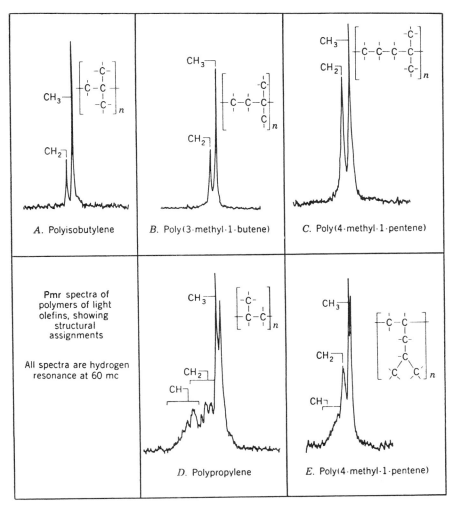

Figure 5.21 PMR peaks for hydrocarbon polymers. (With permission of N. Chamberlain, F. Stehling, K. Bartz, and J. Reed, Esso Research and Engineering Co.)

in Fig. 5.26. The distinctions between DSC and DTA are becoming less clear with the advent of new instrumentation which uses components of both DSC and DTA.

Possible determinations from DSC and DTA measurements include the following: (1) heat of transition, (2) heat of reaction, (3) sample purity, (4) phase diagram, (5) specific heat, (6) sample identification, (7) percentage incorporation of a substance, (8) reaction rate, (9) rate of crystallization or melting, (10) solvent retention, and (11) activation energy. Thus, thermocalorimetric analysis can be a quite useful tool in describing the chemical and physical relationship of a polymer with respect to temperature.

In thermogravimetric analysis, a sensitive balance is used to follow the weight change of a polymer as a function of time or temperature. Usual sample sizes for

Table 5.4 Relationships of Methyl and Methylene Groups in Polymers in Fig. 5.21

Polymer	Structure	$CH_3:CH_2$
Polyisobutylene	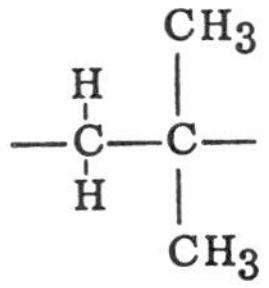 (structure)	2:1
Poly(3-methyl-1-butene)[a]	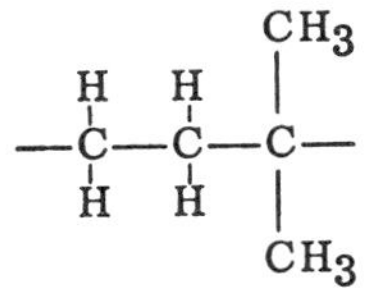 (structure)	1:1
Poly(4-methyl-1-pentene)	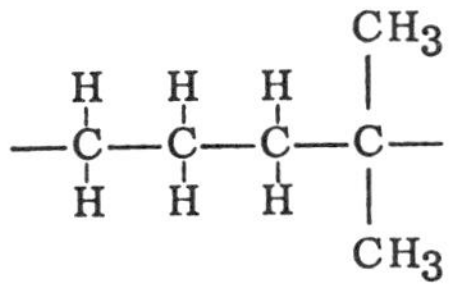 (structure)	2:3
Polypropylene	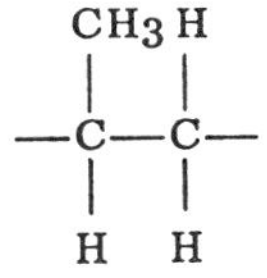 (structure)	1:1
Poly(4-methyl-1-pentene)[a]	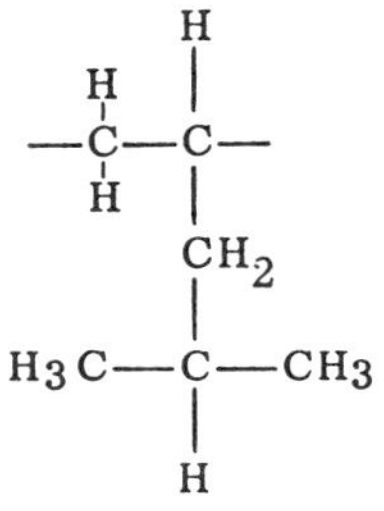 (structure)	2:2

[a]According to Chamberlain, the pmr measurements were used to show an unexpected structure for poly(3-methyl-1-butene) and two different structures for poly(4-methyl-1-pentene).

commercial instruments are in the range of 0.1 mg to 10 g with heating rates of 0.1 to 50°C/min. The most commonly employed heating rates are 10, 15, 20, 25, and 30°C/min. In making both TGA and thermocalorimetric measurements, the same heating rate and flow of gas should be employed to give the most comparable thermograms. TGA can be used to determine the following: (1) sample purity, (2) identification, (3) solvent retention, (4) reaction rate, (5) activation energy, and (6) heat of reaction.

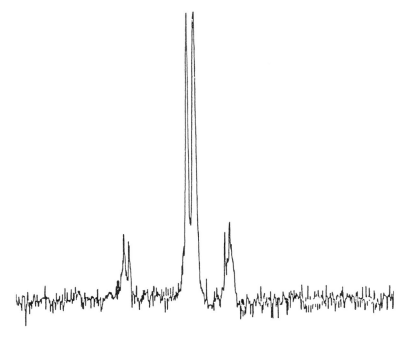

Figure 5.22 Representative ^{13}C-nmr spectra of styrene-acrylonitrile (SAN) alternating co-polymer. [From R. Seymour, G. Stahl, D. Garner, and R. Knapp, Polymer Preprints, *17*(1): 220 (1976). With permission from the Division of Polymer Chemistry, ACS.]

Thermomechanical analysis measures the mechanical response of a polymer as a function of temperature. Typical measurements as a function of temperature include the following: (1) expansion properties, i.e., expansion of a material leading to the calculation of the linear expansion coefficient; (2) tension properties, i.e., the measurement of shrinkage and expansion of a material under tensile stress, e.g., elastic modulus; (3) dilatometry, i.e., volumetric expansion within a confirming medium, e.g., specific volume; (4) single fiber properties, i.e., tensile response of single fibers under a specific load, e.g., single-fiber modulus; and (5) compression properties, such as measuring the softening or the penetration under load.

Compressive, tensile, and single-fiber properties are usually measured under some load, yielding information about softening points, modulus changes, phase transitions, and creep properties. For compressive measurements, a probe is positioned on the sample and loaded with a given stress. A record of the penetration of the probe into the polymer is obtained as a function of temperature. Tensile properties can be measured by attaching the fiber to two fused quartz hooks. One hook is loaded with a given stress. Elastic modulus changes are recorded by monitoring a probe displacement.

In torsional braid analysis, the changes in tensile strength as the polymer undergoes thermal transition is measured as a function of temperature and sometimes also as a function of the applied frequency of vibration of the sample. As thermal

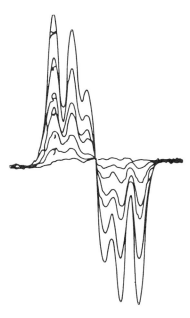

Figure 5.23   EPR spectra for nylon-66 fibers taken under increasing stress.

transitions are measured, irreversible changes such as thermal decomposition or cross-linking are observed, if present. In general, a change in T_g or change in the shape of the curve (shear modulus versus temperature) during repeated sweeps through the region, such as a region containing the T_g, is evidence of irreversible change. The name TBA is derived from the fact that measurements are made on fibers which are "braided" together to give test samples connected between or onto vicelike attachments or hooks.

DSC, DTA, TMA, and TBA analyses are all interrelated, all signaling changes in thermal behavior as a function of heating rate or time. TGA is also related to the others in the assignment of phase changes to observed weight changes.

Recent trends include increased emphasis on coupling thermal techniques and on coupling thermal techniques with other analysis techniques, such as mass spectrometry (MS) of the off-gases produced by thermolysis or pyrolysis of polymers. Useful combinations include TG-MS, GC-MS, TG-GC-MS, and PGC-MS, with intercorrelation of data from IR, nmr, GPC, and EM of analyzed samples at different heating times and temperatures.

The polymer softening range, while not a specific thermodynamic property, is a valuable "use" property. It is normally a simple and readily obtainable property. Softening ranges generally lie between the polymers' T_g and T_m. Some polymers do not exhibit a softening range but rather undergo a solid state decomposition before softening.

Softening ranges are dependent on the technique and procedure used to determine them. Thus, listings of softening ranges should be accompanied by the specific

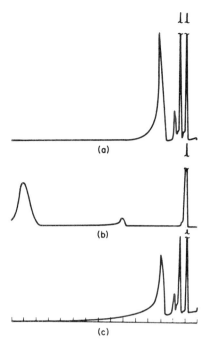

Figure 5.24 Gas chromatography programs of the homopolymer of vinyl acetate (a) and the alternating (b) and random copolymers (c) of vinyl acetate and maleic anhydride. [From R. Seymour, D. Garner, G. Stahl, and L. Sanders, Polymer Preprints, *17*(2):665 (1976). With permission from the Division of Polymer Chemistry, ACS.]

technique and procedure employed for the determination. The following are techniques often used for the determination of polymer softening range.

The capillary technique is analogous to the technique employed to determine melting points of typical organic compounds. The sample is placed in a capillary tube and heated with the temperature recorded from beginning to end of melting. Control of the heating rate lends more importance to the measurements. Instruments such as the Fisher-Johns Melting Point Apparatus are useful in this respect.

Another technique calls for a plug of film (or other suitable form) of the polymer to be stroked along a heated surface whose temperature is being increased until the polymer sticks to the surface. A modification of this utilizes a heated surface containing a temperature gradient between the ends of the surface.

The Vicat needle method (Fig. 5.27) consists of determining the temperature at which a 1-mm penetration of a needle (having a point with an area of 1 mm) occurs on a standard sample (1/8 in. thick, minimum width, 3/4 in.) at a specified heating rate (often 50°C/hr) under specific stress (generally less than 1 kg). This is also referred to as the heat deflection point.

In the ring and ball method the softening range of a sample is determined by noting the temperature at which the sample, held within a horizontal ring, is forced

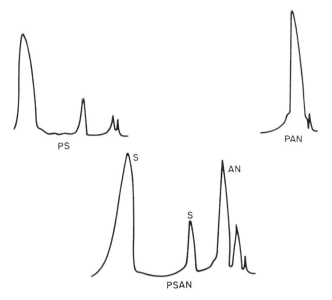

Figure 5.25 Representative PGC pyrograms of polystyrene (PS), polyacrylonitrile (PAN), and poly(styrene-co-acrylonitrile) (PSAN). [From R. Seymour, G. Stahl, D. Garner, and R. Knapp, Polymer Preprints, *17*(1):221 (1976). With permission from the Division of Polymer Chemistry, ACS.]

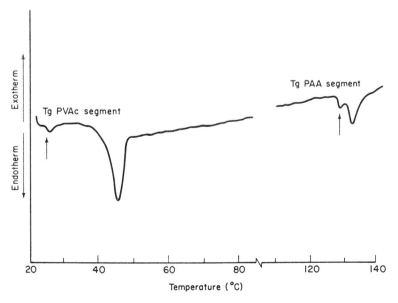

Figure 5.26 Typical DSC thermogram of a block copolymer of vinyl acetate and acrylic acid [P(VAc-*b*-AA)].

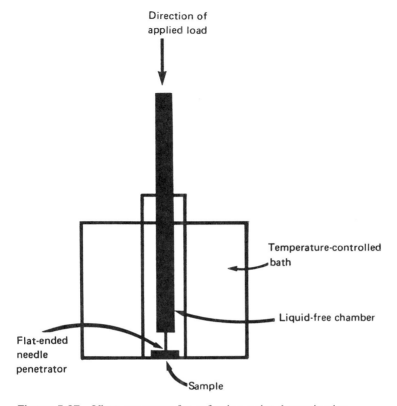

Figure 5.27 Vicat apparatus for softening point determination.

downward by the weight of a standard steel ball supported by the sample. The ball and ring are generally heated by inserting them in a bath.

It is important that softening range data can serve as guides to proper temperatures for melt fabrication, such as melt pressing, melt extruding, and molding. It is also an indication of the product's thermal stability.

5.11 THERMAL PROPERTY TESTS

Thermal Conductivity

As energy—heat, magnetic, or electric—is applied to one side of a material, the energy is transmitted to other areas of the sample. Heat energy is largely transmitted through the increased amplitude of molecular vibrations. The heat flow Q from any point in a solid is related to the temperature gradient dt/dl through the thermal conductivity λ as follows:

$$Q = -\lambda \, (dt/dl) \tag{5.22}$$

The transmission of heat is favored by the presence of ordered lattices and covalently bonded atoms. Thus graphite, quartz, and diamond are good thermal conductors, while less-ordered forms of quartz such as glass have lower thermal conductivities. Most polymeric materials have λ values between 0.1 and 1.0 W/m K, while metals typically have values near and above 100 W/m K.

Practically, thermal conductivity, or K factor (ASTM-C-177-71), is the time rate of heat flow Q required to attain a steady state in the temperature of a sample having a thickness L and an area A. Change in T is the difference in temperature between a hot plate and a cooling plate above and below the test sample. The thermal conductivity K is calculated from the following equation:

$$K = \frac{QL}{A \, \Delta T} \tag{5.23}$$

Thermal Expansion

Coefficients of thermal expansion can refer to differences in length, area, or volume as a function of temperature. Relative to metals such as steel, polymers have large coefficients of thermal expansion (Table 5.5). Polymers also have quite varied coefficients of thermal expansion. Both of these factors are troublesome when different materials are bound together and exposed to wide temperature ranges. Such wide temperature ranges regularly occur in the aerospace industry (aircraft), within computer chips (and many other electrical and visual areas), engines and motors, etc. Thus, it is critical to match the coefficients of thermal expansions of materials that are to be bound through mechanical (such as screws and bolts) and chemical

Table 5.5 Typical Coefficients of Linear Expansion (Multiplied by One Million to Nearest 5)

Material	K range or value (mm/mm)
Ferrous metals	5–20
Iron	10
Steel	10
Nonferrous metals	5–30
Aluminum	25
Copper	15
Thermosets	Near zero to 120
Melamine/formaldehyde	20–60
Phenol/formaldehyde	30–45
Thermoplastics	20–320
(Aromatic) nylons	90–110
Polyethylene	100–250
Poly(vinyl chloride)	185–200
Poly(methyl methacrylate)	50–110
Polystyrene	60–80
Polytetrafluoroethylene	50–100
Glass (window)	10
Wood	5
Concrete	15
Granite	10

(polymer alloys, blends, adhered through use of adhesives) means, or stress will develop between the various materials, possibly resulting in fracture or separation.

For polymeric materials, aspects that restrict gross movement, such as cross-linking, typically result in lowered coefficients of expansion. Thus, the typical range for coefficients of linear expansion for (cross-linked) thermosets is lower than the typical range found for (largely non–cross-linked) thermoplastics. Further, such highly cross-linked polymeric materials such as glass, granite, and concrete (see Chapter 12) also exhibit low coefficients of expansion for the same reason.

Heat Capacity

The thermal conductivity is also related to the specific heat capacity C_p as described in Eq. (5.24), where d is the density of the material and TD is the thermal diffusivity:

$$\lambda = (TD)C_pd \tag{5.24}$$

The amount of heat required to raise the temperature of a material is related to the vibrational and rotational motions thermally excited within the sample. Polymers typically have relatively (compared with metals) large specific heats, with most falling within the range of 1 to 2 kJ kg^{-1} K^{-1}. The values change as materials undergo phase changes (such as that at the T_m) but remain constant between such transitions.

DSC and DTA instruments are able to rapidly give C_p values.

Vicat Softening Point

The softening point of relatively soft polymers (ASTM-D1525) is the temperature at which a flat-ended Vicat needle with a cross-section of 1 mm penetrates a test specimen to a depth of 1 mm under a specified load, which is usually 1 kg. The needle with a specific load is placed on the sample as shown in Fig. 5.27. The temperature of the bath is raised at a specified rate, and the temperature at which the needle penetrates 1 mm is the Vicat Softening Point.

Heat Deflection Temperature

The heat deflection temperature (ASTM-D-648) is determined by noting the temperature at which a simple beam under load deflects a specified amount, typically 0.01 in (0.25 mm) (Fig. 5.28). It is not intended to be a direct measure of high-temperature applications but is generally employed only as a general use indicator.

Glass Transition Temperature

Qualitatively, the glass transition temperature corresponds to the onset of short-range (typically 1 to 5 chain atoms) coordinated motion. Actually, many more (often 10 to 100) chain atoms may attain sufficient thermal energy to move in a coordinated manner at T_g.

The glass transition temperature (T_g; ASTM D-3418) is the temperature at which there is an absorption or release of energy as the temperature is raised or lowered. T_g may be determined from DTA, DSC, TBA, Fourier transform IR (FT-

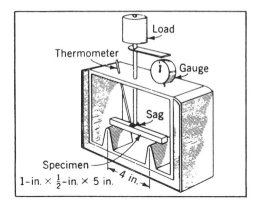

Figure 5.28 Deflection temperature test. (From *Introduction to Polymer Chemistry* by R. Seymour, McGraw-Hill, New York, 1971. Used with permission of McGraw-Hill Book Company.)

IR), nmr, dynamic mechanical spectroscopy (DMS), modulus-dependent techniques, dilatometry (and other associated techniques that measure area, volume, or length), dielectric loss, and TMA.

DMS refers to a group of dynamic techniques where the sample is subjected to repeated small-amplitude strains in a cyclic fashion. The polymer molecules store some of this imparted energy and dissipate a portion in the form of heat. Since the amount of energy stored and converted into heat is related to molecular motion, changes in the ratios of energy stored to energy converted into heat can be employed to measure T_g.

The coefficient of linear thermal expansion (α) (ASTM-D696) is equal to the change in length of a sample (ΔL) divided by its length (L) and the change in temperature (ΔT) during the test, i.e., $\alpha = \Delta L / L \, \Delta T$. The thermal linear expansivity of polymers is usually higher than that of ceramics and metals; polymers have values ranging from 4 to 20×10^{-5} K^{-1}, whereas metals have values of about 1 to 3×10^{-5} K^{-1}. Further, the expansion of polymeric materials, unlike the expansion of metals, is usually not a linear function of temperature.

It must be emphasized that the actual T_g of a sample is dependent on many factors including pretreatment of the sample and the method and conditions of determination. For instance, the T_g for linear polyethylene has been reported to be from about 140 to above 300°K. Calorimetric values centralize about two values—145 and 240°K; thermal expansion values are quite variable within the range of 140 to 270°K; nmr values occur between 220 to 270°K; and mechanical determinations range from 150 to above 280°K. If nothing else, the method of determination and end-property use should be related. Thus, if the area of concern is electrical, then determinations involving dielectric loss are appropriate.

Whether a material is above or below its T_g is important in describing the material's properties and potential end use. Fibers are composed of generally crystalline polymers that contain polar groups. The polymers composing the fibers are generally near their T_g to allow flexibility. Cross-links are often added to prevent wholesale, gross chain movement. An elastomer is cross-linked and composed of essentially nonpolar chains. The use temperature is above its T_g. Largely crystalline plastics may be used above or below their T_g. Coatings or paints must be used near their T_g so that they have some flexibility but are not rubbery. In fact, many coatings behave as "leather" (see Fig. 3.5). Adhesives are generally mixtures where the polymeric portion is above its T_g. Thus the T_g is one of the most important physical properties of an amorphous polymer.

5.12 FLAMMABILITY

Small-scale horizontal flame tests have been used to estimate the flammability of solid (ASTM-D635), cellular (ASTM-D-1692-74), and foamed polymers (ASTM-D 1992), but these tests are useful for comparative purposes only. Large-scale tunnel tests (ASTM-E84) and corner wall tests are more significant, but they are also more expensive than laboratory tests.

One of the most useful laboratory flammability tests is the oxygen index (OI) test (ASTM-D2043 and ASTM-D2863). In this test, the specimen is burned by a candle in controlled mixtures of oxygen and nitrogen. The minimum oxygen concentration that produces downward flame propagation is considered the OI or ignitability of the polymer.

5.13 SURFACE CHARACTERIZATION

There is no exact, universally accepted definition of a surface. Here, the surface will be defined as the outermost atomic layers, including absorbed foreign atoms. The chemical and physical composition, orientation, and properties of surfaces differ from the interior bulk solid.

Current surface characterization techniques fall into two broad categories—those that focus on the outermost few layers and those whose focus includes components present to several thousand angstroms into the solid.

Attenuated total reflectance (ATF) typically employs special cells fitted onto traditional IR, FT-IR, or UV. While some outer surface aspects are gleaned from such techniques, the techniques focus on the bulk surface to several thousand angstroms in depth.

Techniques that analyze the first few atomic layers generally involve low-energy electrons or ions since the incident radiation should penetrate only the top few layers. Normally a combination of techniques is employed to clearly define the outer surface. Special precautions are employed to minimize sample surface contamination. Auger electron spectroscopy (AES) allows the detection of only carbon, oxygen, nitrogen, and heavier atoms. The material surface can be changed by the action of the applied electron beam, often resulting in chemical bonding information being uncertain.

Auger Electron Spectroscopy and X-Ray Photoelectron Spectroscopy

AES and X-ray photoelectron spectroscopy (XPS) are two principal surface analysis techniques. They are employed to identify the elemental composition, to determine the amount and nature of species present at the surface, and to elucidate the properties of the outermost atomic layers (1 nm) of a solid.

In the Auger transition, incident electrons interact with the inner shell electrons (E_i) of the sample. The vacancy created by an ejected inner shell electron is filled by an outer shell electron (E_{o1}), and a second outer shell electron (E_{o2}) is ejected leaving the atom in a doubly ionized state. The electrons ejected from the outer shells are the Auger electrons, named after the Frenchman Pierre Auger, who discovered this effect. Thus, AES is a technique which measures the energies of Auger electrons (E_A) emitted from the first 10 Å of a sample surface. The energy equation can be expressed in the following way:

$$E_A = E_{o1} - E_i + E_{o2} \tag{5.25}$$

AES instrumentation in its simplest form involves a vacuum system, an electron gun for target excitation, and an electron spectrometer for energy analysis of the emitted secondary electrons. Additionally, most AES systems presently in use employ an ion gun for in-depth profiling measurements, a manipulator for positioning the sample, a means for precisely locating the area for analysis, and frequently an attachment for performing in situ fracture or cleavage of specimens.

For XPS, a sample is bombarded by a beam of X-rays with energy $h\nu$ and core electrons are ejected with a kinetic energy (E_k), overcoming the binding energy (E_B) and work function (ϕ). These core electrons are called the X-ray photoelectrons. The energy equation can be expressed as follows:

$$E_k = h\nu - E_B - \phi \tag{5.26}$$

The kinetic energies of these ejected electrons originating within the first 30 Å of the sample surface are measured by XPS.

The functions of an X-ray photoelectron spectrometer are to produce intense X-radiation, to irradiate the sample to photo-eject core electrons, to introduce the ejected electrons into an energy analyzer, to detect the energy-analyzed electrons, and to provide a suitable output of signal intensity as a function of electron-binding energy.

Both techniques are equally well applied to determine the nature of solid surfaces of polymeric materials, ceramics, metals, metallic alloys, and solid small-molecule samples.

Because of the critical nature of solid surfaces and because the nature of the surface may be quite different from the bulk chemical structure, determination of the nature of surfaces is an important endeavor.

Major surface problems addressed by surface analysis techniques include metal-polymer adhesion, gas-polymer interactions, wear, friction, corrosion, sites of deformation, catalysis structure, water content, effect of processing aids and additives, composite structure, and sites of environment and chemical action.

5.14 AMORPHOUS REGION DETERMINATIONS

Experimental tools that have been employed in an attempt to characterize amorphous regions are given in Table 5.6. Techniques such as birefringence and Raman scattering give information related to the short-range (<20 Å) nature of the amor-

Table 5.6 Techniques Employed to Study the Amorphous Regions of Polymers

Short-range interactions	Long-range interactions
Magnetic birefringence	Electron diffraction
Raman scattering	Wide-angle X-ray scattering
Depolarized light scattering	Electron microscopy
Rayleigh scattering	Density
Bruillouin scattering	Small-angle neutron scattering
Nmr relaxation	
Small-angle X-ray scattering	

phous domains, while techniques such as neutron scattering, electron diffraction, and electron microscopy yield information concerning the lower-range nature of these regions.

Birefringence measures order in the axial, backbone direction. The birefringence of a sample can be defined as the difference between the refractive indices for light polarized in two directions 90° apart, i.e.,

$$\Delta n = n_\theta - n_{\theta-90} \tag{5.27}$$

Thus, a polymer sample containing polymer chains oriented in a preferential direction by stretching or some other method will exhibit a different refractive index along the direction of preferred chain alignment compared to that obtained at right angles to this. This change in birefringence as an amorphous material, such as a melt of rubber, is deformed and gives information concerning the amount of order.

The results of utilizing techniques for short-range interactions for vinyl polymers are consistent with at least some amorphous regions consisting of only limited (5 to 10 Å and 1 to 2 units) areas with orientation, similar to that observed for typical low molecular weight hydrocarbons.

Small-angle neutron scattering (SANS) results indicate that vinyl polymers exist in random coils in the amorphous state. Results from electron and X-ray diffraction studies show diffuse halos, meaning that the nearest neighbor spacings are somewhat irregular. The diffuse halos could result from situations where short-range order is present but long-range disorder exists. This is a situation believed to be present in liquid water and may be the result of both short-range order and long-range disorder.

The smooth changes observed with relaxation and annealing studies suggests little order for amorphous polymers. However, the density of an amorphous polymer is generally 85 to 90% of that of the analogous crystalline polymer, and density-related calculations suggest that there may be some short, organized parallel arrangements of vinyl chain segments. The interpretations of such studies are still unclear.

SUMMARY

1. The American Society for Testing and Materials (ASTM) and comparable organizations throughout the world have established meaningful standards for the testing of polymers.

2. The mechanical properties of plastics may be classified by five stress-strain curves, which show the yield point, elongation, and toughness of various classes of plastics.

3. The test for tensile strength, i.e., the measure of the ability of a polymer to withstand pulling stresses, is described in ASTM standard D638-72.

4. The test for flexural strength, or the measure of bending strength of a polymer, is described by ASTM-D790-71.

5. The test for compressive strength, or the measure of the crushing resistance of a polymer, is described by ASTM-D695-60.

6. The test for impact strength, or the measure of toughness of a polymer, is described by ASTM-D256-73.

7. Hardness, which implies resistance to penetration or abrasion and scratching, is described by several ASTM tests.

8. Many electrical tests which are essential for polymers used in electrical applications are described by ASTM tests.

9. ASTM tests have been developed for T_g, softening point, heat deflection temperature, and brittleness at low temperatures.

10. Listed among the ASTM flammability tests is the oxygen index test, which defines the minimum oxygen content in an oxygen-nitrogen mixture that will support combustion of a plastic.

11. ASTM test D-543-67 (1971) measures the changes in weight and dimensions of plastic test specimens immersed in selected standard liquids for 7 days.

12. Most monomers and polymers may be identified qualitatively and quantitatively by IR spectroscopy.

13. Proton and carbon-13 nmr are useful for elucidating polymer structure.

14. The mechanism in fiber rupture and the concentration of primary radicals or macroradicals, may be investigated using epr spectroscopy.

15. PGC, in which products of pyrolysis are analyzed by gas chromatography, is an excellent tool for polymer analysis.

16. DTA and DSC, which provide information on thermal transitions, may also be used for the identification of polymers.

17. The stability of polymers at elevated temperatures may be determined by TGA.

GLOSSARY

A: area

Abbe refractometer: An instrument used for measuring the index of refraction.

alpha (α): Coefficient of expansion.

arc resistance: Resistance to tracking by a high-voltage discharge.

ASTM: American Society for Testing and Materials.

ASTM-D20: Committee responsible for standards for plastics.

Barcol impressor: An instrument used to measure the resistance of a polymer to penetration or indentation.

BSI: British Standards Institution.

BTU: British thermal unit.

Charpy test: An impact resistance test.

chemical shifts: Peaks in pmr spectroscopy.

^{13}C-nmr: Nuclear magnetic resonance spectroscopy based on the effect of neighboring electrons on carbon-13 nuclei.

coefficient of expansion: Change in dimensions per degree Celsius.

compressive strength: Resistance to crushing forces.

ΔT: Difference in temperature.

dielectric constant: Ratio of the capacitance of a polymer to that of air or a vacuum.

dielectric strength: Maximum applied voltage that a polymer can withstand.

differential scanning calorimetry: Measurement of the differences in changes in enthalpy of a heated polymer and a reference standard based on power input.

differential thermal analysis: Measurement of the difference in the temperature of a polymer and a reference standard when heated.

DNA: Deutsche Normenausschuss (German standard tests).

DSC: Differential scanning calorimetry.

DTA: Differential thermal analysis.

elastic range: The range on a stress-strain curve below the yield point.

environmental stress cracking: Cracking of polyolefins in liquid media such as liquid detergents.

epr: Electron paramagnetic resonance.

ESCR: Environmental stress cracking.

esr: Electron spin resonance.

flexural strength: Resistance to bending.

glass transition temperature: Lowest temperature at which segmental motion of polymer chains exists.

heat deflection temperature: The temperature at which a simple loaded beam undergoes a definite deflection.

impact strength: Measure of toughness.

index of refraction: The ratio of the velocity of light in a vacuum to that in a transparent polymer.

infrared spectroscopy: A technique used for the characterization of polymers based on their molecular vibration and vibration-rotation spectra.

IR: Infrared.

ISO: International Standards Organization.

Izod: An impact resistance test.

K factor: A measure of thermal conductivity.

l: Length of a sample.

L: Thickness of test specimen in thermal conductivity test.

loss factor: Power factor multiplied by dielectric constant.

modulus: Stiffness of a polymer.

Mohs scale: A hardness scale ranging from 1 for talc to 10 for diamond.

nmr: Nuclear magnetic resonance spectroscopy.

OI: Oxygen index.

oxygen index: A test for the minimum oxygen concentration in a mixture of oxygen and nitrogen that will support a candlelike flame of a burning polymer.

PGC: Pyrolysis gas chromatography.

phase-contrast microscopy: Measurement of differences in indices of index of refraction.

pmr: Spectroscopy based on proton or hydrogen magnetic resonance.

Poisson's ratio: The ratio of contraction to elongation of a polymer.

power factor: Electrical energy required to rotate the dipoles in a polymer while in an electrostatic field.

Q: Rate of heat flow.

Rockwell hardness: A measure of indentation resistance.

SEM: Scanning electron microscopy.

shear strength: Resistance to shearing forces.

shore durometer: A simple instrument used to measure resistance of a polymer to penetration of a blunt needle.

standard testing conditions: 23°C, 50% humidity.

Sward hardness rocker: A rocking device used as a measure of hardness.

Tabor abrader: A mechanical rotator used to test abrasion resistance.

tensile strength: Resistance to pulling stresses.

T_g: Glass transition temperature.

thermogravimetric analysis (TGA): Measurement of the loss in weight when a polymer is heated.

UV: Ultraviolet.

Vicat needle: One used under load to penetrate polymer surfaces.

volume resistivity: Electrical resistance between opposite forces of a 1-in. cube.

yield point: The point on a stress-strain curve below which there is reversible recovery.

EXERCISES

1. What is the most important standards organization in the United States?

2. What is the tensile strength (TS) of a test sample of PMMA 1.25 cm square with a thickness of 0.32 cm, if failure occurs at 282 kg?

3. What is the tensile strength (TS) of a test sample of PMMA which is 1.27 cm square with a thickness of 0.32 cm, if failure occurs at 282 kg?

4. What is the compressive strength (CS) of a 10-cm-long plastic rod with a cross section of 1.27 cm × 1.27 cm which fails under a load of 3500 kg?

5. If a sample of polypropylene measuring 5 cm elongates to 12 cm, what is the percentage of elongation?

6. If the tensile strength is 705 kg cm^{-2} and the elongation is 0.026 cm, what is the tensile modulus?

7. Define creep.

8. What is the value of Poisson's ratio for hevea rubber at (a) 25°C, and (b) −65°C?

9. What changes occur in a polymer under stress before the yield point?

10. What changes occur in a stressed polymer after the yield point?

11. How can you estimate relative toughness of polymer samples from stress-strain curves?

12. What effect will a decrease in testing temperature have on tensile strength?

13. What effect will an increase in the time of testing have on tensile strength?

14. Is the statement correct that a polymer with a notched impact strength of 2 ft lb per in. of notch is twice as tough as one with a value of 1 ft lb per in. of notch?

15. Why are electrical tests for polymers important?

16. Which is the better insulator: a polymer with (a) a low or (b) a high K factor?

17. Why are the specific heats of polymers higher than those of metals?

18. What are the standard heat deflection loads in metric units?

19. Why is the use of the term flameproof plastics incorrect?

20. Which plastic will be more resistant when immersed in 25% sulfuric acid at 50°C: (a) HDPE, (b) PMMA, or (c) PVAc?

21. What happens to indices of refraction values at T_g?

22. How does the index of refraction change when the molecular weight of a polymer is increased?

23. How many degrees of vibrational freedom are present in polypropylene?

24. What is the ultraviolet region of the spectrum?

25. Which would absorb in the UV region: (a) polystyrene, (b) hevea rubber, or (c) PVC?

26. What technique would you use to determine crystallinity in a polymer?

27. What thermal instrumental technique would you use to determine T_g?

BIBLIOGRAPHY (see also Bibliography, Chap. 3)

Alfrey, T. (1948): *Mechanical Behavior of Polymers*, Interscience, New York.

American Society for Testing and Materials, *Latest Book of ASTM Standards*, Parts 26, 34, 36, Plastics, American Society for Testing and Materials, Philadelphia.

Balke, S. T. (1984): *Quantitative Column Liquid Chromatography*, Elsevier, New York.

Balton-Calleja, Vonk, C. G. (1989): *X-Ray Scattering of Synthetic Polymers*, Elsevier, New York.

Bartener, G. (1993): *Mechanical Strengths and Failure of Polymers*, Prentice-Hall, Englewood Cliffs, New Jersey.

Beamson, G., Briggs, D. (1992): *High Resolution XPS of Organic Polymers*, Wiley, New York.

Belenkii, G., Vilenchik, L. (1983): *Modern Liquid Chromatography of Molecules*, Elsevier, Amsterdam.

Bershtein, V., Egorov, V. (1994): *Differential Scanning Calorimetry of Polymers*, Routledge Chapman and Hall.

Braun, D., Cherdron, H., Kern, W. (1972): *Techniques of Polymer Synthesis and Characterization*, Wiley-Interscience, New York.

Bueche, F. (1962): *Physical Properties of Polymers*, Wiley-Interscience, New York.

Carraher, C. E. (1977): Resistivity measurements, J. Chem. Ed., 54:576.

Carraher, C. E., Sheats, J., Pittman, C. U. (1978): *Organometallic Polymers*, Academic, New York.

Carswell, T. S., Nason, H. K. (1944): Classification of polymers, Mod. Plastics, 21:121.

Chamberlain, N. F., Stehling, F. C., Bartz, K. W., Reed, J. J. R. (1965): *NMR Data for H[1]*, Esso Research and Engineering Company, Baytown, Texas.

Craver, C. D. (1983): *Polymer Characterization*, ACS, Washington, D.C.

Critchley, J. P., Knight, G., Wright, W. W. (1983): *Heat Resistant Polymers: Technologically Useful Materials*, Plenum, New York.

Crompton, T. R. (1993): *Practical Polymer Analysis*, Plenum, New York.

Frisch, K. (1972): *Electrical Properties of Polymers*, Technomic, Lancaster, Pennsylvania.

Galiatsatos, V. (1993): *Molecular Characterization of Networks*, Prentice-Hall, Englewood Cliffs, New Jersey.

Gebelein, C. G., Carraher, C. E. (1985): *Bioactive Polymeric Systems: An Overview*, Plenum, New York.

Goldman, A. (1993): *Prediction of Deformation Properties of Polymeric and Composite Materials*, ACS, Washington, D.C.

Helfinstine, J. D. (1977): Charpy impact test of composites, ASTM Spec. Tech. Publ. STO 617.

Hunt, B. J., Holding, S. (1989): *Size Exclusion Chromatography*, Routledge Chapman & Hall, Boston.

Hwang, C. R., et al. (1989): *Computer-Aided Analysis of the Stress-Strain Response of High Polymers*, Technomic, Westport, Connecticut.

Ibbett, R. N. (1993): *NMR Spectroscopy of Polymers*, Routledge Chapman and Hall.

Ishida, H. (1987): *Fourier Transform-Infrared Characterization of Polymers*, Plenum, New York.

Janca, J. (1984): *Steric Exclusion Liquid Chromatography of Polymers*, Marcel Dekker, New York.

Kausch, H. H. (1987): *Polymer Fracture*, Springer-Verlag, New York.

Keinath, S. E., et al. (1987): *Order in the Amorphous "State" of Polymers*, Plenum, New York.

Kovarskii, A. L. (1993): *High Pressure Chemistry and Physics of Polymers*, CRC Press, Boca Raton, Florida.

Ku, C., Liepins, R. (1987): *Electrical Properties of Polymers: Chemical Principles*, Macmillan, New York.

Labana, S. S., Dickie, R. A. (1984): *Characterization of Highly Cross-Linked Polymers*, ACS, Washington, D.C.

Ladik, J. J. (1987): *Quantum Theory of Polymers as Solids*, Plenum, New York.

Mark, J. E. (1993): *Physical Properties of Polymers*, 2nd ed., ACS, Washington, D.C.

Mathias, L. (1991): *Solid State NMR of Polymers*, Plenum, New York.

Mathot, V. (1994): *Calorimetry and Thermal Analysis of Polymers*, Hauser-Gardner, Cincinnati, Ohio.

McCaffery, E. M. (1970): *Laboratory Preparation for Macromolecular Chemistry*, McGraw-Hill, New York.

Meeten, G. H. (1986): *Optical Properties of Polymers*, Elsevier, New York.

Mitchell, J. (1987): *Applied Polymer Analysis and Characterization*, Oxford University Press, Oxford, England.

Mittall, K. L. (1983): *Physicochemical Aspects of Polymer Surfaces*, Plenum, New York.

Nielsen, L. E. (1966): *Mechanical Properties of Polymers*, Reinhold, New York.

Painter, P. C., Coleman, M. M., Koenig, J. L. (1982): *The Theory of Vibrational Spectroscopy and Its Application to Polymeric Materials*, Wiley, New York.

Provder, T. (1986): *Computer Applications in the Polymer Laboratory*, ACS, Washington, D.C.

Randall, J. C. (1984): *NMR and Macromolecules*, ACS, Washington, D.C.

Richter, D., Springer, T. (1988): *Polymer Motion in Dense Systems*, Springer-Verlag, New York.

Seanor, D. (1982): *Electrical Properties of Polymers*, Academic, New York.

Seymour, R. B., Carraher, C. E. (1984): *Structure-Property Relationships in Polymers*, Plenum, New York.

Siesler, H. W., Holland-Mority, K. (1980): *Infrared and Raman Spectroscopy of Polymers*, Marcel Dekker, New York.

Silverstein, R. M., Bassler, G. C. (Latest Edition): *Spectrometric Identification of Organic Compounds*, Wiley, New York.

Tobolsky, A. V. (1960): *Properties and Structures of Polymers*, Wiley, New York.

Tonelli, A. E. (1989): *NMR Spectroscopy and Polymer Microstructure: Conformational Connection*, VCH Publications, New York.

Tong, H-M., Nguyen, L. T. (1990): *Characterization Techniques for Thin Polymer Films*, Wiley, New York.

Tsvetkov, V. N. (1989): *Rigid-Chain Polymers: Hydrodynamic and Optical Properties in Solution*, Plenum, New York.

Turi, E. A. (1981): *Thermal Characterization of Polymeric Materials*, Academic, New York.

Ward, I. M. (1983): *Mechanical Properties of Solid Polymers*, Wiley-Interscience, New York.

Williams, J. G. (1984): *Fracture Mechanics of Polymers*, Halstead, New York.

Zachariades, A. E., Porter, R. S. (1983): *The Strength and Stiffness of Polymers*, Marcel Dekker, New York.

6

Naturally Occurring Polymers

One of the strongest, most rapidly growing areas of polymers is that of natural polymers. Our bodies are largely composed of polymers: DNA, RNA, proteins, and polycarbohydrates. These are related to aging, awareness, mobility, strength, and so on, i.e., all the characteristics that contribute to our being alive and well. Many medical, health, and biological projects and advances are concerned with materials which are, at least in part, polymeric. There is an ever-increasing emphasis on molecular biology, i.e., chemistry applied to natural systems. Thus, an understanding of polymeric principles is advantageous to those desiring to pursue a career related to their natural environment.

Physically there is no difference in the behavior, study, or testing of natural and synthetic polymers, and information techniques suitable for application to synthetic polymers are equally applicable to natural polymers.

While the specific chemistry and physics dealing with synthetic polymers are complicated, the chemistry and physics of natural polymers are even more complex, complicated by a number of related factors, including (1) the fact that many natural polymers are composed of different, often similar, repeat units; (2) a greater dependency on the exact natural polymer environment; (3) the question of the real structure of the natural polymer in its natural environment; and (4) the fact that polymer shape and size are even more important to natural polymers than to synthetic polymers.

Industrially we are undergoing a reemergence of the use of natural polymers in many new and old areas. Since natural polymers are typically regenerable resources, nature can continue to synthesize as we harvest them. Many natural polymers are also present in large quantities. For instance, cellulose makes up about one-third of the bulk of the entire vegetable kingdom, being present in corn stocks, tree leaves, carrot tops, grass, and so on. With the realization that we must conserve and regulate our chemical resources comes from the awareness that we must find substitutes for resources that are not self-generating, such as oil, gas, and metals—thus, the underlying reason for the increased emphasis in polymer chemistry toward the use and modification of natural, regenerable polymers by industry.

The recognition that the supply of petroleum and coal is limited and relatively costly has led to an emphasis on natural, renewable materials as replacements or substitutes for product materials now derived from petroleum and coal. Also, renewable resources have valuable and sometimes different properties of their own.

Natural feedstocks must serve many human purposes. Carbohydrates as raw materials are valuable due to their actual or potential nutritional value. For example, protein plants are already utilizing rapidly reproducible bacteria that metabolize cellulose wastes. Thus, bacteria are added to a nutrient broth emphasizing cellulose; the bacteria feed on the mixture, converting it to more protein-rich bacteria; the bacteria are harvested and used as a protein feed meal. However, there is potentially available enough renewable carbohydrate to serve both food and polymer needs, and research into the modification of carbohydrates must continue at an increased rate.

When plant or animal tissues are extracted with nonpolar solvents, a portion of the material dissolves. The components of this soluble fraction are called lipids and include fatty acids, triacylglycerols, waxes, terpenes, prostaglandins, and steroids. The insoluble portion contains the more polar plant and animal components and cross-linked materials, including carbohydrates, lignin, proteins, and nucleic acids.

There are numerous natural materials and many ways to partition such materials. Table 6.1 contains one such listing along with suitable general subheadings. Table 6.2 lists a number of natural products as a function of general availability.

Finally, many potential renewable feedstocks are currently summarily destroyed or utilized in a noneconomical manner. Thus leaves and other plantstocks are "ritualistically" burned each fall. A number of these seemingly useless natural materials have already been utilized as feedstock sources for industrial products and more should be included.

6.1 POLYSACCHARIDES

Carbohydrates are the most abundant organic compounds, constituting three-fourths of the dry weight of the plant world. They represent a great storehouse of energy as a food for humans and animals. About 400 billion tons of sugars are produced annually through natural photosynthesis, dwarfing the production of other natural products, with the exception of lignin. Much of this is produced in the oceans, pointing out the importance of harnessing this untapped source of food, energy, and renewable feedstocks.

The potential complexity of even the simple aldohexose monosaccharides is indicated by the presence of five different chiral centers, giving rise to 2^5 or 32 possible stereoisomeric forms of the basic structure, two of which are glucose and mannose. While these sugars differ in specific biological activity, their gross chemical reactivities are almost identical, permitting one to often employ mixtures within chemical reactions without regard to actual structure with respect to most physical properties of the resulting product.

Table 6.1   Renewable Natural Material Groupings

Alkaloids

  Pyrrolidine, pyridine, pyrrolizidine, tropane
  Quinolizidine, isoquinoline, piperidine, indole
  Quinoline, quinazoline, acridone, steroidal, terpenoid

Amino acids

Carbohydrates

  Simple (glucose, sucrose, fructose, lactose, galactose)
  Complex (starch, cellulose, glycogen)

Drying oils and alkyd resins

  Linseed, cottonseed, castor, tung, soybean, oiticica, perilla, menhaden,
    sardine, corn, safflower, vernonia
  Fossil resins—amber, kauri, congo
  Oleoresins—damar, ester gum

Fungus, bacteria, and metabolites

Heme, bile, and chlorophylls

Lignins

Lipids

  Simple (glycerol esters, cholesterol esters)
  Phosphoglycerides
  Sphingolipids (mucolipids, sulfatide, sphingomyelin, cerebroside)
  Complex (lipoproteins, proteolipids, phosphatidopeptides)

Phenolic plant products

  Phenols, resorcinols, anthraquinones, naphthoquinones, hydrangenol,
    stilbenes, coumarins

Polyisoprenes

Proteins

  Enzymes (lysozome, trypsin, chymotrypsin)
  Transport and storage (hemoglobin, myoglobin)
  Antibodies
  Structural (elastin, actin, keratin, myosin, collagen, fibroin)
  Hormones (insulin)

Purines, pyrimidines, nucleotides, nucleic acids

Steroids

  Cholesterol, adrenocortical, bile acids
  Ergosterol, agnosterol, desmosterol

Tannins

Source: Carraher and Sperling, 1983. Used with permission.

Table 6.2 Relative Availability of Assorted Natural Products

Small-scale (biomedical, catalysis)

    Alkaloids
    Heme, bile, and chlorophylls
    Phenolic plant products
    Steroids
    Tannins

Medium- (many with potential for large-) scale

    Amino acids
    Fungus, bacteria
    Lipids
    Proteins (specific)
    Purines, pyrimidines, nucleotides, nucleic acids

Large-scale

    Carbohydrates
    Drying oils, alkyd resins
    Lignins
    Polyisoprenes
    Proteins (general)
    Terpenes and terpenoids

Source: Carraher and Sperling, 1983. Used with permission.

Carbohydrates are diverse with respect to both occurrence and size. Familiar mono- and disaccharides include glucose, fructose, sucrose (table sugar), cellobiose, and mannose. Familiar polysaccharides are listed in Table 6.3 along with their natural sources, purity, molecular weight, amount, and location of source. For instance, cotton is a good source of cellulose, yet the amount of cellulose varies from 85 to 97% depending on the variety of cotton plant, age of plant, and location of growth. Again, the gross chemical reactivity and resulting physical properties are largely independent of the source of cellulose.

The most important polysaccharides are cellulose and starch. These may be hydrolyzed by acids or enzymes to lower molecular weight carbohydrates (oligosaccharides) and finally to D-glucose. The latter is the building block, or mer, for carbohydrate polymers, and since it cannot be hydrolyzed further, it is called a monosaccharide. Cellobiose and maltose, which are the repeat units in cellulose and starch, are disaccharides, consisting of molecules of D-glucose joined together through carbon atoms 1 and 4.

The D-glucose units in cellobiose are joined by a β-acetal linkage while those in maltose are joined by an α-acetal linkage as shown in Fig. 6.1. The hydroxyl groups in the β form of D-glucose are present in the equatorial positions, and the hydroxyl on carbon 1 (the anomeric carbon atom) in the α form is in the axial position. While the chair forms shown for D-glucose, cellobiose, and maltose exist

Table 6.3 Naturally Occurring Polysaccharides

Polysaccharide	Source	Monomeric sugar unit(s)	Structure	Mol. wt.
Amylopectin	Corn, potatoes	D-Glucose	Branched	10^6-10^7
Amylose	Plants	D-Glucose	Linear	10^4-10^6
Chitin	Animals	2-Acetamidoglucose		
Glycogen	Animals (muscles)	D-Glucose	Branched	$>10^8$
Inulin	Artichokes	D-Fructose	Linear (largely)	10^3-10^6
Mannan	Yeast	D-Mannose	Linear	–
Cellulose	Plants	D-Glucose	Linear (2–D)	10^6
Xylan	Plants	D-Xylose	Linear (largely)	–
Lichenan	Iceland moss	D-Glucose	Linear	10^5
Galactan	Plants	D-Galactose	Branched	10^4
Arabinoxylan	Cereal grains	L-Arabinofuranose linked to xylose chain	Branched	$>10^4$
Galactomannans	Seed mucilages	D-Mannopyranose chains with D-galactose side-chains	Linear (largely)	10^5
Arabinogalactan	Lupin, soybean, coffee beans	D-Galactopyranose chain, side-chain galactose and arabinose	Branched	10^5
Carrageenan	Seaweeds	Complex—contains beta-galacto-pyranose linked to 3,6-anhydro-D-galactopyranose	Linear	10^5-10^6
Agar	Red seaweeds	Same as above except for L-galacto-pyranose	Linear	–
Alginic	Brown seaweeds	Beta-D-mannuronic acid and alpha-L-guluronic acid	Linear	–

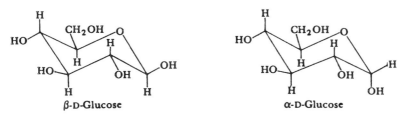

Figure 6.1 Chair forms of α- and β-D-glucose present in an equilibrium of 36% α and 64% β in aqueous solutions.

in all disaccharides and polysaccharides, simple Boeseken-Haworth perspective planar hexagonal rings will be used for simplicity in showing polymeric structures of most carbohydrates.

Accordingly, the molecular structures of cellobiose and maltose are shown in Figs. 6.2 and 6.3. The hydrogen atoms on the terminal carbon 1 and 4 atoms have been deleted to show the bonding present in cellulose and amylose starch. Amylodextrin is a highly branched polysaccharide with branches present on carbon 6.

Cellulose is a polydisperse polymer with a $\overline{\text{DP}}$ that ranges from 3,500 to 36,000. Native cellulose is widely distributed in nature and is the principal constituent of cotton, kapok, flax, hemp, jute, ramie, and wood. Flax has a $\overline{\text{DP}}$ of 36,000 or an average molecular weight of 5,900,000. Regenerated cellulose, such as rayon and cellophane, is produced by precipitating solutions of native cellulose in a nonsolvent.

Cellulose, which comprises more than one-third of all vegetable matter, is the world's most abundant organic compound. Approximately 50 billion tons of this renewable resource are produced annually by land plants, which absorb 4×10^{20} cal of solar energy. Natural cotton fibers, which are the seed hairs from *Gossypium*, are about 1 to 2 cm in length and about 5 to 20 μm in diameter. The molecules in native cellulose are present in threadlike strands or bundles called fibrils.

While the celluloses are often largely linear polymers, they are not soluble in water because of the presence of strong intermolecular hydrogen bonds and sometimes the presence of a small amount of cross-linking. Highly ordered crystalline cellulose has a density as high as 1.63 g cm^{-3}, while amorphous cellulose has a density as low as 1.47 g cm^{-3}. High molecular weight native cellulose, which is insoluble in 17.5% aqueous sodium hydroxide solution, is called α-cellulose. The fraction that is soluble in 17.5% sodium hydroxide solution but insoluble in 8%

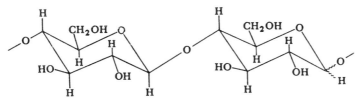

Figure 6.2 Chair form of cellobiose repeat unit in cellulose.

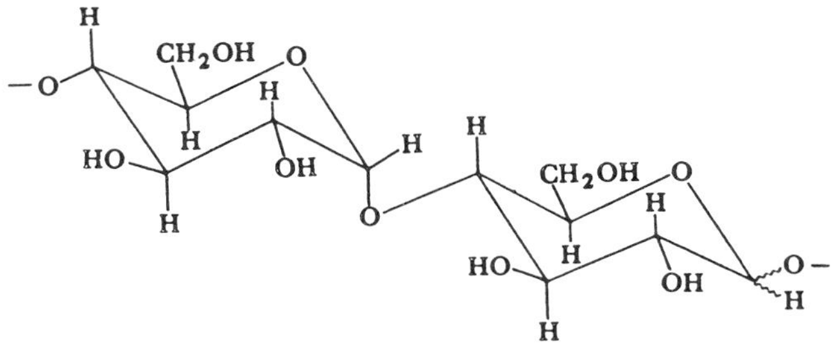

Figure 6.3 Chair form of maltose repeat unit in amylose.

solution is called β-cellulose, and that which is soluble in 8% sodium hydroxide solution is called γ-cellulose.

Strong caustic solutions penetrate the crystal lattice of α-cellulose and produce an alkoxide called alkali or soda cellulose. Mercerized cotton is produced by aqueous extraction of the sodium hydroxide. Cellulose ethers and cellulose xanthate are produced by reactions of alky halides or carbon disulfide, respectively, with the alkali cellulose.

Most linear celluloses may be dissolved in solvents capable of breaking the strong hydrogen bonds. These solvents include aqueous solutions of inorganic acids, calcium thiocyanate, zinc chloride, lithium chloride, dimethyl dibenzyl ammonium hydroxide, iron sodium tartrate, and cadmium or copper ammonia hydroxide (Schweitzer's reagent). Cellulose is also soluble in hydrazine, dimethyl sulfoxide in the presence of formaldehyde, and dimethyl formamide in the presence of lithium chloride. The average molecular weight of cellulose may be determined by measuring the viscosity of these solutions. The product precipitated by the addition of a nonsolvent to these solutions is highly amorphous regenerated cellulose.

6.2 THE XANTHATE VISCOSE PROCESS

The xanthate viscose process which is used for the production of rayon and cellophane is the most widely used regeneration process. The cellulose obtained by the removal of lignin from wood pulp is converted to alkali cellulose. The addition of carbon disulfide to the latter produces cellulose xanthate.

While terminal hydroxyl and aldehyde groups, such as are present in cellobiose, are also present in cellulose, they are not significant because they are present on very long-chain polymeric molecules. For convenience, one may represent cellulose by the semiempirical formula $C_6H_7O_2(OH)_3$. This formula shows the three potentially reactive hydroxyl groups on each repeat unit in the chain.

Presumably, the hydrogen atom of the hydroxyl group on carbon 6 is replaced by the sodium ion in soda cellulose, and, thus, the carbon disulfide reacts with one hydroxyl group only in each anhydroglucose repeat unit as shown by the following equations:

$$[C_6H_7O_2(OH)_3]_n + nNaOH \rightarrow [(C_6H_7O_2(OH)_2O^-, Na^+)]_n + nH_2O \qquad (6.1)$$

$$\text{Cellulose} + \atop \text{Sodium hydroxide} \rightarrow \text{Soda cellulose} + \text{Water}$$

$$[C_6H_7O_2(OH)_2O^-, Na^+]_n + nCS_2 \rightarrow [C_6H_7O_2(OH)_2O\!\!-\!\!\underset{\underset{S}{\|}}{C}\!\!-\!\!S^-, Na^+]_n \qquad (6.2)$$

$$\text{Soda cellulose} + \atop \text{Carbon disulfide} \rightarrow \text{Cellulose xanthate}$$

The orange-colored xanthate solution, or viscose, is allowed to age and is then extruded as a filament through holes in a spinneret. The filament is converted to cellulose when it is immersed in a solution of sodium bisulfite, zinc sulfate, and dilute sulfuric acid. The tenacity, or tensile strength, of this regenerated cellulose is increased by a stretching process which reorients the molecules so that the amorphous polymer becomes more crystalline. Cellophane is produced by passing the viscose solution through a slit die into an acid bath.

Since an average of only one hydroxyl group in each repeating anhydroglucose unit in cellulose reacts with carbon disulfide, the xanthate product is said to have a degree of substitution (DS) of 1 out of a potential DS of 3. Alkyl halides, including chloroacetic acid, may react with soda cellulose to yield ethers with DS average values ranging from 0.1 to 2.9.

The DS of inorganic esters such as cellulose nitrate (CN) may be controlled by the concentration of the esterifying acids. However, in classic esterification with organic acids, an ester with a DS of approximately 3 is obtained. The more polar secondary and primary esters, with DS values of 2 and 1, respectively, are produced by partial saponification of the tertiary ester. The degree of esterification of cellulose solutions in dimethyl-acetamide or dimethylsulfoxide may be controlled by the time of reaction.

Partially degraded cellulose is called hydrocellulose or oxycellulose, depending on the agent used for degradation. The term holocellulose is used to describe the residue after lignin has been removed from wood pulp. Cellulose soluble in 17.5% aqueous sodium hydroxide is called hemicellulose. Tables 6.4 and 6.5 describe a number of important textile fibers, including cellulosic fibers.

6.3 CHITIN

The exoskeletons of shellfish, arachnids, and many insects consist of chitin, which resembles cellulose, whose repeat unit is acetylated D-glucosamine. The hydroxyl group on carbon 2 of glucose is replaced by an amine group in D-glucosamine. Chitin is soluble in Schweitzer's reagent and dilute aqueous acids. It produces a soluble xanthate when reacted with carbon disulfide. Regenerated deacetylated chitin in the form of filaments and fiber is commercially available.

6.4 STARCH

Starch, which is the second most abundant polysaccharide, is widely distributed in plants where it is stored as reserve carbohydrate in seeds, fruits, tubers, roots, and stems. Starch is a polydisperse polymer which exists as a linear polymer amylose

Table 6.4 Noncellulosic Textile Fibers

Fiber name	Definition	Properties	Typical uses	Patent names (assignees)
Acrylic	Acrylonitrile units, 85% or more by weight	Warm; light weight; shape retentive; resilient; quick drying; resistant to sunlight,	Carpeting, sweaters, skirts, baby clothes, socks, slacks, blankets, draperies	Orlon (Du Pont), Acrilan (Monsanto), CHEMSTRAND (Monsanto)
Modacrylic	Acrylonitrile units, 35-85% by weight	Resilient; softenable at low temperatures; easy to dye; abrasion resistant; quick drying; shape retentive; resistant to acids, bases	Simulated fur, scatter rugs, stuffed toys, paint rollers, carpets, hairpieces and wigs, fleece fabrics	Verel (Eastman), Dynel (Union Carbide)
Polyester	Dihydric acid-terephthalic acid ester, 85% or more by weight	Strong, resistant to stretching and shrinking; easy to dye; quick drying; resistant to most chemicals; easily washed, wrinkle resistant; abrasion resistant; retains heat-set pleats and creases (permanent press)	Permanent press wear: skirts, shirts, slacks, underwear, blouses; rope, fish nets, tire cord, sails, thread	Vycron (Beaunit), Dacron (Du Pont), Kodel (Eastman), Fortrel (Fiber Ind., Celanese), CHEMSTRAND (polyester, Monsanto)
Spandex	Segmented polyurethane, 85% or more by weight	Light, soft, smooth, resistant to body oils; stronger and more durable than rubber; can be stretched repeatedly and to 500% without breaking; can retain original form; abrasion resistant; no deterioration from perspirants, lotions, detergents	Girdles, bras, slacks, bathing suits, pillows	Lycra (Du Pont)
Nylon	Recurring amide groups	Exceptionally strong; elastic; lustrous; easy to wash; abrasion resistant; smooth, resilient, low moisture absorbency; recovers quickly from extensions	Carpeting, upholstery, blouses, tent, sails, hosiery, suits, stretch fabrics, tire cord, curtains, rope, nets, parachutes	Caprolan (Allied Chemical), CHEMSTRAND (nylon, Monsanto), Astroturf (Monsanto), Celanese Polyester (Fiber Ind., Celanese), Cantrece (Du Pont)

Table 6.5 Cellulosic Fibers

Fiber name	Definition	Properties	Typical uses	Patent names (asignees)
Rayon	Regenerated cellulose with substitutes no more than 15% of the hydroxyl groups' hydrogens	Highly absorbent; soft; comfortable; soft; easy to dye; good drapability	Dresses, suits, slacks, blouses, coats, tire cord, ties, curtains, blankets	Avril (FMC Corp.), Cuprel (Beaunit), Zantrel (American Enka)
Acetate	Not less than 92% of the hydroxyl groups are acetylated, includes some triacetates	Fast drying; supple; wide range of dyability; shrink resistant	Dresses, shirts, slacks, draperies, upholstery, cigarette filters	Estron (Eastman), Celanese acetate (Celanese)
Triacetate	Derived from cellulose by combining cellulose with acetic acid and/or acetic anhydride	Resistant to shrinking, wrinkling, and fading; easily washed	Skirts, dresses, sportswear (pleat retention important)	Arnel (Celanese)

and a highly branched polymer amylopectin. Starches are usually present in the form of intramolecularly hydrogen-bonded polymer aggregates or granules.

Commercial starch is prepared from corn, white potatoes, wheat, rice, barley, millet, cassava, tapioca, arrowroot, and sorghum. The human digestive tract contains enzymes which are capable of cleaving the α-acetal linkages in starch and producing hydrolysis products, such as dextrins, maltose, and D-glucose.

Amylopectin, which is sometimes called the B fraction, is usually the major type of starch present in grains. However, amylose, which is sometimes called the A fraction, is present exclusively in a recessive strain of wrinkled pea. Since amylopectin serves as a protective colloid, native starch consisting of mixtures of amylose and amylopectin can be suspended in cold water. An opalescent starch paste is produced when this suspension is poured into hot water. In the absence of the amylopectin portion, an amylose solution produces a rigid irreversible gel on standing in a process called retrogradation.

Amylose turns blue when iodine is added and may absorb as much as 20% by weight of this halogen. When iodine is added to dextrins, which are degradation products, a reddish color is produced, depending on $\overline{DP}$ of the dextran. Amylopectin absorbs less than 1% by weight of iodine and yields a violet or pale red color. In addition to its use as food, starch is used as an adhesive for paper and as a textile-sizing agent. Starch forms ethers and esters like cellulose, but these are not used commercially to any great extent.

6.5 OTHER POLYSACCHARIDES

Glycogen, which is the reserve carbohydrate in animals, is a very highly branched polysaccharide similar to amylodextrin. *Polyglycuronic acids* are polysaccharides in which a carboxylic acid group replaces the hydroxyl group on carbon 6 in the anhydrohexose repeat unit. These water-soluble gums include the galactans and mannans found in pectin, algae, seaweed, gum arabic, agar, gum tragacanth, alginic acid, and other plant gums. Pentosans comparable to these hexosans are also naturally occurring polysaccharides.

Dextran is a high molecular weight branched polysaccharide synthesized from sucrose by bacteria. This polymer consists of anhydroglucose repeat units joined by α-acetal linkages through carbons 1 and 6. Partially hydrolyzed dextran is used as a substitute for blood plasma.

6.6 PROTEINS

The many different monodisperse polymers of amino acids, which are essential components of plants and animals, are called *proteins*. This word is derived from the Greek *porteios*, "of chief importance." The 20 different α-amino acids are joined together by *peptide linkages*

$$
\begin{array}{ccc}
O & H & R \\
\parallel & \mid & \mid \\
+\!\!-\!\!C\!\!-\!\!N\!\!-\!\!-\!\!C\!\!-\!\!+ \\
& & \mid \\
& & H
\end{array}
$$

and are *polyamides* or *polypeptides*. The latter term is often used by biologists to denote oligomers or relatively low molecular weight proteins. (Note the structural similarities and differences between proteins and polyamides-nylons.)

All α-amino acids

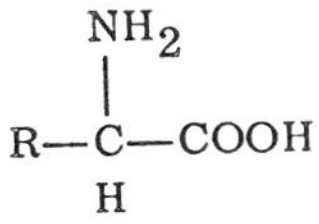

$$
\underset{H}{\overset{NH_2}{R-C-COOH}}
$$

except glycine

$$
\underset{}{\overset{NH_2}{H_2C-COOH}}
$$

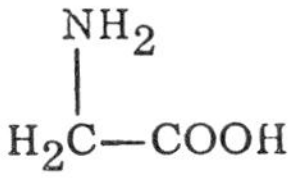

contain a chiral carbon atom and are L-amino acids. The net ionic charge of an amino acid varies with changes in the solution pH. The pH at which an amino acid is electrically neutral is called the isoelectric point. For simple amino acids (containing only one acid and one amine), this occurs at a pH of about 6 with the formation of a *dipolar* or *zwitterion* as shown below.

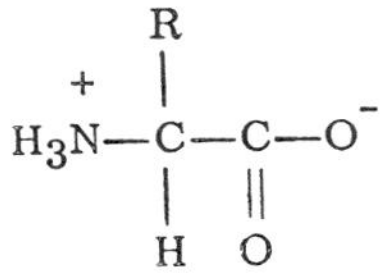

$$
\underset{\underset{O}{\overset{|}{H}}}{\overset{R}{\overset{|}{H_3N^+-C-C-O^-}}}
$$

Hence, α-amino acids, like other salts, are water-soluble, high melting polar compounds which migrate toward an electrode at pH values other than that of the isoelectric point in a process called *electrophoresis*.

Proteins may be classified as intermolecularly hydrogen-bonded *fibrillar*, or hairlike, proteins and as intramolecularly hydrogen-bonded *globular* proteins. Fibrillar proteins, such as keratin of the hair and nails, collagen of connective tissue, and myosin of the muscle, are water-insoluble polymers with relatively good strength. In contrast, globular proteins, such as enzymes, hormones, hemoglobin, and albumin, are water-soluble polymers with relatively poor strength.

Proteins may also be classified as simple polyamides and as conjugated proteins. The latter consist of a protein combined with a prosthetic group. The latter may be a carbohydrate, as in glycoproteins, a nucleic acid, as in nucleoproteins, or a heme, as in hemoglobin.

The amino acids may be neutral, acidic, or basic, in accordance with the relative number of amino and carboxylic acid groups present. For convenience in writing formulas for proteins, the amino acids are represented by abbreviations of three letters, such as Ala for alanine, Leu for leucine, Try for tryptophan, and Glu for glutamic acid. Thus, a dipeptide could be shown as Ala-Try and a tripeptide as Glu-Leu-Ala.

Proteins may be hydrolyzed by dilute acids, and the mixture of amino acids or residues produced may be separated and identified by paper chromatography. The reagent ninhydrin yields characteristic colored products with amino acids, and these may be determined colorimetrically. This chromatographic technique was developed by Nobel laureates Martin and Synge.

The term *primary structure* is used to describe the sequence of amino acid units (configuration) in a polypeptide chain. The sequence for N-terminal amino acids in a chain may be determined by use of a technique developed by Nobel laureate Sanger, who reacted the amino end group with 2,4-dinitrofluorobenzene and characterized the yellow aromatic amino acid produced by hydrolysis. This process is repeated after the end amino acid has been hydrolyzed off.

The C-terminal amino acids may be determined by using hydrazine to form hydrazides from the cleaved amino groups. Since the free carboxyl end group is not affected by hydrazine, the terminal amino acid is readily identified. The sequences of amino acids in several polypeptides, such as insulin and trypsin, have been identified by these techniques.

The term *secondary structure* is used to describe the molecular shape, or conformation, of a protein molecule. Nobel laureate Pauling has shown that a right-handed intramolecularly hydrogen-bonded helical arrangement (α helix; Fig. 6.4) is an important secondary structure when many bulky pendant groups are present on the chain. The distance between the amino acid repeat units along the chain in an α arrangement is 0.15 nm (1.5 Å).

The term *tertiary structure* is used to designate the shape or folding resulting from the presence of sulfur-sulfur cross-links between polymer chains. This structure requires the presence of cysteine units containing mercapto groups which are oxidized to cystine units containing disulfide groups.

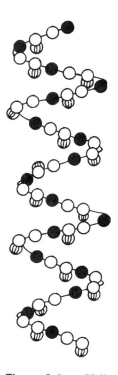

Figure 6.4 α-Helix conformation of proteins.

The specificity of enzymatic catalytic activity is dependent on tertiary structure. When eggs are boiled or ethanol is added, they lose their physiological activity because of a disruption of folded structure. This reorganization of structure is called *denaturation*.

The term *quaternary structure* is employed to describe the overall shape of groups of chains of proteins. For instance, hemoglobin is composed of four distinct but similar protein macromolecules, each with its own tertiary structure that comes together to give the quaternary hemoglobin structure.

A *β arrangement*, or *pleated sheet* conformation, is predominant when small pendant groups are present in the chain, as in silk fibroin. As shown in Fig. 6.5, the distance between amino acid repeating units along the chain in a *β* arrangement is 0.35 nm (3.5 Å), i.e., the repeat unit along the chain is 0.70 nm.

While polypeptides can be synthesized by simply heating *α*-amino acids, the products obtained are random mixtures unless a single amino acid is used. Likewise, nylon-2, which is a homopolypeptide, may be prepared by the Leuch's synthesis from N-carboxy-*α*-amino acid anhydrides. Polypeptides with specific sequences of amino acids can be prepared by protecting the N-terminal amino group by a reactant such as phthalic anhydride and removing the phthalimide group later by reacting with hydrazine.

The most widely used technique is the solid phase technique developed by Nobel prize winner Bruce Merrifield in which all reactions take place on the surface of cross-linked polystyrene beads. Thus, the entire reaction for synthesizing polypeptides with many programmed sequences of amino acids can be carried out automatically in a simple vessel without isolation of any intermediate products.

In nature, extended helical conformations appear to be utilized in two major ways: to provide linear systems for the storage, duplication, and transmission of information (DNA, RNA), and to provide inelastic fibers for the generation and transmission of forces (F-actin, myosin, and collagen).

Examples of the various helical forms found in nature are single helix (messenger and ribosomal DNA), double helix (DNA), triple helix (collagen fibrils),

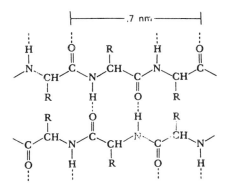

Figure 6.5 *β* Arrangement or pleated-sheet conformation of proteins. (From M. Stevens, *Polymer Chemistry—An Introduction*, Addison-Wesley, Reading, Massachusetts, 1975. With permission of Addison-Wesley Publishing Co.)

Chapter 6

and complex multiple helices (myosin). Generally, these single and double helices are fairly readily soluble in dilute aqueous solution. The triple and complex helices are only soluble if the secondary bonds are broken.

There are a variety of examples in which linear or helical polypeptide chains are arranged in parallel rows joined by covalent cross-links at regular intervals, leading to network structures. These polymers tend to be similar to their synthetic analogues in that they cannot be solubilized without covalent bond breakage and they tend to be good structural materials. The α-keratin of wool consists of parallel polypeptide α helices linked by disulfide bonds. If subjected to tension in the direction of the helix axes, the hydrogen bonds parallel to the axes are broken and the structure can be irreversibly elongated to an extent of about 100%.

A structural protein of skin, elastin, is somewhat similar to α-keratin in that it is composed of polypeptide chains linked covalently to form a network. In this case the cross-linking is formed by the reaction of four lysine side chains from four adjacent polypeptide chains to form desmosine (Fig. 6.6).

In contrast to α-keratin, elastin will stretch in a reversible rubberlike fashion under moderate loads. The tetradentate nature of the desmosine cross-links presumably aids in the reformation of coiled polypeptide chains on the release of tension and thus contributes to the rubberlike properties of the polymer.

In order to understand the physical properties of polymers, use is frequently made of the concept that extensibility depends upon the percentage of amorphous, as opposed to crystalline, material in the polymer. The silk fibroins are monofilament polypeptides with extensive secondary interchain bonding which are spun by various species of silk worms. The crystalline part of the fibroin is a polymer of the hexapeptide $-(Gly-Ser-Gly-Ala-Gly-Ala)-_n$. The polypeptides are arranged in an antiparallel β-pleated sheet, which allows multiple hydrogen bonding at right angles to the polypeptide chains.

Thus in the crystalline segments of silk fibroin there exists directional segregation using three types of bonding: covalent bonding in the first dimension, hydrogen bonding in the second dimension, and hydrophobic bonding in the third dimension. The physical properties of the crystalline regions are in accord with the bonding types. The polypeptide chains are virtually fully extended. There is a little puckering to allow for optimum hydrogen bonding. Thus the structure is inextensible in the direction of the polypeptide chains. On the other hand, the less specific hydrophobic (dispersive) forces between the sheets produces considerable flexibility. The crystalline regions in the polymers are interspersed with amorphous regions in which glycine and alanine are replaced by other amino acids with bulkier pendant groups which prevent the ordered arrangements described above. Furthermore, different silk worm species spin silks with differing amino acid composition and thus with differing degrees of crystallinity. The correlation between the extent of crystallinity and the extension at the break point is shown in Table 6.6.

Table 6.6 also illustrates the effect of changing the relative proportions of monomers on the physical properties of the copolymer.

In natural macromolecules the large number of relatively strong monomer unit-unit interactions results in rigid, only slightly deformable structures, which are usually hydrophobic within themselves but are frequently hydrophilic on their surface. As a consequence, the globular polymers are soluble in aqueous solution but

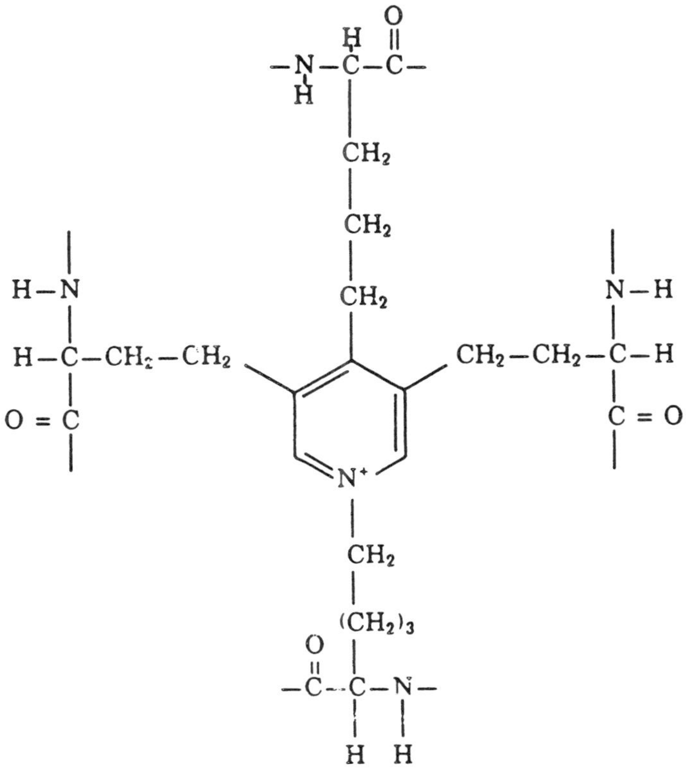

Figure 6.6 Central structure of desmosine.

Table 6.6 Selected Properties as a Function of Silk Worm Species

Silk worm species	Approximate crystallinity (%)	% Extension at the break point
Anaphe moloneyi	95	12.5
Bombyx mori	60	24
Antherea mylitta	30	Flow then extends to 35%

Source: Coates and Carraher, Polymer News, 9(3):77 (1983). Used with permission.

there is no permeation of solvent water within the chains. Water-soluble proteins are regarded as globular polymers which are characterized by the use of large numbers of 20 different α-amino acid monomers linked by peptide bonds in various highly specific orders.

Within a particular globular polymer there may be one or more polypeptide chains folded backwards and forwards through quite distinct structural domains. Each domain is characterized by a particular style of coiling which may be non-repetitive with respect to its peptide chain geometry or may be repetitive, conforming to one of several now well-recognized patterns. The specific chain conformations are determined by the side-chain interactions of the amino acids superimposed on intrapeptide hydrogen bonding along the chain. The form of chain folding is thus ultimately determined by the amino acid sequence and the polymeric nature of the polypeptide chains and is fundamental to the specific geometry of a given protein. The commonly occurring repetitive helical patterns for polypeptides are the α helix, the 2_7 ribbon, the 3_{10} helix, and the π helix. These structures are a consequence of repeated hydrogen bonding between peptide carboxyl and amino groups on the same polypeptide chain, along the direction of the chain, as shown in Fig. 6.7.

In addition to these helical structures, there exist β-sheet structures in which the polypeptide chains lie parallel or antiparallel to each other in a plane, with carbonyl to amino hydrogen bonds approximately at right angles to the directions of the chains. The geometry of the chains and the hydrogen bonds lead to the amino acid side chains lying all on the same side of the sheet, in the case of parallel chains, and on alternate sides of the sheets in the case of antiparallel chains.

As an example of conformational change induced by the binding of a small molecule, we can consider the binding of oxygen to hemoglobin. Hemoglobin is composed of four very similar subunits, each containing a porphyrin ring enclosing a ferrous ion which is capable of binding, reversibly, a dioxygen molecule. Since the ferrous ions are relatively heavy atoms compared with the remainder of the atoms of the hemoglobin, the Fe^{II}–Fe^{II} distances are relatively easily determined

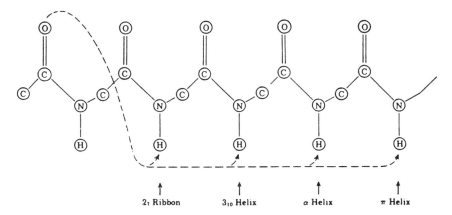

Figure 6.7 Commonly occurring repetitive helical patterns for polypeptides. [From J. Coates and C. Carraher, Polymer News, 9(3):77 (1983). Used with permission.]

by crystallographic techniques. Measurements of the Fe^{II}–Fe^{II} distances in fully oxygenated hemoglobin and in deoxyhemoglobin are given in Fig. 6.8.

As a consequence of the conformation change triggered by the high spin to low spin conversion of one Fe^{II} on binding dioxygen, the affinity for oxygen of the Fe^{II} in the other subunits is changed, a so-called cooperative effect. In addition, it is found that as O_2 is bound, a proton is more readily lost from the subunit—the Haldane effect. Finally, the small molecule 2,3-diphosphoglycerate can bind as an effector to a specific site on the subunit and in so doing distort the polymer towards the deoxy conformation, thus tending to cause O_2 to be released. The complex sets of interactions outlined here would be difficult to envisage in any structure less complex than a macromolecule.

Table 6.7 contains a listing of shapes of selected proteins.

6.7 NUCLEIC ACIDS

Nucleoproteins, which are conjugated proteins, may be separated into nucleic acids and proteins in aqueous sodium chloride. The name "nuclein," which was coined by Miescher in 1869 to describe products isolated from nuclei in pus, was later changed to *nucleic acid*. Pure nucleic acid was isolated by Levene in the early 1900s. He showed that either D-ribose or D-deoxyribose was present in what are now known as ribonucleic acid (RNA) and deoxyribonucleic acid (DNA). These specific compounds were originally obtained from yeast and the thymus gland, respectively.

In 1944, Avery showed that DNA was capable of changing one strain of bacteria into another. It is now known that nucleic acids direct the synthesis of proteins. Thus, our modern knowledge of heredity and molecular biology is based on our knowledge of nucleic acids.

The prosthetic group in a nucleoprotein is a phosphoric acid ester of a nucleoside and is called a *nucleotide*. The nucleoside is a compound consisting of a pentose and a heterocyclic base. The two types of pentose are ribose (in RNA) and deoxyribose (in DNA). There are two classes of heterocyclic bases, namely, *pyrimidines*, which are simple 1,3-diazines, and *purines*, which are more complex fused nitrogenous rings consisting of pyrimidine and imidazole rings.

As shown in Fig. 6.9, the deoxyribose present in DNA differs from the ribose present in RNA by the presence of a hydrogen atom instead of a hydroxyl group

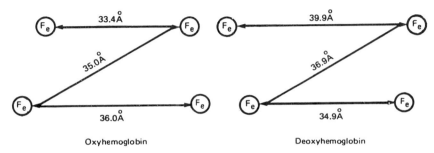

Figure 6.8 Fe^{II}–Fe^{II} distances in oxyhemoglobin and deoxyhemoglobin.

Table 6.7 Shapes of Selected Biologically Important Proteins

Protein	Shape	Molecular weight (Daltons)	Comments
Myoglobin	3-D, oblate spheroid	17,000	Temporary oxygen storage in muscles
Hemoglobin	3-D, more spherical than myoglobin	64,000	Oxygen transport through body
Cytochrome c	3-D, prolate spheroid	12,000-13,000	Heme-containing protein, transports electrons rather than oxygen
Enzymes	Somewhat common features a. Alpha-helix content not as high as myoglobin, but areas of beta-sheeting not unusual. b. Water soluble enzymes have a large number of charged groups on the surface and those not on the surface are involved in the active site. Large parts of the interior are hydrophobic. c. The active site is found either as a cleft in the macromolecule or shallow depression on its surface.		
Lysozyme	3-D, short alpha-helical portions, region of anti-parallel pleated sheet	14,600	Well studied, good illustration of shape-activity
Chymotrypysin and trypsin	3-D, extensive beta-structure		Hydrolysis of peptide bonds on the carboxyl side of certain amino acids
Insulin	3-D, two alpha-helical sections in A-chain, B-chain has alpha-helix and the remainder is extended linear central core	6,000	Regulation of fat, carbohydrate and amino acid metabolism
Somatotropin (human)	3-D, 50% alpha-helix	22,000	Pituitary hormone
Collagen	Tropocollagen, three left-handed helical chains would around each other giving a triple helix	3×10^5	Most abundant protein, major component of skin, teeth, bones, cartilage and tendon
Keratin	Varies with source, 3-D or 2-D, most contain alpha-helix sections	$\sim 10^4$-10^5	Main structural material of skin, feathers, nail, horn, hair—protective coatings of animals
Fibroin	Varies with source, fibrous—linear with cross-links—crystalline regions contain antiparallel, pleated sheets	365,000	A major constituent of silk
Elastin	Varies with source, cross-linked, mostly random coil with some alpha-helix	>70,000	Many properties similar to rubber; gives elasticity to arterial walls and ligaments

Source: Coates and Carraher, 1983. Used with permission.

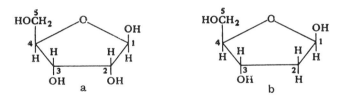

Figure 6.9 Boesekin-Haworth structures of (a) ribose and (b) deoxyribose.

on carbon atom 2. The structural formulas for the pyrimidine and purine bases are shown in Fig. 6.10.

Both DNA and RNA contain both purine bases, namely, adenine and guanine, and the pyrimidine base cytosine. Uracil is found in RNA, and methyluracil or thymine is found in DNA. (It may help to remember that DNA was originally found in the thymus gland.)

The basic nitrogen atoms in the purines and pyrimidines are joined to carbon 1, and the phosphate is on carbon 3 in the nucleotides. The names of the nucleotides and nucleosides correspond to the purine or pyrimidine bases present. Thus, the nucleotide shown in Fig. 6.11 is called adenylic acid, since it is the 3'-phosphate ester of the nucleoside adenosine. The corresponding nucleoside for deoxyribose is deoxyadenosine. The other nucleosides of ribose are called guanosine, cytidine, and uridine. The nucleosides of deoxyribose are called deoxycytidine and, simply, thymidine.

Since there are two hydroxyl groups in each deoxyribose, there are two possible phosphoric acid esters. Adenosine forms esters with more than one phosphate group on carbon 5. These are adenosine monophosphate (AMP), the diphosphate (ADP),

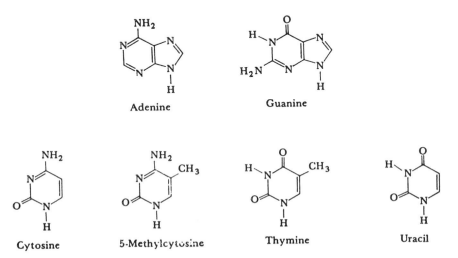

Figure 6.10 Structural formulas of (top) purine and (bottom) pyrimidine bases in nucleic acids.

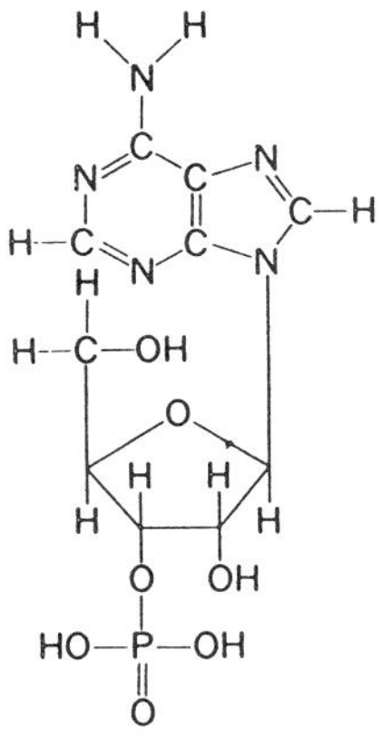

Figure 6.11 Adenylic acid, adenosine-3′-phosphate, nucleotide.

and the triphosphate (ATP). The hydrolysis of ATP to produce ADP is important in biological energy transfer.

As shown in Fig. 6.12, RNA and DNA are polymers in which the backbone consists of phosphate esters on both carbon 3 and carbon 5 of the pentoses.

The bases on the RNA and DNA chains may be either purines or pyrimidines. Nobel laureates Watson and Crick correctly deduced that DNA existed as a double-stranded helix in which a pyrimidine base on one chain or strand was hydrogen bonded to a purine base on the other chain. As illustrated in Fig. 6.13, the total distance for these pairs is 1.07 nm (10.7 Å). As shown, the base pairs are guanine-cytosine and adenine-thymine (which may be remembered from the mnemonic expression *Gee-CAT*). A sketch of the DNA double helix is shown in Fig. 6.14.

It is now known that the molecular weight of RNA is typically less than that of DNA and that there are at least three different types of RNA, namely, high molecular weight ribosomal RNA (r-RNA), moderate molecular weight transfer (t-RNA), and low molecular weight messenger RNA (m-RNA). These are believed to direct the synthesis of proteins which consist of 20 amino acids arranged in specific sequences. Thus, the total number of possible combinations of amino acids is 2.4×10^{18}.

DNA, which is an essential part of the chromosome of the cell, contains the information for the synthesis of a protein molecule. When the double parent strand splits into two single daughter strands, each strand serves as a template for the formation of another strand with the proper base pairs. The single-stranded DNA also serves as a template for m-RNA. The specific amino acids are brought to the cell by t-RNA, to which the amino acids are joined to the phosphoric acid groups.

There are four different nucleotides (or four different bases) in DNA which provide for 4^2, or 16 dinucleotide combinations and 4^3, or 64 trinucleotides, etc. It is believed that these trinucleotides or *codons* are present in all known life forms and that these codons specify the specific amino acids used in protein synthesis. Thus, using U as a symbol for uracil, the codon UUU is specific for phenylalanine (Phe). Most of the code for the 64 combinations of the four bases has been elucidated.

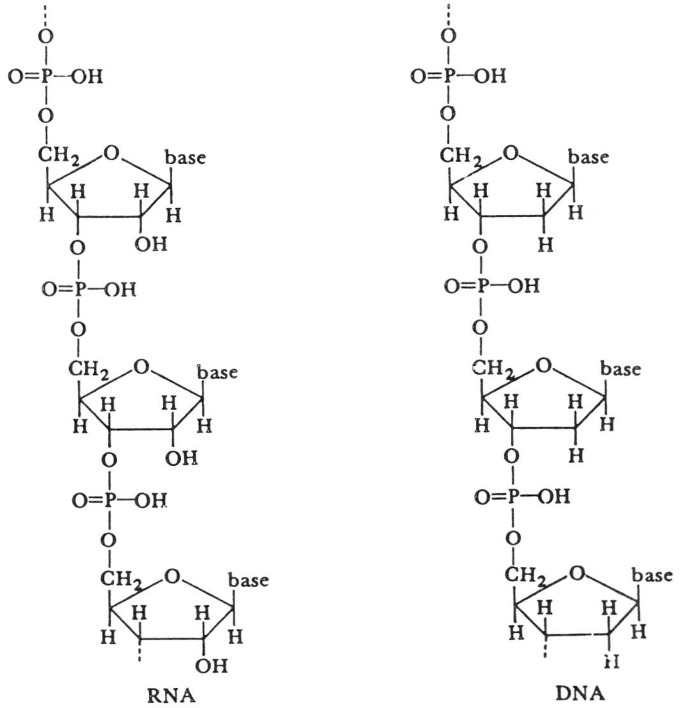

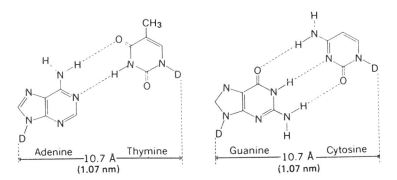

Figure 6.12 Representative segments of nucleic acid polymeric chains.

Other examples of codons are AUG and UAG, which are the initiators and terminators of chains. Of course, with 64 codons and only 20 amino acids, plus start and stop signals, many amino acids are represented by as many as three codons. Thus, in addition to UUU, UUC is also a codon for Phe. Mutations can be caused by mistakes in the code and by specific chemical compounds which affect the base pairs.

Figure 6.13 Allowable base pairs in nucleic acids (DNA).

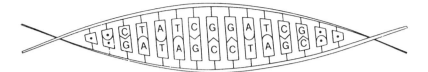

Figure 6.14 A schematic representation of the double helix of DNA.

6.8 NATURALLY OCCURRING POLYISOPRENES

Polyisoprenes

$$[-\underset{\underset{H}{|}}{\overset{\overset{H}{|}}{C}} - \underset{\underset{CH_3}{|}}{\overset{}{C}} = \underset{\underset{H}{|}}{\overset{}{C}} - \underset{\underset{H}{|}}{\overset{\overset{H}{|}}{C}} -]_n$$

occur in nature as hard plastics called gutta percha and balata and as an elastomer or rubber known as *Hevea brasiliensis*, or natural rubber (NR). Approximately 50% of the 500 tons of gutta percha produced annually is obtained from gutta percha trees grown on plantations in Java and Malaya. Balata and about 50% of the gutta percha used are obtained from trees in the jungles of South America and the East Indies. The first gutta-insulated submarine cable was installed between England and France in 1850. Gutta percha (*Palaquium oblongifolium*) continues to be used for wire insulation, and both this polyisoprene and balata (*Mimusops globosa*) are used as the covers for golf balls.

The hardness of these polydisperse naturally occurring crystalline polymers is the result of a *trans configuration* in 1,4-polyisoprene (see Fig. 6.15). The chain extensions on opposite sides of the double bonds in the polymer chain facilitate good fit and cause inflexibility in these chains.

Gutta percha exists in the planar α form, which has an X-ray identity period of 0.87 nm (8.7 Å) and a melting point of 74°C. The α form is transformed to the β form when these trans isomers of polyisoprene are heated above the transition temperature of 68°C. The nonplanar β form has a melting point of 64°C and an identity period of 0.48 nm (4.8 Å). The latter corresponds to the length of the isoprene repeat unit in the polymer chain.

Both trans and cis isomers of polyisoprene are present in chicle, which is obtained from the *Achras sapota* tree in Central America. Since chicle is more flexible than gutta percha, it has been used as a base for chewing gum. Both trans and cis isomers of polyisoprene are also synthesized commercially.

Natural rubber, which is one of the most important biologically inactive, naturally occurring polymers, was used by the Mayan civilization in Central and South America before the twelfth century. In addition to using the latex from the *ule* tree for the waterproofing of clothing, the western hemisphere natives played a game called tlachtli with large rubber balls. The object of this game was to insert the ball into a tight-fitting stone hole in a vertical wall using only the shoulder or thigh. The game ended once a goal was scored, and the members of the losing team were sacrificed to the gods.

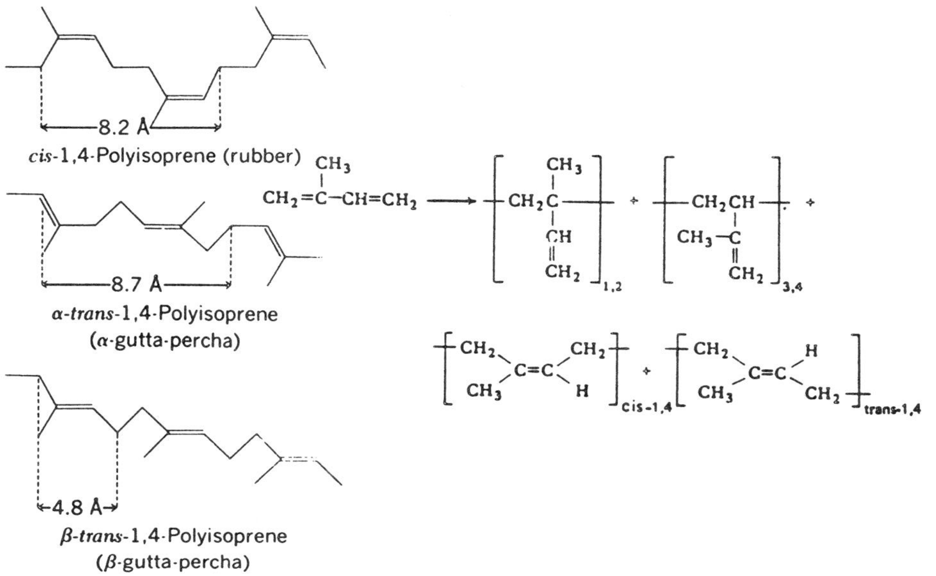

cis-1,4-Polyisoprene (rubber)

α-trans-1,4-Polyisoprene
(α-gutta-percha)

β-trans-1,4-Polyisoprene
(β-gutta-percha)

Figure 6.15 Abbreviated structural formulas for polyisoprenes. (From *Introduction to Polymer Chemistry* by R. Seymour, McGraw-Hill, New York, 1971. Used with permission of McGraw-Hill Book Company.)

While natural rubber, or *caoutchouc*, was brought to Europe by Columbus, little use, beyond erasing pencil marks, was made of this important elastomer until the nineteenth century. Accordingly, Priestly coined the name *India rubber* to describe its major use at that time.

While less than 1% of the world's present supply of natural rubber is obtained from wild rubber trees, most of this elastomer was shipped from Central and South America prior to the twentieth century. The latex, which is an aqueous suspension of 30 to 35% cis-1,4-polyisoprene, occurs in microscopic tubules located between the bark and the cambian layer in the hevea plant. The latex is obtained by collecting the liquid which seeps out after tapping the hevea plants.

Hevea latex is also present in the household decorative rubber plant (*Ficus elastica*), milkweed (*Cryptostegia grandiflora*), goldenrod (*Solidago*), dandelions (*Taraxacum offinale* and *Koksagyhz*), creepers (*Landolphia*), and guayule (*Parthenium argentatum*).

In addition to the cultivation of *H. brasiliensis*, experimental plantations have been established for the cultivation of *F. elastica*, *Funtumia*, *Castilloa*, *Manihot*, *Koksagyhz*, and *P. argentatum*. While *F. elastica* continues to be used as a decorative plant, the cultivation of all nonhevea rubbers, except guayule, has been abandoned.

Much of the interest in the Russian dandelion and guayule resulted from successful growth in cooler temperature zones. As much as 200 kg of rubber per acre has been obtained from the cultivation of dandelions in Turkestan (Uzbekistan), but this project was abandoned after the development of styrene-butadiene synthetic rubber (SBR).

Guayule, which grows in arid regions in southwestern United States and northern Mexico, can be mechanically harvested and is used as a replacement for hevea rubber in pneumatic tires. The guayule plant has a rubber content of about 10%. As much as 1500 kg of rubber per hectare may be obtained by crushing this plant (1 hectare ≃ 2.5 acres).

Approximately 99% of the present supply of natural rubber is obtained from the progenitors of a few thousand seedlings smuggled out of Brazil by Wickham and Cross and cultivated in Ceylon in 1876. The first successful tappings of these trees were made in 1896. The plantations in Malaya and Indonesia now account for over 70% of the world's production of natural rubber.

This widely used elastomer (cis-1,4-polyisoprene) is not produced in nature by the polymerization of isoprene, but by the enzymatic polymerization of isopentenylpyrophosphate:

$$H_2C{=}\overset{\overset{\displaystyle CH_3}{|}}{C}{-}CH_2{-}CH_2O{-}\overset{\overset{\displaystyle OH}{|}}{\underset{\underset{\displaystyle O}{\|}}{P}}{-}O{-}\overset{\overset{\displaystyle OH}{|}}{\underset{\underset{\displaystyle O}{\|}}{P}}{-}OH.$$

While the theoretical yield of hevea rubber is 9000 kg per hectare (ha), the best yields to date are 3300 kg/ha.

Prior to the discovery of the vulcanization or cross-linking of hevea rubber with sulfur by Charles Goodyear in 1838, Faraday had shown that the empirical formula of this elastomer is C_5H_8, and thus rubber is a member of the terpene group. The product obtained by pyrolysis of rubber was named isoprene by Williams in 1860 and converted to a solid (polymerized) by Bouchardat in 1879. Tilden suggested the formula that is now accepted for isoprene:

$$H_2C{=}\overset{\overset{\displaystyle CH_3}{|}}{C}{-}\overset{\overset{\displaystyle H}{|}}{C}{=}CH_2$$

Since levulinic aldehyde

$$H_3C{-}\underset{\underset{\displaystyle O}{\|}}{C}{-}\overset{\overset{\displaystyle H}{|}}{\underset{\underset{\displaystyle H}{|}}{C}}{-}\overset{\overset{\displaystyle H}{|}}{\underset{\underset{\displaystyle H}{|}}{C}}{-}\overset{\overset{\displaystyle H}{|}}{C}{=}O$$

was produced by ozonolysis, Harries suggested the presently accepted structure for the polymer.

The X-ray identity pattern of stretched natural rubber is 0.82 nm (8.2 Å), which represents two isoprene units in a cis configuration in the polymer chain. Rubber has a glass transition temperature of −85°C, but this is increased as the cross-linked density of vulcanized rubber increases. Unvulcanized rubber undergoes typical reactions of olefins, such as hydrogenation, chlorination, hydrohalogenation, epoxidation, and ozonolysis.

Natural rubber crystallizes when stretched in a reversible process. However, the sample remains in its stretched form (racked rubber) if it is cooled below its

T_g. The racked rubber will snap back and approach its original form when it is heated above its T_g. Rubber that has been held in an extended form for long periods of time does not return to its original form immediately when the stress is relieved, since a relaxation process must occur before it decays completely to its original length. The delay in returning to the original form is called hysteresis.

These and other elastic properties of NR and other elastomers above the T_g are based on long-range elasticity. Stretching causes an uncoiling of the polymer chains, but these chains assume the most probable conformations if the stress is removed after a short period of time. Some slippage of chains occurs if the rubber is held in the stretched position for long periods of time.

The absence of strong intermolecular forces, the presence of pendant methyl groups, and the crankshaft action associated with the cis isomer all contribute to the flexibility of the natural rubber molecule. The introduction of a few cross-links by vulcanization with sulfur reduces slippage of the chains, but still permits flexibility in the relatively long chain sections between cross-links (principal sections).

When a strip of natural rubber (NR) or synthetic rubber (SR) is stretched at a constant rate, the tensile strength required for stretching (stress, s) increases slowly until an elongation (strain, γ) of about 500% is observed. This initial process is associated with an uncoiling of the polymer chains in the principal sections.

Considerably more stress is required for greater elongation to about 800%. This rapid increase in modulus (G) is associated with better alignment of the polymer chains along the axis of elongation, crystallization, and decrease in entropy (ΔS). The work done in the stretching process (W_{el}) is equal to the product of the retractile force (f) and the change in length (dl). Therefore, the force is equal to the work per change in length.

$$W_{el} = f\,dl \quad \text{or} \quad f = \frac{W_{el}}{dl} \tag{6.3}$$

W_{el} is equal to the change in Gibbs free energy (dG), which under conditions of constant pressure is equal to the change in internal energy (dE) minus the product of the change in entropy and the Kelvin temperature.

$$f = \frac{W_{el}}{dl} = \frac{dG}{dl} = \frac{dE}{dl} - T\frac{dS}{dl} \tag{6.4}$$

The first term in Eq. (6.4) is important in the initial low-modulus stretching process, and the second term predominates in the second high-modulus stretching process. For an ideal rubber, only the second term is involved.

As observed by Gough in 1805 and confirmed by Joule in 1859, the temperature of rubber increases as it is stretched, and the stretched sample cools as it snaps back to its original condition. (This is easily confirmed by rapidly stretching a rubber band and placing it to your lips, noting that heating has occurred, and then rapidly releasing the tension and again placing the rubber band to your lips.) This effect was expressed mathematically by Kelvin and Clausius in the 1850s. The ratio of the rate of change of the retractive force (df) to the change in Kelvin temperature (dT) in an adiabatic process is equal to the specific heat of the elastomer (C_p) per degree temperature (T) times the change in temperature (dT) with the change in length (dl).

$$\frac{df}{dT} = -\frac{C_p}{T}\frac{dT}{dl} \tag{6.5}$$

Equation (6.5) may be transformed as shown in Eq. (6.6). Unlike most solids, natural rubber and other elastomers contract when heated.

$$\frac{dT}{df} = -\frac{T}{C_p}\frac{dl}{dT} \tag{6.6}$$

In the process of adding various essential ingredients or additives to crude rubber in a process called compounding, the rubber is masticated on a two-roll mill or in an intensive mixer at an elevated temperature in the presence of air. This mechanical action cleaves some of the carbon-carbon covalent bonds and produces macroradicals with lower $\overline{DP}$ values than the original macromolecule. The coupling of these macroradicals may be prevented if chain transfer agents, called peptizers, are present. As shown in the following equation, loosely bonded atoms or groups are abstracted from the chain transfer agent and dead polymers are produced.

$$\text{E}$$
$$\text{\Large\sim}\text{CH}_2\text{—CH}_2\text{—CH}_2\text{—CH}_2\text{\Large\sim} \rightarrow \text{\Large\sim}\text{CH}_2\text{—CH}_2^{\bullet} + {}^{\bullet}\text{CHCH}_2\text{—CH}_2\text{\Large\sim}$$

$$\text{Rubber molecule} \qquad\qquad\qquad \text{Macroradicals}$$

$$\text{\Large\sim}\text{CH}_2\text{—CH}_2^{\bullet} + \text{HSR} \rightarrow \text{\Large\sim}\text{CH}_2\text{—CH}_3 + {}^{\bullet}\text{SR} \tag{6.7}$$
$$\qquad\qquad\qquad\quad \text{Chain}$$
$$\text{Macroradical}\quad \text{transfer agent}\quad \text{Dead polymer}\quad \text{Free radical}$$

$$\text{\Large\sim}\text{CH}_2\text{—CH}_2^{\bullet} + {}^{\bullet}\text{SR} \rightarrow \text{\Large\sim}\text{CH}_2\text{—CH}_2\text{—SR}$$
$$\text{Macroradical}\quad \text{Free radical}\quad \text{Dead polymer}$$

Vulcanization of both natural and synthetic rubber is a cross-linking reaction carried out on an industrial scale. The exact mechanism varies with the vulcanization technique employed. Physical cross-linking occurs through the mechanical cleaving of the carbon-carbon bond as noted above with subsequent rejoining of different chains.

$$\text{\Large\sim}\text{CH}_2\text{—CH}_2^{\bullet} + \text{\Large\sim}\text{CH}_2\text{—CH}_2^{\bullet} \rightarrow \begin{array}{c} \text{\Large\sim}\text{CH}_2\text{—CH}_2 \\ | \\ \text{\Large\sim}\text{CH}_2\text{—CH}_2 \end{array}$$

$$\text{\Large\sim}\text{CH}_2\text{—CH}_2^{\bullet} + \text{\Large\sim}\text{CH}_2\text{—CH}_2\text{\Large\sim} \rightarrow \text{\Large\sim}\text{CH}_2\text{—CH}_3 + \text{\Large\sim}\text{CH}_2\text{—CH}_{\underset{\bullet}{}}\text{\Large\sim} \tag{6.8}$$

$$\begin{array}{c} \text{\Large\sim}\underset{\bullet}{\text{C}}\text{HCH}_2\text{\Large\sim} \\ + \\ \text{\Large\sim}\underset{\bullet}{\text{C}}\text{HCH}_2\text{\Large\sim} \end{array} \rightarrow \begin{array}{c} \text{\Large\sim}\text{CHCH}_2\text{\Large\sim} \\ | \\ \text{\Large\sim}\text{CHCH}_2\text{\Large\sim} \end{array}$$

Peroxidase-initiated cross-linking proceeds by homolytic abstraction of a polymer chain hydrogen atom followed by radical recombination, i.e.,

$$R\text{—}O\text{—}O\text{—}R \rightarrow 2\ RO^{\bullet}$$

$$RO^{\bullet} + \text{\tiny ww}CH_2\text{—}CH_2\text{\tiny ww} \rightarrow ROH + \text{\tiny ww}CH_2\text{—}\underset{\bullet}{C}H\text{\tiny ww}$$

$$\begin{array}{c} \text{\tiny ww}\underset{\bullet}{C}HCH_2\text{\tiny ww} \\ + \\ \text{\tiny ww}\underset{\bullet}{C}HCH_2\text{\tiny ww} \end{array} \rightarrow \begin{array}{c} \text{\tiny ww}CHCH_2\text{\tiny ww} \\ | \\ \text{\tiny ww}CHCH_2\text{\tiny ww} \end{array} \qquad (6.9)$$

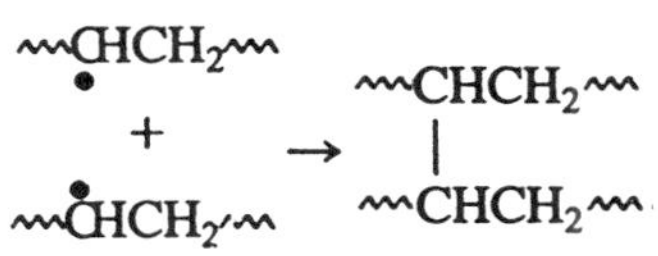

Unsaturated polymer hydrogen abstraction probably occurs largely at the allylic position.

$$\text{\tiny ww}CH_2CH\text{=}CHCH_2\text{\tiny ww} + RO\bullet \rightarrow \text{\tiny ww}\underset{\bullet}{C}HCH\text{=}CHCH_2\text{\tiny ww} + ROH \qquad (6.10)$$

The oldest method of vulcanization utilizing sulfur appears in part to occur through an ionic pathway involving addition to a double bond forming an intermediate sulfonium ion which subsequently abstracts a hydride ion or donates a proton, forming new cations for chain propagation. Other propagation steps may also occur.

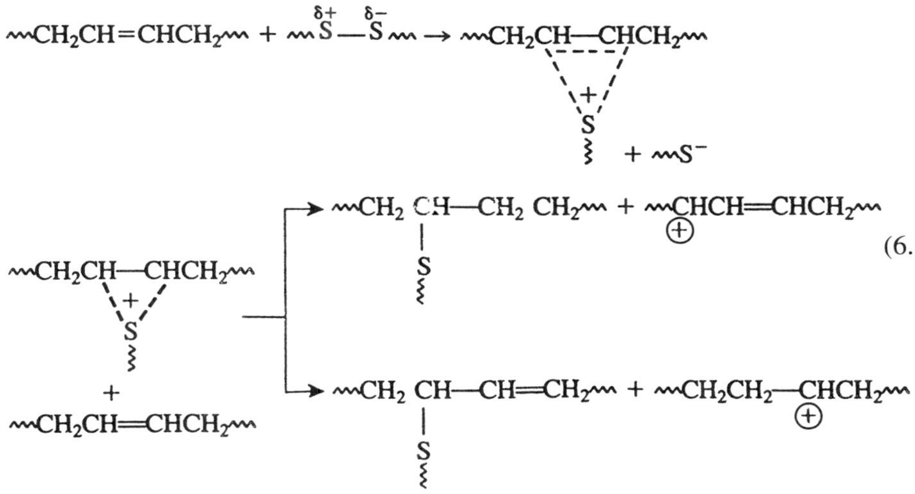

$$\text{\tiny ww}CH_2CH\text{=}CHCH_2\text{\tiny ww} + \text{\tiny ww}\overset{\delta+}{S}\text{—}\overset{\delta-}{S}\text{\tiny ww} \rightarrow \text{\tiny ww}CH_2CH\text{—}CHCH_2\text{\tiny ww}$$

(6.11)

(6.11)

Cold vulcanization can occur by dipping thin portions of rubber into carbon disulfide solutions of S_2Cl_2.

$$2\text{\tiny ww}CH\text{=}CH\text{\tiny ww} + S_2Cl_2 \rightarrow \begin{array}{c} \text{\tiny ww}CH\text{—}S\text{—}CH\text{\tiny ww} \\ |\ \ \ \ \ \ | \\ Cl \ \ \ \ \ Cl \end{array} + S \qquad (6.12)$$

The long time required for vulcanization of rubber by heating with sulfur was shortened drastically by Oenslager, who used organic amines as catalysts, or accelerators, for the vulcanization of rubber in 1904. The most widely used accelerators are derivatives of 2-mercaptobenzothiazole (Captax). Other frequently used accelerators are zinc dimethyldithiocarbamate and tetramethylthiuram disulfide.

2-Mercaptobenzothiazole

Zinc dimethyldithiocarbamate

Tetramethylthiuram
disulfide

Rubber-compounding formulations also include antioxidants, such as phenyl-β-naphthylamine, which retards the degradation of rubber at elevated temperatures, and carbon black, which serves as a reinforcing agent. Amorphous silica is used in place of carbon black in compounding recipes for the white sidewalls of pneumatic tires. Since commercial β-naphthylamine is said to be carcinogenic, relatively pure phenyl-β-naphthylamine should be used. The key to prevention of rubber oxidation by an antioxidant depends on its ability to stop the propagation reaction and the chain reaction, or on its ability to destroy peroxides.

$$R\bullet + O_2 \rightarrow RO_2^{\bullet} \qquad \text{Propagation}$$

$$RO_2^{\bullet} + R'H \rightarrow R\bullet + R'O_2H \qquad \text{Chain reaction} \tag{6.13}$$

$$R'O_2H + AH \text{ (antioxidant)} \rightarrow \text{Stable products} \qquad \text{Peroxide destruction}$$

While natural rubber and cotton account for over 40% of total elastomer and fiber production, only relatively small amounts of natural plastics and resins are used commercially. The principal products of this type are casein, shellac, asphalt or bitumen, rosin, and polymers obtained by the drying (polymerization) of unsaturated oils.

Shellac, which was used by Edison for molding his first phonograph records and is still used as an alcoholic solution (spirit varnish) for coating wood, is a cross-linked polymer consisting to a large extent of derivatives of aleuritic acid (9,10,16-trihydroxyhexadecanoic acid). Shellac is excreted by small coccid insects (*Coccus lacca*) which feed on the twigs of trees in Southeast Asia. Over 20 lakshas (Sanskrit for 100,000), or 2 million insects, must be dissolved in ethanol to produce 1 kg of shellac.

While naturally occurring bitumens were used by the ancients for caulking and waterproofing, they now account for less than 5% of this type of material. Large deposits of natural bitumens are located at Trinidad and Bermudez Lake in the West Indies and in Venezuela. Gilsonite, which was named after its discoverer, is found in Utah and Colorado. Most of today's asphaltic materials are obtained from petroleum-still residues.

Cold molded-filled asphaltic compositions have been used for battery cases and electrical components. Hot-melt asphaltic compositions may be converted to non-

Newtonian fluids by passing air into the melt. Both blown and regular asphalt are used in highway construction, roofing, and flooring construction, and for waterproofing.

Rosin, the residue left in the manufacture of turpentine by distillation, is a mixture of the diterpene, abietic acid, and its anhydride. It is nonpolymeric but is used in the manufacture of synthetic resins and varnishes. Ester gum, a cross-linked ester, is obtained by the esterification of rosin with glycerol or pentaerythritol.

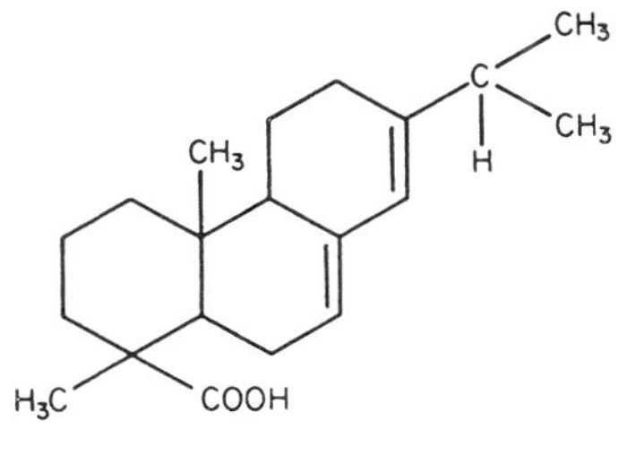

Abietic acid

Lignin is the major noncellulosic constituent of wood. Since its removal is the major step in the production of paper pulp, tremendous quantities of lignin are available as a byproduct of paper manufacture. Lignin is a complex polyphenolic, relatively low molecular weight polymer containing units similar to those described in Fig. 6.16. Its sulfonic acid derivative is used as an extender for phenolic resins, as a wetting agent for applications, such as oil drilling muds, and for the production of vanillin. This last use accounts for less than 0.01%, and all other uses account for less than 1% of this byproduct.

Many natural resins are fossil resins exuded from trees thousands of years ago. Recent exudates are called recent resins, and those obtained from dead trees are called semifossil resins. Humic acid is a fossil resin found with peat, brown coal, or lignite deposits throughout the world. It is a carboxylated phenoliclike polymer used as a soil conditioner, as a component of oil drilling muds, and as a scavenger for heavy metals.

Amber is a fossil resin found in the Baltic Sea regions, and sandarac and copals are found in Morocco and Oceania, respectively. Other copal resins, such as pon-

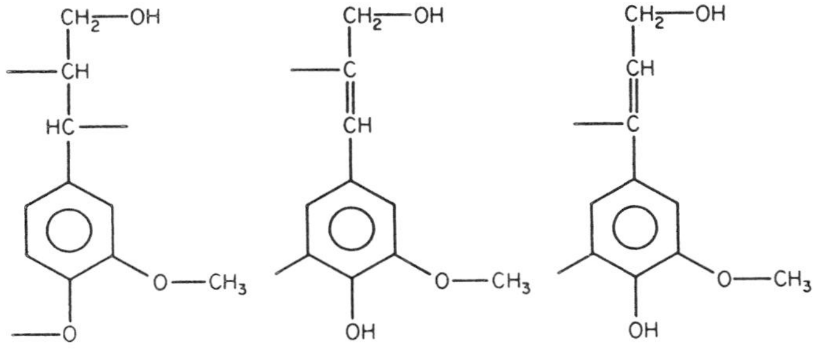

Figure 6.16 Possible units contained within lignin.

tiac, kauri, manila, congo, and batu, are named after the geographical location of the deposit.

Casein, a protein from milk, under the name of Galalith has been used as a molding resin and as an adhesive. Unsaturated oils found in linseed, soybean, safflower, tung, oiticica, and menhaden are used as drying oils in surface coatings at an annual rate of over 300,000 tons. Regenerated proteins include soybeans (glycine max), maize (zein), and ground nuts (arachin).

Synthetic polymeric fibers, plastics, elastomers, and coatings have displaced the natural polymeric products to a large extent. However, the latter are renewable resources independent of the supply of petroleum. Thus, their use will continue to increase as long as there is sufficient land available for the economical cultivation of both food and polymeric crops.

6.9 POLYMER STRUCTURE

In 1954, Linus Pauling received the Nobel Prize for his insights into the structure of materials—in particular, proteins. While the protein chain may assume an infinite number of shapes or conformations, due to essentially free rotation about the various covalent bonds in the chain, Pauling showed that only certain conformations are preferred because of intramolecular and intermolecular hydrogen bonding.

Two major secondary structures are found in nature—the helix (Fig. 6.4) and the sheet (Fig. 6.5). The helix is also a major structure for many synthetic flexible polymers since it can take advantage of both intermolecular secondary bonding and relief of steric constraints. Some materials utilize a combination of helix and sheet structures. Thus, wool consists of helical protein chains connected to give a "pleated" sheet.

Most isotactic polymers and polymers with bulky substituents exist as short-range helices in solution and in the helical conformation in the crystalline phase if sufficient cooling time is allowed. For isotactic vinyl polymers, alternate bond chains take trans and gauche positions. The direction of the twist is such as to relieve stearic hindrance generating a left- or right-handed helix.

The Watson and Crick model for the DNA double helix is only a generalized model to describe what we now know to be a wide variety of DNA structures. The DNA is actually a polyanion with the potential for a single negative charge per each repeat unit. Nature utilizes a combination of DNA structural elements to "code" messages. Much of the DNA structure acts as a "carrier" and "container." For instance, only about one in 10^4 sites are specific binding sites for six base-pair restriction enzymes.

For double-stranded DNA, structural elements include major and minor grooves, super coils, kinks, cruciforms, bends, loops, and triple strands (Fig. 6.17). Each of these structural elements can vary in length, shape, location, and frequency. Further, even the "simple" DNA double helix can vary in pitch per turn of helix, number of units per helix turn, sugar pucker conformation, and helical sense.

Helical sense refers to whether the helix is "left or right handed." The pentose sugar can be present in a wide variety of conformations. Two of these are shown in Fig. 6.18. Such variations can be responsible for variance in the orientation of the base pairs leading to varying pitch of the helix. The orientation of the base with respect to the sugar also varies. Usually the bases lie "away" or anti to the

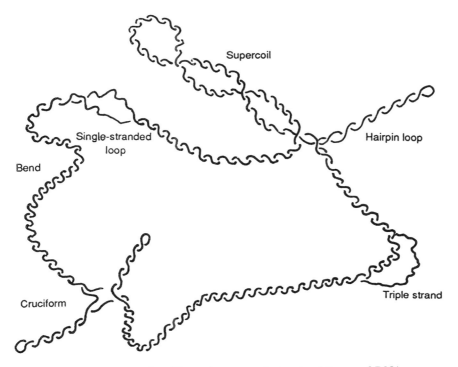

Figure 6.17 Representation illustrating unusual structural forms of DNA.

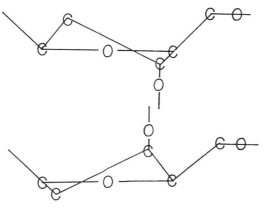

Figure 6.18 Two sugar "pucker" conformations found within DNA. The top form is commonly found in B-DNA, while the bottom form is often found in A-DNA.

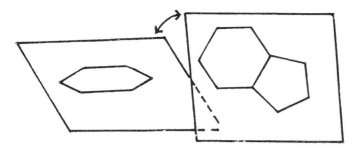

Figure 6.19 Illustration of twisting that occurs between paired bases in DNA.

sugar. Rotation of the base 180° about the glycosidic bond has the base "over" or syn to the sugar. Finally, while the bases pair typically in the mnemonic combination *Gee CAT*—i.e., guanine pairs with cytosine while adenine pairs with thymine—the planes of the paired bases vary to minimize steric requirements and bond angle constraints while maximizing hydrogen bonding (Fig. 6.19).

DNA's helical conformations thus far identified fall into three general groupings—A, B, and Z. Most DNAs are of the B form, which is a regular right-handed helix with the base pairs oriented approximately perpendicular to the helix axis giving well-defined major and minor grooves (Fig. 6.20). The B form has about a 33 Å pitch-per-helix turn with about 10 base pairs per turn and the base

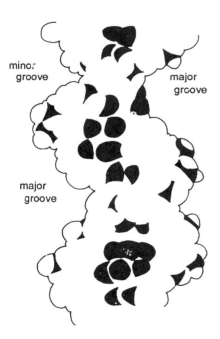

minor groove

major groove

major groove

Figure 6.20 Silhouette-emphasizing major and minor grooves that occur in B-DNA. Phosphorus atoms are darkened.

is situated anti to the sugar. The A form is found in fibers at low hydration and is mainly found in DNA-RNA hybrids and double-stranded RNA segments. Grooves of varying sizes are also present. There are about 10.7 base pairs per turn with about a 25 Å pitch-per-helix turn with the bases anti with respect to the sugar. It is also a right-handed helix. The Z-DNA is a left-handed helix with the cytosine base anti to the sugar, but the guanine base is syn to the sugar. The Z-DNA has about 12 base pairs and a pitch of about 46 Å per helix turn.

As noted before, other unusual conformations also exist, such as the loop, bend, and cruciform. It is now known that certain base sequences encourage each of these unusual conformations. Thus, a series of AT base pairs tends to "bend" the DNA.

This central theme concerning the major secondary structures found in nature is also illustrated with the two major polysaccharides derived from sucrose, i.e., cellulose and the major component of starch—amylose. Glucose exists in one of two forms (Fig. 6.21)—an alpha (α) and a beta (β) form, where the terms α and β refer to the geometry of the oxygen connecting the glucose ring to the fructose ring.

Cellulose is a largely linear homosaccharide of β-D-glucose. Because of the geometry of the β linkage, individual cellulose chains are typically found in nature to exist as sheets, the individual chains being connected through hydrogen bonding (Fig. 6.22). This sheet structure permits the existence of materials with good mechanical strength, allowing them to act as the structural units of most plants.

Amylose, one of the two major components of starch, is a linear polymer of the α-D-glucose. Its usual conformation is as a helix with six residues per turn (Fig. 6.23). Starch is a major energy source occurring in plants, usually as granules.

6.10 GENETIC ENGINEERING

In its broadest sense, genetic engineering refers to any artificial process that alters the genetic composition of an organism. Such alterations can be effected indirectly through chemical methods, through radiation, or through selective breeding. Today, the term usually refers to the processes whereby genes or portions of chromosomes are chemically altered.

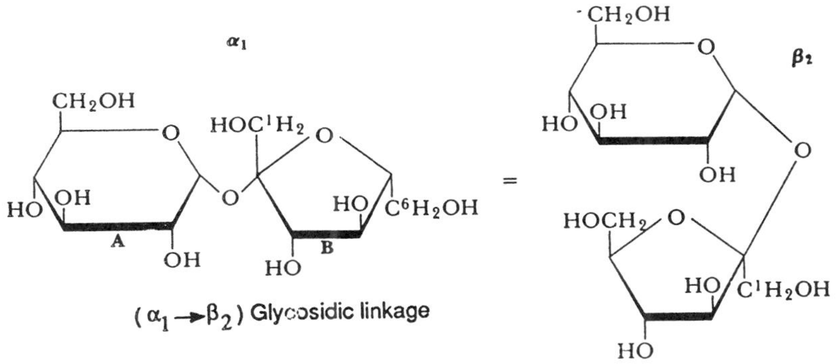

Figure 6.21 α (left) and β (right) forms of D-glucose in sucrose.

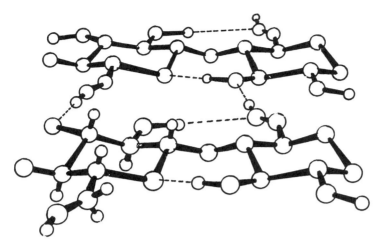

Figure 6.22 Sheetlike structural arrangement of cellulose derived from β-D-glucose units. Hydrogen bonding is noted by -----.

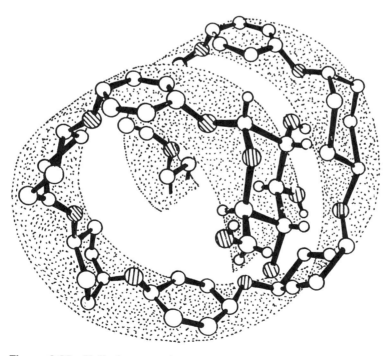

Figure 6.23 Helical structural arrangement of amylose derived from α-D-glucose units.

The term clone comes from the Greek word *klon*, meaning a cutting used to propagate a plant. Cell cloning is the production of identical cells from a single cell. In like manner, gene cloning is the production of identical genes from a single gene, introduced into a host cell. A gene is a chromosomal portion that codes for a single polypeptide or RNA. Gene splicing is currently practiced as the enzymatic attachment of one gene or gene segment to a gene or gene segment.

Enzymes (natural molecular catalysts, protein catalysts) have been found that act as designing tools for the genetic engineer. One of these enzyme groups consists of *restriction endonucleases*. These are highly specific enzymes that recognize specific series of base pairs up to six base pairs in length. They then can split the DNA at specific points within this region. Organisms produce restriction endonucleases that are specific for that organism. A second tool employed by the genetic engineer is the enzyme terminal transferase that adds deoxyribonuclease residues to the 3' end of DNA strands generating 3'-tails of a single type of residue.

The genetic engineer employs special modified plasmid DNAs called vectors or carriers. These circularly shaped vectors can reproduce autonomously in a host cell. Plasmids have two other important properties. First, they can pass from one cell to another allowing a single "infected" bacterial cell to inject neighboring bacterial cells. Second, gene material from other cells can be easily formed into plasmids, allowing ready construction of modified carriers. *Escherichia coli*, better known simply as *E. coli*, is typically employed as the host cell.

Chemically, a gene is but a segment of the DNA chain coded to direct the synthesis of a specific protein. A single DNA molecule often contains many genes. Combinations of these genes along the DNA form chromosomes. The precise steps involved are not difficult to acquire or to understand. Genes are composed of long, linear deoxyribonucleic acid, DNA chains; actually they are composed of two such DNA chains existing in a helical conformation (see Fig. 6.14).

These DNA chains can be manipulated in a variety of ways, but of major interest here is the production of recombinant DNA. For instance, DNA that performs a desired role, such as the production of insulin, is appropriately isolated. The recovered DNA is then spliced using molecular scissors, i.e., specific enzymes. The molecular engineer has a wide variety of these molecular scissors that will "cut" at known sites.

One group of these enzymes is called the restriction endonucleases, briefly described earlier. Certain restriction endonucleases cut double-stranded DNA asymmetrically in regions called palindromes, that is, regions that read (have the identical sequence) the same left to right on one strand as right to left on the other. This produces what is referred to as "sticky ends" that form not only a "cleft" for attachment but also a single-stranded end that has the ability to pair with another similarly produced strand. Both strands of the twin strand have identical coding and a tendency to recombine with complementary strands from another organism.

When mixed under the proper conditions in the presence of DNA-ligase, the lysed ends recombine and rapidly anneal in the correct sequence of base pairing. The hydrogen bonds that form reinforce the recombination reaction.

The resulting recombination products contain a wide variety of recombinant combinations as well as largely unrecombined fragments. This mixture is typically treated in one of two manners. The simplest case requires a chemical-resistant gene, i.e., resistant to an antibiotic such as tetracycline. The recombination product is plated onto a nutrient surface so that colonies can grow. These colonies are then treated with tetracycline. The survivors are transferred into the host organism so the new gene can express its protein.

In cases such as the synthesis of insulin, the recombination mixture is added to a host organism, such as a strain of *E. coli*. This infected mixture is then plated out. The individual colonies are then tested for insulin production. Those colonies that are potential producers are further plated out and grown for mass insulin production. Cells that take up or accept the recombinant DNA are called transformants. More specialized routines have been devised to increase the probability that the desired gene is incorporated into the host.

The steps involved in gene splicing, emphasizing the chemical nature of the individual steps, are as follows:

1.   Lysing, which in reality is simply the hydrolysis of DNA units as follows:

$$\begin{array}{ccc}
\overset{\displaystyle O}{\underset{\displaystyle O^-}{\overset{\displaystyle \|}{\sim\!\!\sim O-\!\!\underset{|}{P}-OH}}} + HO\!\!\sim\!\!\sim & \xrightarrow{-H_2O} & \overset{\displaystyle O}{\underset{\displaystyle O^-}{\overset{\displaystyle \|}{\sim\!\!\sim O-\!\!\underset{|}{P}-O\!\!\sim\!\!\sim}}}
\end{array}$$

2.   Construction of staggered (sticky) ends.
3.   Recombination or lysation.
4.   Chemical recombination of vector. (Insertion into host cell; recombining plasmid genes into the host genetic complement.)

     Steps 2, 3, and 4 can be considered chemically to be the reverse action to lysing. They are dehydration reactions that employ specific enzymes.

$$\begin{array}{ccc}
\overset{\displaystyle O}{\underset{\displaystyle O^-}{\overset{\displaystyle \|}{\sim\!\!\sim O-\!\!\underset{|}{P}-O\!\!\sim\!\!\sim}}} & \xrightarrow[\text{endonuclease}]{\overset{\displaystyle H_2O}{\text{Restriction}}} & \overset{\displaystyle O}{\underset{\displaystyle O^-}{\overset{\displaystyle \|}{\sim\!\!\sim O-\!\!\underset{|}{P}-OH}}} + HO\!\!\sim\!\!\sim
\end{array}$$

This reaction is analogous to esterification reactions including the sequence employed in the synthesis of polyesters.

5.   Replication is identical to plating out of bacteria, fungi, and cell lines.
6.   See step 2.
7.   Annealing of base pairs. Here the major chemical driving force is simply the formation of secondary bonding (hydrogen bonding) between preferred base pairs.
8.   Analysis and selection.

6.11 DNA PROFILING

DNA profiling, also called DNA typing or DNA fingerprinting, relies on technology developed within the past two decades. It is used in paternity identification, classification of plants, criminal cases, etc. We will concentrate on the use of DNA profiling in criminal cases. Here, questions of law become intertwined with statistical arguments and chemical behaviors.

DNA profiling is based on variations that exist in the DNA molecule between each individual's DNA (with the exception of identical siblings). Samples are generally derived from a person's blood, hair roots, semen, skin, and saliva.

Two major types of DNA profiling are the polymer chain reaction (PCR) and the restriction fragment length polymorphism (RFLP) approaches. The PCR approach utilizes a sort of molecular copying process. Here, a specific DNA region is selected for investigation. The PCR sequence requires just a few nanograms of DNA.

The second approach focuses on sections of the DNA called restriction fragment length polymorphisms. Here DNA, isolated from the cell nucleus, is broken into pieces by restriction enzymes that cleave the DNA at specific sites. Thus, each time this base sequence appears in the DNA portion under study, the enzyme will sever the DNA chain. Because of the variability in DNA of individuals, the distance (number of individual phosphate units) between "slices" vary. These spliced fragments are separated largely according to length (size) by gel electrophoresis and transferred to a backing material, where radiolabeled probes that bind only specific DNA sequences are added. The location of the binding is determined using X-ray film that is sensitive to the nuclear breakdown of the radioactive probes. The X-ray film is then developed and analyzed.

Scientists have a number of different restriction enzymes that cut DNA at different specific sites. Thus, the previous sequence can be repeated using a different restriction enzyme giving a different DNA profile for the same DNA sample. Each additional "sequencing," using a different restriction enzyme, decreases the probability that another person will have the same cumulative DNA profile. For instance, for one restriction enzyme the computer data bank may say that the possibility that another person would have the same DNA profile is 1 in 100 (1%), for a second restriction enzyme the probability may be 1 in 1000 (0.1%), and for a third restriction enzyme the probability may be 1 in 500 (0.2%). The probability that the cumulative DNA profile would produce a matching pattern is $0.01 \times 0.001 \times 0.02 = 0.0000002$ or 0.00002% or 1 part in 50,000,000. There is a caution to using the "multiplication rule," in that DNA sequences are not *totally* random. In fact, DNA sequence agreements generally diverge as one's ancestors are less closely related.

The RFLP method requires a sample about 100 times larger than is required for the PCR approach, but with repeated sequences using different restriction enzymes, RFLP is more precise.

It must be noted that factors leading to DNA degradation, such as moisture, chemicals, bacteria, sunlight, and heat, will also impact negatively on DNA profiling since the precise sequences and lengths of the DNA may be changed.

SUMMARY

1. Cellulose and starch, which are the most important polysaccharides, are hydrolyzed by acids to D-glucose, which is a monosaccharide. The disaccharides cellobiose and maltose are precursors of glucose in this hydrolysis. D-glucose exists in two different chair conformations called α- and β-D-glucose. All the hydroxyl groups in the β form are equatorial, and all are equatorial in the α form except the hydroxyl group on carbon 1, which is axial. Cellobiose is a dimer and cellulose is

a polymer of β-D-glucose, and maltose is a dimer and starch is a polymer of α-D-glucose.

2. Cellulose is a linear polydisperse polymer whose molecules are often present as fibrils, because of strong intermolecular hydrogen bonding. Soda cellulose, prepared by the action of strong caustic on cellulose, reacts with carbon disulfide to produce a soluble xanthate and with alkylhalides to produce ethers. Regenerated cellulose in the form of filaments (rayon) or sheets (cellophane) is produced when soluble cellulose xanthate (viscose) is precipitated in aqueous acid solutions.

3. A celluloselike product consisting of repeat units of acetylated D-glucosamine is found in the exoskeleton of shellfish. Amylose is a linear polymer, while amylopectin is a branched polymer. The former forms a blue occlusion compound with iodine. Glycogen is a readily hydrolyzable highly branched polymer of D-glucose.

4. Proteins are polyamides consisting of some 20 different α-amino acids joined through peptide linkages:

$$\begin{array}{c} \quad\; H \qquad\; H \\ \quad\; | \qquad\quad | \\ -N-C-C- \\ \quad\;\; \| \quad\;\; | \\ \quad\;\; O \quad\;\; R \end{array}$$

Since α-amino acids exist as dipolar ions, they are insoluble and high melting and do not migrate to charged poles at their isoelectric points. Since they do migrate at other pH values, the process of electrophoresis may be used to separate amino acids and proteins. Proteins with high intermolecular hydrogen bonding are fibrillar, and those with intramolecular hydrogen bonding are globular. Proteins may be hydrolyzed by weak mineral acids to α-amino acids, and the latter may be separated by paper chromatography and identified by the addition of ninhydrin.

5. The sequence of amino acids in the primary structure (configuration) may be determined by reactions with the amine or carboxyl end groups followed by controlled hydrolysis. The secondary structure or conformation may be a right-handed intramolecularly hydrogen-bonded helix or an intermolecularly bonded pleated sheet. The shape or folding in the tertiary structure is dependent on the presence of sulfur-sulfur cross-links. Polypeptides may be synthesized by protecting one of the functional groups before condensation. The preferred solid-phase polymerization technique takes place on the surface of cross-linked polystyrene beads.

6. The nucleic acid separated from nucleoproteins is a nucleotide, which may be hydrolyzed to phosphoric acid and a nucleoside. The latter consists of ribose (in RNA) or deoxyribose (in DNA) and purine and pyrimidine bases. All nucleic acids contain adenine, guanine, and cytosine. RNA contains uracil, and DNA contains thymine (3-methyluracil). The polymer backbone consists of phosphate esters with the hydroxyl groups in the 3 and 5 positions. The RNA polymer forms a single helix, but DNA forms a double helix capable of replication. The double helix is held together by the hydrogen bonds of purine-pyrimidine pairs from separate chains. The allowable pairs are guanine-cytosine and adenine-thymine. RNA exists in three different molecular weights, namely, messenger, transfer, and ribosomal

RNA. The 64 possible DNA trinucleotides provide 64 codons, which specify the amino acids used in protein synthesis.

7. The hard naturally occurring plastics, balata and gutta percha, are trans isomers of the naturally occurring elastic cis-1,4-polyisoprene, which is natural rubber (*Hevea brasiliensis*). The polymer chain of amorphous natural rubber and other elastomers uncoils during stretching and returns to its original low-entropy form above its T_g. However, if it is held in the stretched position for long periods of time, a relaxation process takes place before it returns to its original length.

8. Slippage of chains is prevented but elasticity is retained when a few cross-links are present between the principal sections of elastomers. The original low-modulus stretching process is related to a change in internal energy per unit length. The more important high-modulus stretching process which follows is related to a change in entropy, resulting from the alignment of chains and crystallization.

9. Unlike other solids, stretched rubber contracts when heated. The long-range elasticity of rubber is dependent on the absence of strong intermolecular forces, the presence of a pendant methyl group, and a cis configuration, which aid the crankshaft-type motion. In addition to sulfur, the compounding ingredients used in rubber include accelerators, antioxidants, and fillers, such as carbon black.

10. The major chemical steps involved in gene splicing are hydrolysis and esterification.

GLOSSARY

A: an abbreviation for adenine.

accelerator: A catalyst for the vulcanization of rubber.

acetal linkage: The linkage between anhydroglucose units in polysaccharides.

adenine: A purine base present in nucleic acids.

adenosine: A nucleoside based on adenine.

adenosine phosphates: Phosphoric acid esters of adenosine.

adenylic acid: A nucleotide based on adenine.

ADP: adenosine diphosphate

alkali cellulose: Cellulose that has been treated with a strong caustic solution.

α-amino acid: A carboxylic acid with an amino group on the carbon next to the carboxyl group.

α arrangement: That present in an α helix.

α-cellulose: Cellulose that is insoluble in a 17.5% caustic solution.

α helix: A right-handed helical conformation.

amorphous silica: A filler used with polymers.

AMP: Adenosine monophosphate.

amylopectin: A highly branched starch polymer with branches or chain extensions on carbon 6 of the anhydroglucose repeating units.

amylose: a linear starch polymer.

antioxidant: A compound that retards the degradation of a polymer.

ATP: Adenosine triphosphate.

axial bonds: Bonds that are perpendicular and above or below the plane of the hexose ring, such as the hydroxyl group on carbon 1 in α-D-glucose.

balata: A naturally occurring plastic (trans-1,4-polyisoprene).

β arrangement: A pleated sheetlike conformation.

β-cellulose: Cellulose soluble in 17.5% but insoluble in 8% caustic solution.

Boeseken-Haworth projections: Planar hexagonal rings used for simplicity, instead of the staggered chain forms, to show the structure of saccharides such as D-glucose.

C: An abbreviation for cytosine.

carbohydrate: An organic compound with an empirical formula of CH_2O.

carbon black: Finely divided carbon used for the reinforcement of rubber.

casein: Milk protein.

cellobiose: A disaccharide consisting of two D-glucose units joined by a β-acetal linkage. It is the repeating unit in the cellulose molecule.

cellophane: A sheet of cellulose regenerated by the acidification of an alkaline solution of cellulose xanthate.

cellulose: A linear polysaccharide consisting of many anhydroglucose units joined by β-acetal linkages.

cellulose xanthate: The reaction product of soda cellulose and carbon disulfide.

chair form: The most stable conformation of a six-membered ring like D-glucose, named because of its slight resemblance to a chair in contrast to the less stable boat conformation.

chitin: A polymer of acetylated glucosamine present in the exoskeletons of shellfish.

chromatography: A separation technique based on the selective absorption of the components present.

codon: A trinucleotide with three different bases which provides the necessary information for protein synthesis.

collagen: A protein present in connective tissue.

compounding: The process of adding essential ingredients to a polymer such as rubber.

conjugated protein: The combination of a protein and a prosthetic group, such as a nucleic acid.

C-terminal amino acid: Amino acid with a carboxylic acid end group.

cytosine: A pyrimidine base present in nucleic acids.

degree of substitution (DS): A number that designates the average number of reacted hydroxyl groups in each anhydroglucose unit in cellulose or starch.

denaturation: The change of conformation of a protein resulting from heat or chemical reactants.

deoxyribonucleic acid (DNA): A nucleic acid in which deoxyribose units are present.

dextran: A high molecular weight branched polysaccharide synthesized from sucrose by bacteria.

dextrin: Degraded starch.

D-glucose: A hexose obtained by the hydrolysis of starch or cellulose.

dipeptide: A dimer formed by the condensation of two amino acids.

disaccharide: A dimer of D-glucose or other monosaccharide, such as cellobiose.

DNA: Deoxyribonucleic acid.

DNA profiling: Identification method based on variations between individual's DNA.

drying: Jargon used to describe the cross-linking of unsaturated polymers in the presence of air and a heavy metal catalyst (drier).

DS: Degree of substitution.

enzyme: A protein with specific catalytic activity.

equatorial bonds: Bonds that are essentially parallel to the plane of the hexose ring, such as those in β-D-glucose.

ester gum: The ester of rosin and glycerol.

fibrillar protein: A hairlike, insoluble, intermolecularly hydrogen-bonded protein.

fibrils: Naturally occurring threadlike strands or bundles of fibers.

fossil resins: Resins obtained from the exudate of prehistoric trees.

G: An abbreviation for guanine.

Galalith: Commercial casein plastics.

γ-cellulose: Cellulose and derivatives soluble in 8% caustic solution.

globular: Nonfibrous soluble proteins containing intramolecular hydrogen bonds.

glycine: The simplest and only nonchiral α-amino acid.

glycogen: A highly branched polysaccharide which serves as the reserve carbohydrate in animals.

guanine: A purine base present in nucleic acids.

guayule: A shrub containing rubber (cis-1,4-polyisoprene).

gutta percha: A naturally occurring plastic (trans-1,4-polyisoprene).

Hevea brasiliensis: Natural rubber.

hexosans: Polysaccharides in which the repeating units are six-membered (pyranose) rings.

hormones: Organic compounds having specific biological effects.

humic acid: A polymeric aromatic carboxylic acid found in lignite.

hydrocellulose: Cellulose degraded (depolymerized) by hydrolysis.

isoelectric point: The pH at which an amino acid does not migrate to either the positive or the negative pole in a cell.

keratin: A fibrillar protein.

latex: A stable dispersion of polymer particles in water.

lignin: The noncellulosic resinous component of wood.

long-range elasticity: The theory of rubber elasticity.

macroradical: A high molecular weight free radical.

maltose: A disaccharide consisting of two D-glucose units joined by an α-acetal linkage; the repeating unit in the starch molecule.

mer: The repeating unit in a polymeric chain.

mercerized cotton: Cotton fiber that has been immersed in caustic solution, usually under tension, and washed with water to remove the excess caustic.

monosaccharide: A simple sugar, such as D-glucose, which cannot be hydrolyzed further.

myosin: A protein present in muscle.

native cellulose: Naturally occurring cellulose.

ninhydrin: A triketohydrindene hydrate which reacts with α-amino acids to produce characteristic blue to purple compounds.

NR: Symbol for natural rubber.

N-terminal amino acid: Amino acid with an amino end group.

nucleoside: The product obtained when phosphoric acid is hydrolyzed off a nucleotide.

nucleotide: A nucleic acid.

oligosaccharide: A relatively low molecular weight polysaccharide.

oxycellulose: Cellulose degraded (depolymerized) by oxidation.

paper chromatography: A technique used for separating amino acids in which paper serves as the stationary phase.

pentosans: Polysaccharides in which the repeating units are five-membered (furanose) rings.

polyglycuronic acids: Also known as uronic or glycuronic acids, naturally occurring polysaccharides in which the hydroxyl on carbon 6 has been oxidized to form a carboxylic acid group.

polypeptide: A protein; often used for low molecular weight proteins.

polysaccharide: A polymer consisting of many hexose units, such as anhydroglucose.

primary structure: A term used to describe the primary configuration present in a protein chain.

prosthetic group: A nonproteinous group conjugated with a protein. Derived from the Greek *prosthesic*, in addition.

protein: A polyamide in which the building blocks are α-amino acids joined by peptide linkages.

purine base: Compounds consisting of two fused heterocyclic rings, namely, a pyrimidine and an imidazole ring.

pyrimidine: A 1,3-diazine.

racked rubber: Stretched rubber cooled below its T_g.

rayon: Cellulose regenerated by acidification of a cellulose xanthate (viscose) solution.

recent resins: Resins obtained from the exudate of living trees.

regenerated cellulose: Cellulose obtained by precipitation from solution.

retrogradation: A process whereby irreversible gel is produced by the aging of aqueous solutions of amylose starch.

ribonucleic acid (RNA): A nucleic acid in which ribose units are present.

RNA: Ribonucleic acid: m-RNA, messenger RNA; r-RNA, ribosonal RNA; t-RNA, transfer RNA.

rosin: A naval stores product consisting primarily of abietic acid anhydride.

Schweitzer's solution: An ammoniacal solution of copper hydroxide.

secondary structure: A term used to describe the conformation of a protein molecule.

semifossil resins: Resins obtained from the exudate of dead trees.

shellac: A natural polymer obtained from the excreta of insects in Southeast Asia.

soda cellulose: Cellulose that has been treated with a strong caustic solution.

starch: A linear or branched polysaccharide consisting of many anhydroglucose units joined by α-acetal linkages. Amylose starch is a linear polymer, while amylodextrin is a branched polymer.

T: An abbreviation for thymine.

tenacity: A term for the tensile strength of fibers.

terpene: A class of hydrocarbons having the empirical formula C_5H_8.

tertiary structure: The shape or folding of a protein resulting from sulfur-sulfur cross-links.

thymine: A pyrimidine base present in DNA.

U: An abbreviation for uracil.

uracil: A pyrimidine base present in RNA.

viscose: An alkaline solution of cellulose xanthate.

zwitterion: a dipolar ion of an amino acid:

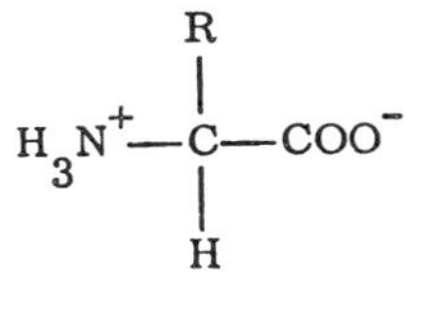

$$H_3N^+ - \underset{\underset{H}{|}}{\overset{\overset{R}{|}}{C}} - COO^-$$

EXERCISES

1. Why is starch digestible by humans? Why is cellulose not digestible by humans?

2. How does cellobiose differ from maltose?

3. Why is cellulose stronger than amylose?

4. How does the monosaccharide hydrolytic product of cellulose differ from that of starch?

5. Which has the higher molecular weight: (a) α-, or (b) β-cellulose?

6. How many hydroxyl groups are present on each anhydroglucose unit in cellulose?

7. Which would be more polar—tertiary or secondary cellulose acetate?

8. Why would you expect chitin to be soluble in hydrochloric acid?

9. Which is more apt to form a helix: (a) amylose, or (b) amylopectin?

10. Why is amylopectin soluble in water?

11. Define a protein in polymer science language.

12. Which α-amino acid does not belong to the L series?

13. To which pole will an amino acid migrate at a pH above its isoelectric point?

14. Why is collagen stronger than albumin?

15. What are the requirements of a strong fiber?

16. Which protein would be more apt to be present in a helical conformation: (a) a linear polyamide with small pendant groups, or (b) a linear polyamide with bulky pendant groups?

17. What is the difference between the molecular weight of (a) ribose, and (b) deoxyribose?

18. What is the repeating unit in the polymer chain of DNA?

19. Which is more acidic: (a) a nucleoside, or (b) a nucleotide?

20. What base found in DNA is not present in RNA?

21. Why would you predict helical conformations for RNA and DNA?

22. If the sequence of one chain of a double helix is ATTACGTCAT, what is the sequence of the adjacent chain?

23. Why is it essential to have trinucleotides rather than dinucleotides as codons for directing protein synthesis?

24. How do the configurations differ for (a) gutta percha, and (b) natural rubber.

25. What is the approximate Poisson's ratio of rubber?

26. Will the tensile force required to stretch rubber increase or decrease as the temperature is increased?

27. Does a stretched rubber band expand or contract when heated?

28. List three requirements for an elastomer.

29. Why is there an interest in the cultivation of guayule?

30. Are the polymerization processes for synthetic and natural cis-polyisoprene (a) similar, or (b) different?

31. What does the production of levulinic aldehyde as the product of the ozonolysis of natural rubber tell you about the structure of NR?

32. Why doesn't cold racked rubber contract readily?

33. Why does a rubber band become opaque when stretched?

34. What is the most important contribution to retractile forces in highly elongated rubber?

35. What is present in so-called vulcanized rubber compounds?

36. What happens to a free radical, such as (·SR), during the mastication of rubber?

37. Why aren't natural plastics used more?

38. What type of solvent would you choose for shellac?

39. What type of solvent would you choose for asphalt?

40. Which is a polymer: (a) rosin, or (b) ester gum?

41. If the annual production of paper is 100 million tons, how much lignin is discarded annually?

42. Is an article molded from Galalith valuable?

BIBLIOGRAPHY

Abelson, J., Butz, E. (eds.) (1980): Recombinant RNA, Science, *209*.

Anderson, W., Diacumakos, E. (1981): Genetic engineering in mammalian cells, Sci. Am., *245*:106.

Andrade, J. D. (1985): *Surface and Interfacial Aspects of Biomedical Polymers*, Plenum, New York.

Asimov, I. (1962): *The Genetic Code*, New American Library, New York.

Atkins, E. D. (1986): *Polysaccharides*, VCH Pubs.

Carraher, C. E., Moore, J. A. (1983): *Modification of Polymers*, Plenum, New York.

Carraher, C. E., Sperling, L. (1983): *Polymer Applications of Renewable-Resource Materials*, Plenum, New York.

Carraher, C. E., Tsuda, M. (1980): *Modification of Polymers*, ACS Symposium Series, Washington, D.C.

Cohn, D., Kost, J. (1993): *Biomedical Polymers: Molecular Design to Clinical Applications*, Elsevier, New York.

Comper, W. D. (1981): *Heparin (and Related Polysaccharides)*, Gordon and Breach, New York.

Dubin, P. (1993): *Macromolecular Complexes in Chemistry and Biology*, Springer-Verlag, New York.

Dusek, K. (1982): *Polymer Networks*, Springer-Verlag, New York.

Edelman, P., Wang, J. (1992): *Biosensors and Chemical Sensors: Optimizing Performance through Polymeric Materials*, ACS, Washington, D.C.

Gebelein, C. G. (1992): *Biotechnology and Polymers*, Plenum, New York.

Gebelein, C. G. (1993): *Biotechnological Polymers: Medical, Pharmaceutical, and Industrial Applications*, Technomic,

Gebelein, C. G., Carraher, C. E. (1994): *Biotechnology and Bioactive Polymers*, Plenum, New York.

Hastings, G., Ducheyne, P. (1983): *Macromolecular Biomaterials*, CRC Pub., Boca Raton, Florida.

Haward, R. N. (1982): *Network Formation and Cyclization in Polymer Reactions*, Applied Science Pubs., London.

Helene, C. (1983): *Structure, Dynamics, Interactions, and Evolution of Biological Macromolecules*, Reidel Pub. Co., Dordrecht, The Netherlands.

Hon, D. (1982): *Graft Copolymerization of Lignocellulosic Fibres*, ACS, Washington, D.C.

Jurnak, F., McPherson, A. (1988): *Biological Macromolecules Assemblies*, 3 vols., Wiley, New York.

Kaplan, D. (1994): *Silk Polymers: Materials Science and Biotechnology*, ACS, Washington, D.C.

Kennedy, J., Mitchell, J., Sandford, P. (1995): *Carbohydrate Polymers*, Elsevier, Tarrytown, New York.

Knorre, D. G., et al. (1989): *Reactive Oligonucleotide Derivatives as Tools for Site Specific Modification of Biopolymers*, Gordon & Breach, New York.

Kramer, O. (1986): *Biological and Synthetic Polymer Networks*, Elsevier, New York.

Lane, D. (1992): *Heparin and Related Polysaccharides*, Plenum, New York.

Luzio, J. P., Thompson, R. J. (1990): *Macromolecular Aspects of Medical Biochemistry*, Cambridge University Press, Cambridge, England.

Mazzarelli, R. A. A. (1974): *Natural Chelating Polymers*, Pergamon, New York.

Merrifield, R. B. (1975): Solid phase peptide synthesis, Polymer Prep., *16*:135.

Meyer, K. H. (1959): *Natural and Synthetic High Polymers*, 2d ed., Interscience, New York.

Mullis, K., Gibbs, R. (1994): *The Polymerase Chain Reaction*, Birkhauser,

Setlow, J., Hollaender, A. (1979): *Genetic Engineering: Principles and Methods*, Plenum, New York.

Seymour, R. B. (1972): *Introduction to Polymer Chemistry*, McGraw-Hill, New York.

———. (1975): *Modern Plastics Technology*, Chap. 14, Reston, Reston, Virginia.

Seymour, R. B., Johnson, E. J. (1977): Solutions of cellulose in organic solvents, Chap. 19 in *Structure-Solubility Relations in Polymers* (F. W. Harris and R. B. Seymour, eds.), Academic, New York.

———. (1978): Acetylation of cellulose solutions. J. Polymer Sci. Chem. Ed., *16*:1.

Soane, D. S. (1992): *Polymer Applications for Biotechnology: Macromolecular Separation and Identification*, Prentice-Hall, Englewood Cliffs, New Jersey.

Watson, J., Tooze, J. (1981): *The DNA Story: A Documentary History of Gene Cloning*, Freeman, San Francisco.

Weaver, R. (1984): Changing life's genetic blueprint, *National Geographic*, Dec., Washington, D.C.

Whistler, R. L., BeMiller, J., Paschall, E. (1984): *Starch*, Academic, New York.

Wyman, J., Gill, S. (1990): *Binding and Linkage: Functional Chemistry of Biological Macromolecules*, University Science Books, Mill Valley, California.

Yalpani, M. (1988): *Polysaccharides*, Elsevier, N.Y.

7
Step-Reaction Polymerization or Polycondensation Reactions

7.1 COMPARISON BETWEEN POLYMER TYPE AND KINETICS OF POLYMERIZATION

There is a large but not total overlap between the terms condensation polymers and stepwise kinetics and the terms addition (or sometimes the term vinyl) polymers and chain kinetics. In this section we will describe briefly each of these four terms and illustrate their similarities and differences.

The terms addition and condensation polymers were first proposed by Carothers and are based on whether the repeat unit of the polymer contains the same atoms as the monomer. An addition polymer has the same atoms as the monomer in its repeat unit.

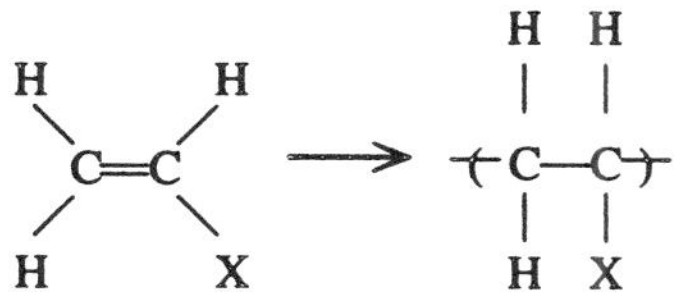

$$
\begin{array}{c}
\text{H} \qquad\quad \text{H} \\
\diagdown \qquad \diagup \\
\text{C}\!=\!\text{C} \\
\diagup \qquad \diagdown \\
\text{H} \qquad\quad \text{X}
\end{array}
\quad\longrightarrow\quad
\begin{array}{c}
\text{H} \quad \text{H} \\
| \quad\; | \\
\text{+C}\!-\!\text{C+} \\
| \quad\; | \\
\text{H} \quad \text{X}
\end{array}
\qquad\qquad (7.1)
$$

The atoms in the polymer backbone are usually carbon atoms. Condensation polymers contain fewer atoms within the polymer repeat unit than the reactants because of the formation of byproducts during the polymerization process, and the polymer backbone usually contains atoms of more than one element.

$$
\text{X—A—R—A—X} + \text{Y—B—R}'\text{—B—Y} \rightarrow \text{(A—R—A—B—R}'\text{—B)}_n \qquad (7.2)
$$

where A—X can be

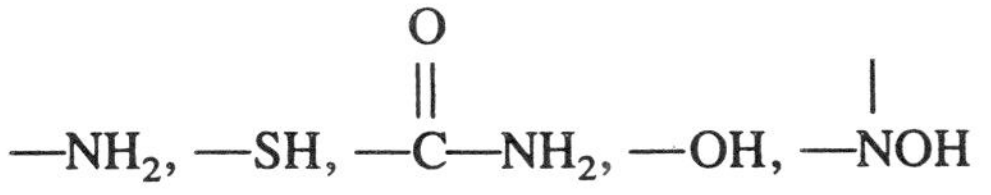

$$
\text{—NH}_2, \text{—SH}, \overset{\displaystyle \overset{O}{\|}}{\text{—C}}\text{—NH}_2, \text{—OH}, \overset{\displaystyle |}{\text{—NOH}}
$$

211

and Y—B can be

$$
-\overset{\overset{\displaystyle O}{\|}}{\underset{\underset{\displaystyle O}{\|}}{C}}-OH, \quad -\overset{\overset{\displaystyle O}{\|}}{\underset{\underset{\displaystyle O}{\|}}{C}}-Cl, \quad -\overset{\overset{\displaystyle O}{\|}}{\underset{\underset{\displaystyle R}{|}}{P}}-Cl, \quad -\overset{\overset{\displaystyle O}{\|}}{\underset{\underset{\displaystyle O}{\|}}{S}}-Cl
$$

The corresponding reactions can then be called addition polymerizations and condensation polymerizations.

The term stepwise kinetics, or step-growth kinetics, refers to polymerizations in which the polymer's molecular weight increases in a slow, steplike manner as reaction time increases.

Polyesterification in a bulk polymerization process will be utilized to illustrate stepwise growth. Polymer formation begins with one dialcohol (diol) molecule reacting with one diacid molecule, forming what we will call one repeat unit of the eventual polyester.

$$
\begin{array}{c}
O \quad\quad O \\
\| \quad\quad \| \\
HO-C-R-C-OH \;+\; HO-R'-OH \;\xrightarrow{-H_2O}\; HO-C-R-C-O-R'-OH
\end{array} \tag{7.3}
$$

This ester unit can now react with either an alcohol or acid group producing chains ending with either two active alcohol functional groups or two active acid groups.

$$
\begin{array}{l}
O \quad O \\
\| \quad\; \| \\
HO-C-R-C-O-R'-OH \;+ \\
\\
\quad\quad\quad \overset{O\;\;\;O}{\overset{\|\;\;\|}{HO-C-R-C-OH}} \xrightarrow{-H_2O} HO-C-R-C-O-R'-O-C-R-C-OH \\
\\
\quad\quad\quad HO-R'-OH \xrightarrow{-H_2O} HO-R'-O-C-R-C-O-R'-OH
\end{array} \tag{7.4}
$$

The chain with the two alcohol end groups can now condense with a molecule containing an acid end group, while the molecule with two acid end groups can react with a molecule containing an alcohol functional group. This reaction continues through the monomer matrix wherever molecules of the correct functionality having the necessary energy of activation and correct geometry collide. The net effect is that dimer, trimer, tetramer, etc., molecules are formed as the reaction progresses.

$$
\begin{array}{l}
O \quad\quad O \quad\quad\quad\quad\quad\quad O \quad\quad O \\
\| \quad\;\; \| \quad\quad\quad\quad\quad\quad \| \quad\;\; \| \\
HO-R'-O-C-R-C-O-R'-OH + HO-C-R-C-OH \xrightarrow{-H_2O} \\
\\
HO-R'-O-C-R-C-O-R'-O-C-R-C-OH \xrightarrow{etc.}
\end{array} \tag{7.5}
$$

$$
\begin{array}{l}
O \quad\quad O \quad\quad\quad\quad O \quad\quad O \\
\| \quad\;\; \| \quad\quad\quad\quad \| \quad\;\; \| \\
HO-C-R-C-O-R'-O-C-R-C-OH + HO-R'-OH \xrightarrow{-H_2O} \\
\\
HO-C-R-C-O-R'-O-C-R-C-O-R'-OH \xrightarrow{etc.}
\end{array} \tag{7.6}
$$

Thus the reactants are consumed with very few long chains being formed throughout the system until the reaction progresses toward total reaction of the chains with themselves. The molecular weight of the total system increases slowly in a stepwise manner. Considering A molecules to be diacid molecules and B to represent dialcohol molecules (diols), we can construct a system containing 10 each of A and B, and assuming a random number of condensations of unlike functional groups, we can calculate changes in DP, maximum DP, and the percentage of unreacted monomer (Fig. 7.1). The maximum DP for this system is 10 AB units. For this system, while the percentage of reacted monomer increases rapidly, both system $\overline{DP}$ and highest DP increase slowly. Figures 7.1 and 7.3 graphically show this stepwise growth as a function of both reaction time and reaction temperature T. It should be noted that the actual plot of molecular weight as a function of reaction time depends on the particular kinetic dependency for the particular system in question and need not be linear (Fig. 7.2).

Chain growth reactions require initiation to begin chain growth. Here, the initiation of a styrene molecule in a bulk reaction system occurs by means of a free radical initiator R•. This free radical quickly adds to a styrene monomer, shifting the unpaired, free electron away from the adding free radical. This new active chain, which contains an unpaired electron, adds to another styrene monomer, with the unpaired electron shifting toward the chain end. This continues again and again, eventually forming a long polystyrene chain.

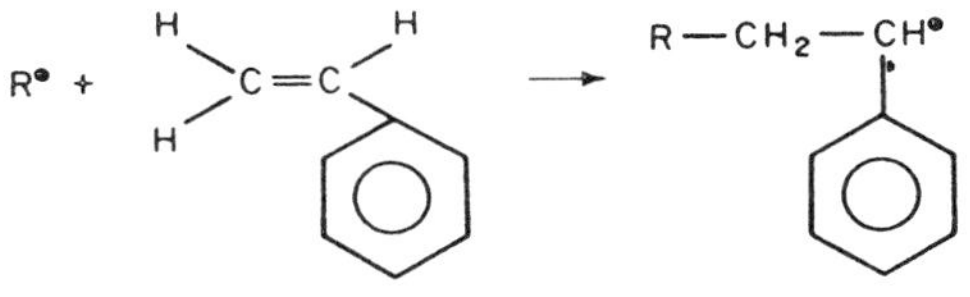

Initiation (7.7)

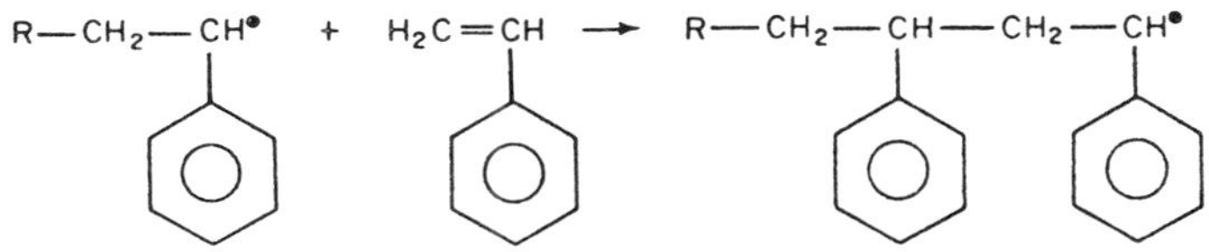

(7.8)

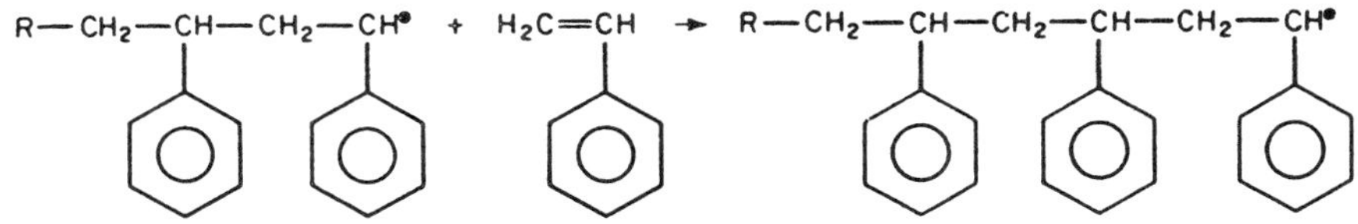

(7.9)

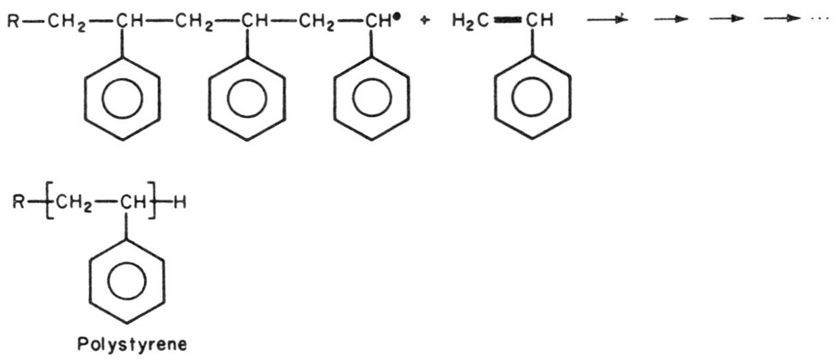

(7.10)

Polystyrene

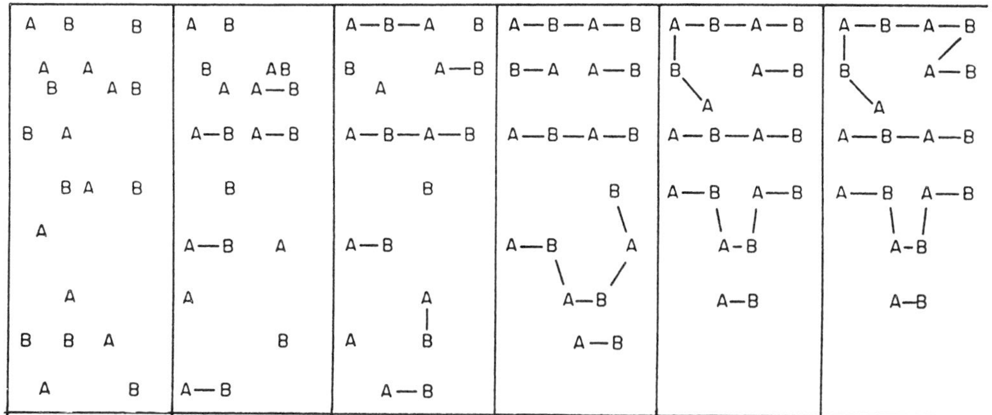

A B	A B	A—B—A B	A—B—A—B	A—B—A—B	A—B—A—B
A A	B AB	B A—B	B—A A—B	A—B	A—B
B A	A A—B			A	A
B A	A—B A—B	A—B—A—B	A—B—A—B	A—B—A—B	A—B—A—B
B A	A—B	B		A—B	A—B
			B A	A—B	A—B
A—B	B	A A—B—A	A—B A	A—B A—B	A A—B
A	A	A—B	A—B		A—B
A	A	B	A—B	A—B	A—B
B B A	A	A			
		A—B	A—B	A—B	A—B
A B	A—B				

% Monomer Unreacted	100	50	25	0	0	0
% Functional Groups Unreacted	100	75	55	30	25	20
System Average DP*	0	1	1.25	1.67	2	2.5
Highest DP	0	1	2	3	3	4

Figure 7.1 Chain-length dependency on reaction time and extent of monomer reaction for a stepwise kinetic model. Since DP refers to reacted units that compose a polymer and since each repeat unit is composed of one A and one B, the repeat unit is AB. By comparison, Fig. 7.4 represents the homopolymerization of vinyl units, and here each repeat unit is represented by one reacted A unit. *Includes only reacted portion.

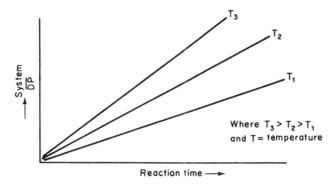

Figure 7.2 System molecular weight for stepwise kinetics as a function of reaction time and reaction temperature.

This growth continues until some termination reaction renders the chain inactive. Since polymerization occurs only with active chains (i.e., chains containing an unpaired electron), and since the concentration of growing chains is maintained at a low concentration at any given time, long chains are formed rapidly, and most of the styrene monomer remains unreacted. Again, we can construct a system illustrating the dependency of chain length on reaction extent and reaction time. Let us consider a system containing 20 styrene molecules (since styrene can react with itself the system can be constructed using only styrene molecules) (Fig. 7.4). While the $\overline{DP}$ for the entire system does not increase at a markedly different rate than for the stepwise kinetic model, the DP for the longest chain increases rapidly. Further, the number of unreacted monomers remains high throughout the system, and in actual bulk and solution polymerizations of styrene, the number of unreacted styrene monomers far exceeds the number of polymer chains until very high conversion. For chain-growth polymerizations it is usual to plot molecular weight only

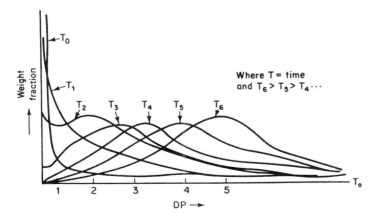

Figure 7.3 Molecular weight distribution for stepwise kinetics as a function of extent of reaction.

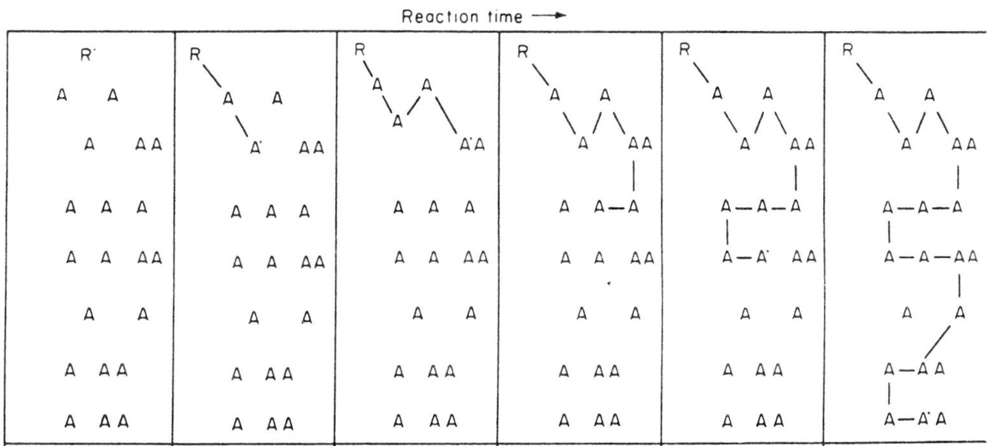

% Unreacted	100	90	80	70	55	25
Highest DP	0.5	2	4.0	6.0	9.0	15.0

Figure 7.4 Molecular weight for chain-growth kinetics as a function of reaction time and extent of reaction. Please note that each repeat unit is represented by one reacted A unit. R is not included in the calculations since it represents the initiator radical in this system.

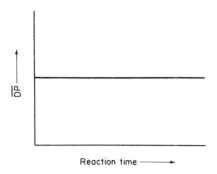

Figure 7.5 Idealized average molecular weight of formed polymers as a function of reaction time for chain-growth kinetics.

of materials that had become free radicals as a function of reaction time, giving plots such as those in Figs. 7.5 and 7.6. Both average chain length and molecular weight distribution remain approximately constant throughout much of the polymerization.

Most addition polymers are formed from polymerizations exhibiting chain-growth kinetics. This includes the typical polymerizations of the vast majority of vinyl monomers, such as ethylene, styrene, vinyl chloride, propylene, methyl acrylate, and vinyl acetate. Further, most condensation polymers are formed from systems exhibiting stepwise kinetics. Industrially this includes the formation of polyesters and polyamides. Thus, there is a large overlap between the terms addition polymers and chain-growth kinetics and the terms condensation polymers and stepwise kinetics. The following are examples illustrating the lack of adherence to the above.

1. The formation of polyurethanes and polyureas typically occurs in the bulk solution through kinetics that are clearly stepwise, and the polymer backbone is heteroatomed, yet there is no byproduct released through the condensation of the isocyanate with the diol or diamine because condensation occurs through internal

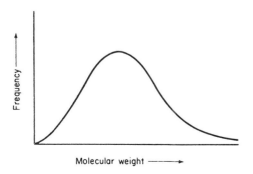

Figure 7.6 Molecular weight distribution for chain-growth kinetics.

rearrangement and shift of the hydrogen—neither necessitating expulsion of a byproduct.

$$OCN\text{-}(CH_2)_6NCO + HO\text{-}(CH_2)_5OH \longrightarrow$$

$$\text{-}(\overset{O}{\overset{\|}{C}}\text{—}\overset{H}{\overset{|}{N}}(CH_2)_6\overset{H}{\overset{|}{N}}\text{—}\overset{O}{\overset{\|}{C}}\text{—}O\text{-}(CH_2)_5O)_n \qquad (7.11)$$

Polyurethane

$$OCN\text{-}(CH_2)_6NCO + H_2N\text{-}(CH_2)_5NH_2 \longrightarrow$$

$$\text{-}(\overset{O}{\overset{\|}{C}}\text{—}\overset{H}{\overset{|}{N}}\text{-}(CH_2)_6\overset{H}{\overset{|}{N}}\text{—}\overset{O}{\overset{\|}{C}}\text{—}\overset{H}{\overset{|}{N}}\text{-}(CH_2)_5\overset{H}{\overset{|}{N}}\text{-})_n \qquad (7.12)$$

Polyurea

2. The Diels-Alder condensation of a bisdiene and benzoquinone forms an addition polymer, one possessing only carbon atoms in its backbone and where no byproduct is produced, yet the polymer is formed through a stepwise kinetic process.

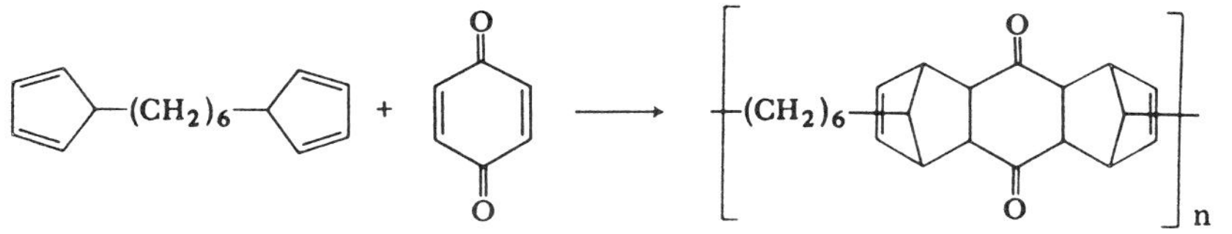

(7.13)

3. Internal esters (lactones) are readily polymerized by chainwise, acid-catalyzed ring openings without expulsion of a byproduct, yet the resulting polyester is a condensation polymer exhibiting a heteroatomed polymeric backbone. Further, the 6-carbon polyester is also formed using typical stepwise polycondensation of ω-hydroxycarboxylic acid.

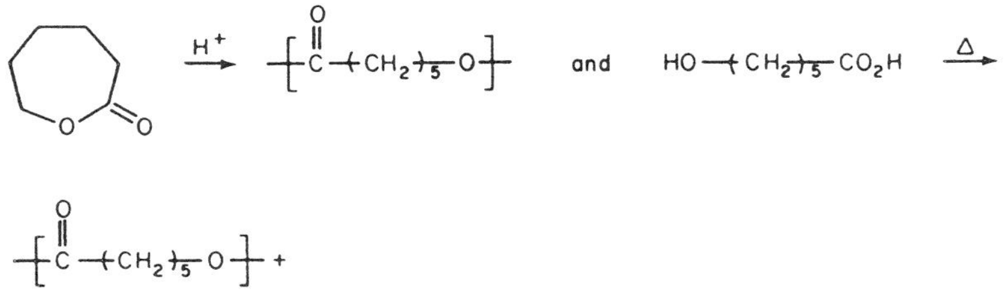

(7.14)

4. Nylon-6, a condensation polymer, is readily formed from either the internal amide (lactam) through a chainwise kinetic polymerization, or from the stepwise reaction of the ω-amino acid.

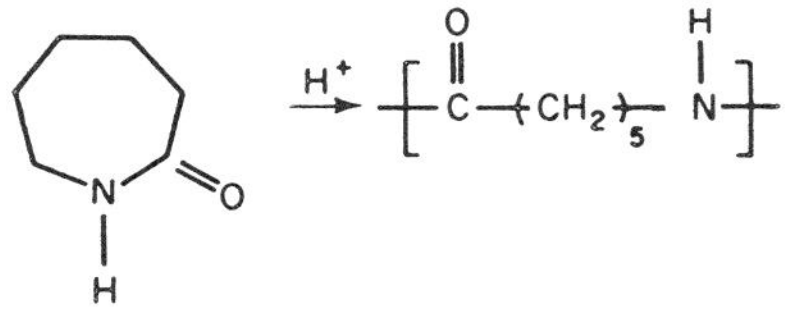

(7.15)

and

$$H_2N \text{+} CH_2 \text{+}_5 CO_2H \xrightarrow{\Delta} \text{+} \overset{\overset{\textstyle O}{\|}}{C} \text{+} CH_2 \text{+}_5 \overset{\overset{\textstyle H}{|}}{N} \text{+} + H_2O \qquad (7.16)$$

5. A hydrocarbon polymer can be made using the typical chainwise polymerization of ethylene, through a steplike polymerization of 1,8-dibromooctane utilizing the Wurtz reaction, or through the chain-growth boron trifluoride–catalyzed polymerization of diazomethane.

$$CH_2 \text{=} CH_2 \longrightarrow \text{+} CH_2 \text{—} CH_2 \text{+}_n \qquad (7.17)$$

$$BrH_2C \text{+} CH_2 \text{+}_6 CH_2 Br \xrightarrow{2Na} \text{+} CH_2 \text{+}_n + 2NaBr \qquad (7.18)$$

$$CH_2N_2 \xrightarrow{BF_3} \text{+} CH_2 \text{+}_n + N_2 \qquad (7.19)$$

6. Polyethylene oxide can be formed from either the usual catalyzed chain polymerization of ethylene oxide or the less usual condensation polymerization by means of a stepwise reaction of ethylene glycol.

$$\overset{\displaystyle H \qquad H}{\underset{\displaystyle H \quad \underset{O}{\diagdown\diagup} \quad H}{\overset{\diagdown\diagup}{C}\text{—}\overset{\diagup\diagdown}{C}}} \xrightarrow{H^+} \text{+} CH_2 \text{—} CH_2O \text{+}_n \qquad (7.20)$$

$$\underset{\underset{\textstyle OH \quad OH}{|\qquad |}}{CH_2 \text{—} CH_2} \longrightarrow \text{+} CH_2 \text{—} CH_2 \text{—} O \text{+}_n + H_2O \qquad (7.21)$$

7. Further interfacially formed condensation polymers, such as polyurethanes, polyesters, polyamides, and polyureas, typically are formed on a microscopic level in a chain-growth manner due to comonomer migration limits and the highly reactive nature of the reactants employed for such interfacial polycondensations.

Thus, a number of examples illustrate the lack of total overlap associated with describing the nature of the polymer backbone and the kinetics of formation for that polymer.

7.2 INTRODUCTION

While condensation polymers account for only a small percentage of all synthetic polymers, most natural polymers are of the condensation type. The first all-synthetic polymer, Bakelite, was produced by the stepwise condensation of phenol and formaldehyde, and many of the other synthetic polymers available before World War II were produced using stepwise polycondensations of appropriate reactants.

As shown by Carothers in the 1930s, the chemistry of condensation polymerizations is essentially the same as classic condensation reactions leading to the formation of monomeric esters, amides, and so on. The principal difference is that the reactants used for polymer formation are bifunctional instead of monofunctional.

The kinetics for stepwise polycondensation reactions and the kinetics for monofunctional aminations and esterifications, for example, are similar. Experimentally, both kinetic approaches are found to be essentially identical. Usual activation energies (30 to 60 kcal/mol) require only about one collision in 10^{12} to 10^{15} to be effective in producing product at 100°C, whereas for the vinyl reactions between R$^\bullet$ and a vinyl-monomer, activation energies are small (2 to 5 kcal/mol), with most collisions of proper orientation being effective in lengthening the chain. This is in agreement with the slowness of stepwise processes compared with radical chain processes. The rate constant of individual polycondensation steps is essentially independent of chain length, being similar to that of the "small-molecule" condensation. This is responsible for the stepwise growth pattern of polycondensations, since addition of a unit does not greatly increase the reactivity of the growing end, whereas for radical and ionic vinyl polymerizations the active radical and ionic site is transmitted to the growing chain end, giving a reactive end group. Table 7.1 contains a listing of a number of industrially important condensation polymers.

7.3 STEPWISE KINETICS

While more complicated situations can occur, we will consider only the kinetics of simple polyesterification. The kinetics of most other common condensations follow an analogous pathway.

For uncatalyzed reactions where the diacid and diol are present in equimolar amounts, one diacid is experimentally found to act as a catalyst. The experimental rate expression dependencies are described in the usual manner as follows:

$$\text{Rate of polycondensation} = -\frac{d[A]}{dt} = k[A]^2[D] \tag{7.22}$$

where [A] represents the diacid concentration and [D] the diol concentration.

Where [A] = [D], we can write

$$-\frac{d[A]}{dt} = k[A]^3 \tag{7.23}$$

Rearrangement gives

$$-\frac{d[A]}{[A]^3} = kdt \tag{7.24}$$

Integration of the above over the limits of $A = A_0$ to $A = A_t$ and $t = 0$ to $t = t$ gives

$$2kt = \frac{1}{[A_t]^2} - \frac{1}{[A_0]^2} = \frac{2}{[A_t]^2} + \text{constant} \tag{7.25}$$

It is convenient to express Eq. (7.25) in terms of extent of reaction p, where p is defined as the fraction of functional groups that have reacted at time t. Thus $1 - p$ is the fraction of groups unreacted. A_t is in turn $A_0 \cdot (1 - p)$, i.e.,

$$A_t = A_0(1 - p) \tag{7.26}$$

Substitution of the expression for A_t from Eq. (7.26) into Eq. (7.25) and re-arrangement gives

$$2A_0^2kt = \frac{1}{(1 - p)^2} + \text{constant} \tag{7.27}$$

A plot of $1/(1 - p)^2$ as a function of time should then be linear with the slope $2A_0^2 k$ from which k is determinable. Determination of k at different temperatures enables the calculation of activation energy.

The number-average degree of polymerization $\overline{DP}_N$ can be expressed as

$$\overline{DP}_N = \frac{\text{number of original molecules}}{\text{number of molecules at a specific time t}} = \frac{N_0}{N} = \frac{A_0}{A_t} \tag{7.28}$$

Thus,

$$\overline{DP}_N = \frac{A_0}{A_t} = \frac{A_0}{A_0(1 - p)} = \frac{1}{1 - p} \tag{7.29}$$

The relationship given in Eq. (7.29) is called the Carothers equation because it was first found by Carothers while working with the synthesis of nylons (polyamides). For an essentially quantitative synthesis of polyamides where p is 0.9999, the $\overline{DP}$ was found to be approximately equal to 10,000, the value calculated using Eq. (7.29).

$$\overline{DP} = \frac{1}{1 - p} = \frac{1}{1 - 0.9999} = \frac{1}{0.0001} = 10,000 \tag{7.30}$$

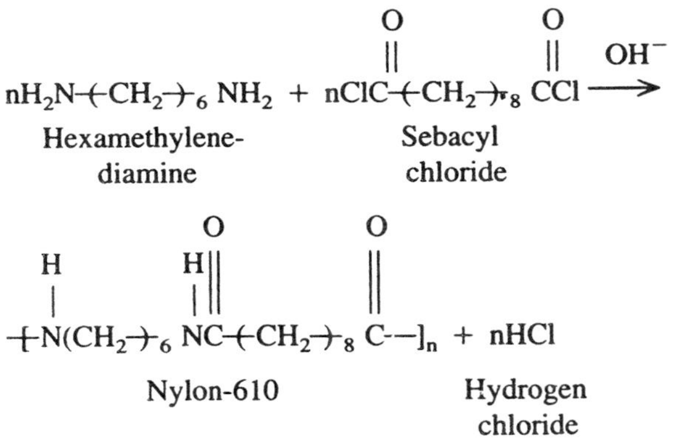

$$nH_2N\!-\!\!(CH_2)_6\,NH_2 \;+\; nClC\!-\!\!(CH_2)_8\,CCl \xrightarrow{\;\;OH^-\;\;}$$

Hexamethylene-   Sebacyl
diamine   chloride

$$\qquad\qquad\qquad\qquad\qquad\qquad\qquad (7.31)$$

$$+\!\!N(CH_2)_6\,NC\!-\!\!(CH_2)_8\,C\!-\!]_n \;+\; nHCl$$

Nylon-610   Hydrogen
                        chloride

The $\overline{DP}$ of 10,000 calculated above for nylon-610 is more than adequate for a strong fiber. Actually, it would be difficult to force such a high molecular weight polymer through the small holes in the spinneret in the melt extrusion process used for fiber production.

The many possible nylons are coded to show the number of carbon atoms in the amine and acid repeat units, respectively. The most widely used nylon fiber is nylon-66.

The high value of p would be decreased and the value of $\overline{DP}$ also decreased accordingly if a competing cyclization reaction occurred.

Table 7.1 Structures, Properties, and Uses of Some Synthetic Condensation-Type Polymers

Type (common name)	Characteristic repeating unit	Typical reactants	Typical properties	Typical uses
Polyamide (nylon)	$$\left(\!\!-N-\overset{O}{\overset{\|}{C}}-R-\overset{O}{\overset{\|}{C}}-N-R\!\!\right)\!\!\!-$$ (H on each N)	$H_2-N-R-\overset{O}{\overset{\|}{C}}-OH$ $H_2NRNH_2 + HO-\overset{O}{\overset{\|}{C}}-R-\overset{O}{\overset{\|}{C}}-OH$ $H_2NRNH_2 + Cl-\overset{O}{\overset{\|}{C}}-R-\overset{O}{\overset{\|}{C}}-Cl$ $(CH_2)_x \overset{C=O}{\underset{N-H}{\big)}}$	Good balance of properties, high strength, good elasticity and abrasion resistance, good toughness, favorable solvent resistance, only fair outdoor weathering properties, fair moisture resistance	Fibers—about 1/2 of all nylon fiber produced goes into tire cord (tire replacement); rope, cord, belting and fiber cloths, thread, hose, undergarments, dresses; plastics—use as an engineering material, substitute for metal in bearings, cams, gears rollers; jackets on electrical wire
Polyurethane	$$-\!\!\left(\!\!N-\overset{O}{\overset{\|}{C}}-O-R-O-\overset{O}{\overset{\|}{C}}-N-R\!\!\right)\!\!\!-$$ (H on each N)	$OCN-R-NCO + HO-R-OH$	Elastomers—good abrasion resistance, hardness, good resistance to grease and good elasticity; fibers—high elasticity, excellent rebound; coatings—good resistance to solvent attack and abrasion, good flexibility and impact resistance; foams—good strength per weight, good rebound, high impact strength	Four major forms utilized—fibers, elastomers, coatings cross-linked foams; elastomers—small industrial wheels, heel lifts; fibers—swimsuits and foundation garments; coatings—floors where high impact and abrasion resistance are required, such as dance floors; bowling pins; foams—pillows, cushions

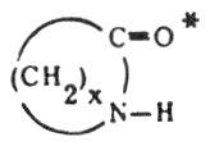

Name	Repeat unit	Monomers	Properties	Uses
Polyurea	$\left[N-C-N-R-N-C-N-R \right]$ (with H, O, H / H, O, H)	$OCN-R-NCO + H_2N-R-NH_2$	High T_g, fair resistance to greases, oils, solvents	Not widely utilized
Polyester	$\left[O-C-R-C-O-R \right]$ (with $O=C$)	$HO-R-OH + Cl-\overset{O}{\overset{\|}{C}}-R-\overset{O}{\overset{\|}{C}}-Cl$ $HO-R-OH + HO-\overset{O}{\overset{\|}{C}}-R-\overset{O}{\overset{\|}{C}}-OH$ * $HO-R-\overset{O}{\overset{\|}{C}}-OH$ **	High T_g, high T_m, good mechanical properties to about 175°C, good resistance to solvent and chemicals; fibers—good crease resistance and rebound, low moisture absorption, high modulus, good resistance to abrasion; film—high—high tensile strength (almost equal to that of some steel), stiff, high resistance to failure on repeated flexing, fair tear strength, high impact strength	Fibers—garments, permanent press and "wash and wear" garments, felts, tire cord; film—magnetic recording tape, high grade films
Polyether	$\left[O-R \right]$	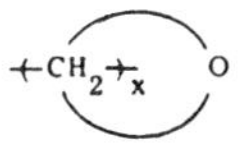 $\left(CH_2 \right)_x$	Good thermoplastic behavior, water solubility, generally good mechanical properties, moderate strength and stiffness (similar to polyethylene)	Sizing for cotton and synthetic fibers; stabilizers for adhesives, binders, and film-formers in pharmaceuticals; thickeners; production of films

* repeat unit $\left[\overset{O}{\overset{\|}{C}} \left(CH_2 \right)_x \overset{H}{\overset{\|}{N}} \right]$

** repeat unit $\left[O-R-\overset{O}{\overset{\|}{C}} \right]$

(continued)

Table 7.1 Continued

Type (common name)	Characteristic repeating unit	Typical reactants	Typical properties	Typical uses
Polycarbonate	$+O-R-O-C+$ (with $C=O$)	$COCl_2 + HO-R-OH$ 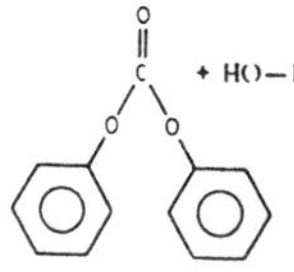 + $HO-R-OH$ (diphenyl carbonate structure)	Crystalline thermoplastic with good mechanical properties, high impact strength, good thermal and oxidative stability, transparent, self-extinguishing, low moisture absorption	Machinery and business
Phenol-formaldehyde resins	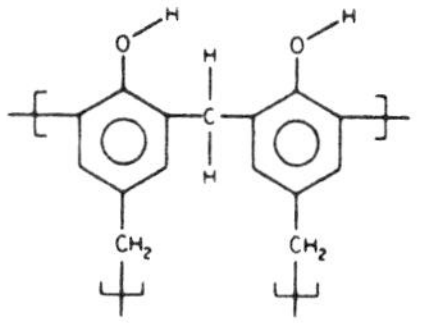 (phenol-formaldehyde repeating unit)	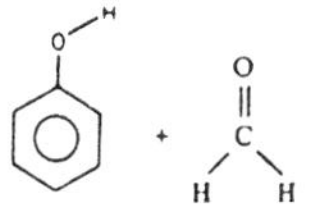 phenol + formaldehyde ($H_2C=O$)	Good heat resistance, dimensional stability and resistance to cold flow; good resistance to most solvents; good dielectric properties	Used in molding applications; phonograph records; electrical, radio, televisions, appliance, and automotive parts where their good dielectric properties are of use; filler; missile nose cones; impregnating paper; varnishes; decorative laminates for wall coverings; electrical parts, such as printed circuits; countertops, toilet seats; coatings for electrical wire; adhesive for plywood, sandpaper, brake linings and abrasive wheels

Name	Structure	Reaction	Properties	Applications
Polyacetal	$-[O-R-O-CH_2]-$	$HO-R-OH + \begin{smallmatrix}RO\\RO\end{smallmatrix}CH_2$	Intermediate physical properties	No large industrial application
Polyanhydride	$-[\overset{O}{\overset{\|\|}{C}}-O-\overset{O}{\overset{\|\|}{C}}-R]-$	$HO-\overset{O}{\overset{\|\|}{C}}-R-\overset{O}{\overset{\|\|}{C}}-OH$	Medium to poor T_g and T_m, medium physical properties	No large industrial application
Polysulfides	$-[S_x-R]-$ $-[\overset{S}{\overset{\|\|}{S}}-S-R]-$ $-[S-S-R]-$	$Cl-R-Cl + Na_2S_x$ $Cl-R-Cl + Na_2S_4$ $Cl-R-Cl$ (oxidation)	Outstanding oil and solvent resistance, good gas impermeability, good resistance to aging and ozone, bad odors, low tensile strength, poor heat resistance	Solvent resistant and gas resistant elastomer—such as gasoline hoses and tanks, gaskets, diaphragms
Polysiloxane	$-[\overset{R}{\underset{R}{Si}}-O]-$	$Cl-\overset{R}{\underset{R}{Si}}-Cl + H_2O \rightarrow HO-\overset{R}{\underset{R}{Si}}-OH \longrightarrow$ polymer	Available in a wide range of physical states—from liquids to greases, to waxes to to resins to rubbers; excellent high and moderate low temperature physical properties; resistant to weathering and lubricating oils	Fluids—cooling and dielectric fluids, in waxes and polishes, as antifoam and release agents, for paper and textile treatment; elastomers—gaskets, seals, cable and wire insulation, hot liquids and gas conduits, surgical and prosthetic devices, sealing compounds; resins—varnishes, industrial paints, encapsulating and impregnating agents
Polyphosphate and polyphosphonate esters	$-[\overset{O}{\overset{\|\|}{P}}-O-R-O]-$ with R	$Cl-\overset{O}{\overset{\|\|}{P}}-Cl + HO-R-OH$ with R	Good fire resistance, fair adhesion, moderate moisture stability, fair temperature stability	Additive to promote flame retardance, adhesive for glass (since refractive index of some esters is about the same as that of glass), certain pharmaceutical applications, surfactant

Since the values of k at any temperature may be determined from the slope $(2kA_0)$ of the line when $1/(1-p)^2$ is plotted against t, one may determine $\overline{DP}_N$ at any time t from the expression

$$(\overline{DP}_N)^2 = 2kt[A_0]^2 + \text{constant} \tag{7.32}$$

Returning again to consider the synthesis of polyesters, generally much longer reaction times are required to effect formation of high polymer in uncatalyzed esterifications than for acid-or base-catalyzed systems. Since the added acid or base is a catalyst, its apparent concentration does not change with time, thus it need not be included in the kinetic rate expression. In such cases the reaction follows the rate expression

$$\text{Rate of polycondensation} = -\frac{d[A]}{dt} = k[A][B] \tag{7.33}$$

For $[A] = [B]$ we have

$$-\frac{d[A]}{dt} = k[A]^2 \tag{7.34}$$

which gives on integration and subsequent substitution

$$kt = \frac{1}{A_t} - \frac{1}{A_0} = \frac{1}{A_0(1-p)} - \frac{1}{A_0} \tag{7.35}$$

Rearrangement gives

$$A_0kt = \frac{1}{1-p} - 1 = \overline{DP}_N - 1 \tag{7.36}$$

which predicts a linear relationship of $1/(1-p)$ with reaction time. This is shown in Fig. 7.7.

For such second-order reactions, $\overline{DP}_N = [A_0]kt + 1$. The effect of time on $\overline{DP}$ can be demonstrated using $A_0 = 2$ mol/liter and $k = 10^{-2}$ liter/mol-sec at the reaction times of 1800, 3600, 5400 sec. Thus, $\overline{DP}$ increases from 37 to 73 to 109.

$$\begin{aligned}
\overline{DP} &= 1 + (10^{-2})\,(2)\,(1800) = 37 \\
\overline{DP} &= 1 + (10^{-2})\,(2)\,(3600) = 73 \\
\overline{DP} &= 1 + (10^{-2})\,(2)\,(5400) = 109
\end{aligned} \tag{7.37}$$

Useful high molecular weight linear polymers are not obtained unless the value for the fractional conversion p is at least 0.990, i.e., a $\overline{DP}$ greater than 100.

It is important to note that the rate constant k for reactions of monofunctional compounds is essentially the same as that for difunctional compounds, and hence these k values, which are essentially unchanged during the reaction, can be used for polycondensation reactions. Likewise, as is the case with reactions of small molecules, the rate constant k increases with temperature in accordance with the Arrhenius equation shown below. The energies of activation (E_a) are also comparable to those for monofunctional reactants.

$$k = Ae^{-E_a/RT} \tag{7.38}$$

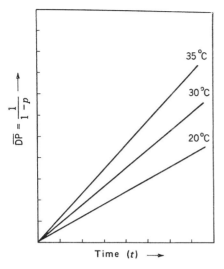

Figure 7.7 Plot of chain length as a function of reaction time for the acid-catalyzed condensation of ethylene glycol with terephthalic acid. (From *Introduction to Polymer Chemistry* by R. Seymour, McGraw-Hill, New York, 1971. Used with permission of McGraw-Hill Book Company.)

7.4 GENERAL STEP-REACTION POLYMERIZATION

From equations, such as Eq. (7.36), it is possible to derive expressions describing the molecular weight distribution of such a stepwise polymerization at any degree of polymerization. The same relationship is more easily derived from statistical considerations. The following statistical treatment assumes the reaction rate to be independent of chain length.

One may write a general equation for the formation of a linear polymer by the step reaction of bifunctional reactants A and B as follows:

$$nA + nB \rightarrow A(BA)_{n-1}B \tag{7.39}$$

The probability of finding a repeating unit AB in the polymer is p, and the probability of finding $n - 1$ of these repeating units in the polymer chain is p^{n-1}. Likewise, the probability of finding an unreacted molecule of A or B is $1 - p$. Thus, the probability (p_n) of finding a chain with n repeating units $(BA)_n$ is

$$p_n = (1 - p)p^{n-1} \tag{7.40}$$

Hence, N_n, the probability of choosing (at random) a molecule with $(BA)_n$ repeating units, where N is equal to the total number of molecules, is

$$N_n = N(1 - p)p^{n-1} \tag{7.41}$$

Since

$$\frac{N_0}{N} = \frac{1}{1 - p} \quad \text{or} \quad N = N_0(1 - p) \tag{7.42}$$

where N_0 is the total number of structural units present, therefore,

$$N_n = N_0(1 - p)^2 p^{n-1} \tag{7.43}$$

The corresponding weight-average molecule weight distribution W_n may be calculated from the relationship $W_n = nN_n/N_0$ as follows:

$$W_n = \frac{nN_0(1 - p)^2 p^{n-1}}{N_0} = n(1 - p)^2 p^{n-1} \tag{7.44}$$

The relationships shown in Eqs. (7.43) and (7.44) demonstrate that high values of p (>0.99) are essential in order to produce either high N_n or high W_n values. The number-average molecular weight M_n and weight-average molecular weight M_w calculated from Eqs. (7.41) and (7.42) are as follows:

$$M_n = \frac{mN_0}{N} = \frac{m}{1 - p} \tag{7.45}$$

where m = the molecular weight of the mer and

$$M_w = \frac{m(1 + p)}{1 - p} = M_n(1 + p) \tag{7.46}$$

Thus, the index of polydispersity M_w/M_n for the most probable molecular weight distribution becomes $1 + p$, as shown below:

$$\frac{M_w}{M_n} = \frac{m(1 + p)/(1 - p)}{m/(1 - p)} = 1 + p \tag{7.47}$$

Thus, when p is equal to 1, the index of polydispersibility for the most probable distribution for step-reaction polymers is 2.

Because the value of p is essentially 1 in some step-reaction polymerizations, the products obtained under normal conditions have very high molecular weights and are difficult to process. It is obvious that the value of p may be reduced by using a slight excess of one of the reactants or by quenching the reaction before completion. Thus, if a reaction is quenched when the fractional conversion p is 0.995, $\overline{DP}$ will be equal to 200.

If more than 1 mol of B is used with 1 mol of A, the ratio of A/B, or r, may be substituted in the modified Carothers equation as shown below:

$$\overline{DP} = \frac{\text{total nA at p}}{\text{total nA at rp}} = \frac{n[(1 + 1/r)]/2}{n[1 - p + (1 - rp/r)]/2} = \frac{1 + (1/r)}{1 - p + [(1 - rp)/r]} \tag{7.48}$$

Therefore, multiplying top and bottom by r gives

$$\overline{DP} = \frac{r + 1}{r(1 - p) + (1 - rp)} = \frac{1 + r}{(1 + r) - 2rp} \tag{7.49}$$

Thus, if r = 0.97 and p ≈ 1, $\overline{DP}$ is equal to

$$\overline{DP} = \frac{1 + r}{(1 + r) - 2rp} = \frac{1 + 0.97}{1 + 0.97 - 2(0.97)} = \frac{1.97}{0.03} = 66 \tag{7.50}$$

The $\overline{DP}$ value of 66 is above the threshold limit of 50 required for nylon-66 fibers.

Since quenching the reaction or adding a stoichiometric excess of one reactant is not economical, the commercial practice is to add a calculated amount of a monofunctional reactant. Acetic acid is used as the monofunctional reactant in the synthesis of polyamides and polyesters. In this case, one may employ a functionality factor (f) which is equal to the average number of functional groups present per reactive molecule. While the value of f in the preceding examples has been 2.0, it may be reduced to lower values and used in the following modified Carothers equation:

$$\overline{DP} = \frac{A_0}{A_0[1 - (pf/2)]} = \frac{2}{2 - pf} \qquad (7.51)$$

Thus, if 0.01 mol of acetic acid is used with 0.99 mol of two difunctional reactants, the average functionality or functional factor, f, is calculated as follows.

$$f = \frac{\text{mol of each reactant} \times \text{functionality}}{\text{total number of moles}}$$

$$= 0.99 \text{ mol} \times 2 + 0.99 \text{ mol} \times 2 + 0.01 \text{ mol} \times \frac{1}{1.99 \text{ mol}} = 1.99$$

Substitution of f = 1.99 and p = 1.00 in Eq. (7.51) gives an average $\overline{DP}$ of 200 representing an upper limit for this situation. The same calculation, except employing p = 0.95, gives an average $\overline{DP}$ of only 20.

7.5 POLYESTERS

In his first experiments after joining the du Pont Company, Carothers attempted to produce fibers from aliphatic polyesters. He condensed adipic acid with ethylene glycol, but since the fractional conversion (p) was less than 0.95, the $\overline{DP}$ of these polyesters was less than 20. He could have solved the problem by use of the Schotten-Baumann reaction or by ester interchange, but he shelved this work in favor of the thermal decomposition of purified ammonium salts in which the fractional conversion (p) was essentially 1.0 (this subject is continued in Sec. 7.6).

Useful commercial polyesters, called *glyptals* and *alkyds*, had been prepared before Carothers conducted his investigations. These processes, like other esterifications, had low fractional conversions, but they were useful as coatings (not fibers) because of high $\overline{DP}$ values resulting from cross-linking.

When the average functionality is greater than 2, cross-linking occurs. Utilization of the modified Carothers equation [Eq. (7.51)] gives large numbers approaching infinity for the average DP as the product of p and f approaches 2. Thus, for the reaction of 0.99 mol of difunctional phthalic anhydride with 0.99 mol of ethylene glycol and 0.01 mol of trifunctional glycerol,

$$f = \frac{0.99 \text{ mol} \times 2 + 0.99 \text{ mol} \times 2 + 0.01 \text{ mol} \times 3}{1.99 \text{ mol}} = 2.02$$

Substitution of r = 2.02 and p = 0.95 into Eq. (7.51) gives an average DP of 25. For p = 0.97, the average DP is 50; for p = 0.99, the average DP increases to 10,000. This march towards infinity as the product of pf approaches 2 is consistent with the formation of high molecular weight cross-linked products.

In the statistical approach to the requirements for incipient gelation, one intro-
duces a branching coefficient α, which is defined as the probability that a reactant
with a value for f of greater than 2.0 is connected to a linear chain segment or to
another multifunctional reactant or branch point. The critical value for incipient
gelation (α_c) is the probability that one or more of the f − 1 chain segments on
the branch unit will be connected to another branch unit, i.e., $\alpha_c = (f − 1)^{-1}$ or
$1/(f − 1)$. Thus, when f = 2.2, $\alpha_c = 0.83$.

Glyptal polyesters were produced in 1901 by heating glycerol and phthalic
anhydride. Since the secondary hydroxyl is less active than the terminal primary
hydroxyls in glycerol, the first product formed at conversions of less than about
70% is a linear polymer. A cross-linked product is produced by further heating:

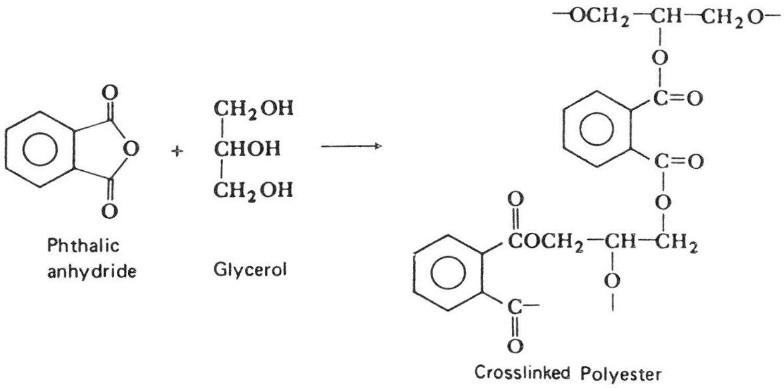

$$(7.52)$$

Alkyds were synthesized by Kienle in the 1920s from trifunctional alcohols
and dicarboxylic acids. Unsaturated oils called drying oils were transesterified with
the phthalic anhydride in the reaction so that an unsaturated polyester was obtained.

The extent of cross-linking or "drying" of these alkyds in the presence of a
soluble lead or cobalt catalyst or drier was dependent on the amount of unsaturated
oil present. The terms short oil, medium oil, and long oil alkyd are used to signify
the "oil length" obtained by use of 30 to 50%, 50 to 65%, and 65 to 80% of
unsaturated oil, respectively.

The term alkyd is sometimes used to describe all polyesters produced from the
condensation of a polybasic acid and a polyhydric alcohol. Thus, the terms nonoil
and oil-free alkyds have been used to distinguish between the principal types of
polyesters. The terms saturated and unsaturated polyesters are also widely used.
The chain reaction mechanism of the curing of these unsaturated polymers will be
discussed in Chap. 8.

Another type of unsaturated polyester is produced by the condensation of ethylene glycol with phthalic anhydride and maleic anhydride [Eq. (7.53)]. These
polyesters may be dissolved in styrene and used as cross-linking resins for the
production of fibrous glass-reinforced plastics.

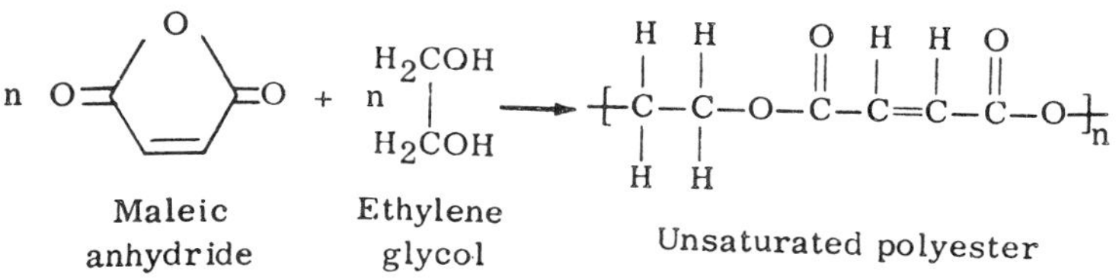

$$(7.53)$$

The difficulties encountered by Carothers in his attempted preparation of crystalline high molecular weight linear fibers were overcome by an ester interchange reaction with ethylene glycol and dimethyl terephthalate [Eq. (7.54)]. More recently, terephthalic acid has been used directly. Polyesters may also be produced from the reaction of terephthalic acid and ethylene oxide. The classic reaction for producing Dacron, Kodel, and Terylene fibers and Dacron film is shown below.

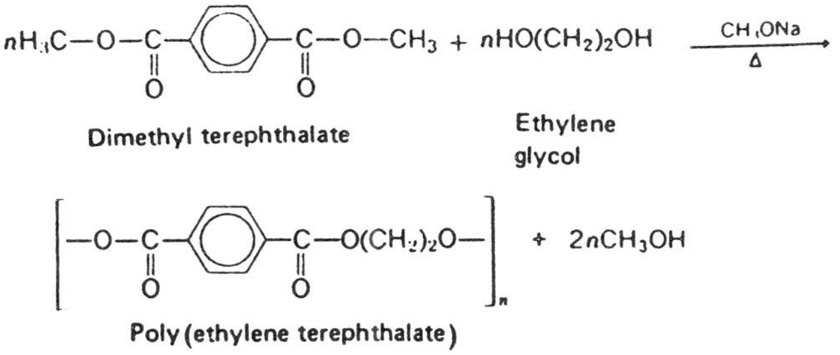

Dimethyl terephthalate   Ethylene glycol

(7.54)

Poly(ethylene terephthalate)

Polyester fibers (poly(ethylene terephthalate), PET), which are now the world's leading synthetic fibers, are produced at an annual rate of over 1.5 million tons in the United States. Biaxially oriented PET film is one of the strongest polymeric films available. Thicker oriented films or sheets are used for containers for carbonated drinks.

The addition of antinucleating agents permits the injection molding of PET (Rynite). Injection-moldable glycol-modified polyesters (PETG, Kodar) have been produced by partial replacement of the ethylene glycol by cyclohexanol dimethylol.

Moldability of aryl polyesters has also been improved by the use of poly-(butylene terephthalate) (PBT) instead of PET or by the use of blends of PET and PBT. PBT under the trade name of Celanex, Valox, Gafite, and Versel is being produced at an annual rate of 25 thousand tons. Copolymers of carbonate and aryl esters, acrylics and aryl esters, and imide and aryl esters as well as physical blends of polyesters and other polymers are available. These aryl polyesters are being used for bicycle wheels, springs, and blow molded containers.

A new high-impact blend of PBT and polybutene (Valox CT) is also available. The melt viscosity of blends of PET and nylon-66 has been reduced by the addition of poly(vinyl alcohol). Self-reinforcing PET has been produced by the addition of p-hydroxybenzoic acid, which forms liquid crystals in the composite.

Since poly(ethylene terephthalate) has a melting point of 240°C, it is difficult to mold. However, poly(butylene terephthalate) produced from butylene glycol has a melting point of 170°C and is more readily molded. It is a strong, highly crystalline engineering plastic.

Polycarbonates (PC), which are polyesters of unstable carbonic acid, are relatively stable polymers which were originally produced by the reaction of phosgene with bisphenol A [2,2-bis(4-hydroxyphenyl) propane] [Eq. (7.55)]. This unusually tough transparent plastic is available under the trade names of Lexan (General Electric) and Merlon (Mobay). Polycarbonates may also be produced by ester interchange between diphenyl carbonate and bisphenol A.

The melting point of polycarbonates is decreased from 225 to 195°C when the methyl pendant groups are replaced by propyl groups. The polycarbonate prepared from bis(4-hydroxyphenyl) ether also has a lower melting point and lower glass transition temperature.

Polycarbonates and polycarbonate-polyester copolymers are used for glazing, sealed beam headlights, door seals, popcorn cookers, solar heat collectors, and appliances. Polycarbonate tends to stress crack in the presence of gasoline, but a 50–50 blend with poly(butyl terephthalate) (PBT, Xenoy) is unusually resistant to gasoline.

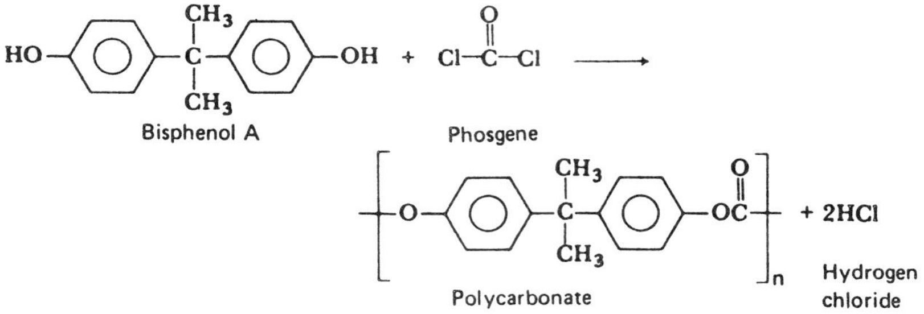

(7.55)

As shown in Eq. (7.56), a highly crystalline, high-temperature–resistant polymer is produced by the self-condensation of an ester of p-hydroxybenzoic acid. This poly-p-benzoate, which is marketed under the trade name of Ekonol, has a melting point greater than 900°C (it probably decomposes before 900°C).

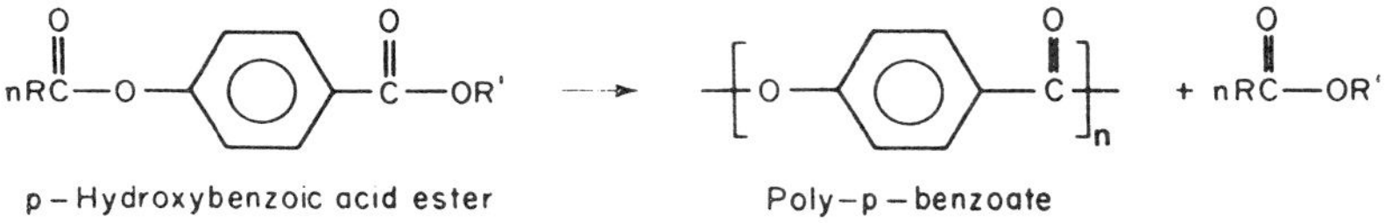

(7.56)

7.6 SYNTHETIC POLYAMIDES

The first polyesters, produced by Carothers, had relatively low molecular weights because of low fractional conversions. Carothers was successful in producing higher molecular weight polymers by shifting the equilibrium by removing the water produced. However, these aliphatic polyesters, which he called "super polymers," lacked stiffening groups in the chain and thus had melting points that were too low for laundering and ironing.

Carothers' next step was to increase the fractional conversion (p) by making salts by the reaction of hexamethylenediamine and adipic acid. These were recrystallizable from ethanol. Thus, a high molecular weight polyamide known generally as nylon, which had a melting point of 265°C, could be produced by the thermal decomposition of this pure, equimolar nylon-66 salt, as shown by

$$n\ HO_2C(CH_2)_4CO_2H\ +\ H_2N(CH_2)_6NH_2\ \rightarrow$$

$$n[^-O_2C(CH_2)_4CO_2^-][H_3\overset{+}{N}(CH_2)_6\overset{+}{N}H_3] \overset{heat}{\rightarrow}$$

$$\begin{bmatrix} O & O \\ \parallel & \parallel \\ C(CH_2)_4CNH(CH_2)_6NH \end{bmatrix}_n + 2nH_2O$$

Adipic acid + 1,6-hexanediamine $\overset{heat}{\rightarrow}$ nylon-66 + water   (7.57)

Since the molecular weight of the original nylon-66 produced by Carothers in

1938 was higher then he desired, he added 1% of acetic acid to the reactants in order to reduce the $\overline{DP}$ value. Because of the stiffening effect of the amide groups, the melting point of nylon-66 is 200°C greater than that of the corresponding polyester. The melting point of nylons (PA) increases as the number of methylene groups between amide groups in the chain are reduced.

Molded nylon-66 is used for lawnmower blades, bicycle wheels, tractor hood extensions, skis for snowmobiles, skate wheels, motorcycle crank cases, bearings, and electrical connections. The radiator in the 1982 model Ford Escort was molded from nylon-66.

The structure of nylons (e.g., nylon-ab) is

$$\begin{array}{cccc} H & H & O & O \\ | & | & \| & \| \\ \end{array}$$
$$+N-(CH_2)_a-N-C-(CH_2)_b-C+_n$$

where a and b are equal to the number of carbon atoms in the repeating units of the diamine and dicarboxylic acid. Mono- and biaxially oriented nylon film is available.

Nylon-610 and nylon-612 produced by the condensation of hexamethylenediamine and sebacic or dodecanoic acid, respectively, are more resistant to moisture and more ductile than nylon-66. The properties of these polyamides may be improved by the formation of polyether blocks (NBC) and by blending with thermoplastics, such as EPDM, PET, PBT, and TPE.

NBC (Nyrim) is more expensive than RIM polyurethane, but it may be heated to 200°C without softening. NBC moldings are produced by the reaction injection molding (RIM) of poly(propylene glycol) and caprolactam in the presence of a catalyst. The tendency for this copolymer to swell in the presence of water is reduced by reinforcing with glass fibers.

Since the chains of nylons having an even number of carbon atoms between the amide groups pack better, their melting points are higher than comparable nylons with odd numbers of carbon atoms. The melting points decrease and the water resistance increases as the number of methylene groups between amide groups is increased.

Nylon-6, which has a heat deflection temperature of 80°C, was produced by the ring-opening polymerization of caprolactam in Germany in the 1940s. Molded articles made of these polymers may be produced in situ in the RIM process. Nylon-11 and nylon-12, produced by the anionic polymerization of 11 and 12 amino acids, are also characterized by good resistance to moisture and superior ductility. The structure of the repeating unit in monadic nylons, such as nylon-6, is:

$$\begin{array}{cc} H & O \\ | & \| \\ \end{array}$$
$$+N-(CH_2)_5C+_n$$
Nylon-6

The copolymer of nylon-6 and nylon-66 is tougher and has a smoother surface than either of the homopolymers.

Injection-moldable, yellow, transparent polyamide (PA7030) is also available.

The aromatic polyamides (aramids) have been produced by the interfacial condensation of aromatic diamines, such as 1,3-phenylenediamine and isophthaloyl chloride in chloroform. Amorphous transparent aramids with heat deflection temperatures of 160°C have been produced from 2,2-bis-4-(aminocyclohexyl) propane.

Aromatic nylons prepared from terephthalic acid (Kevlar) have very high melting points and are called aramids. The solubility and ease of fabrication of aramids is improved by the preparation of ordered copolyamides. For example, a copolymer consisting of meta-benzamide and isophthalamide units is more readily processed than poly(orthophenylene phthalamide).

Because of the presence of the bulky methoxymethyl pendant group, the hydrogen-bonding forces are reduced, the melting point is reduced, and the flexibility is increased in methoxymethylated nylon-66. Comparable results are observed when nylon-66 is condensed with ethylene oxide. The equations for these reactions are shown below.

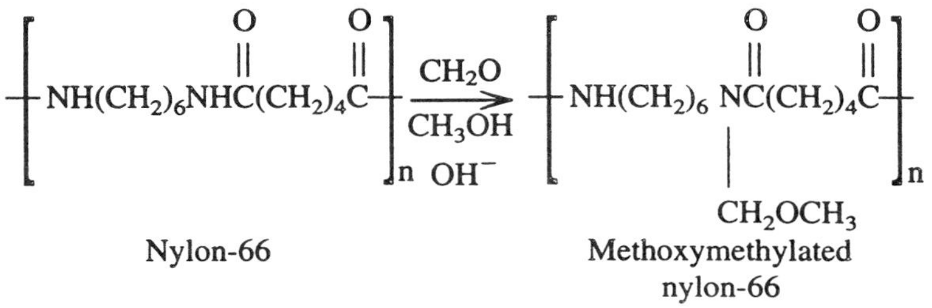

$$\left[NH(CH_2)_6 NHC(CH_2)_4 C \overset{\displaystyle O}{\underset{}{\parallel}} \overset{\displaystyle O}{\underset{}{\parallel}} \right]_n \xrightarrow[\substack{CH_3OH \\ OH^-}]{CH_2O} \left[NH(CH_2)_6 \ \underset{\underset{CH_2OCH_3}{|}}{N} C(CH_2)_4 C \right]_n \tag{7.58}$$

Nylon-66 → Methoxymethylated nylon-66

$$\text{\ensuremath{\sim}\ensuremath{\sim}N}\underset{\underset{\text{Nylon-66}}{\overset{|}{H}}}{-}\overset{\overset{O}{\parallel}}{C}\text{\ensuremath{\sim}\ensuremath{\sim}} + CH_2\overset{\diagup \diagdown}{-}CH_2 \longrightarrow \text{\ensuremath{\sim}\ensuremath{\sim}N}\underset{\underset{\substack{CH_2CH_2OH \\ \text{Ethoxylated} \\ \text{nylon-66}}}{|}}{-}\overset{\overset{O}{\parallel}}{C}\text{\ensuremath{\sim}\ensuremath{\sim}} \tag{7.59}$$

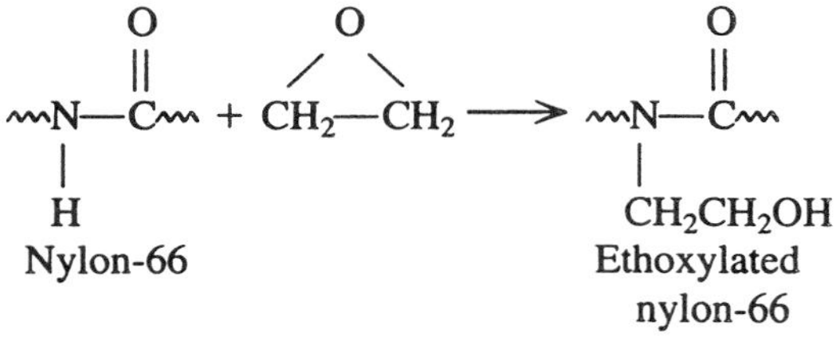

Comparable changes in physical properties are observed when branched dicarboxylic acids and branched diamines are used as the reactants for producing nylons. Thus, nylons produced from α-methyladipic acid and hexamethylenediamine and from adipic acid and 3-methylhexamethylenediamine have melting points that are at least 80°C less than that of nylon-66. These nylons are not suitable for fiber use because of the presence of branches on the chain.

7.7 POLYIMIDES

As shown in the following equation, polyimides (PI) with melting points greater than 600°C are produced by the condensation of an aliphatic diamine and a dianhydride, such as pyromellitic anhydride. It is customary to carry out this reaction stepwise to produce a soluble prepolymer (polyamic acid) which is insolubilized when heated. However, the product of the first step is insoluble when aromatic diamines are used. (Again, decomposition may occur prior to 600°C.)

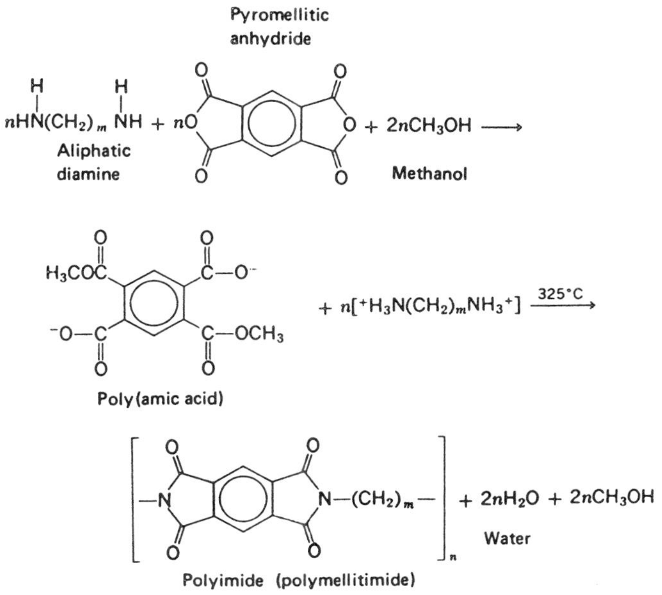

$$(7.60)$$

Heat-resistant intractable thermoset polyimides (Vespel, Kinel, Kapton) have been supplemented by injection-moldable polyamide-imides (Torlon) and modified polyimides (Kanox, Toramid). Thermoplastic polyimide adhesives were cited as one of the top 100 inventions of 1981. Polyimide foam does not ignite at temperatures below 430°C and is being considered as a cushioning material in public conveyances.

The processing properties are improved when polyimides are produced from meta-substituted diamines. A new amorphous polyether-imide (Ultem) with a high heat deflection temperature has been introduced by the General Electric Co.

The ether linkages in polyether-imide (PEI) improve the ease of processing and increase the ductility of these high-performance plastics. PEI reinforced with 30% fiberglass has a heat deflection temperature of 210°C.

The polyamide-imide Torlon (PA-I) has a heat deflection temperature of 285°C. A Ford automobile prototype engine has been built from PA-I.

7.8 POLYBENZIMIDAZOLES AND RELATED POLYMERS

Many heterocyclic polymers have been produced in an attempt to develop high-temperature–resistant polymers for aerospace applications. Among these are the polybenzimidazoles (PBI), which, as shown by the following equation, are prepared from aromatic tetramines and esters of dicarboxylic aromatic acids. In the standardized procedure, the reactants are heated at temperatures below 300°C in order to form a soluble prepolymer, which is converted to the final insoluble polymer by heating at higher temperatures.

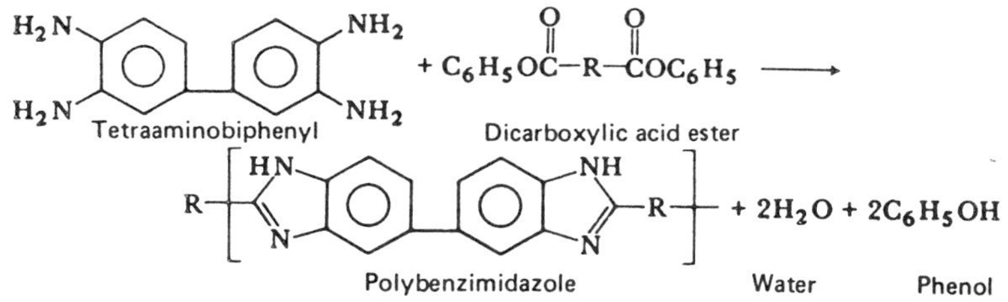

$$(7.61)$$

Polymers such as PBI have a weak link in the single covalent bond connecting the phenyl rings in biphenyl. This weakness is overcome by the synthesis of a ladder polymer, which, as shown by the formula for polyquinoxaline, has two covalent bonds throughout the chain. Thus, the integrity of the polymer is maintained even if one bond is broken. This requirement is also met by spiropolymers, such as the intractable spiroketal polymers produced by the condensation of 1,4-cyclohexanedione and pentaerythritol, as shown below.

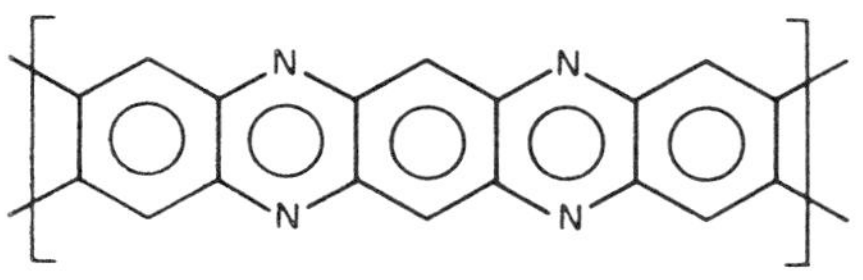

Polyquinoxaline
(a ladder polymer)

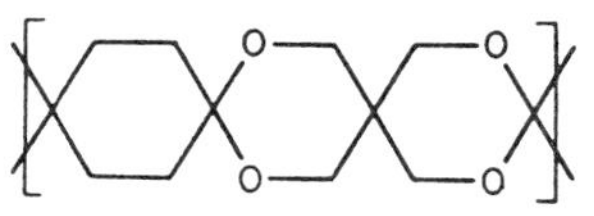

Spiroketal polymer
(a spiropolymer)

7.9 POLYURETHANES AND POLYUREAS

Urethanes, or carbamates, are well-known organic compounds which were formerly used for the characterization of alcohols. Since the fractional conversion (p) of the reaction is relatively high, Bayer was able to prepare numerous useful polyurethanes (PU) by the reaction of dihydric alcohols and diisocyanates. For example, a crystalline polymeric fiber (Perlon U) may be prepared by the reaction of 1,4-butanediol and hexamethylene diisocyanate as shown in the following equation.

$$(7.62)$$

nHO(CH$_2$)$_4$OH + nOCN(CH$_2$)$_6$NCO →

1,4-Butanediol   Hexamethylene
                          diisocyanate

Polyurethane

Reactants with an even number of carbon atoms as used in Eq. (7.62) produce higher-melting polymers than those with an odd number of carbon atoms. The melting point is decreased as the number of methylene groups is increased, and increased by the incorporation of stiffening groups such as phenylene groups.

Isocyanates react with water to produce unstable carbamic acids, which decompose to form diamines and carbon dioxide, which acts as a blowing agent. Hence, polymeric foams are produced when traces of moisture are present in the reactants. Since many of these forms are formed in situ and isocyanates are toxic, it is preferable to use isocyanate-terminated prepolymers. The latter are prepared from flexible or rigid hydroxyl-terminated polyesters or polyethers.

Cross-linked polyurethane coatings, elastomers, or foams may be produced by using an excess of the diisocyanate, which reacts with the urethane hydrogen to produce an allophanate, or by incorporating polyols such as glycerol or pentaerythritol in the reactant mixture. The diamines produced by the decomposition of carbamic acids react with diisocyanates to produce polyureas. Equations for the formation of an allophanate and a polyurea are shown below.

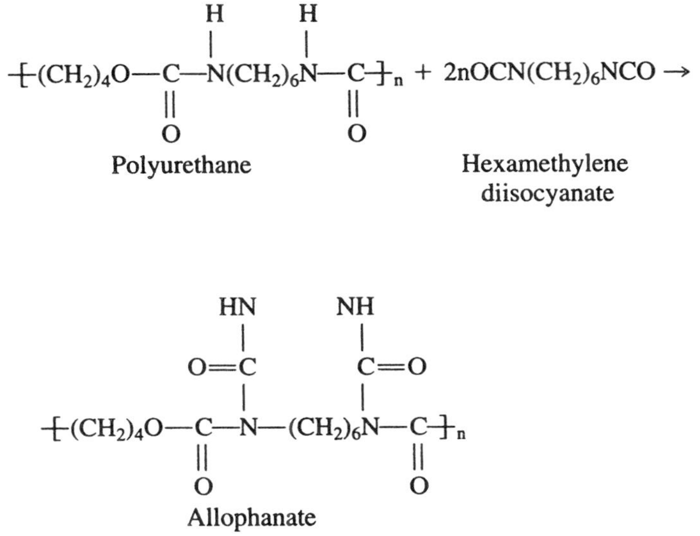

$$\text{---}(CH_2)_4O\text{---}\underset{\underset{O}{\|}}{C}\text{---}\underset{H}{N}(CH_2)_6\underset{H}{N}\text{---}\underset{\underset{O}{\|}}{C}\text{---}]_n + 2n\,OCN(CH_2)_6NCO \rightarrow$$

Polyurethane   Hexamethylene
diisocyanate

$$(7.63)$$

$$\text{---}(CH_2)_4O\text{---}C\text{---}N\text{---}(CH_2)_6\text{---}N\text{---}C\text{---}]_n$$

Allophanate

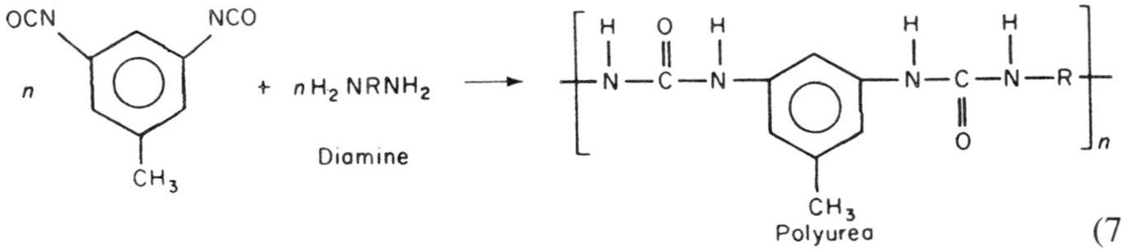

$$(7.64)$$

7.10 POLYSULFIDES

Thiokol, which was the first synthetic elastomer, was synthesized by Patrick by the condensation of alkylene dichlorides and sodium polysulfides in the 1920s. These solvent-resistant elastomers have limited use because of their foul odor. They may be reduced to liquid polymers (LP-2), which may be reoxidized to solid elastomers in caulking material and in some rocket propellant formulations. The equations for the production of these alkylene polysulfides are shown below.

$$nCl\text{---}(CH_2)_2O\text{---}(CH_2)_2CL + nNa_2S_x \longrightarrow$$

bis(2-Chloroethyl) ether   Sodium
polysulfide

$$\text{---}[(CH_2)_2O\text{---}(CH_2)_2S_x]_n + nNaCl$$

Poly(ethylene sulfide)   Sodium
chloride

$$\{(CH_2)_2O\{CH_2\}_2S_x\}_n \xrightarrow{Na_2SO_3} \{\{CH_2\}_2O\{CH_2\}_2S\}H$$

$$+ Na_2S_2O_3$$

$$2\{\{CH_2\}_2O\{CH_2\}_2S\}_nH + PbO_2 \longrightarrow \tag{7.65}$$

Liquid

$$\{\{CH_2\}_2O\{CH_2\}_2S\}_2 + PbO + H_2O$$

Solid

Poly(phenylene sulfide) (PPS; Ryton) is a solvent-resistant plastic that is useful in high-temperature service. This crystalline polymer is synthesized from p-dichlorobenzene as shown by the following equation:

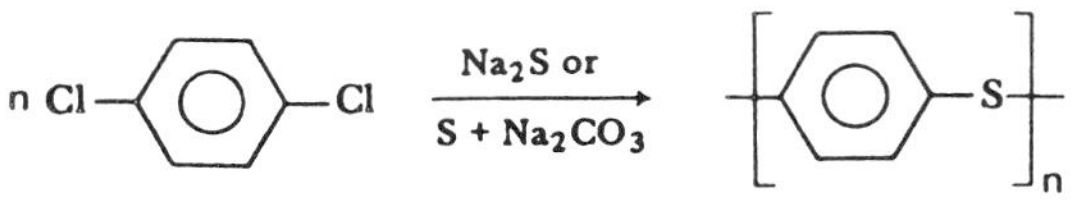

$$n\ Cl-\langle\bigcirc\rangle-Cl \xrightarrow[S + Na_2CO_3]{Na_2S\ or} \left[\{\bigcirc\}-S\right]_n \tag{7.66}$$

PPS is used for pumps, sleeve bearings, cookware, quartz halogen lamp parts, and electrical appliances.

7.11 POLYETHERS

Poly(phenylene oxide) (PPO; Noryl) is a high-temperature–resistant polymer produced by the oxidative coupling of 2,6-disubstituted phenols. As shown by the following equation, this step-growth polymerization is based on a room temperature oxidation, by the bubbling of oxygen through a solution of the phenol, in the presence of copper(I) chloride and pyridine.

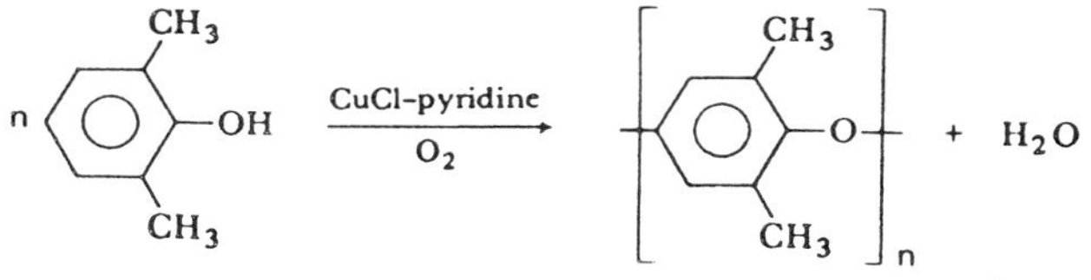

$$n\underset{CH_3}{\overset{CH_3}{\langle\bigcirc\rangle}}-OH \xrightarrow[O_2]{CuCl-pyridine} \left[\underset{CH_3}{\overset{CH_3}{\{\bigcirc\}}}-O\right]_n + H_2O \tag{7.67}$$

PPO extruded sheet is being used for solar energy collectors, life guards on broadcasting towers, airline beverage cases, and window frames.

Polyoxymethylene (POM), which precipitates spontaneously from uninhibited aqueous solutions of formaldehyde, was isolated by Butlerov in 1859 and investigated by Nobel laureate Staudinger in the 1920s. Polyoxymethylene stabilized by acetylation of the hydroxyl end groups (capping) was introduced commercially by du Pont under the trade name of Delrin in the late 1950s. The structure of the repeat unit in the copolymer is

$$\{CH_2-O\}$$

Stable acetal copolymers of formaldehyde, called Hostoform and Celcon, were also produced commercially by the cationic copolymerization of formaldehyde with ethylene oxide. The structure of the repeat unit in the copolymer is

$$\{CH_2-O-CH_2-CH_2-O\}$$

The selling price of polyacetal and other engineering plastics is about one-half that of cast metals. Polyacetals have been approved by the Food and Drug Administration for contact with foods. Some of the uses for molded polyacetals are as follows: valves, faucets, bearings, appliance parts, springs, automobile window brackets, hose clamps, hinges, gears, shower heads, pipe fittings, video cassettes, tea kettles, chains, flush toilet float arms, pasta machines, desktop staplers, and air gun parts.

These unique polymers are resistant to many solvents, aqueous salt and alkaline solutions, and weak acids at a pH greater than 4.5. More complete data on physical properties and resistance to hostile environments of these and other polymers are available.

Epoxy resins (ethoxyline resins), under the trade names of Araldite and Epon, were synthesized in the 1940s by a step-reaction polymerization between epichlorohydrin and bisphenol A. The DP of the prepolymer produced in this reaction is dependent on the ratio of the reactants. The low molecular weight liquid prepolymer is cured, or cross-linked, at room temperature by the addition of alkylene polyamines and at high temperatures by the addition of cyclic carboxylic anhydrides. The equations for the production and curing of these widely used molding, laminating, and surface-coating resins are shown below.

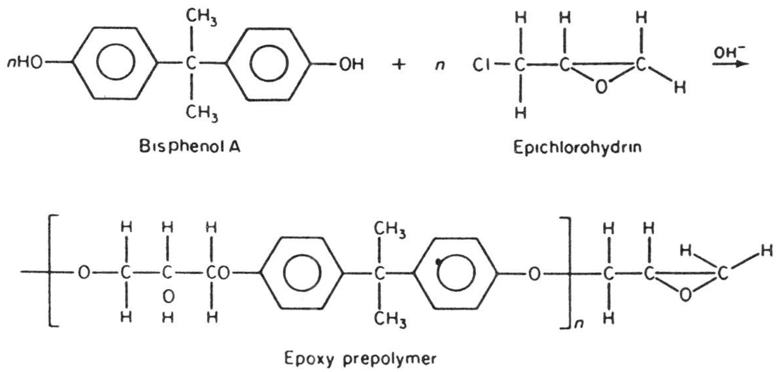

Bisphenol A   +   Epichlorohydrin $\xrightarrow{OH^-}$

$$(7.68)$$

Epoxy prepolymer

$$
\text{Epoxy prepolymer} \quad + \quad \begin{array}{c} \text{alkylene polyamine} \\ \text{or} \\ \text{cyclic anhydride} \end{array} \quad \longrightarrow \quad \text{cured epoxy resin}
$$

$$(7.69)$$

The high molecular weight thermoplastics called phenoxy resins are produced by the hydrolysis of high molecular weight linear epoxy resins. These transparent resins, whose structures resemble epoxy resins, do not contain epoxide groups. They may be molded as such, or cross-linked through the hydroxyl pendant groups by diisocyanates or cyclic anhydrides.

Furan resins are produced by the acid-catalyzed polymerization of furfuryl alcohol or the products obtained by the condensation of this chemurgic product or furfural with acetone or maleic anhydride. The dark-colored resins produced from these unsaturated compounds are characterized by excellent resistance to alkalis, solvents, and nonoxidizing acids.

7.12 POLYSULFONES

While aliphatic polysulfones have been produced from the copolymerization of ethylene and sulfur dioxide, the more important polysulfones are the amorphous aromatic polymers. These high-impact polymers are produced by a Friedal-Crafts condensation of sulfonyl chlorides, as shown in the following equation. The engineering polymers (Astrel or Udel) polysulfone, polyethersulfone, and polyphenylsulfone have heat deflection temperatures of 174, 201, and 204°C, respectively.

$$n \; \langle O \rangle - O - \langle O \rangle - SO_2Cl \xrightarrow{AlCl_3} \left[\langle O \rangle - O - \langle O \rangle - \overset{\overset{O}{\|}}{\underset{\underset{O}{\|}}{S}} \right]_n + HCl \qquad (7.70)$$

    Diphenylene oxide sulfonyl chloride   Poly(aryl sulfone)   Hydrogen chloride

Another polysulfone (Astrel) is produced by the Friedel-Crafts condensation of biphenyl with oxy-bis(benzene sulfonyl chloride). Another polysulfone (Victrex) is produced by the alkaline condensation of bis(chlorophenyl) sulfone. Blends of polysulfones with ABS (Arylon, Mindel) and SAN (Ucardel) are available.

Polysulfones, which are produced at an annual rate of 20,000 tons, are used for ignition components, hair dryers, cookware, and structural foams.

Grumman has developed an automated triangular truss-type beam builder using graphite-reinforced polyether sulfone. This beam can be formed in outer space at the rate of 1.5 m per min from flat stock which is heated and forced continuously around a die. The sections are induction welded and may be protected from deterioration by the application of a coating of poly(ether ether ketone) (PEEK).

7.13 POLY(ETHER ETHER KETONE)

ICI has introduced a new crystalline poly(ether ether ketone) (PEEK), which has a glass transition of 145°C and a heat deflection temperature for reinforced PEEK of 300°C. PEEK is being considered for use as blow-molded containers for nuclear wastes, jet engine components, printed circuits, electrical applications, and wire coatings.

The structure of the repeat unit in PEEK is

$$\left[O - \langle O \rangle - \overset{\overset{O}{\|}}{C} - \langle O \rangle - O - \langle O \rangle - \overset{\overset{CH_3}{|}}{\underset{\underset{CH_3}{|}}{C}} - \langle O \rangle \right]$$

7.14 PHENOLIC AND AMINO PLASTICS

Baekeland showed that a relatively stable resole prepolymer could be obtained by the controlled condensation of phenol and formaldehyde under alkaline conditions. These linear polymers (PF) may be readily converted to infusible cross-linked polymers called resites by heating or by the addition of mineral acids. As shown by the following equation, the first products produced when formaldehyde is condensed with phenol are hydroxybenzyl alcohols. The linear resole polymer is called an A-stage resin, and the cross-linked resite is called a C-stage resin.

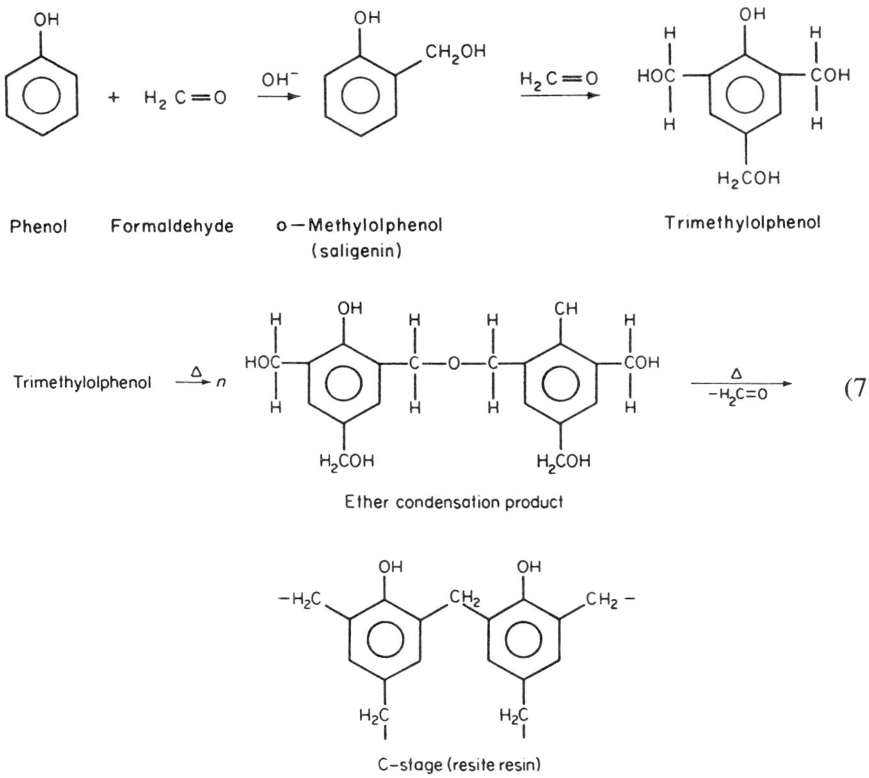

Phenol   Formaldehyde   o—Methylolphenol   Trimethylolphenol
(saligenin)

Ether condensation product

$$(7.71)$$

C—stage (resite resin)

Baekeland recognized that the trifunctional phenol would produce network polymers and hence used difunctional ortho- or para-substituted phenols to produce soluble linear paint resins. As shown by the following equation, linear thermoplastic polymers may be produced by alkaline or acid condensation of formaldehyde with phenolic derivatives such as paracresol.

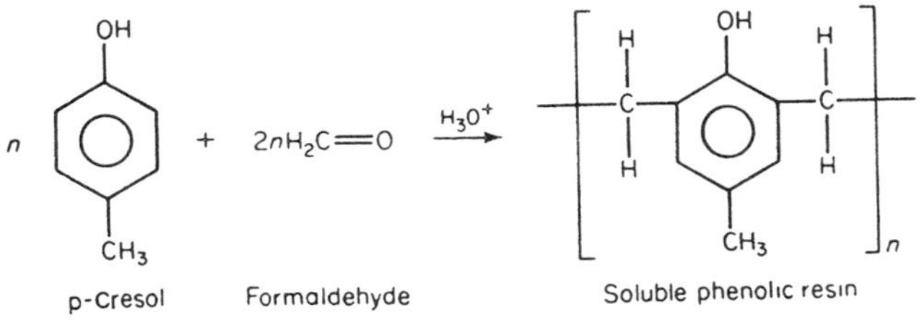

$$(7.72)$$

p-Cresol   Formaldehyde   Soluble phenolic resin

Since the acid condensation of 1 mol of phenol with 1.5 mol of formaldehyde produced infusible C-stage products, Baekeland reduced the relative amount of formaldehyde used and made useful novolac resins in a two-step process. Thus, a stable A-stage novolac resin is produced by heating 1 mol of phenol with 0.8 of formaldehyde in the presence of sulfuric acid.

The A-stage resin thus produced, after the removal of water by vacuum distillation, is cooled to yield a solid, which is then pulverized. The additional formaldehyde required to convert this linear polymer to an infusible thermoset resin is supplied by hexamethylenetetramine. The latter, which is admixed with the pulverized A-stage resin, is produced by the condensation of formaldehyde and ammonia.

Other essential ingredients, such as attrition ground wood (wood flour) filler, pigments, and lubricants, are also admixed with the resin and hexamethylenetetramine. The A-stage resin in this mixture is advanced, or further polymerized, by passing it through heavy heated rolls, through an extruder or a heated heavy-duty mixer. The term phenolic molding compound is applied to the granulated product containing the B-stage novolac resin.

The phenolic molding compound is converted to the infusible C-stage by heating it under pressure in cavities in a compression or transfer molding press. Plywood and other laminates are produced by heating and pressing a series of sheets coated with liquid resole resin. Articles such as automobile distributor heads are molded from novolac molding compounds. The equation for making and curing novolac resins is shown below.

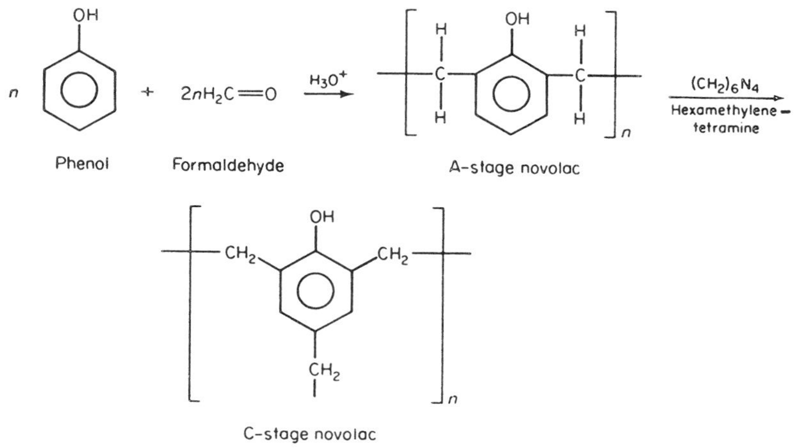

Phenol   Formaldehyde   A-stage novolac

(7.73)

C-stage novolac

While the condensation of urea and formaldehyde had been described in 1884, urea-formaldehyde (UF) resins were not patented until 1918. Comparable products based on the condensation of formaldehyde and melamine (2,4,6-triamino-1,3,5-triazine) were not patented until 1939. The term amino resins is now used to describe both urea and melamine-formaldehyde (MF) resins.

Urea and melamine are tetra- and hexa-functional molecules, respectively. However, the formation of network polymers is prevented by adding alcohols such as n-butanol and by condensing with formaldehyde at low temperature under alkaline conditions. While phenolic resins have better moisture and weather resistance than urea resins, the latter are preferred for light-colored objects.

For example, the interior layers of laminated countertops are bonded together by phenolic resins, but either urea or melamine resins are used for the decorative surface. Melamine plastics are more resistant to heat and moisture than UF and hence are used for the decorative surface and for the manufacture of plastic dinnerware. As shown by the following equations, the intermediate reaction products are methylol derivatives. Also, cyclization occurs when UF resins are cured in the presence of acids.

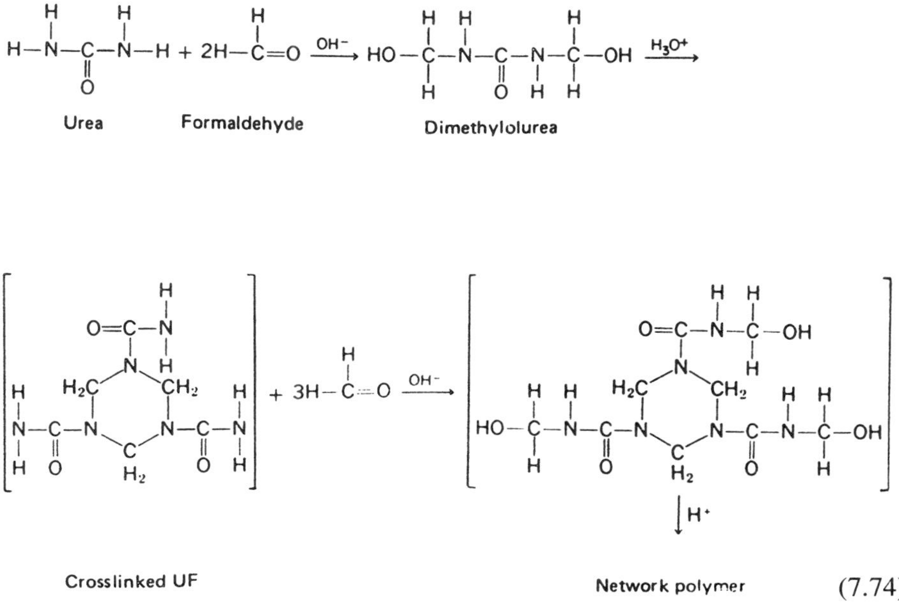

Urea   Formaldehyde   Dimethylolurea

Crosslinked UF   Network polymer   (7.74)

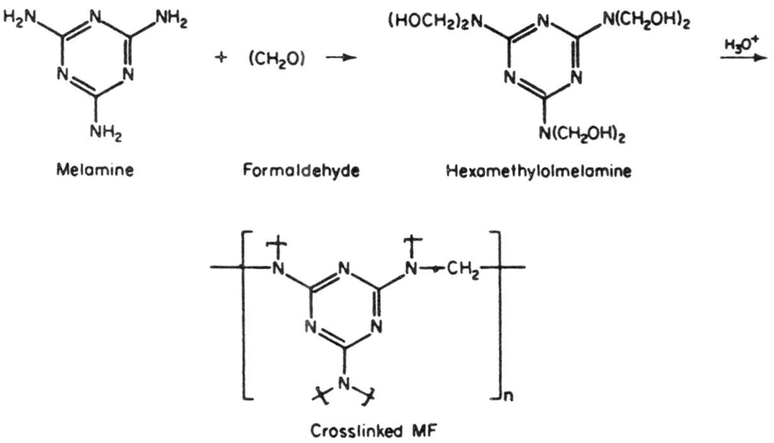

Melamine   Formaldehyde   Hexamethylolmelamine

(7.75)

Crosslinked MF

As shown in Table 7.2, amino resins, phenolic resins, epoxy resins, saturated and unsaturated polyester resins, nylons, and polyurethanes are produced on a relatively large scale by step-reaction polymerization. Synthetic fibers, thermoplastics, thermosetting plastics, adhesives, coatings, and plastic foams are produced by this technique. However, few synthetic elastomers are produced by step-reaction polymerization, and actually most synthetic polymers are produced by chain-reaction polymerization, which is discussed in the next chapter.

7.15 GENERAL INFORMATION ON STEP-REACTION POLYMERIZATION

Since the average molecular weight increases with conversion, useful high molecular weight linear polymers may be obtained by step-reaction polymerization when

Table 7.2 Production of Polymers in the United States by Step-Reaction Polymerization (Millions of lbs.: 1995)

Synthetic fibers (1995)	
Nylon	2,750
Polyester	3,900
Total	6,650
Thermosetting plastics (1995)	
Phenolics	3,230
Polyesters	1,470
Ureas	1,910
Epoxies	600
Melamines	300
Total	7,510
Coatings	
Alkyds	~750
Total for principal synthetic step-reaction polymers	14,810

the fractional conversion (p) is high (>0.99). The concentration of reactants decreases rapidly in the early stages of polymerization, and polymers with many different molecular weights will be present in the final product. The requirement for a linear polymer is a functionality of 2. Network polymers are usually produced when the functionality is greater than 2. The original reactants and all products resulting from their condensation may react to produce higher molecular weight species. The rate constant k is similar to that of corresponding condensation reactions with monofunctional groups and remains essentially unchanged with higher molecular weight species.

7.16 POLYCONDENSATION MECHANISMS

Proposed mechanisms for polycondensations are essentially the same as those proposed in the organic chemistry of smaller molecules. Here we will only briefly consider several examples to illustrate this similarity between proposed mechanisms for reactions involving smaller molecules and those that eventually produce polymers. For instance, the synthesis of polyamides (nylons) can often be envisioned as a simple S_N2 type of Lewis acid-base reaction with the Lewis base nucleophilic amine attacking the electron-poor, electrophilic carbonyl site followed by loss of a proton. A similar mechanism can be proposed for most polyesterifications.

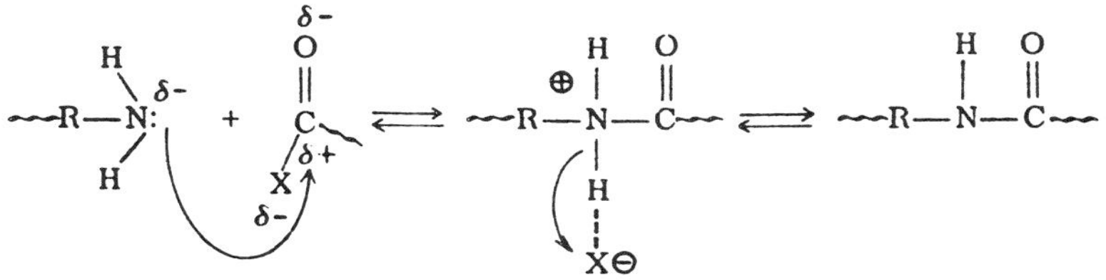

$$(7.76)$$

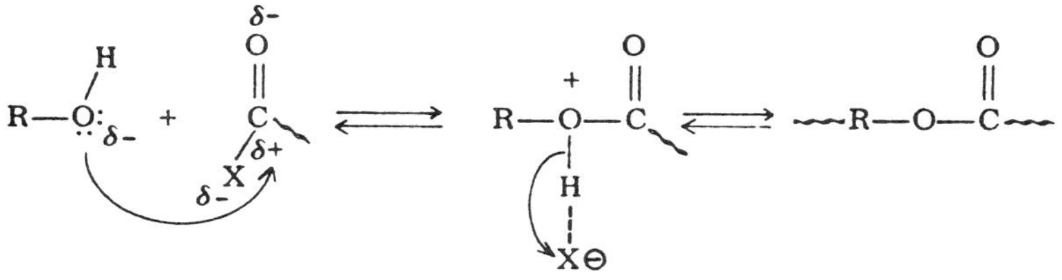

$$(7.77)$$

and a related pathway

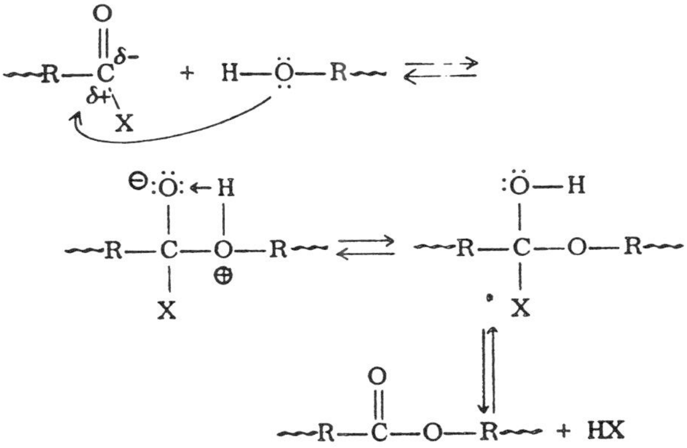

$$(7.78)$$

Below are a number of resonance forms for the isocyanate moiety illustrating the overall electrophilic nature of the carbon atom, giving overall

$$R—\ddot{N}=C=\ddot{O}: \longleftrightarrow R—\overset{\ominus}{\ddot{N}}—\overset{\oplus}{C}=\ddot{O}: \longleftrightarrow R—N=\overset{\oplus}{C}—\ddot{O}:^{\ominus} \longleftrightarrow$$

$$R—\overset{\ominus}{\ddot{N}}—\overset{\oplus}{C}\equiv O: \longleftrightarrow R—\overset{\oplus}{N}\equiv C—\ddot{O}:^{\ominus} \qquad (7.79)$$

an electron arrangement which can be described by

$$R—\overset{\delta-}{\ddot{N}}=\overset{\delta+}{C}=\overset{\delta-}{O}$$

Polyurethene formation occurs with attack of the nucleophilic alcohol at the electron-poor isocyanate carbon with a proton shift followed by a rearrangement to the urethane linkage.

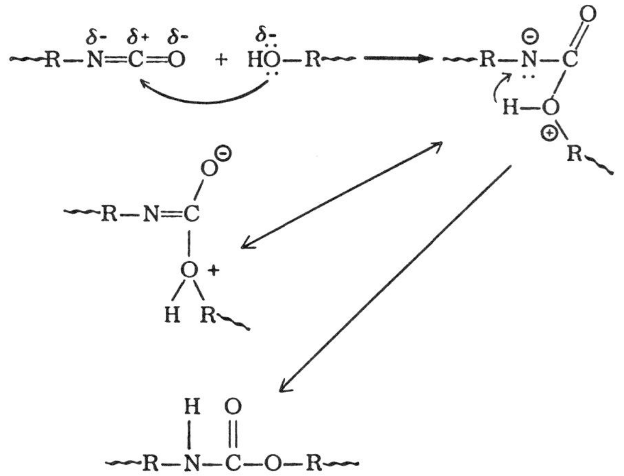

$$(7.80)$$

Polyether formations, such as the formation of poly(ethylene oxide) from ethylene oxide, can occur either through acid or base catalysis as depicted below.

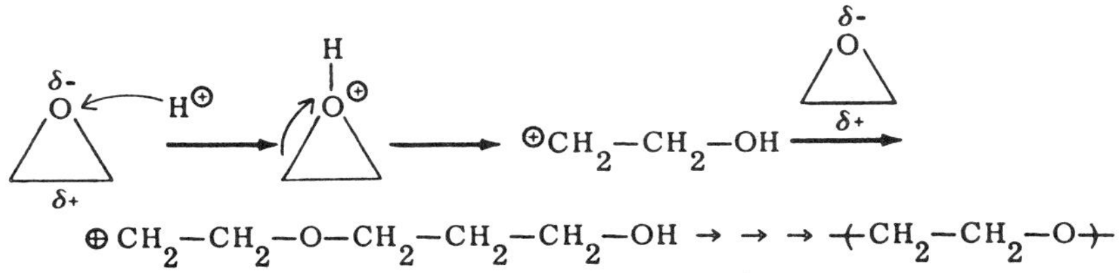

$$(7.81)$$

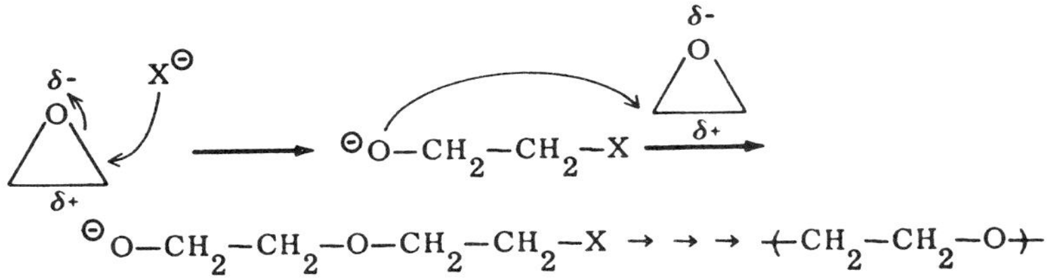

$$\oplus CH_2-CH_2-O-CH_2-CH_2-CH_2-OH \rightarrow \rightarrow \rightarrow +CH_2-CH_2-O+$$

$$(7.82)$$

$$\ominus O-CH_2-CH_2-O-CH_2-CH_2-X \rightarrow \rightarrow \rightarrow +CH_2-CH_2-O+$$

where X = $\ominus$OH, $\ominus$OR. The topic of polyether formation is discussed in Sec. 7.1.

Again, the mechanistic pathways suggested for condensations involving smaller molecules can generally be directly applied to polycondensation processes.

7.17 SYNTHETIC ROUTES

The previous sections contain the general synthesis for a number of important condensation polymers. Here we will consider briefly the three main synthetic techniques utilized in the synthesis of condensation polymers. This will be followed by a discussion of some considerations that must be taken into account when choosing a given synthetic procedure utilizing specific examples to illustrate the points.

The *melt* technique is also referred to by other names to describe the same or similar processes. These names include high melt, bulk melt, and simply bulk or neat. The melt process is an equilibrium-controlled process in which polymer is formed by driving the reaction toward completion, usually by removal of the by-product. For polyesterifications involving the formation of water or HCl, the driving force is the elimination, generally by a combination of reduced pressure and applied heat, of the water or HCl.

$$nCl-\overset{O}{\underset{\|}{C}}-R-\overset{O}{\underset{\|}{C}}-Cl + nHO-R-OH \rightleftarrows \rightleftarrows \rightleftarrows \cdots \cdot (C-R-\overset{O}{\underset{\|}{C}}-O-R-O)_n + HCl \quad (7.83)$$

$$nHO-\overset{O}{\underset{\|}{C}}-R-\overset{O}{\underset{\|}{C}}-OH + nHO-R-OH \rightleftarrows \rightleftarrows \rightleftarrows \cdots \cdot (C-R-\overset{O}{\underset{\|}{C}}-O-R-O)_n + H_2O \quad (7.84)$$

Reactants are introduced along with any added catalyst to the reaction vessel. Heat is applied to melt the reactants, permitting their necessary intimate contact. Heating can be maintained at the reaction melt temperature or increased above it. Pressure is reduced. Typical melt polycondensations take several hours to several days before the desired polymeric product is achieved. Yields are of necessity high.

Solution condensations are also equilibrium processes, with the reaction often driven by removal of the byproduct by distillation or by salt formation with added base. Reactants generally must be more reactive in comparison with the melt technique, since lower temperatures are employed, with a number of solution processes occurring near room temperature. Solvent entrapment is a problem, but since a reaction may occur under considerably reduced temperatures compared to the melt technique, thermally induced side reactions are minimized. Side reactions with the solvent have been a problem in some cases. Because the reactants must be quite energetic, many condensations are not suitable for the solution technique.

The *interfacial* technique, while an old technique, only gained popularity with the work of Morgan and Carraher in the 1960s and 1970s. Many of the reactions can be carried out under essentially nonequilibrium conditions. The technique is heterophasic, with two fast-reacting reactants dissolved in a pair of immiscible liquids, one of which is usually water. The aqueous phase typically contains the Lewis base—a diol, diamine, or dithiol—along with any added base, or other additive. The organic phase consists of a Lewis acid, such as an acid chloride, dissolved in a suitable organic solvent, such as benzene, chloroform, diethyl ether, toluene, octane, or carbon tetrachloride. Reaction occurs near the interface (hence the name). The technique offers the ability to synthesize a wide variety of polymers ranging from modification of cotton, polyester synthesis, synthesis of nucleic acids, and synthesis of polycarbonates, the last being the only polymer produced on a large industrial scale using the interfacial technique. Figure 7.8 describes a simple assembly that can be rapidly put together to form nylon with the interfacial technique. (A note of caution: diamine, carbon tetrachloride, and acid chloride are harmful, and the reaction should be carried out with adequate ventilation and other necessary safety procedures, including rubber gloves.) A few drops of phenolphthalein added to the aqueous phase will give a more colorful nylon material. While aqueous-organic solvent systems are the rule, there are a number of other interfacial liquid-gas and nonaqueous liquid systems. With all the potential that the interfacial system offers, it has not attracted wide industrial use because of the high cost of the necessarily reactive monomers and the added expense of trying to cope with solvent removal and recovery.

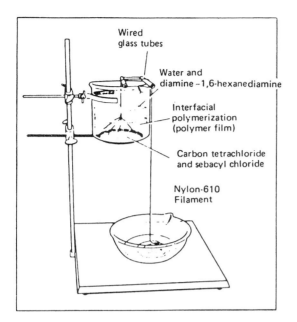

Figure 7.8 Self-propelled interfacial spinning of nylon-610 in the "nylon rope trick." (From R. Seymour and J. Higgins, *Experimental Organic Chemistry*, Barnes and Noble, New York, 1971. With permission of Barnes and Noble Publishing Company.)

In principle, any dibasic acid will condense with any diol or diamine. In practice, few such condensations have been utilized on an industrial scale for a variety of reasons, including availability of inexpensive comonomers in large quantity and undesirable chemical and physical properties of the synthesized polymer. For instance, several companies were interested in the synthesis of aromatic amines for commercial use. Aromatic amines offered much greater tear resistance and strength compared with an aliphatic nylon such as nylon-66. While aromatic amines could be synthesized using several routes, each produced a solid nylon which was only with difficulty soluble in a limited number of undesirable solvents. Solution of the nylon was necessary for fabrication of the polymer. Morgan and others noted that some polymers formed with rapid stirring would remain in solution for different periods of time. Today, aromatic amides, under the name aramids, are synthesized using rapidly stirred systems which permit the aromatic nylon to remain in solution long enough to permit fabrication.

Poly(ethylene terephthalate) is the best known polyester being used as a film (Mylar) and as a fiber (Dacron and Terylene). It can be produced using any of the three major polycondensation procedures. It can be rapidly prepared from the acid chloride using the aqueous interfacial system, giving poor to good yield and short to long chain lengths. Due to the excessive cost of acid chlorides, the interfacial technique of synthesis is presently ruled out industrially. In solution and bulk, polymerization rate is slow, so excess ethylene glycol is employed to increase the esterification rate. The use of excess diol, while producing the polyester at an acceptable rate, effectively decreases molecular weight. This is overcome by subsequent removal of the excess ethylene glycol. The necessity of using two steps in the production of high molecular weight polyester is undesirable in terms of both time and material.

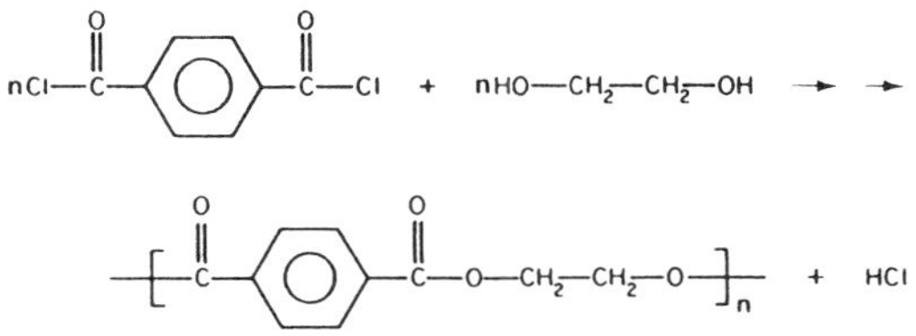

(7.85)

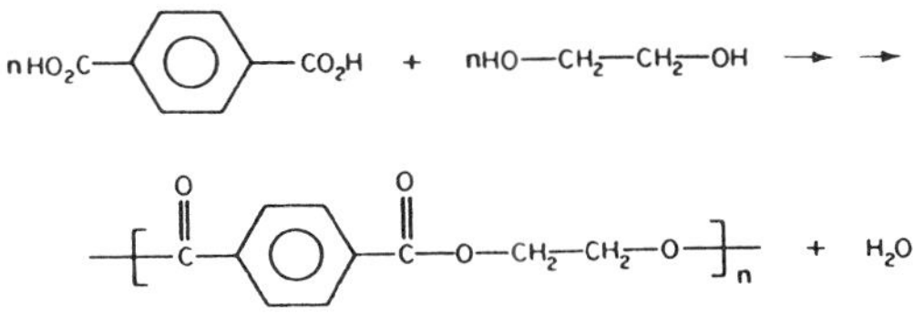

(7.86)

Terephthalic acid is insoluble in most common organic solvents and is also high melting (>300°C), compounding the problem of utilizing it directly for bulk and solution syntheses. Thus, the transesterification reaction utilizing the dimethyl ester of terephthalic acid with removal of methanol and subsequent removal of the "excess" glycol offers an attractive alternative. Yet because the necessary technology, including facilities, are already present to handle terephthalic acid itself, this alternative will probably not become a major part in the industrial synthesis of poly(ethylene terephthalate). Thus, consideration of previous investment of time and facilities are important in deciding the particular polycondensation technique utilized.

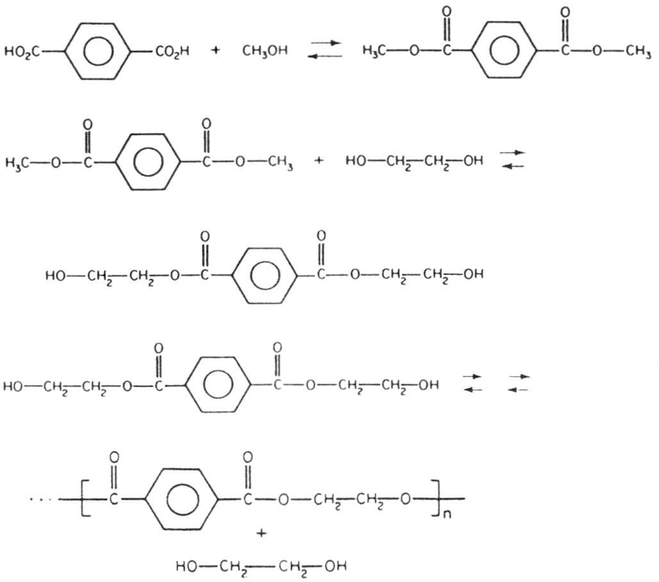

(7.87)

Table 7.3 contains a comparison of the three systems. Other often-noted liabilities, strengths, and comparisons of the systems include the following. A characteristic of melt polycondensations is that they are slow and require that conversion be high before high polymer is formed. Most melt systems utilize less reactive reactants than called for in solution and interfacial systems. Thus, higher temperatures are necessary to achieve a reasonable reaction rate. Undesirable side reactions, depolymerization, and degradation of thermally unstable reactants and products can occur at such elevated temperatures. On the other hand, the less reactive reactants are generally less expensive than those called for in solution and interfacial syntheses. The need for near-equivalence of reactants to achieve high polymers is much greater for the melt process than for the other two processes. Solution and interfacial systems have the added expense of solvent utilized in the reaction system and subsequent solvent removal from the polymer. The solution and interfacial systems are often collectively known as the low-temperature meth-

Table 7.3 Comparison of Requirements Between Different Polycondensation Techniques

Requirement	Melt	Solution	Interfacial
Temperature	High	Limited only by MP and BP of solvents utilized—generally about room temperature	
Stability to heat	Necessary	Unnecessary	Unnecessary
Kinetics	Equilibrium, stepwise	Equilibrium, stepwise	Often nonequilibrium; chainlike on a macroscopic level
Reaction time	1 hr to several days	Several minutes to 1 hr	Several minutes to 1 hr
Yield	Necessarily high	Low to high	Low to high
Stoichiometric equivalence	Necessary	Less necessary	Often unnecessary
Purity of reactants	Necessary	Less necessary	Less necessary
Equipment	Specialized, often sealed	Simple, open	Simple, open
Pressure	High, low	Atmospheric	Atmospheric

ods. Assets of the low-temperature methods often cited include use of cis, trans, or optically active structures without rearrangement; direct polymerization to polymer coatings, fibrous particles, small articles, wires, fibers, and films; and the possible synthesis of thermally unstable products and use of thermally unstable reactants. Most industrial processes utilize the melt and solution techniques, although the interfacial technique is gaining limited use.

7.18   LIQUID CRYSTALS

Every day of our lives we run across liquid crystals. They are commonly found in computer monitors, digital clocks, and other read-out devices, for example.

Reintzer, in 1888, first reported liquid crystal behavior. In working with cholesteryl esters, he observed that the esters formed opaque liquids, which on heating turned clear. We now know that, as a general rule, many materials are clear if they are anisotropic, random, or if the materials are composed of ordered molecules or segments of molecules, whereas they are opaque if they contain a mixture of ordered and disordered regions. Lehmann interpreted this behavior as evidence of a third phase that exists between the solid and isotropic liquid states. This new phase was named by Lehmann the *liquid crystal* phase. Friedel called this phase the *mesophase*, after the Greek word *mesos*, meaning intermediate. The initial molecules investigated as liquid crystals were large monomeric molecules.

Flory, in 1956, predicted that solutions of rodlike polymers could also exhibit liquid crystal behavior. The initial synthetic polymers found to exhibit liquid crystal behavior were concentrated solutions of poly(gamma-benzyl glutamate) and poly(gamma-methyl glutamate). These polymers exist in a helical form that can be oriented in one direction into ordered groupings, giving materials with anisotropic properties.

Liquid crystals (LC) are materials that undergo physical reorganization where at least one of the rearranged structures involve molecular alignment along a preferred direction causing the material to exhibit nonisotropic behavior and associated birefringent properties, i.e., molecular asymmetry.

Liquid crystalline materials can be divided into two large groups: thermotropic and lyotropic. Thermotropic liquid crystals are formed when "pure" molecules such as cholesteryl form ordered structures upon heating. When liquid crystals occur through mixing with solvent, they are called lyotropic liquid crystals.

Thermotropic liquid crystals can be further divided into (1) enantiotropic materials, where the liquid crystal phases are formed on both heating and cooling cycles, and (2) mesotropic materials, where the liquid crystals are stable only on supercooling from the isotropic melt. The mesotropic liquid crystals have been further divided into three groups:

Smectic, meaning "soap"
Nematic, meaning "thread"
Cholesteric, derived from molecules with a chiral center

Liquid crystal polymers are typically composed of materials that are rigid and rodlike with a high length-to-breath ratio or materials that have a disc shape. The smaller groups that give the material liquid crystal behavior are called *mesogens*. These mesogens are simply portions of the overall polymer that are responsible for forming the anisotropic liquid crystal and that, in fact, form the liquid crystal segments. Such mesogens can be composed of segments from the backbone of the polymer, segments from the side chain, or segments from both the backbone and side chain.

The mesogens form the ordered structures necessary to give the overall material anisotropic behavior. A number of different mesogen groupings have been identified. Chains arranged so that the mesogen portions are aligned in one preferred direction with the ordering occurring in a three-dimensional layered fashion compose one group of arrangements called smectic arrangements (Figure 7.9). Here, the lateral forces between the mesogen portions are relatively higher than the lateral forces between the nonmesogen portions, allowing a combination of segments that permit "flowing" (the passage of nonmesogen portions) and segments that retain contact (mesogen portions) as the material flows imparting a "memory"-type behavior of the material. A number of different "packings" of the mesogens have been identified (Fig. 7.9). The most ordered of the mesogenic groupings is called smectic B, which is a hexagonally, close-packed structure present in a three-dimensional arrangement. A much less ordered grouping is called the Smectic A

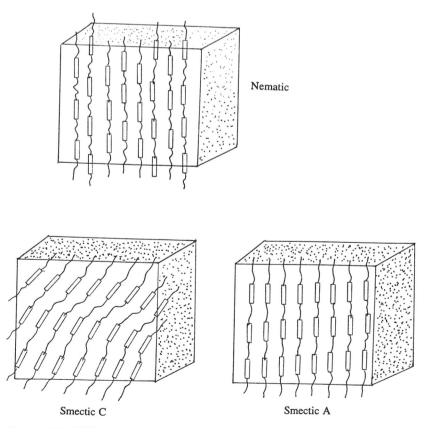

Figure 7.9 Different mesophasic structures where the mesogenic unit is designated as □.

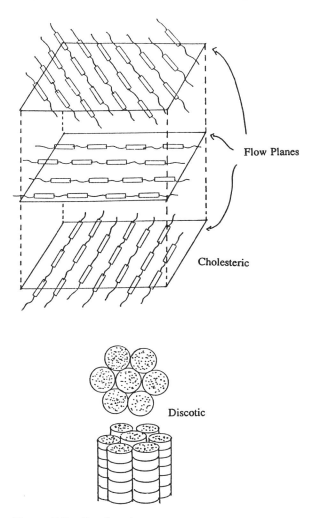

Figure 7.9 Continued.

phase, in which there is a somewhat random distribution of the mesogens between the layers.

Nematic liquid crystals offer much less order in comparison to smectic arrangements. Here, the directional ordering of the mesogen portions along one axis is retained, but the centers of gravity of the mesogen portions are no longer coupled (Fig. 7.9). Thus, the forces between the chains are less, resulting in a generally greater fluidity for nematic liquid crystals in comparison with smectic structures. Nematic liquid crystals still offer nonlinear behavior.

The chiral nematic assembly is formed by materials with chiral centers and that form a nematic phase. Here, a "chiral-imposed twist" is imparted to the linear chains composing each layer, resulting in a three-dimensional helical arrangement.

The molecular asymmetry typically occurs not because of intermolecular interaction, but because two molecules cannot occupy the same space at the same time. Molecular chains can exist in a random arrangement until a given concentration is exceeded, causing the molecules to rearrange in a more ordered fashion to accommodate the larger number of molecules within the same volume. Often, this occurs such that there is an ordered phase and a more random phase. As the concentration of polymer increases, the ordered phase becomes larger at the expense of the disordered phase. This increase in polymer concentration can occur via several routes, such as addition of more polymer, addition of a solution containing a higher concentration of polymer, and evaporation of the solvent.

For crystalline polymer systems, transition from the crystalline structure to a mesophere occurs, whereas in amorphous polymer systems, the mesophase occurs after the Tg has occurred. Some polymer liquid crystal systems form several mesophases, which can be detected using DSC, X-ray diffraction, and polarizing microscopy.

LC polymers include aromatic nylons (aramids), aromatic polyesters, poly(alkyl isonitriles), and polyorganophosphazines. Sample LCs are shown in Fig. 7.10.

Introduction of flexible "spacer" units such as methylene, methylene oxide, and dimethylsiloxane groups lowers the melting point and increases the temperature range within which the mesophase is stable. Often these spacer units are introduced by copolymerization. Thus, preformed p-acetoxybenzoic acid is reacted with poly(ethylene terephthalate), introducing a mesogenic unit in a polymer that has flexible spacer units (from the ethylene glycol) in it.

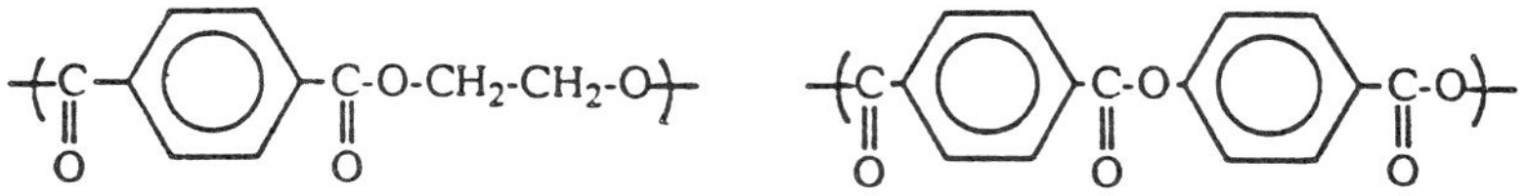

$$(7.88)$$

Poly(ethylene terephthate) **Mesogenic unit**

Along with the mesogen units contained within the polymer backbone, the mesogen units can occur as side chains. These mesogen units can be introduced either through reaction with monomers that contain the mesogen unit or through introduction with already formed polymers.

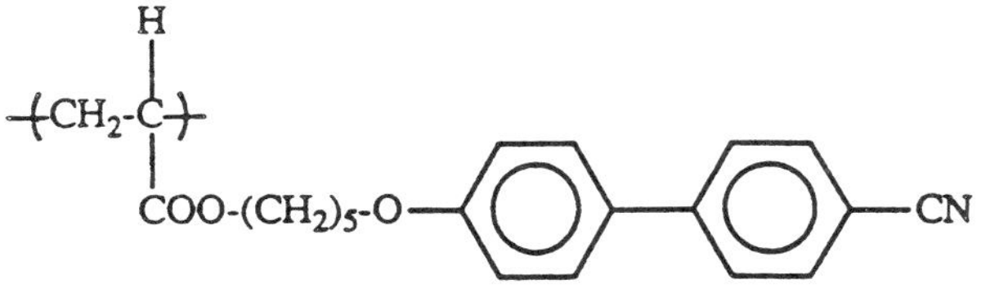

$$(7.89)$$

While most of us connect LCs with watches and PC screens, LC polymers can include all polymers that take on a rigid rod configuration somewhere within the polymer where the "popular" LC behavior is not involved. Aramid fibers are quite strong, with twice the tenacity and almost 10 times the modulus of aliphatic nylons, such as nylon-66. Many of these aramid fibers are stronger than steel wire on a weight basis. Three of the best known aramid fibers are pictured below.

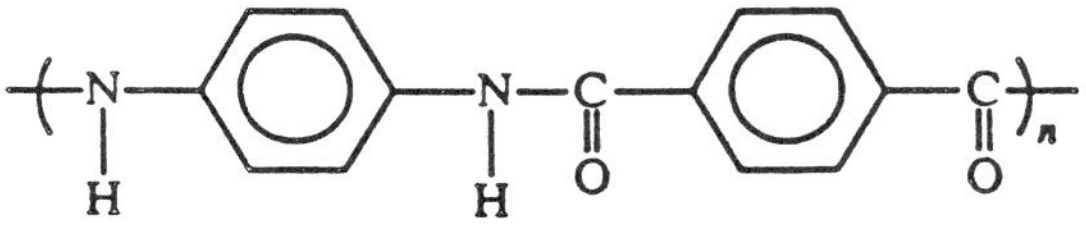

$$(7.90)$$

Poly(p-phenylene terephthalamide), Kevlar

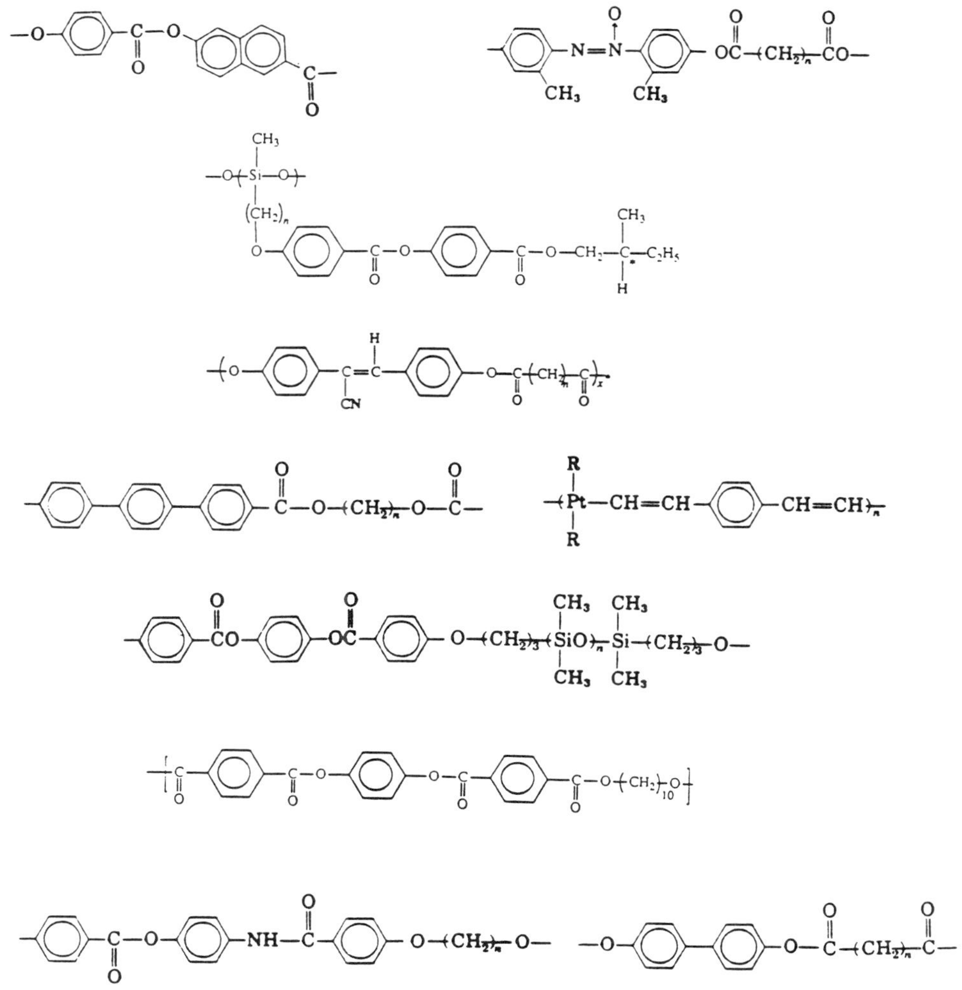

Figure 7.10 Representative structures of polymeric liquid crystals.

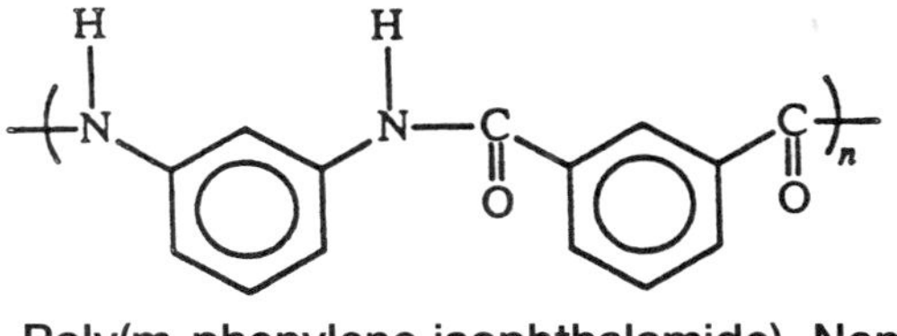

Poly(m-phenylene isophthalamide), Nomex

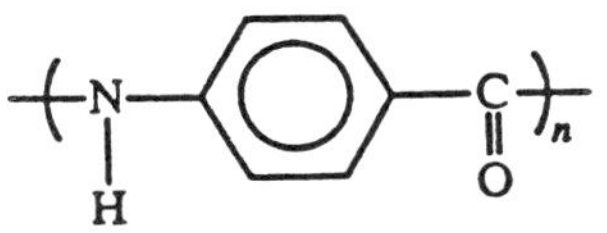

(7.91)

Poly(p-benzamide), Fiber B

(7.92)

These aramid fibers are often employed alone or with carbon or glass fibers for doors, aircraft panels, sporting goods, bullet-proof vests, tires, flooring, and car bodies.

LC materials have also been employed as films, plastics, and resins. Poly(1,4-benzoate) has been marketed under the name Ekonol. It decomposes before it melts, hence it does not form LC melts. Copolymerization with 4,4′-biphenol and terephthalic acid gives Ekkcel, which does melt before it decomposes. Certain forms can be compression-molded and others injection-molded. Reaction of poly(1,4-benzoate) with PET gives a material that can be injection-molded. These LCs are chemical resistant, with high tensile strength.

LC films with mesogenic side chains can be used in information storage devices.

SUMMARY

1. Many naturally occurring and some synthetic polymers are produced by condensation reactions which are described kinetically by the term step-reaction polymerization. Since a high fractional conversion (p) is required, only a relatively few useful linear polymers, such as polyesters, polycarbonates, polyamides, polyurethanes, polyureas, polysulfides, polyoxides, and polysulfones, may be synthesized by step reactions. However, this technique may be used to produce network polymers even when the fractional conversion is relatively low. Phenolic, epoxy, urea, and melamine resins are produced by step-reaction polymerization.

2. Since the fractional conversions are very high, useful polymers may be produced by the stepwise condensation of a bifunctional acyl or sulfonyl chloride with a diamine or glycol. The rate constant k for these second- and third-order reactions is similar to corresponding reactions of monofunctional reactants and is essentially unchanged as the reactions proceed through dimers, tetramers, octamers, oligometers, and higher molecular weight polymers. This rate constant increases with temperature in accordance with the Arrhenius equation.

3. The degree of polymerization $\overline{DP}$ of a reaction of bifunctional reactants may be calculated from the Carothers equation, $\overline{DP} = 1/(1 - p)$. However, this value will be changed if cyclization occurs to form "wasted loops." Since this is a step reaction, DP increases with time.

4. The index of polydispersity $\overline{M}_w/\overline{M}_n$ for the most probable molecular weight distribution is $1 + p$ for the certain stepwise kinetics. When the value of p is very high, the $\overline{DP}$ may be lowered by the inclusion of a small amount of a monofunctional reactant so that the functionality (f) is reduced to below 2.

5. High molecular weight linear polyesters may be produced by ester interchange or by interfacial condensation (Schotten-Baumann reaction). When the functionality of one of the reactants is greater than 2, branching may occur and incipient gelation (α_c) or cross-linking may take place. The critical value for α_c is $1/(f - 1)$.

6. The cross-linking of polyesters produced from phthalic anhydride and glycerol may be controlled using relatively low-temperature conditions under which the secondary hydroxyl group is not esterified until the temperature is increased. The cross-linking of unsaturated polyesters such as alkyds occurs by so-called drying

reactions in which cross-linking occurs in the presence of oxygen and a heavy metal catalyst.

7. High molecular weight polyamides, such as nylon-66, may be produced by the thermal decomposition of pure salts of the diamine and dicarboxylic acid reactants. The melting point of these polyamides is decreased as the number of methylene groups in the reactants is increased or if pendant groups are present. The introduction of stiffening groups such as phenylene groups increases the melting point. Polyamides, polybenzimidazoles, polyquinoxaline, and spiroketal polymers have many stiffening groups in the chains and are useful at high temperatures.

8. Polyurethanes and polyureas are produced by the room temperature reaction of a diisocyanate with a dihydric alcohol or a diamine, respectively. When water is present in the reactants, unstable carbamic acids are produced, which decompose to form carbon dioxide which serves as a blowing agent for foam production.

9. Flexible and rigid polysulfides are produced by a Williamson condensation of aliphatic reactants and a Wurtz condensation of aromatic reactants, respectively. Rigid polyoxides are produced by the low-temperature oxidative coupling of hindered phenols. Stable rigid aromatic polysulfones are produced by the Friedel-Crafts condensation of phenyl sulfonyl chlorides.

10. Low molecular weight epoxy resins are produced by the condensation of epichlorohydrin and dihydroxy compounds such as bisphenol A. These prepolymers, which contain hydroxyl pendant groups and epoxy end groups, may be cured at room temperature by the reaction of polyamines and at elevated temperature by the reaction with cyclic anhydrides.

11. Linear phenolic and amino resins may be produced by the condensation of formaldehyde with phenol, urea, or melamine under alkaline conditions at moderate temperatures. These A-stage resole resins may be advanced almost to incipient gelation (B-stage) by heating and cured to the C-stage by heating at higher temperatures or by the addition of acids. Novolac resins may be produced under acid conditions by condensing formaldehyde with an excess of phenol. These A-stage linear resins are advanced to the B-stage by heating with hexamethylenetetramine under moderate conditions and cured to the C-stage in a mold at higher temperatures.

12. When the functionality of phenol, urea, or melamine is reduced to 2 by the incorporation of substituents, the condensation with formaldehyde yields soluble thermoplastics.

GLOSSARY

advancing: Polymerizing further.

alkyds: Originally used to describe oil-modified polyesters, but now used for many polyester plastics and coatings.

allophanates: The reaction product of a urethane and an isocyanate.

α: Symbol for branching coefficient.

α_c: Critical value for incipient gelation.

amino resins: A term used to describe urea and melamine-formaldehyde resins.

Araldite: Trade name for an epoxy resin.

aramides: Aromatic polyamides.

A-stage: A linear prepolymer of phenol and formaldehyde.

Baekeland, Leo: The inventor of phenol-formaldehyde resins.

Bakelite: A polymer produced by the condensation of phenol and formaldehyde.

bifunctional: A molecule with two active functional groups.

bisphenol A: 2,2′-Bis(4-hydroxyphenyl)propane.

B-stage: An advanced A-stage resin.

C: Concentration at a temperature (T).

C, A: Concentration at a time T.

C_0, A_0: Original concentration.

carbamate: A urethane.

carbamic acids: Unstable compounds which decompose spontaneously to produce amines and carbon dioxide.

Carothers equation: $1/(1 - p)$.

Carothers, W. H.: The inventor of nylon, who also developed the kinetic equations for the step-reaction polymerization.

compression molding press: A press that uses external pressure to force a heat-softened molding compound into a die to produce a molded article.

condensation reaction: A reaction in which two molecules react to produce a third molecule and a byproduct such as water.

cyclization: Ring formation.

Dacron: Trade name for a polyester fiber.

drier: A catalyst for cross-linking by oxygen. Driers are soluble heavy metal salts such as cobalt naphthenate.

drying: Cross-linking of an unsaturated polymer in the presence of oxygen.

drying oil: An unsaturated oil like tung oil.

E_a: Energy of activation.

Ekanol: Trade name for poly-p-benzoate.

engineering plastics: Those with physical properties good enough for use as structural materials.

Epon: Trade name for an epoxy resin.

epoxy resin: A polymer produced by the condensation of epichlorohydrin and a dihydric alcohol or by the epoxidation of an unsaturated compound.

ester interchange: The reaction between an ester of a volatile alcohol and a less volatile alcohol in which the lower boiling alcohol is removed by distillation.

filament: The extrudate when a polymer melt is forced through a hole in a spinneret.

functionality: The number of active functional groups in a molecule.

functionality factor: The average number of functional groups present per reactive molecule in a mixture of reactants.

furan esin: Resin produced from furfuryl alcohol or furfural.

gel point: The point at which cross-linking begins.

glyptals: Polyesters, usually cross-linked by heating at elevated temperatures.

hexamethylenetetramine: A crystalline solid obtained by the condensation of formaldehyde and ammonia.

incipient gelation: The point at which $\overline{DP}$ equals infinity.

interfacial polymerization: One in which the polymerization reaction takes place at the interface of two immiscible liquids.

k: Symbol for rate constant.

Kodel: Trade name for a polyester fiber.

ladder polymer: A double-chained, temperature-resistant polymer.

laminate: Layers of sheets of paper or wood adhered by resins and pressed together like plywood.

long oil alkyd: One obtained in the presence of 65 to 80% of an unsaturated oil.

MF: Melamine-formaldehyde resin.

medium oil alkyd: An alkyd obtained in the presence of 50 to 65% of an unsaturated oil.

melamine-formaldehyde resin: A resin produced by the condensation of melamine and formaldehyde.

methylol: —CH_2OH.

molding compound: A name used to describe a mixture of a resin and essential additives.

N: Number of molecules.

network polymer: Infusible cross-linked polymer.

nonoil alkyd: An oil-free alkyd containing no unsaturated oils.

novolac: Polymers prepared by the condensation of phenol and formaldehyde under acidic conditions.

nucleophilic substitution: A reaction in which a nucleophilic reagent (Greek, nucleus loving) displaces a weaker nucleophile or base and the latter becomes the leaving group.

nylon: A synthetic polyamide.

nylon-610 (PA): A polyamide synthesized from a 6-carbon diamine and a 10-carbon dicarboxylic acid.

nylon rope trick: The preparation of a polyamide by the Schotten-Baumann reaction of a diacyl chloride and a diamine.

nylon salt: A salt of a diamine and a dicarboxylic acid used as the precursor of nylon.

oil length: A term used to indicate the relative percentage of unsaturated oils used in the production of alkyds.

ordered copolyamides: Polymers produced from a mixture of diamine and dicarboxylic acid reactants of different types.

p: Fractional conversion or fractional yield.

PA: Symbol for polyamide.

PBI: Symbol for polybenzimidazole.

Perlon U: A trade name for polyurethane.

PF: A phenolic resin.

phenoxy resin: A polymer with hydroxyl pendant groups resembling an epoxy resin without epoxy groups.

PI: Polyimide.

polyamide: A polymer with repeat units of

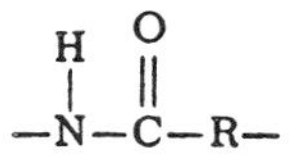

$$
\begin{array}{cc}
\text{H} & \text{O} \\
| & \| \\
\text{—N—C—R—}
\end{array}
$$

polybenzimidazole (PBI): A temperature-resistant heterocyclic polymer.

polycarbonate (PC): A polymer with the repeat unit of

$$
\begin{array}{c}
\text{O} \\
\| \\
\text{—O—C—OR—}
\end{array}
$$

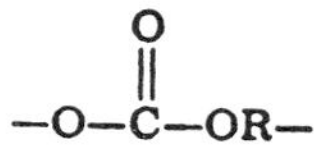

polyester: A polymer with the repeat unit of

$$
\begin{array}{c}
\text{O} \\
\| \\
\text{—COR—}
\end{array}
$$

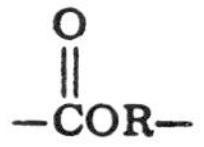

poly(ethylene terephthalate) (PET): A linear polyester used for fibers and for blow molding soft drink bottles.

polyimide (PI): A temperature-resistant heterocyclic polymer.

poly(phenylene oxide): A polymer with the repeat unit

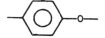

poly(phenylene sulfide): A polymer with the repeat unit

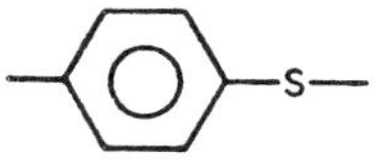

polysulfide: A polymer with the repeat unit —(RS)$_x$—.

polysulfone: A polymer with sulfone groups (SO$_2$) in its backbone.

polyurea: A polymer with the repeat unit of

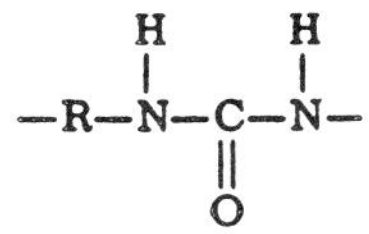

$$-R-\overset{H}{\underset{}{N}}-\overset{H}{\underset{\underset{O}{\parallel}}{C}}-\overset{H}{\underset{}{N}}-$$

polyurethane (PU): A polymer with the repeat unit of

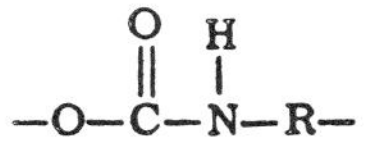

$$-O-\overset{O}{\underset{}{\overset{\parallel}{C}}}-\overset{H}{\underset{}{N}}-R-$$

prepolymer: A low molecular weight material (oligomer) capable of further polymerization produced by step-reaction polymerization.

PU: Polyurethane.

r: Molar ratio of reactants.

resite: A cross-linked resole.

resole: A linear polymer prepared by the condensation of phenol and formaldehyde under alkaline conditions.

Ryton: Trade name for poly(phenylene sulfide).

Schotten-Baumann reaction: A reaction between an acryl chloride and an alcohol or amine in the presence of sodium hydroxide or pyridine.

second-order reaction: A reaction in which the rate of proportional to the concentration of one reactant to the second power or to the product of two reactants to the first power.

short oil alkyd: An alkyd obtained in the presence of 30 to 50% of an unsaturated oil.

spiropolymer: Polymer having a structure resembling a spiral and thus consisting of a double chain.

step-reaction polymerization: Polymerization in which polyfunctional reactants react to produce larger units in a continuous stepwise manner.

t: Time in seconds.

T: Temperature (K).

TDI: 2,4-Tolylene diisocyanate.

Thiokol: Trade name for a polysulfide elastomer.

UF: Symbol for urea-formaldehyde resin.

unsaturated polyester: A term used to describe alkyds with unsaturated chains, particularly those produced by the condensation of maleic anhydride with ethylene glycol.

urea-formaldehyde resin: A resin produced by the condensation of urea and formaldehyde.

wasted loops: The formation of cyclic compounds instead of polymers.

wood flour: A filler produced by the attrition grinding of wood.

EXERCISES

1. Which of the following will yield a polymer when condensed with adipic acid: (a) ethanol, (b) ethylene glycol, (c) glycerol, (d) aniline, or (e) ethylenediamine?

2. Could Carothers have produced strong polyester fibers by ester interchange or Schotten-Baumann reactions using aliphatic reactants?

3. Which would be more useful as a fiber: (a) poly(ethylene terephthalate), or (b) poly(hexylene terephthalate)?

4. If the fractional conversion in an ester interchange reaction is 0.99999, what would be the $\overline{DP}$ of the polyester produced?

5. Use the logarithmic form of the Arrhenius equation to show that the value of the rate constant k increases as the temperature increases.

6. What is the first product produced when a molecule of sebacyl chloride reacts with a molecule of ethylene glycol?

7. What is the next product formed in question 6?

8. How would you improve the strength of the filament produced in the nylon rope trick without changing the reactants?

9. Name the product produced by the condensation of adipic acid and tetra-methylenediamine.

10. In which reaction would you expect the more "wasted loops": the reaction of oxalyl chloride with (a) ethylenediamine, or (b) hexamethylenediamine?

11. Which system would be more apt to produce "wasted loops": (a) a dilute solution, or (b) a concentrated solution?

12. If the values of A_0 and k are 10 mol liter^{-1} and 10^{-3} liter mol sec^{-1}, respectively, how long would it take to obtain a $\overline{DP}$ of 37?

13. Which will yield the lower index of polydispersity: (a) p = 0.999, or (b) p = 0.90?

14. If you used a 2% molar excess of bisphenol A with TDI, what would be the maximum $\overline{DP}$ obtainable assuming p = 1.

15. Why would the product obtained in question 14 be a useful fiber assuming $\overline{DP}$ = 100?

16. Assuming a value of 0.999 for p, what would be the $\overline{DP}$ of a polyester prepared from equimolar quantities of difunctional reactants in the presence of 1.5 mol% of acetic acid? Let the mol% difunctional reactants both be 98.5.

17. What is the functionality of a mixture consisting of 0.7 mol of ethylene glycol, 0.05 mol of ethanol, and 0.25 mol of glycerol?

18. What is the critical value of f for incipient gelation in the mixture described in question 17?

19. What is the functionality of a mixture consisting of 0.4 mol of pentaerythritol and 0.6 mol of diethylene glycol?

20. Which would be the better or stronger fiber: one made from an ester of (a) terephthalic acid, or (b) phthalic acid?

21. What would be the deficiency of a nylon film that was stretched in one direction only?

22. Which would be more flexible: (a) poly(butylene terephthalate), or (b) poly(hexylene terephthalate)?

23. Which would be more apt to deteriorate in the presence of moisture: (a) lexan molding powder, or (b) Lexan sheet?

24. How could you flexibilize Ekanol?

25. How would you prepare a nylon with greater moisture resistance than nylon-66?

26. How would you prepare a nylon that would be less "clammy" when used as clothing?

27. Which would be higher melting: (a) a polyamide, or (b) a polyester with similar numbers of methylene groups in the repeat units?

28. Why is a methoxymethylated nylon more flexible than nylon?

29. Which would perform better at high temperatures: (a) a polyimide, or (b) polyquinoxaline?

30. Isn't it wasteful to decompose a diisocyanate by hydrolysis to produce foams?

31. How would you prepare a hydroxyl-terminated polyester?

32. How would you prepare a more flexible polyalkylene sulfide than the one shown in Sec. 7.10?

33. Why is Ryton stiffer than Thiokol?

34. Why do polyurethanes and epoxy resins have good adhesive properties?

35. Why are furan resins relatively inexpensive?

36. Why it is necessary to add hexamethylenetetramine to a novolac molding compound?

37. Could you produce a soluble novolac resin from resorcinol?

38. Can you explain why there are so many terms used, such as novolac, resole, etc., in phenolic resin technology?

39. Why isn't Bakelite used for dinnerware?

40. Which of the following could be a nonpetrochemical plastic: (a) Bakelite, (b) urea plastics, or (c) melamine plastics?

41. Which would produce the better fiber: the reaction product of phthalic acid and (a) 1,4-butanediol, or (b) 2-hydroxybutanol?

BIBLIOGRAPHY

Aharoni, S. M. (1992): *Synthesis, Characterization and Theory of Polymer Networks and Gels*, Plenum, New York.

Arshady, R. (1994): *Polymer Synthesis*, Springer-Verlag, New York.

Baeyer, A. (1878): Phenol-formaldehyde condensates, Ber. Bunsenges. Phys. Chem., 5:280, 1094.

Bayer, O. (1941): Polyurethanes, Ann., 549:286.

Carothers, W. H. (1929): An introduction to the general theory of condensation polymers, J. Am. Chem. Soc., 51:2548.

———. (1938): Nylon, U.S. Patent 2,130,947.

Carothers, W. H., Arvin, J. A. (1929): Polyesters, J. Am. Chem. Soc., 51:2560.

Carraher, C. E. (1972): Group IVA polymers by the interfacial technique, Inorg. Macromol. Rev., 1:271.

Carraher, C. E., Moore, J. A. (1983): *Modification of Polymers*, Plenum, New York.

Carraher, C. E., Preston, J. (1982): *Interfacial Synthesis*, Vol. III, *Recent Advances*, Marcel Dekker, New York.

Evans, R. (1993): *Polyurethane Sealants: Technology and Applications*, Technomic, Lancaster, PA.

Hill, J. W., Carothers, W. H. (1932, 1933): Polyanhydrides, J. Am. Chem. Soc., *54*:1569, *55*:5023.

Hill, R. (1953): *Fibers from Synthetic Polymers*, Elsevier, Amsterdam.

Inderfurth, K. H. (1953): *Nylon Technology*, McGraw-Hill, New York.

Ivin, K. J., Saegusa, T. (1983): *Ring-Opening Polymerization*, Applied Science Publications, Essex, England.

Kienle, R. H. (1930, 1936): Alkyds, Ind. Eng. Chem., *22*:590, *55*:229T.

Mark, H. F. (1943): The mechanism of polymerization, Chap. 1 in *The Chemistry of Large Molecules* (R. E. Burke and O. Gummit, eds.), Interscience, New York.

Mark, H., Tobolsky, A. V. (1950): *Physical Chemistry of High Polymeric Systems*, Interscience, New York.

Mark, H. F., Whitby, G. S. (eds.) (1940): *The Collected Papers of Wallace Hume Carothers*, Interscience, New York.

Martin, S. M., Patrick, J. C. (1936): Thiokol, Ind. Eng. Chem., *28*:1144.

Marvel, C. S. (1959): *An Introduction to the Organic Chemistry of High Polymers*, John Wiley, New York.

Mathias, L. J., Carraher, C. E. (1984): *Crown Ethers and Phase Transfer Catalysis in Polymer Science*, Plenum, New York.

Millich, F., Carraher, C. E. (1977): *Interfacial Synthesis*, Vols. I and II, Marcel Dekker, New York.

Mijs, W. J. (1992): *New Methods for Polymer Synthesis*, Plenum, New York.

Morgan, P. W. (1965): *Condensation Polymers by Interface and Solution Methods*, Wiley, New York.

Morgan, P. W., Kwolek, S. L. (1959): The nylon rope trick, J. Chem. Ed., *36*:182.

Paleos, C. M. (1992): *Polymerization in Organized Media*, Gordon and Breach, New York.

Russell, D., Smiley, R. (1993): *Practical Chemistry of Polyurethanes and Diisocyanates*, Technomic, Lancaster, PA.

Sandler, S., Karo, W. (1993): *Polymer Synthesis No. 2*, 2nd ed., Academic Press, New York.

Schork, F. (1993): *Control of Polymerization Reactors*, Marcel Dekker, New York.

Soloman, D. H. (1974): *Step Growth Polymerizations*, Marcel Dekker, New York.

Sorenson, W. R., Campbell, T. W. (1968): *Preparative Methods of Polymer Chemistry*, 2nd ed., Wiley-Interscience, New York.

8

Ionic Chain-Reaction and Complex Coordinative Polymerization (Addition Polymerization)

In contrast to the slow step-reaction polymerization discussed in Chap. 7, chain-reaction polymerization is usually rapid, and the initiated species continue to propagate until termination. Thus, in the extreme case, one initiating species could be produced which would produce one high molecular weight polymer molecule, leaving all the other monomer molecules unchanged. In any case, the concentration of the monomer, which is usually a derivative of ethylene, decreases continuously throughout the reaction. In contrast to stepwise polymerization, the first species produced is a high molecular weight polymer.

A kinetic chain reaction usually consists of at least three steps, namely, initiation, propagation, and termination. The initiator may be an anion, a cation, a free radical, or a coordination catalyst. While coordination catalysts are the most important commercially, the ionic initiators will be discussed first in an attempt to simplify the discussion of *chain-reaction polymerization*.

8.1 CATIONIC POLYMERIZATION

The art of cationic polymerization, like many other types of polymerization, is at least a century old. However, the mechanisms for the early reactions were not understood, and, of course, the early scientists did not understand the modern concept of macromolecules. Nevertheless, it is of interest to note that turpentine, styrene, isobutylene, and ethyl vinyl ether were polymerized over a century ago by the use of cationic initiators such as sulfuric acid, tin(IV) chloride, boron trifluoride, and iodine.

The first species produced in these reactions are carbocations, and these were unknown as such prior to World War II. It is now known that pure Lewis acids, such as boron trifluoride or aluminum chloride, are not effective as initiators. A trace of a proton-containing Lewis base, such as water, is also required. As shown in Eq. (8.1), the Lewis base coordinates with the electrophilic Lewis acid, and the proton is the actual initiator. Since cations cannot exist alone, they are always accompanied by a *counterion*, also called a *gegenion*.

$$BF_3 \quad + \quad H_2O \quad \rightleftharpoons H^{\oplus}, \ (BF_3OH^-)$$

Lewis	Lewis	Catalyst-	
acid	base	cocatalyst	(8.1)
(boron	(cocatalyst)	complex	
trifluoride)			

Since the required activation energy for ionic polymerization is small, these reactions may occur at very low temperatures. The carbocations, including the macrocarbocations, repel each other, and, hence, chain termination cannot take place by combination, but is usually the result of reaction with impurities.

Both the initiation step and the propagation step are dependent on the stability of the carbocations. Isobutylene (the first monomer to be polymerized commercially by ionic initiators), vinyl ethers, and styrene may be polymerized by this technique. The order of activity for olefins is $(CH_3)_2C=CH_2 > CH_3(CH=CH_2) > CH_2=CH_2$, and for para-substituted styrenes, the order for the substituents is $OCH_3 > CH_3 > H > Cl$. The mechanism is also dependent on the solvent as well as the electrophilicity of the monomer and the nucleophilicity of the gegenion. Rearrangements may occur in ionic polymerizations.

The rate of initiation (R_i) for typical reactions, as shown in the following equation, is proportional to the concentration of the monomer [M] and the concentration of the catalyst-cocatalyst complex [C].

$$\underset{\substack{\text{Isobutylene} \\ (M)}}{H_2C=\overset{\displaystyle CH_3}{\underset{\displaystyle CH_3}{C}}} + \underset{\substack{\text{Catalyst-} \\ \text{cocatalyst} \\ \text{complex (C)}}}{(H^+, BF_3OH^-)} \xrightarrow{k_i} \underset{\substack{\text{Carbocation} \\ (M^+)}}{H_3C\overset{\displaystyle CH_3}{\underset{\displaystyle CH_3}{C}}\oplus} + \underset{\text{Gegenion}}{BF_3OH^-} \qquad (8.2)$$

$$R_i = k_i[C][M]$$

Propagation, or chain growth, takes place in a head-to-tail configuration as a result of carbocation (M^+) addition to another monomer molecule. The rate constant (k_p) is essentially the same for all propagation steps and is affected by the dielectric constant of the solvent. The rate is fastest in solvents with high dielectric constants, promoting separation of the carbocation-gegenion pairs. The chemical and kinetic equations for propagation are as follows:

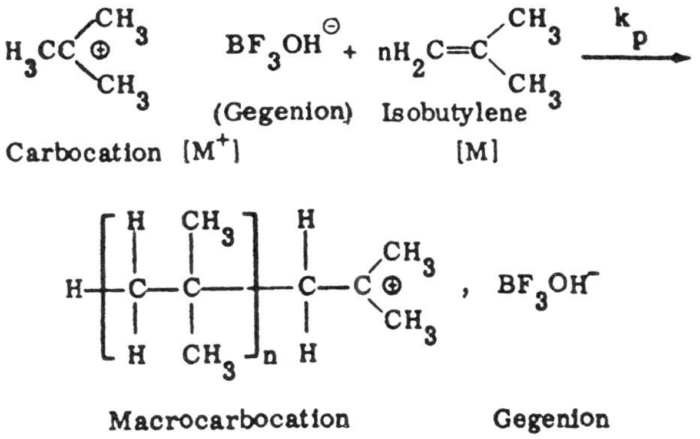

$$R_p = k_p [M][M^+]$$

The termination rate R_T, assumed to be a first-order process, is simply the dissociation of the macrocarbocation-gegenion complex forming BF_3 and H_2O and the now neutral polymer chain. This may be expressed as follows:

$$R_T = k_T[M^+] \tag{8.4}$$

Termination may take place by chain transfer, in which a proton is transferred to a monomer molecule [M], leaving a cation which may serve as an initiator. The $\overline{DP}$ is equal to the kinetic chain length (ν) when chain transfer occurs. The chemical and kinetic equations for chain transfer are shown below.

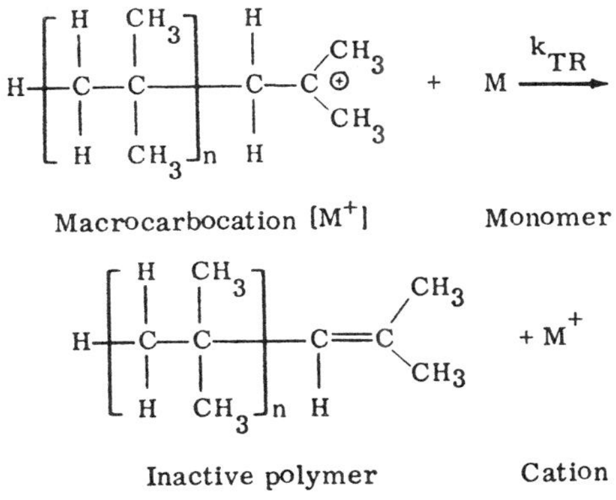

$$R_{Tr} = k_{Tr}[M][M^+]$$

Since it is difficult to solve these equations, which include [M$^+$], one assumes a steady state in which the rate of initiation equals the rate of termination, giving $R_i = R_T$, and solves for [M$^+$] as shown below:

$$k_1[C][M] = k_T[M^+], \text{ therefore, } [M^+] = \frac{k_1[C][M]}{k_T} \tag{8.6}$$

This expression for [M$^+$] may then be substituted in the propagation rate equation [Eq. (8.3)] and the overall rate for cationic polymerization:

$$R_p = k_p[M][M^+] = \frac{k_pk_1[C][M]^2}{k_T} = k'[C][M]^2 \tag{8.7}$$

We may also determine the value for $\overline{DP}$ when termination, via internal dissociation [Eq. (8.4)], is the dominant step, as follows:

$$\overline{DP} = \frac{R_p}{R_T} = \frac{k_p[M][M^+]}{k_T[M^+]} = \frac{k_p}{k_T}[M] = k''[M] \tag{8.8}$$

However, if chain transfer is the dominant step in the termination of a growing chain,

$$\overline{DP} = \frac{R_p}{R_{Tr}} = \frac{k_p[M][M^+]}{k_{Tr}[M][M^+]} = \frac{k_p}{k_{Tr}} = k''' \tag{8.9}$$

It is important to note that regardless of how termination takes place, the molecular weight of a polymer synthesized by the cationic process is independent of the concentration of the initiator. However, the rate of ionic chain polymerization is dependent on the dielectric constant of the solvent, the resonance stability of the carbonium ion, the stability of the gegenion, and the electropositivity of the initiator.

Polyisobutylene (Vistanex, IM) is a tacky polymer with a very low T_g ($-70°C$) used as an adhesive, a caulking compound, a chewing gum base, and an oil additive. Its use as an oil additive is related to its change in shape with increasing temperature. Since lubricating oil is not a good solvent for polyisobutylene, polyisobutylene is present as a coil at room temperature when mixed with an oil. However, the chain tends to uncoil as the temperature increases and as the oil becomes a better solvent. This effect tends to counteract the decrease in viscosity of the oil as the temperature is increased.

Butyl rubber (IIR), widely used for inner tubes and as a sealant, is produced by the cationic low-temperature copolymerization of isobutylene in the presence of a small amount of isoprene (10%), as shown in Scheme 8.1. Thus, the random copolymer chain contains a low concentration of widely isolated double bonds which assure a low cross-linked density as a result of the formation of large "principal sections" when the butyl rubber is vulcanized or cured. (Copolymers are discussed in Chap. 10.)

When the gegenion and the carbocation present in the polymerization of vinyl isobutyl ether form an ion pair in propane at $-40°C$, stereoregular polymers are produced. The carbocations of vinyl alkyl ethers are stabilized by the delocalization of p electrons in the oxygen atom, and thus these monomers are readily polymerized by cationic initiators. Poly(vinyl isobutyl ether)

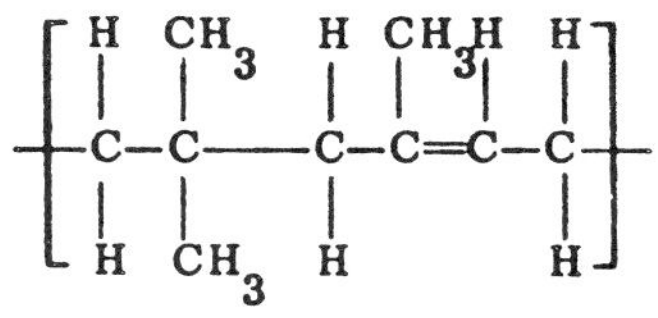

Scheme 8.1 A typical repeat unit in the butyl rubber chain.

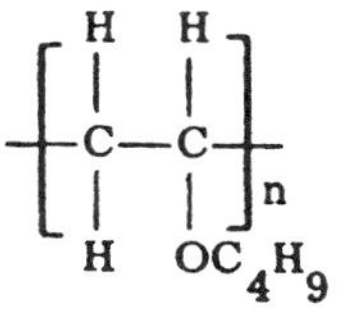

$$\left[\begin{array}{c} \underset{|}{\overset{H}{\underset{|}{C}}} - \underset{|}{\overset{H}{\underset{|}{C}}} \\ H \quad OC_4H_9 \end{array}\right]_n$$

has a low T_g and is used as an adhesive and as an impregnating resin.

The value of the propagation rate constant (k_p) for vinyl isobutyl ether is 6.5 liters mol^{-1} sec^{-1}. This value decreases as one goes from vinyl ethers to isoprene, isobutylene, butadiene, and styrene. The k_p value for styrene in 1,1-dichloroethane is 0.0037 liter mol^{-1} sec^{-1}.

Commercial polymers of formaldehyde may also be produced by cationic polymerization using boron trifluoride etherate as the initiator. As shown in Eq. (8.10), the polymer is produced by ring opening of trioxane. Since the polyacetal is not thermally stable, the hydroxyl end groups are esterified (capped) by acetic anhydride. The commercial polymer is a strong engineering thermoplastic. Engineering plastics usually have higher modulus and higher heat resistance than general-purpose polymers. The commercial polymer Delrin is produced by anionic polymerization.

Another stable polyacetal [polyoxymethylene (POM); Celcon] is produced commercially by the cationic copolymerization of a mixture of trioxane and dioxolane.

$$\frac{n}{3}\;\substack{O\\ \diagup\!\diagdown\\ O} \; + \; H^+ \;\longrightarrow\; HO\!-\!\!\left[CH_2O\right]_n\!\!-\!H \;\xrightarrow{Ac_2O}\; AcO\!-\!\!\left[CH_2O\right]_{n-1}\!\!-\!CH_2OAc \tag{8.10}$$

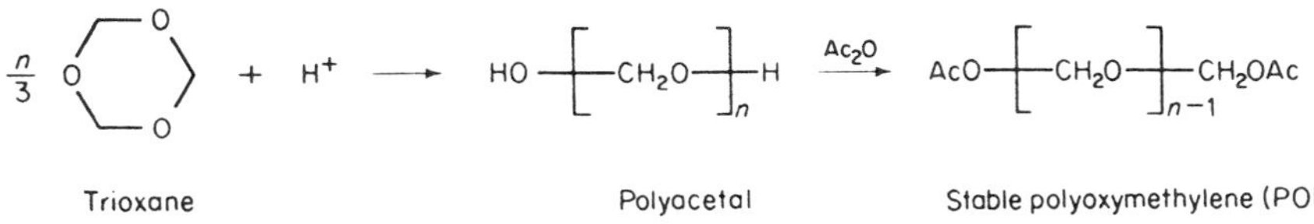

Trioxane   Polyacetal   Stable polyoxymethylene (POM)

As shown in Scheme 8.2, this copolymer contains repeat units from both reactants in the polymer chain. It is believed that the irregularities in the composition of the copolymer hinder the "unzipping" degradation pathway of the polymeric chain.

Polychloral is a flame-resistant strong polymer that can be produced by cationic polymerization. The uncapped polymer decomposes at a ceiling temperature of 58°C. The polymer does not exist above this temperature, which is called the ceiling temperature, T_c. Thus, one may produce solid castings by pouring a hot mixture of trichloroacetaldehyde and initiator into a mold and allowing the polymerization to take place in situ as the mixture cools below the ceiling temperature.

In addition to the production of polyacetals by the ring-opening polymerization of trioxane, this technique was also investigated by Staudinger for the synthesis of ethers and is still used for the production of polymers of ethylene oxide (oxirane). Other homologous, cyclic ethers, such as oxetane and tetrahydrofuran, may be polymerized by cationic ring-opening polymerization techniques. Since the ten-

$$HO\!-\!\!\left[CH_2OCH_2O(CH_2)_2OCH_2O\right]_n\!\!-\!H$$

Scheme 8.2 A segment of a copolymer chain produced from the cationic copolymerization of trioxane and dioxolane.

dency for ring cleavage decreases as the size of the ring increases, it is customary to include some oxirane with the reactants as a promotor. The six-membered ring oxacyclohexane is so stable that it does not polymerize even in the presence of a promotor.

As shown in the following equations, an initiator such as sulfuric acid produces an oxonium ion and a gegenion. The oxonium ion then adds to the oxirane, and the macrooxonium ion produced by propagation may then be terminated by chain transfer with water.

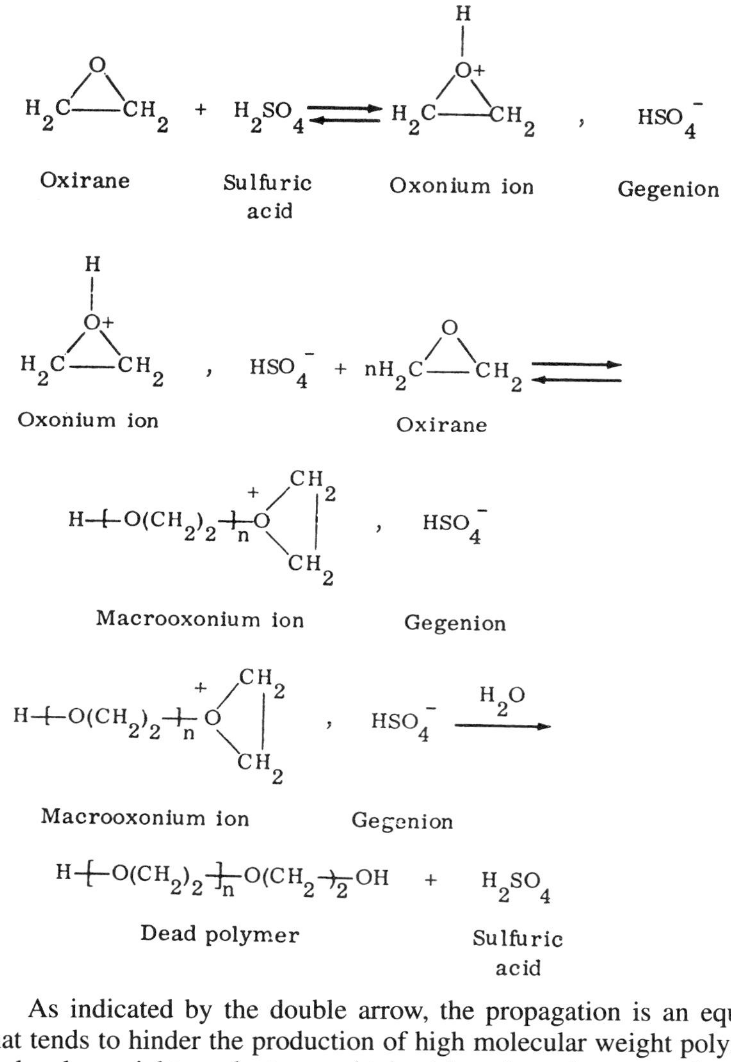

$$(8.11)$$

As indicated by the double arrow, the propagation is an equilibrium reaction that tends to hinder the production of high molecular weight polymers. The highest molecular weight products are obtained in polar solvents, such as methylene chloride, at low temperatures (-20 to $-100°C$).

These water-soluble polymers, which may also be produced by anionic polymerization techniques, are available commercially in several molecular weight ranges under the trade names of Carbowax and Polyox. In addition to their use as water-soluble bases for cosmetics and pharmaceuticals, these polymers may be added to water to increase its flow rate.

The oxacyclobutane derivative, 3,3-bis-chloromethyloxacyclobutane, may be polymerized by cationic ring-opening polymerization techniques to yield a water-insoluble, crystalline, corrosion-resistant polymer. As shown in Scheme 8.3, this polymer (Penton) has two regularly spaced chloromethylene pendant groups on the polymer chain.

An acid-soluble polymer, Montrek, has also been produced by the ring-opening polymerization of ethyleneimine (aziridine). This monomer has been classified as a carcinogen and should be used with extreme caution.

While lactams are usually polymerized by anionic ring-opening reactions, N-carboxyl-α-amino acid anhydrides (NCA) may be polymerized by either cationic or anionic techniques. These polypeptide products, which are now called nylon-2, were first produced by Leuchs in 1908 and are called Leuchs' anhydrides. The synthesis may be used to produce homopolypeptides that can be used as model compounds for proteins. As shown in the following equation, carbon dioxide is eliminated in each step of the propagation reaction.

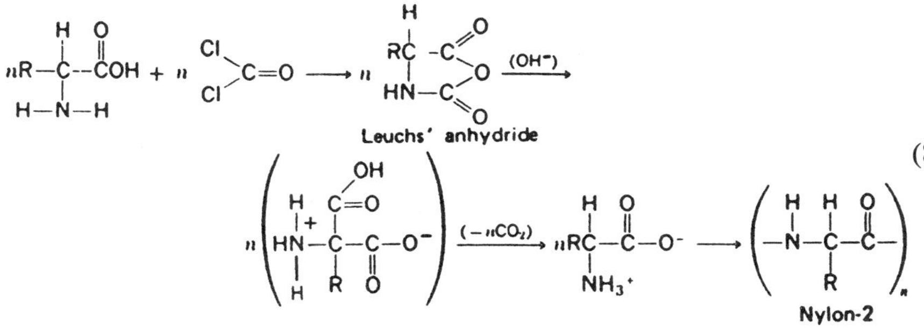

$$(8.12)$$

Polyterpenes, coumarone-indene resins, and the so-called petroleum resins are produced commercially in relatively large quantities by the cationic polymerization of unsaturated cyclic compounds. These inexpensive resinous products are used as additives for rubber, coatings, floor coverings, and adhesives.

It has been known for some time that cationic polymerizations can produce polymers with stereoregular structures. While a number of vinyl monomers have been evaluated in this regard, much of the work has centered about vinyl ethers.

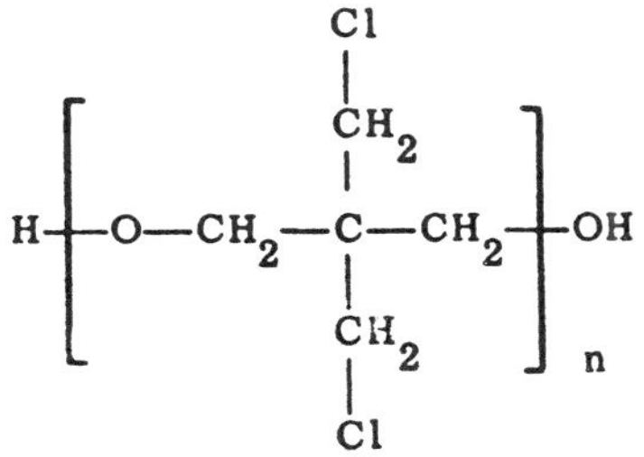

Scheme 8.3 Poly(3,3-bis-chloromethyloxybutylene).

Several general observations have been noted, namely: (1) the amount of stereo-regularity is dependent on the nature of the initiator; (2) stereoregularity increases with a decrease in temperature; and (3) the amount and type of polymer (isotactic or syndiotactic) is dependent on the polarity of the solvent. For instance, t-butyl vinyl ether (Scheme 8.4) has the isotactic form preferred in nonpolar solvents, but the syndiotactic form is preferred in polar solvents.

8.2 ANIONIC POLYMERIZATION

Anionic polymerization was used to produce synthetic elastomers from butadiene at the beginning of the twentieth century. Early investigators used alkali metals in liquid ammonia as initiators, but these were replaced in the 1940s by metal alkyls such as n-butyllithium. In contrast to vinyl monomers with electron-donating groups polymerized by cationic initiators, vinyl monomers with electron-withdrawing groups are more readily polymerized by anionic initiators. Accordingly, acrylonitrile is readily polymerized by anionic techniques, and the order of activity with an amide ion initiator is as follows: acrylonitrile > methyl methacrylate > styrene > butadiene. As might be expected, methyl groups on the α carbon decrease the rate of anionic polymerization, and chlorine atoms on the α carbon increase that activity.

As shown by the following chemical and kinetic equations, potassium amide may be used to initiate the polymerization of acrylonitrile. The propagating species in anionic polymerization are carbanions instead of carbonium ions, but the initiation, propagation, and chain transfer termination steps in anionic polymerization are similar to those described for cationic polymerization.

$$
:NH_2^- + H_2C{=}CH\underset{\displaystyle \text{CN}}{\big|} \xrightarrow{k_i} H_2NC\underset{\substack{| \\ \text{H}}}{\overset{\substack{\text{H} \\ |}}{C}}{-}\overset{\substack{\text{CN} \\ |}}{\underset{\substack{| \\ \text{H}}}{C}}{:}^-
$$

Amide

ion   Acrylonitrile   Carbanion   (8.13)

$$R_i = k_i\, C[M]$$

where C is equal to $[:NH_2^-]$.

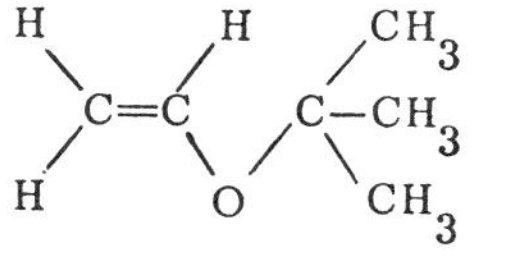

Scheme 8.4 t-Butyl vinyl ether.

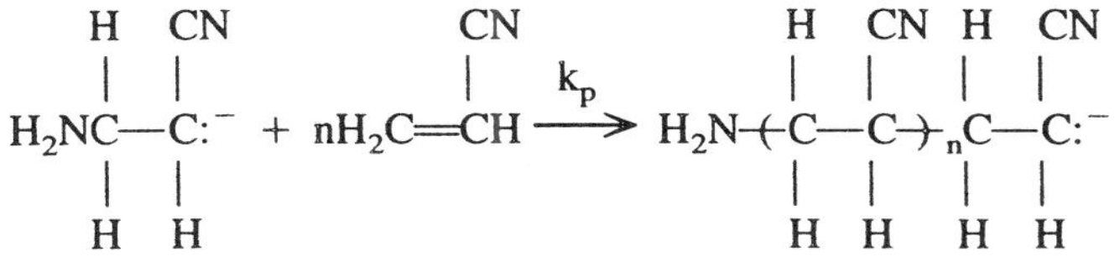

$$
\begin{array}{c}
\text{H} \quad \text{CN} \qquad\qquad \text{CN} \qquad\qquad\qquad \text{H} \quad \text{CN} \ \text{H} \quad \text{CN} \\
| \quad\ | \qquad\qquad\ | \qquad\quad k_p \qquad | \quad\ | \quad | \quad\ | \\
\text{H}_2\text{NC}-\text{C:}^- \ + \ n\text{H}_2\text{C}{=}\text{CH} \ \longrightarrow \ \text{H}_2\text{N}\!\!-\!\!(\text{C}-\text{C})_n\text{C}-\text{C:}^- \\
| \quad\ | \qquad\qquad\qquad\qquad\qquad | \quad\ | \quad | \quad\ | \\
\text{H} \quad \text{H} \qquad\qquad\qquad\qquad\qquad \text{H} \quad \text{H} \quad \text{H} \quad \text{H}
\end{array}
\tag{8.14}
$$

Carbanion   Acrylonitrile   Macrocarbanion

$$R_p = k_p[M][M^-]$$

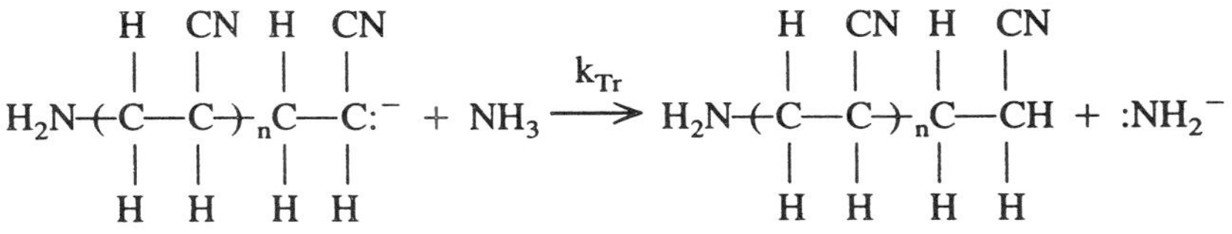

$$
\begin{array}{c}
\text{H} \quad \text{CN} \ \text{H} \quad \text{CN} \qquad\qquad\qquad \text{H} \quad \text{CN} \ \text{H} \quad \text{CN} \\
| \quad\ | \quad | \quad\ | \qquad\quad k_{Tr} \qquad | \quad\ | \quad | \quad\ | \\
\text{H}_2\text{N}\!\!-\!\!(\text{C}-\text{C})_n\text{C}-\text{C:}^- \ + \ \text{NH}_3 \ \longrightarrow \ \text{H}_2\text{N}\!\!-\!\!(\text{C}-\text{C})_n\text{C}-\text{CH} \ + \ :\text{NH}_2^- \\
| \quad\ | \quad | \quad\ | \qquad\qquad\qquad\qquad | \quad\ | \quad | \quad\ | \\
\text{H} \quad \text{H} \quad \text{H} \quad \text{H} \qquad\qquad\qquad \text{H} \quad \text{H} \quad \text{H} \quad \text{H}
\end{array}
\tag{8.15}
$$

Macrocarbanion   Ammonia   Dead polymer
(solvent)

$$R_{Tr} = k_{Tr}[NH_3][M^-]$$

Since it is difficult to determine the concentration of carbanion $[M^-]$, one assumes a steady state in which $R_i = R_{Tr}$ and solves for $[M^-]$, as shown below.

$$
k_iC[M] = k_{Tr}[NH_3][M^-] \quad \text{therefore,} \quad [M^-] = \frac{k_i}{k_{Tr}}\frac{C[M]}{[NH_3]}
\tag{8.16}
$$

Thus,

$$
R_p = k_p[M][M^-] = [M]\frac{k_pk_i}{k_{Tr}}\frac{C[M]}{[NH_3]} = \frac{[M]^2C}{[NH_3]}\frac{k_ik_p}{k_{Tr}} = k'\frac{[M]^2C}{[NH_3]}
\tag{8.17}
$$

Therefore,

$$
\overline{DP} = \frac{R_p}{R_{Tr}} = \frac{k_p[M][M^-]}{k_{Tr}[M^-][NH_3]} = \frac{k_p}{k_{Tr}}\frac{[M]}{[NH_3]} = k''\frac{[M]}{[NH_3]}
\tag{8.18}
$$

Thus, the rate of propagation and the molecular weight are both inversely related to the concentration of ammonia. The activation energy for chain transfer is larger than the activation energy for propagation. The overall activation energy is approximately $+38$ kcal mol^{-1}. The reaction rate increases and molecular weight decreases as the temperature is increased.

The reaction rate is dependent on the dielectric constant of the solvent, the electronegativity of the initiator, the resonance stability of the carbanion, and the degree of solvation of the gegenion. Weakly polar initiators such as Grignard's reagent may be used when strong electron-withdrawing groups are present on the monomer, but monomers with weak electron-withdrawing groups require more highly polar initiators, such as n-butyllithium.

Synthetic cis-1,4-polyisoprene is produced at an annual rate of about 76,000 tons by the polymerization of isoprene in a low dielectric solvent, such as hexane, using n-butyllithium as the initiator. It is assumed that an intermediate cisoid conformation assures the formation of a cis elastomer.

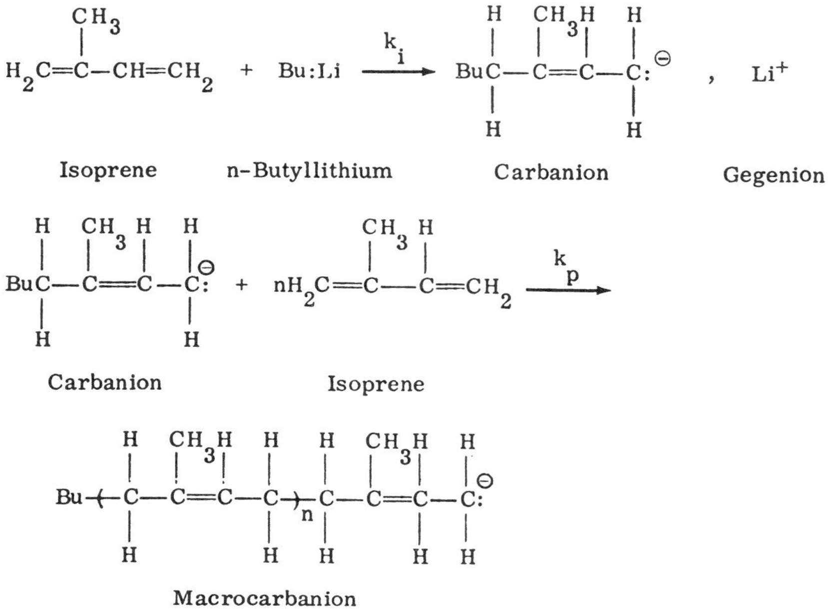

When isoprene is polymerized in a stronger dielectric solvent, such as ethyl ether using butyllithium or sodium, equal amounts of trans-1,4-polyisoprene and cis-3,4-poly-isoprene are produced.

No formal termination step was shown in the previous equations, since in the absence of contaminants, the product is a stable macroanion. Szwarc has used the term "living polymers" to describe these active species. Thus, these macroanions or macrocarbanions may be used to produce a block copolymer, in which, as shown in Scheme 8.5, there are long sequences of similar repeat units. Kraton is an ABA block copolymer of styrene (A) and butadiene (B). Termination may be brought about by the addition of water, ethanol, carbon dioxide, or oxygen.

In addition to the thermal dehydration of ammonium salts, nylons may also be produced by the anionic ring-opening polymerization of lactams. As shown in Eq. (8.20), the polymerization of caprolactam may be initiated by sodium methoxide. This polymer contains six carbon atoms in each repeating unit and is called nylon-6. The term monadic is used to describe nylons such as nylon-6 that have been

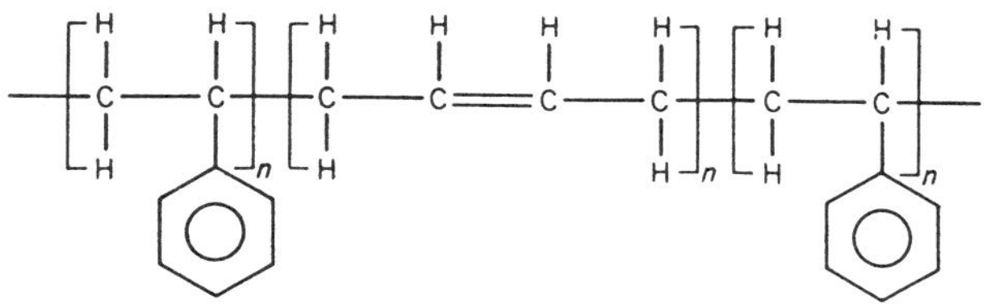

Scheme 8.5   An ABA block copolymer of styrene and butadiene.

produced from one reactant. The term diadic is used to describe nylons such as nylon-66 that have been produced from two reactants.

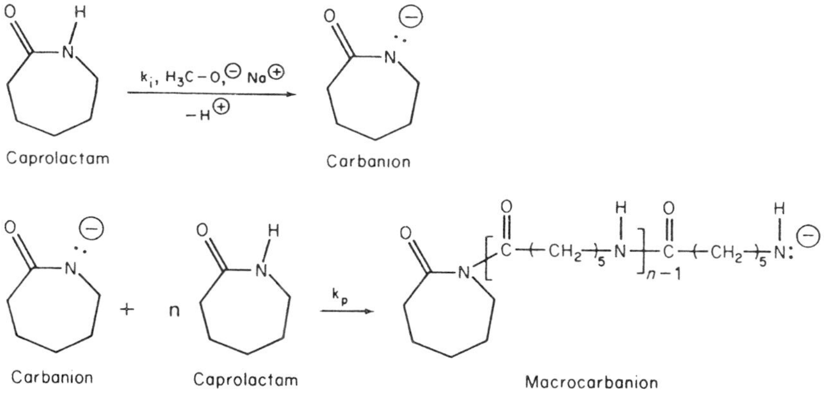

(8.20)

The induction period in lactam ring-opening polymerization may be shortened by the addition of an activator, such as acetyl chloride. Nylon-4, nylon-8, and nylon-12 are commercially available and are used as fibers and coatings.

Lactones may also be polymerized by ring-opening anionic polymerization techniques. While the five-membered ring (γ-butyrolactone) is not readily cleaved, the smaller rings readily polymerize to produce linear polyesters. These polymers are used commercially as biodegradable plastics and in polyurethane foams. A proposed general reaction for the ring-opening polymerization of lactones is shown below:

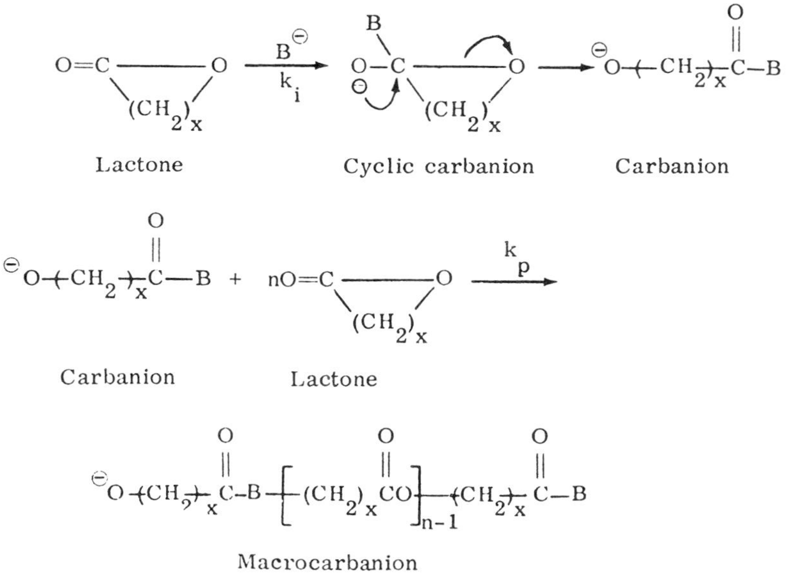

(8.21)

The sterochemistry associated with anionic polymerization is similar to that observed with cationic polymerization. For soluble anionic initiators at low temperatures, syndiotactic formation is favored in polar solvents, whereas isotactic formation is favored in nonpolar solvents. Thus, the stereochemistry of anionic polymerization appears to be largely dependent on the amount of association the growing chain has with the counterion—as it does for cationic polymerizations.

The stereochemistry of diene polymerization is also affected by solvent polarity. For instance, the proportion of cis-1,4 units is increased by using organolithium or lithium itself as the initiator in the polymerization of isoprene or 1,3-butadiene in nonpolar solvents. One can obtain a polymer similar to natural hevea rubber using the anionic polymerization of isoprene. With sodium and potassium initiators the amount of cis-1,4 units decreases and trans-1,4- and trans-3,4 units predominate.

$$CH_2{=}CH{-}CH{=}CH_2 \longrightarrow \left[\begin{array}{c} CH_2CH \\ | \\ CH \\ \| \\ CH_2 \end{array}\right] + \left[\begin{array}{c} CH_2 \qquad CH_2 \\ \diagdown C{=}C \diagup \\ H \qquad H \end{array}\right] \tag{8.22}$$

1,3 - Butadiene   1,2 -   cis - 1,4 -

$$+ \left[\begin{array}{c} CH_2 \qquad H \\ \diagdown C{=}C \diagup \\ H \qquad CH_2 \end{array}\right]$$

trans - 1,4 -

$$\begin{array}{c} CH_3 \\ | \\ CH_2{=}C{-}CH{=}CH_2 \end{array} \longrightarrow \left[\begin{array}{c} CH_3 \\ | \\ CH_2C \\ | \\ CH \\ \| \\ CH_2 \end{array}\right] + \left[\begin{array}{c} CH_2CH \\ | \\ CH_3{-}C \\ \| \\ CH_2 \end{array}\right] \tag{8.23}$$

Isoprene   1,2 -   3,4 -

$$+ \left[\begin{array}{c} CH_2 \qquad CH_2 \\ \diagdown C{=}C \diagup \\ CH_3 \qquad H \end{array}\right] + \left[\begin{array}{c} CH_2 \qquad H \\ \diagdown C{=}C \diagup \\ CH_3 \qquad CH_2 \end{array}\right]$$

cis - 1,4 -   trans - 1,4 -

8.3 POLYMERIZATION WITH COMPLEX COORDINATION CATALYSTS

Prior to 1950, the only commercial polymer of ethylene was a highly branched polymer called high-pressure polyethylene (extremely high pressures were used in the polymerization process). The technique for making a linear polyethylene was discovered by Marvel and Hogan and Banks in the 1940s and by Nobel laureate Karl Ziegler in the early 1950s. Ziegler prepared high-density polyethylene by polymerizing ethylene at low pressure and ambient temperatures using mixtures of triethylaluminum and titanium tetrachloride. Another Nobel laureate, Giulio Natta, used Ziegler's complex coordination catalyst to produce crystalline polypropylene. These catalysts are now known as Ziegler-Natta catalysts. Hogan and Banks also produced crystalline polyethylene in the 1950s.

In general, a Ziegler-Natta catalyst may be described as a combination of a transition metal compound from groups IV to VIII and an organometallic compound of a metal from groups I to III of the periodic table. It is customary to refer to the transition metal compound, such as $TiCl_4$, as the catalyst, and the organometallic compound, such as diethylaluminum chloride, as the cocatalyst.

Several exchange reactions between catalyst and cocatalyst take place, and some of the Ti(IV) is reduced to Ti(III). It is customary to use either the α, γ, or δ form, but not the β crystalline form, of $TiCl_3$ as the catalyst for the production of stereoregular polymers. Both the extent of stereoregularity and the rate of polymerization are increased by the addition of triethylamine and Lewis acids. At least 98% of the isotactic polymer is produced when propylene is polymerized in the presence of triethylamine, γ-titanium(III) chloride, and diethylaluminum chloride.

It is generally agreed that a monomer molecule ($H_2C=CHCH_3$) is inserted between the titanium atom and the terminal carbon atom in the growing chain and that this propagation reaction takes place on the catalyst surface at sites activated by the ethyl groups of the cocatalyst. The monomer molecule is always the terminal group on the chain.

The formation of a π complex is assumed in both the mono- and bimetallic mechanisms. The latter, favored by Natta, involves a cyclic electron-deficient transition complex as shown below:

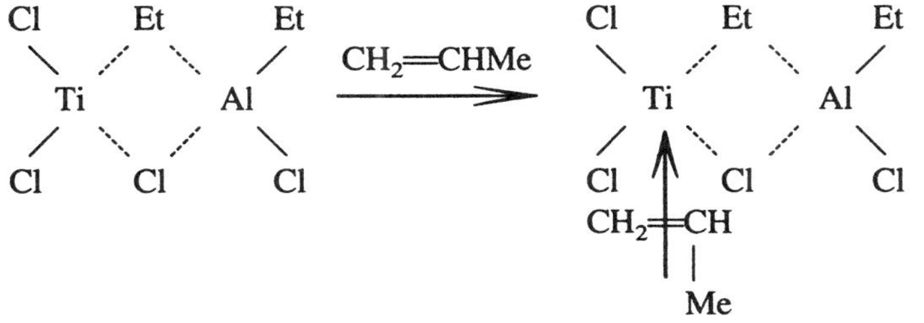

$$ \text{(8.24)} $$

In the more generally accepted monometallic mechanism, shown in the following equation, triethylaluminum reacts to produce ethyltitanium chloride as the active site for polymerization of a nonpolar vinyl monomer such as propylene.

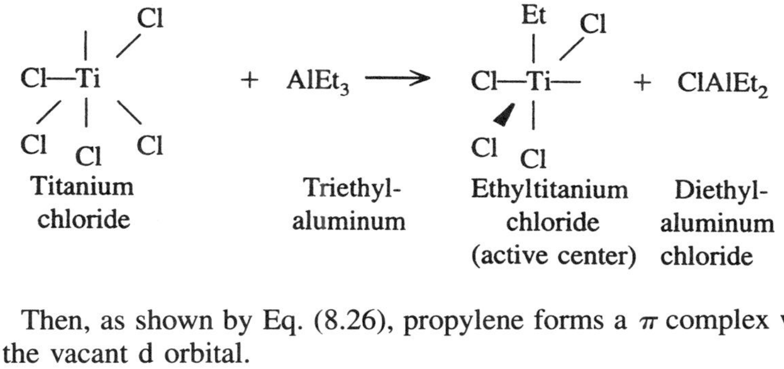

$$ \text{(8.25)} $$

Then, as shown by Eq. (8.26), propylene forms a π complex with the titanium at the vacant d orbital.

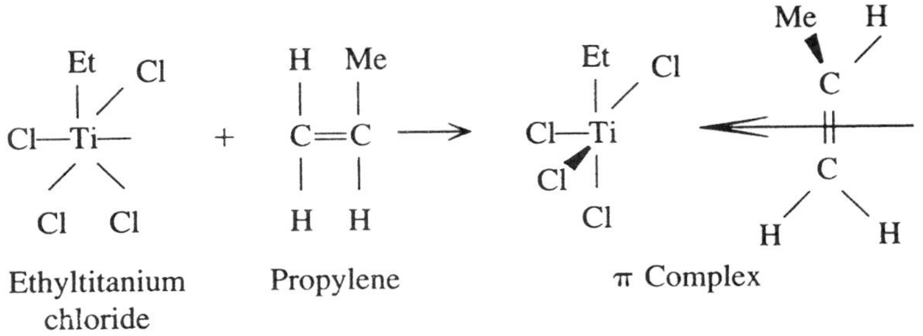

Et   Cl   H Me   Et   Cl   Me   H
 | / | | | /   C
Cl—Ti—   +   C=C   →   Cl—Ti   ‖   (8.26)
 / \ | | Cl | C
Cl   Cl   H   H   Cl   H   H

Ethyltitanium   Propylene   π Complex
chloride

Insertion of the monomer takes place with the formation of a transition state complex, insertion of the monomer, and reformation of an active center in which the ethyl group is now at the end of a propylene group, as shown in the following equation.

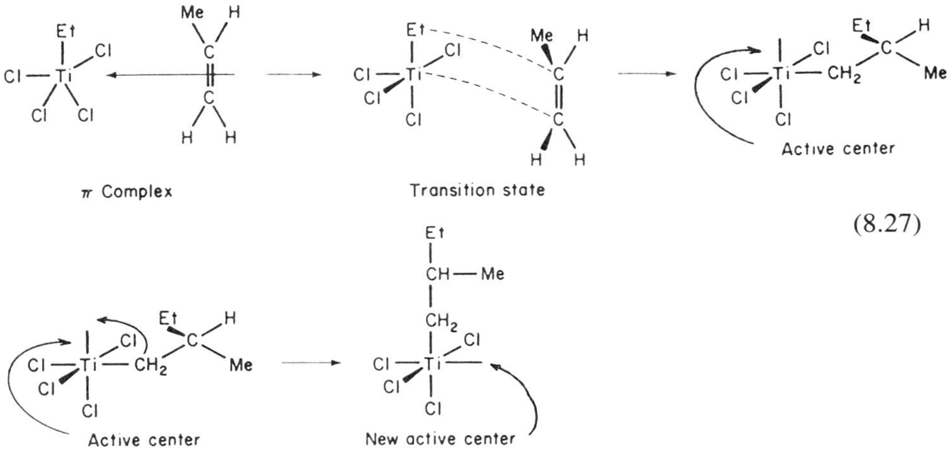

(8.27)

π Complex   Transition state   Active center

Active center   New active center

As shown in Eq. (8.28), the process outlined for initiation is repeated for propagation, and stereoregularity is maintained.

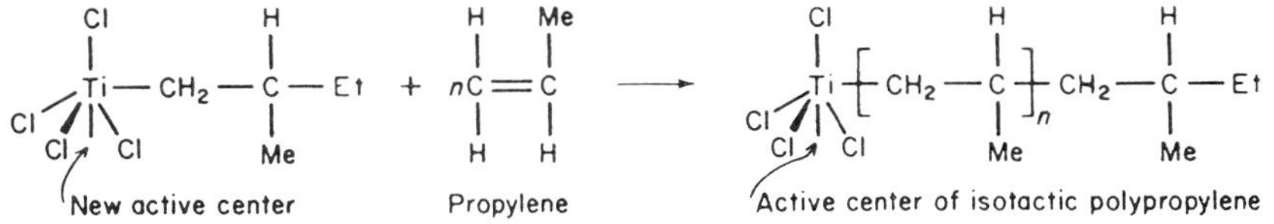

New active center   Propylene   Active center of isotactic polypropylene   (8.28)

For most vinyl monomers, Ziegler-Natta catalysts polymerize to give polymers emphasizing the isotactic form. The degree of stereoregulation appears to be dependent on the amount of exposure of the active site—which is probably a combination of the solid surface and the corners. Typically, the more exposed the catalytic site, the less the isotactic fraction in the resulting chains.

The potential versatility is clearly demonstrated in the polymerization of conjugated dienes, such as 1,3-butadiene, where any of the four possible forms—isotactic 1,2; syndiotactic 1,2; trans-1,4; and cis-1,4—can be synthesized in relatively pure form using different Ziegler-Natta catalysis systems.

Molecular weight is regulated to some degree by chain transfer with monomer and with the cocatalyst, plus internal hydride transfer. However, hydrogen is added in the commercial process to terminate the reaction. Low temperatures, at which the alkyl shift and migration are retarded, favor the formation of syndiotactic polypropylene. Commercial isotactic polymer is produced at ambient temperatures. The percentage of polymer insoluble in n-hexane is called the isotactic index.

High-density polyethylene (HDPE) is produced at an annual rate of 5.7 million tons, but most of this is produced using a chromia catalyst supported on silica, i.e., a Phillips catalyst. Some HDPE and polypropylene (PP) are produced commercially using a Ziegler-Natta catalyst. This initiator is also used for the production of polybutene and poly(-4-methylpentene-1) (TPX). Because of their regular structure, both of these polymers are useful at relatively high temperatures. TPX has a melting point of 300°C and because of its large bulky groups has a low specific gravity of 0.83.

PP, TPX, and LDPE are less resistant to oxidation than HDPE because of the tertiary carbon atoms present in the chain. Their deterioration by weathering and other factors is retarded by incorporation of antioxidants (discussed in Chap. 13).

8.4 POLYMERS OF 1,3-BUTADIENE

cis-Polyisoprene, cis-poly-1,4-butadiene (IR), and ethylene propylene copolymer (EP) elastomers can be produced by use of a Ziegler-Natta catalyst. The annual rate of production of IR and EP elastomers is 444,000 and 181,000 tons, respectively. trans-Polyisoprene can be produced using titanium and vanadium catalysts with an alkylaluminum cocatalyst.

1,3-Butadiene may also be polymerized by a heterogeneous catalyst called alfin, derived from an alchol and olefin. Alfin, which consists of allylsodium, sodium isopropoxide, and sodium chloride, serves as an initiator for the production of very high molecular weight trans-polybutadienes.

1,3-Dienes, such as 2-methyl-1,3-butadiene (isoprene), may polymerize to produce cis- or trans-1,4 or 1,2- and 3,4-isotactic and syndiotactic polybutadienes. The Ziegler-Natta catalyst, consisting of a titanium trichloride catalyst, an alkyl aluminum cocatalyst, and a tertiary amine, produces essentially 100% cis-poly-1,4-isoprene, while tetra-alkoxy titanates produce essentially 100% poly-3,4-isoprene.

When the ratio of Ziegler-Natta catalyst to cocatalyst is greater than 1, the product is trans-poly-1,4-isoprene. When chromium hexacyanobenzene $[Cr(C_6H_5(CN)_6]$ is used as the catalyst for the polymerization of 1,3-butadiene, stereospecific polymers are obtained. A ratio of cocatalyst to catalyst of 2:1 yields syndiotactic (st) poly-1,2-butadiene, and a ratio of 10:1 yields isotactic (it) poly-1,2-butadiene.

8.5 STEREOREGULARITY

One of the more outstanding areas of research involves the synthesis and characterization of stereoregular polymers. Polymers differing in stereoregularity (tacticity) generally vary with respect to such properties as infrared spectra, X-ray diffraction patterns, solubilities, rate and extent of solubility, density, and thermal and mechanical transitions, among others. For instance, the T_g for poly(methyl methacrylate) is about 105°C for atactic forms, 150°C for isotactic forms, and 115°C for syndiotactic forms. Many of these physical differences are due to the ability, or possibility, of more ordered materials to achieve crystalline orientations.

The stereogeometry of 1- and 1,1-disubstituted vinyl addition polymers has been divided into the three conformations illustrated in Eq. (8.29). The isotactic form features a configuration where all the substituents of one kind would all lie on one side, if the molecule were arranged in a linear chain and viewing were done by looking down the "barrel." The syndiotactic form features an alternating arrangement of substituents, whereas a random sequence of substituent placement leads to the atactic configuration. These arrangements assume an adherence to the head-to-tail addition of monomers.

Generally, each substituted vinyl carbon represents a site of asymmetry; it is, in fact, a site of potential optical activity, when included in a polymer chain. Thus, the number of optically active carbons is 2n, where n is the number of asymmetric carbons. Since a polymerizing system can introduce a variety of chain lengths, the total possible number of geometric and chain-length combinations is extremely large. It is quite possible that each polymerization of even common monomers such as styrene might produce many chains as yet not synthesized.

Other geometric possibilities exist for situations involving conjugated double-vinyl compounds. For instance, 1,3-dienes, containing one residual double bond

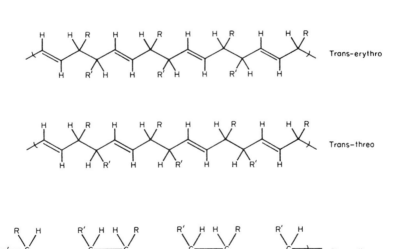

Scheme 8.6 Possible toutadiene-derived polymers.

per repeating unit after polymerization, can contain different configurations [Eq. (8.30)].

Those derivatives of butadiene which are 1- and 4-disubstituted can be polymerized to produce a polymer having two asymmetric carbon atoms and one double bond per repeating unit. Possible geometric forms are illustrated in Scheme 8.6.

The requirements for obtaining measurable optical activity in polymers are still items of active debate. It is known that whole-chain conformations can contribute to the overall optical activity of polymers. For instance, helical conformations offer a major source of optical rotatory power. In addition to the direct use of stereoregular polymers, such products are currently being used (and further investigated) as "templates" to form other products, some of these possessing stereoregularity.

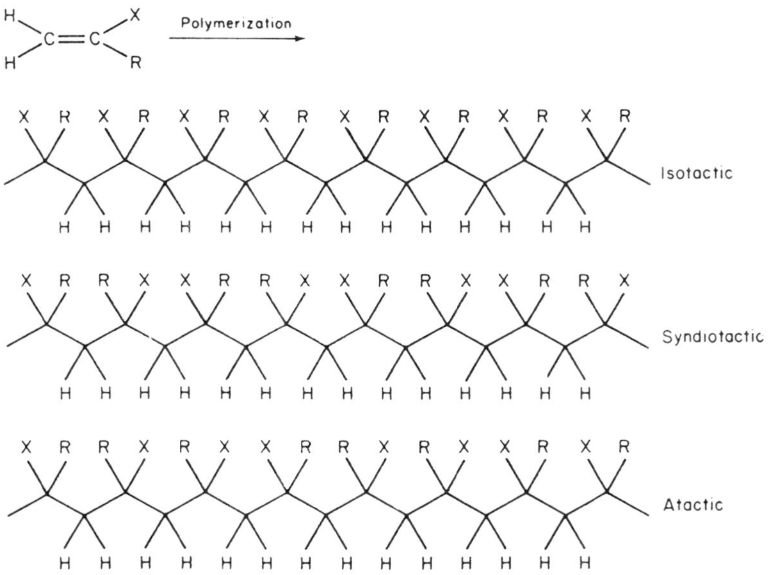

$$(8.29)$$

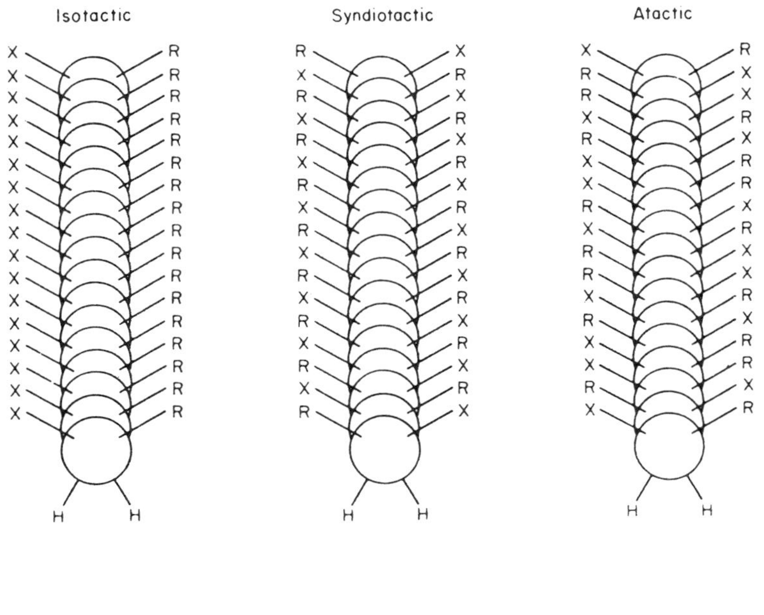

(8.30)

8.6   SOLUBLE STEREOREGULATING CATALYSIS

The 1940s were a time of studying the kinetics and mechanisms of production of the vinyl polymers that would take center stage in the 1950s. The 1950s produced the solid-state stereoregulating catalysis described in Sec. 8.3, which spawned a chemical revolution with the synthesis of stereoregular vinyl polymers in the 1960s. The 1980s and early 1990s witnessed the introduction of soluble stereoregulating catalysis, spawning another revolution related to the production of vinyl polymers with enhanced properties.

The solid-state stereoregulating catalysts suffered from at least three problems. First, while stereoregular polymers were formed with good control of the stereogeometry, polymer properties still fell short of predicted (upper-limit) values. This was probably due to the presence of the associated solid catalyst structure that accompanied the active catalytic site. This "excess baggage" restricted the motion of the growing chains. Second, in many cases the solid-state catalysts were incorporated, as contaminants, within the growing polymer, making an additional purification step necessary in the polymer processing to rid the polymer of this undesired material. Third, many solid-state catalysts offered several active polymerization sites due to differences in the precise structure at and about the active sites.

The new soluble catalysts offer a solution to these three problems. First, the smaller size of the active site and associated molecules allows the growing chains to take advantage of a natural tendency to form a regular helical structure (in comparison to polymers formed from solid-state catalysts). Second, the soluble catalysts allow the synthesis of polymers that contain little or no catalytic agents, allowing the elimination of the typical additional clean-up steps necessary for polymers produced from solid-state catalysts. Third, the newer soluble catalytic sites are homogeneous, offering the same electronic and stereo structure and allowing the synthesis of more homogeneous polymers.

The new soluble stereoregulating polymerization catalysts require three features: a metal atom (active) site, a cocatalyst or counterion, and a ligand system. While the major metal site is zirconium, other metals have been successfully used, including Ti, Hf, Sc, Th, and rare earths (such as Nd, Yb, Y, Lu, and Sm). Cyclopentadienyls (Cp) have been the most commonly used ligands, although a number of others have been successfully employed including substituted Cp and bridged Cp. The most widely used metal-ligand grouping is zirconocene dichloride.

Methylalumoxane (MAO) is the most widely utilized counterion. MAO is an oligomeric material with the following approximate structure.

$$H_3C \diagdown \qquad\qquad \underset{|}{CH_3} \qquad\qquad CH_3 \diagup$$
$$\underset{\diagup}{Al-O-[-Al-O-]_n-Al} \diagdown$$
$$H_3C \qquad\qquad\qquad\qquad CH_3$$
$$\text{Methylalumoxane}$$

where n = 4 − 20

It is believed that MAO plays several roles. MAO maintains the catalyst complex as a cation, but without strongly coordinating to the active site. It also alkylates the metallocene chloride, replacing one of the chloride atoms with an alkyl group and removing the second chlorine, thus creating a coordinately unsaturated cation complex, Cp_2MR^+.

It is believed that as an olefin approaches the ion pair containing the active metal, a metallocene-alkyl-olefin complex forms. This complex is the intermediate stage for the insertion of the monomeric olefin into a growing polymer chain.

The structure of the catalyst complex controls activity, stereoselectivity, and selectivity towards monomers. The catalyst structure is sensitive to Lewis bases such as water and alcohols encouraging the use of strongly oxyphilic molecules, such as MAO, to discourage the inactivation (poisoning) of the catalyst.

These soluble catalysts are able to give vinyl polymers that have increased stereogeometry with respect to tacticity as well as allowing the growing chains to form more precise helical structures. Further, the homogeneity of the catalytic sites also allows for the production of polymers with narrow molecular weight spreads.

The summation of these effects is the production of polymers with increased strength and tensile properties. For polyethylene, the use of these soluble catalysts allows the synthesis of PE chains with less branching compared to those produced using solid-state catalysts such as the Ziegler-Natta catalysts. Table 8.1 gives some comparisons of polyethylenes produced using Ziegler-Natta catalysts with those produced with soluble catalysts.

8.7 POLYETHYLENES

Today there exist a number of polyethylenes that vary in the extent and length of branching as well as molecular weight and molecular weight distribution and amount of crystallinity. Some of these are pictured in Fig. 8.1. Commercial LDPE typically has between 40 and 150 short alkyl branches for every 1000 ethylene units. It is produced employing high pressure (15,000 to 50,000 psi and temperatures to 350°C). It has a density of about 0.912 to 0.935. Because of the branching, LDPE is amorphous (about 50%) and sheets can allow the flow-through of liquids and gasses. Because of the branching and low amount of crystallinity, LDPE has a low melting point of about 100°C, making it unsuitable for use with materials requiring sterilization through use of boiling water. LDPE has a ratio of about 10 short branches to every long branch.

Table 8.1 Comparison of Properties of Polyethylene Using Solid (Ziegler-Natta) Catalysts and Soluble Catalysts

Property	Unit	Soluble	ZNC
Density	g/cc	0.967	0.964
Melt index		1.3	1.1
Haze		4.2	10.5
Tensile yield	psi	800	750
Tensile brake	psi	9400	7300
Elongation break	%	630	670

Values of M_w/M_n of 2 or less are common for the soluble catalyst systems, whereas values of 4–8 are usual for ZNC systems.

The soluble catalyst systems also are able to polymerize a larger number and greater variety of vinyl monomers to form homogeneous polymers and copolymers than solid catalyst systems.

From C. F. Pain, Proceedings Worldwide Metallocene Conference May 26–28, 1993, Catalyst Consultant Inc., Houston, Texas.

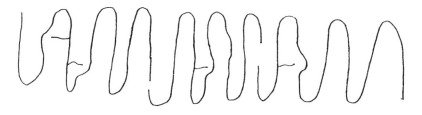

HDPE

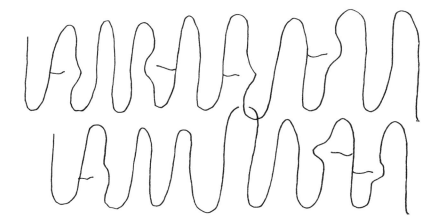

UHMWPE

Figure 8.1 Simulated skeletal structural formulas for various commercially available forms of polyethylene.

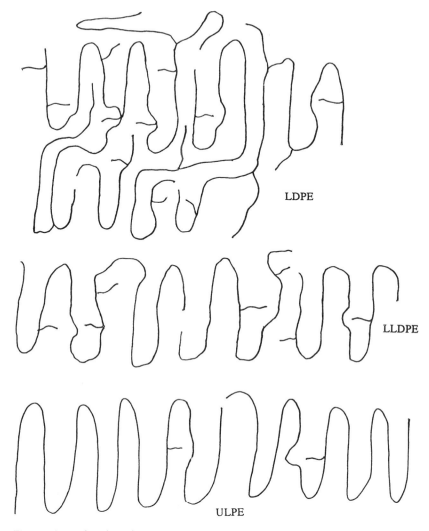

LDPE

LLDPE

ULPE

Figure 8.1 Continued.

HDPE produced using organometallic catalysts, such as the Ziegler-Natta or Phillips catalysts, have less than 15 (normally within the range of 1 to 6) short alkyl branches (and essentially no long branches) per 1000 ethylene units. Because of the regular structure of the ethylene units themselves and the low extent of branching, HDPE chains can pack more efficiently, resulting in a material with greater crystallinity (generally up to 90%), higher density (0.96), with increased chemical resistance, hardness, stiffness, barrier properties, melting point (about 130°C), and tensile strength. Low molecular weight (chain lengths in the hundreds) HDPE is a wax, while typical HDPE is a tough plastic.

Linear low density polyethylene (LLDPE) can be produced with less than 300 psi and at about 100°C. It is actually a copolymer of ethylene with about 8 to 10% of an α-olefin such as 1-butene, 1-pentene, 1-hexene, or 1-octene produced by solution or gas phase polymerization. Through control of the nature and amount of α-olefin, it is possible to produce materials with densities and properties between those of LDPE and HDPE. LLDPE does not contain the long branches found in LDPE.

Ultrahigh molecular weight polyethylene (UHMWPE) is a high-density polyethylene with chain lengths of over 100,000 ethylene units. Because of the long chain lengths, they "intertangle" causing physical crosslinks, increasing the tensile strength and related properties of these materials. (By comparison, HDPE rarely is longer than 2000 ethylene units.) UHMWPE is about 45% crystalline and offers outstanding resistance to corrosion and environmental stress-cracking, outstanding abrasion resistance and impact toughness, and good resistance to cyclical fatigue, radiation failure with a low surface friction. It is produced utilizing catalyst systems similar to those employed for the production of HDPE (i.e., Ziegler-Natta and Phillips catalysts). It has a density of about 0.93.

Ultralinear polyethylene (ULPE) has recently become available through the use of soluble stereoregulating catalysts described in Sec. 8.6. Along with a decreased amount of short-chain alkyl branching, ULPE has a narrower molecular weight spread.

Polymethylene can be produced through several routes including the use of diazomethane or a mixture of carbon monoxide and hydrogen. This polymer has only a little branching.

Types of applications for the various polyethylenes include:

1. UHMWPE: battery separators, lightweight fibers, permanent solid lubricant materials in rail car manufacture, automobile parts, truck liners; liners to hoppers, bins, and chutes; farm machinery such as sprockets, idlers, wear plates, and wear shoes; sewage-treatment bearings, sprockets, wear shoes; lumbering-chute, sluice, and chain-drag liners; neutron shields
2. "Typical" HDPE: blow-molded products—bottles, cans, trays, drums, tanks, and pails; injection-molded products—housewares, toys, food containers, cases, pails, and crates; films, pipes, bags, conduit, wire and cable coating, foam, insulation for coaxial and communication cables
3. Low molecular weight HDPE: spray coatings, emulsions, printing inks, wax polishes and crayons
4. LDPE: packaging products, bags, industrial sheeting, and piping and tubing, films, garbage cans, industrial containers, house-hold items
5. LLDPE: telephone jacketing, wire and cable insulation, piping and tubing, drum liners, bottles, films

SUMMARY

1. Chain reactions, including ionic chain polymerization reactions, consist of at least three steps, namely, initiation, propagation, and termination. Because of the repulsion of similarly charged species, termination by coupling seldom occurs. Instead, termination may take place by chain transfer to produce a new ion and a dead polymer. The $\overline{DP}$ of the latter is equal to the kinetic chain length.

2. Sulfuric acid and Lewis acids with a cocatalyst of water or ether serve as possible initiators in cationic polymerizations, producing a carbocation and a gegenion. Monomers with electron-donating groups, such as isobutylene and vinyl alkyl ethers, may be polymerized at low temperatures in solvents with high dielectric constants.

3. The degree of polymerization is proportional to the concentration of monomer in cationic polymerization, and the overall rate of polymerization is proportional to the square of the concentration of monomers.

4. In general, the rate of cationic polymerization is dependent on the dielectric constant of the solvent, the resonance stability of the carbocation, the degree of solvation of the gegenion, and the electropositivity of the initiator.

5. Stereoregular polymers are produced at low temperatures in solvents that favor the formation of ion pairs between the carbocation and the gegenion.

6. The stability of formaldehyde polymers is improved by capping the hydroxyl end groups or by copolymerizing with other monomers, such as dioxolane.

7. Water-soluble polymers of ethylene oxide are readily formed, but those of more stable cyclic ethers require a promotor, such as ethylene oxide, for this formation.

8. Nylon-2, a polyamide with a wide variety of pendant groups, may be prepared from N-carbosyl amino acid anhydrides.

9. Monomers with electron-withdrawing groups, such as acrylonitrile, may be polymerized in the presence of anionic initiators, such as butyl lithium.

10. The rate of anionic polymerization is dependent on the dielectric constant of the solvent, the stability of the carbanion, the electronegativity of the initiator, the degree of solvation of the gegenion, and the strength of the electron-withdrawing groups in the monomer.

11. Monadic nylons are produced by the anionic ring opening of lactams, such as caprolactam, yielding nylon-6.

12. Stereospecific polymers of nonpolar monomers may be produced by polymerization with a Ziegler-Natta catalyst. The most widely used system consists of a titanium trichloride catalyst and an alkylaluminum cocatalyst.

13. This mechanism (cf. 12) involves a reaction on the surface of $TiCl_3$, activated by the addition of the ethyl group from the cocatalyst. The monomer adds to this active site to produce a π complex, which forms a new active center by insertion of the monomer between the titanium and carbon atoms. This step is repeated in the propagation reactions, in which the alkyl group from the cocatalyst is always the terminal group.

14. Stereospecific polymers are also produced using alfin and chromia on silica initiators. The former, which consists of allylsodium, sodium isopropoxide, and sodium chloride, yields high molecular weight transpolydienes.

15. The production of cis- and trans-polydienes, as well as stereospecific poly-1,2-dienes, is influenced by the proper choice of Ziegler-Natta catalysts and the polymerization temperature.

GLOSSARY

alfin: A complex catalyst system consisting of allylsodium, sodium isopropoxide, and sodium chloride.

anionic polymerization: A polymerization initiated by an anion.

aziridine: Ethyleneimine.

block copolymer: A macromolecule consisting of long sequences of different repeat units $(A_nB_nA_n)$.

butyl rubber (IIR): A copolymer of isobutylene and isoprene.

C: Catalyst-cocatalyst complex.

capping: Reacting the end groups to produce a stable polymer.

carbanion: A negatively charged organic ion.

carbocation: A positively charged organic ion, i.e., one lacking an electron pair on a carbon atom.

cationic polymerization: Polymerization initiated by a cation and propagated by a carbonium ion.

ceiling temperature: A threshold temperature above which a specific polymer cannot exist.

Celcon: A trade name for a copolymer of formaldehyde and dioxolane.

chain-reaction polymerization: A rapid polymerization based on initiation, propagation, and termination steps.

chain transfer: A process in which a growing chain becomes a dead polymer by abstracting a group from some other compound, thereby generating another active site.

cocatalyst: The alkylaluminum compound in the Ziegler-Natta catalyst system.

copolymer: A polymer chain containing repeat units from more than one monomer.

copolymerization: A polymerization of a mixture of monomers.

coupling: Joining of two active species.

Delrin: Trade name for a polyacetal.

diadic: A polyamide produced from more than one reactant.

dioxolane: A five-membered cyclic ether containing three carbon and two oxygen atoms in the ring.

electrophilic: Electron seeking.

EP: Ethylene-propylene copolymer.

gegenion: A counterion.

IIR: Butyl rubber.

initiation: The start of a chain reaction.

it: Isotactic.

k: Rate constant.

Kraton: A trade name for an ABA block copolymer of styrene-butadienestyrene.

lactam: A heterocyclic amide with one nitrogen atom in the ring.

lactone: A heterocyclic ester with one oxygen atom in the ring.

Leuchs' anhydride: A cyclic anhydride which decomposes to carbon dioxide and an amino acid.

living polymers: Macroanions or macrocarbanions.

[M]: Monomer concentration.

[M^+]: Carbonium ion concentration.

macroions: Charged polymer molecules.

monadic: A polyamide produced from one reactant.

Natta, Giulio: The discoverer of stereospecific polymers.

nylon-4: A polymer with the following repeat unit:

$$-\overset{\overset{\text{O}}{\|}}{\text{C}}\!\!-\!\!(\text{CH}_2)_{3}\!\!-\!\!\overset{\overset{\text{H}}{|}}{\text{N}}-$$

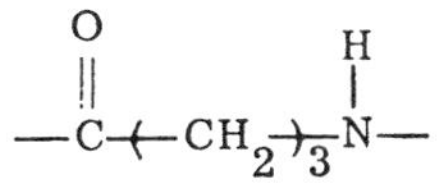

nylon-6: A polymer with the following repeat unit:

$$-\overset{\overset{\text{O}}{\|}}{\text{C}}\!\!-\!\!(\text{CH}_2)_{5}\!\!-\!\!\overset{\overset{\text{H}}{|}}{\text{N}}-$$

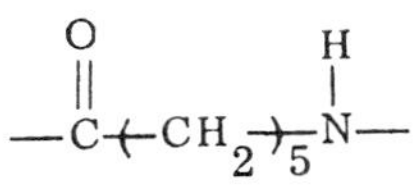

oxacycloalkane: A cyclic compound containing an oxygen atom in the ring.

oxirane: Ethylene oxide.

Penton: A trade name for a polychloroether.

π Complex: A complex formed by a metal with an empty orbital, such as titanium, over-lapping with the p orbitals in an alkene.

polyacetal: Polyoxymethylene.

polychloral: A polymer of trichloroacetaldehyde.

polyisobutylene: A polymer with the following repeat units:

$$-\!\!\left(\!\!\begin{array}{cc} \overset{\text{H}}{\underset{\text{H}}{|}} & \overset{\text{CH}_3}{\underset{\text{CH}_3}{|}} \\ \text{C}\!\!-\!\!\text{C} \end{array}\!\!\right)_{\!n}\!\!-$$

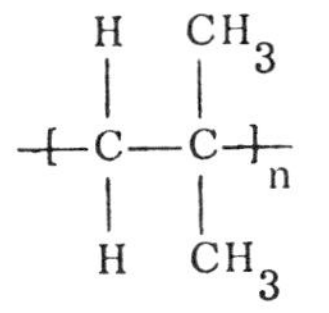

polyoxymethylene: A polymer with the following repeat units:

$$-\!\!\left(\!\!\begin{array}{c} \overset{\text{H}}{\underset{\text{H}}{|}} \\ \text{C}\!\!-\!\!\text{O} \end{array}\!\!\right)_{\!n}\!\!-$$

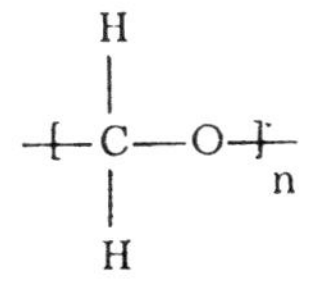

poly(vinyl isobutyl ether): A polymer with the following repeat units:

$$-\!\!\left(\!\!\begin{array}{cc} \overset{\text{H}}{\underset{\text{H}}{|}} & \overset{\text{H}}{\underset{\text{OC}_4\text{H}_9}{|}} \\ \text{C}\!\!-\!\!\text{C} \end{array}\!\!\right)_{\!n}\!\!-$$

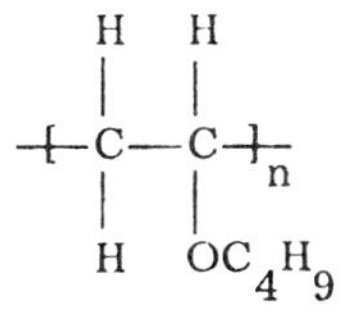

POM: Polyoxymethylene.

PP: Polypropylene.

promotor: A term used for strained cyclic ethers that are readily cleaved.

propagation: The continuous successive chain extension in a chain reaction.

R: Rate.

soluble stereoregulating catalysts: Soluble catalysts requiring a metal active site, cocatalyst or counterion, and a ligand system; capable of producing polymers with high stereoregularity and a minimum of branching.

st: Syndiotactic.

termination: The destruction of active species in a chain reaction.

TPX: Poly-4-methylpentene.

trioxane: A trimer of formaldehyde.

Ziegler, Karl: The discoverer of complex coordination catalysts.

Ziegler-Natta catalyst: $TiCl_3$-AlR$_3$.

EXERCISES

1. Describe the contents of the reaction flask 10 min after the polymerization of (a) reactants in step polymerization, such as dimethyl terephthalate and ethylene glycol, and (b) monomer in chain reaction, such as isobutylene.

2. What is the initiator in the polymerization of isobutylene?

3. What is the general name of the product produced by cationic initiation?

4. What reactant besides the monomer is present in cationic chain propagation reactions?

5. What name is used to describe the negatively charged counterion in cationic chain-reaction polymerization?

6. Is a Lewis acid (a) an electrophile, or (b) a nucleophile?

7. Is a Lewis base (a) an electrophile, or (b) a nucleophile?

8. Why isn't coupling a preferred termination step in the cationic chain polymerization of pure monomer?

9. Is the usual configuration of polymers produced by ionic chain polymerization (a) head to tail, or (b) head to head?

10. Which condition would be more apt to produce stereoregular polymers in ionic chain polymerizations: (a) high temperatures, or (b) low temperatures?

11. Name (a) a thermoplastic, (b) an elastomer, and (c) a fiber that is produced commercially by ionic chain polymerization.

12. Which technique would you choose for producing a polymer of isobutyl vinyl ether: (a) cationic, or (b) anionic?

13. Which technique would you choose for producing a polymer of acrylonitrile: (a) cationic, or (b) anionic?

14. Which of the following could be used to initiate the polymerization of isobutylene: (a) sulfuric acid, (b) boron trifluoride etherate, (c) water, or (d) butyl lithium?

15. Which of the following could be polymerized by cationic chain polymerization?

(a)
$$H_2C=\overset{\displaystyle\overset{H}{|}}{C}-CN$$

(b) $H_2C=C(CH_3)_2$

(c)

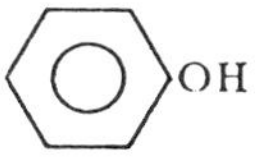

(d) 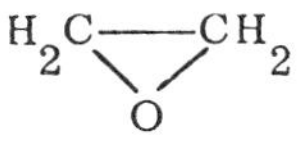 $H_2C\overset{\displaystyle\diagdown\ \diagup}{\underset{\displaystyle O}{}}CH_2$

(e)
$$H_2C=\overset{\displaystyle\overset{H}{|}}{C}-O-C_4H_9$$

16. Which polymer is more susceptible to oxidation: (a) HDPE, or (b) PP?

17. When termination is by chain transfer, what is the relationship of $\overline{DP}$ and the kinetic chain length?

18. What would be the composition of the product obtained by the cationic low-temperature polymerization of a solution of isobutylene in ethylene?

19. What is the relationship of the rate of initiation (R_i) to the monomer concentration [M] in ionic chain polymerizations?

20. What effect will the use of a solvent with a higher dielectric constant have on the rate of propagation (R_p) in ionic chain polymerization?

21. How does the rate constant k_p change as the yield of polymer increases?

22. Which will have the higher T_g value: (a) polystyrene, or (b) polyisobutylene?

23. Which of the following could serve as an initiator for an anionic chain polymerization (a) $AlCl_3 \cdot H_2O$, (b) $BF_3 \cdot H_2O$, (c) butyllithium, or (d) sodium metal?

24. What species, in addition to a dead polymer, is produced in a chain transfer reaction with a macrocarbocation in cationic chain polymerization?

25. What is the relationship of R_i and R_T in a steady-state condition?

26. What is the relationship of $\overline{DP}$ to R_p and R_T?

27. Which would yield the higher molecular weight aldehyde: ozonolysis of (a) natural rubber, or (b) butyl rubber?

28. What percentage of polymer is usually found when a polymer produced by chain-reaction polymerization is heated above its ceiling temperature?

29. What is the relationship of $\overline{DP}$ to initiator concentration in cationic chain polymerization?

30. Can the polymers found in the bottom of a bottle of insolubilized formaldehyde solution be useful?

31. How would you prepare stable polymers from formaldehyde?

32. Why is the thermal decomposition of polymers of formaldehyde called unzipping?

33. Can chloral be polymerized at 60°C?

34. How would you promote the ring-opening polymerization of tetrahydrofuran?

35. How would you increase the rate of flow of water in a fire hose?

36. Why is poly-3,3-bis-chloromethyloxybutylene crystalline?

37. Why must care be used in the polymerization of aziridine?

38. What byproduct is produced when Leuchs' anhydride is polymerized?

39. How could you remove unsaturated hydrocarbons from petroleum or coal tar distillates?

40. What species is produced by the reaction of an anionic chain polymerization initiator and the monomer?

41. What are the propagating species in anionic chain polymerization?

42. Why are polymers produced by the anionic polymerization of pure monomers called "living polymers"?

43. Using the symbols A and B for repeating units in the polymer chain, which of the following is a block copolymer: (a) —ABABAB—, (b) AABABBA—, (c) —(A)$_n$B$_n$—?

44. What is the most widely used monadic nylon?

45. What is the repeating unit in nylon-4?

46. What is the catalyst and cocatalyst in the most widely used Ziegler-Natta catalyst?

47. Why is β-TiCl$_3$ not used as a polymerization catalyst?

48. What is the principal difference between propagation reactions with butyllithium and a Ziegler-Natta catalyst?

49. In addition to good strength, clarity, and good resistance to heat and corrosives, what is a unique feature of poly-4-methylpentene-1?

50. Show the skeletal structures of cis- and trans-polyisoprene.

51. Write formulas for repeating units in the chains of (a) poly-1,4-isoprene, and (b) poly-1,2-isoprene.

52. What is the most widely used catalyst for the production of HDPE?

53. What elastomer is produced by anionic chain polymerization?

54. What elastomer is produced by use of a Ziegler-Natta catalyst?

APPENDIX

Production of polymers in the United States and Canada in 1978 by ionic chain reaction and complex coordination polymerization (thousands of metric tons) is listed below.

By cationic polymerization	
butyl rubber	146
By anionic polymerization	
polyisoprene	76
By coordination polymerization,	
including chromia on silica	
(Phillips catalyst)	
high-density polyethylene (HDPE)	1852
polypropylene plastics	1341
polybutadiene	444
ethylene-propylene copolymer	181
polypropylene fibers	332

BIBLIOGRAPHY

Bailey, F. E. (1983): *Initiation of Polymerization*, ACS, Washington, D.C.

Chien, J. C. (1993): *Homogeneous Polymerization Catalysis*, Prentice-Hall, Englewood Cliffs, New Jersey.

Gaylord, N. G., Mark, H. F. (1959): *Linear and Stereoregular Addition Polymers*, Interscience, New York.

Goodman, M. (1967): *Concepts of Polymer Stereochemistry*, Wiley-Interscience, New York.

Hogan, J. P., Banks, R. L. (1955): (Philips Process), U.S. Patent 2,717,888.

Hogen-Esch, T. E., Smidgd, M. (1987): *Recent Advances in Anionic Polymerization: Carbanions, Mechanism and Synthesis*, Elsevier, New York.

Hummel, K., Schurz, D. (1989): *Disperse Systems, Interfaces and Membranes*, Springer-Verlag, New York.

Kaminsky, W., Sinn, H. (1988): *Transition Metals and Organometallics as Catalysts for Olefin Polymerization*, Springer-Verlag, New York.

Kennedy, J. P., Ivan, B. (1992): *Designed Polymers by Carbocationic Macromolecular Engineering: Theory and Practice*, Hanser-Gardner, Cincinnati, OH.

Kissin, Y. V. (1985): *Isospecific Polymerization of Olefins*, Springer-Verlag, New York.

Lenz, R. W. (1975): *Coordination Polymerization*, Academic, New York.

Marvel, C. S. (1959): *An Introduction to Organic Chemistry of High Polymers*, Wiley, New York.

Morton, A., Bolton, F. H., et al. (1952): Alfin catalysis, Ind. Eng. Chem., *40*:2876.

Morton, A., Lanpher, E. J. (1960): Alfin catalyst, J. Polymer Sci., *44*:233.

Morton, M. (1983): *Anionic Polymerization*, Academic, New York.

Natta, G. (1958): Ziegler catalysts, J. Inorg. Nucl. Chem., *8*:589.

Natta, G., Danusso, F. (1967): *Stereoregular Polymers and Stereospecific Polymerization*, Pergamon, New York.

Price, C. C., Vandenberg, E. J. (1983): *Co-ordination Polymerization*, Plenum, New York.

Szwarc, M. (1968): *Carbanions, Living Polymers and Electron-Transfer Processes*, Wiley, New York.

Thomas, R. M., Sparks, W. J., et al. (1940): Polyisobutylene, J. Am. Chem. Soc., *62*:276.

Willert, H. G. (1990): *Ultra High Molecular Weight Polyethylene*, Hogrefe and Huber Pubs., Kirkland, WA.

Ziegler, K. (1952, 1955): Ziegler catalysts, Angew. Chem., *64*:323; *67*:541.

9

Free-Radical Chain Polymerization (Addition Polymerization)

Since most synthetic plastics and elastomers and some fibers are prepared by free-radical polymerization, this method is obviously most important from a commercial viewpoint. Table 9.1 contains a listing of a number of commercially important addition polymers. Many of the concepts discussed in Chap. 8 on ionic chain polymerization also apply to free-radical polymerization. However, because of the versatility of this polymerization technique, several new concepts will be introduced in this chapter.

As is the case with other chain reactions, free-radical polymerization is a rapid reaction which consists of the characteristic chain-reaction steps, namely, initiation, propagation, and termination. Unlike the cationic and anionic initiators produced from the hetrolytic cleavage of ion pairs such as H^+,HSO_4^- and Bu^-,Li^+, respectively, free-radical initiators are produced by homolytic cleavage of covalent bonds. The formation of free radicals is dependent on high-energy forces or the presence of weak covalent bonds.

9.1 INITIATORS FOR FREE-RADICAL CHAIN POLYMERIZATION

The rate of decomposition of initiators usually follows first-order kinetics and is dependent upon the solvent present and the temperature of polymerization. The rate is usually expressed as a half-life time ($t_{1/2}$) where $t_{1/2} = \ln 2/k_d = 0.693/k_d$. The rate constant ($k_d$) changes with temperature in accordance with the Arrhenius equation as shown below:

$$k_d = Ae^{-E_a/RT} \tag{9.1}$$

The rate constants for several common initiators are listed in Table 9.2. Typical equations for the dissociation of 2,2'-azo-bis-isobutyronitrile (AIBN) and benzoyl peroxide (BPO) are shown below. It should be pointed out that because of recombination, which is solvent dependent, and other side reactions of the free radicals (R·), the initiator efficiency is seldom 100%. Hence, an efficiency factor (f) is employed to show the percentage of effective free radicals produced.

$$(CH_3)_2C-N{=}N-C(CH_3)_2 \xrightarrow[\substack{or \\ 3600\text{Å}}]{\Delta} 2(CH_3)_2C\bullet + N_2 \tag{9.2}$$
$$\underset{\substack{| \\ CN}}{} \qquad \underset{\substack{| \\ CN}}{} \qquad \qquad \underset{\substack{| \\ CN}}{}$$

AIBN   Free radical

$$C_6H_5-\underset{\substack{\| \\ O}}{C}-OO-\underset{\substack{\| \\ O}}{C}C_6H_5 \longrightarrow 2C_6H_5-\underset{\substack{\| \\ O}}{C}-O\bullet \longrightarrow 2C_6H_5\bullet + 2CO_2 \tag{9.3}$$

BPO   Free radical

The rate of decomposition of peroxides such as benzoyl peroxide may be increased by the addition of small amounts of tertiary amines such as N,N-dimethylaniline. The rate of decomposition of initiators may also be increased by exposure to ultraviolet (UV) radiation. For example, AIBN may be decomposed at low temperatures by UV radiation at a wavelength of 360 nm.

9.2 MECHANISM FOR FREE-RADICAL CHAIN POLYMERIZATION

In general, the decomposition of the initiator (I) may be expressed by the following equations in which k_d is the rate or decay constant.

$$I \xrightarrow{k_d} 2R\cdot \tag{9.4}$$

$$R_d = \frac{-d[I]}{dt} = k_d[I] \tag{9.5}$$

where R_d is the rate of decomposition.

Initiation of a free-radical chain takes place by addition of a free radical (R·) to a vinyl molecule. Polystyrene (PS) is produced by free-radical polymerization at an annual rate of 2.9 million metric tons. The polymerization of styrene (S) will be used as an illustration. Styrene, like many other aromatic compounds, is toxic, and concentrations in the atmosphere should not exceed 10 ppm. It is important to note that the free radical (R·) is a companion of all polymerizing species and hence should not be called a catalyst, even though it often is referred to as a catalyst.

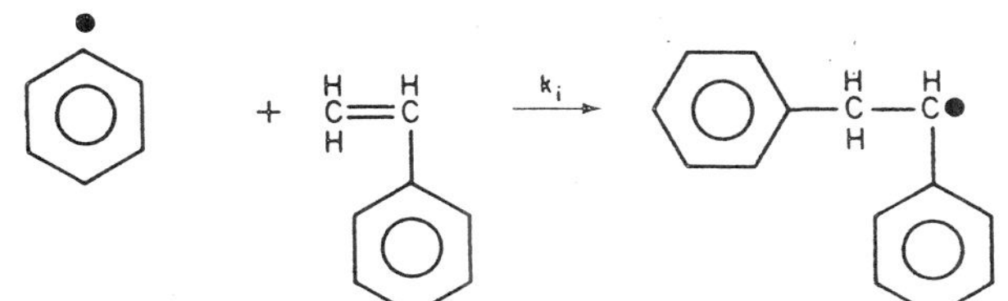

$$\tag{9.6}$$

Free radical   Styrene   Styrene–free radical
from benzoyl
peroxide

Table 9.1　Industrially Important Addition Polymers

Name	Repeating unit	Typical properties	Typical uses
Polyacrylonitrile (including acrylic fibers)	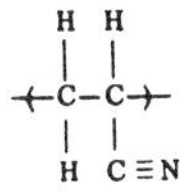	High strength; good stiffness; tough; abrasion resistant; resilient; good flex life; relatively good resistance to moisture and stains, chemicals, insects, and fungi; good weatherability	Carpeting, sweaters, skirts, socks, slacks, baby garments
Poly(vinyl acetate)	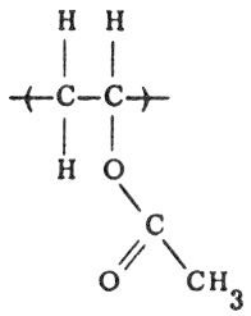	Water sensitive with respect to physical properties such as adhesion and strength; generally good weatherability, fair adhesion	Lower molecular weight used in chewing gums, intermediate in production of poly(vinyl alcohol), water-based emulsion paints
Poly(vinyl alcohol)	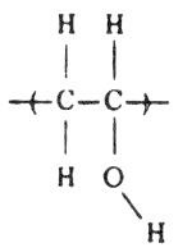	Water soluble, unstable in acidic or basic aqueous systems; fair adhesion	Thickening agent for various suspension and emulsion systems, packaging film, wet-strength adhesive
Poly(vinyl butyral)	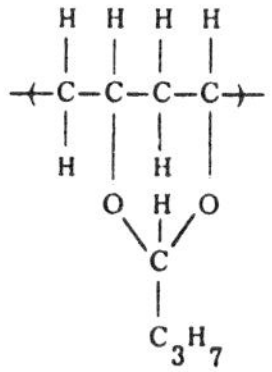	Good adhesion to glass; tough; good stability to sunlight; good clarity; insensitive to moisture	Automotive safety glass as the interlayer

Polymer	Structure	Properties	Uses
Poly(vinyl chloride) and poly(vinylidene chloride) (called "the vinyls" or "vinyl resins")	$\begin{array}{cc} H & H \\ -C-C- \\ H & Cl \end{array}$ $\begin{array}{cc} H & Cl \\ -C-C- \\ H & Cl \end{array}$	Relatively unstable to heat and light, fire resistant; resistant to chemicals, insects, fungi; resistant to moisture	Calendered products such as film sheets and floor coverings; shower curtains, food covers, rainwear, handbags, coated fabrics, insulation for electrical cable and wire, phonograph records
Polytetrafluoroethylene (Teflon)	$\begin{array}{cc} F & F \\ -C-C- \\ F & F \end{array}$	Insoluble in most solvents, chemically inert, low dielectric loss, high dielectric strength, uniquely nonadhesive, low friction properties, constant electrical and mechanical properties from 20 to about 250°C, high impact strength, not hard, outstanding mechanical properties	Coatings for frying pans, etc.; wire and cable insulation; insulation for motors, oils, transformers, generators; gaskets; pump and valve packings; nonlubricated bearings
Polyethylene (low-density, branched)	$\begin{array}{cc} H & H \\ -C-C- \\ H & H \end{array}$	Dependent on molecular weight, branching, molecular weight distribution, etc.; good toughness and pliability over a wide temperature range, outstanding electrical properties, good transparency in thin films, inert chemically, resistant to acids and bases, ages on exposure to light and oxygen, low density, flexible without plasticizer, resilient, high tear strength, moisture resistant	Films; sheeting used in bags, pouchers, produce wrapping, textile materials, frozen foods, etc.; drapes, table cloths; covers for construction, ponds, greenhouses, trash can liners, etc.; electrical wire and cable insulator; coating of foils, papers, other films; squeeze bottles

(continued)

Table 9.1 (continued)

Name	Repeating unit	Typical properties	Typical uses
Polyethylene (high-density, linear)		Most of the differences in properties between branched and linear concerns the high crystallinity of the latter; linear polyethylene has a high T_g, T_m, softening range, greater hardness and tensile strength.	Bottles, housewares, toys, films, sheets, extrusion coating, pipes, conduit, wire and cable insulation
Polypropylene	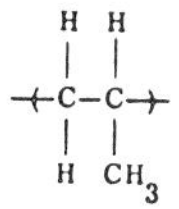 $\begin{array}{c} H \quad H \\ \mid \quad \mid \\ +C-C+ \\ \mid \quad \mid \\ H \quad CH_3 \end{array}$	Lightest major plastic; its high crystallinity imparts to it high tensile strength, stiffness, and hardness, good gloss, high resistance to marring; high softening range permits polymer to be sterilized; good electrical properties, chemical inertness, moisture resistance	Filament—rope, webbing, cordage; carpeting; injection molding applications in appliance, small houseware, and automotive fields
Polyisoprene (cis-1,4-polyisoprene)	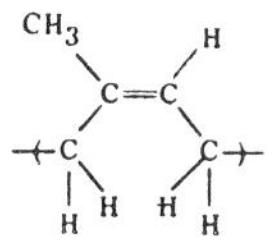 $\begin{array}{c} H \\ \mid \\ CH_3 \quad C+ \\ \backslash \quad \mid \\ C=C \quad H \quad H \\ / \quad \backslash \\ +C \quad \\ \mid \\ H \quad H \end{array}$	Structure closely resembling that of natural rubber; properties similar to those of natural rubber	Replacement of natural rubber; often preferred because of its greater uniformity and cleanliness
SBR (styrene-butadiene rubber)	Random copolymer	Generally slightly poorer physical properties than those of natural rubber	Tire treads for cars, inferior to natural rubber with respect to heat buildup and resilience, thus not used for truck tires; belting; molded goods, gum, flooring, rubber shoe soles, electrical insulation, hoses

Polymer	Structure	Properties	Uses
Butyl rubber (copolymer of isobutylene with small amounts of isoprene added to permit vulcanization)	Amorphous, isoprene—largely 1,4 isomer	Good chemical inertness, low gas permeability, high viscoelastic response to stresses, less sensitive to oxidative aging than most other elastomers, better ozone resistance than natural rubber, good solvent resistance	About 70 to 60% used for inner tubes for tires
Polychloroprene (Neoprene)	Mostly 1,4 isomer	Outstanding oil and chemical resistance; high tensile strength, outstanding resistance to oxidative degradation and aging; good ozone and weathering resistance; dynamic properties are equal or greater than those of most synthetic rubber and only slightly inferior to those of natural rubber	Can replace natural rubber in most applications; gloves, coated fabrics, cable and wire coatings, hoses, belts, shoe heels, solid tires
Polystyrene	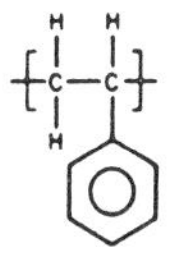	Clear; easily colored; easily fabricated; transparent; fair mechanical and thermal properties; good resistance to acids, bases, oxidizing and reducing agents; readily attacked by many organic solvents; good electrical insulator	Used for the production of ion-exchange resins, heat- and impact-resistant copolymers, ABS resins, etc., foams, plastic optical components, lighting fixtures, housewares, toys, packaging, appliances, home furnishings
Poly(methyl methacrylate)	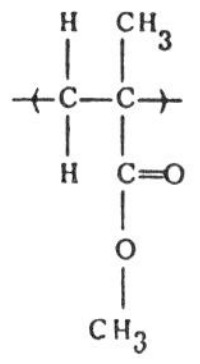	Clear, transparent, colorless, good weatherability, good impact strength, resistant to dilute basic and acidic solutions, easily colored, good mechanical and thermal properties, good fabricability, poor abrasion resistance compared to glass	Available in cast sheets, rods, tubes, and molding and extrusion compositions; applications where light transmission is needed, such as tail- and signal-light lenses, dials, medallions, brush backs, jewelry, signs, lenses, skylight "glass"

Table 9.2 Rate Constants for Common Initiators in Various Solvents[a]

Initiator	Solvent	Temp. (°C)	k_d (sec^{-1})	E_a (kcal/mol^{-1})
2,2'-Azo-bis-isobutyronitrile (AIBN)	Benzene	40	5.4×10^{-7}	30.7
Phenyl-azo-triphenylmethane	Benzene	25	4.3×10^{-6}	26.8
tert-Butyl peroxide (TBP)	Benzene	80	7.8×10^{-8}	34
Cumyl peroxide	Benzene	115	1.6×10^{-5}	40.7
Acetyl peroxide	Benzene	35	9.5×10^{-5}	32.3
Benzoyl peroxide (BPO)	Benzene	30	4.8×10^{-8}	27.8
Lauroyl peroxide	Benzene	30	2.6×10^{-7}	30.4
tert-Butyl hydroperoxide	Benzene	154	4.3×10^{-6}	40.8
tert-Butyl perbenzoate	Benzene	100	1.1×10^{-5}	34.7

[a]All initiators are unstable compounds and should be handled with extreme caution! Data from *Polymer Handbook* (Brandrup and Immergut, 1975).

$$R\bullet + M \xrightarrow{k_i} RM\bullet$$
$$\text{Free radical} \quad \text{Monomer} \quad \text{New free radical}$$

$$R_i = \frac{d[RM\cdot]}{dt} = k_i[R\cdot][M] \tag{9.7}$$

where R_i is the rate of initiation.

The rate of decomposition of I [Eq. (9.4)] is the rate-controlling step in the free-radical polymerization. Thus, the overall expression describing the rate of initiation can be given as

$$R_i = 2k_d f[I] \tag{9.8}$$

where f is the efficiency factor and is a measure of the fraction of initiator radicals that produce growing radical chains, i.e., are able to react with monomer.

A 2 is inserted in Eq. (9.8) because, in this presentation, for each initiator molecule that decomposes, two radicals are formed. The 2 is omitted from Eq. (9.5) because this rate expression describes the rate of decomposition of the initiator, but not the rate of formation of free radicals (R·). (Similarly, in Eqs. (9.14) and (9.16), each termination results in the loss of two growing chains, thus a 2 appears in the descriptions.)

Propagation is a bimolecular reaction [Eq. (9.9)], which takes place by the addition of the new free radical (RM·) to another molecule of monomer (M), and by many repetitions of this step. While there may be slight changes in the propagation rate constant (k_p) in the first few steps, the rate constant is generally considered to be independent of chain length. Hence, the symbols M·, RM·, and RM_nM· may be considered equivalent in rate equations for free-radical chain-reaction polymerization.

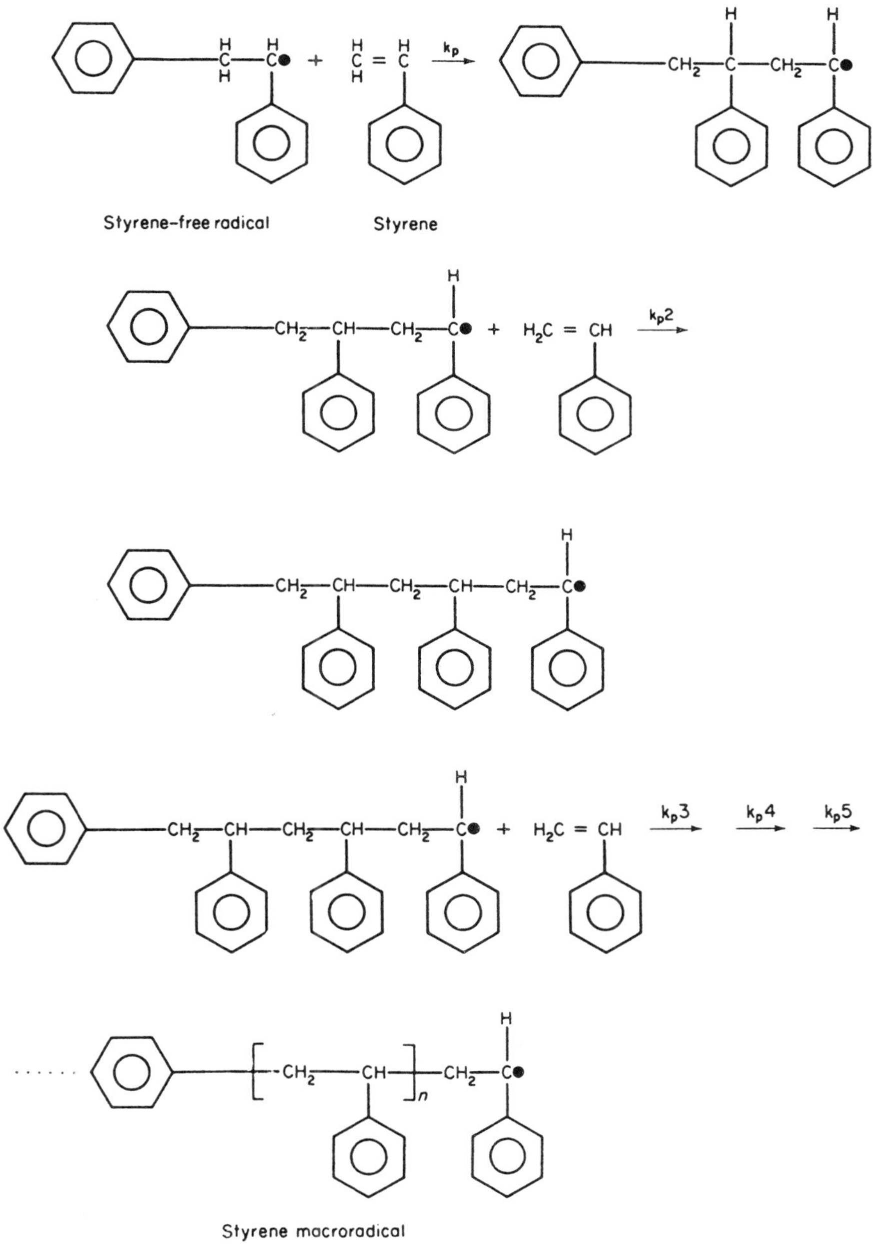
Styrene-free radical
Styrene
k_p
k_p2
k_p3
k_p4
k_p5
Styrene macroradical

(9.9)

Experimentally it is found that the specific rate constants associated with propagation are approximately independent of chain length; thus, the specific rate constants for each propagation step are considered to be the same, permitting all the propagation steps to be described by a single specific rate constant k_p. Thus, Eq. (9.9) can be summed giving the overall propagation expression as

$$\sim M\cdot + nM \xrightarrow{k_p} \sim M—M_{n-1}—M\cdot$$

or simply

$$M\cdot + nM \xrightarrow{k_2} M—M_{n-1}—M\cdot \tag{9.10}$$

The rate of demise of monomer with time is described as:

$$-\frac{d[M]}{dt} = k_p[M\cdot][M] + k_i[R\cdot][M] \tag{9.11}$$

i.e., monomer consumption occurs only in the steps 2 and 3 described by Eqs. (9.6) and (9.9).

For long chains the amount of monomer consumed by step 2 [Eq. (9.6)] is small compared with that consumed in step 3 [Eq. (9.9)], permitting Eq. (9.11) to be rewritten as:

$$R_p = \frac{-d[M]}{dt} = k_p[M][M\cdot] \tag{9.12}$$

The polarity of the functional group in the monomers polymerized by free-radical chain polymerization is between the positively inclined monomers, characteristic of undergoing cationic polymerization, and the negatively inclined monomers, characteristic of undergoing anionic polymerizations. However, as was true for the configuration of growing chains in ionic propagations, head-to-tail arrangement is the customary sequence in free-radical propagation. The functional groups in the vinyl monomers are better stabilizers than the hydrogen atom that would be present as the macroradical end group in a head-to-head arrangement.

Unlike ionic polymerizations, the termination of the growing free-radical chains usually takes place by coupling of two macroradicals. Thus, the kinetic chain length (v) is equal to DP/2. The chemical and kinetic equations for bimolecular termination are shown below.

RM● + ●MR $\xrightarrow{K_t}$ RMMR

Macroradicals   Dead polymer

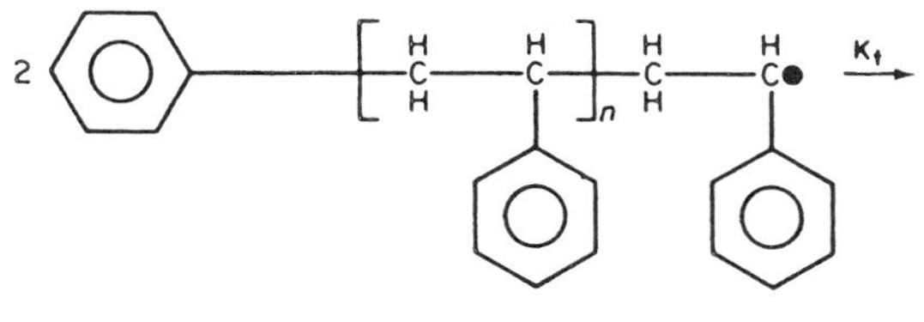

Styrene macroradical

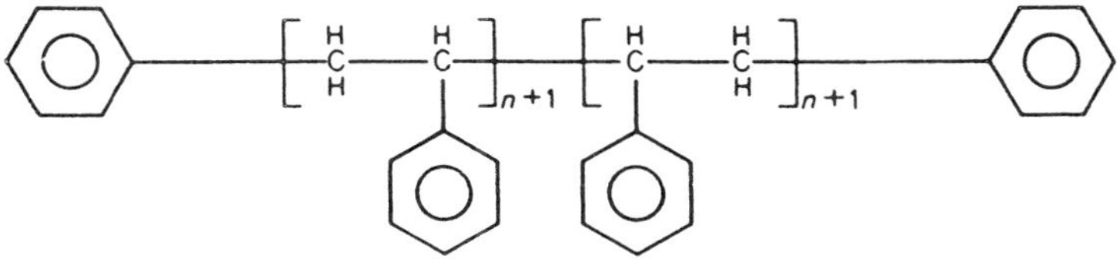

$$(9.13)$$

Dead polystyrene

It should be noted that there is a head-to-head configuration at the juncture of the two macroradicals in the dead polymer. The kinetic equation for coupling termination is shown below.

$$R_t = \frac{-d[M\cdot]}{dt} = 2k_t[M\cdot][M\cdot] = 2k_t[M\cdot]^2 \qquad (9.14)$$

Termination of free-radical chain polymerization may also take place by disproportionation. This termination process involves chain transfer of a hydrogen atom from one chain end to the free radical chain end of another growing chain, resulting in one of the "dead" polymers having an unsaturated chain end. Thus, the type and/or proportion of each type of chain termination can be obtained by determining the number of head-to-head configurations and number of end groups containing unsaturation. The kinetic description for chain termination by disproportionation is given in Eq. (9.16).

The kinetic chain length v is the number of monomer molecules consumed by each primary radical and is equal to the rate of propagation divided by the rate of initiation as described in Eq. (9.17) for termination by disproportionation. The kinetic chain length is independent of the type of termination, whereas the actual degree of polymerization or chain length depends on the mode of termination. For coupling, $\overline{DP} = 2v$ since the coupling acts to double the actual chain length. For disproportionation, $\overline{DP} = v$.

Styrene macroradicals are usually terminated by coupling. However, while methyl methacrylate macroradicals terminate by coupling at temperatures below 60°C, they tend to terminate by disproportionation at higher temperatures.

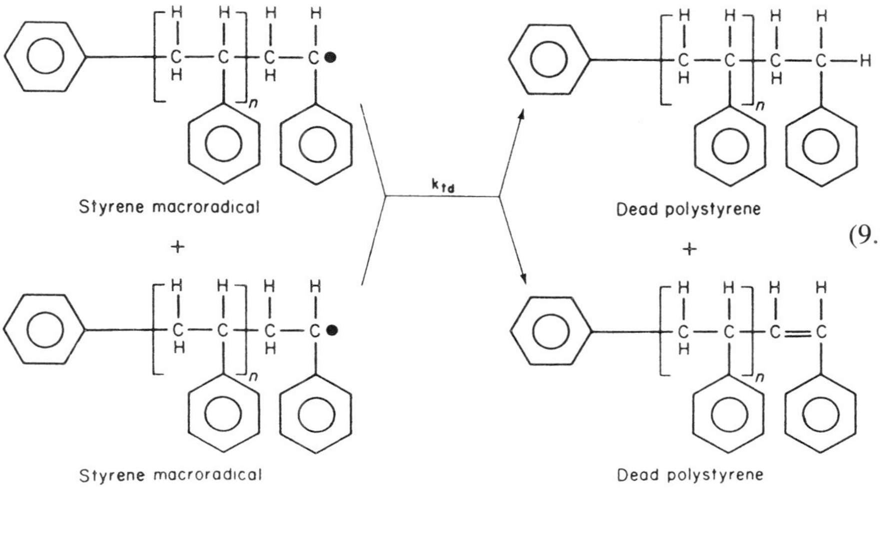

Styrene macroradical

+

Styrene macroradical

k_{td}

Dead polystyrene

+

Dead polystyrene

$$(9.15)$$

$$R_{td} = 2k_{td}[M\cdot]^2 \tag{9.16}$$

$$\nu_t = \frac{R_p}{R_i} = \frac{R_p}{R_{td}} = \frac{k_p[M][M\cdot]}{2k_{td}[M\cdot]^2} = \frac{k_p}{2k_{td}}\frac{[M]}{[M\cdot]} = \overline{DP} = k'''\frac{[M]}{[M\cdot]} \tag{9.17}$$

However, while Eqs. (9.12) and (9.16) are theoretically important, they contain [M·], which is difficult to experimentally determine and practically of little use. The following is an approach to render such equations more useful by generation of a description of [M·] involving more experimentally accessible terms.

The rate of monomer-radical change is described by:

$$\frac{d[M\cdot]}{dt} = k_i[R\cdot][M] - 2k_t[M\cdot]^2 \tag{9.18}$$

$$= \text{(monomer-radical formed)} - \text{(monomer radical utilized)}$$

It is experimentally found that the number of growing chains is approximately constant over a large extent of reaction. This situation is referred to as a "steady state." For Eq. (9.18), this results in $d[M\cdot]/dt = 0$ and

$$k_i[R\cdot][M] = 2k_t[M\cdot]^2 \tag{9.19}$$

Additionally, a steady-state value for the concentration of R· exists, yielding

$$\frac{d[R\cdot]}{dt} = 2k_d f[I] - k_i[R\cdot][M] = 0 \tag{9.20}$$

Solving for [M•] from Eq. (9.19) and [R•] from Eq. (9.20) gives

$$[M\cdot] = \left(\frac{k_i[R\cdot][M]}{2k_t}\right)^{1/2} \tag{9.21}$$

and

$$[R\cdot] = \frac{2k_d f[I]}{k_i[M]} \tag{9.22}$$

Substituting into Eq. (9.21) the expression for [R•] from Eq. (9.22), we obtain an expression for [M•] [Eq. (9.23)], which contains readily determinable variables.

$$[M\cdot] = \left(\frac{k_d f[I]}{k_t}\right)^{1/2} \tag{9.23}$$

We then obtain useful rate and kinetic chain-length equations by using the relationship shown above for [M•].

$$R_p = k_p[M][M\cdot] = k_p[M]\left(\frac{k_d f[I]}{k_t}\right)^{1/2}$$

$$= [M][I]^{1/2}\left(\frac{k_p^2 k_d f}{k_t}\right)^{1/2} = k'[M][I]^{1/2} \tag{9.24}$$

where $k' = (k_p^2 k_d f/k_t)^{1/2}$

$$R_T = 2k_t[M\cdot]^2 = \frac{2k_t k_d f[I]}{k_t} = 2k_d \, f[I] \tag{9.25}$$

$$\overline{DP} = \frac{R_p}{R_i} = \frac{k_p[M](k_df[I]/k_t)^{1/2}}{2k_df[I]} = \frac{k_p[M]}{2(k_dk_tf[I])^{1/2}} \tag{9.26}$$

$$\overline{DP} = \frac{[M]}{[I]^{1/2}} \frac{k_p}{(2k_dk_tf)^{1/2}} = \frac{[M]}{[I]^{1/2}} k''$$

where $k'' = k_p/(2k_dk_tf)^{1/2}$.

Thus, one may make the following conclusions about free-radical chain polymerizations using a chemical initiator:

1. The rate of propagation is proportional to the concentration of the monomer and the square root of the concentration of the initiator.
2. The rate of termination is proportional to the concentration of the initiator.
3. The average molecular weight is proportional to the concentration of the monomer and inversely proportional to the square root of the concentration of initiator.
4. The first chain that is initiated rapidly produces a high molecular weight polymer.
5. The monomer concentration decreases steadily throughout the reaction and approaches zero at the end.
6. The increases in rates of initiation, propagation, and termination with increases in temperature are in accord with the Arrhenius equation. The energies of activation of initiation, propagation, and termination are approximately 35, 5, and 3 kcal/mol, respectively. Data for typical energies of activation are given in Tables 9.3 and 9.4.
7. Increasing the temperature increases the concentration of free radicals, and thus increases the rate of reactions, but decreases the average molecular weight.
8. If the temperature exceeds the ceiling temperature (T_c), the polymer will decompose and no propagation will take place at temperatures above the ceiling

Table 9.3 Energies of Activation for Propagation (E_p) and Termination (E_T) in Free-Radical Chain Polymerization

Monomer	E_p (kcal/mol)	E_t (kcal/mol)
Methyl acrylate	7.1	5.3
Acrylonitrile	4.1	5.4
Butadiene	9.3	—
Ethylene	8.2	—
Methyl methacrylate	6.3	2.8
Styrene	7.8	2.4
Vinyl acetate	7.3	5.2
Vinyl chloride	3.7	4.2

Data from Brandrup and Immergut, 1975.

Table 9.4 Typical Free-Radical Kinetic Values

k_d	10^{-3} sec^{-1}	E_{ad}	30 to 50 kcal/mol
k_i	10^3 liter/mol sec	E_{ai}	5 to 7 kcal/mol
k_p	10^3 liter/mol sec	E_{ap}	4 to 10 kcal/mol
k_t	10^7 liter/mol sec	E_{at}	0 to 6 kcal/mol

temperature. The ceiling temperature for styrene is 310°C and is only 61°C for α-methylstyrene.

It is interesting to note that because of the great industrial importance of free-radical polymerizations, they are the most studied reactions in all of chemistry. Further, the kinetic approaches described in this chapter are experimentally verified for essentially all typical free-radical vinyl polymerizations studied.

There is some tendency for the formation of stereoregular segments, particularly at low temperatures, but ionic and coordination catalyst techniques are preferred for the production of stereoregular polymers. For instance, the amount of trans-1,4 is increased from 71% at 100°C to 94% at −46°C for the free-radical polymerization of the diene chloroprene. In fact, the trans-1,4 form (Scheme 9.1) appears to be favored for most free-radical polymerizations of dienes. For simple vinyl monomers, no clear trend with respect to stereoregular preference has yet emerged.

Chapters 8 and 9 are concerned mainly with the polymerization of vinyl monomers. Experimentally, only a few vinyl monomers can undergo polymerization by way of anionic, cationic, and free-radical pathways. As would be expected, anionic polymerizations occur mainly with vinyl monomers containing electron-withdrawing substituents, leaving the resulting vinyl portion electron deficient, whereas cationic polymerizations occur mainly with vinyl monomers which contain electron-donating groups. Free-radical polymerizations occur for vinyl monomers that are typically intermediate between electron poor and electron rich. It must be noted that these general tendencies are just that—general tendencies—and that variations do in fact occur. For instance, vinylidene chloride, which contains two chlorine atoms typically considered to be electron withdrawing, does not undergo homopolymerization by an anionic mechanism. Figure 9.1 contains a listing for some of the more common vinyl monomers as a function of type of chain initiation.

Scheme 9.1 trans-1,4-Polychloroprene.

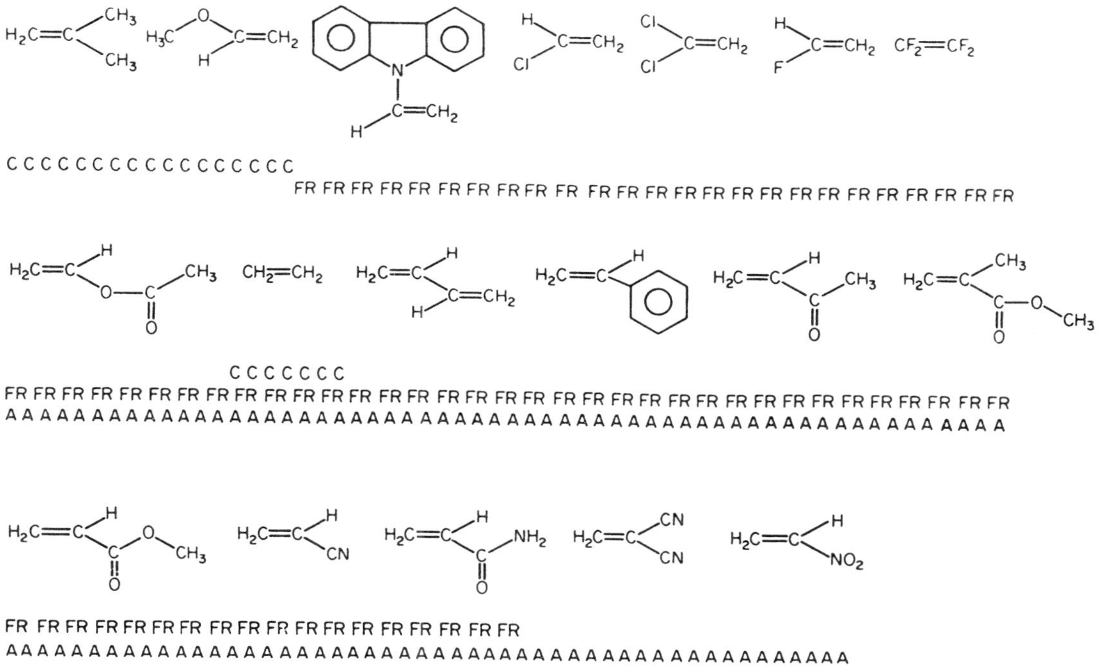

C C C C C C C C C C C C C C C C C

FR FR FR FR FR FR FR FR FR FR FR FR FR FR FR FR FR FR FR FR FR FR FR FR FR FR

C C C C C C

FR FR FR FR FR FR FR FR FR FR FR FR FR FR FR FR FR FR FR FR FR FR FR FR FR FR FR FR FR FR FR FR FR FR
A A A A A A A A A A A A A A A A A A A A A A A A A A A A A A A A A A A A A A A A A A A A A A A A A A

FR FR FR FR FR FR FR FR FR FR FR FR FR FR FR FR FR FR
A A A A A A A A A A A A A A A A A A A A A A A A A A A A A A A A A A A A A A A A A A A A A A A

Figure 9.1 Type of chain initiation suitable for some common monomers in order of general decrease in electron density associated with the double bond: A = anionic, C = cationic, FR = free-radical initiation possible.

9.3 CHAIN TRANSFER

As shown in Eq. (9.15), two macroradicals may terminate by chain transfer of a hydrogen atom at one chain end to a free-radical end of another chain in a chain-transfer process called disproportionation. When the abstraction takes place intra-molecularly or intermolecularly by a hydrogen atom, some distance from the chain end, branching results. Thus, low-density polyethylene (LDPE), which is produced by a free-radical chain polymerization at extremely high pressure, is highly branched because both types of chain transfer take place during the polymerization of ethylene at high pressure.

Each of these chain-transfer processes causes termination of one macroradical and produces another macroradical. In either case, the unpaired electron is no longer on the terminal carbon atom. The new radical sites serve as branch points for additional chain extension or branching.

As shown by the following equation, short-chain branching is the result of "backbiting" as the chain end assumes a preferred conformation resembling a stable hexagonal ring. The new active center, or branch point, is the result of an abstraction of a hydrogen atom on carbon 6 by the free radical on carbon 1.

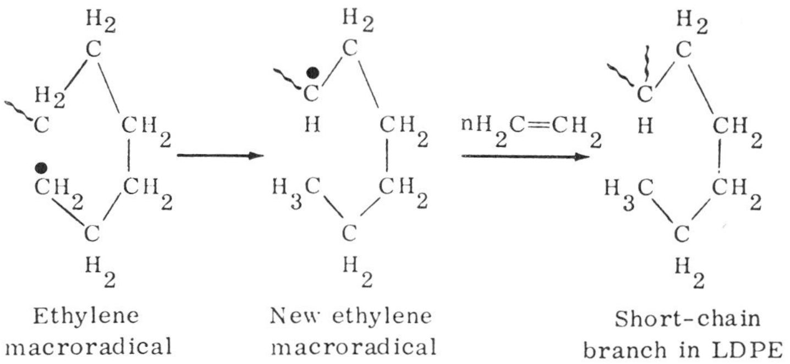

$$ \text{(9.27)} $$

Ethylene macroradical   New ethylene macroradical   Short-chain branch in LDPE

As shown in the following equation, long-chain branching takes place as the result of the formation of a branch point by intermolecular chain transfer. Since it is more probable that the new active site will be at a considerable distance from the chain end, the result is a long branch.

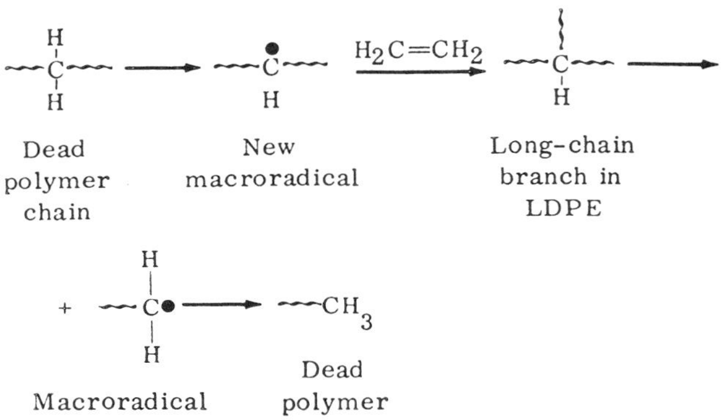

$$ \text{(9.28)} $$

Dead polymer chain   New macroradical   Long-chain branch in LDPE

Macroradical   Dead polymer

Chain transfer may also take place with monomer, initiator, solvent, or any other additives present in the polymerization system. Chain transfer to polymer is usually disregarded in the study of chain-transfer reactions, and the emphasis is on chain transfer with solvents or other additives. Thus, while the average chain length is equal to R_p divided by the sum of all termination reactions, it is customary to control all termination reactions except the chain-transfer reaction under investigation. The chain transfer to all other molecules, except solvent or additive, is typically negligible.

This chain-transfer reaction decreases the average chain length in accordance with the concentration of chain transfer agent S. The resulting degree of polymerization ($\overline{DP}$) is equal to that which would have been obtained without the solvent or additive plus a factor related to the product of the ratio of the rate of propagation (R_p) to the rate of chain transfer (R_{tr}) and the ratio of the concentration of the monomer [M] to the concentration of chain-transfer agent [S].

The Mayo equation, which yields positive slopes when the data are plotted, is the reciprocal relationship derived from the previously cited expression. The ratio of the rate of cessation or termination by transfer to the rate of propagation is called the chain-transfer constant (C_s). The latter is related to relative bond strengths in the solvent or additive molecule and the stability of the new free radical produced. The Mayo equation is shown below.

$$\frac{1}{\overline{DP}} = \frac{1}{\overline{DP}_0} + C_s \frac{[S]}{[M]} \tag{9.29}$$

As shown in Fig. 9.2, the molecular weight of polystyrene is reduced when it is polymerized in solvents, and the reduction or increase in slope is related to the chain-transfer efficiency of the solvent. The slopes in this graph are equal to C_s.

Chain-transfer agents have been called regulators (of molecular weight). When used in large proportions, they are called telogens, since they produce low molecular weight polymers (telomers) in these telomerization reactions. As shown in Table 9.5, alkyl mercaptans are effective chain transfer agents for the polymerization of styrene.

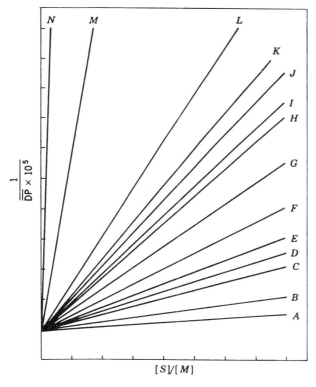

Figure 9.2 Molecular weight of polystyrene as a function of solvent and solvent concentration; A = benezne, B = toluene, C = n-heptane, D = chloroform, E = ethylbenzene, F = cumene, G = sec-butylbenzene, H = phenol, I = m-cresol, J = p-cresol, K = o-cresol, L = carbon tetrachloride, M = carbon tetrabromide, N = n-butylmercaptan, [S] = concentration of chain transfer agent, [M] = concentration of styrene monomer. (From *Introduction to Polymer Chemistry* by R. Seymour, McGraw-Hill, New York, 1971. Used with permission of McGraw-Hill Book Company.)

Table 9.5 Chain-Transfer Constants of Solvents to Styrene in Free-Radical Chain Polymerization at 60°C

Transfer agent	$C_S \times 10^4$
Acetic acid	2.0
Benzene	0.01
Butyl alcohol	0.06
tert-Butyl alcohol	6.7
Butyl disulfide	0.24
Carbon tetrabromide	18,000
Carbon tetrachloride	84
Chloroform	0.5
O-Chlorophenol	6.0
2,6-ditert-Butylphenol	49
1-Dodecanethiol	148,000
Hexane	0.9
N,N-Dimethylaniline	12
1-Naphthalenethiol	1500
1-Octanethiol	190,000
p-Methoxyphenol	260
Phenol	8.1
Triethylamine	1.4
Toluene	0.105
Water	0

The hydrogen atom is abstracted from many telogens, including the phenolic hydrogen of phenol and the hydrogen of mercaptans. However, the hydrogen on the α carbon atom is abstracted from carboxylic acids, and the hydroxyl group is abstracted from alkanols. Halogens are abstracted from many halogen compounds, such as carbon tetrabromide.

It is important to note that the new free radical produced by chain transfer may or may not initiate another polymer chain formation, depending on its activity. Retarders, chain stoppers, and many antioxidants produce new free radicals with low polymerization activity.

9.4 POLYMERIZATION TECHNIQUES

Many monomers such as styrene, acrylonitrile, and vinyl chloride are toxic, and the polymerization reaction is highly exothermic. Hence, precautions must be taken to minimize exposure to these compounds and to control the temperature of the polymerization reaction. The principal methods are bulk, solution, and emulsion polymerization (Table 9.6). Each has characteristic advantages and disadvantages (Table 9.7).

Bulk polymerization of a liquid monomer such as methyl methacrylate is relatively simple in the absence of oxygen when small bottles or test tubes are used. As shown by the following equation, one may heat this monomer in the presence of an initiator and obtain a clear plastic, shaped like the container but a little smaller because of shrinkage. Most monomers shrink during polymerization, and thus the density of the polymers is greater than that of the monomers. Poly(methyl methacrylate) (PMMA), as an atactic amorphous polymer, is sold under the trade names Lucite and Plexiglas.

$$
R\cdot + n\underset{\substack{H \\ \\ H}}{\overset{\substack{CH_3 \\ \\ C=O \\ OCH_3}}{C=C}} \xrightarrow{\Delta} R\left(\underset{\substack{H \\ \\ H}}{\overset{\substack{CH_3 \\ \\ C=O \\ OCH_3}}{C-C}}\right)_n
$$

Free radical   Methyl methacrylate   Poly(methyl methacrylate)

(9.30)

The rate of polymerization of liquid monomers such as methyl methacrylate may be followed by monitoring the change in volume by dilatometry or the increase in viscosity. The latter has essentially no effect on rate of polymerization or molecular weight unless it is relatively high. When the viscosity is high, the termination reaction is hindered, since the macroradicals are unable to diffuse readily in the viscous medium. In contrast, the monomer may diffuse quite readily and high molecular weight macroradicals are produced as the result of propagation in the absence of termination.

This autoacceleration, called the Norris-Trommsdorff, or gel effect, causes the formation of unusually high molecular weight polymers. The additional heat of polymerization may be dissipated in a small test tube, but special design of equipment is necessary for large-scale bulk (mass, or bulk) polymerizations. Fortunately, monomers such as methyl methacrylate may be polymerized without difficulty in sheets up to 5 cm in thickness either in static or continuous systems. The very high

Table 9.6 Types of Polymerization Systems

Monomer-polymer phase relationship	Monomer location	
	Continuous	Dispersed
Homogeneous (same phase)	Bulk, solid–state solution	Suspension
Heterogeneous (different phases)	Bulk with polymer precipitating	Emulsion Suspension with polymer precipitating

Table 9.7 Comparison of Polymerization Systems

Type	Advantages	Disadvantages
Homogeneous		
Bulk–batch	Simple equipment	May require solution and subsequent precipitation for purification and/or fabrication May require reduction to usable particle size Heat control important Broad molecular weight distribution
Bulk-continuous	Easier heat control Narrower molecular weight distribution	Requires reactant recycling May require solution and subsequent ppt. for purification and/or fabrication Requires more complex equipment May require reduction to usable particle size
Solution	Easy agitation May allow longer chains to be formed Easy heat control	Requires some agitation Requires solvent removal and recycling Requires polymer recovery Solvent chain transfer may be harmful (i.e., reaction with solvent)
Heterogeneous		
Emulsion	Easy heat control Easy agitation Latex may be directly usable High polymerization rates possible Molecular weight control possible Usable, small-particle size possible Usable in producing tacky, soft, and solid products	Polymer may require additional cleanup and purification Difficult to eliminate entrenched coagulants, emulsifiers, surfactants, etc. Often requires rapid agitation
Precipitation	Molecular weight and molecular weight distribution controllable by control of polymerization environment	May require solution and reprecipitation of product to remove unwanted material Precipitation may act to limit molecular weight disallowing formation of ultrahigh molecular weight products
Suspension	Easy agitation Higher purity product when compared to emulsion	Sensitive to agitation Particle size difficult to control

molecular weight product, produced because of increased viscosity that results in autoacceleration, is advantageous for cast plastics, but not for those that must be molded or extruded.

It is standard practice to polymerize liquid monomers with agitation in appropriate vessels as long as the system is liquid. In some instances, the unreacted monomer is removed by distillation and recycled. In most cases, the polymerization is continued in special equipment in which the viscous material is forced through extruderlike equipment under controlled temperature conditions. The product is polydisperse, i.e., it consists of a polymer with a broad distribution of molecular weights.

As shown in the following equation, ethyl acrylate may be polymerized by a free radical process. Poly(ethyl acrylate) does not have a pendant methyl group on the α-carbon atom like poly(methyl methacrylate), and it has a lower T_g and is flexible at room temperature. Polyacrylates with larger alkyl groups have lower T_g values and are more soluble in hydrocarbon solvents. However, the value of T_g increases when more than 10 carbon atoms are present in the alkyl group because of side-chain crystallization.

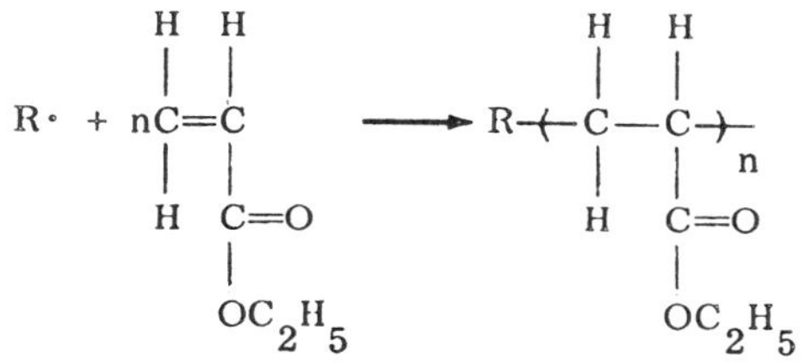

$$R\cdot + nC=C \longrightarrow R\left(C-C\right)_n \qquad (9.31)$$

Ethyl acrylate   Poly(ethyl acrylate)

Water-insoluble monomers such as vinyl chloride may be polymerized as suspended droplets (10 to 1000 nm in diameter) in a process called suspension (pearl) polymerization. Coalescing of the droplets is prevented by use of small amounts of water-soluble polymers, such as poly(vinyl alcohol). The suspension process is characterized by good heat control and ease of removal of the discrete polymer particles.

Since poly(vinyl chloride) (PVC) is insoluble in its monomer (VCM), it precipitates as formed in the droplets. This is actually advantageous, since it permits ready removal of any residual carcinogenic monomer from the solid beads by stripping under reduced pressure.

PVC, which is produced at an annual rate of 6.0 million tons, is characterized by good resistance to flame and corrosives. While the rigid product is used for some articles in which flexibility is not required, most PVC is plasticized or flexibilized by the addition of relatively large amounts of a compatible high-boiling liquid plasticizer such as dioctyl or didecyl phthalate. It has been recently reported that nonylphenols used sometimes in PVC are endocrine-system disrupters. The equation for the polymerization of vinyl chloride is shown below:

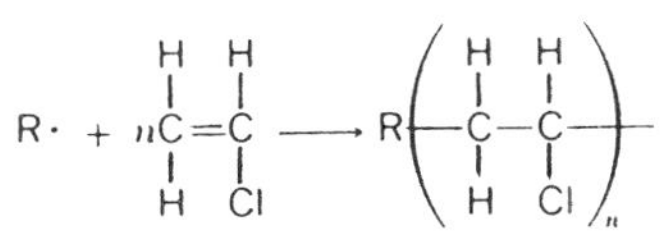

$$R\cdot + nC=C \longrightarrow R\left(C-C\right)_n \qquad (9.32)$$

Vinyl chloride   Poly(vinyl chloride)

Monomers may also be polymerized in solution using good or poor solvents for homogeneous and heterogeneous systems, respectively. Solvents with low chain-transfer constants should be used whenever possible to minimize reduction in molecular weight. While telogens decrease molecular weight, the molecular weight and the rate of polymerization are independent of the polarity of the solvent in homogeneous solution systems.

Poly(vinyl acetate) (PVAc) may be produced by the polymerization of vinyl acetate in the presence of an initiator in a solution such as benzene. The viscosity of the solution continues to increase until the reaction is complete, but the concentration of the solution is usually too dilute to exhibit autoacceleration because of the gel effect. The solution may be used as prepared, the solvent may be stripped off, or the polymer may be recovered by pouring the solution into an agitated poor solvent, such as ethanol.

PVAc is used in adhesives and coatings and may be hydrolyzed to produce water-soluble poly(vinyl alcohol) (PVA). The PVA, which is produced at an annual rate of 56,000 tons, may be reacted with butyraldehyde to produce poly(vinyl butyral) (PVB) (used as the inner lining in safety glass). The equations for the production of these polymers are as follows:

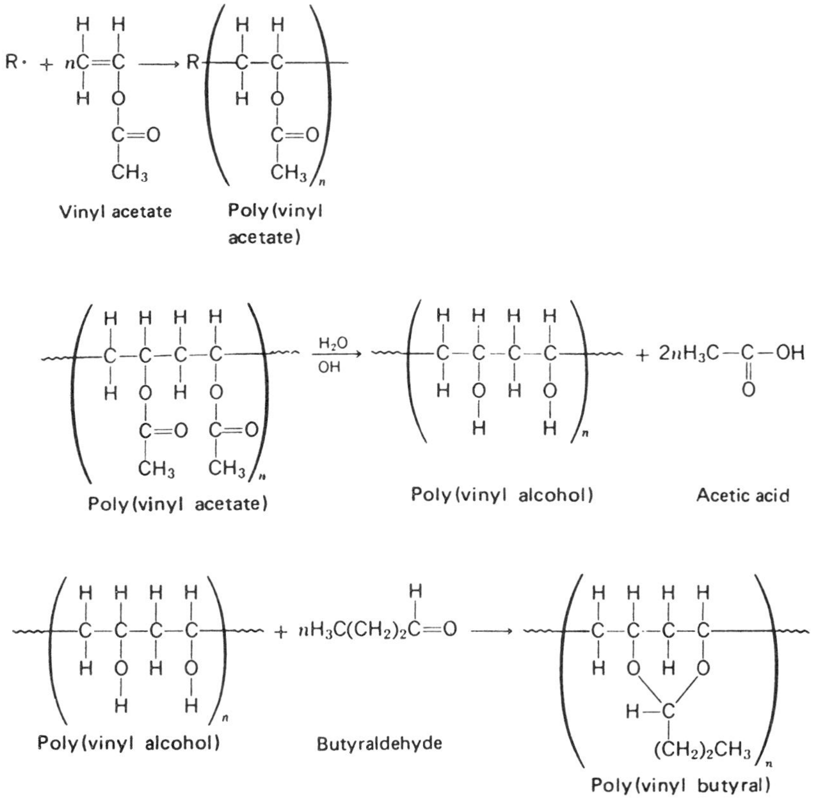

Vinyl acetate — Poly(vinyl acetate)

Poly(vinyl acetate) — Poly(vinyl alcohol) — Acetic acid

$$(9.33)$$

Poly(vinyl alcohol) — Butyraldehyde — Poly(vinyl butyral)

When a monomer such as acrylonitrile is polymerized in a poor solvent such as benzene, macroradicals precipitate as they are formed. Since these are "living polymers," the polymerization continues as more acrylonitrile diffuses into the precipitated particles. This heterogeneous solution type of polymerization has been called precipitation polymerization. Acrylic fibers (Acrilan; produced at an annual rate of 440 million lbs.) are based on polyacrylonitrile (PAN).

Since the acrylonitrile is carcinogenic, precautions must be taken to avoid contact with this monomer. The monomer must not be present in high concentrations in acrylic fibers or plastics. It is of interest to note that polyacrylonitrile forms a heat-resistant ladder polymer when heated at elevated temperatures. The equations for these reactions are as follows:

Acrylonitrile   Polyacrylonitrile

(9.34)

Polyacrylonitrile   Ladder polymers

The black ladder polymer produced by pyrolysis of PAN is sometimes called black orlon or fiber AF. Polyacrylonitrile is a hydrogen-bonded polymer with a high solubility parameter value, on the order of 15 H, and hence is soluble only in polar solvents such as N,N-dimethylformamide (DMF).

Many water-insoluble vinyl monomers may also be polymerized by the emulsion polymerization technique. This technique, which differs from suspension polymerization in the size of the suspended particles and in the mechanism, is widely used for the production of many commercial plastics and elastomers. While the particles in the suspension range from 10 to 1000 nm, those in the emulsion process range from 0.05 to 5 nm. The small beads produced in the suspension process may be separated by filtering, but the latex produced in emulsion polymerization is a stable system in which the charged particles cannot be removed by ordinary separation procedures.

Since relatively stable macroradicals are produced in the emulsion process, the termination rate is decreased and a high molecular weight product is produced at a rapid rate. It is customary to use a water-soluble initiator such as potassium persulfate, and an anionic surfactant such as sodium stearate and to stir the aqueous mixture of monomer, initiator, and surfactant in the absence of oxygen at 40 to 70°C.

A typical recipe for emulsion polymerization includes 100 g of monomer, such as styrene, 180 g of water, 5 g of sodium stearate (a soap), and 0.5 g of potassium persulfate. When the concentration of soap exceeds the critical micelle concentration (CMC), these molecules are present as micelles in which the hydrophilic carboxylic acid ends are oriented toward the water-micelle interface, and the lyophilic hydrocarbon ends are oriented toward the center of the micelle. The micelles are present as spheres with a diameter of 5 to 10 nm when the soap concentration is less than 2%. However, with the higher concentrations customarily used, the micelles resemble aggregates of rods which are 100 to 300 nm in length.

As shown in Fig. 9.3, the water-insoluble monomer is attracted to the lyophilic ends in the micelles, causing the micelles to swell. The number of swollen micelles per milliliter of water is on the order of 10^{18}. However, at the initial stages of polymerization (phase I), most of the monomer is present at globules which resemble those observed in suspension polymerization.

Since the initiation of polymerization takes place in the aqueous phase, essentially no polymerization takes place in the globules. Thus, they serve primarily as a reservoir of monomer supplied to the micelles to replace the monomer converted to polymer. The number of droplets per milliliter of water is on the order of 10^{11}. Hence, since there are 10 million times as many micelles as droplets, the chance of initiation of monomer in a droplet is very, very small.

As shown in the following equations, the persulfate ion undergoes homolytic cleavage to produce two sulfate ion radicals. These serve as initiators for the few water-soluble monomer molecules present in the aqueous phase.

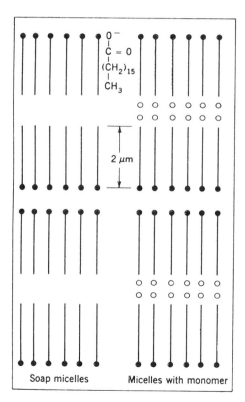

Soap micelles   Micelles with monomer

Figure 9.3 Micelles swollen with solubilized styrene monomer. (From *Introduction to Polymer Chemistry* by R. Seymour, McGraw-Hill, New York, 1971. Used with permission of McGraw-Hill Book Company.)

$$S_2O_8^{-2} \longrightarrow 2SO_4^{\bullet -}$$

Persulfate ion / Sulfate ion radical

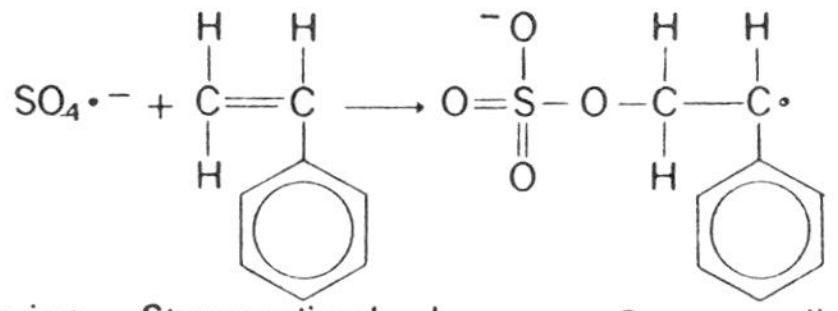

(9.35)

Sulfate ion radical   Styrene dissolved in water   Styrene radical

According to a theory proposed by Harkins and refined by Smith and Ewart, the first stages of propagation in an emulsion system also take place in the aqueous phase to produce a more lyophilic surface-active oligoradical, as shown below.

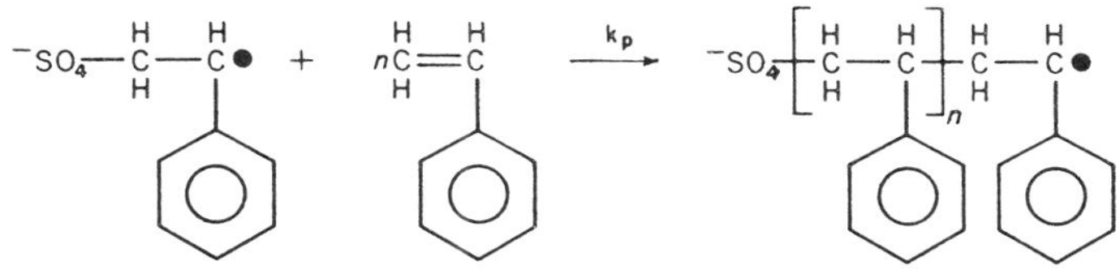

(9.36)

Styrene radical   Styrene   Oligoradical

When the $\overline{DP}$ of the styrene oligoradical is 3 to 5, its solubility is much like that of styrene, and it migrates to the swollen micelle where propagation continues with the styrene molecules already present. According to accepted theories, each micelle can accommodate only one free radical, and until a second one enters and terminates the propagation reaction by coupling, propagation continues to take place in the micelles. From a statistical point of view, only one-half of the micelles (N/2) will contain growing chains at any one time. It should also be noted that since propagation occurs in the micelles, the rate of polymerization will be proportional to the number of micelles present; i.e., the rate is proportional to the soap concentration.

As the micelles grow by absorption of more monomer and formation of polymer, they become relatively large particles which absorb soap from micelles that have not been inoculated or stung by oligoradicals. Thus, in stage II, when about 20% of the monomer has been converted to polymer, the micelles disappear and are replaced by larger, but fewer, monomer-polymer particles.

Polymerization continues in stage II, and monomer continues to be supplied to the particles by the droplets in the aqueous phase. These droplets disappear when about 30% of the monomer has been converted to polymer. Polymerization continues in stage III after about a 60% conversion, but all monomer must now be supplied to the macroradicals by a diffusion process within the micelles.

As shown below, the rate of initiation in the aqueous phase of emulsion polymerization is the same as that described for other free-radical chain initiations.

$$R_d = k_d[S_2O_8^{2-}]$$

(9.37)

$$R_i = k_i [SO_4 \cdot {}^-][H_2C = \overset{\overset{\displaystyle H}{\displaystyle |}}{C}] = 2k_d \, f \, [S_2O_8^{2=}]$$

$$\tag{9.38}$$

The rate of propagation in the micelles is similar to that described for other free-radical chain propagations, but since the free-radical concentration is equal to the number of active micelles, the value of N/2 is used instead of [M·]. Thus, as shown by the following equation, the rate of propagation is dependent on the number of micelles present.

$$R_p = k_p \, [M][M \cdot] = k_p[M] \, \frac{N}{2} \tag{9.39}$$

The rate of production of free radicals at 50°C is about 10^{13} radicals/ml/sec. Thus, since there are 100,000 micelles for every free radical produced in a second, inoculation of any of the 10^{18} micelles per milliliter is infrequent. Hence, since propagation is a very fast reaction, long chains are produced before termination by coupling, which takes place as the result of the entrance of a new oligoradical in the active micelle. As shown by the following equation, the degree of polymerization is also proportional to the number of active micelles (N/2).

$$\overline{DP} = \frac{R_p}{R_i} = \frac{k_p}{R_i} \, [M] \, \frac{N}{2} \tag{9.40}$$

The rate of polymerization in emulsion systems may be increased by the addition of reducing agents such as iron(II) salts. The presence of high molecular weight polymers is advantageous when the latex is used directly as a coating, adhesive, or film. However, the high molecular weight solid polymer obtained when the emulsion is coagulated may make subsequent processing difficult. This difficulty is overcome by the addition of chain transfer agents, such as dodecyl mercaptan (1-dodecanethiol).

In addition to LDPE, PS, PVC, PMMA, PVAc, and PAN, many other commercial polymers are produced by free-radical chain-reaction polymerization. Among these are the polyfluorocarbons, polyvinylidene chloride, neoprene elastomer, and SBR rubber. The latter, which is a copolymer of butadiene (75%) and styrene (25%), will be discussed in Chap. 10.

Polytetrafluoroethylene (Teflon, PTFE) was discovered accidentally in 1938 by Plunkett who found a solid in his cylinder of gaseous tetrafluoroethylene. As evident by its structural formula, PTFE contains no hydrogen atoms. Because of the stability of the carbon-fluorine bond, the closeness of the carbon atoms to each other, the crowding by the fluorine atoms, and the regularity in its structure, PTFE has outstanding resistance to heat. It is a crystalline polymer which does not melt below a temperature of 327°C. Teflon is one of the most expensive polymers with extensive commercial use. At least some of its high cost is due to difficulty in its processability.

The processability of this type of polyfluorocarbon is improved by replacing one of the eight fluorine atoms by a trifluoromethyl group. The product, called FEP or Viton, is a copolymer of tetrafluoroethylene and hexafluoropropylene. Polytrifluoromonochloroethylene (CTFE, Kel F), in which one fluorine atom has been replaced by a chlorine atom, has a less regular structure than FEP and is also more easily processed.

Poly(vinylidene fluoride) (Kynar) and poly(vinyl fluoride) (Tedlar) are also more readily processable and less resistant to solvents and corrosives than PTFE. The former has piezoelectric properties, i.e., it generates electric current when compressed. Polymers and copolymers of vinylidene chloride (PVDC, Saran) are used as film (Saran Wrap).

Production data for commercial polymers produced by free-radical chain polymerizations are shown in Table 9.8.

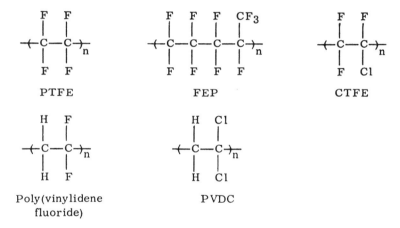

SUMMARY

1. Peroxides, such as benzoyl peroxide, and diazo compounds, such as azo-bis-isobutyronitrile, are readily homolytically cleaved by heat or ultraviolet light to produce free radicals, which serve as initiators for chain-reaction polymerization. Each initiator has its characteristic half-life. Their decomposition into free radicals may be accelerated by heat or by reducing agents.

2. The rate of initiation of free-radical chain polymerization, which is the rate-controlling step, is proportional to the product of the concentrations of the free radical (R·) and the monomer.

3. The rate of propagation is proportional to the concentrations of the monomer and the macroradicals. The additional stability of the macroradical with the functional group or the terminal carbon favors the head-to-tail configuration.

4. Termination of propagating macroradicals may take place by coupling or disproportionation. The kinetic chain length v is equal to $\overline{DP}$ for the latter and $\overline{DP}/2$ for the former. The rate of termination is proportional to the square of the concentration of macroradicals.

Table 9.8 Production of Homopolymers in the United States by Free-Radical Chain Polymerization in 1985 (Thousands of Metric Tons)

Thermoplastics	
Low–density polyethylene (LDPE)[a]	3825
Poly(vinyl chloride)	3551
Polystyrene	1798
Poly(vinyl alcohol), from PVAc	56
Total thermoplastics	9230
Elastomers	
Neoprene	126
Fibers	
Acrylic fibers (PAN)	339
Total polymers via free-radical polymerization	9695

[a]Includes LLDPE, which is a copolymer of ethylene and α-olefins.

5. The degree of polymerization $\overline{DP}$ is inversely proportional to the square root of the concentration of initiator and decreases as temperture increases.

6. Since growing chains continue to propagate until a high molecular weight product is formed, the concentration of monomer decreases steadily and approaches zero at the end of the reaction. At all times, prior to the end of the polymerization, the composition of the system consists of monomer and high molecular weight polymers.

7. Polymerization will not take place above the characteristic ceiling temperature.

8. Chain transfer, in which the macroradical abstracts a weakly bonded atom, causes branching when chain transfer with polymer occurs. Chain transfer with solvent or some other additive results in a dead polymer and a new free radical. If the latter does not serve as an initiator for further propagation, it is called a retarder or inhibitor.

9. The efficiency of a chain-transfer agent (chain-transfer constant) is the slope of the line with $1/\overline{DP}$ plotted against $[S]/[M]$ when S is the solvent or telogen.

10. Vinyl monomers may be polymerized without other additives except free radicals in bulk polymerization. When the monomer is polymerized while suspended in a stirred aqueous medium, the process is called suspension, or pearl, polymerization. Polymerization may also take place in good or poor solvents. In the former case, the viscosity continues to increase, and in the latter, the macroradicals precipitate.

11. When the system is viscous, the termination step is hindered but propagation continues. Thus, this so-called Trommsdorff or gel effect produces high molecular weight polymers.

12. Monomers may also be polymerized by a water-soluble initiator while dispersed, by agitation, in a concentrated soap solution. In this emulsion polymerization process, initiation takes place in the aqueous phase and propagation takes place in the soap micelles. Since the growing macroradicals are not terminated until a new free radical enters the micelle, high molecular weight products are obtained in a relatively short time.

13. The rate of polymerization and the degree of polymerization in the emulsion process are proportional to the number of activated micelles.

14. The polyfluorocarbons are resistant to heat, solvents, and corrosives. The resistance is greatest in the regularly structured polytetrafluoroethylene and decreases as the geometry is upset by substitution of larger or smaller groups for the fluorine atoms.

15. Low-density polyethylene, poly(vinyl chloride), polystyrene, and neoprene are made in large quantities by free-radical chain polymerization. Over 9 million tons of homopolymers are produced annually in the United States by this mechanism.

GLOSSARY

AIBN: Abbreviation for 2,2'-azo-bis-isobutyronitrile.

Acrilan: Trade name for fibers based on polymers of acrylonitrile (PAN).

backbiting: the hydrogen atom abstraction that occurs when a chain end of a macroradical doubles back on itself to form a more stable hexagonal conformation.

bimolecular reaction: A reaction involving two reactants.

BPO: Abbreviation for benzoyl peroxide.

branch point: The point on the polymer chain where additional chain extension takes place to produce a branch.

bulk polymerization: The polymerization of monomer without added solvents or water.

[]: Concentration.

C_s: Chain-transfer constant.

ceiling temperature (T_c): A characteristic temperature above which polymerization does not take place and polymers decompose.

chain stopper: A chain-transfer agent that produces inactive free radicals.

chain transfer: A process in which a free radical abstracts an atom or group of atoms from a solvent, telogen, or polymer.

chain-transfer constant (C_s): The ratio of cessation or termination of transfer to the rate of propagation.

CMC: Critical micelle concentration.

critical micelle concentration: The minimum concentration of soap in water that will produce micelles.

dead polymer: A polymer in which chain growth has been terminated.

dilatometer: An instrument which measures changes in volume.

disproportionation: A process by which termination takes place as the result of chain transfer between two macroradicals to yield dead polymers, one of which has an ethylenic end group.

DMF: Abbreviation for N,N-dimethylformamide.

E_a: Activation energy.

f: The efficiency factor in the decomposition of initiators.

FEP: A copolymer of tetrafluoroethylene and hexafluoropropylene.

first-order reaction: A reaction in which the rate is proportional to the concentration of the reactant to the first power.

half-life time ($t_{1/2}$): The time required for half the reactants to be consumed in a first-order reaction.

heterolytic cleavage: A cleavage of a covalent bond or ion pairs which leaves the two electrons on one of the atoms. The products are a carbonium ion and a carbanion, or a cation and an anion.

homogeneous cleavage: A cleavage of a covalent bond which leaves one of the two electrons on each atom. The products are free radicals.

homopolymer: A polymer made up of only one repeating unit, in contrast to a copolymer, which is made up of more than one repeat unit.

I: Initiator.

k_d: Decay constant or rate of decomposition constant.

Kel F: Trade name for polytrifluorochloroethylene (TFE).

kinetic chain length: The length of the polymer chain (DP) initiated by one free radical, which can be described as follows:

$$\nu = \frac{R_p}{R_i} = \frac{R_p}{R_t}$$

Kynar: Trade name for poly(vinylidene fluoride).

latex: A stable dispersion of a polymer in water.

M: Monomer.

M·: A monomer radical or a macroradical.

macroradicals: Electron-deficient polymers, i.e., those having a free radical present on the chain.

Mayo equation:

$$\frac{1}{\overline{DP}} = \frac{1}{\overline{DP_0}} + C_s \frac{[S]}{[M]}$$

micelles: Ordered groups of soap molecules in water.

N: The number of active micelles present.

nm: Nanometers (1×10^{-9} m).

v: Symbol for the kinetic chain length.

Norris-Trommsdorff effect: Same as Trommsdorff effect.

oligoradical: A low molecular weight macroradical.

PAN: Polyacrylonitrile.

piezoelectric: The conversion of mechanical force such as pressure into electrical energy.

plasticizer: A high-boiling compatible liquid which lowers the T_g and flexibilizes a stiff polymer like PVC.

PMMA: Poly(methyl methyacrlate).

ppm: Parts per million.

PS: Polystyrene.

PTFE: Polytetrafluoroethylene.

PVA: Poly(vinyl alcohol).

PVAc: Poly(vinyl acetate).

PVB: Poly(vinyl butyral).

PVDC: Poly(vinylidene chloride).

R: Rate of reaction.

R·: A free radical.

retarder: An additive which acts as a chain-transfer agent to produce less active free radicals.

RM: A macroradical.

S: SOlvent or telogen.

S: Styrene.

Saran: Trade name for polymers of vinylidene chloride.

SBR: A rubbery copolymer of styrene and butadiene.

stage I: The first stage in emulsion polymerization when up to 20% of the monomer is being polymerized.

stage II: The intermediate stage in emulsion polymerization when 20 to 60% of the monomer is being polymerized at a steady rate.

stage III: The last stage in emulsion polymerization when the last 40% of monomer is being polymerized.

suspension polymerization: A process in which liquid monomers are polymerized as liquid droplets suspended in water.

t: Time.

T_c: Ceiling temperature.

TBP: tert-Butyl peroxide.

Tedlar: Trade name for poly(vinyl fluoride).

telogen: An additive that readily undergoes chain transfer with a macroradical.

telomer: A low molecular weight polymer resulting from chain transfer of a macroradical with a telogen.

telomerization: The process in which telomers are produced by chain-transfer reactions.

TFE: Polytrifluoromonochloroethyelen.

Trommsdorff effect: The decrease in termination rate in viscous media which results in higher molecular weight polymers.

vinyl groups:

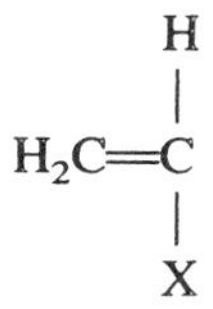

$$H_2C{=}\underset{\underset{X}{|}}{\overset{\overset{H}{|}}{C}}$$

where X may be R, X, $\overset{\overset{O}{\|}}{C}{-}OR$, OR, etc.

Viton: Trade name for FEP.

EXERCISES

1. Use a slanted line to show the cleavage of (a) boron trifluoride-water, (b) sodamide, and (c) AIBN in cationic, anionic, and free-radical initiations, respectively.

2. Which type of chain-reaction polymerization is most likely to terminate by coupling?

3. If an initiator has a half-life of 4 hr, what percentage of this initiator will remain after 12 hr?

4. If some head-to-head configuration is detected in a polymer chain known to propagate by head-to-tail addition, what type of termination has occurred?

5. Which is the better way to increase polymer production rates: (a) increasing the temperature, or (b) increasing the initiator concentration?

6. Name three widely used thermoplastics produced by free-radical chain polymerization.

7. What effect does the increase of polarity of the solvent have on free-radical polymerization rates in solution?

8. Show the repeating units for (a) PS, (b) PVC, and (c) PMMA.

9. Can you think of any advantage of the Trommsdorff effect?

10. What is the limiting step in free-radical chain polymerization?

11. In general, which is more rapid: (a) free-radical chain reaction, or (b) step-reaction polymerization?

12. If one obtained a yield of 10% polymer after 10 min of polymerizing styrene by a free-radical mechanism, what would be the composition of the other 90%?

13. Why is $t_{1/2}$ for all first-order reactions equal to $0.693/k_d$?

14. How could you follow the rate of decomposition of AIBN without measuring the rate of polymerization?

15. What is the usual value for the energy of activation of free-radical initiator?

16. What is the advantage of producing free radicals by use UV radiation?

17. Which is the better catalyst for the polymerization of styrene: (a) AIBN, or (b) BPO?

18. If [M·] is equal to 1×10^{-11} mol/liter under steady-state conditions, what will [M·] equal after (a) 30, (b) 60, and (c) 90 min?

19. In general, what is the activation energy in free-radical chain propagation of polymer chains?

20. What is the relationshiip of the rate of propagation to the concentration of initiators [I]?

21. When chain transfer with solvent occurs, what effect does this have on $\overline{DP}$?

22. Name a plasticizer for PVC.

23. What monomer is used to produce PVA?

24. How can you reconcile the two different equations for $R_i : R_i \propto [R \cdot]$ and $R_i \propto 2[I]$?

25. Does k_p increase or decrease when $\overline{DP}$ goes from 10 to 10^4?

26. Why is ethylene more readily polymerized by free-radical chain polymerization than isobutylene?

27. What is the termination mechanism in free-radical polymerization if $\overline{DP} = v$?

28. The value of v increases as the polymerization temperature of a specific monomer is increased. What does this tell you about the termination process?

29. In general, what is the activation energy of termination?

30. Why wouldn't you recommend the use of poly-α-methylstyrene for a handle of a cooking utensil?

31. When backbiting occurs, a long branch forms at the branch point. Why is this called "short-chain branching"?

32. Which would you expect to have the higher chain-transfer constant: (a) carbon tetrafluoride, or (b) carbon tetrachloride?

33. While the addition of dodecyl mercaptan to styrene causes a reduction in $\overline{DP}$, the rate of polymerization is essentially unchanged. Explain.

34. Would it be safe to polymerize styrene by bulk polymerization in a 55-gal drum?

35. How do the kinetics of polymerization differ in the bulk and suspension of polyermization methods?

36. Since the monomers are carcinogenic, should the polymerization of styrene, acrylonitrile, and vinyl chloride be banned?

37. What happens when a filament of polyacrylonitrile is pyrolyzed?

38. Why doesn't polymerization take place in the droplets, instead of in the micelles in emulsion polymerization?

39. Why doesn't initiation occur in the micelles in emulsion polymerization?

40. What would happen if one added a small amount of an inhibitor to styrene before bulk polymerization?

41. What would be the effect on the rate of polymerization if one used AIBN as the initiator and an amount of soap that was less than CMC in emulsion polymerization?

42. Why is the T_g of PTFE higher than that of FEP?

43. Why does an increase in soap concentration increase the $\overline{DP}$ and R_p of emulsion polymerization?

44. Which will have the higher specific gravity: (a) PVC, or (b) PVDC?

BIBLIOGRAPHY

Allen, P. E. M., Patrick, D. R. (1900): Kinetics and mechanisms of polymerization reactions, Makromol. Chem., *47*: 154.

Bamford, C. H., Barb, W. G., Jenkins, A. D., Onyon, P. F. (1958): *Kinetics of Vinyl Polymerization by Radical Mechanisms*, Butterworths, London.

Barton, J. (1993): *Radical Polymerization in Disperse Systems*, Prentice-Hall, Englewood Cliffs, New Jersey.

Blackley, D. C. (1975): *Emulsion Polymerization*, Halsted, New York.

Boundy, R. H., Boyer, R. F. (1952): *Styrene*, Reinhold, New York.

Brandrup, J., Immergut, E. H. (1975): *Polymer Handbook*, Wiley, New York.

Carraher, C. E., Moore, J. A. (1983): *Modification of Polymers*, Plenum, New York.

Chien, J. (1984): *Polyacetylenes*, Academic, New York.

El-Aaser, M. S., Vanderhoff, J. W. (1981): *Emulsion Polymerization of Vinyl Acetate*, Applied Science, Essex, England.

Fawcett, E. W., Gibson, R. O., et al. (1937): British Patent 471,590, September 6.

Gaylord, N. G., Mark, H. F. (1958): *Linear and Stereoaddition Polymers*, Interscience, New York.

Gertel, G. (1985): *Polyurethane Handbook*, MacMillan, New York.

Ham, G. E. (1967): *Vinyl Polymerization*, Vol. I, Interscience, John Wiley, New York.

Harkins, W. D. (1947; 1950): Emulsion polymerization, J. Am. Chem. Soc., *69*:1429; J. Polymer Sci., *5*:217.

Ivin, K. J. (1983): *Olefin Metathesis*, Academic, New York.

Mayo, F. R. (1943): Chain transfer, J. Am. Chem. Soc., *65*:2324.

Piirma, I. (1982): *Emulsion Polymerization*, Academic, New York.

Plunkett, R. J. (1941): Polytetrafluoroethylene, U.S. Patent 2,230,654, February 4.

Schildknecht, C. E. (ed.) (1956): *Polymer Processes*, Interscience, New York.

Seymour, R. B. (1975): *Modern Plastics Technology*, Reston Publishing, Reston, Virginia.

————. (1987): Polyurethanes, in *Encyclopedia of Physical Science and Technology*, Academic, New York.

Seymour, R. B., Cheng, T. (1985): *History of Polyolefins*, Riedal, Dortrecht, The Netherlands.

————. (1987): *Advances in Polyolefins*, Plenum, New York.

Seymour, R. B., Patel, V. (1973): Chain transfer, J. Macromol. Sci. Chem., *7*(4):961.

Seymour, R. B., Sosa, J. M., Patel, V. (1971): Chain transfer, J. Paint Technol., *43*(563):45.

Small, P. A. (1975): *Long Chain Branching in Polymers*, Springer-Verlag, New York.

Smith, W. V., Ewart, R. H. (1948): Emulsion polymerization, J. Chem. Phys., *16*:592.

Sorenson, W. R., Campbell, T. W. (1968): *Preparative Methods of Polymer Chemistry*, 2nd ed., Wiley-Interscience, New York.

Trommsdorff, E., Kohle, H., Lagally, P. (1948): Viscous polymerization, Makromol. Chem., *1*:169.

Walling, C. (1957): *Free Radicals in Solution*, Wiley, New York.

Willbecker, E. L. (1974): *Macromolecular Synthesis*, Wiley, New York.

Yokum, R. H., Nyquist, E. G. (1974): *Functional Monomers: Their Preparation, Polymerization, and Application*, Marcel Dekker, New York.

10

Copolymerization

While the mechanism of copolymerization is similar to that discussed for the polymerization of one reactant (homopolymerization), the reactivities of monomers may differ when more than one is present in the feed, i.e., reaction mixture. Copolymers may be produced by step-reaction or by chain-reaction polymerization. It is important to note that if the reactant species are M_1 and M_2, the composition of the copolymer is not a mixture or blend $[M_1]_n + [M_2]_n$.

Most naturally occurring polymers are largely homopolymers, but both proteins and nucleic acids are copolymers. While many synthetic polymers are homopolymers, the most widely used synthetic rubber (SBR) is a copolymer of styrene (S) and butadiene (B). A widely used plastic (ABS) is a copolymer or blend of polymers of acrylonitrile, butadiene, and styrene. A special fiber called Spandex is a block copolymer of a stiff polyurethane and a flexible polyester.

Copolymers may be alternating copolymers, in which there is a regular order of M_1 and M_2 in the chain, i.e., $(M_1M_2)_n$; random copolymers, in which the sequences of M_1 and M_2 are arranged in a random fashion, i.e., $M_1M_1M_2M_1M_2M_2$ . . . ; block copolymers, in which there are long sequences of the same repeating unit in the chain, i.e., $(M_1)_n(M_2)_n$; or graft copolymers, in which the chain extension of the second monomer are as branches, i.e.,

$$
\begin{array}{l}
| \\
M_1 \\
M_1\!\!-\!\!(M_2)_n\!\!- \\
M_1 \\
M_1 \\
|_m
\end{array}
$$

It is interesting to note that block copolymers may be produced from one monomer only if the arrangements around the chiral carbon atom change sequentially. The copolymers in which the tacticity of the monomers in each sequence differs are called stereoblock copolymers.

10.1 KINETICS OF COPOLYMERIZATION

Because of a difference in the reactivity of the monomers, expressed as reactivity ratios (r), the composition of the copolymer (n) may be different from that of the reactant mixture or feed (x). When x equals n, the product is said to be an azeotropic copolymer.

In the early 1930s, Nobel laureate Staudinger analyzed the product obtained from the copolymerization of equimolar quantities of vinyl chloride (VC) and vinyl acetate (VAc). He found that the first product produced was high in VC, but as the composition of the reactant mixture changed because of a depletion of VC, the product was higher in VAc. This phenomenon is called the composition drift.

Wall showed that n was equal to rx when the reactivity ratio r was equal to the ratio of the propagation rate constants. Thus, r was the slope of the line obtained when the ratio of monomers in the copolymer (M_1/M_2) was plotted against the ratio of monomers in the feed (m_1/m_2). The Wall equation shown below is not a general equation.

$$n = \frac{M_1}{M_2} = r \frac{m_1}{m_2} = rx \tag{10.1}$$

The copolymer equation that is now generally accepted was developed in the late 1930s by a group of investigators including Wall, Dostal, Lewis, Alfrey, Simha, and Mayo. These workers considered the four possible chain extension reactions when M_1 and M_2 were present in the feed. As shown below, two of these equations are homopolymerizations, or self-propagating steps, and the other two are heteropolymerizations or cross-propagating steps. The ratio of the propagating rate constants are expressed as monomer reactivity ratios, where $r_1 = k_{11}/k_{12}$ and $r_2 = k_{22}/k_{21}$. $M_1\cdot$ and $M_2\cdot$ are used as symbols for the macroradicals with M_1 and M_2 terminal groups, respectively.

Reactions	Rate expressions	
$M_1\cdot + M_1 \xrightarrow{k_{11}} M_1M_1\cdot$	$R_{11} = k_{11}[M_1\cdot][M_1]$	(10.2)
$M_1\cdot + M_2 \xrightarrow{k_{12}} M_1M_2\cdot$	$R_{12} = k_{12}[M_1\cdot][M_1]$	(10.3)
$M_2\cdot + M_2 \xrightarrow{k_{22}} M_2M_2\cdot$	$R_{22} = k_{22}[M_2\cdot][M_2]$	(10.4)
$M_2\cdot + M_1 \xrightarrow{k_{21}} M_2M_1\cdot$	$R_{21} = k_{21}[M_2\cdot][M_1]$	(10.5)

Experimentally it is found that the specific rate constants for the reaction steps described above are essentially independent of chain length, with the rate of monomer addition primarily dependent only on the adding monomer unit and the growing end. Thus, the four copolymerizations between two comonomers can be described using only four equations.

The rate of consumption of M_1 and M_2 in the feed or reactant mixture during the early stages of the reaction can be then described by the following equations:

Disappearance of M_1: $\dfrac{-d[M_1]}{dt} = k_{11}\,[M_1 \cdot][M_1] + k_{21}\,[M_2 \cdot][M_1]$　　　(10.6)

Disappearance of M_2: $\dfrac{-d[M_2]}{dt} = k_{22}\,[M_2 \cdot][M_2] + k_{12}\,[M_1 \cdot][M_2]$　　　(10.7)

Since it is experimentally observed that the number of growing chains remains approximately constant throughout the duration of most copolymerizations (that is there is a steady state in the number of growing chains), the concentrations of $M_1 \cdot$ and $M_2 \cdot$ are constant, and the rate of conversion of $M_1 \cdot$ to $M_2 \cdot$ is equal to the conversion of $M_2 \cdot$ to $M_1 \cdot$. Solving for $M_1 \cdot$ gives

$$[M_1 \cdot] = \frac{k_{21}[M_2 \cdot][M_1]}{k_{12}[M_2]}$$　　　(10.8)

The ratio of disappearance of monomers M_1/M_2 is described by Eq. (10.9a) from Eqs. (10.6) and (10.7).

$$\frac{d[M_1]}{d[M_2]} = \frac{[M_1]}{[M_2]}\left(\frac{k_{11}[M_1 \cdot] + k_{21}[M_2 \cdot]}{k_{12}[M_1 \cdot] + k_{22}[M_2 \cdot]}\right)$$　　　(10.9a)

Substitution of $[M_1 \cdot]$ into Eq. (10.9a) gives

$$\frac{d[M_1]}{d[M_2]} = \frac{[M_1]}{[M_2]}\left(\frac{\dfrac{k_{11}k_{21}[M_2 \cdot][M_1]}{k_{12}[M_2]} + k_{21}[M_2 \cdot]}{\dfrac{k_{12}k_{21}[M_2 \cdot][M_1]}{k_{12}[M_2]} + k_{22}[M_2 \cdot]}\right)$$　　　(10.9b)

Division by k_{21} and cancellation of the appropriate k's gives

$$\frac{d[M_1]}{d[M_2]} = \frac{[M_1]}{[M_2]}\left(\frac{\dfrac{k_{11}[M_2 \cdot][M_1]}{k_{12}[M_2]} + [M_2 \cdot]}{\dfrac{[M_2 \cdot][M_1]}{[M_2]} + \dfrac{k_{22}[M_2 \cdot]}{k_{21}}}\right)$$　　　(10.9c)

Substitution of $r_1 = k_{11}/k_{12}$ and $r_2 = k_{22}/k_{21}$ and cancellation of $[M_2 \cdot]$ gives

$$\frac{d[M_1]}{d[M_2]} = \frac{[M_1]}{[M_2]}\left(\frac{\dfrac{r_1[M_1]}{[M_2]} + 1}{\dfrac{[M_1]}{[M_2]} + r_2}\right)$$　　　(10.9d)

Multiplication by $[M_2]$ yields what are generally referred to as the "copolymerization equations" [Eq. (10.10)], which gives the copolymer composition without the need to know any free-radical concentration.

$$n = \frac{d[M_1]}{d[M_2]} = \frac{[M_1]}{[M_2]}\left(\frac{r_1[M_1] + [M_2]}{[M_1] + r_2[M_2]}\right)$$　　　(10.10)

or

$$n = \frac{d[M_1]}{d[M_2]} = \frac{r_1([M_1]/[M_2])}{r_2([M_2]/[M_1]) + 1} = \frac{r_1x + 1}{(r_2/x) + 1}$$

The copolymer equations may be used to show the effect of the composition of the feed (x) on the composition of the copolymer (n). While the values of these two parameters are equal in azeotropic copolymerization, they are different in most copolymerizations. Hence, it is customary to make up for this difference in reactivity ratios by adding monomers continuously to the feed in order to produce copolymers of uniform composition.

For example, from Table 10.1, the reactivity ratios for butadiene and styrene are $r_1 = 1.39$ and $r_2 = 0.78$, respectively. Since the rate of consumption of butadiene (M_1) is faster than that of styrene (M_2), the compositions of the feed (x) would change rapidly if butadiene were not added to prevent a change in its composition, i.e., composition drift.

As shown by the following equation, an equimolar ratio of butadiene and styrene would produce a butadiene-rich copolymer in which there would be four molecules of butadiene to every three molecules of styrene in the polymer chain.

$$n = \frac{r_1x + 1}{(r_2/x) + 1} = \frac{1.39(1) + 1}{(0.78/1) + 1} = \frac{2.39}{1.78} = 1.34 \tag{10.11}$$

The reactivity ratios may be determined by an analysis of the change in composition of the feed during the very early stages of polymerization. Typical free-radical chain copolymerization reactivity ratios are listed in Table 10.1.

If the reactivity ratio for r_1, shown in Table 10.1, is greater than 1, then monomer M_1 tends to produce homopolymers, or block copolymers. Preference for reaction with the unlike monomer occurs when r_1 is less than 1. When both r_1 and r_2 are approximately equal to 1, the conditions are said to be ideal, and a random (not alternating) copolymer is produced, in accordance with the Wall equation. Thus, a perfectly random copolymer (ideal copolymer) would be produced when chlorotrifluoroethylene is copolymerized with tetrafluoroethylene.

When r_1 and r_2 are approximately equal to zero, as is the case with the copolymerization of maleic anhydride and styrene, an alternating copolymer is produced. In general, there will be a tendency toward alternation when the product of r_1r_2 approaches zero. In contrast, if the values of r_1 and r_2 are similar and the product r_1r_2 approaches 1, the tendency will be to produce random copolymers. The value of r_1r_2 for most copolymerizations is between 1 and 0, and thus this value may be used with discretion for estimating the extent of randomness in a copolymer.

In the absence of steric or polar restrictions, the assumptions used in the development of the copolymerization equation are valid. Temperature conditions below T_c and the polarity of solvents have little effect on free-radical copolymerization. However, steric effects can be important. For example, as shown by the data in Table 10.1, styrene should tend to produce an alternating copolymer with fumaronitrile. However, since there is not enough space for the fumaronitrile mer (M_1) in alternating sequences, it is necessary that two styrene mers (M_2) be in the chain after each regular alternating sequence, i.e., $-M_1M_2M_2M_1M_2M_2M_1M_2-$.

While 1,2-disubstituted vinyl monomers are not readily polymerized, they may form alternating copolymers. For example, stilbene and maleic anhydride produce

Table 10.1 Typical Free-Radical Chain-Copolymerization Reactivity Ratios at 60°C[a]

M_1	M_2	r_1	r_2	$r_1 r_2$
Acrylamide	Acrylic acid	1.38	0.36	0.5
	Methyl acrylate	1.30	0.05	0.07
	Vinylidene chloride	4.9	0.15	0.74
Acrylic acid	Acrylonitrile (50°C)	1.15	0.35	0.40
	Styrene	0.25	0.15	0.04
	Vinyl acetate (70°C)	2	0.1	0.2
Acrylonitrile	Butadiene	0.25	0.33	0.08
	Ethyl acrylate (50°C)	1.17	0.67	0.78
	Maleic anhydride	6	0	0
	Methyl methacrylate	0.13	1.16	0.15
	Styrene	0.04	0.41	0.16
	Vinyl acetate	4.05	0.06	0.24
	Vinyl chloride	3.28	0.02	0.07
Butadiene	Methyl methacrylate	0.70	0.32	0.22
	Styrene	1.39	0.78	1.08
Chlorotrifluoroethylene	Tetrafluoroethylene	1.0	1.0	1.0
Isoprene	Styrene	1.98	0.44	0.87
Maleic anhydride	Methyl acrylate	0	2.5	0
	Methyl methacrylate	0.03	3.5	0.11
	Styrene	0	0.02	0
	Vinyl acetate (70°C)	0.003	0.055	0.0002
Methyl acrylate	Acrylonitrile	0.67	1.26	0.84
	Styrene	0.18	0.75	0.14
	Vinyl acetate	9.0	0.1	0.90
	Vinyl chloride	5	0	0
Methyl isopropenyl ketone	Styrene (80°C)	0.66	0.32	0.21
Methyl methacrylate	Styrene	0.50	0.50	0.25
	Vinyl acetate	20	0.015	0.30
	Vinyl chloride	12.5	0	0
α-Methylstyrene	Maleic anhydride	0.038	0.08	0.003
	Styrene	0.38	2.3	0.87
Styrene	p-Chlorostyrene	0.74	1.025	0.76
	Fumaronitrile	0.23	0.01	0.002
	p-Methoxystyrene	1.16	0.82	0.95
	Vinyl acetate	55	0.01	0.55
	Vinyl chloride	17	0.02	0.34
	2-Vinylpyridine	0.56	0.9	0.50
Vinyl acetate	Vinyl chloride	0.23	1.68	0.39
	Vinyl laurate	1.4	0.7	0.98
Vinyl chloride	Diethyl maleate	0.77	0.009	0.007
	Vinylidene chloride	0.3	3.2	0.96
N-Vinylpyrrolidone	Styrene (50°C)	0.045	15.7	0.71

[a]Data from Brandrup and Immergut, 1975. Temperatures other than 60°C shown in parentheses.

an alternating copolymer but have little tendency to form homopolymers. It is now believed that a charge-transfer complex is the active species in alternating copolymerization.

The formation of these charge-transfer complexes is enhanced by the presence of salts such as zinc chloride. These complexes can be detected at low temperatures by ultraviolet or nuclear magnetic resonance spectrometry. Since the equilibrium constants for the formation of these charge complexes decrease as the temperature increases, random, instead of alternating copolymers, may be produced at elevated temperatures.

The resonance stability of the macroradical is an important factor in free-radical propagation. Thus, a conjugated monomer such as styrene is at least 30 times as apt to form a resonance-stabilized macroradical as vinyl acetate, resulting in a copolymer rich in styrene. Providing there is not excessive steric hindrance, a 1,1-disubstituted ethylene monomer will polymerize more readily than the unsubstituted vinyl monomer.

Strongly electrophilic or nucleophilic monomers will polymerize exclusively by anionic or cationic mechanisms. However, monomers such as styrene or methyl methacrylate, which are neither strongly electrophilic nor nucleophilic, will polymerize by ionic and free-radical chain polymerization mechanisms. As shown in Table 10.1, the values of r_1 and r_2 for these monomers are identical in free-radical chain polymerization. Thus, the formation of a random copolymer would be anticipated when equimolar quantities of these two monomers are present in the feed.

In contrast, the reactivity ratios for styrene and methyl methacrylate are $r_1 = 0.12$ and $r_2 = 26.4$ in an anionic system. Thus, a copolymer rich in methyl methacrylate would be predicted. However, since $r_1 = 10.5$ and $r_2 = 0.1$ in cationic systems, the reverse would be true.

Butyl rubber is a random copolymer prepared by the cationic copolymerization of isobutylene ($r_1 = 2.5$) and isoprene ($r_2 = 0.4$) at $-100°C$. The elastomeric copolymer of ethylene and propylene is prepared with a homogeneous Ziegler-Natta catalyst in chlorobenzene using $VO(OR)_3$ as the catalyst and $(C_2H_5)_2AlCl$ as the cocatalyst. In this case, r_1r_2 is approximately 1, so the product tends to be a random copolymer. It is customary to add a diene monomer so that the product may be cured through cross-linking of the double bonds in the copolymer (EPDM).

The contrast between anionic, cationic, and free-radical methods of addition polymerization is clearly illustrated by the results of copolymerization utilizing the three modes of initiation. The composition of initial copolymer formed from a feed of styrene and methyl methacrylate is shown in Fig. 10.1. Such results illustrate the variations of reactivities and copolymer composition that are possible from employing the different initiation modes. The free-radical "tie-line" resides near the middle since free-radical polymerizations are less dependent on the electronic nature of the comonomers relative to the ionic modes of chain propagation.

10.2 THE Q-e SCHEME

A useful scheme for predicting r_1 and r_2 values for free-radical copolymerizations was proposed by Alfrey and Price in 1947. The Alfrey-Price Q-e scheme is similar to the Hammett equation, except that it is not limited to substituted aromatic compounds. In the semiempirical Q-e scheme, the reactivities or resonance effects of

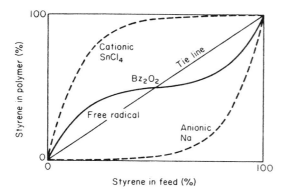

Figure 10.1 Instantaneous copolymer composition as a function of monomer composition and initiator employed for the comonomer system of styrene and methyl methacrylate utilizing different modes of initiation. From Y. Landler, Comptes Rendus, *230*:539 (1950). With permission of the Academie des Sciences, Paris, France; D. Pepper, Chem. Soc. Q. Rev., *8*:88 (1954). With permission of The Chemical Society (Great Britain).

the monomers and macroradicals are evaluated empirically by Q and P values, respectively. The polar properties of both monomers and macroradicals are designated by arbitrary e values.

Thus, as shown in Table 10.2, Q_1 and Q_2 are related to the reactivity, and e_1 and e_2 are related to the polarity of monomers M_1 and M_2, respectively. Styrene, with Q and e values of 1.00 and -0.80, is used as the comparative standard. Higher Q values indicate greater resonance stability or reactivity, and higher e values (less negative) indicate greater electron-withdrawing power of the α substituents on the vinyl monomer.

Thus, as shown in Table 10.2, butadiene and isoprene with higher Q values are more reactive than styrene. Likewise, acrylonitrile and maleic anhydride with positive e values are more polar than styrene. While the Q-e scheme neglects steric factors, it is a useful guide when data for r_1 and r_2 are not available. The Alfrey-Price equation, in which P_1 relates to macroradical $M_1 \cdot$, Q_2 relates to monomer M_2, and e_1 and e_2 relate to polarity, is shown below.

$$k_{11} = P_1 Q_1 e^{-e_1^2}$$

$$k_{12} = P_1 Q_2 e^{-e_1 e_2}$$

therefore,

$$r_1 = \frac{k_{11}}{k_{12}} = \frac{Q_1}{Q_2} e^{-e_1(e_1 - e_2)} \tag{10.12}$$

$$k_{22} = P_2 Q_2 e^{-e_2^2}$$

$$k_{21} = P_2 Q_1 e^{-e_1 e_2}$$

$$r_2 = \frac{k_{22}}{k_{21}} = \frac{Q_2}{Q_1} e^{-e_2(e_2 - e_1)}$$

$$r_1 r_2 = e^{-(e_1 - e_2)^2} \qquad \text{or} \qquad r_1 r_2 = \text{exponential } -(e_1 - e_2)^2 \tag{10.13}$$

Table 10.2 Typical Q and e Values for Monomers[a]

Monomers	Q	e
Benzyl methacrylate	3.64	0.36
Methacrylic acid	2.34	0.65
2-Fluoro-1,3-butadiene	2.08	-0.43
p-Cyanostyrene	1.86	-0.21
p-Nitrostyrene	1.63	0.39
2,5-Dichlorostyrene	1.60	0.09
Methacrylamide	1.46	2.24
p-Methoxystyrene	1.36	-1.11
2-Vinylpyridine	1.30	-0.50
p-Methylstyrene	1.27	-0.98
2-Vinylnaphthalene	1.25	-0.38
Isopropyl methacrylate	1.20	-0.15
Methacrylonitrile	1.12	0.81
p-Bromostyrene	1.04	-0.32
Styrene	1.00	-0.80
m-Methylstyrene	0.91	-0.72
n-Amyl methacrylate	0.82	-0.32
Methyl methacrylate	0.74	0.40
Acrylonitrile	0.60	1.20
Methyl acrylate	0.42	0.60
Vinylidene chloride	0.23	0.36
Vinyl chloride	0.044	0.20
Vinyl acetate	0.026	-0.22

[a]Data from *Polymer Handbook.*

It is important to note that while the reactivity is related to resonance stability of the macroradical $M_1 \cdot$, the composition of the copolymer is related to the relative polarity of the two monomers M_1 and M_2.

10.3 COMMERCIAL COPOLYMERS

One of the first commercial copolymers was the copolymer of vinyl chloride (87%) and vinyl acetate (13%) (Vinylite, VYHH) which was introduced in 1928. As shown in Table 10.1, the r_1 value for vinyl acetate is 0.23 and r_2 for vinyl chloride is 1.68. Hence, a 4:1 molar ratio of VC to VAc in the feed produced VYHH as the initial copolymer, and it was necessary to continue to add the monomers in the correct proportions in order to obtain the same composition throughout the polymerization.

Because the presence of vinyl acetate mers disrupted the regular structure of PVC, the copolymer was more flexible and more soluble than the homopolymer. A copolymer of vinyl chloride and vinyl isobutyl ether (Vinoflex) has been produced in Germany and a copolymer of vinyl chloride and acrylonitrile (Vinyon, Dynel) has been used as a fiber in the United States. Copolymers of styrene and acrylonitrile are sometimes classified as high-performance or engineering plastics.

Copolymers of vinyl chloride and vinylidene chloride were introduced in the 1930s. The copolymer with a high vinyl chloride content was used as a plastic film (Pliovic), and the copolymer with a high vinylidene chloride content as been used as a film and filament (Saran). Three-component copolymers of vinyl chloride, vinyl acetate, and maleic anhydride have superior adhesive properties because of the polar anhydride groups present on the polymer chain.

Copolymers of ethylene and methacrylic acid (ionomers) also have good adhesive properties and transparency. When salts such as those of zinc, lithium, or sodium are added to these ionomers, they act like thermosetting plastics at room temperature but are readily molded since the ion pairs dissociate at processing temperatures.

Polybutadiene, produced in emulsion polymerization, is not useful as an elastomer. However, the copolymers with styrene (SBR) and acrylonitrile (Buna-N) are widely used as elastomers. Over 8 million tons of SBR were produced worldwide in 1995.

While most fibers are homopolymers, polyacrylonitrile fibers cannot be readily dyed. However, this difficulty is overcome when acrylonitrile is copolymerized with a small amount of acrylic acid, acrylamide, vinylpyridine, or vinylpyrrolidone. The dyability of polypropylene fibers is improved by graft copolymerization with vinylpyrrolidone.

Ethylene-propylene copolymers show good resistance to ozone, heat, and oxygen. Two general types of ethylene-propylene elastomers are commercially available. Ethylene-propylene copolymers (EPM) are saturated and require vulcanization employing free-radical processes. They are used in a wide variety of automotive applications, including as body and chassis parts, bumpers, radiator and heater hoses, seals, mats, and weatherstrips. EPM are produced by solution polymerization employing Ziegler-Natta–type catalysts.

The second type of ethylene-propylene copolymer is the ethylene-propylene-diene terpolymers (EPDM). EPDM are made by polymerizing ethylene, propylene, and a small amount (3 to 9%) of nonconjugated diolefin employing Ziegler-Natta–type catalysts. The side chains permit vulcanization with sulfur. They are employed in the production of appliance parts, wire and cable insulation, coated fabrics, gaskets, hoses, seals, and high-impact polypropylene.

10.4 BLOCK COPOLYMERS

While block copolymers do not occur naturally, synthetic block copolymers have been prepared by all known polymerization techniques. Block copolymers consisting of long sequences of mers are produced inadvertently when hevea rubber is milled with other polymers or when mixtures of different polyesters or mixtures of polyesters and polyamides are melted.

The first commercial block copolymer was a surfactant (Pluronics) prepared by the addition of propylene oxide to polycarbanions of ethylene oxide. While neither water-soluble poly(ethylene oxide) nor water-insoluble poly(propylene oxide) exhibits surface activity, the ABA block copolymer consisting of hydrophilic and lyophilic segments, shown below, is an excellent surfactant.

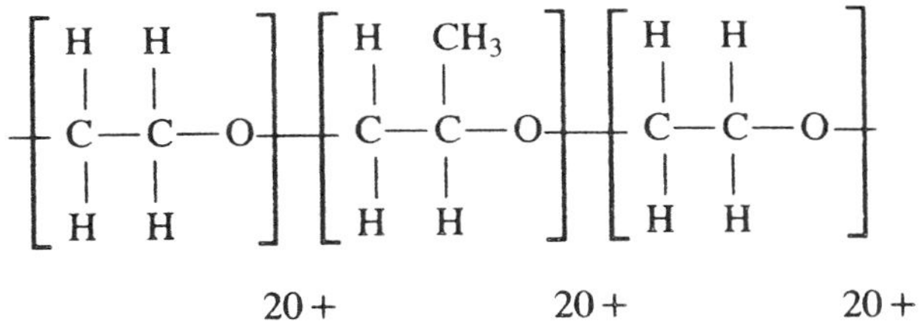

$$\left[\begin{matrix} & H & H \\ & | & | \\ +\!\! & C\!-\!C\!-\!O \\ & | & | \\ & H & H \end{matrix}\right]\!\!\left[\begin{matrix} & H & CH_3 \\ & | & | \\ & C\!-\!C\!-\!O \\ & | & | \\ & H & H \end{matrix}\right]\!\!\left[\begin{matrix} & H & H \\ & | & | \\ & C\!-\!C\!-\!O\!+ \\ & | & | \\ & H & H \end{matrix}\right]$$

$$20+ \qquad\qquad 20+ \qquad\qquad 20+$$

ABA copolymer of ethylene oxide and propylene oxide

Elastomeric polyurethane fibers consisting of at least 85% segmented polyurethane, which are commercially available under the generic name of Spandex, are block copolymers. As shown by the following equation, these fibers, as well as some thermoplastic elastomers (TPE), are prepared by the reaction of a stiff polymeric dissocyanate with a flexible polymeric diol. Polyethers with terminal hydroxyl groups are used for the production of flexible polyurethane foams. TPE (Hytrel) is a block copolymer of a polyether and a polyester.

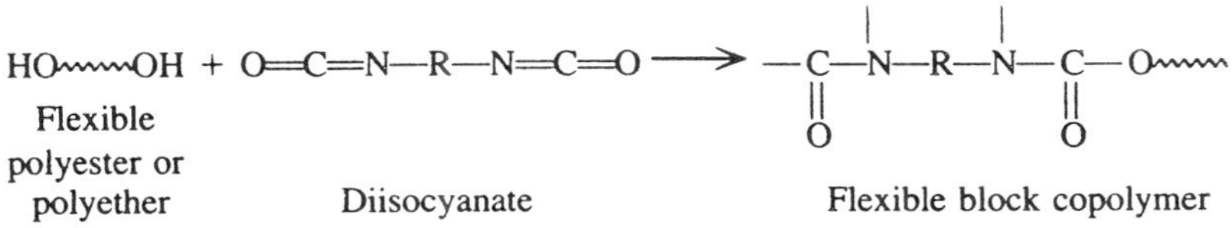

$$HO\text{\small\char`\~\char`\~}OH + O\!=\!C\!=\!N\!-\!R\!-\!N\!=\!C\!=\!O \longrightarrow \overset{|}{\underset{\underset{O}{\|}}{-\!C}}\!-\!N\!-\!R\!-\!N\!-\!\overset{|}{\underset{\underset{O}{\|}}{C}}\!-\!O\text{\small\char`\~\char`\~}$$

Flexible
polyester or
polyether   Diisocyanate   Flexible block copolymer (10.14)

ABA block copolymers of dimethylsiloxane(A)-diphenylsiloxane(B)-dimethylsiloxane(C) are synthesized by the sequential polymerization of reactants such as hexaethylcyclotrisiloxane and hexaphenylcyclotrisiloxane in the presence of an initiator such as dilithiodiphenylsilanolate and a promotor such as tetrahydrofuran. As shown in the following structure, ABCCBA siloxane block copolymers may be prepared in the same way.

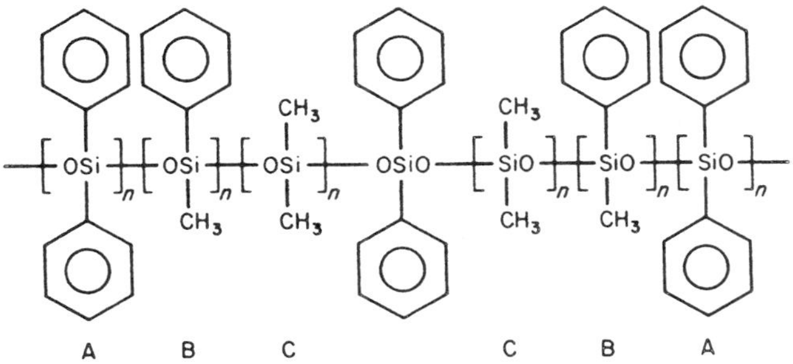

As shown in the following equation, linear polysilarylenesiloxanes are prepared by the condensation of silphenylenesiloxanes, such as p-bis(dimethylhydroxy-silyl)benzene and polydimethylsiloxane in the presence of n-hexylamine-2-ethylhexoate.

Other siloxane block copolymers have been made from polysiloxanes and polyalkylene ethers, polysiloxanes and polyarylene ethers, and polysiloxanes and polyvinyl compounds. The latter are obtained by the coupling of polysiloxanes with polycarbanions of vinyl monomers such as styrene, α-methylstyrene, and isoprene. The chemistry of silicones is discussed in Chap. 12.

$$(10.15)$$

Block copolymers with segments or domains of random length have been produced by the mechanical or ultrasonic degradation of a mixture of two or more polymers such as hevea rubber and poly(methyl methacrylate) (Heveaplus). As shown by the following general equation, the products should have different properties than mixtures of the separate polymers, since segments of more than one polymer are present in the new chain.

$$\underset{n}{M_1\,M_1\cdot} \quad + \quad \underset{n}{M_2\,M_2\cdot} \quad \rightarrow \quad \underset{n+1\quad n+1}{M_1{-}M_2}$$

| Macroradical from elastomer molecule | Macroradical from plastic molecule | Block copolymer |

$$(10.16a)$$

The impact resistance of polypropylene has been increased by the preparation of block copolymers (polyallomers) with small segments of polyethylene. These polyallomers, and those with larger domains of polyethylene, are produced by Ziegler-Natta polymerization. Block copolymers have also been prepared from the same polymer with domains of different tacticity and with different configurations. For example, block copolymers with domains formed by the 1,2 and 1,3 polymerization of 4-methylpentene have been reported.

The most widely used chain-reaction block copolymers are those prepared by the addition of a new monomer to a macroanion. The latter have been called "living polymers" by Szwarc. AB and an ABA "block" copolymers called Soprene and Kraton, respectively, are produced by the addition of butadiene to styryl macroanions or macrocarbanions. As shown by the following equation, a more complex block copolymer is produced by the hydrogenation of poly(styrene-b-butadiene-b)styrene. The produce is actually a saturated copolymer block of styrene (A), butene (B), and styrene (A).

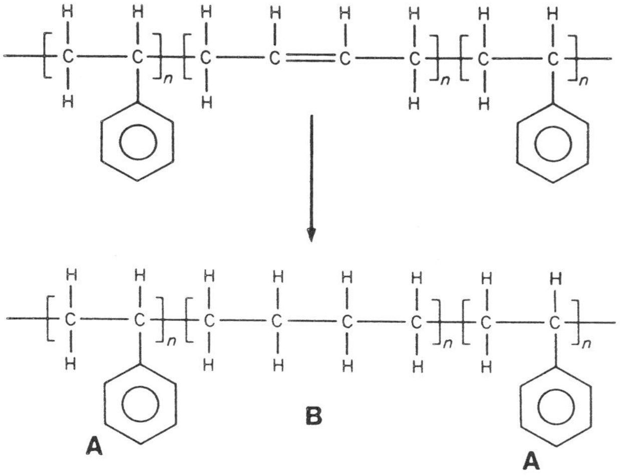

$$(10.16b)$$

Block copolymers have been prepared by free-radical chain polymerization by introducing free-radical ends, by the cleavage of weak links such as peroxy links, by using polyperoxides, such as phthaloyl polyperoxide as initiators, by introducing an active group by telomerization, or by using a complex compound such as azobiscyanopentanoic acid as the initiator. When only one of the isopropyl groups in diisopropylbenzene is oxidized to form the peroxide, this monofunctional initiator will produce a polymer that may be oxidized to produce a new isopropyl peroxide, which can cause chain extension with a new monomer.

The stable macroradicals produced in heterogeneous solution polymerization or those formed in viscous homogeneous solutions may be used to produce a wide variety of block copolymers. These systems must be oxygen free and the new vinyl monomer must be able to diffuse into the macroradical. This diffusion is related to the difference in the solubility parameters between the new monomer and the macroradical. When this value is less than 3H, diffusion will take place. The rate of diffusion is inversely related to the difference in solubility parameters ($\Delta\delta$) between the monomer and macroradical.

It is of interest to note that while acrylonitrile ($\delta = 10.8$ H) will diffuse into acrylonitrile macroradicals ($\delta = 12.5+$d H), styrene ($\delta = 9.2$ H) will not diffuse into these macroradicals. However, acrylonitrile will diffuse into styryl macroradicals ($\delta = 9.2$ H) so that block copolymers of styrene and acrylonitrile are readily produced. Likewise, charge transfer complexes of styrene and acrylonitrile and styrene and maleic anhydride will diffuse into acrylonitrile macroradicals.

When soluble organic initiators are used in emulsion polymerization in the absence of droplets, initiation takes place in the micelles and stable macroradicals are produced. Block copolymers may be formed by the addition of a second monomer after all the primary radicals have been used up.

10.5 GRAFT COPOLYMERS

Since the major difference between block and graft copolymers is the position of chain extension, much of the information on block copolymerization may be applied to graft copolymers. The chain extensions in the latter are at branch points along the chain. Of course, branch polymers are typically homograft polymers.

Graft copolymers of nylon, protein, cellulose, or starch or copolymers of vinyl alcohol may be prepared by the reaction of ethylene oxide with these polymers. Graft copolymers are also produced when styrene is polymerized by Lewis acids in the presence of poly-p-methoxystyrene. The Merrifield synthesis of polypeptides is also based on graft copolymers formed from chloromethylated polystyrene.

Isobutylene and butadiene will also form cationic graft copolymers with chloromethylated polystyrene. Styrene has been grafted onto chlorinated butyl rubber and PVC in the presence of Lewis acids. The latter reaction is as follows:

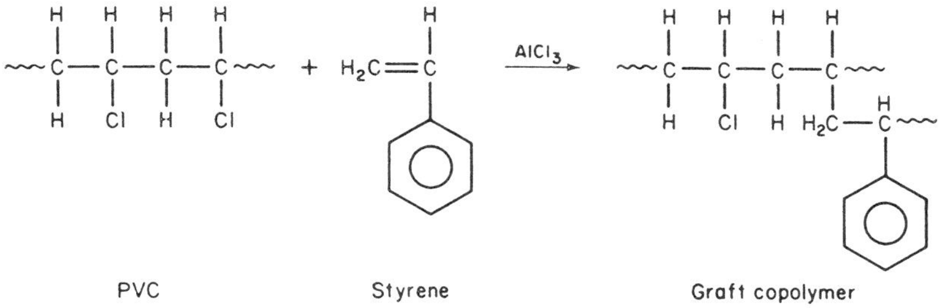

$$(10.17)$$

As shown in Eq. (10.18), acrylonitrile graft copolymers of poly-p-chlorostyrene may be produced by anionic polymerization techniques.

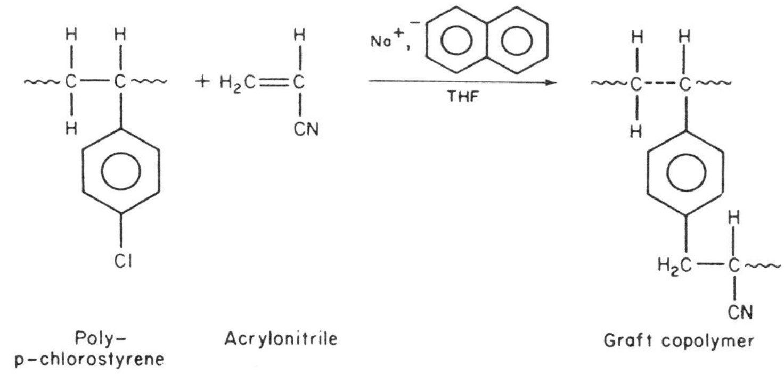

$$(10.18)$$

As shown in Eq. (10.19), graft copolymers are also produced from the reaction of vinyl compounds and unsaturated polymers in the presence of Ziegler-Natta catalysts.

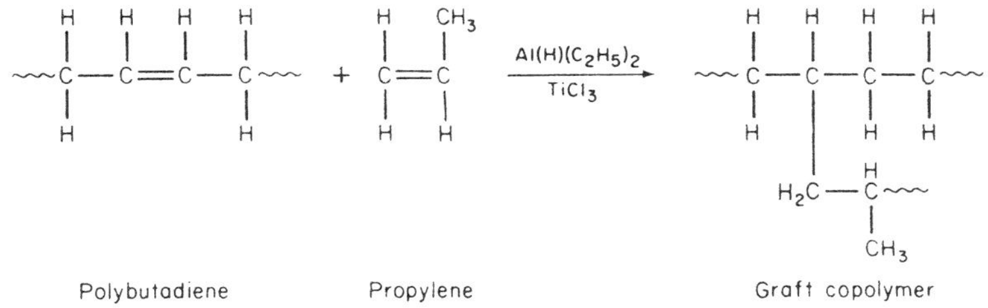

$$(10.19)$$

Graft copolymers of many polymers have been prepared by irradiation with visible light in the presence of a photosensitizer or ionization radiation and a vinyl monomer. While the normal procedure involves direct radiation of the mixture of polymer and monomer, better results are obtained when the polymer is preradiated. High-energy radiation often causes cross-linking.

The most widely used graft copolymer is the styrene–unsaturated polyester copolymer. This copolymer, which is usually reinforced by fibrous glass, is prepared by the free-radical chain polymerization of a styrene solution of an unsaturated polyester. Since oxygen inhibits the polymerization, it is preferable to polymerize in an oxygen-free atmosphere. As is the case with many other graft copolymers, considerable homopolymer is also produced. The general reaction is as follows:

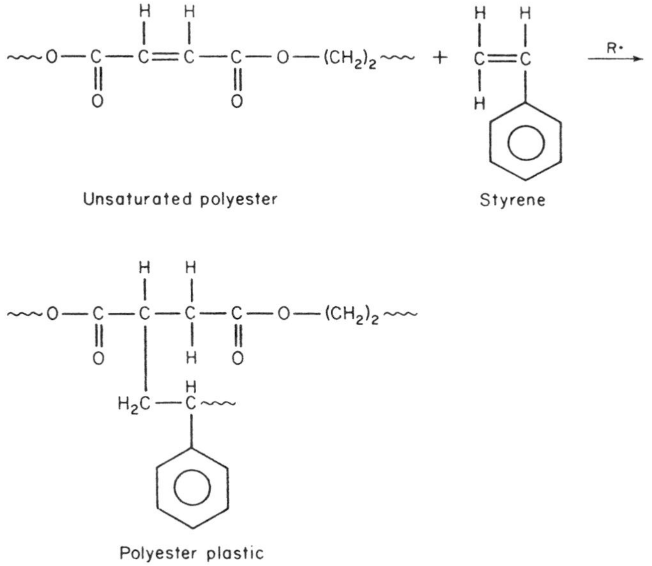

Unsaturated polyester   Styrene

(10.20)

Polyester plastic

High-impact polystyrene (HIP) and some ABS copolymers may be produced by the free-radical chain polymerization of styrene in the presence of an unsaturated elastomer. While azobisisobutyronitrile (AIBN) is a useful initiator for the polymerization of vinyl monomers, including styrene, it is not a satisfactory grafting initiator presumably because of the higher resonance stability of

$$(CH_3)_2C\cdot$$
$$|$$
$$CN$$

as compared to $C_6H_5COO\bullet$ or $C_6H_5\bullet$ from benzoyl peroxide. AIBN is not used as an initiator for styrene with unsaturated polyesters for the same reason.

The graft copolymers of acrylamide and cellulose or starch are of particular interest. These products, which are usually produced in the presence of cerium(IV) ions, have potential use as water absorbents and in enhanced oil recovery systems. The graft copolymers with acrylic acid or the hydrolyzed graft copolymers with acrylonitrile have exceptionally good water absorbency.

10.6 ELASTOMERS

As noted in Sec. 2.5, elastomers, or rubbers, are high polymers that typically contain chemical and/or physical cross-links. While there are thermoplastic elastomers such as styrene-butadiene-styrene, SBS, and block copolymers, most elastomers are thermosets. Elastomers are characterized by a disorganized (high entropy) structure in the rest or nonstressed state. Application of stress is accompanied by a ready distortion requiring (relative to fibers and plastics) little stress to effect the distortion. This distortion brings about an aligning of the chains forming a structure with greater order. The driving force for the material to return to its original shape is largely a return to the original less organized state. While entropy is the primary driving force for elastomers to return to the original rest state, the cross-links allow the material to return to its original shape giving the material what is often termed memory. Materials that allow easy distortion generally have minimal interactions between the same or different chains. This qualification is filled by materials that do not bond through the use of dipolar (or polar) or hydrogen bonding. Thus, the intermolecular and intramolecular forces of attraction are small relative to those required of fibers. Hydrocarbon polymers are examples of materials that have small inter- and intramolecular attractive forces.

As noted in Table 1.4, traditional synthetic rubbers were produced at the rate of about 5,500 million pounds in 1994. Production of elastomeric materials typically requires the initial synthesis of the polymeric backbone, insertion of cross-links through a process called vulcanization, addition of fillers and other additives such as carbon black, and finally the processing of the complex mixture.

The introduction of cross-links to inhibit chain slippage is called vulcanizing (curing) in the rubber industry. Goodyear, in 1839, was the first to recognize the importance of introducing cross-links. He accomplished this through the addition of sulfur to natural rubber (NR). Shortly thereafter, an accelerator [zinc(II) oxide] was added to the mixture to increase the rate of cross-linking. Other additives were discovered, often through observation of a problem and a somewhat trial-and-error approach towards a solution.

In about 1915, Mote found that a superior abrasion-resistant elastomer was produced through the use of carbon black. Carbon black can be considered to simply be the result of the thermal decomposition of hydrocarbons. Today it is recognized that factors such as surface area, structure, and aggregate size are important factors in allowing the carbon black to impart particular properties to the rubber. For instance, high surface areas (small particle size) increase the reinforcement and consequently the tensile strength and improve the resistance to tearing and abrasion. Large aggregates give elastomers that have improved strength before curing, high modulus, and an improved extrusion behavior.

Rubbers typically have low hysteresis. Hysteresis is a measure of the energy absorbed when the elastomer is deformed. A rubber that absorbs a great amount of energy as it is deformed is said to have a high hysteresis. The absorbed energy is equivalent to the reciprocal of resilience such that a material with a low hysteresis has a high resilience and a material with a high hysteresis conversely has a low resilience. Rubbers with a particularly low hysteresis are used where heat build-up is not wanted, such as in tire walls. Rubbers with a high hysteresis are used where heat build-up is desirable, such as in tires to give the tread a better grip on the road and the tire a smoother ride.

Table 10.3 contains a listing of elastomers and sample applications.

Recycling

Typical thermosetting elastomers are difficult to recycle because their cross-linking prevents them from being easily solubilized when added to liquids and because the cross-linking prevents them from being easily reformed through application of heat or heat and pressure. Recycling can be accomplished through the particalizing (grinding into small particles) of the elastomeric material followed by a softening-up by application of a suitable liquid and/or heat and, finally, addition of a binder that physically or chemically allows the particles to be bound together in a desired shape.

Thermoplastic Elastomers

A number of thermoplastic elastomers have been developed since the mid-1960s. The initial thermoplastic elastomers were derived from plasticized PVC and are called plastisols. Plastisols are formed from the fusing together of PVC with a compatible plasticizer through heating to about 200°C. A homogeneous matrix is formed. The plasticizer acts to lower the glass transition temperature of the PVC to below room temperature. Conceptually, this can be though of as the plasticizer acting to put additional distance between the PVC chains, thus lowering the inter- and intrachain attractive forces. The resulting rubbery materials are utilized in the construction of many products, including boot soles.

Styrene-butadiene-styrene (SBS) block copolymers differ from the structurally related random copolymer of styrene and butadiene (SBR). Because the styrene and butadiene blocks are incompatible, they form separate phases joined at the junctures where the various blocks are connected. This gives an elastomeric material where the butadiene blocks form what are called soft segments and the styrene blocks form hard blocks. The junctures act as cross-links between the phases, but these cross-links can be mobilized through application of heat.

Thermoplastic elastomeric polyurethanes are made from the inclusion of polyether or polyester soft segments connected to hard segments through urethane linkages (Sec. 10.4) forming what is often called segmented polyurethanes. Again, these are block copolymers where the various blocks act to allow the material to have elastomeric properties as in the case of the SBS copolymers.

The copolymer of poly(butylene terephthalate) (PBT) and poly(tetramethylene ether glycol) (PTMEG) is marketed as a thermoplastic elastomer under trade names such as Hytrel (duPont) and Arnitel (Akzo) where the PBT acts as the hard segment and the PTMEG acts as the soft segment.

Ionomers are thermosets that can be processed as thermoplastics. They are another example of cross-linked materials that can be made to behave as thermoplastics (see Sec. 10.10).

Table 10.3 Common Elastomers and Their Uses

Common name (chemical composition)	Abbreviation	Uses and properties
Natural rubber (polyisoprene)	NR	General purpose tires, bushings, and couplings, seals footwear, belting; Good resilience
Styrene-butadiene-rubber (random copolymer)	SBR	Tire tread, footwear, wire and cable covering, adhesives; High hysteresis
Butadiene rubber	BR	Tire tread, hose, belts; Very low hysteresis, high rebound
Butyl rubber (from isobutene and 0.5–3% isoprene)	IIR	Inner tubes, cable sheathing tank liners, roofing, seals, coated fabrics; Very low rebound, high hysteresis
Silicons (typically polydimethylsiloxane)		Medical applications, flexible molds, gaskets, seals; Extreme use temperature range
Polyurethanes		Sealing and joining, printing rollers, fibers, industrial tires, footwear, wire and cable covering
Nitrile rubber (random copolymer of butadiene) and acrylonitrile	NBR	Seals automotive parts that are in contact with oils and gas, footwear, hose; Outstanding resistance to oils and gas, little swelling in organic liquids
Ethylene-propylene rubbers (random copolymers with 60–80% ethylene)	EP or EPM	Cable insulation, window strips; Outstanding insulative properties
Ethylene-propylene-diene (random ter-polymer)	EPDM	Good resistance to weathering, resistant to ozone attack
Chloroprene rubber (polychloroprene)	CR	Wire and cable insulation, hose, footwear, mechanical automotive products; Good resistance to oil and fire, good weatherability
Acrylonitrile-butadiene-styrene (ter-polymers)	ABS	Oil houses, fuel tanks, gaskets, pipe and fittings, appliance and automotive housings; Resistant to oils and gas
Ionomers (largely copolymers of ethylene and acid-containing monomer reacted with metal ions)		Golf ball covers, shoe soles, weather stripping; Tough, flame resistant, good clarity, good electrical properties, abrasion resistant
Fluoroelastomers (fluorine-containing copolymers)		Wire and cable insulation, aerospace applications; Outstanding resistance to continuous exposure to high temperatures, chemicals, and fluids
Epichlorohydrine (epoxy copolymers)		Seals, gaskets, wire and cable insulation; Good resistance to chemicals
Polysulfide		Adhesive, sealants, hose binders; Outstanding resistance to oil and organic solvents

10.7 GENERAL INFORMATION ON BLOCK AND GRAFT COPOLYMERS

Unlike mixtures that may be separated, these copolymers that cannot be separated into components may consist of sequences of polymers with diverse properties. For example, block or graft copolymers of styrene and methacrylic acid are soluble in both polar and nonpolar solvents. The acrylic acid chains are extended in the former but are coiled in nonpolar solvents. Each segment of the chain or each long branch exhibits its own characteristic properties. Thus, the copolymer will not only exhibit different conformations in solvents but will have characteristic glass transition temperatures.

Elastomeric fibers and thermoplastic elastomers have been produced by controlling the size and flexibility of the domains in block and graft copolymers. These techniques have opened up new fields for the tailor-making of functional polymers.

ABS copolymers, based on a wide variety of formulations, are produced at an annual rate of over 500,000 tons. Some of these high-performance plastics are graft copolymers, but most are empirically developed blends of polystyrene and nitrile rubber [poly(butadiene-co-acrylonitrile)].

10.8 POLYMER MIXTURES—IPNs, COMPOSITES, BLENDS, AND ALLOYS

Copolymers are polymers composed of two or more different mer units. When such different mers occur in a random or alternating fashion, where the sequence of like mer units is small (generally less than six), the resulting physical properties are some average of the properties of each mer and the influence of surrounding like and unlike mers on each other. Such copolymers offer single physical responses such as having a single glass transition temperature. Examples of such copolymers include statistical, random, periodic, alternating, and block (for short blocks) copolymers.

There are also a number of polymer combinations where larger groupings of like mers or similar sequences of mers are chemically connected to one another. In these cases the various chain segments may act somewhat independent of one another such that the bulk material exhibits two glass transition temperatures or responds as a brittle material and as a tough material under different physical influences. While there is no common length of sequence that will "guarantee" this dualistic behavior, mer sequences of greater than about 10 are typically required. Terms such as star, block, graft, and starblock are common for such copolymers (Fig. 10.2). These copolymers may also exist as three-dimensional networks.

There also exist a number of polymer combinations that are linked together through secondary forces. This grouping includes interpenetrating polymer networks (IPNs) and polymer blends. These may be called physical copolymers.

On solidification, most small molecules crystallize with "their own," obeying the saying "like likes like" (see Sec. 17.2). Thus a solution containing benzoic acid and para-dichlorobenzene upon evaporation will produce mainly crystals of benzoic acid and crystals of para-dichlorobenzene and not crystals containing a mix of the two. Polymers, because of their large size and high viscosity, can be

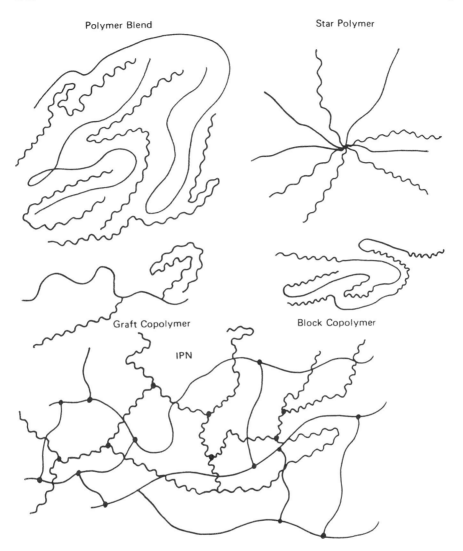

Figure 10.2 Polymer mixtures.

mixed and allowed to solidify such that solids consisting of combinations of different polymers can be achieved. This mixing or blending can give materials with properties similar to short sequenced copolymers or quite different, depending upon the particular blend or property. Compatible blends exist as a single phase, whereas incompatible blends exist as two phases. Compatibilizing agents and other additives have been used to assist in the formation of blends.

Nylon-phenolic blends are used as ablative heat shields on space vehicles during reentry when surface temperatures of several thousand degrees are attained.

The toughness of high-impact polystyrene (HIP) is dependent on a dispersion of flexible elastomer molecules in a rigid polystyrene matrix. These blends may be made by mixing solutions or lattices of the components and by melt-mixing a mixture of the finely divided polymers. Thus HIP is a two-phase blend in which a graft copolymer is formed at the interface between the PS matrix and dispersed SBR or BR elastomer particles.

ABS thermoplastics have a two-phase morphology consisting of elastomer particles dispersed in a styrene-acrylonitrile (SAN) copolymer matrix. Today the term polymer alloy is often applied to such rubber-toughened materials in which the matrix can be a mixture of polymer types.

ABS alloys with PVC, polyurethane (PU), and polycarbonate (PC) are widely used. The processing difficulties of poly(phenylene oxide) (PPO) are lessened by blending with polystyrene (PS). Elastomers and thermoplastic elastomers (TPE) have been used to flexibilize polyolefins and nylons. Many useful blends have also been developed empirically because of material shortages or the need to meet unusual specifications. For example, the use of blends of natural, synthetic, and reclaimed elastomers and thermoplastics is standard practice in the rubber industry.

IPNs are described by Sperling to be an intimate combination of two polymers, both in network form, where at least one of the two polymers is synthesized, and /or cross-linked in the immediate presence of the other. This is similar to taking a sponge cake and soaking in it warm ice cream and then refreezing the ice cream, resulting in a dessert (or mess) that has both spongy and stiff portions.

Such IPNs, grafts, blocks, and blends offer synergistic properties that are being widely exploited. Applications include impact-resistant plastics, specialty coating, adhesives, and as thermoplastic elastomers. New combinations are being studied.

Composites consist of a continuous phase and a discontinuous phase. The difference between a blend and a composite is the size scale. Blends are mixtures of different polymer chains on a molecular level, whereas composites can be considered as mixtures on a macroscopic level. Fiber-reinforced materials are the major type of commercial composite. Examples include plywoods which are wood laminates bonded with a resin, usually PF and amino resins; particle boards made from waste-wood chips; and fiberglass. The topic of composites is further discussed in Secs. 13.1 and 13.2.

There continue to emerge new materials to meet new needs. Following is a brief description of some of these. Simultaneous interpenetrating networks (SINs) are products where the connection of two differing materials occurs simultaneously.

A plastic or polymer alloy is a physical mixture of two or more polymers in a melt. While some cross-linking may occur, they are generally not viewed as copolymers, though they are presented in this chapter. Alloys using ABS are common.

Another new area of work involves "fuzzy cotton-ball–like" polymers with sizes between those of small molecules and very large super macromolecules. These new materials have been given many names, including dendrites, superatoms, and mesoscopic materials. The size, shape, flexibility, and length of the "arms," hardness of the core, etc., can be varied depending upon the need.

"Smart materials" (see Sec. 17.4) are materials whose response varies depending upon a specific input. These include specifically compounded, blended, etc., materials such as HIPs, IPNs, and Silly Putty that react to external stress and strain dependent upon the "interaction time." Today's smart material efforts are focused upon piezoelectronic-embedded ceramics and organic polymers where electricity is generated as they are compressed and bent and where they will bend and twist as an electrical charge is passed through them.

10.9 DENDRITES

Along with the structures shown in Fig. 10.2, there exist other structurally complex molecules called dendrites. These molecules can act as spacers, ball-bearings, and building blocks for other structures. Usually, they are either wholly organic or they may contain metal atoms. They may or may not be copolymers depending on the particular synthetic route employed in their synthesis.

Dendrites are highly branched, usually curved, structures. The name comes from the greek word for tree, *dendron*. Another term often associated with these structures is dendrimers, describing the oligomeric nature of many dendrites. Because of the structure, dendrites can contain many terminal functional groups for each molecule that can be further reacted. Also, most dendrites contain unoccupied space that can be used to carry drugs, fragrances, adhesives, diagnostic molecules, cosmetics, catalysts, herbicides, and other molecules.

The dendrite structure is determined largely by the functionality of the reactants. The dendrite pictured in Fig. 10.3 can be considered as being derived from a tetra-functional monomer formed from the reaction of 1,4-diaminobutane and acrylonitrile. The resulting polypropylenimine dendrimer has terminal nitrile groups that can be further reacted extending the dendrimer or terminating further dendrimer formation. The resulting molecule is circular with some three-dimensional structure. The dendrimer shown in Fig. 10.4 is derived from difunctional reactants

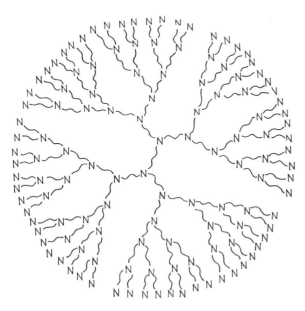

Figure 10.3 Dendrite structure derived from the reaction of 1,4-diaminobutane and acrylonitrile.

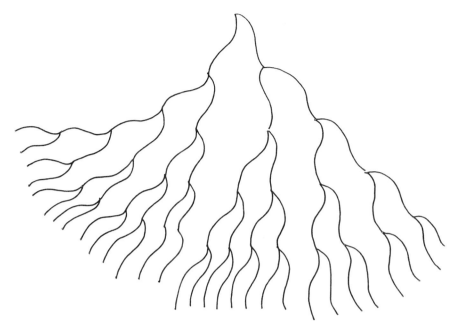

Figure 10.4 Dendrite structure derived from bent difunctional reactants.

that are bent so as to encourage fanlike expansion rather than the typical linear growth.

10.10 IONOMERS

Inonomers are ion-containing copolymers typically containing over 90% (by number) ethylene units, with the remaining being ion-containing units such as acrylic acid. These ionic sites are connected through metal atoms. Ionomers are often referred to as processable thermosets. They are thermosets because of the cross-linking introduced through the interaction of the ionic sites with metal ions. They are processable or exhibit thermoplastic behavior because they can be reformed through application of heat and pressure.

As with all polymers, the ultimate properties are dependant upon the various processing and synthetic procedures that the material is exposed to. This is especially true for ionomers where the location, amount, nature, and distribution of the metal sites strongly determine the properties. Many industrial ionomers are made where a significant fraction of the ionomer is unionized and where the metal-containing reactants are simply added to the preionomer followed by heating and agitation of the mixture. These products often offer superior properties to ionomers produced from fully dissolved preionomers.

Bonding sites are believed to be of two different grouping densities. One of these groupings involves only a few or individual bonding between the acid groups and the metal atoms as pictured in Fig. 10.5. The second bonding type consists of

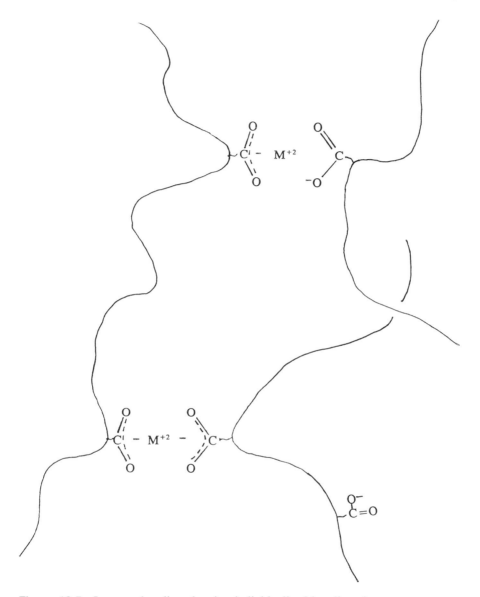

Figure 10.5 Ionomer bonding showing individualized bonding sites.

large concentrations of acid groups with multiple metal atoms (clusters) as pictured in Fig. 10.6 This metal-acid group bonding (salt formation) constitutes sites of cross-linking. It is believed that the processability is a result of the combination of the movement of the ethylene units and the metal atoms acting as ball bearings. The sliding and rolling is believed to be a result of the metallic nature of the acid-metal atom bonding. (Remember that most metallic salts are believed to have a high degree of ionic, nondirectional bonding as compared with typical organic

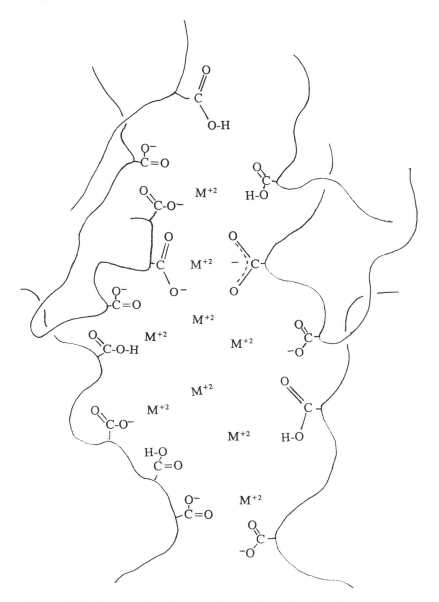

Figure 10.6 Ionomer bonding showing a cluster bonding site.

bonds where there is a high amount of covalent, directional bonding.) Carraher and coworkers have shown that the ethylene portions alone are sufficient to allow ionomers to be processed through application of heat and pressure.

Ionomers are generally tough and offer good stiffness and abrasion resistance. They offer good visual clarity, high melt viscosities, superior tensile properties, and oil resistance and are flame retarders. They are used in the automotive industry in the formation of exterior trim and bumper pads, in the sporting goods industry as bowling pin coatings and golf ball covers, and in the manufacture of roller skate wheels and ski boots. Surlyn (duPont; poly(ethylene-co-methacrylic acid)] is used in vacuum packaging for meats, in skin packaging for hardware and electronic items (e.g., seal layers and foil coatings of multiwall bags), and in shoe soles.

Sulfonated EPDM are formulated to form a number of rubbery products including adhesives for footwear and garden hoses. Perfluorinated ionomers marketed as Naflon (duPont) are used for membrane applications including chemical processing separations, spent acid regeneration, electrochemical fuel cells, ion-selective separations, electrodialysis, and in the production of chlorine. They are employed as solid state catalysts in chemical synthesis and processing, and are also used in blends with other polymers.

SUMMARY

1. Unlike homopolymers, which consist of chains with identical repeating units, copolymers are macromolecules with more than one repeating unit in the same molecule. These repeating units may be randomly arranged or they may alternate in the chain. Block copolymers are linear polymers which consist of long sequences of repeating units in the chain, and graft copolymers are branch polymers in which the branches consist of sequences of repeating units which differ from those in the backbone.

2. In the special case of azeotropic copolymerization, the composition of the copolymer is identical to that of the feed. However, copolymer compositions usually differ because of the difference in reactivity ratios or rates at which the different monomers add to macroradicals with similar end groups. The composition of the copolymer may be estimated from the copolymer equation

$$n = \frac{r_1 x + 1}{(r_2/x) + 1}$$

where n and x are the molar ratios of the monomer in the copolymer and feed, respectively, and r_1 and r_2 are the reactivity ratios.

3. The product of the reactivity ratios $r_1 r_2$ may be used to estimate the relative randomness of the copolymer. When $r_1 r_2 \simeq 1$, the copolymer arrangement is random, and when $r_1 r_2 \simeq 0$, the arrangement is alternating. The latter is the result of the homopolymerization of charge-transfer complexes which are unstable at elevated temperatures.

4. Strongly electrophilic or nucleophilic monomers will polymerize exclusively by anionic and cationic mechanisms. Those like styrene and methyl methacrylate, which are neither strong electrophiles nor nucleophiles, also polymerize by free-radical mechanisms. Since these two monomers also polymerize by ionic mechanisms, the reactivity ratios for all three systems may be compared.

5. The Q-e scheme, in which Q is related to reactivity and e to polarity, may be used to predict reactivity ratios:

$$r_1 = \frac{k_{11}}{k_{12}} = \frac{Q_1}{Q_2} e^{-e_1(e_1 - e_2)}$$

Styrene, which is assigned a Q value of 1.00 and an e value of -0.80, is used as the comparative reference for determining Q-e values for other monomers.

6. The principal copolymers are SBR, butyl rubber, poly(vinyl chloride-co-vinyl acetate), Saran, ionomers, and Acrilan fibers.

7. Block and graft copolymers differ from mixtures but having the properties of each component. Thus, block copolymers may be used as thermoplastic elastomers and graft copolymers with a flexible backbone may be used for high-impact plastics.

8. Block and graft copolymers may be produced by step-reaction polymerization and by chain-reaction polymerization.

9. The principal block copolymers are thermoplastic elastomers and elastic fibers. The principal graft copolymers are ABS, HIP, polyester plastics, and grafted starch and cellulose.

GLOSSARY

AB: A block copolymer with two sequences of mers.

ABA: A block copolymer with three sequences of mers in the order shown.

ABS: A three-component copolymer of acrylonitrile, butadiene, and styrene.

AIBN: Azobisisobutyronitrile.

alloy: Rubber-toughened materials in which the matrix can be a mixture of polymer types.

alternating copolymer: An ordered copolymer in which every other building unit is different $(—M_1M_2—)_n$.

azeotropic copolymer: Copolymer in which the composition is the same as that of the feed.

blends: Mixtures of different polymers on a molecular level; may exist as one or two phases.

block copolymer: A copolymer of long sequences or runs of one mer followed by long sequences of another mer $(M_1)_2(M_2)_n$.

buna-N: An elastomeric copolymer of butadiene and acrylonitrile.

buna-S: An elastomeric copolymer of butadiene and styrene.

butyl rubber: An elastomeric copolymer of isobutylene and isoprene.

charge transfer complex: A complex consisting of an electron donor (D) and an electron acceptor (A) in which an electron has been transferred from D to A, i.e., (D^+, A^-).

composites: Mixtures, on a macroscopic level, of different polymers, one forming a continuous phase and another a discontinuous phase.

composition drift: The change in composition of a copolymer that occurs as copolymerization takes place with monomers of different reactivities.

copolymer: A macromolecule consisting of more than one type of building unit or mer.

copolymer equation:

$$n = \frac{d[M_1]}{d[M_2]} = \frac{r_1x + 1}{(r_2/x) + 1}$$

$\Delta\delta$: Difference in solubility parameters.

dendrites: Complex molecules that are highly branched, usually with a curved structure.

domains: Sequences in block copolymers.

Dynel: Trade name for a copolymer of vinyl chloride and acrylonitrile.

e: Polarity of monomers.

EMP: Copolymer of ethylene and propylene.

EPDM: Abbreviation for a curable (vulcanizable) terpolymer of ethylene, propylene, and diene.

graft copolymer: A branched copolymer in which the backbone and the branches consist of different mers.

GRS: A name used for SBR during World War II.

Heveaplus: Trade name for block copolymers of methyl methacrylate and hevea rubber.

HIP: High-impact polystyrene.

homopolymer: A macromolecule consisting of only one type of building unit or mer.

Hytrel: Trade name for a commercial TPE.

ideal copolymer: A random copolymer, $r_1 = r_2$ and $r_1 r_2 = 1$.

interpenetrating polymer networks (IPNs): Intimate combination of two polymers, both in network form, where at least one of the two polymers is synthesized and/or cross-linked in the immediate presence of the other.

ionomers: Copolymers typically containing mostly ethylene units, with the remaining units being ion containing, such as acrylic acid neutralized through reaction with metals.

k_{11}: Rate constant for the addition of M_1 to M_1.

Kraton: Trade name for an ABA block copolymer of styrene and butadiene.

living polymers: A name used by Szwarc for macrocarbanions.

M_1: A macroradical with a chain end of M_1.

M_2: A macroradical with a chain end of M_2.

n: The molar ratio of mers M_1 and M_2 in a copolymer chain, i.e., M_1/M_2.

P: Reactivity or resonance effect of monomers.

Pliovic: Trade name for copolymers of vinyl chloride and vinylidene chloride.

Pluronics: Trade name for block copolymers of ethylene oxide and propylene oxide.

polyallomers: Block copolymers of ethylene and propylene.

polyester plastic: A name used to describe polymers produced by the polymerization of a solution of an unsaturated polyester in styrene.

Q: Reactivity or resonance effect of monomers.

Q-e scheme: A simiempirical method for predicting reactivity ratios.

r: Reactivity ratios.

r_1: k_{11}/k_{12}.

r_2: k_{22}/k_{21}.

r_1r_2: The product of the reactivity ratios which predict the extent of randomness in a copolymer chain. When r_1 and r_2 are similar and their product equals 1, the macromolecule is an ideal random copolymer.

random copolymer: A copolymer in which there is no definite order for the sequence of the different mers or building blocks ($—M_1M_2M_1M_1M_2M_1M_2M_2—$).

reactivity ratio (r): The relative reactivity of one monomer compared to another monomer.

Saran: Trade name for copolymers of vinylidene chloride and vinyl chloride.

SBR: The elastomeric copolymer of styrene and butadiene.

siloxane: $+O—SiR_2—O+_n$

Solprene: Trade name for an AB block copolymer of styrene and butadiene.

Spandex: An elastic fiber consisting of a block copolymer of a polyurethane and a polyester.

TPE: Thermoplastic elastomer.

Vinoflex: Trade name for copolymers of vinyl chloride and vinyl isobutyl ether.

Vinylite: Trade name for copolymers of vinyl chloride and vinyl acetate.

Vinyon: Trade name for copolymers of vinyl chloride and acrylonitrile.

Wall equation: The predecessor to the copolymer equation.

x: Symbol for the molar of monomers in the feed, i.e., $[M_1]/[M_2]$.

EXERCISES

1. Draw representative structures for (a) homopolymers, (b) alternating copolymers, (c) random copolymers, (d) AB block copolymers, and (e) graft copolymers of styrene and acrylonitrile.

2. If equimolar quantities of M_1 and M_2 are used in an azeotropic copolymerization, what is the composition of the feed after 50% of the copolymer has formed?

3. Define r_1 and r_2 in terms of rate constants.

4. Do the r_2 and r_2 values increase or decrease during copolymerization?

5. What is the effect of temperature on r_1 and r_2?

6. What will be the composition of copolymers produced in the first part of the polymerization of equimolar quantities of vinylidene chloride and vinyl chloride?

7. What monomer may be polymerized by anionic, cationic, and free-radical chain techniques?

8. Which chain polymerization technique would you select to polymerize (a) isobutylene, (b) acrylonitrile, and (c) propylene?

9. If $r_1r_2 \simeq 0$, what type of copolymer would be formed?

10. Show a structure for an AB block copolymer.

11. What is the value of r_1r_2 for an ideal random copolymer?

12.   Which would polymerize more readily: (a) 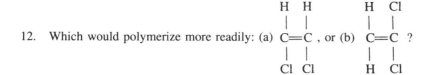, or (b) ?

13.   An equimolar mixture of styrene and maleic anhydride in decalin produces an alternating copolymer at 60°C and a random copolymer at 140°C; explain.

14.   What is the composition of the first copolymer chains produced by the copolymerization of equimolar quantities of styrene and methyl methacrylate in (a) free-radical, (b) cationic, and (c) anionic copolymerization?

15.   What is the composition of the first copolymer butyl rubber chains produced from equimolar quantities of the two monomers?

16.   What is the composition of the first copolymer butyl rubber chains produced from a feed containing 9 mol of isobutylene and 1 mol of isoprene?

17.   How would you insure that production of butyl rubber of uniform composition in question 16?

18.   Which would be more reactive in free-radical polymerization: (a) p-methylstyrene, or (b) m-methylstyrene?

19.   What is the composition of the first polymer chains produced by the copolymerization of equimolar quantities of vinyl chloride and vinyl acetate?

20.   What are the advantages, if any, of the vinyl chloride–vinyl acetate copolymer over PVC?

21.   Why are ionomers superior to LDPE?

22.   What is the difference between buna-S, GRS, and SBR?

23.   What do the acrylonitrile comonomers, acrylic acid, acrylamide, vinylpyridine, and vinylpyrrolidone have in common?

24.   Which sequence in the ABA block copolymer of ethylene oxide and propylene oxide is lyophilic?

25.   What is the advantage of TPE, if any, over hevea rubber?

26.   What is the advantage, if any, of the hydrogenated over the unhydrogenated ABA block copolymer of styrene-butadiene-styrene?

27.   What product is obtained if 1.5 mol of styrene is copolymerized with 1 mol of maleic anhydride of benzene?

28.   What precautions must be taken when making articles such as boats from fibrous glass-reinforced polyester plastics?

29.   Are so-called polyester plastics cross-linked?

30.   How could you use graft copolymerization techniques to reduce the water solubility of starch.

31.   What is the end group when azobiscyanopentanoic acid is used as an initiator?

BIBLIOGRAPHY

Aggarawl, S. L. (1970): *Block Copolymers*, Plenum, New York.

Alfrey, Jr., T. Bohrer, J. J., Mark, H. (1952): *Copolymerization*, Interscience, New York.

Brandrup, J., and Immergut, E. H. (1975): *Polymer Handbook*, Wiley, New York.

Butler, G. C. (1992): *Cyclopolymerization and Cyclocopolymerization*, Marcel Dekker, New York.

Carraher, C. E., Moore, J. A. (1984): *Modification of Polymers*, Plenum, New York.

Carraher, C. E., Sperling, L. H. (1983): *Polymer Applications of Renewable-Resource Materials*, Plenum, New York.

Casale, A., Porter, R. S. (1975): *Mechanical Synthesis of Block and Graft Copolymers*, Springer-Verlag, New York.

Ceresa, R. J. (1976): *Block and Graft Copolymerization*, Vol. 2, Wiley, New York.

Dole, M. (1972): *The Radiation Chemistry of Macromolecules*, Academic, New York.

Ham, G. E. (1964): *Copolymerization*, Interscience, New York.

―――. (1967): *Vinyl Polymerization*, Marcel Dekker, New York.

Houtz, R. C., Adkins, H. (1933): Graft copolymers, J. Am. Chem. Soc., *55*: 1609.

Lunsted, L. G. (1931): Block copolymer surfactants, J. Am. Oil Chem. Soc., *28*:294.

Meier, D. J. (1983): *Block Copolymers: Science and Technology*, Harwood, New York.

Osada, Y., Ross-Murphy, S. B., Siegel, R. A. (1995): *Polymer Gels and Networks*, Elsevier, Tarrytown, New York.

Seymour, r. B., and Stahl, G. A. (1975): Block copolymers of styrene, Appl. Polymer Sci. Symp., *26*:249.

―――. (1986): *Block Copolymers*, Tamkang University Press, Taipei, Taiwan.

Seymour, R. B., Stahl, G. A., Owen, D. R., Wood, H. (1975): Block copolymers of methyl methacrylate, Adv. Chem., *142*:309.

Sweeney, F. M. (1988): *Polymer Blends and Alloys—Guidebook to Commercial Products*, Technomic, Lancaster, Pennsylvania.

Utracki, L. A. (1990): *Polymer Alloys and Blends: Thermodynamics and Rheology*, Oxford University Press, Oxford, England.

Utracki, L. A., Weiss, R. A. (1989): *Multiphase Polymers*: *Blends and Ionomers*, ACS, Washington, D.C.

Wall, F. T. (1944): The structure of copolymers, II, J. Am. Chem. Soc., *66*:2050.

Walling, C. (1957): *Free Radicals in Solution*, Wiley, New York.

Woodward, A. E. (1989): *Atlas of Polymer Morphology*, Oxford University Press, Oxford, England.

11

Inorganic–Organic Polymers

11.1 INTRODUCTION

Classical polymer chemistry emphasizes materials derived from about a dozen elements (including C, H, O, N, S, P, Cl, and F). The following two chapters deal with polymers containing additional elements.

The present chapter will focus on inorganic polymers containing organic portions, which are called inorganic–organic polymers.

Elements such as silicon, sulfur, and phosphorus catenate similar to the way carbon does, but such catenation generally does not lead to (homo) chains with degrees of polymerization greater than 10. Further, such products might be expected to offer lesser thermal stabilities and possibly lesser strengths than carbon-based polymers, since their bond energies are generally less (Table 11.1). The alternative of utilizing heteroatomed backbones is attractive, since the resultant bond can exhibit greater bond energies.

One frequent misconception concerns the type of bonding that occurs between the inorganic and organic unit. With the exception of metals present in metallocene "sandwich" compounds, the organic to inorganic bonding is of the same general nature as that present in organic compounds. The percentage contribution of the organic–inorganic bonding due to covalent contributions is typically well within that found in organic acid, alcohol, and thio and nitro moieties (for instance, usual limits are about 5% ionic bonding character for the B—C bond to 55% for the Sn—O bond). Thus the same natural spatial, geometric laws that apply to the bonding in organic compounds apply equally well to the polymers covered in this chapter.

Table 11.1 General Magnitudes of Bonds

Bond	General bond energy[a] (Kcal/mol)	Ionic character[b] (%)
Al—O	140	60
B—C	90	5
B—N	110	20
B—O	115	45
Be—O	125	65
C—C	85	0
C—H	100	5
C—N	75	5
C—O	85	20
C—S	65	5
P—P	50	0
P—N	140	20
P—O	100	40
P—S	80	5
S—S	60	0
Si—Si	55	0
Si—C	75	10
Si—N	105	30
Si—O	110	50
Si—S	60	10
Sn—Sn	40	0
Sn—O	130	55
Ti—O	160	60

[a]Given to nearest 5 kcal/mol.
[b]Based on Pauling electronegativity values. The percentage of ionic bonding should be less where pi-bonding occurs. Given to nearest 5%.

The number of potential inorganic–organic polymers is great. The inorganic portions can exist as oxides, salts, etc., in differing oxidation states and differing geometries.

The importance of inorganic polymers can be readily appreciated by considering the following. First, photosynthesis, the conversion of carbon dioxide and water by sunlight in the synthesis of saccharides, is based on a metal-containing polymer—chlorophyll. Also a number of critical enzymes, such as hemoglobin, contain a metal site as the key site for activity. Second, the inorganic–organic polymers produced thus far exhibit a wide range of applications not common to most organic polymers, including electrical conductivity, specific catalytic operations, wide operating temperatures, greater strengths, and geater thermal stabilities (Table 11.2). Third, inorganic polymers form the basis for many insulators and building materials. Fourth, inorganic elements are present in high abundance in the Earth's crust (Table 11.3).

The topic of metal-containing inorganic–organic polymers can be divided by many means. Here the topic will be divided according to the type of reaction employed to incorporate the inorganic and/or inorganic-containing moiety into the polymer chain—addition, condensation, and coordination. Emphasis is given to unifying factors and to metal-containing materials.

11.2 INORGANIC REACTION MECHANISMS

Many of the polymerizations and monomer syntheses are simple extensions of known inorganic, organometallic, and organic reactions. The types and language used to describe inorganic reaction mechanisms are more diversified than those employed by classical organic chemists.

The majority of inorganic reactions can be placed into one of two broad classes—oxidation-reduction reactions, including atom and electron transfer reactions, and substitution reactions.

Terms such as inner sphere, outer sphere, and photo-related reactions are employed to describe more reduction-oxidation (redox) reactions. Such reactions are

Table 11.2 Actual and Potential Smaller-Scale Applications for Inorganic–Organic Polymers

Biological	Anticancer, antiviral, treatment of arthritis, antibacterial, antifungal, antifouling, treatment of Cooley's anemia, algicides, molluscicides, treatment of vitamin deficiency
Electrical	Conductors, semiconductors, piezoelectronic, pyroelectronic, solar energy conversion, electrodes, computer chip circuitry
Analytical, catalytic, photo	UV absorption, laser, nonlinear optics
Building	Sealants, caulkants, lubricants, gaskets

Table 11.3 Relative Abundance of Selected Elements in Earth's Upper (10-Mile) Crust

Element	Weight (%)	Element	Weight (%)
Oxygen	50	Titanium	0.4
Silicon	26	Fluorine	0.3
Aluminum	7.3	Chlorine	0.2
Iron	4.2	Carbon	0.2
Calcium	3.2	Sulfur	0.1
Sodium	2.4	Phosphorus	0.1
Potassium	2.3	Barium	0.1
Magnesium	2.1	Manganese	0.1
Hydrogen	0.4		

important in the synthesis of polymers and monomers and in the use of metal-containing polymers as catalysts and in applications involving transfer of heat, electricity, and light. They will not be dealt with to any appreciable extent in this chapter.

Terms such as lability, inertness, ligand, associative, interchange, and dissociative are important when discussing substitution reactions. The ligand is simply (typically) the Lewis base that is substituted for and is also the agent of substitution. Thus in the reaction between tetrachloroplatinate and diamines to form the anticancer and antiviral platinum II polyamines [Eq. (11.1)], the chloride is the leaving group or departing ligand, while the amine-functional group is the ligand that is the agent of substitution.

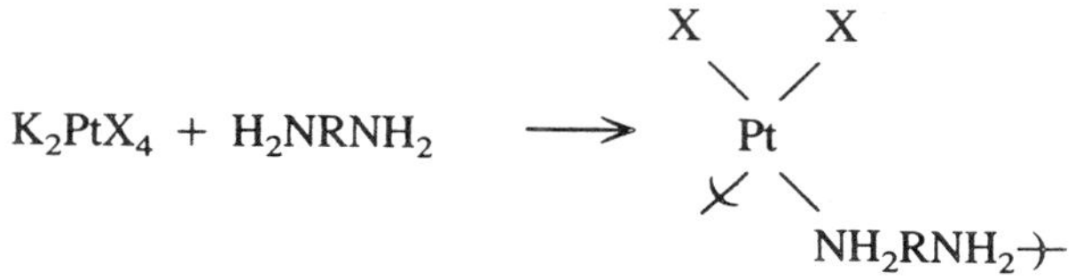

$$K_2PtX_4 + H_2NRNH_2 \longrightarrow \begin{array}{c} X \quad X \\ \diagdown \diagup \\ Pt \\ \diagup \diagdown \\ NH_2RNH_2 \end{array} \rightarrow \qquad (11.1)$$

where X = Cl, Br, or I.

There is a difference between the thermodynamic terms stable and unstable and the kinetic terms labile and inert. Further, the difference between the terms stable and unstable and the terms labile and inert are relative. Thus $Ni(CN)_4^{-2}$ and $Cr(CN)_6^{-3}$ are both thermodynamically stable in aqueous solution, yet kinetically, the rate of exchange of radiocarbon-labeled cyanide is quite different. The half-life for exchange is about 30 sec for the nickel complex and one month for the chromium complex. Taube has suggested that those complexes that react completely within about 60 sec at 25°C should be considered labile while those taking longer times should be called inert. This "rule of thumb" is often given in texts, but is not in general use in the literature. Actual rates and conditions are superior tools for the evaluation of the kinetic/thermodynamic stability of complexes.

The term "D mechanism" (dissociation) is loosely comparable to S_N1-type reaction mechanisms, but it does not imply the observed rate law. Here a transient intermediate is assumed to live long enough to be able to differentiate between various ligands, including the one just lost, and between solvent molecules. Thus the overall rate expression may be dependent on the nature of L,L', solvents, or some combination.

$$ML_4 \rightleftharpoons ML_3 + L \tag{11.2}$$

$$ML_3 + L' \rightarrow ML_3L' \tag{11.3}$$

$$ML_3 + S \rightleftharpoons ML_3S \tag{11.4}$$

In the I_d mechanism, dissociative interchange, the transition state involves extension of the M . . . L bond, but not rupture.

$$ML_4 + L' \rightarrow ML_4L' \rightarrow [L{\cdots}ML_3{\cdots}L'] \rightarrow ML_3L' + L \tag{11.5}$$

The ML_4L' species is called an outer-sphere complex, or, if ML_4 is a cation and L an anion, the couple is called an ion pair.

For the I_a mechanism, associative interchange, the interaction between M and L' is more advanced in the transition state than in the case of the I_d. The M$\cdots$L' bonding is important in defining the activated complex. Both of these interchange mechanisms are loosely connected to the S_N2-type reaction mechanism.

For the A mechanism, associative, there is a fully formed intermediate complex, ML_4L', which then dissociates, being roughly analogous to the E_1 type of reaction mechanism.

It is important to remember that the same factors that operate in regard to smaller molecules are in operation during a polymerization process. The same electronic, steric, mechanistic, kinetic, and thermodynamic directives present in the addition, condensation, etc., reactions occurring between small molecules are present as such molecule interactions lead to the formation of macromolecules.

11.3   CONDENSATION ORGANOMETALLIC POLYMERS

Condensation reactions exhibit several characteristics such as (typically) expulsion of a smaller molecule (often H_2O or HX) on reaction leading to a repeat unit containing fewer atoms than the sum of the two reactants, and most reactions can be considered in terms of polar (Lewis acid-base, nucleophilic-electrophilic) mechanisms. The reaction site can be at the metal atom adjacent to the metal atom [Eq. (11.6)] or somewhat removed from the metal atom [Eq. (11.7)].

$$R_2MX_2 + H-\overset{\overset{\displaystyle H}{|}}{N}-R-\overset{\overset{\displaystyle H}{|}}{N}-H \longrightarrow (\overset{\overset{\displaystyle R}{|}}{\underset{\underset{\displaystyle R}{|}}{M}}-\overset{\overset{\displaystyle H}{|}}{N}-R-\overset{\overset{\displaystyle H}{|}}{N})- \; + HX \tag{11.6}$$

$$Cl-\overset{\overset{O}{\|}}{C}-R-\overset{\overset{O}{\|}}{C}-Cl \ + \ HO-\underset{\underset{H}{|}}{\overset{\overset{H}{|}}{C}}\langle\bigcirc\rangle Fe \langle\bigcirc\rangle \underset{\underset{H}{|}}{\overset{\overset{H}{|}}{C}}-OH \longrightarrow \text{+}(\overset{\overset{O}{\|}}{C}-R-\overset{\overset{O}{\|}}{C}-O-CH_2\langle\bigcirc\rangle Fe \langle\bigcirc\rangle CH_2O\text{+} \quad (11.7)$$

Research involving condensation organometallic polymers was catalyzed by the observation that many organometallic halides possess a high degree of covalent character within their composite structure and that they can behave as organic acids in many reactions, such as hydrolysis [Eqs. (11.8, 11.9)] and polyesterifications [Eqs. (11.10, 11.11)].

$$R-\overset{\overset{O}{\|}}{C}-Cl + H_2O \longrightarrow R-\overset{\overset{O}{\|}}{C}-OH + HCl \qquad (11.8)$$

$$-\overset{|}{\underset{|}{M}}-Cl + H_2O \longrightarrow -\overset{|}{\underset{|}{M}}-OH + HCl \qquad (11.9)$$

$$Cl-\overset{\overset{O}{\|}}{C}-R-\overset{\overset{O}{\|}}{C}-Cl + HO-R'-OH \longrightarrow \text{+}(\overset{\overset{O}{\|}}{C}-R-\overset{\overset{O}{\|}}{C}-O-R'-O\text{+} + HCl \qquad (11.10)$$

$$Cl-\overset{\overset{R}{|}}{\underset{\underset{R}{|}}{M}}-Cl + HO-R'-OH \longrightarrow \text{+}(\overset{\overset{R}{|}}{\underset{\underset{R}{|}}{M}}-O-R'-O\text{+} + HCl \qquad (11.11)$$

Thus many of the affected organometallic polycondensations can be considered as extensions of organic polyesterification, polyamination, etc., reactions.

The most important organometallic polymers are the polysiloxanes based on the same Si—O linkage present in glass and quartz (Chap. 12). The polysiloxames were incorrectly named silicones by Kipping in the 1920s, and this name continues to be widely used.

The production of silicate glass is believed to be a transcondensation of the siloxane linkages in silica in the presence of calcium and sodium oxides. Thus, some calcium and sodium silicates are produced in glass making, but the siloxane backbone remains intact. A comparable polysilicic acid is produced when silicon tetrachloride is hydrolyzed, as shown in the following equation:

$$n\ Cl\overset{\overset{Cl}{|}}{\underset{\underset{Cl}{|}}{Si}}Cl \xrightarrow{4nH_2O} n\left[HO-\overset{\overset{OH}{|}}{\underset{\underset{OH}{|}}{Si}}-OH\right] \longrightarrow \left[\text{+}O-\overset{\overset{OH}{|}}{\underset{\underset{OH}{|}}{Si}}-O\text{+}\right]_n \qquad (11.12)$$

Silicon · · · · · · · · · · Silicic acid · · · · · · · · · · Polysilicic
tetrachloride · acid

The polysilicic acid condenses further to produce a cross-linked gel. This cross-linking may be prevented by replacing the hydroxyl groups in silicic acid by alkyl groups. Thus, Ladenburg prepared the first silicone polymer in the nineteenth century by the hydrolysis of diethyldiethoxysilane. Kipping recognized that these siloxanes were produced by the hydrolysis of dialkyldichlorosilanes in the early 1940s. In both cases, two of the hydroxyl groups shown in the formula for silicic acid in Eq. (11.12) were replaced by alkyl groups.

The instability of silicones is overcome by capping the hydroxyl end groups with a monofunctional trialkylchlorosilane or trialkylalkoxysilane. The monoalkyltrichlorosilane is used as a trifunctional cross-linking agent for silicones.

In 1945 Rochow discovered that a silicon-copper alloy reacted with organic chlorides forming a new class of compounds called organosilanes.

$$CH_3Cl + Si(Cu) \rightarrow (CH_3)_2SiCl_2 + Cu \tag{11.13}$$

Silicon-	Dimethyldichloro-
copper	silane
alloy	

These compounds react with water, forming dihydroxylsilanes [Eq. (11.14)] that in turn condense, splitting out water, eventually forming polysiloxanes [Eqs. (11.15, 11.16)].

$$(CH_3)_2SiCl_2 + 2H_2O \rightarrow (CH_3)_2Si(OH)_2 + 2HCl \tag{11.14}$$

$$(11.15)$$

$$(11.16)$$

Organosiloxanes are characterized by combinations of chemical, mechanical, and electrical properties which taken together are not common to any other commercially available class of polymers. They exhibit relatively high thermal and oxidation stability, low power loss, high dielectric strength, and unique rheological properties and are relatively inert to most ionic reagents.

As with alkanes, the oligomeric polysiloxanes are oils while high molecular weight polysiloxanes are solids. Silicone oils are prepared in the presence of hexamethyldisiloxane which produces trimethylsiloxyl-capped ends. The fluidity and viscosity of silicone oils is controlled by the ratio of hexamethyldisiloxane to dimethyldimethoxysilane (or other suitable silane) employed.

Room-temperature-vulcanizing (RTV) silicone rubbers make use of the room temperature hydrolyzability of chloro, alkoxyl, and ester-silicone moieties. These high polymer silicones contain a specific amount of such easily hydrolyzable groups. These products are exposed to water-forming Si—OH products that subsequently form cross-links resulting in a silicone rubber as follows:

$$
\begin{array}{c}
\text{CH}_3 \quad \text{CH}_3 \quad \text{CH}_3 \\
\mid \qquad \mid \qquad \mid \\
-\text{O}-\text{Si}-\!\!-\text{O}-\text{Si}-\text{O}-\text{Si}-\text{O}- \\
\mid \qquad \mid \qquad \mid \\
\text{CH}_3 \quad \text{O} \quad \text{CH}_3 \\
\mid \\
\text{O}{=}\text{C}-\text{CH}_3
\end{array}
\xrightarrow{\text{H}_2\text{O}}
\begin{array}{c}
\text{CH}_3 \quad \text{CH}_3 \quad \text{CH}_3 \\
\mid \qquad \mid \qquad \mid \\
-\text{O}-\text{Si}-\text{O}-\text{Si}-\text{O}-\text{Si}-\text{O}- \\
\mid \qquad \mid \qquad \mid \\
\text{CH}_3 \quad \text{OH} \quad \text{CH}_3
\end{array}
+ \text{CH}_3\text{COOH}
$$

$$\downarrow$$

$$
\begin{array}{c}
\text{CH}_3 \quad \text{CH}_3 \quad \text{CH}_3 \quad \text{CH}_3 \quad \text{CH}_3 \\
\mid \qquad \mid \qquad \mid \qquad \mid \qquad \mid \\
-\text{O}-\text{Si}-\text{O}-\!\!-\text{Si}-\text{O}-\text{Si}-\text{O}-\text{Si}-\text{O}-\text{Si}-\text{O}- \\
\mid \qquad \mid \qquad \mid \qquad \mid \qquad \mid \\
\text{CH}_3 \quad \text{CH}_3 \quad\;\; \text{CH}_3 \quad \text{CH}_3 \\
\qquad\qquad\qquad \text{O} \\
\text{CH}_3 \quad \text{CH}_3 \quad\;\; \text{CH}_3 \quad \text{CH}_3 \\
\mid \qquad \mid \qquad \mid \qquad \mid \qquad \mid \\
-\text{O}-\text{Si}-\text{O}-\!\!-\text{Si}-\text{O}-\text{Si}-\text{O}-\text{Si}-\text{O}-\text{Si}-\text{O}- \\
\mid \qquad \mid \qquad \mid \qquad \mid \qquad \mid \\
\text{CH}_3 \quad \text{CH}_3 \quad \text{CH}_3 \quad \text{CH}_3 \quad \text{CH}_3
\end{array}
\qquad + \text{H}_2\text{O}
$$

(11.17)

Silicones can also be cured (vulcanized, cross-linked) by free-radical reactions.

Currell and Parsonage recently reported the production of oligomeric silicones from mineral silicates. Products suitable for use as greases and cements and masonry water-proofing agents have already been produced.

The silicon reactants easily form six- and eight-membered cyclic compounds resulting in "wasted-loop" side reactions. Fortunately, the cyclic siloxanes can be removed by distillation and reintroduced in the feed.

Six-membered compound   Eight-membered compound

Siloxanes find many uses. The initial footprints on the moon were made by silicone elastomer boots. Silicones are employed as high-performance caulks, lubricants, anti-foaming agents, window gaskets, O-rings, and sealants. Bouncing or Silly Putty is produced when polydimethylsiloxanes with capped ends are heated with boric acid.

Carraher and coworkers have produced a wide variety of organometallic condensation polymers based on the Lewis acid-base concept. Polymers have been produced from the condensation of traditional Lewis bases such as diols (Table 11.4) with Lewis acids such as Cp_2TiCl_2 (Table 11.5, where Cp represents a cyclopentadienyl moiety). These polymers show promise in a wide variety of areas including electrical, catalytic, and biomedical applications.

Table 11.4 Lewis Bases Employed in the Synthesis of Condensation Organometallic Polymers

Oxides ... Amidoximes ... Thiols

Diols ... Hydrazines ... Hydrazides and thiohydrazides

(Primary)amines ... (Secondary)amines ... Acid salts

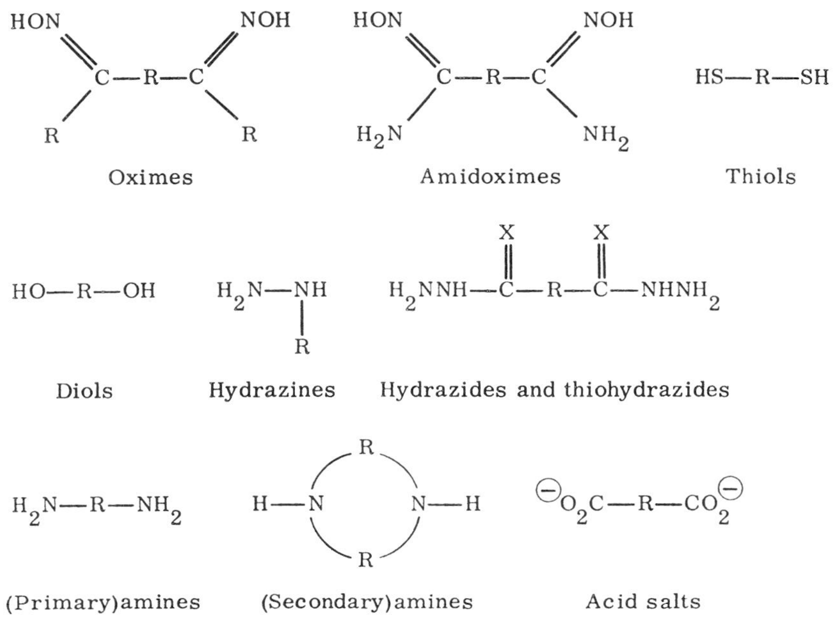

One requirement for large-scale reclaiming will be ready identification of select materials. Organometallic polydyes offer one solution to the problem of content identification since such polydyes can be "permanently" incorporated into polymeric materials such as plastics, paper-based products, coatings, rubbers, and fibers. Further, due to the variety of metals, organometallic materials and dyes, a large array of combinations are readily available. Further, the presence of the metal allows for both optical and metal-detection techniques to be employed. Organometallic polydyes is the name associated with products of metal-containing compounds and dyes. One such dye [Eq. (11.18)] is based on mercurochrome and

Table 11.5 Lewis Acids Employed in the Synthesis of Organometallic Polymers

R_2TiX_2	R_2SiX_2	R_3SbX_2	R_3Sb
R_2ZrX_2	R_2GeX_2	R_3As	R_3BiX_2
R_2HfX_2	R_2SnX_2	R_3AsX_2	R_2MnX_2
R_2PbX_2			

dibutyltin dichloride. The resulting products are typically both colorful and anti-fungal. Such polydyes may also be useful in photosensitive media for photocopy-ing; as permanent coloring agents in cloth, paper, sealants, coatings, and plastics; in biological applications such as specialty bandages and antifungal and antibac-terial topical drugs; as specialty stains and toxins; and as radiation enhancers and targets for coating-curing applications.

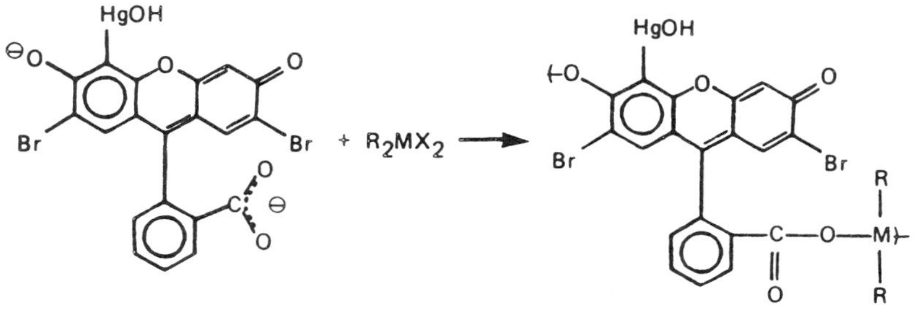

$$(11.18)$$

Condensation organometallic polymers have also been employed as biologi-cally active agents, where the biological activity is derived from the polymer chain in total, oligomeric portions of the chain or through control release of monomerlike fragments. The desired drug may be either or both comonomers. In the case of Scheme 11.1a, derived from reacting Cp_2MCl_2 with the dioxime of vitamin K, the drug may be only the vitamin K moiety where M is Zr or both moieties where M is Ti since Cp_2 Ti-containing compounds are known to be antitumoral agents. The product in Scheme 11.1b is employed to deliver the dibutyltin moiety whereas in scheme 11.1c the pyrimidine moiety is included to encourage host acceptance and subsequent demise through release of the arsenic moiety. Thus, there exists a wide latitude within which to tailor-make products with desired biological activities.

A number of polymers (mostly containing the Cp_2Ti moiety) exhibit a phe-nomenon called "anomalous fiber formation," reminiscent of "metallic whiskers."

The Lewis acid-base condensations can also occur employing a polymer as one of the reactants. Thus, reactions analogous to those reported before have been carried out employing both synthetic and natural polymers. One illustration follows.

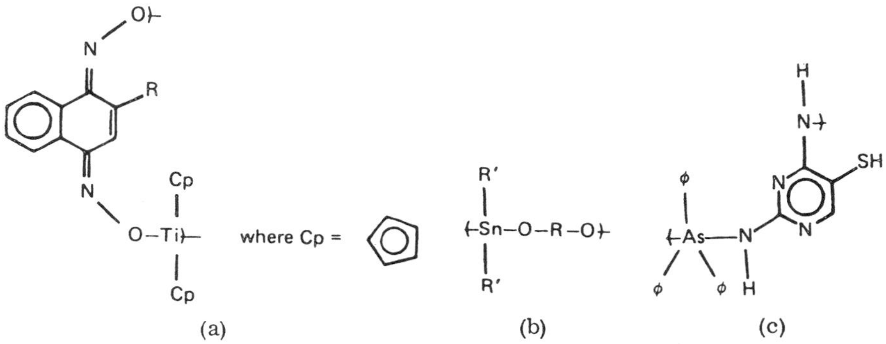

(a) (b) (c)

Scheme 11.1

There exists a need to conserve our slowly renewable resources such as coal, oil, and gas. As the price of these natural resources increases and the supply dwindles, there is an increased awareness that alternative sources of inexpensive, large-quantity, readily available feedstocks must be pursued. The use of polyhydroxylic materials such as sugars and polysaccharides fulfills these requirements and is particularly desirable since they are continually replenished. The modification of carbohydrates such as sucrose, dextrose, dextran, and cellulose has been effected using a wide variety of reaction systems and metal-containing reactants. Potential uses range from specialty bandages to treated wood for improved hydrolytic and biological properties, to paint and plastic additives and for use in commercial insulation.

11.4 COORDINATION POLYMERS

Coordination polymers have served humankind since before recorded history. The tanning of leather and generation of select colored pigments depend on the coordination of metal ions. A number of biological agents, including humans, owe their existence to coordinated polymers such as hemoglobin. Many of these coordination polymers have unknown and/or irregular structures.

The drive for the synthesis and characterization of synthetic coordination polymers was catalyzed by work supported and conducted by the United States Air Force in a search for materials which exhibited high thermal stabilities. Attempts to prepare stable, tractable coordination polymers which would simulate the exceptional thermal and/or chemical stability of model monomeric coordination compounds such as copper ethylenediamino-bis-acetylacetate (II) or copper phthalocyanine (I) have been disappointing at best. Typically only short chains were formed and the thermally stable "monomers" lost most of their stability when linked by the metals into polymeric units.

Bailar listed a number of principles which can be considered in designing coordination polymers. Briefly these are: (a) little flexibility is imparted by the metal ion or within its immediate environment; thus, flexibility must arise from the organic moiety; flexibility increases as the covalent nature of metal-ligand bond increases; (b) metal ions only stabilize ligands in their immediate vicinity; thus the chelates should be strong and close to the metal atom; (c) thermal, oxidative, and hydrolytic stability are not directly related; polymers must be designed specifically for the properties desired; (d) metal-ligand bonds have enough ionic character to permit them to rearrange more readily than typical "organic bonds"; (e) polymer structure (such as square planar, octahedral, planar, linear, network) is dictated by the coordination number and stereochemistry of the metal ion or chelating agent; and (f) employed solvents should not form strong complexes with the metal ion or chelating agent or they will be incorporated into the polymer structure and/or prevent reaction from occurring.

Coordination polymers can be prepared by a number of routes, the three most common being: (a) preformed coordination metal complexes polymerized through functional groups where the actual polymer-forming step may be a condensation or addition reaction [Eq. (11.19)]; (b) metal chelation ligands [Eq. (11.20)]. (The ionomers produced by duPont consist of copolymers of ethylene with a small portion of methacrylic acid. The polymer chains are cross-linked by addition of metal ions [Eq. (11.20)]. These polymers exhibit good thermal stability and flame retardance. The ionomers are believed to contain domains of crystalline polyethylene bound together by the metal ion. This permits a favorable combination of high strength with some flexibility); and (c) polychelation through reaction of the metal-containing moiety with an appropriate Lewis base-containing reactant [Eq. (11.21)].

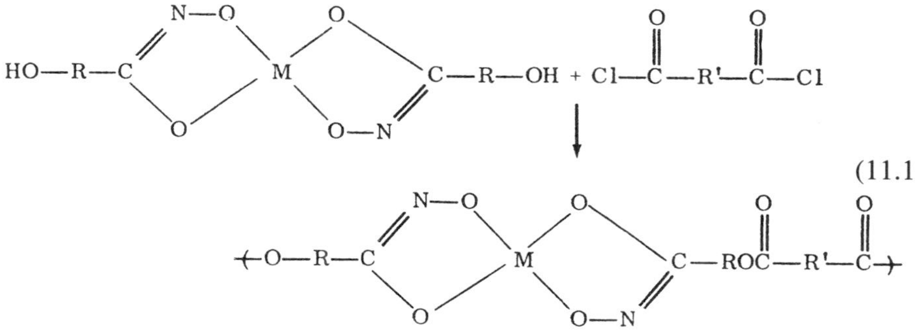

$$
\text{HO—R—C} \overset{\displaystyle N—O}{\underset{\displaystyle O}{\diagdown}} \text{M} \overset{\displaystyle O}{\underset{\displaystyle O—N}{\diagdown}} \text{C—R—OH} + \text{Cl—}\overset{O}{\overset{\|}{C}}\text{—R'—}\overset{O}{\overset{\|}{C}}\text{—Cl}
$$

(11.19)

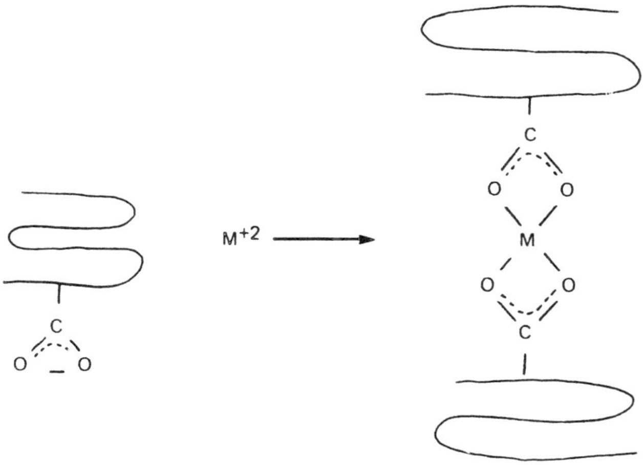

$$
\text{+O—R—C} \overset{\displaystyle N—O}{\underset{\displaystyle O}{\diagdown}} \text{M} \overset{\displaystyle O}{\underset{\displaystyle O—N}{\diagdown}} \text{C—RO}\overset{O}{\overset{\|}{C}}\text{—R'—}\overset{O}{\overset{\|}{C}}\text{+}
$$

$$
M^{+2} \longrightarrow
$$

(11.20)

$$
\overset{\ominus}{O}\text{—}\overset{O}{\overset{\|}{C}}\text{—R—}\overset{O}{\overset{\|}{C}}\text{—}\overset{\ominus}{O} + \left[\begin{array}{c} O \\ \| \\ U \\ \| \\ O \end{array}\,\overset{OH_2}{\diagup}\,\,H_2O\right]^{2+} \longrightarrow
$$

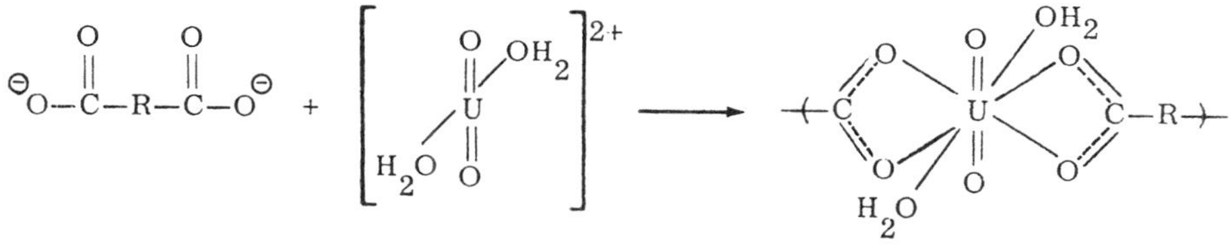

Uranyl ion   Uranyl polyester   1.21)

Carraher and coworkers employed the last process for the recovery of the uranyl ion. The uranyl ion is the natural water-soluble form of uranium oxides. It is also toxic, acting as a heavy metal toxin. Through the use of salts of dicarboxylic acids and polyacrylic acids, uranyl ion was removed to 10^{-5} molar with the resulting product being much less toxic.

Many of the organometallic polymers are semiconductors with bulk specific resistivities for the range of 10^3 to 10^{12} ohm cm suitable for specific semiconductor activities. Further, some exhibit interesting photoproperties. Some polymers degrade on heating to give metal oxides with the process suitable for the homogeneous doping of other polymers and matrices with metal sites.

In 1964, Rosenberg and coworkers found that bacteria failed to divide, but continued to grow. The major cause of this inhibition to cell division is cis-dichlorodiamineplatinum II (c-DDP). It is now licensed under the name Platinol and employed extensively in combination with other drugs in the treatment of a wide variety of cancers. The use of c-DDP is complicated due to a number of negative side effects. Carraher and Allcock recognized that many of these side effects might be overcome if the platinum is present in a polymer that can act as a long-acting controlled-release agent.

Carraher and coworkers synthesized many polymeric derivatives of c-DDP [Eq. (11.1)] employing nitrogen-containing Lewis bases. Some of the products show good inhibition to a wide variety of cancer cell lines, prolong the life expectancies of terminally ill test animals, are more toxic to cancer-like cells compared with analogous healthy cells, and are much less toxic. One of these polymers is also able to prevent the onset of virally related juvenile diabetes in test animals.

11.5 ADDITION POLYMERS

A large number of vinyl organometallic monomers has been prepared, homopolymerized, and copolymerized with other classic vinyl monomers by Pittman and others. These include polymers containing Mo, W, Fe, Cr, Ir, Ru, and Co.

The effect that the presence of organometallic functions exert in vinyl polymerizations is beginning to become understood in some instances. A transition metal may be expected, with its various readily available oxidation states and large steric bulk, to exert unusual electronic and steric effects during polymerization. The polymerization of vinyl-ferrocene will be employed as an example. Its homopolymerization has been initiated by radical, cationic, coordination, and Ziegler-Natta initiators. Unlike the classic organic monomer styrene, vinylferrocene undergoes oxidation at the iron atom when peroxide initiators are employed. Thus, azo initiators (such as AIBN) are commonly used. Here we see one difference between an organic and an organometallic monomer in the presence of peroxide initiators. The stability of the ferricinium ion makes ferrocene readily oxidizable by peroxides, whereas styrene, for example, undergoes polymerization.

Unlike most vinyl monomers, the molecular weight of polyvinylferrocene does not increase with a decrease in initiator concentration. This is the result of vinylferrocene's anomalously high chain-transfer constant ($C_m = 2 \times 10^{-3}$ versus 6×10^{-5} for styrene at 60°C). Finally, the rate law for vinylferrocene homopolymerization is first order in initiator in benzene. Thus, intramolecular termination occurs. Mossbauer's studies support a mechanism involving electron transfer from iron to the growing chain radical to give a zwitterion which terminates and further results in a high spin Fe(III) complex.

The great electron richness of vinylferrocene as a monomer is illustrated in its copolymerizations with maleic anhydride, where 1:1 alternating copolymers are obtained over a wide range of M_1/M_2 feed ratios and $r_1 \cdot r_2 = 0.003$. Subsequently, a large number of detailed copolymerization studies were carried out between vinylferrocene and classic organic monomers such as styrene, methacrylates, N-vinylcarbazole, and acrylonitrile. The relative reactivity ratios (r_1 and r_2) were obtained, and from them the values of the Alfrey-Price Q and e parameters were obtained. The value of e is a semiempirical measure of the electron richness of the vinyl group. The best value of e for vinylferrocene is about -2.1, which, when compared to the e values of maleic anhydride ($+2.25$), p-nitrostyrene ($+0.39$), styrene (-0.80), p-N,N-dimethylaminostyrene (-1.37), and 1,1'-dianisylethylene (-1.96), illustrates the exceptional electron richness of vinylferrocene's vinyl group.

Remarkably, the presence of different metals or the presence of electron-withdrawing carbonyl groups on metal atoms attached to the N^5-vinyl-cyclopentadienyl group does not markedly diminish the electron richness of the vinyl group as measured by the Alfrey-Price e value.

Other inorganic polymers are formed from addition reactions. These include polyphosphazenes, polyphosphonitriles, and poly(sulfur nitride). Phosphonitrilic polymers have been known for centuries but since they lacked resistance to water, they were not of interest as commercial polymers. However, when the pendant chlorine groups are replaced by fluorine atoms, amino, alkoxy, or phenoxy groups, these polymers are much more resistant to moisture. Phosphonitrile fluoroelastomers (PNF-200) are useful throughout a temperature range of -56 to $180°C$. Phosphazenes are produced by the thermal cleavage of a cyclic trimer obtained from the reaction of phosphorus pentachloride and ammonium chloride [Eq. (11.22)]. Similar reactions occur when the chloro group is replaced by fluoro, bromo, or isothiocyano groups.

$$PCl_5 + NH_4Cl \xrightarrow{170°C} HCl + \text{(Hexachlorocyclo-triphosphazene)} \xrightarrow[\substack{ROH \\ RCOOH \\ Na}]{catalyst} \left(-\underset{\underset{Cl}{|}}{\overset{\overset{Cl}{|}}{P}}=N- \right)_n$$

Phosphorus pentachloride

Poly(phosphonitrilic chloride)

(11.22)

As shown in the following equations, amorphous elastomers are obtained when phosphazene is refluxed with nucleophiles, such as sodium trifluoroethoxide or sodium cresylate, and secondary amines. Copolymers are produced when mixtures of reactants are employed.

$$\left[-N=\underset{\underset{Cl}{|}}{\overset{\overset{Cl}{|}}{P}}- \right]_n + 2n\ NaOCH_2CF_3 \xrightarrow[-2nNaCl]{\Delta} \left[-N=\underset{\underset{OCH_2CF_3}{|}}{\overset{\overset{OCH_2CF_3}{|}}{P}}- \right]_n$$

(11.23)

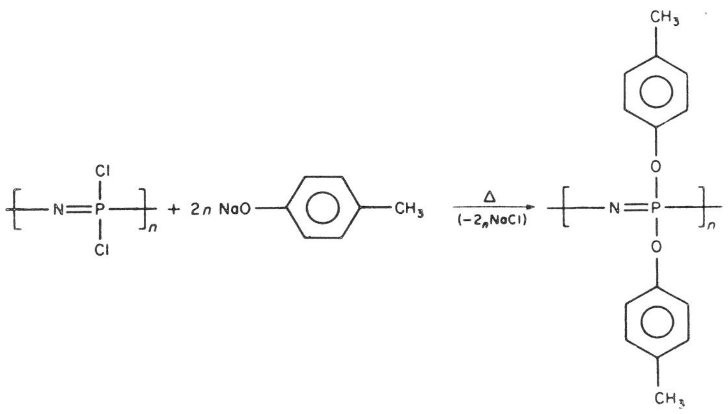

(11.24)

(11.25)

Difunctional reactants such as dihydroxybenzenes (hydroquinone) produce cross-linked phosphazenes. Fibers may be made from polyphosphates and sheets may be pressed from polymeric black phosphorus. The latter is a semiconductor which reverts to white or red phosphorus when heated.

The observed low T_gs for most polyphosphazenes are consistent with the barriers to internal rotation being low and indicate the potential these polymers offer for elastomer applications. In fact, theoretical calculations based on a rotational isomeric model assuming localized pi-bonding predict the lowest (100 cal/mol repeating unit) known polymer barrier to rotation for the skeletal bonds for polydifluorophosphazene.

Temperature intervals between T_g and T_m are unusually small and generally fall outside the frequently cited empirical relation $0.5 < T_g/T_m(A) < 0.67$. This behavior could be related to complications in the first-order transition generally found for organo-substituted phosphazenes are not common to other semicrystalline polymers. Two first-order transitions are usually observed for organo-substituted phosphazenes with a temperature interval from about 150 to 200°C. The lower first-order transition can be detected using DSC, DTA, and TMA. Examination by optical microscopy reveals that the crystalline structure is not entirely lost but persists throughout the extended temperature range to a higher temperature transition which appears to be T_m, the true melting temperature. The nature of this transition behavior resembles the transformation to a mesomorphic state similar to that observed in nematic liquid crystals. It appears from the relationship between the equilibrium melting temperature (heat and entropy of fusion; $T_m = H_m/S_m$) and the low value of H_m at T_m compared with the lower transition temperature that the upper transition, T_m, is characterized by a quite small entropy change. This may be due to an onset of chain motion between the two transitions leading to the small additional gain in conformational entropy at T_m.

The absence of an undetected endothermic transition at T_m, for aryloxy polymers and polyarylaminophosphazenes, indicates that the gain in conformational entropy at T_m for these products is indeed small.

The lower transition is sensitive to structural changes and usually parallels T_g changes. Further, these are commonly noted as T_g values. This will be done here.

From a practical point of view, polyphosphazenes are usually soft just above the lower transition, so that compression molding of films can be carried out. This suggests that the lower transition represents the upper temperature for most useful engineering applications of polyphosphazenes in an unmodified form.

The results indicate that organization in the mesomorphic state, as influenced by thermal history, has a great effect on the transition from a crystalline state in an unmodified form.

Allcock and others have employed polyphosphazenes in a variety of uses including the broad areas of electrical and biomedical applications.

Sulfur nitride polymers $[(SN)_n]$, which have optical and electrical properties similar to metals, were synthesized in 1910. These crystalline polymers, which are superconductive at $0.25°K$, may be produced by the room-temperature solid-state polymerization of the dimer (S_2N_2). Amorphous polymers are also obtained from the tetramer.

A dark, blue-black, amorphous paramagnetic form of poly(sulfur nitride) may be produced by quenching the gaseous tetramer in liquid nitrogen. The tetramer is produced when the polymer is heated at 145°C. Golden metallic films of poly(sulfur nitride) are obtained when the gaseous tetramer is condensed on a glass surface at room temperature. Films of bromine-containing derivatives of poly(sulfur nitride) are superconductors at $0.31°K$.

$$\xrightarrow{300°C} \; (-S-N=)_{4n} + n\,SOF_4 + n\,NH_3 \longrightarrow \quad (11.26)$$

SUMMARY

1. There is a wide variety of inorganic polymers. The potential uses are many and include the broad areas of biological, electrical, analytical, catalytic, building, and photo applications.

2. The bond strength for many combinations is higher than that of many carbon-intensive organic polymers leading to products with superior strength and thermal stabilities.

3. The majority of inorganic reactions are of two general groupings—redox (oxidation-reduction) and substitution.

4. The majority of the condensation polymerizations can be considered extensions of typical organic-types of Lewis acid-base reactions.

5. Polysiloxanes (silicones) offer a good combination of properties not found in organic polymers. Silicones are employed in a number of applications, including antifoaming agents, lubricants, caulks, sealants, O-rings, and gaskets.

6. Polyphosphazenes offer unique thermal properties and have shown a number of uses in the field of electronics.

7. Polymers of sulfur nitride decompose into tetramers when heated at 145°C, and interesting polymers with electrical conductivity are obtained when the gaseous tetramer is cooled.

8. The number and variety of organometallic polymers and potential applications for organometallic polymers are great. Because of the high cost of many of the metal-containing reactants, uses will probably be limited to applications employing minute quantities of the polymers.

GLOSSARY

amorphous: Noncrystalline.

borazoles: Molecules made up of boron and nitrogen atoms.

capping: Protecting the end groups.

carboranes: Molecules made up of carbon and boron atoms.

coordination polymers: Polymers based on coordination complexes.

Dexsil: Trade name for copolymers of carborane and siloxane.

ferrocene: A sandwichlike molecule of cyclopentadiene and iron.

inorganic polymers: Those containing elements other than typically carbon, nitrogen, and oxygen in their backbone or pendant groups.

M: Any metal.

metallocenes: Sandwichlike molecules of cyclopentadiene and metals.

PNF: Poly(phosphonitrilic fluoride).

polyphosphonitrile: Polymer with the repeat unit $+\!\!\overset{\displaystyle |}{\underset{\displaystyle |}{P}}\!\!=\!\!N\!\!+$.

RTV: Room-temperature vulcanization.

silanes: A homologous series, like the alkanes, based on silicon instead of carbon.

SN: Sulfur nitride.

EXERCISES

1. What is meant by "lost loops" in the production of silicones?

2. How could you produce a silicone with a low DP?

3. What would you estimate the solubility parameters of silicones to be?

4. Sodium silicate is water soluble (forming water glass), but silicones are water repellents. Can you explain this difference?

5. How could you polymerize an aqueous solution of sodium silicate?

6. How could you explain the good temperature resistance of silicones?

7. Show the repeat unit for polydiethylsiloxane.

8. What are the reactants used to make phosphazenes?

9. Why would you predict that the chloro groups in phosphonitrilic polymers would be attacked by water?

10. Which phosphazene would be more flexible—one made by a reaction of poly-(phosphonitrilic chloride) with (a) sodium trifuloroethoxide, or (b) sodium trifluorobutoxide?

11. Show the structure of borazole.

12. Since tin-containing organometallic polymers are used in marine antifouling coatings, what would you predict about their water resistance?

13. In addition to high cost, name another disadvantage of coordination polymers.

14. What is the ceiling temperature of sulfur nitride polymers?

BIBLIOGRAPHY

Allcock, H. R. (1972): *Phosphorus-Nitrogen Compounds*, Academic, New York.

Carraher, C. E. (1977): Organometallic polymers, Coatings Plastics Preprints, *37*(1):59.

Carraher, C. E., Preston, J. (1982): *Interfacial Synthesis*, Vol. III, *Recent Advances*, Marcel Dekker, New York.

Carraher, C. E., Reese, D. R. (1977): Lead polyesters, Coatings Plastics Preprints, *37*(1): 162.

Carraher, C. E., Sheats, J., Pittman, C. U. (eds.) (1978): *Organometallic Polymers*, Academic, New York.

———. (1981): *Metallo-Organic Polymers*, Mer, Moscow.

———. (1982): *Advances in Organometallic and Inorganic Polymer Science*, Marcel Dekker, New York.

Clarson, S. J., Semlyen, J. A. (1993): *Silozane Polymers*, Prentice-Hall, Englewood Cliffs, New Jersey.

Kipping, F. S. (1927): Silicones, J. Chem. Soc., *130*:104.

Ladenburg, A. (1872): Silicones, Ann. Chem., *164*:300.

Lee, J., Mykkanen, D. L. (1987): *Metal and Polymer Matrix Composites*, Noyes, Park Ridge, New Jersey.

Neuse, E. W. (1968): Ferrocene polymers. In *Advances in Macromolecular Chemistry*, Vol. 1 (W. M. Pasika, ed.), Academic, New York.

Neuse, E. W., Rosenberg, H. (1970): *Metallocene Polymers*, Marcel Dekker, New York.

Pittman, C. U., Carrahu, C. E., Sheats, J., Culberton, B., Zeldin, M. (1996): *Metal-Containing Polymeric Materials*, Plenum, New York.

Pomogailo, A., *Materials*, Plenum, New York. Savostyanov, V. S. (1994): *Synthesis and Polymerization of Metal-Containing Monomers*, CRC Press, Boca Raton, Florida.

Rochow, E. G. (1951): *An Introduction to the Chemistry of the Silicones*, Wiley Interscience, New York.

Roesky, H. W. (1989): *Clusters and Polymers of Main Group and Transition Elements*, Elsevier, New York.

Sheats, J. E., Carraher, C. E., Pittman, C. U. (1985): *Metal-Containing Polymeric Systems*, Plenum, New York.

Sheats, J. E., Carraher, C. E., Pittman, C. E., Zeldin, M., Currell, B. (1991): *Inorganic and Metal-Containing Polymeric Materials*, Plenum, New York.

Zeigler, J. M., Fearon, F. W. (1989): *Silicon-Based Polymer Sciences*, ACS, Washington, D.C.

Zeldin, M., Wynne, K. J., Allcock, H. R. (1987): *Inorganic and Organometallic Polymers*, ACS, Washington, D.C.

12

Inorganic Polymers

12.1 INTRODUCTION

Just as polymers abound in the world of organics, so do they abound in the world
of inorganics. Inorganic polymers are the major components of soil, mountains,
and sand. Inorganic polymers are also extensively employed as abrasives and cutting materials [diamond, boron carbide, silicon carbide (carborundum), aluminum
oxide], fibers (fibrous glass, asbestos, boron fibers), coatings, flame retardants,
building and construction materials (window glass, stone, Portland cement, brick,
tiles), and lubricants and catalysts (zinc oxide, nickel oxide, carbon black, graphite,
silica gel, alumina, aluminum silicate, chromium oxide, clays, titanium chloride).

The first somewhat manmade, synthetic polymer was probably inorganic in
nature. Alkaline silicate glass was used in the Badarian period in Egypt (about
12,000 B.C.) as a glaze, which was applied to steatite after it has been carved into
various animal, etc., shapes. Faience, a composite containing a powdered quartz or
steatite core covered with a layer of opaque glass, was employed from about 9000
B.C. to make decorative objects. The earliest known piece of regular (modern-
day–type) glass, dated to 3000 B.C., is a lion's amulet found at Thebes and now
housed in the British Museum. This is a blue opaque glass partially covered with
a dark green glass. Transparent glass appeared about 1500 B.C. Several fine pieces
of glass jewelry were found in Tutankhamen's tomb (ca. 1300 B.C.) including two
bird's heads of light blue glass incorporated into the gold pectoral worn by the
Pharaoh.

Portions of this chapter are loosely based on columns appearing in *Polymer News*, Gordon and
Breach, and used with permission of the editor, Gerald Kirschenbaum.

As with the organic-containing inorganic polymers covered in the previous chapter, there are many methods of presenting the topic of inorganic polymers. Because of the wide variety and great number of inorganic polymers, this chapter will focus on only a few of the more well-known inorganic polymers.

Table 12.1 contains a partial listing of common inorganic polymers.

12.2 PORTLAND CEMENT

Portland cement is the least expensive, most widely used synthetic inorganic polymer. It is employed as the basic nonmetallic, nonwoody material of construction. Concrete highways and streets span our countryside and concrete skyscrapers silhouette the urban skyline. Less spectacular uses are found in everyday life as sidewalks, fence posts, and parking bumpers.

The name "Portland" is derived from the cement having the same color as the natural stone quarried on the Isle of Portland, a peninsula on the south of Great Britain. The word cement comes from the Latin word *caementum*, which means

Table 12.1   Important Inorganic Polymers

Agate	Dickite	Phosphorus oxynitride
Alumina	Edingtonite	Polyphosphates (many)
Aluminum oxide	Enstatite	Polyphosphazenes
Amosite	Epistilbite	Polysilicates (many)
Amphiboles	Faujasite	Polysulfur
Anorthite	Feldspars	(polymeric sulfur)
Anthophyllite	Flint	Pyrophyllite
Arsenic selenide	Fuller's Earth	Pyroxmangite
Arsenic sulfide	Garnet	Quartz
Asbestos	Germanium selenide	Rhodonite
Berlinite	Gibbsite	Scolecite
Beryllium oxide	Glasses	Serpentine
Boehmite	(many kinds)	Silicon (many)
Boron nitride	Graphite	Silicon carbide
Boron oxides	Halloysite	Spodumene
Boron phosphate	Heulandite	Stilbite
Calcite	Hiddenite	Stishorite
Carbon black	Imogolite	Sulfur nitride
Cerium phosphate	Jasperite	Talc
Chabazite	Kaolinite	Thomsonite
Chalcedony	Keatite	Titanium
Chalcogens	Mesolite	Tremolite
Chert	Mica	Tridymite
Chrysotile	Montasite	Valentinite
Coesite	Montmorillonite	Vermiculite
Concrete	Mordenite	Wollastonite
Cristobalite	Muscovite	Xonotlite
Crocidolite	Nacrite	Ziolites
Diamond	Natrolite	Zirconia

pieces of rough, uncut stone. Concrete comes from the Latin word *concretus*, meaning to grow together.

Common cement consists of anhydrous crystalline calcium silicates (the major ones being tricalcium silicate, Ca_3SiO_5, and beta-dicalcium silicate, Ca_2SiO_4), lime (CaO, 60%), and alumina (a complex aluminum-containing silicate, 5%). While cement is widely used and has been studied in good detail, its structure and the process whereby it is formed are not completely known. This is due to at least two factors. First, its three-dimensional arrangement of various atoms has a somewhat ordered array when any small (molecular level) portion is studied, but as larger portions are viewed, less order is observed. This arrangement is referred to as short-range order—long-range disorder and is a good description of many three-dimensional, somewhat amorphous inorganic and organic polymers. Thus, there exists only an average structure for the cement which varies with amount of water, etc., added, time after application (that is, age of cement), and source of concrete mix and location (surface or internal). Second, three-dimensional materials are insoluble in all liquids, therefore tools of characterization and identification that require materials to be in solution cannot be employed to assist in the structural identification of cement.

When anhydrous cement mix is added to water, the silicates react, forming hydrates and calcium hydroxide. Hardened Portland cement contains about 70% cross-linked calcium silicate hydrate and 20% crystalline calcium hydroxide.

$$2Ca_3SiO_5 + 6H_2O \rightarrow Ca_3Si_2O_7 \cdot 3H_2O + 3Ca(OH)_2 \quad (12.1)$$

$$2Ca_2SiO_4 + 4H_2O \rightarrow Ca_3Si_2O_7 \cdot 3H_2O + Ca(OH)_2 \quad (12.2)$$

A typical cement paste contains about 60 to 75% water by volume and only about 40 to 25% solids. The hardening occurs through at least two major steps (Fig. 12.1). First a gelatinous layer is formed on the surface of the calcium silicate particles. The layer consists mainly of water with some calcium hydroxide. After about two hours the gel layer sprouts fibrillar outgrowths that radiate from each calcium silicate particle. The fibrillar tentacles increase in number and length, becoming enmeshed and integrated. The lengthwise growth slows, with the fibrils now joining up sideways, forming striated sheets that contain tunnels and holes. During this time, calcium ions are washed away from the solid silicate structure by water molecules and react further, forming additional calcium hydroxide. As particular local sites become saturated with calcium hydroxide, calcium hydroxide itself begins to crystallize, occupying once vacant sites and carrying on the process of interconnecting about and with the silicate "jungle."

In spite of the attempts by the silicate and calcium hydroxide to occupy all of the space, voids are formed, probably from the shrinkage of the calcium hydroxide as it forms a crystalline matrix (generally crystalline materials have higher densities than amorphous materials, thus a given amount will occupy less volume leaving some unfilled sites). Just as a chain is no stronger than its weakest link, so also is cement no stronger than its weakest sites, its voids. Much current research concerns attempts to generate stronger cement with a focus toward filling these voids. Interestingly enough, two of the more successful cement void-fillers are also polymers—dextran, a polysaccharide, and polymeric sulfur.

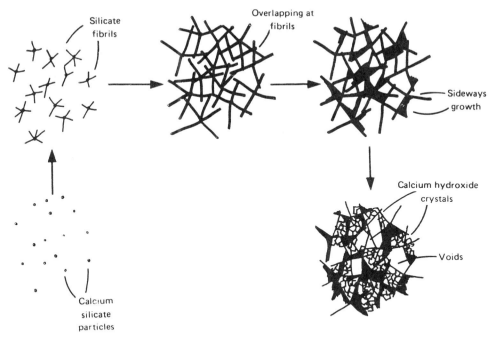

Figure 12.1 Steps in the hardening of Portland cement.

Table 12.2 shows a typical concrete mix. The exact amounts may vary as much as 50% depending on the intended use and preference of the concrete maker.

The manufacture of Portland concrete consists of three basic steps—crushing, burning, and finish grinding. As noted before, Portland cement contains about 60% lime, 25% silicates, and 5% alumina with the remainder being iron oxide and gypsum. Most cement plants are located near limestone ($CaCO_3$) quarries since

Table 12.2 Sample Concrete Mix

Material	Amount	
	By volume	By weight
Portland cement	90–100 lb (1 cubic foot)	90–100 lb (40–45 kg)
Water	5 1/2 gallons	45 lb (20 kg)
Sand	2 cubic feet	200 lb (90 kg)
Gravel (small rocks)	3 cubic feet	250 lb (120 kg)

this is the major source of lime. Lime may also come from oyster shells, chalk, and a type of clay called marl. The silicates and alumina are derived from clay, silicon sand, shale, and blast-furnace slag.

The initial step in the manufacture of cement after the limestone rock is quarried is the powdering of the rock. The powdered rock is mixed together giving a material with a somewhat uniform composition. Two major systems are now employed to obtain the finished cement. In the dry process, the grinding and mixing are done on dry materials. As one would guess, this results in excessive particulate dust formation. In the wet process certain operations are carried out in a water slurry. In both processes, the material is further ground to a fine consistency through a series of pounding, shaking, and recombining process. Alumina, silicates, and iron oxide are added during this time.

In the dry process the mixture is fed directly into kilns for burning. In the wet process, preheating for removal of much of the water precedes introduction into the kiln. The material is typically placed into kilns (rotating at a rate of about one rotation per minute), which are heated to about 1500°C. Water and organics are burned away and the carbonates liberate carbon dioxide.

$$CaCO_3 \rightarrow CaO + CO_2 \qquad (12.3)$$

These cement kilns are the largest pieces of moving machinery used in industry and can be 25 feet in diameter and 750 feet long. The heat changes the mixture into particles called clinkers, about the size of a marble. The clinkers are cooled and reground, the final grinding producing Portland cement finer than flour. The United States produces over 60 million metric tons of Portland cement a year.

12.3 OTHER CEMENTS

There are a number of cements specially formulated for specific uses.

Air-entrained concrete contains small air bubbles formed by the addition of soaplike resinous materials to the cement or to the concrete when it is mixed. The bubbles permit the concrete to expand and contract (as temperature changes) without breaking (since the resistance of air to changes in volume is small).

Lightweight concrete may be made through use of lightweight fillers such as clays and pumice in place of sand and rocks or through the addition of chemical foaming agents that produce air pockets as the concrete hardens. These air pockets are usually much larger than those found in air-entrained concrete.

Reinforced concrete is made by casting concrete about steel bars or rods. Most large cement-intense structures such as bridges and skyscrapers employ reinforced concrete.

Prestressed concrete is typically made by casting concrete about steel cables stretched by jacks. After the concrete hardens, the tension is released, resulting in the entrapped cables compressing the concrete. Steel is stronger when tensed, and concrete is stronger when compressed. Thus prestressed concrete takes advantage of both of these factors. Archways and bridge connections are often made from prestressed concrete.

Concrete masonry is simply the name given to the cement building blocks employed in the construction of most houses, and it is simply a precast block of cement.

Precast concrete is concrete that is cast and hardened before it is taken to the site of construction. Concrete sewer pipes, wall panels, beams, girders, and spillways are all examples of precast concrete.

The aforementioned cements are all typically derived from Portland cement. Following are non-Portland cements.

Calcium-aluminate cement has a much higher percentage of alumina than does Portland cement. Further, the active ingredients are lime, CaO, and alumina. In Europe it is called melted or fused cement. In the United States it is manufactured under the trade name of Lumnite. Its major advantage is its rapidity of hardening, developing high strength within a day or two.

Magnesia cement is largely composed of magnesium oxide (MgO). In practice, the magnesium oxide is mixed with fillers and rocks and an aqueous solution of magnesium chloride. This cement sets up (hardens) within 2 to 8 hours and is employed for flooring in special circumstances.

Gypsum, hydrated calcium sulfate ($CaSO_4 \cdot 2H_2O$), serves as the basis of a number of products including plaster of Paris (also known as molding plaster, wall plaster, and finishing plaster), Keen's cement, Parisian cement, and Martin's cement.

The ease with which plaster of Paris and the other gypsum cements can be mixed and cast and the rapidity with which they harden contribute to their importance in the construction field as a major component for plaster wall boards. Plaster of Paris's lack of shrinkage in hardening accounts for its use in casts. Plaster of Paris is also employed as a dental plaster, pottery plaster, and as molds for decorative figures. Unlike Portland cement, plaster of Paris requires only about 20% water and dries to the touch in 30 to 60 minutes, giving maximum strength after 2 or 3 days. Portland cement requires several weeks to reach maximum strength.

12.4 SILICON DIOXIDE (AMORPHOUS)

Silicon dioxide (SiO_2) is the repeating general structural formula for the vast majority of rock, sand, and dirt about us and for the material we refer to as glass.

The term glass can refer to many materials, but here we will use the ASTM definition that glass is an inorganic product of fusion which has been cooled to a rigid condition without crystallization. In this section silicate glasses, the common glasses for electric light bulbs, window glass, drinking glasses, glass bottles, glass test tubes and beakers, and glass cookware, will be emphasized.

Glass has many useful properties, as listed in Table 12.3. It ages (changes chemical composition and/or physical property) slowly, typically retaining its fine optical and hardness-related properties for centuries. Glass is referred to as a supercooled liquid or a very viscous liquid. Indeed it is a slow-moving liquid, as attested to be sensitive measurements carried out in many laboratories. Concurrent with this is the observation that the old stained glass windows adorning European cathedrals are a little thicker at the bottom of each small, individual piece than at the top of the piece. For most purposes though, glass can be treated as a brittle solid that shatters on sharp impact.

Glass is mainly silica sand (SiO_2) and is made by heating silica sand and powdered additives together in a specified manner and proportion much as a cake is baked, following a recipe which describes the items to be included, amounts,

Table 12.3   General Properties of Silicate Glasses

High transparency to light	Good electrical insulator
Permanent (long time) transparency	Wide range of colors possible
Hard	High sparkle
Scratch resistant	Good luster
Chemically inert	Low porosity
Low thermal expansion coefficients	Ease of reforming

mixing procedure (including sequence), oven heat and heating time. The amounts, nature of additives, etc., all affect the physical properties of the final glass.

Typically cullet, recycled or waste glass (5 to 40%), is added along with the principal raw materials. The mixture is thoroughly mixed and then added to a furnace where the mixture is heated to near 1500°C to form a viscous syruplike liquid. The size and nature of the furnace corresponds to the glasses' intended uses. For small, individual items the mixture may be heated in small clay (refractory) pots.

Most glass is melted in large (continuous) tanks that can melt 400 to 600 metric tons a day for the production of other glass products. The process is continuous with raw materials fed into one end as molten glass is removed from the other. Once the process (called a campaign) is begun it is continued indefinitely, night and day, often for several years until the demand is met or the furnace breaks down.

A typical window glass will contain 95 to 99% silica sand with the remainder being soda ash (Na_2CO_3), limestone ($CaCO_3$), Feldspar, and borax or boric acid along with the appropriate coloring and decolorizing agents, oxidizing agents, etc. As noted previously, 5 to 40% by weight of crushed cullet is also added. The soda ash, limestone, Feldspar, and borax or boric acid all form oxides, as they are heated, that become integrated into the silicon structure:

$$Na_2CO_3 \xrightarrow{\Delta} Na_2O + CO_2 \qquad (12.4)$$
Soda ash
(sodium carbonate)

$$CaCO_3 \xrightarrow{\Delta} CaO + CO_2 \qquad (12.5)$$
Limestone
(calcium carbonate)

$$R_2OAl_2O_3\ 6SiO_2 \xrightarrow{\Delta} R_2O + Al_2O_3 + 6SiO_2 \qquad (12.6)$$
Feldspar   Alumina
                                   (aluminum oxide)

$$H_3BO_4 \xrightarrow{\Delta} B_2O_3 \qquad (12.7)$$
Boric   Boron
acid   oxide

The exact structure varies according to the ingredients and actual processing conditions (as how long heated at what temperature). As in the case of cement, glass is a three-dimensional array which offers short-range order and long-range disorder—it is amorphous offering little or no areas of crystallinity. The structure is based on the silicon atoms existing in a tetrahedral geometry with each silicon atom attached to four oxygen atoms, generating a three-dimensional array of inexact tetrahedrals. Thus structural defects occur, due in part to the presence of impurities such as Al, B, and Ca, intentionally introduced, and numerous other impurities not intentionally introduced. These impurities encourage the glass to cool to an amorphous structure since the different-sized impurity metal ions, etc., disrupt the rigorous space requirements necessary to exact crystal formation. Figure 12.2 is an illustration of this situation emphasizing the presence of the impurities.

Processing includes shaping and retreatments of glass. Since shaping may create undue sites of amorphous structure, most glass objects are again heated to near their melting point. This process is called *annealing*. Since many materials tend to form more ordered structures when heated and recooled slowly, the effect of annealing is to "heal" these sites of major dissymmetry. It is important to "heal" these sites since they represent preferential sites for chemical and physical attack such as fracture.

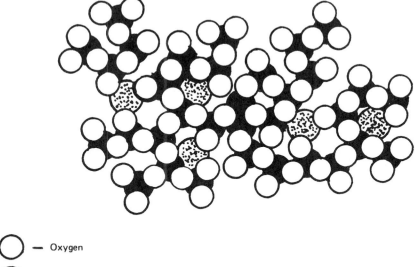

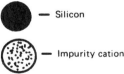

○ — Oxygen

● — Silicon

◉ — Impurity cation

Figure 12.2 Two-dimensional representation of a multicomponent silicon dioxide intensive glass.

Four main methods are employed for shaping glass. They are drawing, pressing, casting, and blowing. Drawing is employed for shaping flat glass, glass tubing, and fiberglass. Most flat glass is shaped by drawing a sheet of molten (heated so it can be shaped but not so it freely flows) glass onto a tank of molten tin. Since the glass literally floats on the tin, it is called "float glass." The temperature is carefully controlled. The glass from a "float bath" typically has both sides quite smooth with a brilliant finish that requires no polishing.

Glass tubing is made by drawing molten glass around a rotating cylinder of appropriate shape and size. Air can be blown through the cylinder or cone giving glass tubing used in the laboratory. Fiberglass is made by drawing molten glass through tiny holes, the drawing process helping to align the tetrahedral clusters.

Pressing is accomplished by simply dropping a portion of molten glass into a form and then applying pressure to ensure that the glass takes the form of the mold. Lenses, glass blocks, baking dishes, and ashtrays are examples of objects commonly press-processed.

The casting process involves filling molds with molten glass much the same way cement and plaster of Paris molded objects are produced. Art glass pieces are typical examples of articles produced by casting.

Glass blowing is one of the oldest arts known to man. For art or tailor-made blown objects, the working and blowing of the glass is done by a skilled worker who blows into a pipe intruded into the molten glass. The glass must be maintained at a temperature that permits working but not free flow, and the blowing must be at such a rate and controlled as to give the desired result. Mass-produced items are made using mechanical intense blowers, often having the glass blown to fit a mold much as the blow molding of plastics.

As noted above, annealing encourages the removal of stress and strain locations through raising the temperature to give nearly molten glass, which allows the glass to "heal" these stress and strain locations. Slow cooling will result in a glass containing larger areas of crystallinity resulting in stronger but more brittle glass. Tempering is the name given when the glass is rapidly cooled, resulting in a glass that is more amorphous as well as weaker but less brittle than more slowly cooled glass. The correlation between crystallinity, rate of cooling, and brittleness is readily demonstrated by noting that older window glass exposed to periods of full sun is quite brittle and easily shattered, since the sunlight raises the temperature sufficiently to permit small molecular movements (even in full sunlight the glass is not near the temperature to permit ready movement), and over a period of years this results in localized crystalline formation.

There is a wide variety of uses and kinds of glass. Silicon-based glasses account for almost all of the glasses manufactured. Silica is finely ground sand. Yet most sand is unsuitable for general glasses making due to the presence of excessive impurities. Thus while sand is plentiful, sand useful in the production of glass is much less common. In fact, the scarcity of large deposits of glass sand is one major reason for the need to recycle glass items. The second major reason is associated with the lowered energy requirements for glass to be made molten again for reshaping compared with a virgin glass mixture, i.e., glass becomes molten at lower temperatures than are necessary to form molten glass from original feedstock.

Kinds of Glass

The types and properties of glass can be readily varied by changing the relative amounts and nature of ingredients. Soda-lime glass is the most common of all glasses, accounting for about 90% of glass made. Window glass, glass for bottles and other containers, glass for light bulbs, and many art glass objects are all soda-lime glass. Soda-lime glass typically contains 72% silica, 15% soda (sodium oxide, NaO), 9% lime (calcium oxide, CaO), and the remaining 4% minor ingredients. Its relatively low softening temperature and low thermal shock resistance limit its high-temperature applications.

Vycor, or 96% silicon glass, is made using silicon and boron oxide. Initially the alkali-borosilicate mixture is melted and shaped using conventional procedures. The article is then heat-treated, resulting in the formation of two separate phases—one high in alkalis and boron oxide and the other containing 96% silica and 3% boron oxide. The alkali-boron oxide phase is soluble in strong acids and is leached away by immersion in hot acid. The remaining silica-boron oxide phase is quite porous. This porous glass is then again heated to about 1200°C, resulting in a 14% shrinkage due to the remaining portion filling the porous voids. The best variety is "crystal" clear and called fused quartz. The 96% silica glasses are more stable and exhibit higher melting points (1500°C) than soda-lime glass. Crucibles, ultraviolet filters, range-burner plates, induction-furnace linings, optically clear filters and cells, and super heat-resistant laboratoryware are often 96% silicon glass.

Borosilicate glass contains about 80% silica, 13% boric oxide, 4% alkali, and 2% alumina. It is more heat-shock resistant than most glasses due to its unusually small coefficient of thermal expansion [typically between 2 to 5×10^{-6} cm/cm/°C (or simply 1/°C); for soda-lime glass 8 to 9×10^{-6} cm/cm/°C]. It is better known by such trade names as Kimax and Pyrex. Bakeware and glass pipelines are often borosilicate glass.

Lead glasses (often also called heavy glasses) are made by replacing some or all of the calcium oxide by lead oxide (PbO). Very high amounts of lead oxide can be incorporated—up to 80%. Lead glasses are more expensive than soda-lime glasses, and they are easier to melt and work with. They are more easily cut and engraved, giving a product with high sparkle and luster (due to higher refractive indexes). Fine glass and tableware are often lead glass.

Silicon glass is made by fusing pure quartz crystals or glass sand (impure crystals), and it is typically about 99.8% SiO_2. Silica glass has a high melting point (1750°C), making it hard to melt and difficult to fabricate.

Colored glass has been made for several thousands of years, first by Egyptians and later by Romans. Color is typically introduced by addition of transition metals and oxides. Table 12.4 contains a listing of selected inorganic colorants and the resulting colors. Because of the high clarity of glass, a small amount of coloring agent "goes a long way." One part of cobalt oxide in 10,000 parts of glass gives an intense blue glass. The most well-known use for colored glass is in the construction of stained-glass windows. In truth there are numerous other uses such as industrial color filters and lenses.

Coloring is usually one of three basic types: (1) solution colors where the coloring agent such as oxides of transition metals (titanium, vanadium, chromium, manganese, iron, cobalt, nickel, and copper) are dissolved in the glass; (2) colloidal

Table 12.4 Colorants for Stained Glass

Additive	Color(s)
Nickel oxide (NiO)	Yellow to purple
Calcium fluoride (CaF_2)	(Milky) white
Cobalt oxide (CoO)	Blue
Iron II compounds (salts and oxides)	Green
Iron III compounds (salts and oxides)	Yellow
Copper oxide (Cu_2O)	Red, blue, or green
Tin IV oxide (SnO_2)	Opaque
Manganese IV oxide (MnO_2)	Violet
Gold oxide (Au_2O_3)	Red

particle colors that are produced by the presence of certain small particles sus-pended in the glass. (Most ruby glass is made by introducing selenium, cadmium sulfide (CdS), and zinc oxide (ZnO) into the glass. This combination is believed to form small colloidal particles that result in the glass appearing to be ruby red-colored.); and (3) color caused by larger particles that can be seen under a regular microscope by the naked eye. Milky-colored opal and alabaster glasses are often of this variety.

Glazes are thin, transparent coatings (colored or colorless) fused on ceramic materials. Vitreous enamels are thin, normally opaque or semi-opaque, colored coatings fused on metals, glasses, or ceramic materials. Both are special glasses but can contain little silica. They are typically low melting and often are not easily mixed in with more traditional glasses.

There also exist special glasses for special applications. Laminated automotive safety glass is a sandwich made by combining alternate layers of poly(vinyl butyral) (containing about 30% plasticizer) and soda-lime glass. This sticky organic polymer layer acts to both absorb sudden shocks (like hitting another car) and to hold broken pieces of the glass together. Bullet-proof (or more correctly stated, bullet-resistant) glass is typically a thicker, multilayer form of safety glass.

Tempered safety glass is a single piece of specially heat-treated glass often used for industrial glass doors, laboratory glass, lenses, and side and rear auto-motive windows. Due to the tempering process, the material is much stronger than normal soda-lime glass. Optical fibers can be glass fibers that are coated with a highly reflective polymer coating such that light entering one end of the fiber is transmitted through the fiber (even around curves and corners as within a person's stomach) emerging from the other end with little loss of light energy. These optical fibers can also be made to transmit sound and serve as the basis for transmission of television and telephone signals over great distances.

Foam glass is simply soda-lime glass filled with tiny cells of gas. It is light-weight and a good heat insulator. Optical glass for eyeglass lenses and camera lenses is again typically soda-lime glass, but it is highly purified so as to be highly transmissive of light. This special processing increases greatly the cost of this glass. There exist numerous other specialty glasses that are important in today's society—laser glasses, photosensitive glass, photochromic window and eyeglass glass, invisible glasses, and radiation absorbing glasses are only a few of these.

There are a number of silica-intensive fibers, many of them lumped together under the terms fibrous glass and fiberglass. A general-purpose soda-lime glass fiber may contain silica (72%), calcium oxide (9.5%), magnesium oxide (3.5%), aluminum oxide (2%), and sodium oxide (13%). Borosilicate fibers, synthesized for electrical applications where there is a danger of water, may contain silica (52%), calcium oxide (17%), magnesium oxide (5%), aluminum oxide (14%), sodium and potassium oxides (1%), and boric oxide (11%). Thus, through practice, various recipes have been developed for making fiberglass suitable for specific uses.

The fibers are produced by melting the "glass mixture" with the molten glass drawn through a number of orifices. The filaments are passed through a pan containing sizing solution onto a winding drum. The take-up rate of the filament is more rapid than the rate of flow from the orifices acting to align the molecules and draw the fibers into thinner filaments. Thus a fiber forced through a 0.1-cm orifice may result in filaments of 0.0005-cm diameter. This drawing increases the strength and flexibility of the fiberglass.

Applications of fiberglass include as insulation (thermal and electrical) and in the reinforcing of plastics.

Table 12.5 contains a listing of major glass-producing companies in the United States.

12.5 SILICON DIOXIDE (CRYSTALLINE FORMS)—QUARTZ FORMS

Just as silicon dioxide forms the basis of glass, so also does it form the basis of many of the rocks, grains of sand, and dirt particles that make up the Earth's crust. Most rocks are inorganic polymers, but here we will deal with only a few of those containing goodly amounts of silicon.

Silicon crystallizes in mainly three forms—quartz, tridymite, and cristobalite. After the feldspars, quartz is the most abundant mineral in the Earth's crust, being a major component of igneous rocks and one of the most common sedimentary materials, in the form of sandstone and sand. Quartz can occur as large (several

Table 12.5 Leading U.S. Glass Companies in Decreasing Order of Output

Owens–Illinois, Inc.

PPG Industries, Inc.

Corning Glass Works

Owens–Corning Fiberglass Corporation

Libbey–Owens–Ford Company

pounds) single-crystals but is normally present as much smaller components of many of the common materials about us. The structure of quartz (Fig. 12.3) is a three-dimensional network of six-membered Si—O rings (three SiO_4 tetrahedra) connected such that every six rings enclose a 12-membered Si—O (six SiO_4 tetrahedra) ring.

Quartz is found in several forms in all three major kinds of rocks—igneous, metamorphic, and sedimentary. Is is one of the hardest minerals known. Geologists often divide quartz into two major groupings—coarse crystalline and cryptocrystalline quartz.

Coarse crystalline quartz includes six-sided crystals and massive granular clumps. Some colored varieties of coarse crystalline quartz crystals, amethyst and citrine, are cut into gem stones. Others include pink (rose) and purple and milky quartz, but most coarse crystalline quartz is colorless and transparent. Sandstone is a ready example of granular quartz. Color is a result of the presence of small amounts of metal cations such as calcium, iron, magnesium, aluminum, and iron.

Cryptocrystalline forms contain microscopic quartz crystals and include the chalcedony grouping of rocks, including chert, agate, jasper, and flint.

Quartz exhibits an important property called the piezoelectric effect. When pressure is applied to a slice of quartz, it develops a net positive charge on one side of the quartz slice and a negative charge on the other. This phenomenon is piezoelectric generation of a voltage difference across the two sides of the quartz crystal. Further, the same effect is found when pressure is applied not mechanically, but through application of an alternating electrical field with only certain frequencies permitted to pass through the crystal. The frequencies allowed to pass vary with the crystal shape and thickness. Such crystals are used in radios, televisions, and radar. This effect also forms the basis for quartz watches and clocks. Voltage applied to a quartz crystal causes the crystal to expand and contract at a set rate, producing vibrations. The vibrations are then translated into a uniform measure time.

While quartz crystals are suitable for the production of optical lenses, most lenses are manufactured from synthetically produced quartz due to the scarcity of good grade large quartz crystals.

The feldspars are the most abundant minerals in the Earth's crust, accounting for about 60% of all igneous rocks. They are derivatives of silica, where about one-half or one-quarter of the silicon atoms have been replaced by aluminum. The minerals can be divided into two groupings—alkali and plagioclase feldspars. All alkali feldspars contain potassium and most contain sodium. The most common alkaline feldspars are orthoclase, microcline, and sanidine. Most plagioclase feldspars contain calcium and sodium and include andesine and labradorite. Feldspar is used in the manufacture of certain glass and pottery. Some feldspar crystals such as moonstone (white perthilte), Amazon stone (green microcline), and multicolored labradorite are used as gemstones and in architectural decorations. Some feldspar is also used as a coating and filler in the production of paper.

The three-dimensional network in a typical feldspar is composed of eight-membered rings containing four tetrahedra (AlO_4 and/or SiO_4) linked through additional tetrahedra into chains.

Granite is a hard crystalline rock chiefly composed of quartz and feldspar. Many of the quartz and feldspar crystals are large enough to be seen by the naked

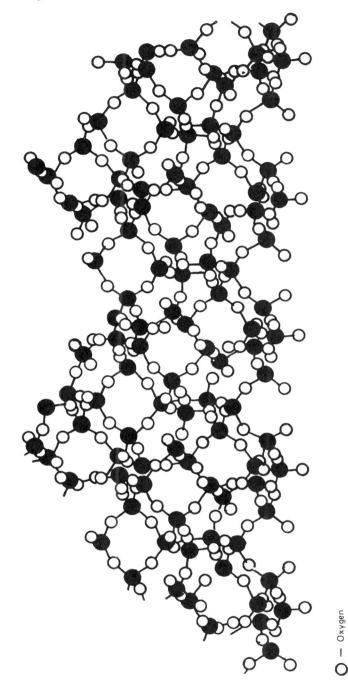

Figure 12.3 Structure of crystalline SiO_2 tetrahedra as found in quartz.

O — Oxygen

● — Silicon

eye. Quartz is transparent, much like glass. Feldspar is typically colored a dull white, grey, or pink. On a broken, cleaved rock surface, the feldspar crystals are smooth and reflect sunlight, being seen as "sparkling." The dark-colored minerals in granite are mainly black mica and hornblende, also polymeric.

Granite is used in building bridges and buildings where great strength is needed. It is also employed in the construction of monuments and gravestones since it can be polished to give a lasting luster and because of its tendency to resist wear by the natural elements.

Granite was once melted rock like lava, but it remained deep underground where the Earth's pressure acted to both encourage crystal formation and to encourage a slow cooling rate, which also encouraged crystal formation. Geologic upheavals brought the granite from deep in the Earth's mantle to the surface. At times, the melted pregranite rock was forced to the surface in its molten state where it cooled more rapidly and without benefit of high pressure, resulting in smaller crystals that can only be seen through a microscope. Such lava is called rhyolite.

Sand is loose grains of minerals or rocks, larger than silt but smaller than gravel. Sand can be divided by kind or size. Size differential is normally accomplished by simply placing the raw material on top of a series of wire screens of decreasing size. Most grains of sand are parts of rocks and have been weathered—thus they are usually smooth. The most common sand is largely quartz, although the sands of White Sands, New Mexico are largely gypsum grains. Most grains of sand are hues of brown and white with a few being clear quartz and others small red-colored rubies.

Soil contains mineral and organic particles, the majority by weight of the mineral particles being simply sand. The contents of soil change constantly. The organic particles bring organic nutrients to plants while the inorganic sand particles supply the mineral, trace element nutrients necessary to life. Plants demand a delicate balance of nutrients. Fortunately as one plant dies, it deposits both organic and inorganic nutrients for later generations. Soil erosion, flooding, and fires all result in drastic changes in the composition of soil, and thus present local plants with new soil features that must be coped with if survival is to occur.

As noted in previous chapters, much organic decay material is polymeric. Thus soil is a complex, variable mixture of organic and inorganic polymers, small molecules, gases (nitrogen, oxygen, carbon monoxide, carbon dioxide, and methane), and water.

Micas are also composed of silicon-oxygen tetrahedra. The anionic charge on the silicate sheet is the result of the replacement of silicon by aluminum. Cations such as potassium are interspaced between these negatively charged sheets. Muscovite, $[KAl_2(OH)_2Si_3AlO_{10}]$ and phlogopite $[KMg_3(OH)_2Si_3AlO_{10}]$ are naturally occurring micas employed in constructional and electrical engineering applications.

Synthetic mica is manufactured on a large scale for industrial consumption. The China clay or kaolin minerals are sheet polymers with a basic formula of $Al_2(OH)_4Si_2O_5$. Kaolinite, dickite, and nacrite all have this empirical composition, differing in the number of kaolin layers—keolinite with one layer per unit cell, dickite two, and nacrite six. Mica, claylike silica, and fibrous glass are used as fillers and reinforcements in plastics and elastomers.

12.6 ASBESTOS

Asbestos has been known and used for over 2000 years. Egyptians used asbestos cloth to prepare bodies for burial. The Romans called it *aminatus* and used it is a cremation cloth and for lamp wicks. Marco Polo described its use in the preparation of fire-resistant textiles in the thirteenth century. Asbestos is not a single mineral but rather a grouping of materials that give soft, threadlike fibers (Table 12.6). These materials are examples of two-dimensional sheet polymers containing two-dimensional silicate $(Si_4O_{10})^{-4}$ anions bound on either one or both sides by a layer of aluminum hydroxides [$Al(OH)_3$, gibbsite] or magnesium hydroxide [$Mg(OH)_2$, brucite]. The aluminum and magnesium are present as positively charged ions (cations). These cations can also contain a varying number of water molecules (waters of hydration) associated with them. The spacing between silicate layers varies with the nature of the cation and amount of its hydration.

Amphibole asbestos consists of a number of mineral types, including the cummingtonite-grunerite series, actinolite-tremolite series, anthophyllite, and crocidolite. Amphibole asbestos is generally coarser than serpentine asbestos. The amphibole asbestos contains silicate sheets with a wider number of cations, including iron, magnesium, calcium, sodium, and aluminum. Amphibole asbestos is noted for its good resistance to both heat and acids. Kaolinite is the most important member of this group. It is actually a mixture of a number of asbestoses with the natural aggregates known as China clay or kaolin.

Due to its fibrous nature and ability to resist elevated temperature (compared with most organic-based fibers), chrysotile is used to make fabrics for the production of fire-resistant fabric, including laboratory glove-wear and theater curtains. Shorter fibers are used in electrical insulation and in automotive brake linings. The shortest fibers are employed in the production of cement pipe, floor tile, reinforced plastics, and reinforced caulking agents. A summary of uses is given in Table 12.6.

Table 12.6 Major Uses of Chrysotile Asbestos as a Function of Fiber Length

Description[a]	Uses
Crude (3/8 to 1 inch and unsorted)	Textiles, theater curtains, brake linings
2 M (medium; milled fibers)	Textiles, theater curtains, packings, brake linings
8 M (medium; milled fibers)	Shingles, cement pipe, insulating jackets, packings, gaskets, asbestos paper, millboard, plastic fireproof cements, asbestos-asphalt roof coatings, electrical panels, insulation, plastic fillers
10 M (shortest; milled fibers)	Asbestos-cement sheets, asbestos-cement shingles, brake linings, asbestos paper, asphalt roof coatings, plastic fillers, welding rods, floor tiles, boiler insulation

[a]Sized according to ability to pass through standardized screen meshes with the screen mesh sizes decreasing as screen number increases.

Though asbestos has been known and used for thousands of years, it has only been in the last 60 to 80 years that the potential hazards of asbestos have begun to be recognized. For instance, asbestos miners and manufacturing personnel who work with asbestos for 20 years or more are 10 times more likely than the general public to contract asbestosis. Families of asbestos workers and those living near the mines also have a greater than average incidence of asbestosis.

Asbestosis is a disease that blocks the lungs with thick fibrous tissue, causing shortness of breath and swollen fingers and toes. Bronchogenic cancer (cancer of the bronchial tubes) is prevalent among asbestos workers who also smoke cigarettes. Asbestos also causes mesothelioma, a fatal cancer of the lining of the abdomen or chest. These diseases may lie dormant until many years after exposure.

The exact origin of these diseases is unknown, but they appear to be caused by particles (whether asbestos or other particulates) about 5 to 20 micrometers in length (2×10^{-4} inches) corresponding to the approximate sizes of the mucous openings in the lungs. Further, the sharpness of asbestos fibers intensifies their toxicity, since these fibers actually cut the lung walls. When the walls heal, the deposited asbestos, if not flushed from the lungs, again cuts the lung walls as the individual coughs, with more scar tissue being formed. The cycle continues until the lungs are no longer able to function properly.

Because of its widespread use (see Table 12.6), small amounts of airborne asbestos are probably present throughout most of the United States. Hopefully the amounts are small enough as to not be harmful.

12.7 POLYMERIC CARBON—DIAMOND

Just as carbon serves as the basic building element for organic materials, so also does it form a building block in the world of inorganic materials. Elemental carbon exists in two distinct crystalline forms—graphite and diamond. Graphite is the more stable allotrope of carbon, with graphite readily formed from heating diamonds.

Natural diamonds (Fig. 12.4) were formed millions of years ago when concentrations of pure carbon were subjected by the Earth's mantle to great pressure and heat. The majority of diamonds (nongems) are now manmade. The majority of synthetic diamonds are no larger than a grain of common sand. By 1970, General Electric was manufacturing diamonds of gem quality and size through compressing pure carbon under extreme pressure and heat. These diamonds, available at a cost much higher than natural diamonds, are used for research. For instance, it was discovered that the addition of small amounts of boron to diamonds causes them to become semiconductors. Today such doped diamonds are used to make transistors.

Imitation diamonds are prevalent in the marketplace. These include spinel, strontium titanate glass, and yttrium aluminate. While they are less expensive than true diamonds, they are able to give good sparkle and be fashioned into diamond shapes. They are not, however, as hard as real diamonds, and their luster is more easily marred through clouding and formation of surface scratches.

The major use of diamonds is as industrial shaping and cutting agents to cut, grind, and bore (drill) holes in metals and ceramics. Many record players employ a diamond needle to transmit differences in the record grooves into sound.

India was the earliest supplier of diamonds. In 1867 the first diamonds were discovered in South Africa, which has since become the largest supplier of dia-

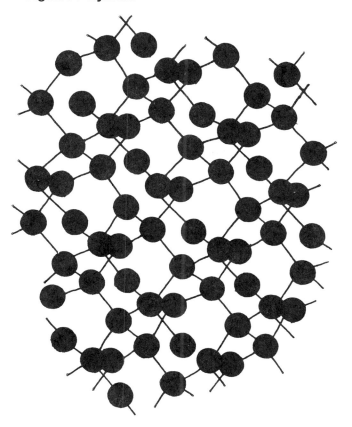

Figure 12.4 Representation of diamond. Where each carbon is at the center of a tetrahedron composed of four other carbon atoms. It is the hardest known natural material.

monds. Diamonds have been discovered in many parts of Africa, including the Gold Coast, Angola, and Sierra Leone. While diamonds were discovered in Brazil in 1725, South America has only accounted for a small fraction of the diamonds found. Small deposits have also been found in Arkansas, Australia, the Ural Mountains (Russia), and Mexico. Diamonds are few in number even in the richest mines. For instance, it takes an average of about 15 metric tons of ore to find one gram equivalent of diamond from the diamond fields of South Africa.

While diamonds can be cut, shaping has been done in the past by a trained gem cutter by striking the rough diamond on one of its cleavage planes. These cleavage plates are called faces and represent sites of preferential cleavage and reflection of light. Thus a gem cutter will seek out these cleavage planes in combinations that will result in a spectacular gem stone where the maximum amount of diamond is included in the gem stone. Unlike food or other commodities, the price of a diamond increases as the bulk increases.

12.8 POLYMERIC CARBON—GRAPHITE

While diamond is the hardest naturally occurring material, the most common form of crystalline carbon is the much softer graphite. Graphite occurs as sheets of hexagonally fused benzene rings (Fig. 12.5) or "hexachicken wire." The bonds holding the fused hexagons together are traditionally primary, covalent bonds. The bonding between the sheets of fused hexagons consists of a weak overlapping of pi-electron orbitals and is considerably weaker than the bonding within the sheet. Thus, graphite exhibits many properties that are dependent on the angle at which they are measured. Further, they show some strength when measured along the sheet but very little strength if the layers are allowed to "slide over one another." Also, the fused hexagons are situated such that the atoms in each layer lie opposite to the centers of the six-membered rings in the next layer. This arrangement further weakens the overlapping of pi-electrons between layers such that the magnitude of layer-to-layer attraction is on the order of ordinary secondary van der Waals forces.

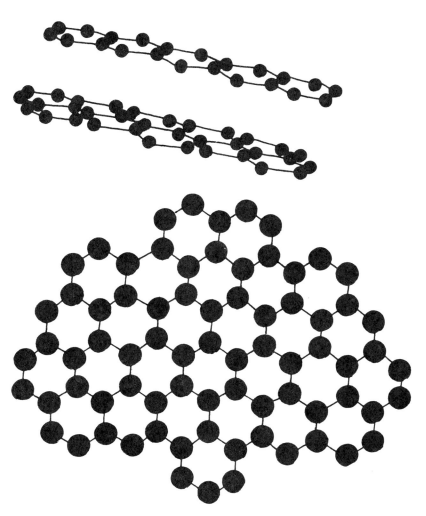

Figure 12.5 *Pictorial* representation of graphite emphasizing the layered (top) and sheet (bottom) nature of graphite.

The "slipperiness" of the layers accounts for graphite's ability to be a good lubricant.

The variance of property with angle of applied force, light, magnetism, etc., is called anisotropic behavior. Calcite is anisotropic in its crystal structure resulting in a dependency of its interaction with light with the angle of incidence of the light. This is employed in the construction of laboratory polariscopes. Certain organic polymers can also be made to be anisotropic in their reaction to light, electricity, magnetism, or other applied forces. This serves as the basis for many polarized sunglasses.

As with diamonds and most other natural materials, graphite's discovery and initial usage is lost in antiquity. It was long confused with other minerals such as molybdenite (MoS_2). At one time it was known as plumbago ("like lead"), crayon noir, silver lead, black lead, and carbo mineralis. Werner, in 1789, first named it graphite, meaning (in Greek) "to write."

While graphite has been extensively mined in China, Mexico, Austria, North and South Korea, and Russia, the majority of graphite used in the United States is manufactured from coke.

The Acheson process for graphite production begins by heating a mixture of charcoal, or coke, and sand. The silica is believed to be reduced to silicon that combines with carbon-forming silicon carbide, which subsequently dissociates into carbon and silicon. The silicon vaporizes and the carbon condenses, forming graphite. Graphite may also be produced employing other processes. The precise properties of a graphite sample is dependent on the specific conditions employed in its production. For instance, certain processes will give graphite that contains a large number of defects that allow a high permeability. Other processes, such as the high-temperature pyrolysis of hydrocarbon vapors, gives a dense layer of graphite with few defects and a low gas permeability.

Its properties led directly to its many uses in today's society. Because of its tendency to mark, hardened mixtures of clay and graphite are the "lead" in today's lead pencils. Graphite conducts electricity and is not easily burned. Thus, many industrial electrical contact points (electrodes) are made of graphite. Graphite is a good conductor of heat and is chemically quite inert, even at high temperatures. Thus many crucibles for melting metals are graphite lined. Graphite has good stability to even strong acids, thus it is employed to coat acid tanks. It is also effective at slowing down neutrons and thus composite bricks and rods (often called carbon rods) are employed in some nuclear generators to regulate the progress of the nuclear reaction. As already noted, its good "slipperiness" accounts for its use as a lubricant for clocks, door locks, and hand-held tools. Graphite is also the major starting material for the synthesis of synthetic diamonds. Graphite is sometimes used as a component of industrial coatings (paints). Dry cells and some types of alkali storage batteries also employ graphite. Graphite fibers are used for the reinforcement of thermosets and thermoplastics.

At ordinary pressures and temperatures both graphite and diamond are stable. At high temperatures (1700 to 1900°C), diamond is readily transformed to graphite. The reverse transformation of graphite to diamond occurs only with application of great pressure and high temperatures. Thus, as noted before, the naturally more stable form of crystalline carbon is not diamond but rather graphite.

12.9   HIGH-TEMPERATURE SUPERCONDUCTORS

Discovery of the 123-Compound

In early 1986, George Bedorz and K. Alex Muller reported a startling discovery: a ceramic material, La-Ba-Cu-O, lost its resistance to electrical current at about 30°K. This was the first report of a so-called high-T_c superconductor. Intensive efforts were then concentrated on substituting the component ions with similar elements on both of the La and Ba sites. The first success was reported by Kishio et al. with a $(La_{1-x}Sr_x)_2CuO_4$ system that exhibited a T_c higher than the former by about 7°K. Then the substitution on the La sites led Wu et al. to find another superconductor, the Y-Ba-Cu-O system, with a T_c of 93°K, in February 1987. This finally broke the technological barrier of liquid nitrogen temperature. The superconducting phase was identified as $Y_1Ba_2Cu_3O_7$ (generally referred to as the *123-compound*).

The 123-compound, therefore, was the first of the 90°K plus superconductors to be discovered, and it has been the most thoroughly studied.

Structure of the 123-Compound

The structure of the 123-compound is related to that of an important class of minerals called *perovskites*. These minerals contain three oxygen atoms for every two metal atoms. The new compound, with six metal atoms in its unit cell, would thus be expected to have nine oxygens if it were an ideal perovskite. In fact, it has, in most samples, between 6.5 and 7 oxygens. In other words, by comparison to an ideal perovskite, about one quarter of the oxygens are missing. The unit cell of the 123-compound can be thought of as a pile of three cubes. Each cube has a metal atom at its center: barium in the bottom cube, yttrium in the middle one, and barium in the top one. At the corners of each cube, a copper would be surrounded by six oxygens in an octahedral arrangement linked at each oxygen in an ideal perovskite. Each barium and yttrium would then be surrounded by 12 oxygens. But X-ray and neutron diffraction studies have shown that the unit cell doesn't conform to this simple picture because certain oxygen positions are vacant. All oxygen positions in the horizontal plane containing yttrium are vacant. The other vacancies are located in the top and bottom Cu-O planes. Several studies of the unit-cell parameters indicate that the unit cell is orthorhombic with b slightly larger than a. The perpendicular square planar arrangements of copper and oxygen ions are linked together by sharing their corners forming "linear chains" along the b axis of the structure. Linking of horizontal CuO_4 results in two-dimensional puckered sheets in the a-b plane of the structure.

As shown in Fig. 12.6, there are two nonequivalent Cu sites: Cu(1) in the chains, and Cu(2) in the planes. Above each Cu(2) there is an O(4) site (in a Ba plane); and in the Cu(2) planes there are O(2) and O(3) sites along the a and b axes, respectively. The oxygen atom site in the chains is labeled O(1). It was believed, early on, that the occurrence of Cu-O chains was crucial for obtaining high transition temperatures, but the recent discovery of the new Bi and Tl layered oxides, which do not have chains, have shown that this is not likely to be the key.

The two copper-oxide layers can be considered as polymeric since the covalent character is in the same range as the carbon-fluoride bond in teflon. Thus, the 123-

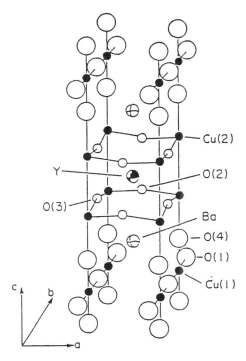

Figure 12.6   The structure of the 123-compound.

superconductive materials consist of two types of polymeric copper oxide layers held together by ionic bonding metals such as barium and yttrium. This theme of polymeric layers held together by ionic bonding to metals is often found in inorganic materials including many of the silicates and asbestos.

SUMMARY

1. Inorganic polymers are widely employed in the construction and building businesses, as abrasives and cutting materials, as fibers, as coatings, as lubricants, and as catalysts. They also serve as the basis of rocks and soils.

2. Portland cement is the least expensive, most widely used synthetic polymer. It has a complex (short-range order with long-range disorder), three-dimensional structure. A typical cement paste contains Portland cement, sand, gravel, and about 75% water.

3. There are many specialty cements including reinforced concrete, lightweight concrete, prestressed concrete, calcium-aluminate cement, magnesia cement, and gypsum.

4. Silicon dioxide is present in nature in both amorphous and crystalline forms.

5. Amorphous silicon dioxide–intensive materials include a wide variety of glasses such as fiberglass, window glasses, Vycor glasses, borosilicate glasses, lead (or heavy) glasses, colored glasses, glazes, and tempered safety glass. Glass can be shaped by drawing, pressing, casting, and blowing.

6. Silicon dioxide also exists in crystalline forms that form the basis of many rocks, sands, and soils.

7. Asbestos is a widely employed, silicon oxide–intensive group of minerals. While these materials offer good thermal stabilities, great strength, and good flexibility, certain ones are responsible for selected illnesses.

8. Polymeric carbon exists in two common forms—as diamonds and as graphite, the latter being thermodynamically more stable.

9. Diamonds are almost entirely pure carbon, where each carbon is at the center of a tetrahedron composed of four other carbon atoms. It is the hardest known natural material. Diamonds are commercially synthesized for industrial use as cutting and shaping agents.

10. Graphite occurs as sheets of hexagonally fused benzene rings. The bonds holding the fused benzene rings together are covalent bonds, while the bonding between the sheets results from the weak overlapping of pi-electron orbitals. Thus, many of the properties of graphite are anisotropic. The weak forces holding the graphite sheets together are responsible for its "slipperiness." Graphite is commercially made from charcoal or coke.

11. 123-superconductive materials contain two polymeric copper oxide layers held together by ionic bonding to metal atoms. This theme of polymeric layers held together by ionic bonding to metal ions is often found in inorganic materials such as silicates and asbestos.

GLOSSARY

alumina: Complex aluminum-containing silicate.

anisotropic: Dependent on direction; directionally dependent.

annealing: Subjecting materials to heat near their melting point.

asbestos: Grouping of silica-intensive materials containing aluminum and magnesium that gives soft, threadlike fibers.

asbestosis: Disease that blocks lung sacks with thick fibrous tissue.

borosilicate glass: Relatively heat-shock–resistant glass with a small coefficient of thermal expansion (e.g., Kimax and Pyrex).

calcium-aluminate cement: Contains more alumina than Portland cement.

Chrysotile: Most abundant and widely used type of asbestos.

colored glass (stained glass): Glass containing coloring agents such as metal salts and oxides.

concrete: Combination of cement, water, and filler material such as rocks and sand.

diamond: Polymeric carbon where the carbon atoms are at centers of tetrahedra composed of four other carbon atoms; hardest known natural material.

feldspars: Derivatives of silica where one half to one quarter of the silicon atoms are replaced by aluminum atoms.

fibrous glass (fiberglass): Fibers of drawn glass.

float glass: Glass made by cooling sheets of molten glass in a tank of molten tin; most common window glass is of this type.

glass: Inorganic product of fusion which has been cooled to a rigid condition without crystallization; most glasses are based on amorphous SiO_2.

glass sand: Impure quartz crystals.

glaze: Thin, transparent coatings fused on ceramic materials.

granite: Hard crystalline rock containing mainly quartz and feldspar.

graphite: Polymeric carbon consisting of sheets of hexagonally fused rings where the sheets are held together by weak overlapping pi-electron orbitals; anisotropic in behavior

gypsum: $CaSO_2 \cdot 2H_2O$; serves as the basis of plaster of Paris, Martin's cement, Keen's cement, and Parisian cement; shrinks very little on hardening; rapid crying.

High-T_c superconductors: Polymeric copper oxide–containing materials that exhibit superconductivity above the boiling temperature of liquid nitrogen.

inorganic polymers: Polymers containing no organic portions.

kaolinite: Important type of asbestos.

lead glasses (heavy glass): Glass where some or all of the calcium oxide is replaced by lead oxide.

lime: $CaCO_3$ from oyster shells, chalk, and marl.

magnesia cement: Composed mainly of magnesium oxide; rapid hardening.

optical fibers: Glass fibers coated with highly reflective polymer coatings; allows light entering one end of the fiber to pass through to the other end with little loss of energy.

piezoelectric materials: Materials that develop net electronic charges when pressure is applied; sliced quartz is piezoelectric.

Portland cement: Major three-dimensional inorganic polymer construction material consisting of calcium silicates, lime, and alumina.

precast concrete: Portland concrete cast and hardened prior to it being taken to the site of use.

prestressed concrete: Portland concrete cast about steel cables stretched by jacks.

quartz: Crystalline forms of silicon dioxide; basic material of many sands, soils, and rocks.

recipe: Description of content, sequence, and amounts of materials that, when followed, produces a desired material.

reinforced concrete: Portland concrete cast about steel rods or bars.

safety glass: Laminated glass; sandwich containing alternate layers of poly(vinyl butyral) and soda-lime glass.

sand: Loose grains of minerals or rocks larger than silt but smaller than gravel.

sandstone: Granular quartz.

silica: Based on SiO_2; finely ground sand.

silicon glass: Made by fusing pure quartz crystals or glass sand; high melting.

soda: Na_2O.

soda ash: Na_2CO_3.

soda-lime glass: Most common glass; based on silica, soda, and lime.

soil: Contains mineral and organic particles; majority by weight is sand.

tempered safety glass: Single piece of specially heat-treated glass.

tempering: Process of rapidly cooling glass resulting in an amorphous glass that is weaker but less brittle.

vitreous enamels: Thin, normally somewhat opaque-colored inorganic coatings fused on materials.

Vycor: 96% silicon glass; made from silicon and boron oxide; best variety is called fused quartz.

EXERCISES

1. What properties of glass correspond to those of organic polymers?

2. Why is Portland cement an attractive large-bulk building material?

3. Name five important synthetic inorganic polymers. Name five important natural inorganic polymers.

4. Describe what is meant by three-dimensional polymers. Name five important three-dimensional inorganic polymers. Name two general properties typical of three-dimensional polymers.

5. Why are specialty cements and concretes necessary?

6. What is meant by the comment that "glass is a supercooled liquid"?

7. Briefly describe techniques employed to shape glass.

8. Why are specialty glasses important in today's society?

9. Name two important inorganic fibers employed with resins to form useful materials. Describe briefly what fibers impart to the overall structure of these materials.

10. Which would you predict to be more brittle—quartz, fibrous glass, or window glass?

11. Briefly describe the piezoelectric effect.

12. We do not live a a risk-free society. Discuss this statement in light of asbestos.

13. What does anisotropic behavior mean? Why does graphite exhibit anisotropic behavior?

14. Briefly compare the structures of diamond and graphite.

BIBLIOGRAPHY

Adams, D. W. (1974): *Inorganic Solids*, Wiley, New York.

Clarke, J. L. (1993): *Alternative Materials for the Reinforcement and Prestressing of Concrete*, Routledge Chapman & Hall, London.

Dagani, R. (1990): C & En, January 1, 24.

Dagani, R. (1987): C & En, May 11, 8.

Deer, W., Howie, R., Zussman, J. (1982): *Rock-Forming Minerals*, Longman, New York.

Ellis, A. B. (1987): J. Chem. Ed., *64*(10):838.

Eitel, W. (1964): *Silicate Science*, Academic, New York.

Harris, D. C., Hills, M. E., Hewston, T. A. (1987): J. Chem. Ed., *64*(10):847.

Holliday, L. (1975): *Ionic Polymers*, Halsted Press, New York.

Klein, C., Hurlbut, C. (1985): *Manual of Mineralogy*, Wiley, New York.

MacGregor, E., Greenwood, C. (1980): *Polymers in Nature*, Wiley, New York.

Mineralogical Society of America, *Reviews in Mineralogy*, ongoing series, Mineralogical Society of America, Washington, D.C.

Ray, N. H. (1978): *Inorganic Polymers*, Academic, New York.

Saifullin, R. S. (1993): *Physical Chemistry of Inorganic Polymer and Composite Materials*, Prentice-Hall, Englewood Cliffs, New Jersey.

Tinkham, M. (1975): *Introduction to Superconductivity*, McGraw-Hill, New York.

Zoltai, T., Stout, J. (1984): *Mineralogy*, Burgess, Minneapolis, Minnesota.

13

Fillers and Reinforcements for Polymers

Some films, fibers, and plastics are used as unfilled polymers, but the strength and cost of most elastomers and plastic composites are dependent on the presence of appropriate fillers or reinforcements. Some rubber articles such as crepe rubber shoe soles, rubber bands, inner tubes, and balloons are unfilled. However, the tread stock in pneumatic tires would not be serviceable without the addition of carbon black or amorphous silica. For example, addition of these fillers increases the tensile strength of SBR from 100 to 4000 psi. Likewise, most high-performance plastics are composites of polymers reinforced by fibrous glass.

13.1 THEORY OF THE EFFECT OF FILLERS

Many naturally occurring functional materials, such as wood, bone, and feathers, are composites consisting of a continuous resinous phase and a discontinuous phase. The first synthetic plastics such as celluloid and Bakelite were also composites. Wood flour was used to reinforce the pioneer phenolic resins and is still used today for the same purpose.

According to the American Society for Testing and Materials standard ASTM-D-883, a filler is a relatively inert material added to a plastic to modify its strength, permanence, working properties, or other qualities or to lower costs, while a reinforced plastic is one with some strength properties greatly superior to those of the base resin resulting from the presence of high-strength fillers embedded in the composition. According to ASTM, plastic laminates are the most common and strongest type of reinforced plastics.

According to one widely accepted definition, fillers are comminuted spherical or spheroidal solids. Glass beads which meet the requirements of this definition are used to reduce mold wear and to improve the quality of molded parts. The word extender, sometimes used for fillers, is not always appropriate since some fillers are more expensive than the resin.

Current theories describing the action of spherical fillers in polymers are based on the Einstein equation shown below. Einstein showed that the viscosity of a viscous Newtonian fluid (η_0) was increased when small, rigid, noninteracting spheres were suspended in the liquid. According to the Einstein equation, the viscosity of the mixture (η) is related to the fractional volume (c) occupied by the spheres, and η is independent of the size of the spheres or the polarity of the liquid.

$$\eta = \eta_0(1 + 2.5c) \qquad \text{and} \qquad \eta_{sp}/c = 2.5 \qquad (13.1)$$

Providing that c is less than 0.1, good agreement with the Einstein equation is noted when glass spheres are suspended in ethylene glycol. Maximum packing of spheres (c = 90%) is attained when the composition equals 40% each of 20 and 325 mesh and 10% each of 35 and 100 mesh spheres.

The Einstein equation has been modified by including a hydrodynamic or crowding factor (β) which is equal to 1.35 and 1.91 for closely packed and loosely packed spheres, respectively. The modified Mooney equation shown below resembles the Einstein equation when $\beta = 0$.

$$\eta = \eta_0 \frac{2.5c}{1 - \beta c} \qquad \text{and} \qquad \eta_{rel}/c = 2.5 \qquad (13.2)$$

Many other empirical modifications of the Einstein equation have been made to predict actual viscosities of resinous composites. Since the modulus (M) is related to viscosity, these empirical equations such as the Einstein-Guth-Gold (EGG) equation [Eq. (13.3)], may be used to predict changes in modulus when spherical fillers are added.

$$M = M_0(1 + 2.5c - 14.1c^2) \qquad (13.3)$$

Since carbon black and amorphous silica tend to form clusters of spheres (graping effect), an additional modification of the Einstein equation has been made to account for the nonspherical shape or aspect ratio (1/D). This factor (f) is equal to the ratio of the length (l) to the diameter (D) of nonspherical particles:

$$\eta = \eta_0(1 + 0.67fc + 1.62f^2c^2) \qquad (13.4)$$

It is generally recognized that the segmental mobility of a polymer is reduced by the presence of a filler. Small particles with a diameter of less than 10 mm increase the cross-linked density, and active fillers increase the glass transition temperature (T_g) of the composite.

The high strength of composites is dependent on strong van der Waals interfacial forces. These forces are enhanced by the presence of polar functional groups on the polymer and by the treatment of filler surfaces with silanes, titanates, or other surface-active agents. Composites have been produced from almost every available polymer using almost every conceivable comminuted material as a filler.

13.2 FILLERS

Among the naturally occurring filler materials are cellulosics—such as wood flour, α-cellulose, shell flour, starch—and proteinaceous fillers such as soybean residues.

Approximately 40,000 tons of cellulosic fillers are used annually by the American polymer industry. Wood flour, which is produced by the attrition grinding of wood wastes, is used as a filler for phenolic resins, dark-colored urea resins, polyolefins, and PVC. Shell flour, which lacks the fibrous structure of wood flour, has been made by grinding walnut and peanut shells. It is used as a replacement for wood flour.

α-Cellulose, which is more fibrous than wood flour, is used as a filler for urea and melamine plastics. Melamine dishware is a laminated structure consisting of molded resin-impregnated paper. Presumably, the formaldehyde in these thermosetting resins reacts with the hydroxyl groups in cellulose to produce a more compatible composite. Starch and soybean derivatives are biodegradable, and the rate of disintegration of resin composites may be controlled by the amount of these fillers present.

As was discussed in Chap. 10, many incompatible polymers are added to increase the impact resistance of other polymers such as polystyrene. Other comminuted resins such as silicones or polyfluorocarbons are added to increase the lubricity of other plastics. For example, a hot melt dispersion of polytetrafluoroethylene in poly(phenylene sulfide) is used as a coating for antistick cookware.

Since cellulose acetate butyrate is compatible with uncured polyester resins, but incompatible with the cured resin, it is added to the premix to reduce shrinkage during curing. Finely divided polyethylene, which is also incompatible with polyester resins, is added to reduce surface roughness in the so-called low profile resin technique.

Carbon black, which was produced by the smoke impingement process by the Chinese over a thousand years ago, is now the most widely used filler for polymers. Much of the 1.5 million tons produced annually in the United States is used for the reinforcement of elastomers. The most widely used carbon black is furnace carbon black. The particle size of this carbon black is about 0.08 mm. Its hardness on the Mohs' scale is less than 1.

Carbon-filled polymers, especially those made from acetylene black, are fair conductors of heat and electricity. Polymers with fair conductivity have also been obtained by embedding carbon black in the surfaces of nylon or polyester filament reinforcements. The resistance of polyolefins to ultraviolet radiation is also improved by the incorporation of carbon black.

While glass spheres are classified as nonreinforcing fillers, the addition of 40 g of these spheres to 60 g of nylon-66 increases the flexural modulus, compressive strength, and melt index. The tensile strength, impact strength, creep resistance, and elongation of these composites are less than those of the unfilled nylon-66. A greater increase in flexural modulus is noted when the glass surface is altered by surface-active agents.

Milled glass fibers, multicellular glass nodules, and glass flakes have also been used as fillers. When added to resins, hollow glass and carbon spheres, called microballoons or microspheres, produce syntactic foams with varying specific gravities depending on the ratio of filler to resin. Superior high-performance composites are obtained when glass or graphite fibers are used as the additives.

Conductive composites are obtained when powdered metal fillers, metal flakes, or metal-plated fillers are added to resins. These composites have been used to produce forming tools for the aircraft industry and to overcome electromagnetic interference in office machines. Powdered lead-filled polyolefin composites have been used as shields for neutron and γ radiation. Zinc-filled plastics have been used as sacrificial composite electrodes, and magnetizable composites have been obtained by the incorporation of powdered aluminum-nickel alloys or barium ferrite.

Zinc oxide, which has a hardness of 2.5 on the Mohs' scale, is used to a large extent as an active filler in rubber and as a weatherability improver in polyolefins and polyesters. Anatase and rutile titanium dioxide are used as white pigments and as weatherability improvers in many polymers. Ground barites $(BaSO_4)$ yield X-ray–opaque plastics with controlled density.

The addition of finely divided calcined alumina, corundum, or silicon carbide produces abrasive composites. However, alumina trihydrate (ATH), which has a Mohs' hardness of less than 3, serves as a flame-retardant filler in plastics. Zirconia, zirconium silicate, and iron oxide, which have specific gravities greater than 4.5, are used to produce plastics with controlled high densities.

Calcium carbonate, which has a Mohs' hardness of 3, is available both as ground natural limestone and as synthetic chalk. This filler is widely used in paints, plastics, and elastomers. The volume relationship of calcium carbonate to resin or the pigment volume required to fill voids in the resin composite is called the pigment-volume concentration (PIVC). The critical PIVC (CPIVC) is the minimum required to satisfy the resin demand.

In contrast to the cubically shaped, naturally occurring calcium carbonate filler, the synthetic product produced by the addition of carbon dioxide to a slurry of calcium hydroxide is acicular, or needle shaped. Calcium carbonate is used at an annual rate of 700 million tons as a filler in PVC, polyolefins, polyurethane foams, and in epoxy and phenolic resins (see Table 13.1).

Silica, which has a specific gravity of 2.6, is used as naturally occurring and synthetic amorphous silica, as well as in the form of large crystalline particulates such as sand and quartz.

Diatomaceous earth, also called infusorial earth and fossil flour, is a finely divided amorphous silica consisting of the skeletons of diatoms. This filler has a Mohs' hardness of less than 1.5 in contrast to sharp silica sand, which has a value of 7. Diatomaceous earth is used to prevent a roll of film from sticking to itself (antiblocking) and to increase the compressive strength of polyurethane foams. Tripoli, or rotten stone, is porous silica formed by the decomposition of sandstone.

Pyrogenic, or fumed, silica is a finely divided filler obtained by reacting silicon tetrachloride in an atmosphere of hydrogen and oxygen. This filler is used as a thixotrope to increase the viscosity of liquid resins. Finely divided silicas are also produced by the acidification of sodium silicate solutions and by the evaporation of alcoholic solutions of silicic acid.

Sharp silica sand is used as a filler in resinous cement mortars. Reactive silica ash produced by burning rice hulls and a lamellar filler, novaculite, from the novaculite uplift in Arkansas, are also used as silica fillers in polymers.

Hydrated finely divided silicas which contain surface silanol groups are used for the reinforcement of elastomers. Extreme care must be taken in handling silica to prevent silicosis. Approximately 25 million tons of silica fillers are used annually by the polymer industry.

Table 13.1 Types of Fillers for Polymers

I. Organic materials
 A. Cellulosic products
  1. Wood products
   a. Kraft paper
   b. Chips
   c. Coarse flour
   d. Ground flour
    (1) Softwood flour
    (2) Hardwood flour
    (3) Shell flour
  2. Comminuted cellulose products
   a. Chopped paper
   b. Diced resin board
   c. Crepe paper
   d. Pulp preforms
  3. Fibers
   a. α-Cellulose
   b. Pulp preforms
   c. Cotton flock
   d. Textile byproducts
   e. Jute
   f. Sisal
   g. Rayon
 B. Lignin-type products
  1. Ground bark
  2. Processed lignin
 C. Synthetic fibers
  1. Polyamides (nylon)
  2. Polyesters (Dacron)
  3. Polyacrylonitrile (Orlon, Acrilan)
 D. Carbon
  1. Carbon black
   a. Channel black
   b. Furnace black
  2. Ground petroleum coke
  3. Graphite filaments
  4. Graphite whiskers
II. Inorganic materials
 A. Silica products
  1. Minerals
   a. Sand
   b. Quartz
   c. Tripoli
   d. Diatomaceous earth
  2. Synthetic materials
   a. Wet-processed silica
   b. Pyrogenic silica
   c. Silica aerogel
 B. Silicates
  1. Minerals
   a. Asbestos
    (1) Chrysotile
    (2) Amosite

    (3) Anthophyllite
    (4) Crocidolite
    (5) Tremolite
    (6) Actinolite
   b. Kaolinite (China clay)
   c. Mica
   d. Nepheline syenite
   e. Talc
   f. Wollastonite
  2. Synthetic products
   a. Calcium silicate
   b. Aluminum silicate
 C. Glass
  1. Glass flakes
  2. Solid glass spheres
  3. Hollow glass spheres
  4. Milled fibers
  5. Fibrous glass
   a. Filament
   b. Rovings
   c. Woven roving
   d. Yarn
   e. Mat
   f. Fabric
 D. Metals
 E. Boron filaments
 F. Metallic oxides
  1. Ground material
   a. Zinc oxide
   b. Alumina
   c. Magnesia
   d. Titania
  2. Whiskers
   a. Aluminum oxide (sapphire)
   b. Beryllium oxide
   c. Magnesium oxide
   d. Thorium oxide
   e. Zirconium oxide
 G. Calcium carbonate
  1. Chalk
  2. Limestone
  3. Precipitated calcium carbonate
 H. Polyfluorocarbons
 I. Other fillers
  1. Whiskers (nonoxide)
   a. Aluminum nitride
   b. Beryllium carbide
   c. Boron carbide
   d. Silicon carbide
   e. Silicon nitride
   f. Tungsten carbide
  2. Barium ferrite
  3. Barium sulfate

Both naturally occurring and synthetic silicates are also widely used as fillers. Hydrated aluminum silicate, or kaolin, has a specific gravity of 2.6 and a hardness value of 2.5 on the Mohs' scale. Kaolin and other clays may be dissolved in sulfuric acid and regenerated by the addition of sodium silicate. Clays are used as fillers in synthetic paper, rubber, and bituminous products.

Mica, which has a specific gravity of 2.8 and a Mohs' hardness value of 3, is a naturally occurring lamellar or platelike filler with an aspect ratio below 30. However, much higher aspect ratios are obtained by ultrasonic delamination.

Talc is a naturally occurring fibrouslike hydrated magnesium silicate with a Mohs' hardness of 1 and a specific gravity of 2.4. Since talc-filler polypropylene is much more resistant to heat than polypropylene (PP), it is used in automotive accessories subject to high temperatures. Over 40 million tons of talc are used annually as fillers.

Nepheline syenite, a naturally occurring sodium potassium aluminum silicate filler, and wallastonite, an acicular calcium metasilicate filler, are used for the reinforcement of many plastics. Fibers from molten rock or slag (PMF) are also used for reinforcing polymers.

Asbestos is a naturally occurring magnesium silicate which, in spite of its toxicity, has been used for over 250 years as a flame-resistant fiber. Approximately 180 million tons of asbestos are used annually as an additive for polymers, but this use is decreasing (see also Sec. 12.6).

13.3 REINFORCEMENTS

According to the ASTM definition, fillers are relatively inert while reinforcements improve the properties of plastics. Actually, few fillers are used that do not improve properties, but reinforcing fibers produce dramatic improvements in the physical properties of the composites. Many fibrous reinforcements are available, but most theories have been developed as a result of investigations of fibrous glass, which is the most widely used reinforcement for polymers.

That filaments could be produced from molten glass was known for centuries, but fibrous glass was not produced commercially until the mid-1930. Unlike the previously cited isotropic filler-resin composites, the stress in a fibrous glass–reinforced composite is concentrated at the fiber ends. Therefore, providing the strength of the fiber is greater than that of the resin, the properties of the composite are anisotropic and dependent on the direction of the stress.

The transverse modulus (M_T) and many other properties of a long fiber-resin composite may be estimated from the law of mixtures. The longitudinal modulus (M_L) may be estimated from the Kelly Tyson equation shown below; the longitudinal modulus is proportional to the sum of the fiber modulus (M_F) and the resin matrix modulus (M_M). Each modulus is based on the fractional volume (c). The constant k is equal to 1 for paralle continuous filaments and decreases for more randomly arranged shorter filaments.

$$M_L = kM_Fc_F + M_Mc_M \tag{13.5}$$

Since the contribution of the resin matrix is small in a strong composite, the second term in the Kelly Tyson equation may be disregarded. Thus, the longitudinal modulus is dependent on the reinforcement modulus, which is independent of the diameter of the reinforcing fiber. The full length of filament reinforcement is utilized in filament-wound and pultruded composites. In each case, the impregnated resin becomes part of the finished composite.

While unsuccessful attempts were made to use cotton threads as reinforcements, the first successful reinforcements were with fibrous glass. The latter may be spun from low-cost soda-lime, type E, or type C glass. In the spinning process, filaments are produced by passing the molten glass through orifices in the bushings.

In addition to the filament winding and pultrusion processes, chopped fibrous glass may be used as a reinforcement for spraying resin-impregnated chopped strand and for bulk molding compounds (BMC) and sheet molding compounds (SMC). The SMC is the most widely used form of molded fibrous glass resin composite.

In many processes, such as SMC, the fibers are not continuous. When they are longer than the critical length (l_c), it is necessary to modify the first term in the Kelly Tyson equation by multiplying by a factor $(1 - l_c)/2$ in which l is equal to the actual fiber length. The constant k approaches 0.5 for two-directionally oriented fibers.

Fibrous glass–reinforced resinous composites were introduced in 1940, and their use has increased steadily since then. Their use was originally confined to thermosetting resins, but fibrous glass–reinforced thermoplastics such as nylon are now important composites. Over 500,000 tons of fibrous glass–reinforced composites are produced annually in the United States.

Short, randomly oriented glass fibers and those from nylon, aramids, poly(vinyl alcohol), polyacrylonitrile, and polyesters have been used successfully in the preparation of strong composites. Since these organic fibers crystallize, they serve as nucleating agents when used to reinforce crystallizable polymers such as polypropylene.

Polyester resin–impregnated fibrous glass roving or mat is used for SMC and BMC. The former is used like a molding powder and the latter is hot pressed in the shape of the desired object, such as one half of a suitcase. Chopped fibrous glass roving may be impregnated with resin and sprayed, and glass mats may be impregnated with resin just prior to curing.

The strongest composites are made from continuous filaments impregnated with resin before curing. These continuous filaments are wound around a mandrel in the filament-winding process and gathered together and forced through an orifice in the pultrusion molding process.

The first continuous filaments were rayon, and these as well as polyacrylonitrile fibers have been pyrolyzed to produce graphite fiber. High-modulus reinforcing filaments have also been produced by the deposition of boron atoms from boron trichloride vapors on tungsten or graphite filaments.

Small single crystals, such as potassium titanate, are being used at an annual rate of over 10,000 tons for the reinforcement of nylon and other thermoplastics. These PMRN composites are replacing die-cast metals in many applications. Another microfiber, sodium hydroxycarbonate (called Dawsonite), also improved the physical properties and flame resistance of many polymers. Many other single crystals, called whiskers, such as alumina, chromia, and boron carbide, have been used for making high-performance composites.

13.4 COUPLING AGENTS

The first commerical glass filaments were protected from breakage by a starch sizing which was removed before use in resin composites. Since poly(vinyl acetate) sizing, now in use, is more compatible with resins, it does not have to be removed from the glass surface. However, the interfacial attraction between fibrous glass and resins is poor, and strong composites require the use of coupling agents to increase the interfacial bond between the resin and reinforcing agent.

The original coupling agents, which were called promotors, were used to assure a good bond between rubber and the carbon black filler. The first commercial promotors were N-4-dinitroso-N-methylaniline and N-(2-methyl-2-nitropropyl)-4-nitrosoaniline. Presumably, coupling took place between carbon black and the N-nitroso group and between the elastomer and the p-nitroso group.

These promotors increased the tensile strength, modulus, and bound rubber content of rubber. While natural rubber is soluble in benzene, it becomes less soluble when carbon black or amorphous silica is added. The insoluble mixture of filler and rubber is called bound rubber.

Some of the original promotors or coupling agents for fibrous glass and polyester resins were silanes. It is assumed that the trimethoxy groups in a coupling agent such as γ-mercaptopropyltrimethoxysilane couple with the silanol groups on the surface of fibrous glass and that the mercapto groups couple with the polymer.

Many silane zirconate and titanate coupling agents have been developed for the treatment of different reinforcing agents and fillers used with specific polymers. While it is the custom to treat the surface of the filler, coupling may also occur when the silane or titanate derivative is added to the resin or the mixture of resin and filler.

It is believed that the continuous resin matrix in a composite transfers applied stress to the reinforcing discontinuous phase through the interface between the two components. This interfacial attraction is often weakened in the presence of moisture, but a strong interfacial bond is maintained when coupling agents are used.

The interfacial bond between calcium carbonate fillers and resins has been improved by surface treatment with stearic acid. The bond between silica and resins has been strengthened by the addition of O-hydroxybenzyl alcohol or ethylene glycol.

Titanate coupling agents such as triisostearylisopropyl titanate (TTS) are effective in reducing energy requirements for processing mixtures of fillers and resins. It has been proposed that monoalkyl titanates form titanium oxide monomolecular layers on the filler surface and modify the surface energy so that the viscosity of the resin-filler mixture is reduced.

Thus, ferric oxide loadings as high as 90% in nylon-66 and calcium carbonate loadings of 70% in PP are possible when appropriate titanates are added. The melt flow of the filled PP is similar to that of the unfilled polymer, but the impact strength of the composite is much higher than that of PP itself.

13.5 COMPOSITES

Composites (see also Sec. 17.5) are materials that contain strong fibers embedded in a continuous phase. The fibers are called ''reinforcement'' fibers and the contin-

Table 13.2 Fibers Frequently Employed in Composites

Alumina (aluminum oxide)	Polyolefin
Aromatic nylons	Silicon nitride (Si_3N_4)
Boron	Titanium carbide (TiC)
Carbon and graphite	Tungsten carbide (WC)
Glass	Zirconia (ZrO_2)

uous phase is called the matrix. While the continuous phase can be a metallic alloy or inorganic material, the continuous phase is typically an organic polymer that is termed a "resin." Because of the use of new fibers and technology, most of the composites discussed in this section are referred to as "space-age" and "advanced materials" composites. Composites can be fabricated into almost any shape, and after hardening, they can be machined, painted, etc., as desired.

Tables 13.2 and 13.3 contain a partial listing of the main materials employed in the fabrication of composites. It is important to note that many of the entries represent whole families of materials. Thus, there are many possible combinations, but not all combinations perform in a satisfactory manner. Generally good adhesion between the matrix and fiber is needed. Table 13.4 contains a listing of some of the more utilized combinations.

While there is a lot of science and space-age technology involved in the construction of composites, many composites have been formulated through a combination of this science and trial and error, giving recipes that contain the nature and form of the fiber and matrix materials, amounts, additives, and processing conditions.

Composites have high tensile strengths (on the order of thousands of MPa), high Young's modulus (on the order of hundreds of GPa), and good resistance to weathering exceeding the bulk properties of most metals. The resinous matrix, by itself, is typically not particularly strong relative to the composite. Further, the overall strength of a single fiber is low. In combination, the matrix-fiber composite becomes strong. The resin acts as a transfer agent, transferring and distributing applied stresses to the fibers. Generally, the fibers should have aspect ratios (ratio of length to diameter) exceeding 100, often much larger. Most fibers are thin (less

Table 13.3 Polymer Resins Employed in the Fabrication of Composites

Thermosets	Thermoplastics
Epoxys	Nylons
Melamine-formaldehyde	Polycarbonates
Phenol-formaldehyde	Poly(ether ether ketone)
Polybenzimidazoles	Poly(ether ketone)
Polyimides	Poly(ether sulfones)
Polyesters	Poly(phenylene sulfide)
Silicones	

Table 13.4 Typically Employed Fiber/Resin Pairs

Alumina/Epoxy	Carbon/Nylon/Epoxy
Alumina/Polyimide	Carbon/Polyimides
Boron/Carbon/Epoxy	Glass/Epoxy
Boron/Epoxy	Glass/Carbon/Polyester
Boron/Polyimide	Glass/Polyester
Boron/Carbon/Epoxy	Glass/Polyimide
Carbon/Acrylic	Glass/Silicon
Carbon/Epoxy	Nylon/Epoxy

than 20 μm thick, about a tenth the thickness of a human hair). Fibers should have a high tensile strength and most have a high stiffness, i.e., low strain for high stress or little elongation as high forces are applied.

There exists a relationship between the ideal length of a fiber and the amount of adhesion between the matrix and the fiber. For instance, assume that only the tip, one end, of a fiber is placed in a resin (Fig. 13.1, top). The fiber is pulled. The adhesion is insufficient to hold the fiber, and it is pulled from the resin. The experiment is repeated until the fiber is broken (outside the matrix) rather than being pulled (without breaking) from the resin (Fig. 13.1, bottom). Somewhere between the two extremes, there is a length where there exists a balance between the strength of the fiber and the adhesion between the fiber and matrix. Most modern composites utilize fiber/matrix combinations that exploit this balance.

Fiber failure is usually of the catastrophic type, where the failure is sudden. This is typical in polymeric materials where the material is broken at the weak link.

Fibers

Most (up to 98%) of the fibers employed in composites today are of three general varieties—glass (Sec. 12.4), carbon (graphite; Sec. 5.2), and aromatic nylons (often referred to as aramids; Sec. 7.6). Asbestos, a major fiber of choice years ago, holds less than 1% of the fiber-composite market, with even this small amount on the decrease.

Glass

Fiberglass is manufactured from a number of materials that are largely composed of silicon dioxide that are cooled below their melting points (supercooled liquids) without crystallizing (Sec. 12.4). Other oxides are included giving glass fibers with differing characteristics. Table 13.5 contains a brief description of the most important glass fiber types.

The glass fibers are "pulled" from the melted glass, forming fibers that typically range from 2 to 25 μm in diameter. This pulling acts to orient the overall three-dimensional structure, producing a material with greater strength and stiffness along the axis of the pull.

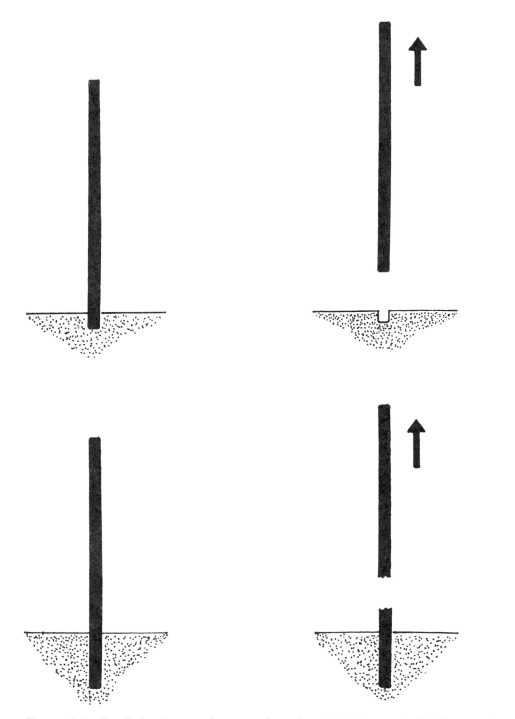

Figure 13.1 Tensile loading experiments performed on single fibers embedded in a matrix. The illustrations on the left are preapplication of the tensile loading, and those on the right are postapplication of the tensile loading.

Table 13.5 Types of Glass Fibers

Designation	General properties
C-glass	Chemical resistant
E-glass	"Typical" glass fiber
R-glass and S-glass	Stiffer and stronger than E-glass

As with other three-dimensional materials, the limits of the strength are due to the presence of voids. In the case of glass fibers, these voids largely occur on the surface of the fiber. Interestingly, adherence of water moisture, grease from handling, or other foreign materials can severely reduce the strength of the fibers. Thus, great care is taken to protect the surface by application of surface agents such as methacrylatochromic chloride, vinyl trichlorosilanes, and other silanes (Sec. 11.3). It is believed that these surface agents chemically react with the fiber surface, acting to repel and protect the surface from harmful agents such as water.

Carbon

Carbon can be made in many allotropic forms ranging from diamond (Sec. 12.4), graphite (Sec. 12.8), carbon black, and as high strength fibers and whiskers known as carbon and graphite fibers (Sec. 5.2). Carbon whiskers are sheets of hexagonal carbon atoms layered like a laminate, one on top of the other in an ordered array. Leslie Phillips, one of the inventors of carbon fibers, describes them as bundles of oriented crystalline carbon held in a matrix of amorphous carbon (Fig. 13.2). Order is a key to the strength of these allotropic carbon forms.

As in glass fibers, carbon voids are present in the carbon fibers. The carbon fibers are treated by application of a surface treatment such as low molecular weight epoxy resin. Additionally, carbon fibers are also normally treated to improve adhesion to the resin.

Aromatic Nylons

Two general types of aromatic nylon fibers are produced. The less stiff variety is utilized in situations where flexibility is required. The second type is stiffer with a higher Young's modulus and is employed for applications where greater strength

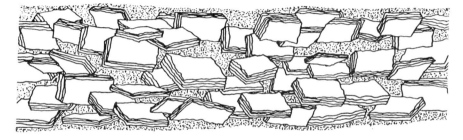

Figure 13.2 Idealized drawing of crystalline carbon bundles in a "sea of amorphous carbon."

is required and little flexibility is needed. Poor adhesion to the resin is desirable in some applications of aromatic nylon materials such as in body-armor (bullet-resistant material), where the "delamination" is a useful mode for absorbing an impact.

Boron

Boron fibers (Sec. 17.5) are larger in diameter (often 50 to 150 μm) than many of the other fibers employed in composite manufacture. They are made from the high-temperature reduction of boron trichloride.

$$2\ BCl_3 + 3\ H_2 \rightarrow 2\ B + 6\ HCl$$

Polyolefins

The flow properties and consequences of flow in polymers are becoming better understood. It is known that elongational flow through orifices can result in the stretching of polymer chains. Such polymer chains can become entangled, bringing about additional orientation as they flow. Finally, polymer solutions may be stable at rest, but under high rates of extrusion they may be removed from solution, forming a gel phase. These phenomena have allowed the production of a number of new polyolefin fibers, with an emphasis on the production of new polyethylene fibers (Sec. 8.7). Ultrahigh-modulus polyethylene (UHMPE) fibers have been produced which have a low density (about 0.96 g/cc) with a tensile strength of 1.5 GPa and a Young's modulus of 70 GPa. They have a higher elongation at break, over two times greater than fiberglass and aromatic nylons.

Alumina

As in the case of carbon, aluminum oxide, or alumina, is allotropic, existing is different forms. Polycrystalline alumina produces fibers. Two commercially available forms have about 95% aluminum oxide and 5% silica. Both have a diameter of about 3 μm. One is referred to as low-density alumina fiber exhibiting a tensile strength of 2 GPa and a Young's modulus of 200 GPa with a use temperature to about 900°C. The higher-density material has a use temperature to 1600°C with a similar tensile strength and Young's modulus.

Single fibers can be spun from a single aluminum oxide crystal. Alumina whiskers (Sec. 5.2) are elongated crystals of alpha-alumina that are stable to temperatures of 1650°C for about 2 hours, allowing their use in ceramic sintering and firing and in metal casting processes. Such whiskers have aspect ratios to only about 10.

Resins

Resin systems for composites can be divided into thermosetting and thermoplastic resins (Table 13.3). Most of the actual resin formations are complex, requiring many additives and supplemental treatments. Following is a brief discussion of the main resin components.

Thermosets

Polyimides.   The formation of polyimide resins (Sec. 7.7) typically involves two reaction steps. First, a polyamic acid is formed through reaction of a diamine and dianhydride. Cross-linking then occurs, generally through application of heat, with the loss of water accompanying the cyclization of the polyamic acid forming the thermoset polyimide resin. Other reaction systems are similarly employed in the formation of polyimide resins. An important subgroup is the maleimide resins formed from bismaleimides. Polymerization occurs through reaction with the double bonds of the maleimides.

Like the polyimides themselves, the polyimide resins are known for their thermal stability.

Unsaturated polyesters. Typically, the precured (pre–cross-linked) polyesters are formed by reaction of an anhydride (usually maleic anhydride) or diacid with diols such as ethylene and propylene glycol (Sec. 7.5). These polyesters contain sites of unsaturation that are subsequently employed for the purposes of introducing cross-links. Cross-linking can be accomplished through heating or through introduction of free-radical initiators utilizing a comonomer such as a diallyl phthalate or styrene.

Epoxys. Most epoxy resins (Sec. 7.11) employed in composites are derived from the epichlorhydrin with aromatic amines and phenols. An intermediate chlorhydrin is formed that cyclizes forming an epoxy resin containing epoxy or glycidyl groups. These are then reacted with amines, anhydrides, carboxylic acids, and phenols resulting in an addition reaction that does not produce side products resulting in a three-dimensional network with few voids.

Phenol-formaldehyde and amino-formaldehyde. The phenol-formaldehyde resins (Sec. 7.14) and the amino-formaldehyde resins (Sec. 7.14) are among the oldest resin systems. Reaction of formaldehyde with more than equimolar amounts of phenol, in the presence of acid, gives pre-resins called novolacs that contain mainly aromatic phenol units connected by methylene bridges. These are subsequently cross-linked through reaction with reactants such as hexamethylenetetramine.

Resole prepolymers are made by reaction of formaldehyde with less than an equimolar amount of phenol, in the presence of base, forming hydroxymethyl phenols. These prepolymers are less viscous in comparison with the novolac prepolymers, allowing their use in fiber situations where high flow is necessary for a good fiber-matrix mixing. Cross-linking occurs through heating with the elimination of water and formaldehyde from the reaction of hydroxymethyl groups forming "methylene" cross-links.

Thermoplastics

Thermoplastic resins typically use simply the polymers (Table 13.3) involved in the fabrication of the composites. Some of these are only soluble in liquids that are difficult to remove because they are dipolar and high-boiling. Further, many of the polymers have high melting ranges requiring high temperatures and pressures for processing. Finally, many of these polymers are rigid and adhere to the resin phase only with difficulty. In spite of these difficulties, these thermoplastic resins can offer good storage life and good damage tolerance.

Applications

Many of the applications for composite materials involve their (relative) light-weight, resistance to weathering and chemicals, and ability to be easily fabricated and machined. Bulk applications employ composites that are relatively inexpensive. Combinations of rigorous specifications, low volume, specific machining and fabrication specifications, and comparable price to alternative materials and solutions allow more expensive specialized composites to be developed and utilized. Applications are increasing but can be roughly divided into seven areas.

Marine Craft

One of the largest and oldest applications of composites is in the construction of watergoing vessels from rowboats, sailboats, racing boats, and motor craft to large seagoing ships.

The use of fresh and salt water–resistant composites allowed the boating industry to grow from almost a "mom-and-pop" operation to the use of large boatyards producing craft in an assembly line–like fashion. Most boats are composed of fiberglass and fiberglass/carbon combination composites. Other fibers are also being utilized in the boating industry, including aromatic nylons and fiberglass/aromatic nylon combinations.

Outer Space

Because of the large amount of fuel required to propel spacecraft into outer space, weight reduction, offered by composites, is an essential ingredient in the construction of materials utilized in outer space.

Some of the solid propellant tanks are made from composites. The tanks are often composed of fiberglass and glass/carbon fiber–containing composites. In fact, the development of S-glass (Table 13.5) was a result of the space effort. Carbon fibers are employed in the construction of some of the nose fairings.

The reusable space shuttles contain various composites. The cargobay doors are sandwich composites composed of carbon/epoxy/honeycomb materials. The "manipulator arm" used for loading the payload bay is composed of a number of composites including carbon/epoxy composite laminates and aromatic nylon laminates and sandwich materials. Composites are also used for the construction and mounting of mirrors, telescopes, solar panels, and antennae reflectors.

Biomaterials

Bones and skin are relatively light compared with metals. Composite structures can approach the densities of bone and skin and offer necessary inertness and strength to act as body-part substitutes.

Power-assisted arms have been made by placing hot-form strips of closed-cell PE foam over the cast of an arm. Grooves are cut into these strips prior to application and carbon/resin added to the grooves. The resulting product is strong and light, and the cushioned PE strips soften the attachment site of the arm to the living body. Artificial legs can be fashioned in glass/polyester and filled with polyurethane foam adding strength to the thin-shelled glass/polyester shell. Artificial legs are also made from carbon/epoxy composite materials. Some of these contain a strong interior core with a soft, flexible, skin.

Carbon/epoxy plates are now used in bone surgery, replacing the titanium plates previously employed. Usually a layer of connective tissue forms about the composite plate.

Rejection of composite materials typically does not occur, but as is the case of all biomaterials, compatibility is a major factor. Often, lack of biocompatibility has been found to be the result of impurities (often additives) found in the materials. Removal of these impurities allows the these materials to be used.

Sports

Carbon and carbon/glass composites are being used to make advanced-material fishing rods, bicycle frames, golf clubs, baseball bats, racquets, skis and ski poles, basketball backboards, etc. These come in one color—black—because the carbon fibers are black. Even so, they can be coated with about any color desired.

Automobiles

Composites are being employed in a number of automotive applications. These include racing car bodies as well as "regular" automobiles. Most automobiles have the lower exterior panels composed of rubbery and/or composite materials. Other parts such as drive shafts and leaf springs, antennas, and bumpers in private cars and heavy trucks are being made from composite materials.

Industry

Industrial storage vessels, pipes, reaction vessels, and pumps are now made from composite materials. They offer needed resistance to corrosion, acids and bases, oils and gases, and salt solutions and the necessary strength and ease of fabrication to allow their continued adoption as a major industrial building material.

Aerospace

The Gulf War spotlighted the use of composite materials in new-age aircraft. The bodies of both the Stealth fighter and bomber are mainly carbon composites. The versatility is apparent when one realizes that the Gossamer Albatross, the first plane to cross the English Channel with human power, consisted largely of composite materials, including a carbon/epoxy and aromatic nylon composite body and propellers containing a carbon composite core.

The increase in use of composite materials by the aerospace industry is generally due to their outstanding strength and resistance to weathering and fraction and their light weight, allowing fuel reduction savings. Succeeding families of Boeing aircraft have used ever greater amounts of fiberglass composite material in their manufacture, from about 20 square yards for the 707, to 200 square yards for the 727, to 300 square yards for the 737 and over 1000 square yards for the 747. This amount is increased again in the Boeing 767 and includes other structural applications of other space-age composites. Thus, the Boeing 767 uses carbon/aromatic nylon/epoxy landing gear doors and wing-to-body fairings.

As noted above, most applications of composites in the aerospace industry involve decreased weight, resulting in increased (potential) payloads and decreased fuel consumption. Interestingly, the lack of even limited flexibility for many composites limits their use (currently) for large commercial aircraft where normal wing "flapping" amplitudes may be several feet during a flight. Even so, composite use is increasing in the construction of small aircraft such as the McDonnell Douglas F-18, where roughly 50% of the outer body surface is composite, with the remainder being largely a mixture of titanium, aluminum, and steel.

Other areas of increased composite use include helicopter blades (giving about a twofold increase in life expectancy compared to metal blades) and jet engines (e.g., turbo fans, cowling, container rings).

SUMMARY

1. According to Einstein's equation, spherical fillers of any size increase the viscosity of a fluid in accordance with the partial volume of the filler. The Einstein equation, which relates to moduli of composites, has been modified to account for filler aggregates and nonspherically shaped fillers.

2. Many comminuted materials, such as wood flour, shell flour, α-cellulose, starch, synthetic polymers, carbon black, glass spheres and flakes, powdered metals, metallic oxides, calcium carbonate, silica, mica, talc, clay, and asbestos, have been used as fillers in polymers.

3. When added to polymers, fibrous reinforcements are much more effective than spherical fillers in improving strength properties. The most widely used reinforcing fiber is fibrous glass. Its effect and that of other reinforcing fibers is dependent on fiber length and the interfacial bond between it and the continuous resin matrix.

4. Fibrous glass may be used in the form of resin-impregnated mat or roving in the SMC and BMC processes. Resin-impregnated glass filament may be wound on a mandrel of forced through an orifice and cured in the filament-winding and pultrusion processes, respectively.

5. Many sophisticated reinforcements based on graphite fibers, aramid fibers, boron-coated tungsten, and single crystals, such as those of potassium titanate and sapphire, have been used as the discontinuous phase in high-performance composites.

6. The interfacial bond between the resin and fibrous glass, as well as other fillers, may be improved by surface treatment with surface-active agents such as silanes and titanates. These coupling agents not only improve physical properties of the composite but are effective in reducing the energy required in processing.

GLOSSARY

acicular: Needle shaped.

α-Cellulose: Cellulose insoluble in 17.5% NaOH.

anisotopic: Properties vary with direction.

aramid: Nylon produced from aromatic reactants.

asbestos: Fibrous magnesium silicate.

aspect ratio: Ratio of length to diameter of particles.

ASTM: American Society for Testing and Materials.

ATH: Alumina trihydrate.

barites: Barium sulfate.

β: Hydrodynamic or crowding factor.

blocking: The sticking of sheets of film to each other.

BMC: Bulk molding compound; resin-impregnated bundles of fibers.

bound rubber: Rubber adsorbed on carbon black which is insoluble in benzene.

c: Fractional volume occupied by a filler.

carbon black: Finely divided carbon made by the incomplete combustion of hydrocarbons.

comminuted: Finely divided.

composites: Materials that contain strong fibers embedded in a continuous phase called a matrix or resin.

continuous phase: The resin in a composite.

coupling agents: Products such as silanes or organic titanates which improve the interfacial bond between filler and resin.

CPIVC: Critical pigment volume concentration.

D: diameter.

diatomaceous earth: Siliceous skeletons of diatoms.

discontinuous phase: The discrete filler additive in a composite.

EGG: Einstein-Guth-Gold equation: $M = M_0(1 + 2.5c + 14.1c^2)$.

Einstein equation: $\eta = \eta_0(1 + 2.5c)$.

η: Coefficient of viscosity of a mixture of a solid and liquid.

η_0: Coefficient of viscosity of a liquid.

extender: A term sometimes applied to an inexpensive filler.

f: Aspect ratio (l/D).

Fiberglass: Trade name for fibrous glass.

fibrous filler: One in which the aspect ratio is at least 150:1.

fibrous glass: Filaments made from molten glass.

filament winding: A process in which resin-impregnated continuous filaments are wound on a mandrel and the composite is cured.

filler: Usually a relatively inert material used as the discontinuous phase of a resinous composite.

Fuller's earth: Diatomaceous earth.

graphite fibers: Fibers made by the pyrolysis of polyacrylonitrile fibers.

isotropic: Identical properties in all directions.

kaolin: Clay.

Kelly Tyson equation: $M_L = kM_FC_F + M_MC_M$.

l_c: Critical fiber length.

lamellar: Sheetlike.

laminate: A composite consisting of layers adhered by a resin.

low profile resins: Finely divided incompatible resins that make a rough surface smooth.

lubricity: Slipperiness.

M: Modulus of a composite.

M_0: Modulus of an unfilled resin.

M_F: Modulus of a fiber.

M_L: Longitudinal modulus.

M_M: Modulus of resin matrix.

M_T: Transverse modulus.

mesh size: Size of screens used to classify finely divided solids.

mica: Naturally occurring lamellar silicate.

microballoons: Hollow glass spheres.

microspheres: Hollow carbon spheres.

Mohs' scale: A scale of hardness from 1 for talc and 10 for diamonds.

Mooney equation: $\eta = \eta_0(2.5c/1 - \beta c)$.

novaculite: Finely ground quartzite rock.

PMRN: Composites reinforced by potassium titanate.

PP: Polypropylene.

promotor: Coupling agent.

pultrusion: A process in which bundles of resin-impregnated filaments are passed through an orifice and cured.

PIVC: Pigment-volume concentration.

reinforced plastic: A composite whose additional strength is dependent on a fibrous additive.

roving: A bundle of untwisted strands.

silanes: Silicon compounds corresponding to alkanes.

silanol group: SiOH.

SMC: Sheet molding compound; resin-impregnated mat.

strand: A bundle of filaments.

syenite: Igenous rock similar to feldspar.

syntactic foam: Composite of resin and hollow spheres.

talc: Naturally occurring hydrated magnesium silicate.

thixotrope: An additive which yields thixotropic liquids.

tripoli: Rotten stone; porous decomposed sandstone.

TTS: Triisostearylisopropyl titanate.

type C glass: Acid-resistant glass.

type E glass: Electrical grade glass.

van der Waals forces: Intermolecular attractions.

wallastonite: Acicular calcium metasilicate.

whiskers: Single crystals used as reinforcements.

wood flour: Attrition-ground, slightly fibrous wood particles.

EXERCISES

1. Name three unfilled polymers.

2. What is the continuous phase in wood?

3. What filler is used in Bakelite?

4. Name three laminated plastics.

5. How would you change a glass sphere from an extender to a reinforcing filler?

6. If one stirs a 5-ml volume of glass beads in 1 liter of glycerol, which will have the higher viscosity, small or large beads?

7. When used in equal volume, which will have the higher viscosity: (a) a suspension of loosely packed spheres, or (b) a suspension of tightly packed spheres?

8. Why is the segmental mobility of a polymer reduced by the presence of a filler?

9. What effect does a filler have on T_g?

10. Which would yield the stronger composite: (a) peanut shell flour, or (b) wood flour?

11. What is the advantage and disadvantage, if any, of α-cellulose over wood flour?

12. What filler is used in decorative laminates such as Formica table tops?

13. Which is the filler (discontinuous phase) and which is the resin (continuous phase) in a cookware coating produced from (a) polytetrafluoroethylene (Teflon), and (b) poly(phenylene sulfide) (Ryton)?

14. Would finely divided polystyrene make a good low-profile resin for reinforced polyesters?

15. How would you make a conductive syntactic foam?

16. Which would be stronger: (a) a chair made from polypropylene, or (b) one of equal weight made from cellular polypropylene?

17. What advantage would a barium ferrite–filled PVC strip have over an iron magnet?

18. How would you make X-ray–opaque PVC?

19. How would you make an abrasive foam from polyurethane?

20. How could you explain the flame-retardant qualities of ATH?

21. How would you explain the improvement in strength of composites with stearic acid–treated calcium carbonate?

22. How could you justify the high cost of pyrogenic silica?

23. In what elastomer would hydrated silica be most effective as a filler?

24. Why is a good interfacial bond between the filler surface and the resin essential?

25. Providing the volumes of the fibers are similar, which will yield the stronger composite: fibers with (a) small, or (b) large cross sections?

26. How would you make a strong, corrosion-resistant pipe?

27. What is the advantage of BMC and SMC over hand lay-up techniques such as those used in boat building?

28. What technique would you use to incorporate large amounts of filler (e.g., 70% calcium carbonate) in a resin?

BIBLIOGRAPHY

Jones, F. R. (1994): *Handbook of Polymer-Fibre Compositer*, Halsted Press, New York.

Katz, H. S., Milewski, J. (1978): *Handbook of Plastic Fillers and Reinforcements*, Van Nostrand-Reinhold, New York.

Langley, M. (1973): *Carbon Fibers in Engineering*, McGraw-Hill, New York.

Manson, J. H., Sperling, L. H. (1975): *Polymer Blends and Composites*, Plenum, New York.

Nielson, L., Landel, X. (1993): *Mechanical Properties of Polymers and Composites*, Marcel Dekker, New York.

Paul, D. R., Sperling, L. H. (1986): *Multicomponent Polymer Materials*, ACS, Washington, D.C.

Platzer, N. A. (1975): *Copolymers, Polyblends, and Composites*, ACS, Washington, D.C.

Serafini, T. T. (1987): *High Temperature Polymer Matrix Composites*, Noyes, Park Ridge, New Jersey.

Seymour, R. B. (1975): Plastics composites, in *Polymer Plastics Technology and Engineering* (L. Natureman, ed.), Marcel Dekker, New York.

————. (1976): Additives for plastics—fillers and reinforcements, Plast. Eng., *32*(8):29.

————. (1976): Fibrous reinforcements, Modern Plastics, *53*(10A):169.

————. (1976): Fillers for plastics, Modern Plastics, *53*(10A):172.

————. (1976): The role of fillers and reinforcements in plastics technology, Polym. Plast. Technol. Eng., *7*(1):49.

Summerscales J., Short, D. (1988): *Fibre Reinforced Polymers*, Technomic, Lancaster, Pennsylvania.

Sweeney, F. M. (1988): *Polymer Blands and Alloys*, Technomic, Lancaster, Pennsylvania.

14

Plasticizers, Stabilizers, Flame Retardants, and Other Additives

Additives are added to modify properties, assist in processing, and introduce new properties to a material. Additives can be added as solids, liquids, or gasses. They are usually added mechanically (with or without subsequent chemical reaction), during formulation, or as the material is in the fluid state. Many additives have become parts of general formulations developed as much as an art as a science. Thus, a general paint formulation for a water-based paint has titanium dioxide as the white pigment, China clay as an extender, a fungicide, defoaming agent, a coalescing liquid, a surfactant-dispersing agent, and calcium carbonate as the extender. A typical tire tread recipe has a processing aid, accelerator activator, antioxidant, antiozonate, softener, finishing aid, retarder, vulcanizing agent, and accelerator as additives.

Typical additives include:

Antiblocking agents
Antifoaming agents
Antifogging agents
Antimicrobial agents
Antioxidants
Antistatic agents
Blowing agents
Coloring agents
Coupling agents
Curing agents
Fillers
Flame retardants
Foaming agents

Impact modifiers
Low profile materials
Lubricants
Mold release agents
Odorants or fragrances
Plasticizers
Preservatives
Reinforcements
Slip agents
Stabilizers, including
        Radiation (UV/VIS)
        Heat
Viscosity modifiers
        Flow enhancers
        Thickening agents
        Antisag materials

While the modulus of polymers is usually increased by the addition of fillers and reinforcements, it may be decreased by the addition of moderate amounts of plasticizers. Other essential additives such as antioxidants, heat stabilizers, ultraviolet stabilizers, and flame retardants may reduce the modulus and other physical properties. Thus, it may be necessary to add reinforcing agents to counteract the weakening effect of some other additives. Nevertheless, additives are essential functional ingredients of polymers, and whenever possible, each should be used in optimum amounts for the attainment of high-quality products.

14.1 PLASTICIZERS

According to the ASTM-D-883 definition, a plasticizer is a material incorporated in a plastic to increase its workability and flexibility or distensibility. The addition of a plasticizer may lower the melt viscosity, elastic modulus, and glass transition temperature (T_g) of a plastic. Thus, the utility of cellulose nitrate (CN) produced by Schonbein in 1846 was limited until Parkes added castor oil to CN in 1865 and Hyatt added camphor to plasticize CN in 1870. Another plasticizer, tricresyl phosphate (TCP), was used to replace part of the camphor and reduce the flammability of celluloid in 1910.

Waldo Semon patented the use of tricresyl phosphate as a plasticizer for poly(vinyl chloride) (PVC) in 1933. This was later replaced by the less toxic di-2-ethylhexyl phthalate (DOP), which is now the most widely used plasticizer. The annual worldwide production of plasticizers is 3.2 million tons, and the U.S. production is in excess of 1 million tons. In fact, plasticizers are major components of a number of polymer-containing products. For instance, the adhesive in automobile safety glass is typically composed mainly of poly(vinyl butyral) and about 30% plasticizer.

The effect of plasticizers may be explained by the lubricity, gel, and free volume theories. The first states that the plasticizer acts as an internal lubricant and permits the polymer chains to slip by each other. The gel theory, which is applicable to amorphous polymers, assumes that a polymer such as PVC has many intermolecular attractions which are weakened by the presence of a plasticizer such as

DOP. It is assumed that the addition of a plasticizer increases the free volume of a polymer and that the free volume is identical for polymers at T_g.

Since plasticizers are essentially nonvolatile solvents, compatibility requires that the difference in the solubility parameter of the plasticizer and polymer ($\Delta\delta$) be less than 1.8 H (see Table 14.1). It is of interest to note that δ for PVC is 9.66 H and for DOP is 8.85 H. Thus, $\Delta\delta = 0.81$ H for this widely used plasticizer-resin system. When present in small amounts, plasticizers generally act as antiplasticizers, i.e., they increase the hardness and decrease the elongation of polymers.

Inefficient plasticizers require relatively large amounts of these additives to overcome the initial antiplasticization. However, good plasticizers such as DOP change from antiplasticizers to plasticizers when less than 10% of the plasticizer is added to PVC.

The development of plasticizers has been plagued with toxicity problems. Thus, the use of highly toxic polychlorinated biphenyls (PCB) has been discontinued. Phthalic acid esters, such as DOP, may be extracted from blood stored in plasticized PVC blood bags and tubing. These aromatic esters are also distilled from PVC upholstery in closed automobiles in hot weather. These problems have been solved by using oligomeric polyesters as nonmigrating plasticizers, instead of DOP.

Table 14.1   Solubility Parameters of Typical Plasticizers

Plasticizer	δ Solubility parameter (H)
Paraffinic oils	7.5
Dioctyl phthalate	7.9
Dibutoxyethyl phthalate	8.0
Tricresyl phosphate	8.4
Dioctyl sebacate	8.6
Triphenyl phosphate	8.6
Chlorinated biphenyl (Arochlor 1248)	8.8
Dihexyl phthalate	8.9
Hydrogenated terphenyl (HB-40)	9.0
Dibutyl sebacate	9.2
Dibutyl phthalate	9.3
Dipropyl phthalate	9.7
Diethyl phthalate	10.0
Dimethyl phthalate	10.7
Santicizer 8	11.9
Glycerol	16.5

Many copolymers, such as poly(vinyl chloride-co-vinyl acetate), are internally plasticized because of the flexibilization brought about by the change in structure of the polymer chain. In contrast, DOP and TCP are said to be external plasticizers. The presence of bulky groups on the polymer chain increases segmental motion. Thus, the flexibility increases as the size of the pendant group increases. However, linear bulky groups with more than 10 carbon atoms will reduce flexibility because of side-chain crystallization when the groups are regularly spaced.

Water is a widely utilized plasticizer in nature, permitting flexibility of much of the human body as well as the "bendability" of flowers, leaves, tree branches, and so on. Fats and many proteins also act as plasticizers in animals.

Plasticizer containment still remains a major problem, particularly for periods of extended use. For instance, most plastic floor tiles become brittle with extended use, mainly due to the leaching out of plasticizer. This is being overcome through many routes including surface treatment of polymer product surfaces affecting less porous surface features and use of branched polymers which can act as plasticizers to themselves. Being polymers themselves, the highly branched polymers are slow to leach because of physical entanglements within the total polymer matrix.

14.2 ANTIOXIDANTS

Antioxidants retard oxidative degradation. Heat, mechanical shear, UV radiation, and heat can be responsible for the formation of free radicals, which in turn can act to shorten polymer chains and increase crosslinking, both leading to a deterioration in material properties. Free-radical production often begins a chain reaction. Primary antioxidants donate active hydrogen atoms to free radical sites, thereby quenching or stopping the chain reaction. Secondary antioxidants or synergists act to decompose free radicals to more stable products.

Polymers such as polypropylene (PP) are not usable outdoors without appropriate stabilizers, because of the presence of readily removable hydrogen atoms on the tertiary carbon atoms. PP and many other polymers (RH) are attacked during processing or outdoor use in the absence of stabilizers because of a chain degradation reaction, as shown in the following equations.

By initiation:

$$R{\text{---}}H \;\rightarrow\; R\cdot + H\cdot$$

Polymer   Free
                     radicals

$$(14.1)$$

propagation:

$$R\cdot \;+\; O_2 \;\rightarrow\; ROO\cdot$$

Free   Oxygen   Peroxy
radical   free
                                    radical

$$(14.2)$$

$$ROO\cdot \;+\; R{:}H \;\rightarrow\; ROOH \;+\; R\cdot$$

Peroxy   Polymer   Dead   Free
 free   polymer   radical
radical

termination:

$$R\cdot + R\cdot \rightarrow \quad R{:}R$$
Dead
polymer

$$R\cdot + ROO\cdot \rightarrow \quad ROOR$$
Dead
polymer

$$ROO\cdot + ROO\cdot \rightarrow \quad ROOR \; + O_2$$
Dead
polymer

(14.3)

The rate of the free-radical chain reactions shown above is accelerated by the presence of heavy metals, such as cobalt(II) ions, as follows.

$$ROOH + Co^{2+} \rightarrow Co^{3+} + RO\cdot + OH^-$$
$$ROOH + Co^{3+} \rightarrow ROO\cdot + H^+ + Co^{2+}$$

(14.4)

In contrast, the rate of chain-reaction degradation is retarded by the presence of small amounts of antioxidants. Naturally occurring antioxidants are present in many plants, including hevea rubber trees. The first synthetic antioxidants were synthesized independently by Caldwell and by Winkelman and Gray by the condensation of aromatic amines with aliphatic aldehydes. While unpurified commercial products such as phenyl-β-naphthyl-amine are toxic, they are still used as antioxidants for rubber tires.

Many naturally occurring antioxidants are derivatives of phenol and hindered phenols, such as di-tert-butyl-para-cresol. As shown by the following equation, the antioxidant acts as a chain-transfer agent to produce a dead polymer and a stable free radical that does not initiate chain-radical degradation. However, the phenoxy free radical may react with other free radicals to produce a quinone derivative.

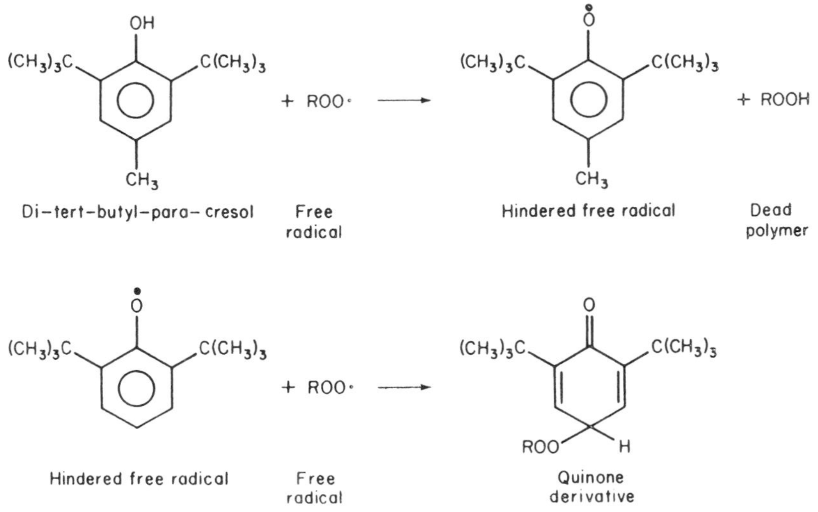

Di–tert–butyl–para– cresol   Free radical   + ROO· ⟶ Hindered free radical   + ROOH   Dead polymer

(14.5)

Hindered free radical   Free radical   + ROO· ⟶ Quinone derivative

Since carbon black has many free electrons, it may be added to polymers such as polyolefins to retard free-radical degradation of the polymer. It is customary to add small amounts of other antioxidants, such as aliphatic thiols or disulfides, to enhance the stabilization by a so-called synergistic effect. The latter term is used to explain the more effective stabilization by a mixture of antioxidants. Over 3000 tons of antioxidants are used annually by the polymer industry in the United States.

14.3 HEAT STABILIZERS

Heat stabilizers are added to materials to impart protection against heat-induced decomposition. Such stabilizers are needed to protect a material when it is subjected to a thermal-intense process (such as melt extrusion) or when the material is employed under conditions where increased heat stability is needed.

In addition to the free-radical chain degradation described for polyolefins, another type of degradation (dehydrohalogenation) also occurs with chlorine-containing polymers such as PVC. As shown by the following equation, when heated, PVC may lose hydrogen chloride and form a chromophoric conjugated polyene structure. Since the allylic chlorides produced are very unstable, the degradation continues as an unzipping type of chain reaction.

$$
\begin{array}{c}
\underset{\text{PVC}}{
\begin{array}{c}
\text{H} \quad \text{H} \quad \text{H} \quad \text{H} \quad \text{H} \quad \text{H} \quad \text{H} \quad \text{H} \\
| \quad\ | \quad\ | \quad\ | \quad\ | \quad\ | \quad\ | \quad\ | \\
-\text{C}-\text{C}-\text{C}-\text{C}-\text{C}-\text{C}-\text{C}-\text{C}- \\
| \quad\ | \quad\ | \quad\ | \quad\ | \quad\ | \quad\ | \quad\ | \\
\text{H} \quad \text{Cl} \quad \text{H} \quad \text{Cl} \quad \text{H} \quad \text{Cl} \quad \text{H} \quad \text{Cl}
\end{array}}
\xrightarrow[-2\text{HCl}]{\Delta}
\underset{\text{Conjugated structure}}{
\begin{array}{c}
\text{H} \quad \text{H} \quad \text{H} \quad \text{H} \quad \text{H} \quad \text{H} \quad \text{H} \quad \text{H} \\
| \quad\ | \quad\ | \\
-\text{C}-\text{C}-\text{C}-\text{C}=\text{C}-\text{C}=\text{C}-\text{C}- \\
| \quad\ | \quad\ | \qquad\qquad\qquad\ | \\
\text{H} \quad \text{Cl} \quad \text{H} \qquad\qquad\qquad\ \text{Cl}
\end{array}}
\end{array}
\quad (14.6)
$$

This type of degradation is accelerated in the presence of iron salts, oxygen, and hydrogen chloride. Toxic lead and barium and cadmium salts act as scavengers for hydrogen chloride and may be used as heat stabilizers in some applications, such as wire coating. Mixtures of magnesium and calcium stearates are less toxic. In spite of their toxicity, alkyl tin mercaptides and alkyl tin derivatives of thio acids have also been used. Dioctyltin salts are less toxic and produce clear PVC films.

Organic phosphites, such as mixed aryl and alkyl phosphites or triphenyl phosphite, form complexes with free metallic ions and prevent the formation of insoluble metal chlorides. Less toxic, epoxidized, unsaturated oils such as soy bean oil act as HCl scavengers as shown below.

$$
\underset{\substack{\text{Epoxy}\\\text{group}}}{
\begin{array}{c}
\text{O} \\
/ \ \backslash \\
+\text{C}\!-\!-\!\text{C}\!+ \\
| \quad\ | \\
\text{H} \quad \text{H}
\end{array}}
+ \text{HCl} \longrightarrow
\underset{\substack{\text{Chlorohydrin}\\\text{derivative}}}{
\begin{array}{c}
\text{H} \\
\backslash \\
\text{O} \quad \text{Cl} \\
| \qquad | \\
+\text{C}\!-\!\text{C}\!+ \\
| \quad\ | \\
\text{H} \quad \text{H}
\end{array}}
\quad (14.7)
$$

14.4 ULTRAVIOLET STABILIZERS

While much of the sun's high-energy radiation is absorbed by the atmosphere, some radiation in the 280 to 400 nm (ultraviolet) range reaches the Earth's surface. Since the energy of this radiation is 100 to 72 kcal, it is sufficiently strong to cleave covalent bonds and cause yellowing and embrittlement of organic polymers.

Polyethylene, PVC, polystyrene, polyesters, and polypropylene are degraded at wavelengths of 300, 310, 319, 325, and 370 nm, respectively. The bond energy required to cleave the tertiary carbon hydrogen bond in polypropylene is 90 kcal/mol, corresponding to a wavelength of 318 nm.

Since the effect of ultraviolet radiation on synthetic polymers is similar to its effect on the human skin, it is not surprising that ultraviolet stabilizers such as phenyl salicylate have been used for many years in suntanning lotions. As shown in Eq. (14.8), phenyl salicylate rearranges in the presence of high-energy radiation to form a 2,2'-dihydroxybenzophenone. The latter and other 2-hydroxybenozophenones act as energy-transfer agents, i.e., they absorb energy to form chelates which release energy at longer wave-lengths by the formation of quinone derivatives.

Phenyl salicylate   2,2'-Dihydroxybenzophenone

$$(14.8)$$

Chelate   ⟶   Quinone + $h\nu$

Many commercial ultraviolet stabilizers have alkoxyl groups on carbon 4 of the phenyl group. 2-Alkoxybenzophenones and those with bulky groups on carbon 6 are not useful as stabilizers. Other ultraviolet stabilizers are benzotriazoles, such as 2-(2'-hydroxyphenyl) benzotriazole; substituted acrylonitriles, such as ethyl-2-cyano-3,3'-diphenyl acrylate; metallic complexes, such as nickel dibutyldithiocarbamate; and pigments, such as carbon black.

The metal complexes function as energy-transfer agents, free-radical scavengers, and decomposers of hydroperoxides. The pigments absorb UV radiation and act as screening agents. Over 100,000 tons of UV stabilizers are used annually by the U.S. polymer industry.

14.5 FLAME RETARDANTS

While some polymers such as PVC are not readily ignited, most organic polymers, like other carbonaceous materials, will burn at elevated temperatures, such as those present in burning buildings, Polyolefins, SBR, EPDM, and wood, of course, will support combustion when ignited with a match or some other source of flame. In addition to burning, thermoplastics such as polyester fibers will melt, and other plastics such as PVC, polyurethanes, and proteins, when ignited, will produce smoke and toxic gases such as CO, HCl, and HCN.

Since some polymers are used as shelter and clothing and in household furnishing, it is essential that they have good flame resistance. Combustion is a chain reaction that may be initiated and propagated by free radicals like the hydroxyl free radical. As shown in Eq. (14.9), hydroxyl radicals may be produced by the reaction of oxygen with macroalkyl radicals. Halogen radicals produced by the reaction of hydroxyl radicals with halides, such as HX, may serve as terminators for the chain reaction.

$$\text{~RCH}_2\cdot \; + \; \text{O}_2 \; \rightarrow \; \text{~RCHO} + \; \cdot\text{OH}$$

Macroalkyl   Oxygen   Dead   Hydroxyl   (14.9)
radical   polymer   radical

$$\cdot\text{OH} \; + \; \text{~RCH}_2\text{H} \rightarrow \; \text{~RCH}_2\cdot \; + \; \text{HOH}$$

Hydroxyl   Polymer   Macroradical   Water
radical

$$\text{HX} \; + \; \cdot\text{OH} \; \rightarrow \text{HOH} + \; \text{X}\cdot$$

Hydrogen   Hydroxyl   Water   Halogen
halide   radical   radical

$$\text{H}\cdot \; + \; \text{~RCH}_2\cdot \; \rightarrow \text{RCH}_2\text{X}$$

Halogen   Macroradical   Dead
radical   polymer

14.6 FLAME-RETARDANT MECHANISMS

Since halogen and phosphorus radicals couple with free radicals produced in the combustion process and terminate the reaction, many flame retardants are halogen or phosphorus compounds. These may be (a) additives, (b) external retardants, such as antimony oxide and organic bromides, or (c) internal retardants, such as tetrabromophthalic anhydride, which can become part of the polymer. Over 100,000 tons of flame retardants are used annually in the United States.

Fuel, oxygen, and high temperature are essential for the combustion process. Thus, polyfluorocarbons, phosphazenes, and some composites have flame-retardant properties since they are not good fuels. Fillers such as alumina trihydrate (ATH) release water when heated, and hence reduce the temperature of the combustion reaction. Compounds such as sodium carbonate, which release carbon dioxide, shield the reactants from oxygen.

Char, formed in some combustion processes, also shields the reactants from oxygen and retards the outward diffusion of volatile combustible products. Aromatic polymers tend to char, and some phosphorus and boron compounds catalyze char formation.

Synergistic flame retardants such as a mixture of antimony trioxide and an organic bromo compound are much more effective than single flame retardants. Thus, while a polyester containing 11.5% tetrabromophthalic anhydride burned without charring at high temperatures, charring but no burning was noted when 5% antimony oxide was added.

Since combustion is subject to many variables, tests for flame retardancy may not predict flame resistance under unusual conditions. Thus, a disclaimer stating that flame-retardant tests do not predict performance in an actual fire must accompany all flame-retardant polymers. Flame retardants, like many other organic compounds, may be toxic or they may produce toxic gases when burned. Hence, extreme care must be exercised when using fabrics or other polymers treated with flame retardants.

14.7 COLORANTS

Color is a subjective phenomenon whose esthetic value has been recognized for centuries. Since it is dependent on the light source, the object, and the observer, color is not subject to direct measurement. Colorants which provide color in polymers may be soluble dyes or comminuted pigments.

Some polymeric objects, such as rubber tires, are black because of the presence of high proportions of carbon black filler. Many other products, including some paints, are white because of the presence of titanium dioxide, the most widely used inorganic pigment. Over 50,000 tons of colorants are used annually by the American polymer industry.

Pigments are classified as organic or inorganic. The former are brighter, less dense, and smaller in particle size than the more widely used, more opaque inorganic colorants. Iron oxides or ochers, available as yellow, red, black, brown, and tan, are the second most widely used pigments.

Other pigments, such as yellow lead chromate, molybate orange, yellow cadmium pigments, and green zinc chromate, are toxic. Green zinc chromate is a blend of yellow lead chromate and iron blue–iron (II) ferrocyanide, or Prussian blue. Ultramarine blue is also widely used as a pigment.

Carbon black is the most widely used organic pigment, but phthalocyanine blues and greens are available in many different shades and are also widely used. Other organic pigments are the azo dyestuffs such as the pyrazolone reds, diarylide yellows, dianisidine orange, and tolyl orange; quinacridone dyestuffs, such as quinacridone violet, magenta, and red; the red perylenes; acid and basic dyes, such as rhodamine red and victoria blue; anthraquinones, such as flavanthrone yellow; dioxazines, such as carbazole violet; and isindolines, available in the yellow and red range.

14.8 CURING AGENTS

The use of curing agents began with the serendipitous discovery of vulcanization of hevea rubber with sulfur by Charles Goodyear in 1838. The conversion of an A- or B-stage phenolic novolac resin with hexamethylenetetramine in the early 1900s was another relatively early example of the use of a curing (cross-linking) agent. Organic accelerators, or catalysts, for the sulfur vulcanization of rubber were discovered by Oenslager in 1912. While these accelerators are not completely innocuous, they are less toxic than aniline, used previously to the discovery of accelerators. Sample accelerators are thiocarbanilide and 2-mercaptobenzothiazole (Captax).

Captax is used to the extent of 1% with hevea rubber and accounts for the major part of the 30,000 tons of accelerators used annually in the United States. Other accelerators, whose structural formulas are shown below, are 2-mercaptobenzothiazole sulfonamide (Santocure), used for the vulcanization of SBR, dithiocarbamates, and thiuram disulfides. The last, called ultraaccelerators, catalyze the curing of rubber at moderate temperatures and may be used in the absence of sulfur.

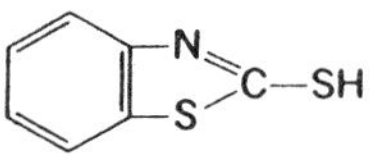

2-Mercaptobenzothiazole
(Captax)

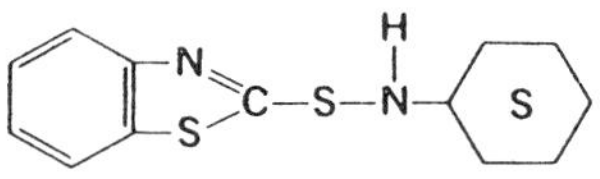

2-Mercaptobenzothiazole sulfenamide
(Santocure)

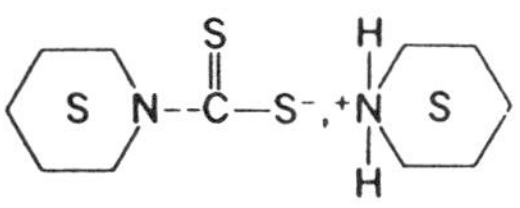

Piperidinium pentamethylene
dithiocarbamate (pip-pip)

(14.10)

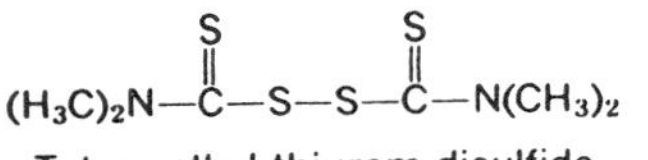

Tetramethyl thiuram disulfide
(Tuads)

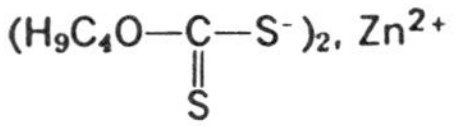

Zinc butyl xanthate

Initiators such as benzoyl peroxide are used not only for the initiation of chain-reaction polymerization but also for the curing of polyesters and ethylene-propylene copolymers and for the grafting of styrene on elastomeric polymer chains. Peroxides such as 2,5-dimethyl-2,5-di(t-butylperoxy)hexyne-3 are used for cross-linking HDPE. Since these compounds contain weak covalent bonds, precautions must be taken in their storage and use to prevent explosions. Free radicals for cross-linking may also be produced by electrons, γ rays, or UV irradiation. The latter is more effective in the presence of additives such as the methyl ether of benzoin.

Unsaturated polymers such as alkyd resins may be cured or "dried' in the presence of oxygen, a salt of a heavy metal and an organic acid called a drier. The commonly used metals are cobalt, lead, and manganese, and the most common organic acids are linoleic, abietic, naphthenic, octoic, and tall oil fatty acids.

As shown in Eq. (14.11), oxygen is now believed to form a peroxide in the presence of a drier, yielding a macroradical capable of cross-linking.

$$
\begin{array}{c}
\underset{\text{Unsaturated oil}}{-\text{C}=\text{C}-\text{C}-\text{C}-} \;+\; O_2 \;\longrightarrow\; \underset{\text{Peroxide}}{-\text{C}=\text{C}-\text{C}-\overset{\text{H}}{\underset{\text{OOH}}{\text{C}}}-}
\end{array}
$$

$$
\underset{\substack{\text{Peroxide}}}{-\text{C}=\text{C}-\text{C}-\overset{\text{H}}{\underset{\text{OOH}}{\text{C}}}-} \;+\; Co^{2+} \;\longrightarrow\; Co^{3+} \;+\; OH^- \;+\; \underset{\text{Macroradical}}{-\text{C}=\text{C}-\text{C}-\overset{\text{H}}{\underset{\overset{\bullet}{\text{O}}}{\text{C}}}-}
$$

(14.11)

14.9 ANTISTATIC AGENTS (ANTISTATS)

Antistatic agents (antistats) dissipate static electrical charges. Insulating materials including most organic plastics, fibers, films, and elastomers can build up electrical charge. Because these largely organic materials are insulators, they are not able to dissipate the charge. Such charge build-up is particularly noticeable in cold, dry climates and leads to dust attraction and sparking. These charges, resulting from an excess or deficiency of electrons, may be counteracted by the use of air-ionizing bars during processing or by the addition of antistatic agents.

Antistatic agents may reduce the charge by acting as lubricants, or they may provide a conductive path for the dissipation of the charge. Most antistats are hygroscopic and attract a thin film of water to the polymer surface. Internal antistats are admixed with the polymer, while external antistats are usually sprayed on the polymer surface. Some examples of antistatic agents are quaternary ammonium compounds, hydroxyalkylamines, organic phosphates, derivatives of polyhydric alcohols such as sorbitol, and glycol esters of fatty acids.

14.10 CHEMICAL BLOWING AGENTS

Chemical blowing agents (CBA) are employed to create lighter-weight material through formation of a foam. Physical CBA are volatile liquids and gases that expand and volatilize during processing through control of the pressure and temperature.

Cellular polymers not only provide insulation and resiliency but are usually stronger on a weight basis than solid polymers. Fluid polymers may be foamed by the addition of low-boiling liquids such as pentane or fluorocarbons, by blowing with compressed nitrogen gas, by mechanical heating, and by the addition of foaming agents. While some carbon dioxide is produced when polyurethanes are produced in the presence of moisture, auxiliary propellants are also added to the prepolymer mixture.

The most widely used foaming agents are nitrogen-producing compounds such as azobisformamide (ABFA). Other foaming agents, which decompose at various temperatures, are available. These may be used in extrusion, rotational molding, injection molding, and slush molding of plastisols. Plastisols consist of suspensions of polymer particles in a liquid plasticizer. These products, such as PVC plastisols, solidify when the temperature reaches a point at which the plasticizer penetrates the polymer particles.

14.11 COMPATIBILIZERS

Compatibilizers are compounds that provide miscibility or compatibility to materials that are otherwise immiscible or only partially miscible, yielding a homogeneous product that does not separate into its components. Typically compatibilizers act to reduce the interfacial tension and are concentrated at phase boundaries. Reactive compatibilizers chemically react with the materials they are to make compatible. Nonreactive compatibilizers perform their task by physically making the various component materials compatible.

14.12 IMPACT MODIFIERS

Impact modifiers improve the resistance of materials to stress. Most impact modifiers are elastomers such as ABS, BS, methacrylate-butadiene-styrene, acrylic, ethylene-vinyl acetate, and chlorinated polyethylene.

14.13 PROCESSING AIDS

Processing aids are added to improve the processing characteristics of a material. They may increase the rheological and mechanical properties of a melted material. Acrylate copolymers are often utilized as processing aids.

14.14 LUBRICANTS

Lubricants are added to improve the flow characteristics of a material during its processing. They operate by reducing the melt viscosity or by decreasing adhesion between the metallic surfaces of the processing equipment and the material being processed. Internal lubricants reduce molecular friction, consequently decreasing the material's melt viscosity and allowing it to flow more easily. External lubricants act by increasing the flow of the material by decreasing the friction of the melted material as it comes into contact with surrounding surfaces. In reality, lubricants such as waxes, amides, esters, acids, and metallic stearates act as both external and internal lubricants.

14.15 MICROORGANISM INHIBITORS

While most synthetic polymers are not directly attacked by microorganisms such as fungi, yeast, and bacteria, they often allow growth on their surfaces. Further, naturally occurring polymeric materials such as cellulosics, starch, protein, and vegetable oil–based coatings are often subject to microbiological deterioration. Finally, some synthetics that contain linkages "recognized" by enzymes contained within the microorganism (such as amide and ester linkages) may also be susceptible to attack.

One major antimicrobial grouping is the organotin-containing compounds. These monomeric organotin-containing compounds are now outlawed because of the high "leaching" rates of these materials affecting surrounding areas. Even so, polymeric versions are acceptable and can be considered "nonleaching" or slowly leaching.

Organic fungistatic and bacteriostatic additives are currently employed, but in all cases, formation of resistant strains and the toxicity of the bioactive additive must be considered.

PVC is also subject to pink staining as a result of the diffusion of byproducts of attack by microorganisms. Quaternary ammonium carboxylates and tributyltin compounds are effective preservatives against pink staining. Other effective preservatives or biocides are esters of p-hydroxybenzoic acid, N-(trichloromethylthio)-4-phthalimide, bis(tri-n-butyltin), and bis(8-quinolinato) copper.

SUMMARY

1. Stiff polymers such as PVC may be flexibilized by the addition of a nonvolatile compatible liquid or solid which permits slippage of polymer chains and thus reduces the T_g and modulus of the polymer. Comparable effects may be accomplished by introducing randomness into the polymer by copolymerization.

2. The rate of degradation of polymers may be retarded by the addition of chain-transfer agents, called antioxidants, which produce inactive free radicals.

3. The rate of decomposition of polymers such as PVC at elevated temperatures may be decreased by the addition of heat stabilizers which react with the decomposition products, like HCl. Soluble organic metal compounds, phosphites, and epoxides act as thermal stabilizers or scavengers for HCl.

4. The degradative effects of high-energy photolytic decomposition of polymers by ultraviolet radiation may be lessened by the addition of compounds such as 2-hydroxybenzophenones which serve as energy-transfer agents, i.e., they absorb radiation at low wavelengths and reradiate it at lower energy at longer wavelengths.

5. Since fuel, oxygen, and high temperature are essential for the combustion of polymers, the removal of any one of these prerequisites will retard combustion. Thus, additives which when heated produce water or carbon dioxide are effective flame retardants.

6. The rapid combustion of organic polymers at elevated temperatures is also retarded by the presence of flame retardants which terminate the free-radical combustion reaction.

7. A wide variety of organic and inorganic pigments is used as additives to color polymers.

8. The rate of cross-linking of natural rubber by sulfur is accelerated by catalysts like Captax, which are called accelerators. Initiators like BPO which produce free radicals and heavy metal salts (driers) will also promote cross-linking.

9. Antistats reduce the electrostatic charge on the surface of polymers.

10. Gas-producing additives are essential for the formation of cellular products.

11. Biocides are used as additives to prevent attack on polymers by microorganisms.

12. Lubricants serve as processing aids which prevent the sticking of polymers to metal surfaces during processing.

GLOSSARY

accelerator: A catalyst for the vulcanization of rubber.

antioxidant: An additive which retards the degradation of polymers.

antiplasticization: The hardening and stiffening effect observed when small amounts of a plasticizer are added to a polymer such as PVC.

antistat: An additive that reduces static charges on polymers.

biocide: An additive which retards attack by microorganisms.

Captax: 2-Mercaptobenzothiazole.

cellular polymers: Foams.

char: Carbonaceous residue produced by burning.

chelate: A cyclic complex resembling pincerlike claws.

chemical blowing agents: Volatile liquids and gases that expand and/or volatilize during processing of a polymeric material creating pockets leading to lighter-weight materials.

CN: Cellulose nitrate.

colorant: A dye or pigment.

compatibilizers: Chemicals that provide miscibility or compatibility to materials that are otherwise immiscible or only partially miscible, giving a homogeneous product.

curing agent: An additive that causes cross-linking.

$\Delta\delta$: Difference in solubility parameters.

dehydrohalogenation: Loss of HX.

DOP: di-2-Ethylhexyl phthalate.

drier: A catalyst which aids the reaction of polymers with oxygen.

drying: The cross-linking of an unsaturated polymer chain.

energy-transfer agent: A molecule which absorbs high energy and reradiates it in the form of lower energy, i.e., longer wavelength.

flame retardant: An additive that increases the flame resistance of a polymer.

foaming agent: A gas producer.

free volume: Holes not occupied by polymer chains.

gel theory: The theory that assumes the presence of a pseudo-three-dimensional structure in PVC and that the intermolecular attractions are weakened by the presence of a plasticizer.

h: Planck's constant.

HDPE: High-density polyethylene.

heat stabilizers: Additives which retard the decomposition of polymers at elevated temperatures.

Hyatt, J. W.: The inventor of celluloid.

impact modifiers: Materials that improve the resistance of materials to stress.

internal plasticization: The flexibilization resulting from the introduction of bulky groups in a polymer by copolymerization.

lubricants: Materials that improve the flow characteristics of materials during processing.

lubricity theory: The theory that explains plasticization on the basis of an increase in polymer chain slippage.

microorganism inhibitors: Compounds that impart the ability to resist habitation and destruction by microorganisms.

mold release agent: A lubricant which prevents the polymer from sticking to the mold cavity.

nm: Nanometer (1×10^{-9} m); 1 nm = 10 Å.

v: Frequency.

ocher: An iron oxide pigment.

oligomeric polyesters: Low molecular weight polyesters.

PCB: Polychlorinated biphenyls.

pink staining: Discoloration of PVC resulting from attack by microorganisms.

plasticizer: A compatible nonvolatile liquid or solid which increases the flexibility of hard polymers.

plastisol: A suspension of finely divided polymer in a liquid plasticizer. The latter penetrates and plasticizes the polymer when heated.

PP: Polypropylene.

processing aid: A lubricant.

side-chain crystallization: A stiffening effect noted when long, regularly spaced bulky pendant groups are present on a polymer chain.

synergistic effect: The enhanced effect of a mixture of additives.

T_g: Glass transition temperature.

TCP: Tricresyl phosphate.

ultraaccelerator: A catalyst which cures rubber at low temperatures.

ultraviolet stabilizer: An additive that retards degradation caused by ultraviolet radiation.

UV: Ultraviolet.

vulcanization: Cross-linking with heat and sulfur.

EXERCISES

1. Cellulose nitrate explodes when softened by heat. What would you add to permit processing at lower temperatures?

2. PVC was produced in the 1830s but not used until the 1930s. Why?

3. What is the source of "fog" on the inside of the windshield on pre-1976 automobiles?

4. Can you propose a mechanism for antiplasticization?

5. Why is plasticized PVC said to be toxic?

6. The T_g decreases progressively as the size of the alkoxy group increases from methyl to decyl in polyalkyl methacrylates, but then increases. Explain.

7. PP is now used in indoor/outdoor carpets. However, the first PP produced deteriorated rapidly when subjected to sunlight because of tertiary hydrogen atoms present. Explain.

8. Tests showed that crude phenyl-β-naphthylamine was a carcinogen, but the mice used for testing lived longer than normal when injected with di-tert-butyl-paracresol. Explain.

9. Lead stearate is an effective thermal stabilizer for PVC, yet its use in PVC pipe is not permitted. Why?

10. When a PVC sheet fails in sunlight, it goes through a series of colors before becoming black. Explain.

11. What would be the advantage of using epoxidized soybean oil as a stabilizer for PVC?

12. Why do PVC films deteriorate more rapidly when used outdoors?

13. Which of the following is more resistant to UV light: (a) polypropylene, (b) polyethylene, or (c) PVC?

14. Would 2-methoxybenzophenone be useful as a UV stabilizer?

15. Would 3,3'-dihydroxybenzophenone be useful as a UV stabilizer?

16. Polytetrafluoroethylene is considered to be noncombustible. Yet two astronauts burned to death in a space capsule because of burning PTFE. Explain.

17. The disastrous effect of discarded burning cigarettes on polyurethane-"stuffed" upholstery has been lessened by the use of a protective film of neoprene. yet disastrous fires have occurred, e.g., in jail cells. Explain.

18. The addition of "tris" as a retardant was required for children's sleeping garments by a government order in the early 1970s. Nevertheless, the use of "tris" was discontinued in 1977. Why?

19. Antimony oxide is a good flame retardant in the presence of organic halogen compounds. Would you recommend it for textile applications?

20. Is a phosphazene elastomer flame resistant?

21. Cotton, wool, silk, flax, and wood, which are combustible, have been used for many centuries without flame retardants. Yet synthetic polymers which are not any more combustible are banned unless flame retardants are added. Why?

22. Was carbon black always used as a reinforcing filler for tires?

23. How would you explain the different colors of ocher ($Fe_2O_3 \cdot XH_2O$)?

24. What is the formula for Santocure?

25. How would you synthesize pip pip?

26. BPO is explosive. How would you store it?

27. Can you propose a better name for a drier?

28. What is the objection to the presence of static charges on polymer surfaces?

29. Explain how an internal antistat performs.

30. Why are foamed wire coatings preferred over solid coatings?

31. How would you design a foamed plastisol?

32. Why doesn't the liquid plasticizer penetrate a polymer particle such as PVC at ordinary temperatures?

33. Which would be more resistant to attack by microorganisms: (a) PVC, or (b) plasticized PVC?

34. What naturally occurring fiber is more resistant to microbiologic attack than nylon?

35. Sometimes, molded plastics have opaque material called "bloom" on the surface. Explain.

BIBLIOGRAPHY

Allen, N. S., McKellar, J. (1980): *Photochemistry of Dyed and Pigmented Polymers*, Applied Science, Essex, England.

Bhatnger, V. (1979): *Advances in Fire Retardants*, Technomic, Lancaster, Pennsylvania.

Boldus, L. R. (1976): Foaming agents, Mod. Plast., *53*(10A):192.

Copp, J. (1976): Colorants, Mod. Plast., *53*(10A):142.

DiBattisto, A. D. (1976): Antioxidants, Mod. Plast., *53*(10A):138.

Doolittle, A. K. (1954): *The Technology of Solvents and Plasticizers*, Wiley, New York.

Edwards, R. W. (1976): Antistatic agents, Mod. Plast., *53*(10A):139.

Gachter, R., Muller, H. (1983): *Plastics Additives Handbook*, Hanser Publications, Munich, Germany.

Henman, T. J. (1983): *World Index of Polyolefin Stabilizers*, RSC, Herts, England.

Hilado, C. J. (1974): *Flammability of Fabrics*, Technomic, Lancaster, Pennsylvania.

———. (1974): *Flammability of Solid Plastics*, Technomic, Lancaster, Pennsylvania.

Johnson, K. (1976): *Antistatic Compositions for Textiles and Plastics*, Noyes, Park Ridge, New Jersey.

Lewin, M., Atlas, S. M., Pearce, E. M. (1975): *Flame Retardant Polymeric Materials*, Plenum, New York.

Lubin, G. (1982): *Handbook of Composites*, Van Nostrand-Reinhold, New York.

Powell, G. M., Brister, J. E. (1900): Plastigels, U.S. Patent 2,427,507.

Scott, G. (1981): *Developments of Polymer Stabilization—4*, Applied Science, Essex, England.

Semon, W. L. (1936): Plastisol, U.S. Patent 2,188,396.

Seymour, R. B. (1975): *Modern Plastics Technology*, Chap. 3, Reston Publishing, Reston, Virginia.

———. (1976): Additives for polymers, in *Encyclopedia of Chemical Processing and Design*, Vol. 11 (J. McKetta, W. A. Cunningham, eds.), Marcel Dekker, New York.

———. (1978): *Additives for Plastics*, Academic, New York.

———. (1987): *Coatings, Pigments and Paints*, Encyclopedia of Physical Science and Technology, Academic, New York.

———. (1987): Fillers for plastics, in *Developments in Plastics Technology*, Applied Science, Essex, England.

15

Reactions of Polymers

Synthesis and curing (cross-linking) of polymers and telomerization are chemical reactions which have been discussed in previous chapters. Other reactions of polymers, such as hydrogenation, halogenation, cyclization, and degradation, will be described in this chapter. Providing the reaction sites are accessible to the reactants, reactions take place that are similar to classic organic chemical reactions. The rates of reactions of polymers may be enhanced by the presence of neighboring groups. This effect is called anchimeric assistance.

15.1 REACTIONS WITH POLYOLEFINS

Deuterated polyethylene could be produced from HDPE, but it is more readily produced by the polymerization of perdeuterated ethylene. The latter is prepared by the reduction of deuterated acetylene with chromium(II) sulfate in a dimethylformamide (DMF/H_2O) solution.

Polyethylene, like other alkane polymers, is resistant to chemical oxidation, but will burn in the presence of oxygen. The folds in crystalline polyethylene are less resistant to attack, and α,ω-dicarboxylic acids can be produced by the reaction of concentrated nitric acid on crystalline HDPE.

Polyolefins, like simple alkanes, may be chlorinated by chlorine at elevated temperatures or in the presence of ultraviolet light. This free-radical reaction produces HCl and chlorinated polyolefins (Tyrin), which are used as plasticizers and flame retardants. The commercial product is available with various percentages of chlorine. Since a tertiary hydrogen atom is more readily replaced than a secondary hydrogen, polypropylene is more readily chlorinated than HDPE. HDPE is readily converted to polytetrafluoroethylene by direct fluorination using fluorine.

Sulfochlorinated polyethylene (Hypalon) is obtained when a suspension of polyethylene in carbon tetrachloride is chlorinated by a mixture of chlorine and sulfur dioxide in pyridine. The commercial product, which contains 27.5% chlorine and 1.5% sulfur, is soluble in tetralin. It may be used as a coating which may be cured with sulfur and diphenylguanidine.

A chlorinated product called poly(vinyl dichloride) (PVDC) is obtained when PVC is chlorinated. Since the heat resistance of PVDC is superior to that of PVC, it is used for hot water piping systems.

Polyolefins may be cross-linked by heating with peroxides such as di-tert-butyl peroxide or by irradiation. It is advantageous to cross-link these polymers after they have been fabricated, as in wire coatings. The cross-linked products are less soluble and more resistant to heat than are the linear polyolefins.

Reactions comparable to those described for HDPE take place with many organic polymers. The pendant groups may also react with appropriate reactants.

15.2 REACTIONS OF POLYENES

Providing the ethylenic groups are accessible, the reactions of polyenes are similar to the classic reactions of alkenes. Thus, as demonstrated by Nobel laureate Hermann Staudinger, polyenes such as *Hevea brasiliensis* may be hydrogenated, halogenated, hydrohalogenated, and cyclized. In this classic experiment in the early 1900s, Harries produced ozonides of rubber. Epoxides were also obtained by peroxidation.

The hydrogenation of *H. brasiliensis* was investigated by Berthelot in 1869. The rate of this reaction with natural rubber, gutta percha, balata, or SBR may be followed by observing the changes in the glass transition temperature and the degree of crystallinity. The product obtained by the partial hydrogenation of polybutadiene (Hydropol) has been used as a wire coating.

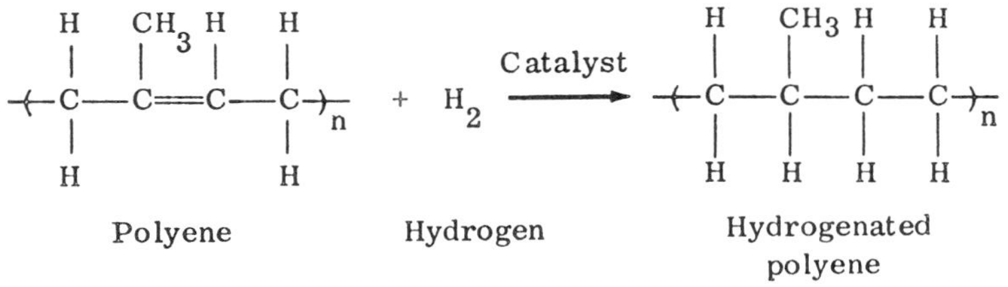

$$(15.1)$$

A saturated ABA block copolymer (Kraton) is produced by the hydrogenation of the ABA block copolymer of styrene and butadiene.

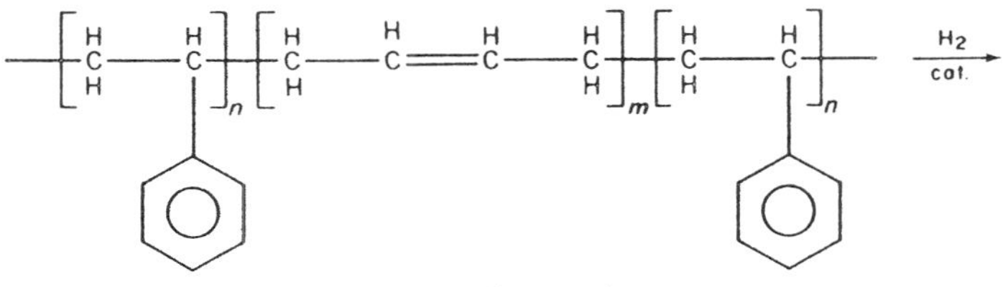

Block copolymer of butadiene and styrene

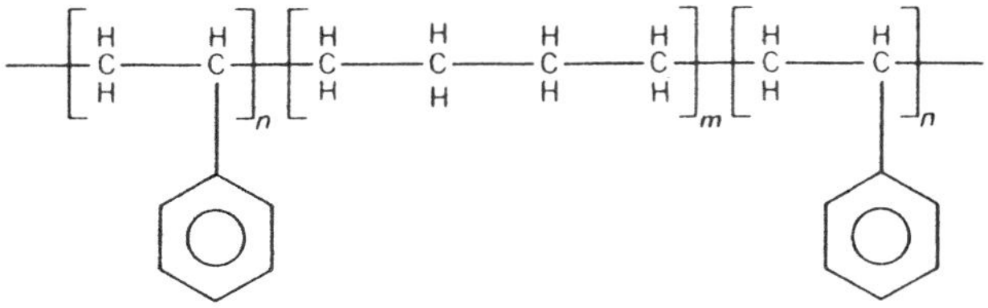

Block copolymer of styrene and butene

$$(15.2)$$

Chlorinated rubber (Tornesit or Parlon), which was produced by Traun in 1859, is available with varying amounts of chlorine and is used for the coating of concrete. While chlorinated NR is soluble in carbon tetrachloride, chlorinated SBR is insoluble, but both are soluble in benzene and chloroform.

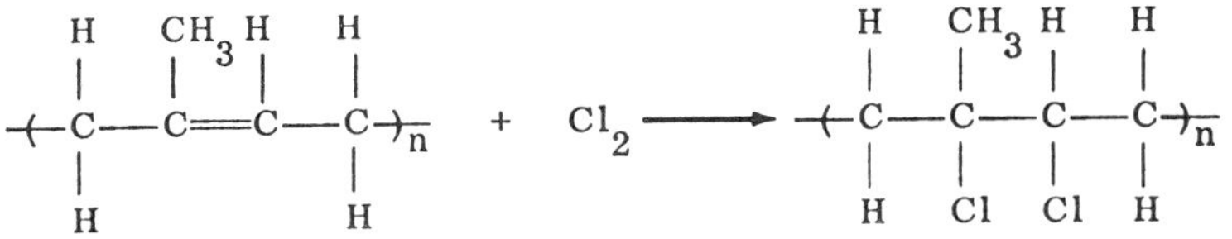

Hevea rubber   Chlorine   Chlorinated rubber

$$(15.3)$$

Hydrohalogenation of *H. brasiliensis* proceeds by an ionic mechanism to yield a Markownikoff addition product with the halogen on the tertiary carbon atom. The principal use of rubber hydrochloride (Pliofilm) is as a packaging film. Some cyclization of the polymer also takes place during hydrohalogenation.

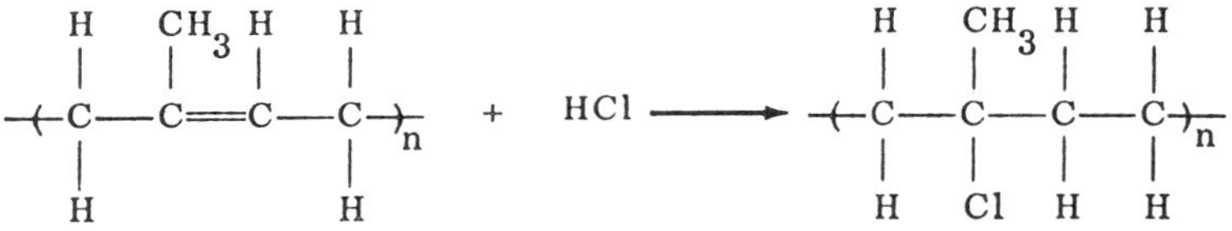

Hevea rubber   Hydrogen chloride   Rubber hydrochloride

$$(15.4)$$

Staudinger prepared cyclized polydienes by heating rubber hydrochloride with zinc dust. Commercial cyclized rubber (Thermoprene and Pliolite) has been produced by heating rubber with sulfuric acid or Lewis acids, respectively. These cyclized products or isomerized polymers, which are soluble plastics, are used as adhesives and coatings.

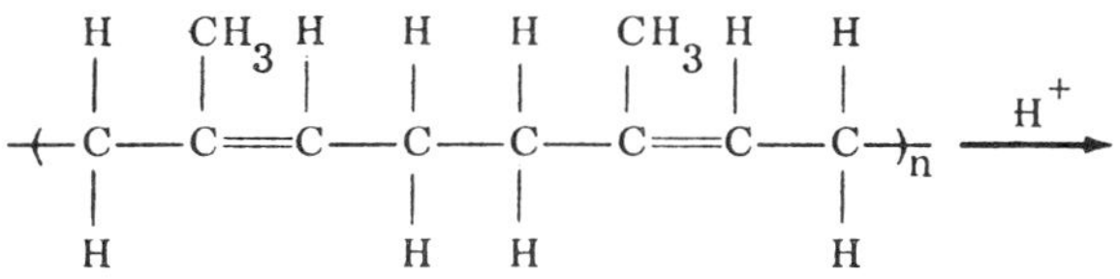

Hevea rubber

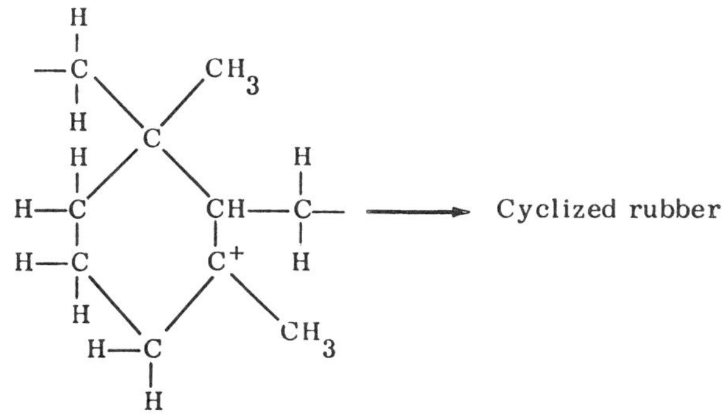

Carbonium ion

(15.5)

Hydrogen peroxide or peracetic acid is added to polyenes such as hevea rubber or unsaturated vegetable oils to produce epoxidized polybutadiene (Oxiron) or epoxidized vegetable oils, respectively. These products react readily with water, alcohols, anhydrides, and amines. The reaction product of hevea rubber and hydrogen peroxide is used as a specialty rubber. The reaction product of unsaturated vegetable oils is used as a PVC stabilizer.

(15.6)

Hevea rubber — Hydrogen peroxide — Epoxidized polyene — Water

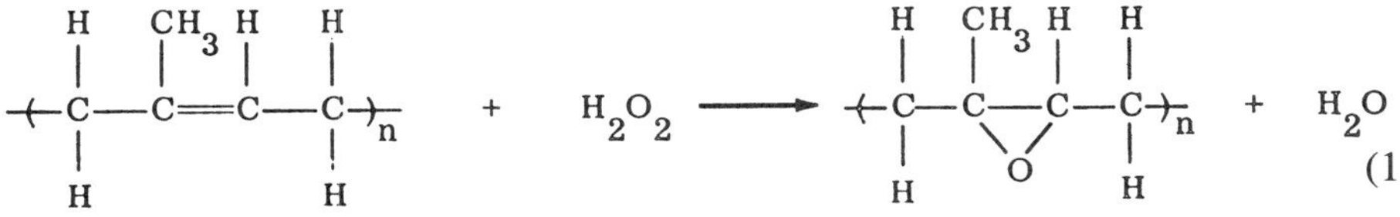

The reaction of ozone with polyenes yields ozonides which cleave to produce aldehydes in the presence of water and zinc dust. Since the ozonides are explosive, they are not usually isolated. As shown by the following equation, ozonolysis can be used as a diagnostic procedure to locate the position of the double bonds in polyenes.

Hevea rubber — Ozone

(15.7)

Aldehyde — Levulinic aldehyde — Aldehyde

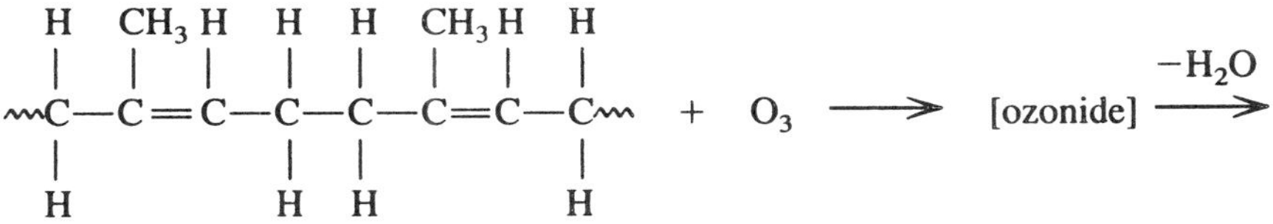

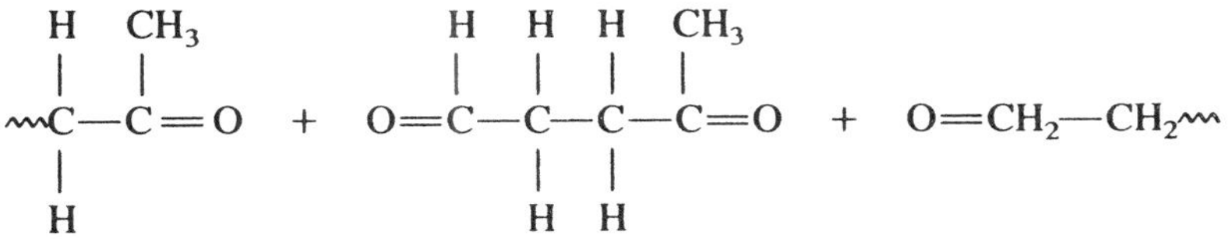

15.3 REACTIONS OF ALIPHATIC PENDANT GROUPS

Esters such as poly(vinyl acetate) (PVAc) may be hydrolyzed to produce alcohols such as poly(vinyl alcohol) (PVA), which should have the same $\overline{DP}$ as the ester. In truth, PVAc contains some branching through the acetate portion that is eliminated upon hydrolysis. While the enol tautomer of acetaldehyde (vinyl alcohol) cannot be isolated, its water-soluble polymer can be obtained by the hydrolysis of any poly(vinyl carboxylic acid ester). The solubility of the product in water is dependent on the extent of hydrolysis.

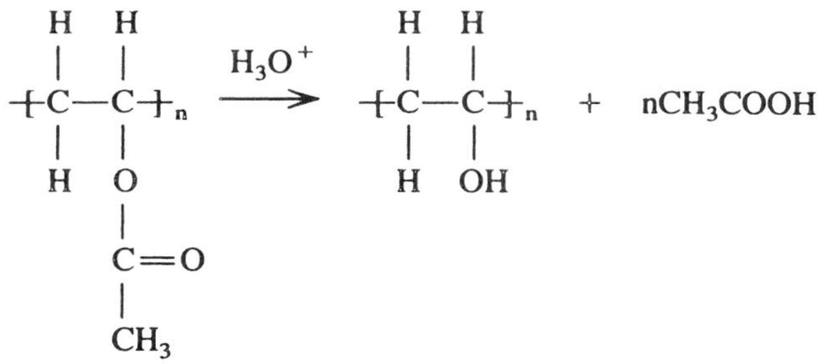

$$
\begin{array}{ccc}
\text{H} & \text{H} & \\
| & | & \\
\begin{matrix} \text{+C—C+} \end{matrix}_n & \xrightarrow{\text{H}_3\text{O}^+} & \begin{matrix} \text{+C—C+} \end{matrix}_n \; + \; n\text{CH}_3\text{COOH} \\
| & | & \\
\text{H} & \text{O} & \text{H} \quad \text{OH} \\
& | & \\
& \text{C=O} & \\
& | & \\
& \text{CH}_3 &
\end{array}
\tag{15.8}
$$

Poly(vinyl acetate)   Poly(vinyl alcohol)   Acetic acid

The hydroxyl pendant groups in cellulose will form alkoxides with sodium hydroxide, and these alkoxides will produce ethers by the classic Williamson reaction. Thus, Suida produced methylcellulose in 1905 by the reaction of soda cellulose and methyl sulfate.

Alkylcellulose may also be produced by the reaction of soda cellulose and alkyl chlorides. Since these alkoxy pendant groups reduce the hydrogen-bonding forces, the partially reacted products are water soluble. This solubility in water will decrease as the degree of substitution (DS) increases.

$$
\begin{array}{lll}
\text{Cellulose—OH} + & \text{NaOH} + & \text{RCl} \rightarrow \\
\text{Cellulose} & \text{Sodium hydroxide} & \text{Alkyl chloride} \\
\text{Cellulose—OR} + & \text{NaCl} + & \text{H}_2\text{O} \\
\text{Cellulose ether} & \text{Sodium chloride} & \text{Water}
\end{array}
\tag{15.9}
$$

Methylcellulose is widely used as a viscosity improver and thickener. Ethylcellulose is used as a melt coating. One of the most widely used cellulosic ethers

is carboxymethylcellulose (CMC). This water-soluble cellulosic derivative is produced by the reaction of soda cellulose and chloroacetic acid.

CMC, with a DS of 0.5 to 0.8, is used in detergents and as a textile-sizing
agent. Products with higher DS are used as thickeners and in drilling mud formulations. Aluminum or chromium ions will produce insoluble gels.

$$
\begin{array}{ll}
\underset{\text{Cellulose}}{\text{Cellulose—OH}} + \underset{\substack{\text{Sodium} \\ \text{hydroxide}}}{\text{NaOH}} + \underset{\substack{\text{Sodium salt} \\ \text{of} \\ \text{chloroacetic} \\ \text{acid}}}{\text{ClCH}_2\text{COONa}} \rightarrow \\[2em]
\underset{\text{CMC}}{\text{Cellulose—OCH}_2\text{COONa}} + \underset{\substack{\text{Sodium} \\ \text{chloride}}}{\text{NaCl}} + \underset{\text{Water}}{\text{H}_2\text{O}}
\end{array}
\tag{15.10}
$$

Water-soluble sodium cellulose xanthate, with a DS of about 0.5, is produced
by the room temperature reaction of carbon disulfide and soda cellulose. Since
xanthic acids are unstable, the cellulose is readily regenerated by passing extrudates
into an acid bath. Rayon and cellophane are produced from cellulose xanthate in
the viscose process.

$$
\begin{array}{l}
\underset{\text{Cellulose}}{\text{Cellulose—OH}} + \underset{\substack{\text{Sodium} \\ \text{hydroxide}}}{\text{NaOH}} + \underset{\substack{\text{Carbon} \\ \text{disulfide}}}{\text{CS}_2} \longrightarrow \\[2em]
\underset{\substack{\text{Cellulose} \\ \text{xanthate}}}{\text{Cellulose—O—}\overset{\displaystyle\overset{\text{S}}{\|}}{\text{C}}\text{—S}^-, \text{Na}^+} \xrightarrow{\text{H}_3\text{O}^+} \underset{\substack{\text{Regenerated} \\ \text{cellulose}}}{\text{Cellulose—OH}}
\end{array}
\tag{15.11}
$$

Hydroxyethylcellulose and hydroxypropylcellulose are obtained from the reaction of soda cellulose and ethylene oxide or propylene oxide, respectively. The
former (Cellosize) is used as a water-soluble sizing agent and descaling agent in
boiler water.

$$
\underset{\text{Cellulose}}{\text{Cellulose—OH}} + \underset{\substack{\text{Ethylene} \\ \text{oxide}}}{\text{H}_2\text{C}\underset{\displaystyle\text{O}}{\diagdown\diagup}\text{CH}_2} \longrightarrow \underset{\substack{\text{Hydroxyethylcellulose} \\ \text{(Cellosize)}}}{\text{Cellulose—O—}\overset{\displaystyle\overset{\text{H}}{|}}{\underset{\displaystyle\underset{\text{H}}{|}}{\text{C}}}\text{—}\overset{\displaystyle\overset{\text{H}}{|}}{\underset{\displaystyle\underset{\text{H}}{|}}{\text{C}}}\text{OH}}
\tag{15.12}
$$

Cyanoethylcellulose is produced by the reaction of soda cellulose and acrylonitrile. This derivative is more resistant to abrasion and biological attack than is

cellulose. Cyanoethylcellulose and graft copolymers of acrylonitrile and cellulose may be hydrolyzed to yield products with excellent water absorbency characteristics.

$$\text{Cellulose—OH} + \text{H}_2\text{C}=\overset{\overset{\text{H}}{|}}{\text{C}}\text{—CN} \longrightarrow \text{Cellulose—O}\overset{\overset{\text{H}}{|}}{\underset{\underset{\text{H}}{|}}{\text{—C—}}}\overset{\overset{\text{H}}{|}}{\underset{\underset{\text{H}}{|}}{\text{C—}}}\text{CN} \quad (15.13)$$

  Cellulose   Acrylonitrile   Cyanoethylcellulose

Cellulose nitrate (CN) (incorrectly called nitrocellulose) was one of the first synthetic plastics. The degree of substitution of CN is dependent on the concentration of the nitric acid used. The commercial product has a DS of about 2.0, corresponding to a nitrogen content of about 11%. Despite its flammability, CN plasticized by camphor is still used for personal accessories, toiletries, and industrial items. However, the principal use of CN is for coatings, in which the CN has a $\overline{\text{DP}}$ of 200 and a DS of about 2.0.

$$\text{Cellulose}\text{—}(\text{OH})_2 + 2\text{HONO}_2 \overset{\text{H}_2\text{SO}_4}{\longrightarrow} \text{Cellulose—}(\text{ONO}_2)_2 + 2\text{H}_2\text{O}$$

  Cellulose   Nitric   Cellulose   Water   (15.14)
                     acid   nitrate

Cellulose triacetate (DS 2.8) was originally prepared by Schutzenberger and Naudine in 1865 by the reaction of cellulose with acetic acid and acetic anhydride in the presence of sulfuric acid. This ester, which has an acetyl content of 43%, is soluble in chloroform or in mixtures of methylene chloride and ethanol. When plasticized by ethyl phthaloyl ethyl glycolate, it is used as a film, a molding, and as an extrusion. The unplasticized cellulose triacetate is used as a specialty fiber.

$$\begin{array}{c} \text{H}_3\text{CCOOH} \\ + \\ \text{Cellulose (OH)}_3 + (\text{H}_3\text{CCO})_2\text{O} \overset{\text{H}_2\text{SO}_4}{\longrightarrow} \text{Cellulose —}(\text{OOCCH}_3)_3 \end{array} \quad (15.15)$$

  Cellulose   Acetylation   Cellulose
                     agents   triacetate

Cellulose diacetate may be produced directly by the acetylation of cellulose in a solution of dimethyl sulfoxide (DMSO) and formaldehyde, but this ester cannot be made directly from the acetylation of cellulose in a conventional process. The commercial product, which is also called secondary cellulose acetate ester, is produced by the partial saponification of cellulose triacetate.

Cellulose diacetate is widely used as extruded tape, in molded toys, electrical appliance housings, and sheet and blister packaging and fiber. Cellulose diacetate fiber, which is extruded from an acetone solution called "dope," is called acetate rayon.

$$\text{Cellulose —}(\text{OOCCH}_3)_3 \overset{\text{NaOH}}{\longrightarrow} \text{Cellulose}\text{—}(\text{OOCCH}_3)_2 + \text{CH}_3\text{COONa}$$

  Cellulose   Cellulose   Sodium   (15.16)
  triacetate   diacetate   acetate

Cellulose propionate and cellulose acetate butyrate are also available commercially. These esters, which are more readily processed and more resistant to moisture than CA, are used as sheets, tool handles, steering wheels, and for packaging.

Polyvinyl(vinyl formal) (PVF) and poly(vinyl butyral) (PVB) are produced by the reaction of poly(vinyl alcohol) with formaldehyde or butyraldehyde, respectively. PVF has been used as a wire coating, and PVB forms the inner layer of safety-glass windshields. These products contain residual acetate and hydroxyl pendant groups.

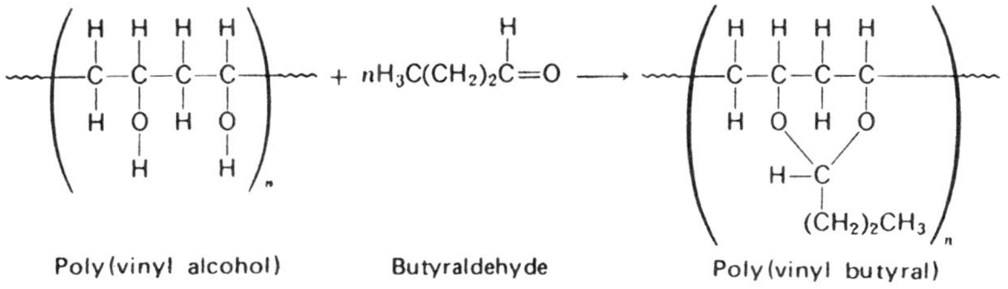

$$\text{(15.17)}$$

Poly(vinyl alcohol)   Butyraldehyde   Poly(vinyl butyral)

Esters of polycarboxylic acids, nitriles, or amides may be hydrolyzed to produce polycarboxylic acids. Thus polyacrylonitrile, polyacrylamide, or poly(methyl acrylate) may be hydrolyzed to produce poly(acrylic acid).

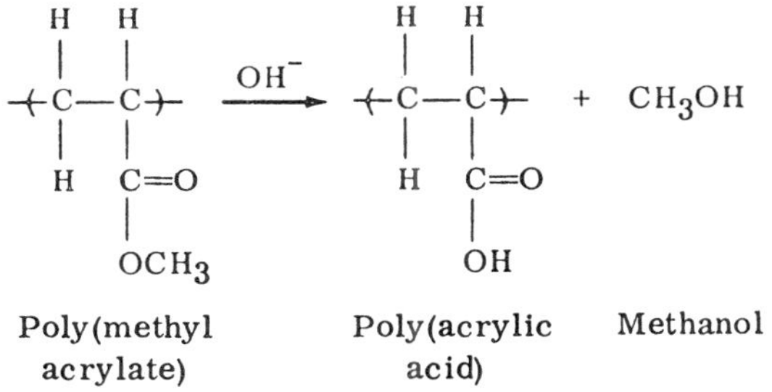

$$\text{(15.18)}$$

Poly(methyl acrylate)   Poly(acrylic acid)   Methanol

Poly(acrylic acid) and partially hydrolyzed polyacrylamide are used for the prevention of scale in water used for boilers and for flocculating agents in water purification. In the presence of aluminum ions, these polymeric acids produce flocs which remove impurities in water as they settle.

When heated, poly(methacrylic acid) loses water and forms a polymeric anhydride. The latter, as well as copolymers of maleic anhydride, such as styrene-maleic anhydride copolymers (SMA), undergo characteristic anhydride reactions with water, alcohols, and amines to produce acids, esters, and amides.

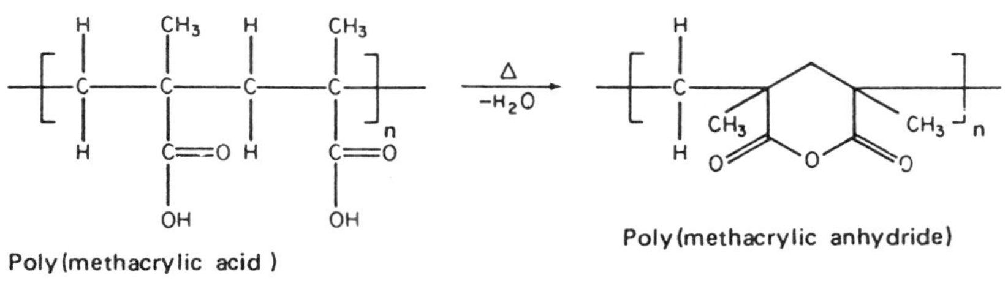

$$\text{(15.19)}$$

Poly(methacrylic acid)   Poly(methacrylic anhydride)

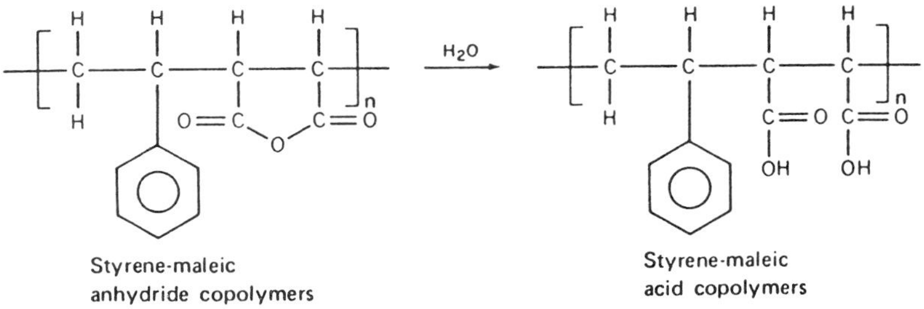

$$\text{(15.20)}$$

Styrene-maleic anhydride copolymers → Styrene-maleic acid copolymers

When heated, polyacrylonitrile forms a complex ladder polymer containing a quniazirinelike structure that approximates the structure shown below.

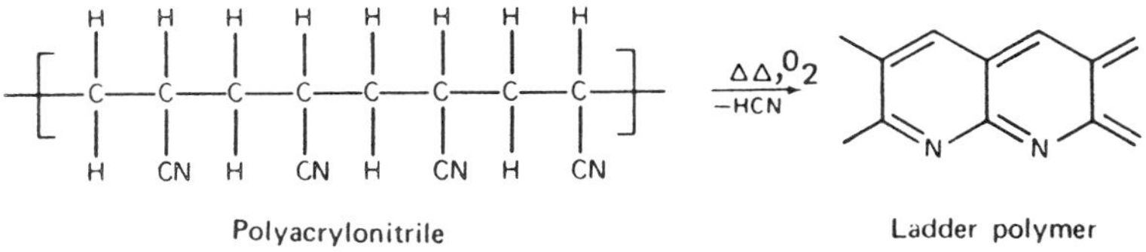

$$\text{(15.21)}$$

Polyacrylonitrile   Ladder polymer

Poly(vinyl amine) is produced by the Hofmann degradation of polyacrylamide. Vinyl amine, like vinyl alcohol, is unstable. Thus its polymer is prepared by an indirect reaction. Poly(vinyl amine) is soluble in aqueous acids.

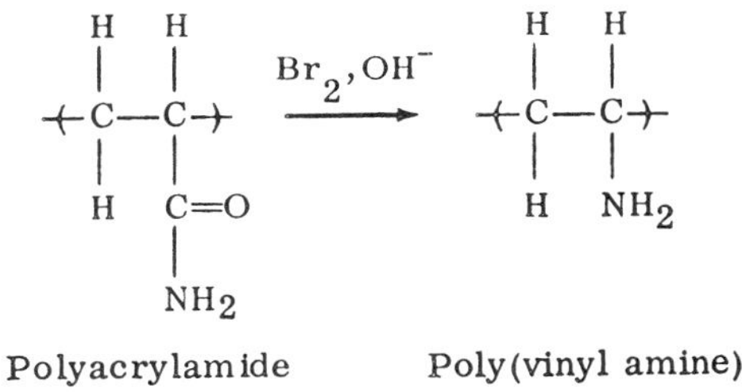

$$\text{(15.22)}$$

Polyacrylamide   Poly(vinyl amine)

15.4   REACTIONS OF AROMATIC PENDANT GROUPS

Polymers with aromatic pendant groups, such as polystyrene, undergo all the characteristic reactions of benzene, such as alkylation, halogenation, nitration, and sulfonation. Thus, oil-soluble polymers used as viscosity improvers in lubricating oils are obtained by the Friedel-Crafts reaction of polystyrene and unsaturated hydrocarbons such as cyclohexene. Poly(vinyl cyclohexylbenzene) is produced in the latter reaction.

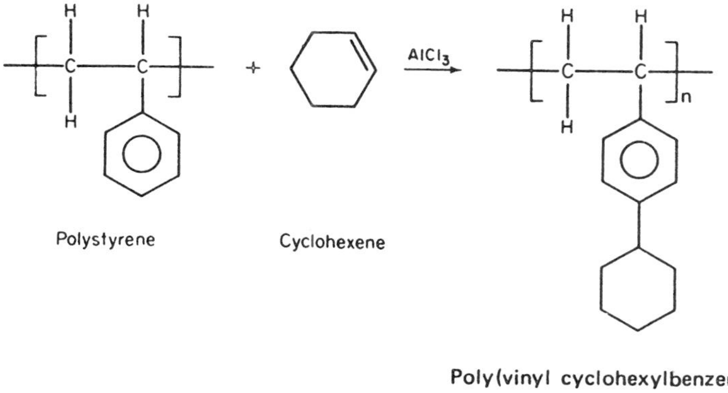

(15.23)

Poly(vinyl cyclohexylbenzene)

In the presence of a Lewis acid, halogens such as chlorine react with polystyrene to produce chlorinated polystyrene. The latter has a higher softening point than polystyrene.

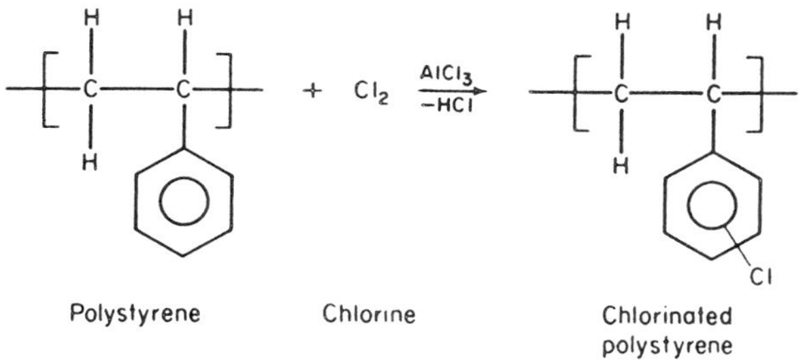

(15.24)

Perfluoropolystyrene is produced by the reaction of fluorine and polystyrene. This reaction may also be run on the surface (topochemical) of polystyrene articles.

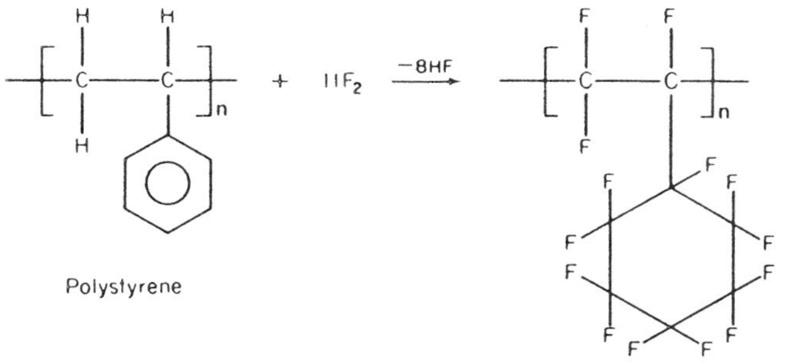

(15.25)

Polynitrostyrene is obtained by the nitration of polystyrene. The latter may be reduced to form polyaminostyrene. Polyaminostyrene may be diazotized to produce polymeric dyes.

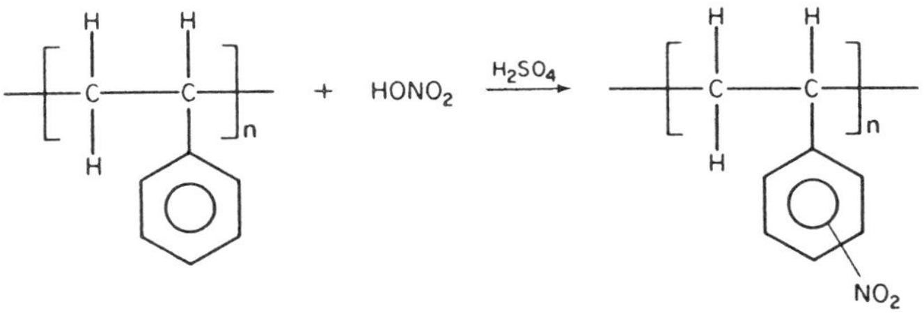

$$(15.26)$$

Polystyrene   Nitric acid   Polynitrostyrene

Polystyrene and other aromatic polymers have been sulfonated by fuming sulfuric acid. Sulfonated cross-linked polystyrene has been used as an ion-exchange resin.

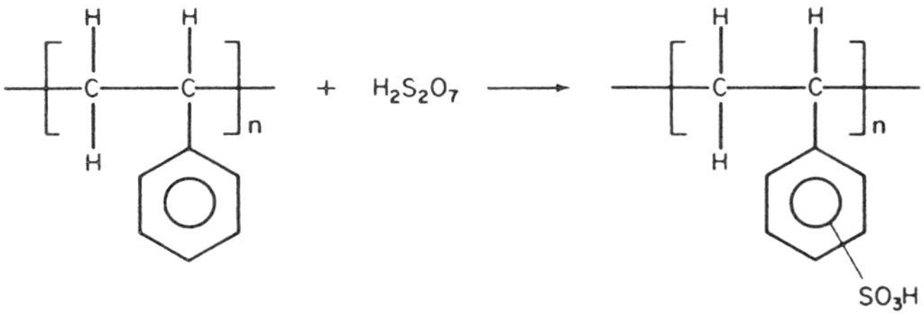

$$(15.27)$$

Polystyrene   Sulfonated polystyrene

15.5 REACTIONS OF POLYAMIDES

Polyamides such as proteins or nylons may be reacted with ethylene oxide or formaldehyde. The latter serves as a cross-linking agent with proteins in embalming, leather production, and the stabilization of regenerated protein fibers.

15.6 POLYMERIZATION REACTIONS

The cross-linking of many thermosetting resins, such as novolac phenolic resins, takes place during the molding operation. Polymerization also occurs during the production of polyurethane (PU) foams and molded products. The latter are produced when liquid reactants are introduced in the injection molding press. One process is called liquid injection molding (LIM) and the other is called reaction injection molding (RIM). The latter has been used to mold automotive fascias, bumpers, and flexible fenders.

15.7 DEGRADATION OF POLYMERS

In many chemical reactions with pendant groups, such as the hydrolysis of ester groups, the polymer is said to be degraded, but since the degree of polymerization is unchanged, the integrity of the polymer chain is maintained. However, the $\overline{DP}$ is decreased when chain-scission degradation occurs, and this type of degradation will be emphasized in this section.

Saturated linear polymers such as HDPE are resistant to degradation, but slow degradation will occur in the presence of oxygen, ultraviolet (UV) light, heat, and impurities. Since tertiary carbon atoms are more readily attacked, polypropylene is less resistant to degradation than HDPE. Unsaturated polymers are even less resistant to degradation as evidenced by the ozonolysis of hevea rubber and the degradation of dehydrohalogenated PVC.

The kinetics of random chain scission are essentially the reverse of those in stepwise propagation. Much information on this type of degradation has been obtained from studies of the acid-catalyzed homogeneous degradation of cellulose in which $\overline{DP}^{-1}$ increases with time. The initial degradation of polymeric hydrocarbons is usually the result of homolytic cleavage at weak points in the polymer chain. As shown in Eq. (15.28), the initial products are macroradicals.

$$
\begin{array}{ccc}
\text{H} & \text{H} & \\
| & | & \\
\sim\!\!\text{C}\!-\!\text{C}\!\sim & \xrightarrow{\ E\ } & \sim\!\!\text{C}\bullet \ + \ \bullet\text{C}\!\sim \\
| & | & \\
\text{H} & \text{H} & \\
\end{array}
\qquad (15.28)
$$

Polymer   Macroradicals

As shown in Eq. (15.29), hydroperoxides are formed in the presence of oxygen. These readily cleaved hydroperoxides also yield macroradicals.

$$
\begin{array}{cccc}
\text{H} & \text{H} & & \text{H}\quad\text{OOH} \\
| & | & & |\qquad| \\
\sim\!\!\text{C}\!-\!\text{C}\!\sim & +\ \text{O}_2 & \longrightarrow & \sim\!\!\text{C}\!-\!\text{C}\!\sim \longrightarrow \\
| & | & & |\qquad| \\
\text{H} & \text{CH}_3 & & \text{H}\quad\text{CH}_3 \\
\end{array}
\qquad (15.29)
$$

Polymer   Oxygen   Hydroperoxide

$$
\begin{array}{ccc}
\text{H} & & \text{O} \\
| & & \| \\
\sim\!\!\text{C}\bullet & +\ \bullet\text{OH}\ + & \text{C}\!\sim \\
| & & | \\
\text{H} & & \text{CH}_3 \\
\end{array}
$$

Macro-   Hydroxyl   Ketone
radical   radical

Oxidizing acids such as concentrated nitric acid will also cause scission at weak links. This hypothesis is verified by the attack by nitric acid on the folds of polyethylene crystals to produce α,ω-dicarboxylic acids. Acids and alkalies will also hydrolyze ester and amide linkages in the polymer chain.

While most polymers undergo random chain scission, 1,1-disubstituted polymers, such as poly(methyl methacrylate) (PMMA), depolymerize or unzip quantitatively when heated above the ceiling temperature (T_c). In contrast, nonsubstituted polymers such as polystyrene degrade by both depolymerization and random chain scission reactions. Since higher temperatures favor the former, more than 85% styrene monomer is produced at 725°C.

Some degradation occurs when polymers are irradiated, but cross-linking is the predominant reaction. Thus, thermoplastic coatings may be applied and cross-linked by radiation after application.

Today, major items related to the topic of polymer degradation involve degradation of polymers for potential reuse of the degradation products and degradation of polymers as a "garbage" material.

Polystyrene gives different thermal degradation products dependent on the exact degradation conditions. Under one set of conditions the major products are benzene, hydrogen, and ethylene, all three of which are usable chemical molecules.

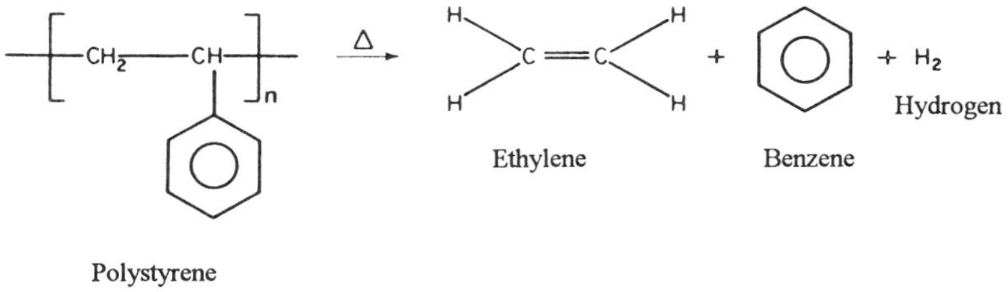

$$\text{Polystyrene} \xrightarrow{\Delta} \text{Ethylene} + \text{Benzene} + H_2 \text{ (Hydrogen)} \tag{15.30}$$

A major problem in the burning of polymers containing halogens, such as poly(vinyl chloride), is the emission of the hydrogen halide (e.g., hydrogen chloride) which is quite dangerous to human and plant life and is the major problem in the deterioration of commercial incinerators.

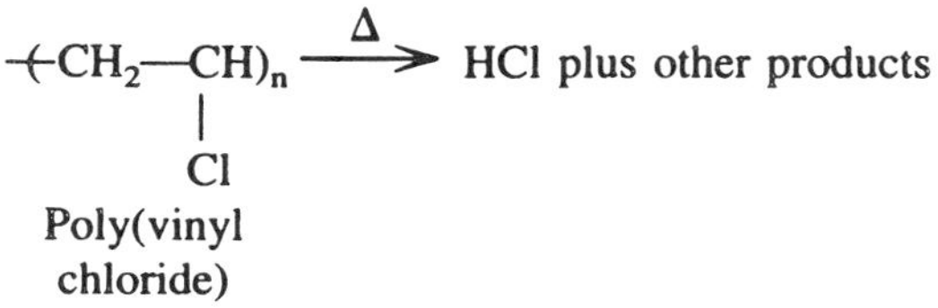

$$-(CH_2-CH)_n \xrightarrow{\Delta} \text{HCl plus other products} \tag{15.31}$$

Poly(vinyl chloride)

Radiation is also responsible for the deterioration of many polymeric materials. Radiation can form macroradicals as noted in Eqs. (15.28) and (15.32). Depending on the fate of the newly formed, energetic free radical, chain scission may occur, resulting in a decreased molecular weight, or cross-linking may occur which increases the stiffness of the material but may eventually lead to the formation of a hard, brittle material with little toughness.

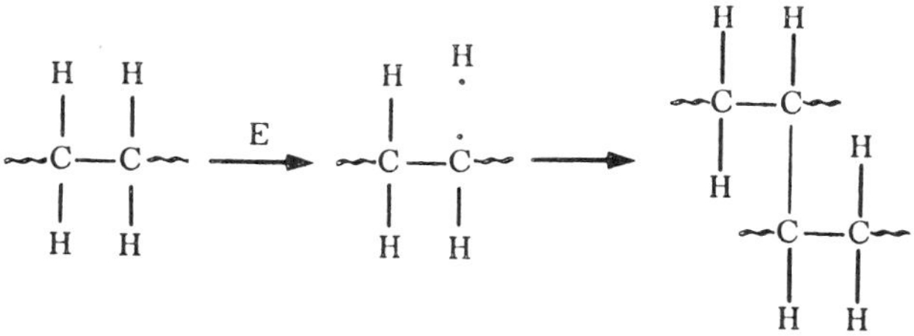

$$\tag{15.32}$$

Radiation-induced reactions occur on and within the body and are particularly important in materials exposed to the outdoors such as human skin and exterior house paints. Specific agents such as UV stabilizers are common additives to exterior paints. Such degradation typically occurs with UV and more energenic radiations but can occur with lower energy radiation (such as microwave) if the energy is quite intense.

For space exploration, high energy radiation and the formation of ionic and free radical species are important causes of polymer degradation. Torre and Pippin recently noted the importance of atomic oxygen in materials' degradation in long-term missions at low Earth orbitals. In general, increases in bond strength and bulk improve stability to atomic oxygen. For instance, the bond strength for C—F is about 20% greater than the C—H bond strength. Complete fluoridation of the ethylene structure (polytetrafluoroethylene) decreases the loss rate by a factor of 12 relative to polyethylene.

15.8 CONDENSATION AND CHELATION REACTIONS

In a broad sense, reactions that occur with smaller molecules also occur with macromolecules. Thus, reactions of acid chlorides with alcohols to give esters can occur whether the acid chloride is contained within a small molecule,

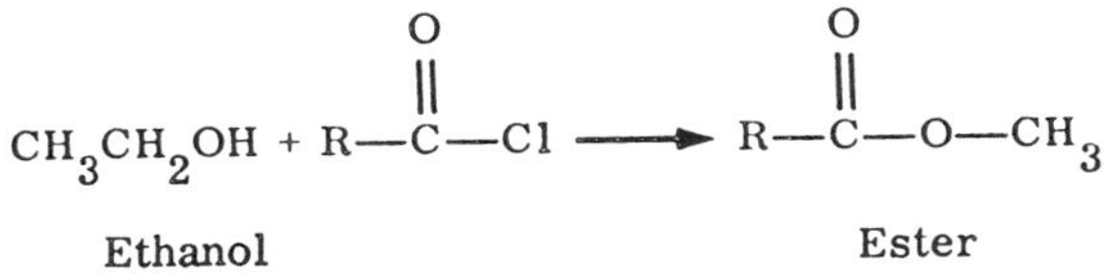

$$
\underset{\text{Ethanol}}{CH_3CH_2OH} + R\overset{\overset{\displaystyle O}{\|}}{-C}-Cl \longrightarrow \underset{\text{Ester}}{R\overset{\overset{\displaystyle O}{\|}}{-C}-O-CH_3} \tag{15.33}
$$

within a polymer,

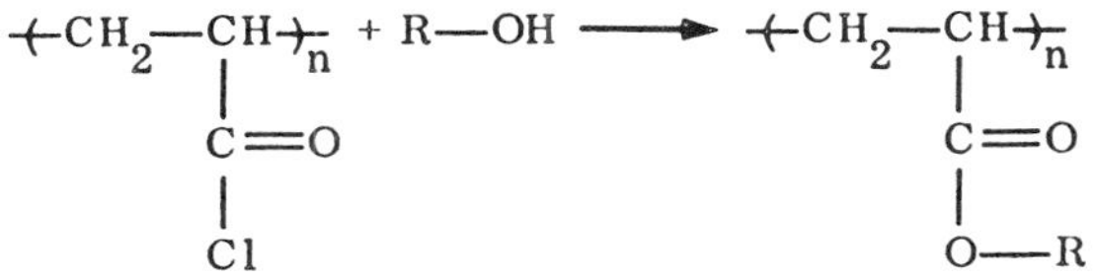

$$
\begin{array}{c} +CH_2-CH\!\!+_n \\ | \\ C=O \\ | \\ Cl \end{array} + R-OH \longrightarrow \begin{array}{c} +CH_2-CH\!\!+_n \\ | \\ C=O \\ | \\ O-R \end{array} \tag{15.34}
$$

or whether the alcohol is contained within a small molecule such as in Eq. (15.33) or within a polymer such as poly(vinyl alcohol).

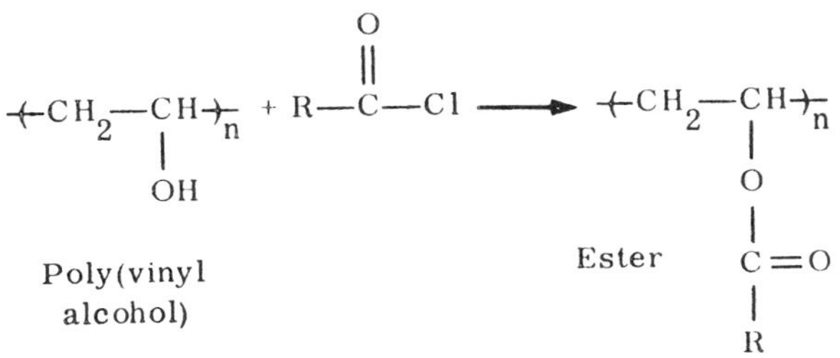

$$
\underset{\substack{\text{Poly(vinyl} \\ \text{alcohol)}}}{\begin{array}{c} +CH_2-CH\!\!+_n \\ | \\ OH \end{array}} + R\overset{\overset{\displaystyle O}{\|}}{-C}-Cl \longrightarrow \underset{\text{Ester}}{\begin{array}{c} +CH_2-CH\!\!+_n \\ | \\ O \\ | \\ C=O \\ | \\ R \end{array}} \tag{15.35}
$$

This is true for both condensation reactions such as those noted above or for chelation reactions such as the following.

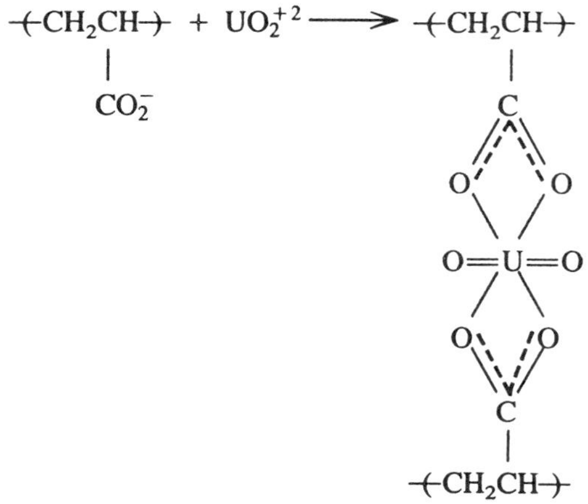

$$\begin{array}{c}
\text{+CH}_2\text{CH+} \ + \ \text{UO}_2^{+2} \longrightarrow \text{+CH}_2\text{CH+} \\
\quad | \qquad\qquad\qquad\qquad\qquad | \\
\quad \text{CO}_2^- \qquad\qquad\qquad\qquad\qquad \text{C} \\
\end{array}$$

(15.36)

These reactions can occur with synthetic polymers and with natural macromolecules such as cellulose, proteins, and nucleic acids.

For most of these chemical modifications, special interactions may be important and can result in the reaction rate being greatly enhanced or decreased. Overberger and others have demonstrated many cases where polymers act as catalysts assisting the modification of the polymer, or specific polymers assisting "small molecule" reactions.

15.9 REACTIVITIES OF END GROUPS

The use of placing specific end groups onto polymers and oligomers is widespread and will be illustrated here for dimethylsiloxanes. These end groups are given various names, often dependent on their intended use. Some of these names are capping agent, blocking agent, chain-length modifier, and coupling agent. For instance, siloxanes containing unreactive trimethylsiloxyl groups are formed from addition of hexamethyldisiloxane to the reaction system. These products can be used as fluids for a variety of applications according to the viscosity of the siloxane, which is dependent on the chain length of the material. Fluid siloxanes are employed as anti-foaming agents, for flow control in coating applications, heat exchangers, in baths and thermostats, as dielectric material in rectifiers and other electronics, and for electronic cooling applications in magnetrons and klystrons. Intermediate molecular weight trimethylsiloxyl-terminated liquids are used as mold release agents, in timing devices, as dielectric fluids, as hydraulic fluids, in inertial guidance systems, in polish formulations, and in grease and oil formulations. High molecular weight materials are employed as internal lubricants and process aids for thermoplastics. They are also used as liquid springs in shock absorbers, as impact modifiers for thermoplastic resins, and as band-ply lubricants in the rubber industry.

$$\text{Polydimethylsiloxane} \quad \underset{\underset{CH_3}{|}}{\overset{\overset{CH_3}{|}}{-\!(\!Si-\!O\!)\!}_x}$$

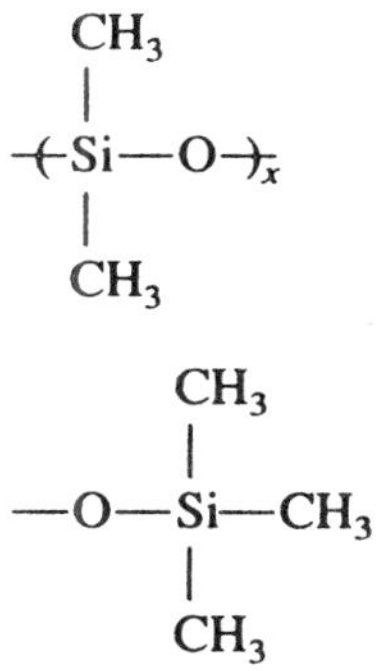

(15.37)

$$\text{trimethylsiloxy group} \quad -O-\underset{\underset{CH_3}{|}}{\overset{\overset{CH_3}{|}}{Si}}-CH_3$$

Reactive end groups such as silanol, alkoxy, vinyl, and hydrogen can be formed from reaction of water, alcohol, divinyltetramethyldisiloxane, or tetramethyldisiloxane, respectively, with chlorosilane end groups.

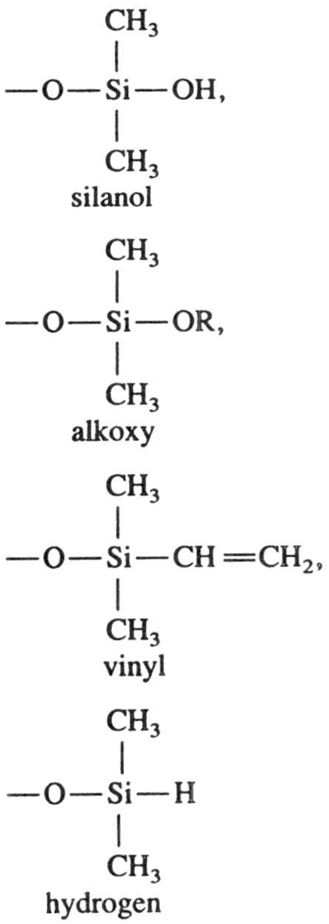

$$-O-\underset{\underset{CH_3}{|}}{\overset{\overset{CH_3}{|}}{Si}}-OH,$$
silanol

$$-O-\underset{\underset{CH_3}{|}}{\overset{\overset{CH_3}{|}}{Si}}-OR,$$
alkoxy

(15.38)

$$-O-\underset{\underset{CH_3}{|}}{\overset{\overset{CH_3}{|}}{Si}}-CH=CH_2,$$
vinyl

$$-O-\underset{\underset{CH_3}{|}}{\overset{\overset{CH_3}{|}}{Si}}-H$$
hydrogen

These reactive end groups can be further reacted to form a wide variety of useful materials. The formation of room temperature vulcanizing liquid rubbers (RTVs) illustrates this point. Most silicone RTV adhesives, sealants, and caulks are moisture curing, that is, they contain a hydroxyl-capped silane that is reacted with acyloxy, amine, oxime, or alkoxyl moisture-sensitive compounds. Reaction with acyloxyl-capped siloxanes results in the formation of acetic acid, producing a "vinegarlike" odor (also see Sec. 11.3).

Acyloxy
$$\equiv Si\text{-}OH \; + \; AcOSi\equiv \quad \longrightarrow \quad \equiv Si\text{-}O\text{-}Si\equiv \; + \; AcOh$$

Amine
$$\equiv Si\text{-}OH \; + \; Me_2NSi\equiv \quad \longrightarrow \quad \equiv Si\text{-}O\text{-}Si\equiv \; + \; Me_2NH$$

Oxime
$$\equiv Si\text{-}OH \; + \; R_2C\equiv NOSi \quad \longrightarrow \quad \equiv Si\text{-}O\text{-}Si\equiv \; + \; R_2C\equiv NOH$$

Alkoxy
$$\equiv Si\text{-}OH \; + \; MeOSi\equiv \quad \longrightarrow \quad \equiv Si\text{-}O\text{-}Si\equiv \; + \; MeOH$$

$$(15.39)$$

15.10 TRANSFER AND RETENTION OF OXYGEN

Today, the polymer chemist should be aware of both synthetic, inorganic, and biological macromolecules. The area of biological macromolecules is large and is one of the most expanding areas of knowledge today. It involves gene splicing and other related biological engineering aspects, neurobiology, many areas of medicine, biological reactions and drugs, neuroreceptors—the very elements of life and death, of thought and action, pain and health, of biological transference, of energy and biological matter. The polymer scientist can learn from these advances but must also contribute to their understanding on a molecular and chain-aggregate level.

The scientist who investigates biological macromolecules works typically with quite complex systems; yet, through persistence, the use of state-of-the-art instrumentation and techniques, and the use of scientific intuition (at times, simply educated guesses), the world of some of these natural macromolecules is yielding information allowing an understanding on a molecular level. A striking example of this involves oxygen transfer and retention of mammals.

Oxygen retention and transfer involves the iron-containing organometallic planer porphyrin-containing structure called *heme* (Fig. 15.1). The iron is bonded through what can be considered classical coordination. The ferrous or iron(II) ion has six coordination sites. Four of these coordination sites are occupied by the nitrogen atoms of the four pyrrole-related rings of the porphyrin shown in Fig. 15.1. A fifth site is occupied by one of the nitrogens of an imidazole side chain found as part of the protein structure and located just opposite the planer porphyrin moiety. The sixth site acts to bind oxygen. The iron remains in the +2 oxidation state, whether oxygen is being bound or the site is vacant. An additional histidine is present, residing in the protein chain opposite the sixth site of the iron atom. This second histidine does not bind iron but serves to stabilize the binding site for oxygen. Experimentally, isolated heme does not bind oxygen. Instead a complex protein wrapping is necessary to both assist binding and protect the binding site from foreign competitor molecules that could render the heme site inactive, either through structural change, change in iron oxidation state, or through occupation of the site, thus preventing oxygen access to the active binding site.

The precise electronic environment of iron deserves special comment. In deoxyhemoglobin, the iron atom has four unpaired electrons, but in oxyhemoglobin iron has no unpaired electrons. The iron in the oxygen-free deoxyhemoglobin is referred to as "high spin" iron, whereas the iron in oxyhemoglobin is called "low spin"

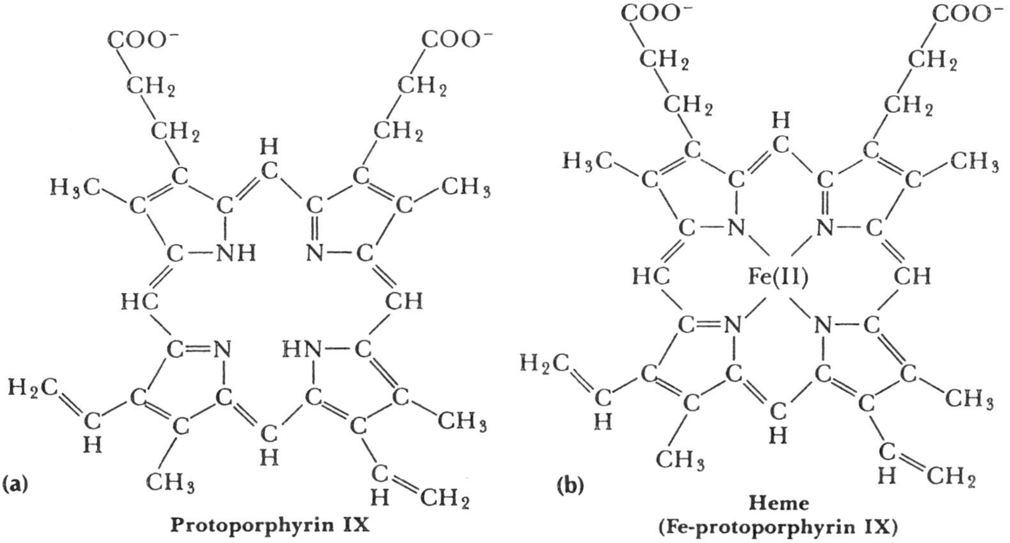

Figure 15.1 (a) Porphyrin structure that serves as the basis of heme. Upon addition of iron, this porphyrin, which is called protoporphyrin IX, forms (b) the heme group.

iron. Hund's rule of maximum multiplicity calls for the most energy favored, lowest energy form to be the structure containing the highest number of unpaired electrons. The binding of oxygen, itself with two unpaired electrons, is probably the result of a favored energy of binding brought about through the coupling of the two sets of unpaired electrons—the favorable energy allowing the violation of Hund's rule.

There are two major protein/heme-binding macromolecules. These are myoglobin (Fig. 15.2), which is used as an oxygen *storage* molecule in mammalian muscle, and hemoglobin, which is active in the *transport* of oxygen. Myoglobin is single-stranded with one heme site per chain, whereas hemoglobin is composed of four protein chains, each one containing a single heme site. There are two sets of equivalent chains composing the quaternary structure of hemoglobin. These two types of chains are referred to as α and β chains. The α chains contain 141 units and the β chains contain 146 units. The myoglobin contains 153 units. Each of these three chains is similar and each forms the necessary environment to allow the heme site to bind oxygen in a reversible manner. The protein segments can be described to be loosely helical with about 60 to 80% of the structure helical. The various segments of these chains are given specific designations for identification purposes. For instance, the histidine that binds iron at the fifth site on the heme moiety resides as a side chain at the F8 location—that is, the eighth amino acid unit in the F segment. The nonbinding histidine is located at the seventh amino acid unit in the E segment, i.e., E7 (Fig. 15.3).

While the three chains are similar in overall structure, there exist somewhat subtle differences. For instance, the quaternary structure of hemoglobin permits interaction between the four chains. Thus, structural movement brought about through binding of oxygen at one of the four heme sites on hemoglobin acts to

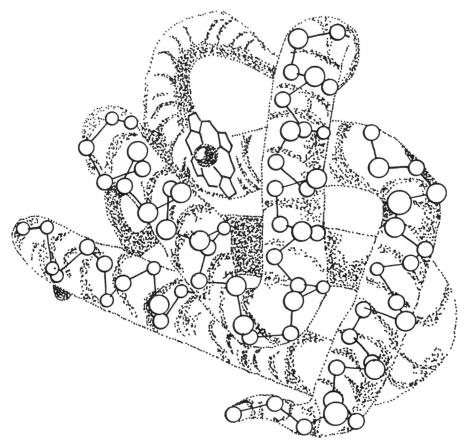

Figure 15.2 Generalized myoglobin structure showing some amino acid units as open circles illustrating the "folded" tertiary structure.

make it easier for subsequent addition of oxygen at other heme sites. Such cooperative binding of oxygen is not possible in the single-chained myoglobin. The result of this cooperative binding is clearly seen in a comparison of oxygen binding by both myoglobin and hemoglobin as a function of oxygen pressure (Fig. 15.4). As seen, the oxygen binding by myoglobin occurs, to a large extent, even at low oxygen pressure, and this behavior is referred to as being hyperbolic. By comparison, the binding by hemoglobin increases more slowly as the pressure of oxygen increases, occurring in what is referred to as a sigmoidal fashion. Thus, the initial binding of oxygen by a heme site on hemoglobin is relatively difficult, becoming increasingly easier as the number of heme sites binding oxygen increase.

On a molecular level, the binding of oxygen is accompanied with the F subunit moving towards the H subunit, forcing the tyrosine amino acid–derived unit located at HC2 to move so that a hydrogen bond is formed with the peptide subunit at FG5. This movement causes a disruption of the ionic bonds that hold the precise structure of the deoxyhemoglobin, i.e., the ionic cross-links. This structural reor-

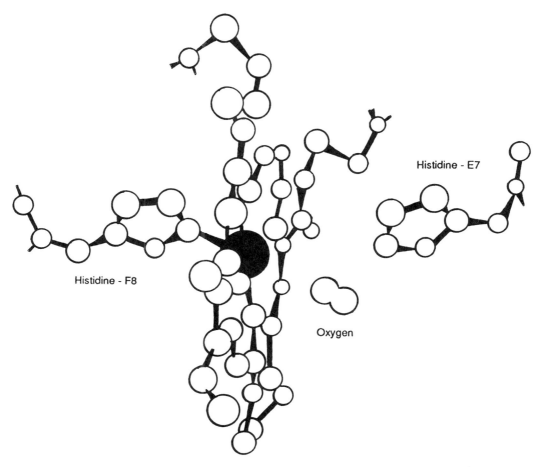

Histidine - E7

Histidine - F8

Oxygen

Figure 15.3 Oxygen-binding site illustrating the positions of the surrounding nitrogen-containing groups of the porphyrin, the binding and nonbinding positions of the histidine groups, and the entrance of oxygen.

ganization is transmitted to the other chains through cooperative interactions, making the remaining heme sites more vulnerable to oxygen binding.

The differences in oxygen-binding characteristics are related to the differing roles of hemoglobin and myoglobin. Thus, myoglobin is employed for the storage of oxygen in muscle. Binding must occur even at low oxygen contents. Hemoglobin is active in the transport of oxygen and becomes saturated only at higher oxygen concentrations. The oxygen content in the alveoli portion of our lungs is of the order of 100 torr (one atmosphere of pressure is 760 torr). Here almost total saturation of the heme-binding sites in hemoglobin occurs. By comparison, the oxygen level in the capillaries of active muscles is of the order of 20 torr, allowing for the hemoglobin to deliver about 75% of its oxygen and for myoglobin to almost reach saturation with respect to oxygen-binding sites.

Conformational changes accompany the binding and release of oxygen. These changes are clearly seen by superimposing the oxygen-containing form of hemo-

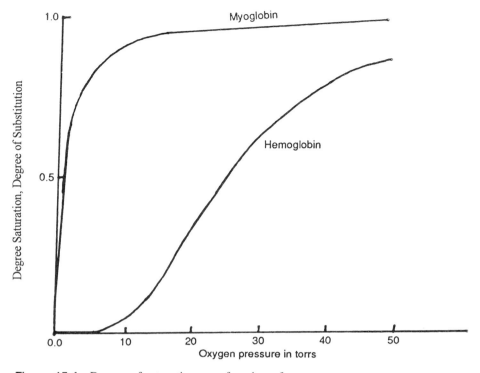

Figure 15.4 Degree of saturation as a function of oxygen pressure.

globin-oxyhemoglobin over the non–oxygen-containing structure of hemoglobin-deoxyhemoglobin (Figs. 15.5 and 15.6).

Bonding changes involving iron also occur as oxygen is bound. For deoxyhemoglobin, the iron is bonded to the four porphyrin nitrogens through electrostatic, ionic bonding, whereas in oxyhemoglobin, the iron is bonded to the nitrogen through covalent bonds.

As noted before, the behavior of hemoglobin involves the interaction between the α and β chains. While these chains are similar, there exist sufficient differences that the presence of four β chains joined similarly to hemoglobin behaves as myoglobin, not hemoglobin.

While typically present in natural deoxyhemoglobin, 2,3-bisphosphoglycerate (BPG) is bound electrostatically to the central core intersecting the four main protein chains of hemoglobin (Fig. 15.7). Figures cited for deoxyhemoglobin behavior are typically for BPG-bound hemoglobin. The presence of BPG has a major affect on the oxygen-binding characteristics of hemoglobin. For instance, for non–BPG-containing hemoglobin, almost all of the heme sites are occupied at a pressure of 10 torr, whereas the bound hemoglobin contains less than 10% of its sites bound by oxygen at 10 torr pressure. As noted before, natural BPG-bound hemoglobin contains less than half its heme sites bound at about 20 torr pressure, allowing it to deliver oxygen to the active muscles.

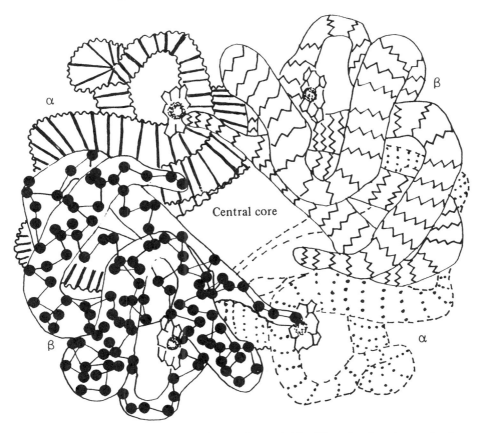

Figure 15.5 Space-filling model of deoxyhemoglobin. Note the four intertwined protein subunits and the four heme sites.

$$
\begin{array}{c}
\underset{\text{O}}{\overset{\text{O}}{\underset{\|}{\text{C}}}}\text{---O}^- \\
\text{H---C---O---}\overset{\overset{\text{O}}{\|}}{\underset{\underset{\text{O}^-}{|}}{\text{P}}}\text{---O}^- \\
\text{H---C---H} \\
\text{O} \\
\text{O}=\overset{}{\underset{\underset{\text{O}^-}{|}}{\text{P}}}\text{---O}^- \\
\end{array}
\qquad (15.40)
$$

Structure of 2,3-bisphosphoglycerate (BPG)

DPG binds to the non–oxygen-binding deoxyhemoglobin influencing favorably its tendency to bind to oxygen. As oxygen is bound, structural changes occur, as noted previously. One pronounced change occurs in the inner core where the BPG

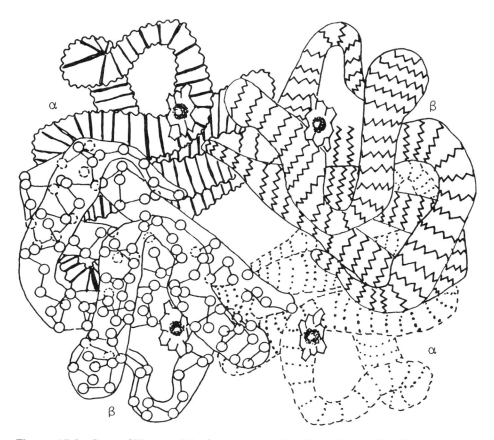

Figure 15.6 Space-filling model of oxyhemoglobin. Note the small shifts in the overall geometry of the various protein chains and the decreased size of the inner core. The top left segment and the lower right segment are known as α chains, while the upper right and lower left chains are referred to as β chains.

resides. The cavity becomes smaller, forcing the exit of DPG. In fact, DPG does not bind fully to oxygenated hemoglobin because the inner core area is too small to accommodate DPG.

Additional factors affect the behavior of hemoglobin. Two major factors are pH and the amount and form of carbon dioxide. We recall that dissolved carbon dioxide resides as a hydrated species that acts as an acid we call carbonic acid, H_2CO_3, or, in reality, H_2OCO_2. In the presence of base, carbon dioxide forms salts, with the combination of carbon dioxide and its salts acting as a buffer preventing or minimizing drastic pH changes. Both carbon dioxide and H^+ bind directly and reversibly to iron at the sixth binding site. At high pH, that is, in highly acidic solutions, the tendency to bind oxygen by hemoglobin is decreased. Increased concentrations of carbon dioxide also act to decrease the oxygen-binding capacity of hemoglobin. Further, the non–oxygen-bound form of hemoglobin has a greater affinity for H^+ than does the bound form. Thus, at high concentrations of H^+ and

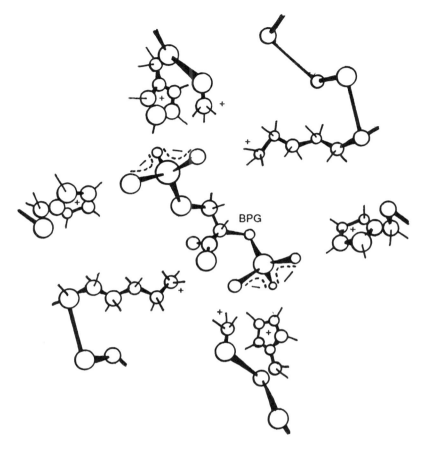

Figure 15.7 Central core binding of 2,3-bisphosphoglycerate (BPG) to deoxyhemoglobin.

carbon dioxide, oxygen binding is discouraged and carbon dioxide binding is favored. High carbon dioxide concentrations are found in active metabolizing muscles where the oxygen concentration is also low. Thus, hemoglobin takes on oxygen in the lungs and transports it to the actively metabolizing sites, exchanging the bound oxygen for carbon dioxide, which is then transferred to the lungs where the carbon dioxide is released and oxygen bound. Of interest is the insensitivity of myoglobin's ability to bind oxygen with respect to pH and the concentration of carbon dioxide.

Changes in overall structure and reactivity are also common for many synthetic polymers. Thus, the neutralization of poly(acrylic acid) causes the polymer chain to become elongated. Associative and electrophilic changes are well known in polymers undergoing modification.

15.11 NATURE'S MACROMOLECULAR CATALYSTS

Probably the most important reaction occurring on polymers involves the catalytic activity of a class of proteins called enzymes. The catalytic action is a result of a lowering of the activation energy for the rate-determining step in the reaction. In general terms, the catalytic action results from the formation of a complex between the enzyme and the molecule undergoing reaction. The decreased activation energy is a result of the reacting molecules(s) being held by the enzyme in such a manner as to favor the appropriate reaction to occur. The two primary models currently employed to describe the formation of the complex between the reacting molecule(s) and the enzyme are the classical "lock-and-key" model and the "induced-fit" model. Briefly, the lock-and-key model calls for an exact or highly similar "complementary fit" between the enzyme and the reacting molecule(s) (Fig. 15.8). Geometry plays an essential part in permitting the electronic (polar and electrostatic) attractions to form the necessary complex. Release is encouraged by the new geometry and electronic distribution of the resulting products of the reaction being dissimilar in comparison to the original reactant molecule(s). The induced-fit model is similar except that the enzyme originally does not fit the required shape. The required shape is achieved upon binding—the binding causing needed "assisting

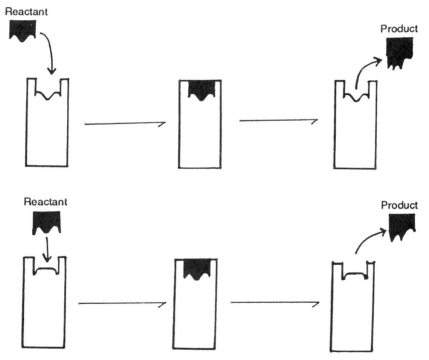

Figure 15.8 The two major models for the binding of reactant molecules to the active sites of nonallosteric enzymes. The top sequence describes the essential steps in the lock-and-key model, where the reactant(s) is attracted to the active site on the enzyme where the active site is a cavity of the same general size, shape, and (complementary) electronic features. Binding occurs and the appropriate reaction(s) occurs resulting in a change in the geometry and electronic configuration of the product, causing it to be released. The second model (below) describes the induced-fit model where the individual steps are similar to the lock-and-key except the reactants "induce" a change in the conformation of the active site on the enzyme, allowing it to accept the reactant(s).

factors" of proximity and orientation to effect a decrease in the energy of the transition state.

Enzyme reactions generally follow one of two kinetic behaviors (see Sec. 15.10). Briefly, the oxygen-binding curve for myoglobin is hyperbolic, while that for hemoglobin is sigmoidal (Fig. 15.9). In general, it is found that similar enzymes such as myoglobin follow a similar hyperbolic relationship between reaction extent and reaction time. More complex enzymes such as hemoglobin follow a sigmoidal relationship between reaction extent and reaction time. The primary difference involves the ability of different portions of the overall hemoglobin structure to affect other, removed reaction sites. Molecules in which various removed sites affect the reactivity of other removed sites are called allosteric enzymes.

The Michaelis-Menten model is commonly employed in describing nonallosteric enzyme reactions. The overall model can be pictured as follows, where E represents the enzyme and M the reacting molecule(s).

$$E + M \underset{k_{-1}}{\overset{k_1}{\rightleftharpoons}} EM \overset{k_2}{\rightarrow} E + P \qquad (15.41)$$

Here EM represents the enzyme complex and P the product(s). The rate of complex formation is described as

$$\text{Rate of complex formation} = \frac{\Delta[EM]}{\Delta t} = k_1 \, [E] \, [M] \qquad (15.42)$$

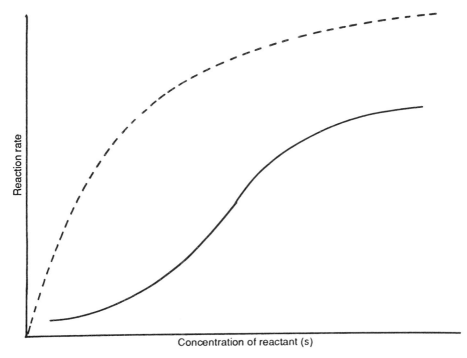

Figure 15.9 Dependence of reaction rate showing sigmoidal (solid line) and hyperbolic (dashed line) behaviors.

The complex then either returns to form the initial reactants or forms the product(s) and the free enzyme. In kinetic terms, the change, or rate of breakdown of the complex is described as

$$\text{Rate of complex change} = -\frac{\Delta[EM]}{\Delta t} = k_1\,[EM] + k_2\,[EM] \qquad (15.43)$$

The negative sign associated with the equation means that the terms are describing the rate of decrease in complex concentration. The rate of complex formation is rapid, and fairly soon the rate at which the complex is formed is equal to the rate at which it breaks down. Such a situation is called a steady state. Mathematically, this is described by

$$\frac{\Delta[EM]}{\Delta t} = -\frac{\Delta[EM]}{\Delta t} \qquad (15.44)$$

and

$$k_1[E]\,[M] = k_{-1}[EM] + k_2\,[EM] \qquad (15.45)$$

Often it is difficult to directly measure the concentration of E as the reaction progresses. Thus, the concentration of E is generally substituted for using the relationship

$$[E] = [E]_0 - [EM] \qquad (15.46)$$

where $[E]_0$ is the initial enzyme concentration.

Substitution of this description of [E] into Eq. (15.45), gives, after collection of the constants,

$$k_1\,([E]_0 - [EM])[M] = k_{-1}\,[EM] + k_2\,[EM] \qquad (15.47)$$

Equation (15.47) can now be solved for [EM], giving

$$[EM] = \frac{[E]_0\,[M]}{K + [M]} \qquad (15.48)$$

where K is a collection of rate constants.

The initial rate of product formation, R_i, for the Michaelis-Menten model depends only on the rate of complex breakdown, that is,

$$R_i = k_2\,[EM] \qquad (15.49)$$

Substitution of Eq. (15.49) into Eq. (15.48) gives

$$\text{Rate}_i = \frac{k_2\,[E]_0\,[M]}{K + [M]} \qquad (15.50)$$

This expression is dependent on the concentration of M and describes the initial part of the plot given in Fig. 15.10.

Generally, the concentration of M far exceeds that of the enzyme sites such that essentially all of the enzyme sites are complexed, that is, $[EM] = [E]_0$. (This is similar to a situation that occurs regularly in south Florida, where four and six lane roads are funnelled into a two-lane section of road. The "rate-determining

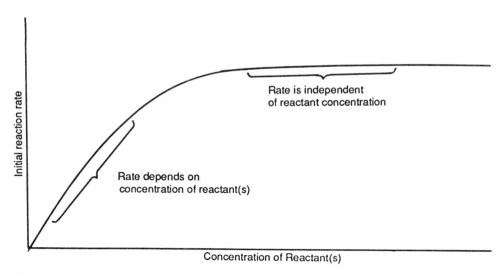

Figure 15.10 Plot of the initial rate of reaction as a function of reactant concentration when the concentration of the enzyme remains constant. The initial reaction rate varies initially until the number of reactants clearly outnumbers the number of reaction sites on the enzyme, at which time the rate becomes zero order, independent of the reactant concentration.

step" is how many cars can get through the "bottleneck," not the number of cars attempting to drive through the two-lane portion. The bottleneck or two-lane portion represents the enzyme active sites, and the automobiles represent the molecules undergoing reaction.) Thus, the rate of product formation is maximized under these conditions. This maximum rate, R_m, allows us to substitute $[E]_0$ for $[EM]$ in Eq. (15.50) to give

$$R_m = k_2 [E]_0 \tag{15.51}$$

Since the enzyme concentration is constant, the rate of product formation under these conditions is independent of $[M]$ and is said to be zero order (Fig. 15.10).

The maximum rate is directly related to the rate at which the enzyme "processes" or permits conversion of the reactant molecule(s). The number of moles of reactants processed per mole of enzyme per second is called the turnover number. Turnover numbers vary widely. Some are high, such as for the scavenging of harmful free radicals by catalase, with a turnover number of about 40 million. Other turnover numbers are small, such as for the hydrolysis of bacterial cell walls by the enzyme lysozyme, with a turnover number of about one-half.

The Michaelis-Menten approach does not describe the behavior of allosteric enzymes, such as hemoglobin, where rate curves are sigmoidal rather than hyperbolic. A more complex model is called for to account for the biofeedback that occurs with allosteroid enzymes. Such affects may be positive, such as those associated with hemoglobin, where binding by one site changes the geometry and electronic environments of the other remaining sites, allowing these additional sites to bind oxygen under more favorable conditions. The affects may also be negative, such as that of cytidine triphosphate, which inhibits ATCase and catalyzes the condensation of aspartate and carbamoyl phosphate-forming carbamoyl aspartate.

Two major models are typically used: the concerted model and the sequential model. In the concerted model, the enzyme has two major conformations—a relaxed form that can bind the appropriate reactant molecule(s) and a tight form that is unable to tightly bind the reactant molecule(s). In this new model, all subunits containing reactive sites change at the same time (Fig. 15.11). An equilibrium exists between the active and inactive structures. Binding at one of the sites shifts the equilibrium to favor the active relaxed form.

The major feature in the sequential model is the induction of a conformational change from the inactive tight form to the active relaxed form as the reacting molecule(s) is bound at one of the sites. This change from an unfavorable to a favorable structure is signaled to other potentially reactive sites bringing about a change to the more favored structural arrangement in these other sites (Fig. 15.12).

Structural changes can be brought about through simple electrostatic and steric events caused by the presence of the reacting molecule(s). Structure changes also result as cross-linking and other primary bonding changes occur. Chymotrypsinogen is formed in the pancreas. It is inactive, fortunately, as it resides in the pancreas. When needed, it is secreted into the small intestine where it is activated by trypsin through cleavage of the peptide bond that resides between the 15 and 16 amino acid units counting from the N-end of the chymotrypsinogen. The rearranged conformation is active and is called pi-chymotrypsin, which folds back on itself to remove two dipeptide fragments—Ser 14–Arg 15 and Thr 147–Asn 148—finally yielding the active form called α-chymotrypsin. In this form, the α-chymotrypsin

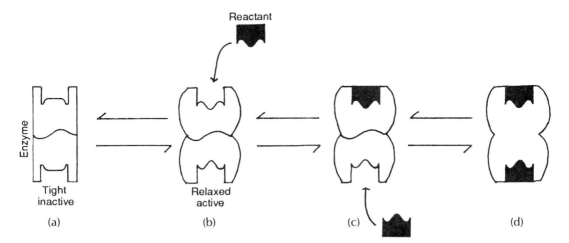

Figure 15.11 Concerted model for allosteric enzymes. The major steps are (a) and (b) an equilibrium exists between the tight and relaxed forms of the allosteric enzyme; (b) the reactant molecule(s) approaches the reactive site of one of the enzyme portions present in the relaxed form; (c) binding occurs, shifting the equilibrium towards the relaxed form(s); (d) the second site is bound.

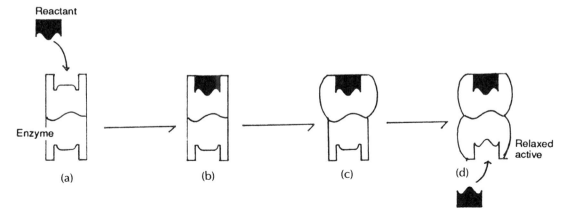

Figure 15.12 Sequential model for allosteric enzymes. The individual steps are: (a) approach of the reactive molecule(s) to the reactive site, which typically has a cavity similar to but not the same as the reactant molecule(s); (b) a conformational change is effected so that the reactive molecule(s) can be bound; (c) the bound portion of the enzyme changes shape; this shape change is transmitted to the other unit(s) containing active sites; (d) the remaining enzyme portion containing active sites undergoes a conformational change that makes binding easier.

catalyzes peptide bonds adjacent to aromatic amino acid–containing sites. While the most important activities of chymotrypsin are those that occur in our bodies, it has also been employed, as have a number of other biologically active agents, to perform selective functions with synthetic molecules. Of interest is the fact that α-chymotrypsin is held together by a set of disulfide cross-links (Fig. 15.13).

Several generalities exist with respect to the nature of the active site. First, the active site typically occupies only a small portion of the enzyme. Some of the remainder of the enzyme acts to create the needed chemical and geometrical environment necessary for the active site to create the needed chemical and geometrical environmental necessary for the active site to function appropriately. Yet, it is still unknown what other functions are carried out by this "extra" bulk. Second, the active site typically exists as a hole, cleft, or crevice. It has a three-dimensionality that allows it to be specific both with respect to the molecules allowed to "approach" it and with respect to providing the needed geometry and electronic arrangement to perform the necessary "catalytic activity." Water is usually not present within the reactive site. Third, the reactive molecule(s) is generally held by numerous secondary forces, including van der Waals forces, hydrogen bonding, and hydrophobic interactions. Fourth, the catalytic site is often mediated by the presence of functional groups such as the side chains of histidine in the case of hemoglobin. Fifth, the activity of enzymes is sensitive to pH and can be sensitive to the presence of other reactive sites. Finally, reactive sites can undergo conformational changes prior to, during, and subsequent to binding.

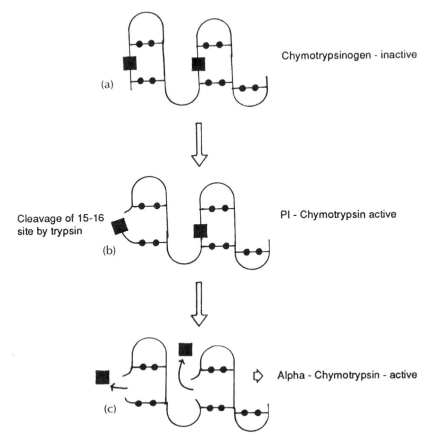

(a) Chymotrypsinogen - inactive

Cleavage of 15-16 site by trypsin

(b) PI - Chymotrypsin active

(c) Alpha - Chymotrypsin - active

Figure 15.13 Description of the activation of chymotrypsinogen. (a) The inactive chymo-trypsinogen with five disulfide cross-links (solid circles); (b) cleavage of the peptide bond between the arginine in the 15 position from the N-terminal end and the isoleucine at the 16 position; (c) the active pi-form of chymotrypsin is now free to move and it carries out three additional cleavages, resulting in the splitting "out" of two dipeptide units and finally the formation of the active α-chymotrypsin.

15.12 MECHANISMS OF ENERGY ABSORPTION

Let us consider a force, stress, acting on a material producing a deformation. The action of this force can be described in terms of two models—a Hookean spring and a Newtonian dashpot (see Secs. 3.1, 5.1, and 5.2). In the Hookean spring, the energy of deformation is stored in the spring and may be recovered by allowing the spring to act on another load or through release of the stress; in either case, the site is returned to zero strain. A Newtonian dashpot is pictorially a frictionless piston and is used to describe chains flowing past one another. The energy of deformation is converted to heat. In actuality, the deformation of most plastics results in some combination of Hookean and Newtonian behavior. The Newtonian behavior results in net energy absorption by stressed material, some of this energy producing the work of moving chains past one another in an irreversible manner while some of the energy is converted to heat.

There are three major mechanisms of energy absorption: shear yielding, crazing, and cracking. The latter two are often dealt with together and called normal stress yielding.

We can distinguish between a crack and a craze. When stresses are applied to polymeric materials, the initial deformation involves shear flow of the macromolecules past one another if it is above T_g, or bond bending, stretching, or breaking for glassy polymers. Eventually a crack will begin to form, presumably at a microscopic flaw, then propagate at high speed, often causing catastrophic failure.

The applied stress results in a realigning of the polymer chains, resulting in greater order but decreased volume taken up by the polymer chains, i.e., an increase in free volume. This unoccupied volume often acts as the site for opportunistic smaller molecules to attack, leading to cracking and crazing and eventual property failure.

A crack is an open fissure, whereas a craze is spanned top to bottom by fibrils that act to resist entrance of opportunistic molecules such as water. Even here some smaller molecular interaction can occur with the void space, and eventually the specimen is weakened.

Crazing and cracking can be induced by stress or combined stress and solvent action. For general polymers they show similar features. To the naked eye, crazing and cracking appear to be a fine, microscopic network of cracks generally advancing in a direction at right angles to the maximum principal stress. Such stress yielding can occur at low stress levels under long-term loading. Suppression of stress yielding has been observed for some polymers by imposition of high pressure.

In shear yielding, oriented regions are formed at 45° angles to the stress. No void space is produced in shear yielding.

Crazing often occurs prior to and in front of a crack tip. As noted before, the craze portion contains both fibrils and small voids that can be exploited after the stress is released or if the stress is maintained, since, while many materials are somewhat elastic, most plastics are not ideal elastomers and additional microscopic voids occur each time a material is stressed.

Aqueous solutions of surface-active materials, such as detergents, can produce brittle cracking, particularly in stressed materials. The term environmental stress cracking (ESC) was introduced to describe such situations. ESC is now more widely applied to describe the promotion of slow, brittle failure in stressed materials by organic substances.

Subjection of polymeric materials to organic liquids and gases can also promote formation of networks of small voids, crazes, particularly in amorphous polymers where voids can be either unfilled or occupied by smaller, somewhat readily removed molecules. These voids are sites of opportunity for organic materials to exploit further.

In both cases, ESC and environmental crazing, direct chemical attack on the polymer chain is minor or not involved. It appears that the organic liquid or other promoting substance is absorbed or locally dissolved at the defects so as to assist further failure, possibly by plasticization of the stressed material or modification of the surface energy.

All three mechanisms result in a difference in the optical properties of the polymeric materials because of the preferential reorientation, with realignment of the polymer chains resulting in a change in optical properties such as refractive index, allowing detection through various optical methods including visual, microscopy and infrared spectroscopy of films. Thus crazed and cracked sites of optically clear materials appear opaque while shear yielded sites may appear to be "wavy" when properly viewed by the naked eye employing even partially refracted light. Thus the three major types of energy absorption can be differentiated though they may appear in combination.

It is important to emphasize that the surface layers of a polymeric material are often different from that of the bulk material and are typically more susceptible to environmental attack. Thus special surface treatments are often employed in an attempt to protect the surface molecules.

15.13 RECENT TRENDS

Today modifications can be roughly grouped into two categories: (1) physical modifications, including entanglement and entrapment and radiation-induced changes, and (2) chemical modifications where chemical reactions on the polymer are emphasized. This distinction is often unclear at best.

Modification through exposure to radiation (thermal, light, and particle) continues to be at the forefront of many areas of polymer modification. A major problem involves use of industrial radiation curing of coatings surfaces because of the present practical limitation in depth of cure penetration.

The problem is common to the application of all industrial coatings. Potential solutions are numerous including: (1) repeatable coatings application (negative features include time, adhesion of the separate coats, and increased energy of radiation (currently largely ruled out due to energy, safety, and cost considerations); (3) formulation of polymer mixtures that can be "set" with radiation but which continue to cure on standing by a slower mechanism; and (4) addition of species which can transfer "captured" radiation to a greater depths.

The advent of computer chips and laser signal and controlling devices will permit more complex modifications to be carried out on an industrial scale.

A remaining problem and one where no real widespread solution has even been (experimentally) proposed is the adequate description of molecular weight of cross-linked materials and the interrelationship(s) of amount and type of cross-linking, polymer molecular weight, and physical and chemical characteristics. Related to this is the need to better control the extent and location (i.e., random, homogeneous, etc.) of modifications on polymers. Some of the good NMR work concerning identification of sequence with copolymers can be utilized in the description of many graph and block copolymers. Mass spectrophotometry utilizing laser excitation of modified polymers may make possible a better description of the actual framework of many cross-linked modified materials since laser excitation allows the examination of both small and large (to greater than 1000 amu) fragments.

The construction of a powerful, continuously variable wavelength laser is approaching reality. Such a laser could be of great use in tailoring polymer modifications through activation of only selected sites for reaction. A number of groups are currently conducting selected reactions utilizing laser energy, so the needed technology is becoming available.

While much of the current and recent research has emphasized modification of synthetic polymers, increasing efforts will undoubtedly focus on the modification of regenerable polymers and on the blending of natural polymers with natural polymers and with synthetic polymers through block, graft, etc., approaches.

The need for replacement of objects currently derived from nonregenerable materials (most plastics, rubbers, elastomers, metals) with objects derived from regenerable materials is critical and must be continually emphasized in our research efforts.

The use of polymers as catalysts may involve the chemical modification of the polymer, with adjacent or neighboring side groups assisting in the catalysis event, or may involve no real modification at all, such as in the case of anchored metal catalysts where the polymer can act as a site for reaction without undergoing permanent chemical modification.

Another area in need of work is on-site grafting—attachment of polymeric materials on biological sites such as particularly badly broken bones. Here the leg is surgically opened and a polymeric material chemically attached after suitable bone activation. The polymeric material degrades after its use period is up. This area is mentioned only to reinforce the notion that interdisciplinary team efforts and polymer chemists with broad training are needed to make the best use of applications of polymer modifications.

The area of delivery of biologically active materials will also involve, in great part, polymer modifications. For instance, Gebelein describes the ideal polymer for good drug delivery as being composed of three parts—one to give the overall polymer the desired solubility, the second containing the drug to be delivered, and the third containing chemical units to direct the overall material only to the site where the drug is to be delivered. It may be possible to combine several of these aspects by a judicious choice of polymeric units, but presently more fruitful approaches include grafting the desired components together forming the needed overall polymeric properties. As a side comment, relatively little work has been done with the generation of "directing groups," and this is an area where much work is needed if the advantages of polymeric drugs are to be realized.

SUMMARY

1. In addition to the reactions occurring during synthesis, telomerization, and cross-linking, polymers, like small molecules, may also react with selected reactants if the reaction sites are accessible.

2. Saturated polymeric hydrocarbons such as HDPE may be chlorinated, sulfochlorinated, or oxidized.

3. Polyenes such as hevea rubber may be hydrogenated, halogenated, hydrohalogenated, cyclized, epoxidized, and ozonized. While many of these plastics are used commercially, ozonide formation is also used to determine the position of double bonds in the polyene.

4. Pendant groups such as the ester groups in poly(vinyl acetate) may be hydrolyzed to yield poly(vinyl alcohol). The latter and other polymeric alcohols, such

as cellulose, undergo typical reactions of alcohols, such as the formation of ethers, xanthates, inorganic and organic esters, and acetals.

5. Esters, amides, or nitriles of polycarboxylic acids may be hydrolyzed to produce polycarboxylic acids. When heated, the latter form polymeric anhydrides, which undergo typical reactions of anhydrides, such as hydrolysis, alcoholysis, and amidation. When heated, polyacrylonitrile forms a dark ladder polymer. Polyamines are produced by the Hofmann degradation of polyamides.

6. Phenyl pendant groups such as those present in polystyrene undergo all the characteristic reactions of benzene, such as alkylation, halogenation, nitration, and sulfonation.

7. Degradation, in which the degree of polymerization is reduced, may occur by random chain scission, by depolymerization, or both. In the former reaction, which may occur in the presence of oxygen, ultraviolet light, heat, and impurities, tertiary carbon atoms are preferentially attacked. The latter is the preferred reaction for 1,1-disubstituted polymers such as PMMA.

8. Biomacromolecular scientists are rapidly understanding more fully biomacromolecules. Much of the work involves investigating reactions involving enzymes where shape, size, and electronic configuration are essential to the catalytic nature of these biomacromolecules. The activity of myoglobin can be described mathematically using the Michaelis-Menten approach. The two major models describing the activity of nonallosteric enzymes such as myoglobin are the lock-and-key model and the induced-fit model. Hemoglobin is an example of an allosteric enzyme where the two popular models are the concerted model and the sequential model.

GLOSSARY

acetate rayon: Cellulose diacetate.

anchimeric reactions: Those enhanced by the presence of neighboring groups.

cellophane: Regenerated cellulose sheet.

Cellosize: Trade name for hydroxyethylcellulose.

cellulose diacetate: Cellulose acetate with a DS of about 2.0.

cellolose triacetate: Cellulose acetate with a DS of about 3.0.

chain scission: Breaking of a polymer chain.

CMC: Carboxymethylcellulose.

CN: Cellulose nitrate.

curing: Cross-linking to produce a network polymer.

cyclized rubber: Isomerized rubber containing cyclohexane rings.

deuterated polyethylene: Polyethylene in which ^1H is replaced by ^2H.

1,1-disubstituted polymers:

DMF: Dimethylformamide.

DMSO: Dimethylsulfoxide.

dope: Jargon for a solution of cellulose diacetate in acetone.

$\overline{DP}$: Average degree of polymerization.

DS: Degree of substitution.

fascias: Panels in the front and rear of an automobile.

halogenation: The reaction of a halogen such as chlorine with a molecule.

heme: Iron containing active site for hemoglobin and myoglobin.

hemoglobin: Allosteric enzymes responsible for the transport of oxygen in our bodies. It contains four "myoglobinlike" protein chains, each containing a single heme active site. The four units are interdependent, working together to give hemoglobin the necessary selectivity and oxygen-binding characteristics.

homolytic cleavage: Breaking of a covalent bond to produce two radicals.

hydrogenation: The addition of hydrogen to an unsaturated molecule.

Hydropol: Trade name for hydrogenated polybutadiene.

Hypalon: Trade name for sulfochlorinated polyethylene.

induced-fit model: One of two basic models employed to describe the enzymatic behavior of nonallosteric molecules. Here the steps are similar to the lock-and-key model except the reactants "induce" a change in the conformation of the active site allowing the active site to bind with the reactant.

isomerization: Term often applied to cyclization reactions of polymers.

Kraton: Trade name for ABA block copolymer of styrene (A) and butadiene (B).

ladder polymer: Polymer having a double-stranded backbone.

LIM: Liquid injection molding.

lock-and-key model: One of two basic models used to describe the selectivity and catalytic nature of nonallosteric enzymes. In the model the reactant(s) is attracted to the active site on the enzyme, which is of the same general size, shape, and complementary electronic nature as the reactant(s).

myoglobin: Nonallosteric enzyme responsible for the storage of oxygen in our bodies. It contains an iron-porphyrin catalytic site responsible of its enzyme activity. The activity follows the kinetic scheme described by Michaelis-Menten.

nitrocellulose: Incorrect name used for cellulose nitrate.

NR: *Hevea brasiliensis*, for natural rubber.

Oxiron: Trade name for epoxidized polybutadiene.

ozonolysis: The reaction of an unsaturated organic compound with ozone followed by cleavage with zinc and water.

Parlon: Trade name for chlorinated rubber.

perfluoropolystyrene: Polystyrene in which all hydrogenatoms have been replaced by fluorine.

Pliofilm: Trade name for rubber hydrochloride.

PMMA: Poly(methyl methacrylate).

polyacrylamide:

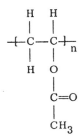

$$\begin{array}{ccc} H & H & \\ | & | & \\ -\!\!\left[-C\!-\!C\!-\right]\!\!-_n & & \\ | & | & \\ H & C\!=\!O & \\ & | & \\ & NH_2 & \end{array}$$

poly(acrylic acid):

$$\begin{array}{ccc} H & H & \\ | & | & \\ -\!\!\left[-C\!-\!C\!-\right]\!\!-_n & & \\ | & | & \\ H & C\!=\!O & \\ & | & \\ & OH & \end{array}$$

poly(vinyl acetate):

$$\begin{array}{ccc} H & H & \\ | & | & \\ -\!\!\left[-C\!-\!C\!-\right]\!\!-_n & & \\ | & | & \\ H & O & \\ & | & \\ & C\!=\!O & \\ & | & \\ & CH_3 & \end{array}$$

poly(vinyl alcohol):

$$\begin{array}{ccc} H & H & \\ | & | & \\ -\!\!\left[-C\!-\!C\!-\right]\!\!-_n & & \\ | & | & \\ H & OH & \end{array}$$

PU: Polyurethane.

PVA: Poly(vinyl alcohol).

PVAc: Poly(vinyl acetate).

PVB: Poly(vinyl butyral).

PVDC: Poly(vinyl dichloride), or chlorinated PVC.

PVF: Poly(vinyl formal).

rayon: Regenerated cellulose filaments.

RIM: Reaction injection molding.

SMA: Copolymer of styrene and maleic anhydride.

soda cellulose: The reaction product of cellulose and sodium hydroxide.

telomerization: Abstraction of an atom by a macroradical.

topochemical reactions: Reactions on the surface of a polymer.

Tornesit: Trade name for chlorinated rubber.

Tyrin: Trade name for chlorinated polyethylene.

viscose process: The production of regenerated cellulose from cellulose xanthate.

Williamson reaction: The reaction of an alkoxide and an alkyl chloride.

xanthate:

$$RO-\underset{\underset{S}{\overset{\parallel}{C}}}{C}-S^-, Na^+$$

EXERCISES

1. What is the general mechanism for the curing of step-reaction polymers?

2. What is the general mechanism for the curing of unsaturated polymers?

3. Write the formula for perdeuterated polyethylene.

4. Explain why good yields of α,ω-dicarboxylic acids can be obtained by the reaction of concentrated nitric acid on crystalline HDPE.

5. Write the formula for diphenylguanidine.

6. How would you cross-link a polyethylene coating after it is applied to a wire?

7. How would you prepare a block copolymer of styrene and an alternating copolymer of ethylene and propylene?

8. What is the difference between completely hydrogenated *Hevea brasiliensis* and completely hydrogenated gutta percha?

9. Explain why the use of rubber hydrochloride film is limited.

10. Which would be more resistant to ozone: (a) *H. brasiliensis*, or (b) cyclized rubber?

11. Name a use for epoxidized vegetable oil.

12. What product would be produced by the ozonolysis of polybutadiene?

13. Write the structural formula for the polymeric hydrolytic products from (a) poly(vinyl acetate), and (b) poly(methyl methacrylate).

14. Why is commercial methylcellulose more soluble in water than cellulose?

15. Why is CMC used in detergent formulations?

16. Propose a mechanism for the use of hydroxyethylcellulose or poly(acrylic acid) as a descaling agent.

17. What is the DS of cellulose nitrate when it is used as an explosive?

18. Why is the DS of cellulose triacetate only 2.8?

19. Which is more polar: (a) cellulose triacetate, or (b) cellulose diacetate?

20. Why is it not possible to prepare pure poly(vinyl butyral)?

21. Why is poly(acrylic acid) an effective flocculating agent?

22. Propose a use for pyrolyzed filaments of polyacrylonitrile.

23. What monomer would be obtained by the decomposition of PVA?

24. Which would be more resistant to nitric acid: (a) polystyrene, or (b) perfluoropolystyrene?

25. What ions would be removed from water by sulfonated polystyrene: (a) cations, or (b) anions?

26. What reaction occurs when tannic acid is added to proteins such as those present in cowhide?

27. What chain-reaction polymerization reactions take place in the molding operation?

28. Propose a procedure for recovering monomeric methyl methacrylate from scrap PMMA.

29. Which will produce the larger yield of monomer when heated at moderate temperatures: (a) polystyrene, or (b) poly-α-methylstyrene?

BIBLIOGRAPHY

Allen, N. S. (1983): *Degradation and Stabilization of Polyolefins*, Applied Science, Essex, England.

Aseeva, R. M., Zaikov, G. E. (1986): *Combustion of Polymer Materials*, Oxford University Press, Oxford, England.

Bamford, C. H., Tipper, C. F. H. (1975): *Degradation of Polymers*, Elsevier, Amsterdam.

Berthelot, P. E. M. (1869): Hydrogenation of rubber, Bull. Soc. Chem. France, *11*:33.

Bovey, F. A. (1958): *The Effects of Ionizing Radiation on Natural and Synthetic High Polymers*, Wiley-Interscience, New York.

Carraher, C. E., Moore, J. A. (1983): *Modification of Polymers*, Plenum, New York.

Carraher, C. E., Sperling, L. H. (1983): *Polymer Applications of Renewable-Resource Materials*, Plenum, New York.

Carraher, C. E., Tsuda, M. (eds.) (1980): *Modification of Polymers*, ACS Symposium Series, New York.

Charlesby, A. (1960): *Atomic Radiation and Polymers*, Pergamon, New York.

Davis, A., Sims, D. (1983): *Weathering of Polymers*, Applied Science, Essex, England.

Dole, M. (1973): *The Radiation Chemistry of Macromolecules*, Academic, New York.

Engelhard, G. A., Day, H. H. (1859): Chlorinated rubber, British Patent 2734.

Fettes, E. M. (ed.): (1964): *Chemical Reactions of Polymers*, Wiley-Interscience, New York.

Geuskens, G. (1975): *Degradation and Stabilization of Polymers*, Halsted, New York.

Grassie, N. (1981): *Developments in Polymer Degradation*, Applied Science, Essex, England.

Grassie, N., McNeill, I. C. (1958): Cyclization of polyacrylonitrile, J. Polymer Sci., *27*:207.

Grot, W. G. F. (1972): Nafions, sulfonated fluorocarbons, Chem. Ing. Tech., *44*:167.

Guilett, J. (1984): *Polymer Photophysics and Photochemistry*, Cambridge University Press, New York.

Hawkins, W. L. (1983): *Polymer Degradation and Stabilization*, Springer-Verlag, New York.

Jellinek, H. H., Kachi, H. (1989): *Degradation and Stabilization of Polymers*, Elsevier, New York.

Kelen, T. (1983): *Polymer Degradation*, Van Nostrand Reinhold, New York.

Kinloch, A. J. (1983): *Durability of Structural Adhesives*, Applied Science, Essex, England.

Klemchuk, P. P. (1985): *Polymer Stabilization and Degradation*, ACS, Washington, D.C.

Liatov, Y. S. (1988): *Colloid Chemistry of Polymers*, Elsevier, New York.

Moiseev, Y., Zaikov, G. E. (1987): *Chemical Resistance of Polymers in Aggressive Media*, Plenum, New York.

Moore, J. A. (1974): *Reactions of Polymers*, D. Reidal Publishing, Boston.

Patsis, A. V. (1989): *Advances in the Stabilization and Controlled Degradation of Polymers*, Technomic, Lancaster, Pennsylvania.

Powers, P. O. (1943): *Synthetic Resins and Rubber*, Part 5, Wiley, New York.

Pummerer, R., Burkhard, P. A. (1922): Hydrogenated rubber, Ber. Bunsenges. Phys. Chem., *55*:3458.

Rabek, J. F. (1990): *Photostabilization of Polymers: Principles and Applications*, Elsevier, New York.

Ranby, B. G., Rabek, J. (1974): *Photodegradation, Photo-oxidation, and Photostabilization of Polymers*, Wiley-Interscience, New York.

Reich, L., Stivala, S. S. (1974): *Elements of Polymer Degradation*, McGraw-Hill, New York.

Seymour, R. B., Branum, I., Hayward, R. W. (1949): Surface reaction of polymers, Ind. Eng. Chem., *41*(7):1479, 1482.

Seymour, R. B., Steiner, R. S. (1955): *Plastics for Corrosion Resistant Applications*, Reinhold, New York.

Starks, C. M. (1974): *Free Radical Telomerization*, Academic, New York.

Staudinger, H., Geiger, E. (1926): Cyclized rubber, Helv. Chim. Acta, *9*:549.

Thies, H. R., Clifford, A. M. (1934): Isomerized rubber, Ind. Eng. Chem., *26*:123.

Wilson, J. E. (1975): *Radiation Chemistry of Monomers, Polymers, and Plastics*, Marcel Dekker, New York.

Wypych, J. (1985): *Polyvinyl Chloride Degradation*, Elsevier, New York.

16

Synthesis of Reactants and Intermediates for Polymers

Many of the difunctional reactants used for the production of step-reaction polymers are standard organic chemicals. However, because the degree of polymerization ($\overline{DP}$) is dependent on high purity, these reactants must be adequately pure.

Impurities in the vinyl monomer may reduce the degree of polymerization through telomerization reactions. Less active impurities may not affect the rate or degree of polymerization, but may be present as contaminants in the polymer.

16.1 REACTANTS FOR STEP-REACTION POLYMERIZATION

Adipic acid (1,4-butanedicarboxylic acid), is used for the production of nylon-66 and may be produced by the oxidation of cyclohexane using air or nitric acid, from furfural or from 1,3-butadiene. As shown by Eq. (16.1), cyclohexane is obtained by the Raney nickel-catalytic hydrogenation of benzene. Both the cyclohexanol and cyclohexanone produced by the cobalt(II) acetate-catalyzed air oxidation of cyclohexane are oxidized to adipic acid by heating with 50% nitric acid; 1.8 billion lbs. of adipic acid are produced annually in the United States.

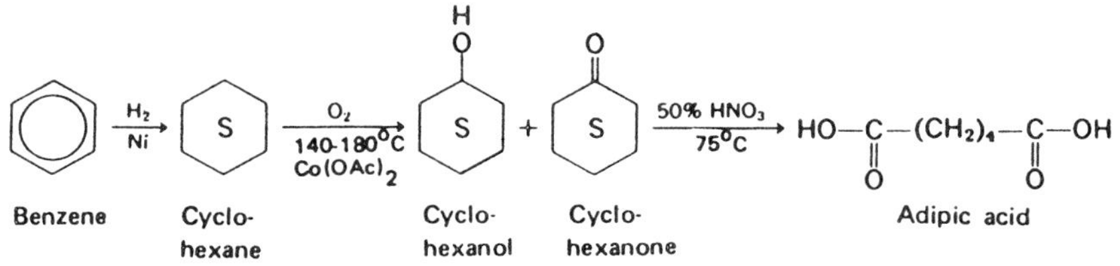

As shown in Eq. (16.2), tetrahydrofuran (THF), which is obtained from furfural, may be carbonylated in the presence of nickel carbonyl–nickel iodide catalyst at a pressure of 2000 atm and at a temperature of 270°C. Furfural is a chemurgic product obtained by the steam-acid digestion of corn cobs, oat hulls, bagasse, or rice hulls.

$$\text{Tetrahydro-} \quad + 2CO + H_2O \xrightarrow[Ni(CO_4)-NiI_2]{270°C} HO-\underset{\underset{O}{\|}}{C}-(CH_2)_4-\underset{\underset{O}{\|}}{C}-OH \qquad \text{Adipic acid} \tag{16.2}$$

Adiponitrile may be produced by the hydrodimerization of acrylonitrile or from 1,3-butadiene via 1,4-dicyanobutene-2.

$$2\underset{\underset{H}{|}}{C}=\underset{|}{\overset{H}{C}}-CN \xrightarrow{(H_2)} NC(CH_2)_4CN \xrightarrow{(H_2)} H_2N(CH_2)_6NH_2 \tag{16.3}$$

Acrylonitrile   Adiponitrile   Hexamethylene-diamine

As shown in Eq. (16.3), hexamethylenediamine (1,6-diaminohexane), which is used for the production of nylon-66, is obtained by the liquid-phase catalytic hydrogenation of adiponitrile or adipamide. As shown in Eq. (16.4), the latter is obtained by the ammonation of adipic acid.

$$HO-\underset{\underset{O}{\|}}{C}-(CH_2)_4-\underset{\underset{O}{\|}}{C}-OH \xrightarrow[350°C]{NH_3} H_2N-\underset{\underset{O}{\|}}{C}-(CH_2)_4-\underset{\underset{O}{\|}}{C}-NH_2 \xrightarrow[125°C \ 600 \ atm]{Co/Cu(H_2)} H_2N(CH_2)_6NH_2 \tag{16.4}$$

Adipic acid   Adipamide   Hexamethylene-diamine

Sebacic acid (decanedioic acid, or 1,8-octane dicarboxylic acid), which has been used for the production of nylon-610, has been produced from 1,3-butadiene and by the dry distillation of castor oil (ricinolein) with sodium hydroxide at 250°C. The cleavage of the ricinoleic acid produces 2-octanol (iso-capryl alcohol) and the sodium salt of sebacic acid.

$$\text{Castor oil} \xrightarrow{OH^-} H_3C(CH_2)_5\underset{\underset{H}{\underset{|}{O}}}{\overset{\overset{H}{|}}{C}}-CH_2-\overset{H}{C}=\overset{H}{C}-(CH_2)_7\underset{\underset{O}{\|}}{C}-OH \xrightarrow[\Delta, \ H_2O]{NaOH}$$

Ricinoleic acid

$$H_3C(CH_2)_5\underset{\underset{H}{\underset{|}{O}}}{\overset{\overset{H}{|}}{C}}CH_3 + Na^+, \ ^-O-\underset{\underset{O}{\|}}{C}-(CH_2)_8-\underset{\underset{O}{\|}}{C}-O^-, \ Na^+ + H_2 \tag{16.5}$$

2-Octanol   Sodium sebacate   Hydrogen

Phthalic acid (1,2-benzene dicarboxylic acid), isophthalic acid (1,3-benzene dicarboxylic acid), and terephthalic acid (1,4-benzene dicarboxylic acid) are made by the selective catalytic oxidation of the corresponding xylenes. Terephthalic acid may also be produced by the isomerization of its isomers by heating the potassium salts at 400°C in the presence of cadmium iodide.

About 8 billion lbs. of terephthalic acid are produced annually in the United States. Phthalic acid is converted to phthalic anhydride when heated. This acid may also be produced by the classic oxidation of naphthalene and by the hydrolysis of terephthalonitrile.

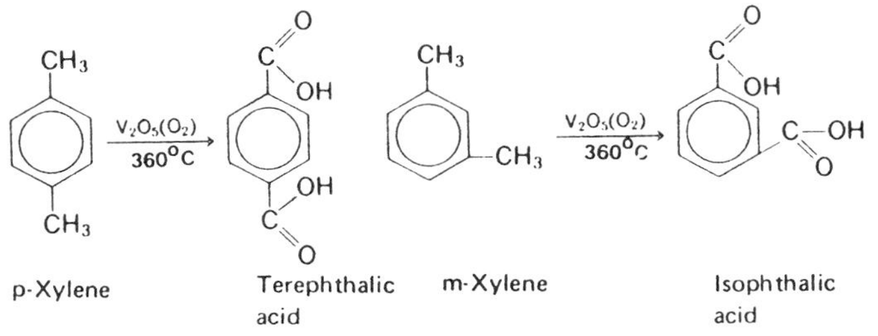

$$(16.6)$$

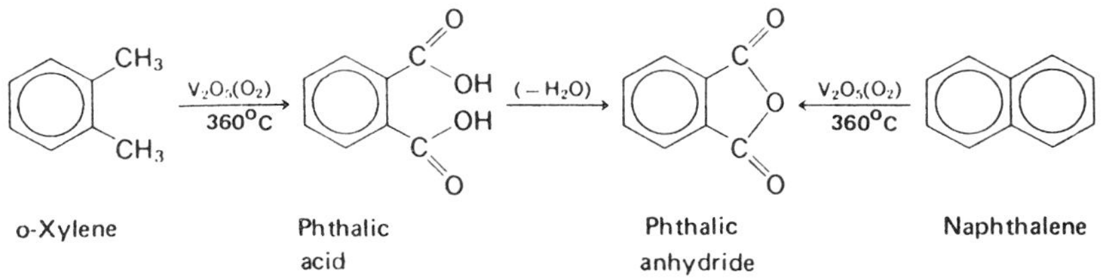

| p-Xylene | Terephthalic acid | m-Xylene | Isophthalic acid |

$$(16.7)$$

| o-Xylene | Phthalic acid | Phthalic anhydride | Naphthalene |

Maleic anhydride (2,5-furandione, toxilic anhydride) is obtained as a byproduct (about 6%) in the production of phthalic anhydride, and by the vapor-phase oxidation of butylene or crotonaldehyde. It is also obtained by the dehydration of maleic acid and by the oxidation of benzene. Maleic anhydride is used for the production of unsaturated polyester resins. This reactant, like most reactants, including benzene, is fairly toxic.

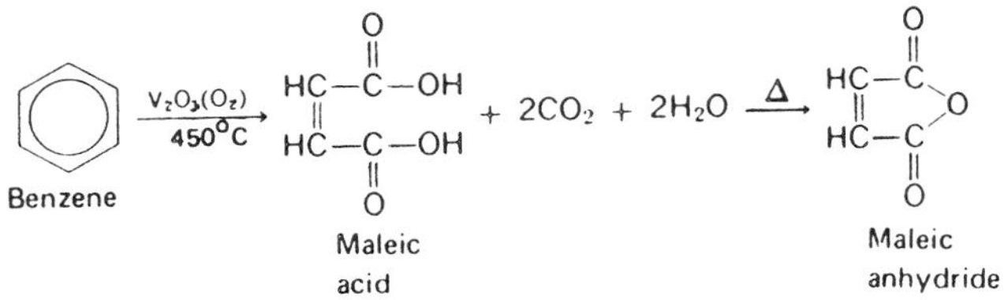

$$(16.8)$$

| Benzene | Maleic acid | Maleic anhydride |

2-Pyrrolidone is a lactone which is used for the production of nylon-4. This reactant may be produced by reductive ammonation of maleic anhydride. ϵ-Caprolactam, which is used for the production of nylon-6, may be produced by the Beckman rearrangement of cyclohexanone oxime [Eq. (16.11)]. The oxime may be produced by the catalytic hydrogenation of nitrobenzene, the photolytic nitrosylation of cyclohexane [Eq. (16.9)], or the classic reaction of cyclohexanone and hydroxylamine [Eq. (16.10)]. ϵ-Caprolactam is also produced by the nitrosylation of cyclohexane carboxylic acid. The latter is obtained by the hydrogenation of benzoic acid, which is obtained by the oxidation of toluene. Nearly one-half of the production of caprolactam is derived from phenol.

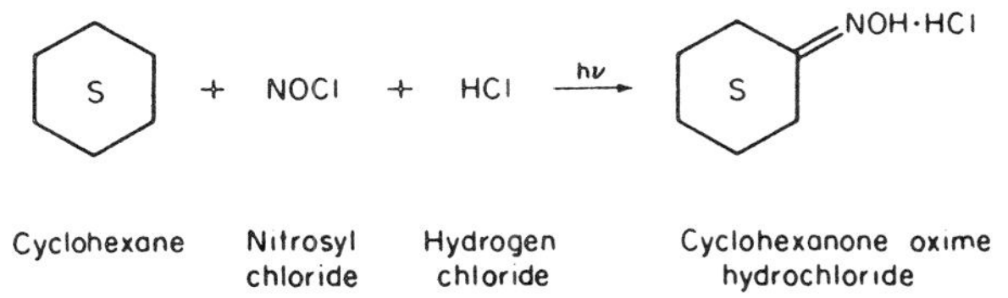

(16.9)

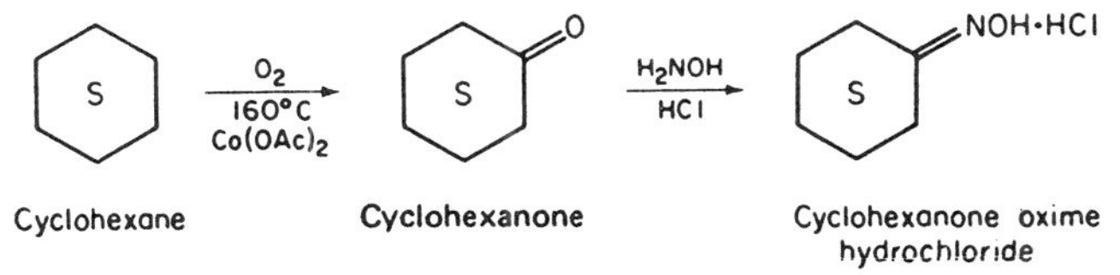

Cyclohexane   Nitrosyl   Hydrogen   Cyclohexanone oxime
         chloride   chloride   hydrochloride

(16.10)

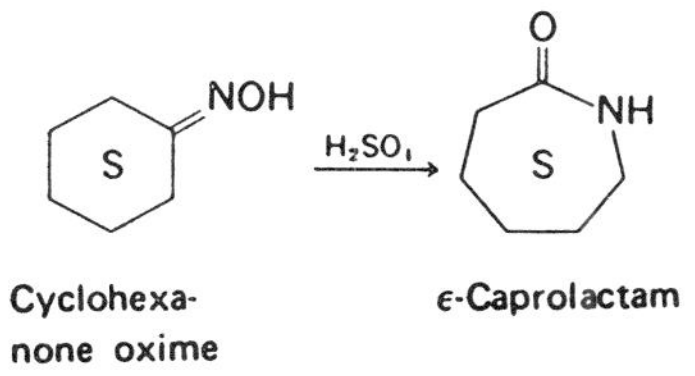

Cyclohexane   Cyclohexanone   Cyclohexanone oxime
                                           hydrochloride

(16.11)

Cyclohexa-   ε-Caprolactam
none oxime

Ethylene oxide (oxirane), used for the production of ethylene glycol and poly(ethylene oxide), is obtained by the catalytic oxidation of ethylene [Eq. (16.12)]. Ethylene glycol, used for the production of polyesters, is produced by the hydrolysis of ethylene oxide [Eq. (16.12)]. Ethylene oxide and ethylene glycol are produced at an annual rate of 7.6 billion lbs. and 5.2 billion lbs., respectively, in the United States.

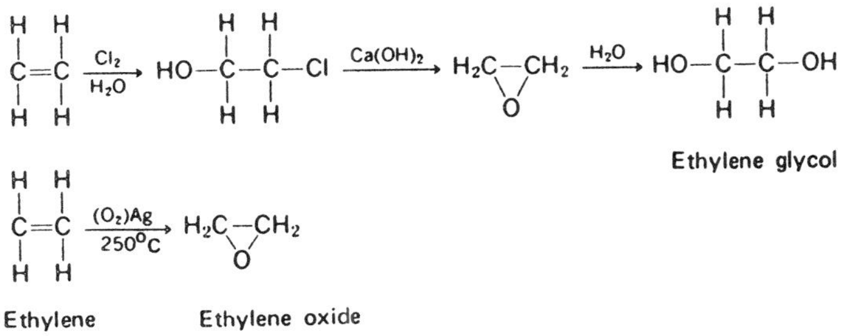

Ethylene glycol

(16.12)

Ethylene   Ethylene oxide

Propylene oxide is obtained by the oxidation of propylene in a similar manner.

Glycerol (glycerin), used for the production of alkyds, is produced by the catalytic hydroxylation or the hypochlorination of allyl alcohol. The latter is produced by the reduction of acrolein, which is obtained by the oxidation of propylene [Eq. (16.13)].

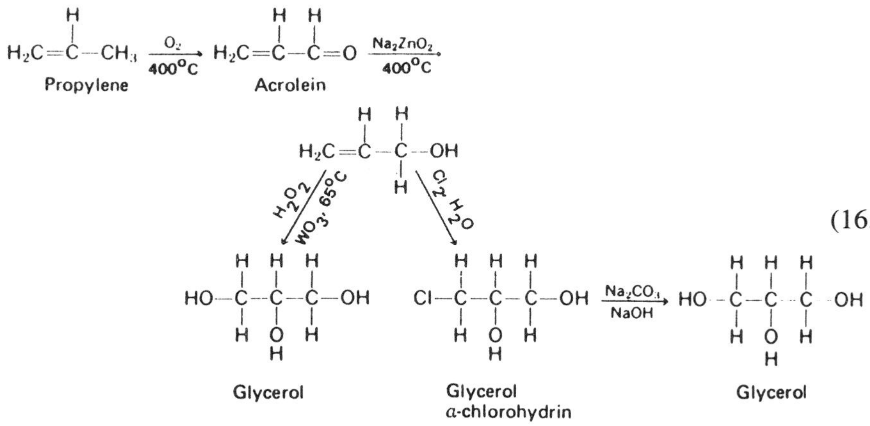

(16.13)

Pentaerythritol, used in the production of alkyds, is produced by a crossed Cannizzaro reaction of the aldol condensation product of formaldehyde and acetaldehyde. The byproduct, calcium formate, serves as the major source of formic acid.

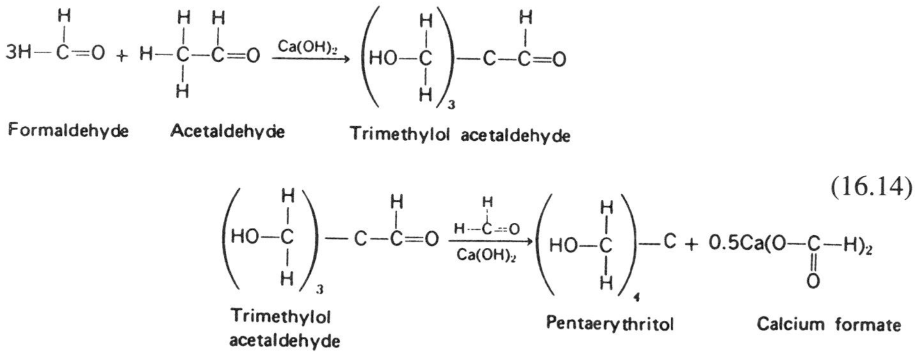

(16.14)

2,4-Toluene diisocyanate (TDI), used for the production of polyurethanes and polyureas, is obtained by the phosgenation of 2,4-toluenediamine. Phosgene is obtained by the reaction of chlorine and carbon monoxide. TDI may also be produced by the catalytic liquid-phase reductive carbonylation of dinitrotoluene.

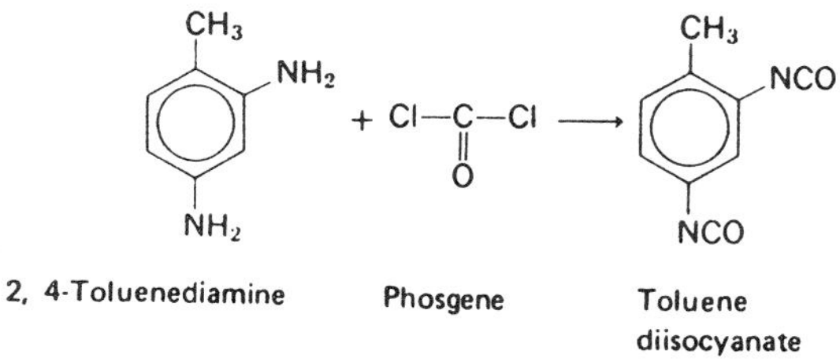

(16.15)

Formaldehyde, which is used for the production of phenolic and amino resins, is produced at an annual rate of over 7 billion lbs. by the catalytic hot-air oxidation of methanol. Hexamethylenetetramine (hexa) is produced by the condensation of ammonia and 30% aqueous formaldehyde (formalin).

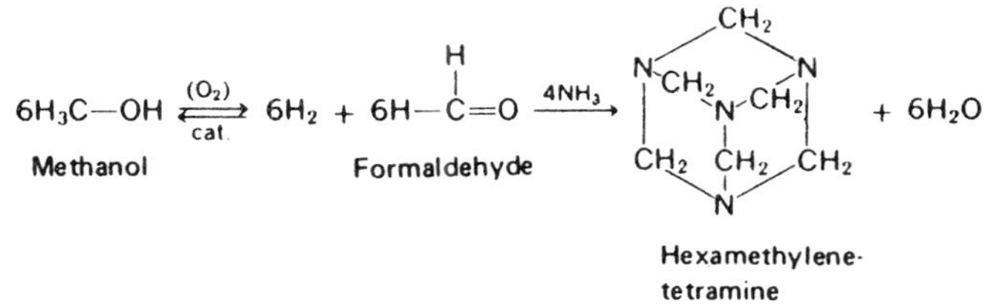

(16.16)

While Dow continues to produce some of its phenol by the nucleophilic substitution of chlorine in chlorobenzene by the hydroxyl group [Eq. (16.17)], most synthetic phenol (carbolic acid, phenylic acid) is produced by the acidic decomposition of cumene hydroperoxide [Eq. (16.18)]. The cumene is obtained by the oxidation of cumene hydroperoxide.

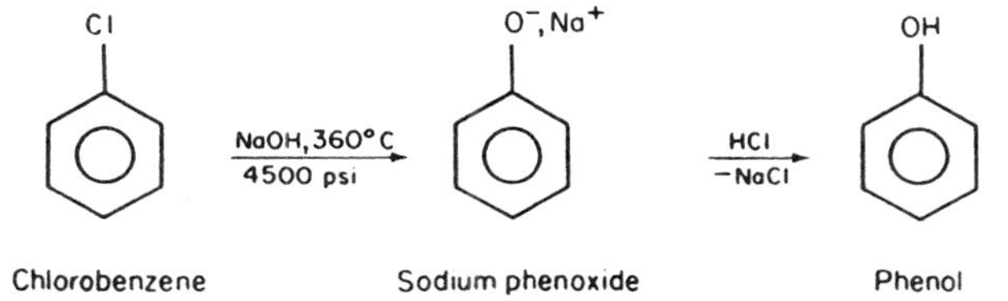

(16.17)

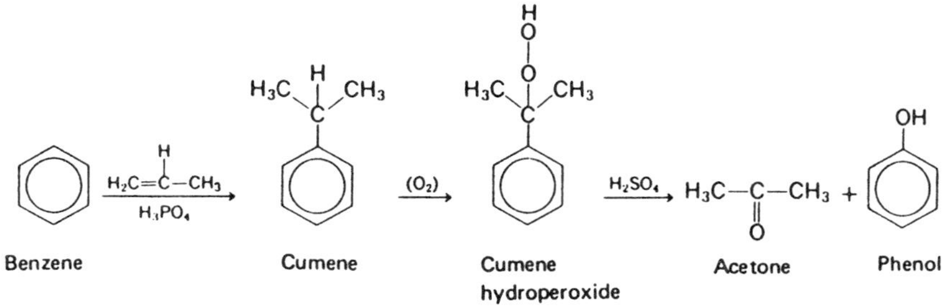

(16.18)

Some of the newer processes for synthesizing phenol are the dehydrogenation of cyclohexanol, the decarboxylation of benzoic acid, and the hydrogen peroxide hydroxylation of benzene. Phenol, which is used for production of phenol-formaldehyde resins as well as a starting material for the synthesis of many important materials, is produced at an annual rate of over 4 billion lbs.

Urea (carbamide), used for the production of urea-formaldehyde resins, is produced at an annual rate of about 15 billion lbs. by the in situ decomposition of ammonium carbamate at 5 atm pressure. The latter is obtained by the condensation of liquid ammonia and liquid carbon dioxide at a temperature of 200°C and at a pressure of 200 atm.

$$\underset{\text{Ammonia}}{2H_3N} + \underset{\text{Carbon dioxide}}{CO_2} \longrightarrow \underset{\text{Ammonium carbamate}}{H_2N{-}\overset{\overset{\displaystyle O}{\|}}{C}{-}ONH_4} \longrightarrow H_2O + \underset{\text{Urea}}{H_2N{-}\underset{\underset{\displaystyle O}{\|}}{C}{-}NH_2} \qquad (16.19)$$

Melamine (cyanuramide) is obtained by heating dicyanodiamide (dicy) at 209°C [Eq. (16.20)]. The latter is obtained by heating cyanamide at 80°C. Melamine, which is used for the production of melamine-formaldehyde resins, is also obtained by heating urea [Eq. (16.21)].

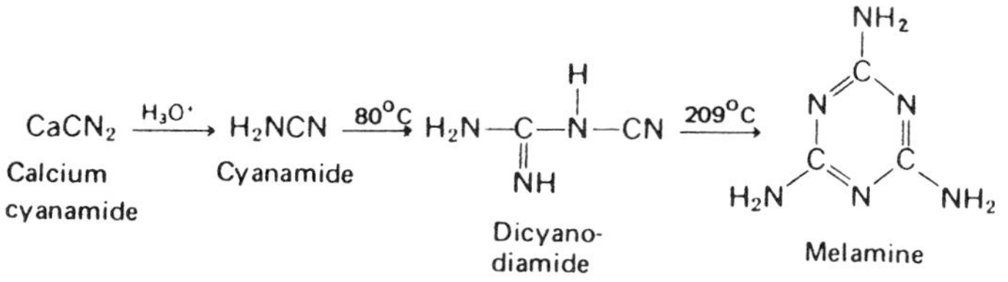

$$\text{(16.20)}$$

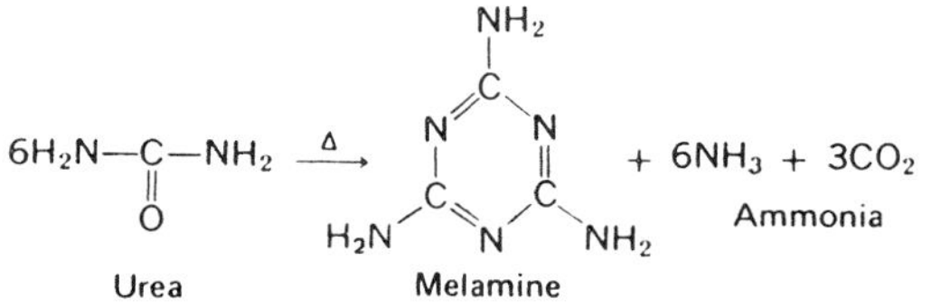

$$\text{(16.21)}$$

Bisphenol A [(bis-4-hydroxyphenyl)dimethylmethane], used for the production of epoxy resins and polycarbonates, is obtained by the acidic condensation of phenol and acetone. In this reaction, the carbonium ion produced by the protonation of acetone attacks the phenol molecule at the para position to produce a quinoidal oxonium ion which loses water and rearranges to a p-isopropylphenol carbonium ion. The water attacks another phenol molecule in the para position to produce a quinoidal structure which rearranges to bisphenol A. Recently, it has been found that bisphenol A is an endocrine-system disrupter. The consequences of this are still being determined.

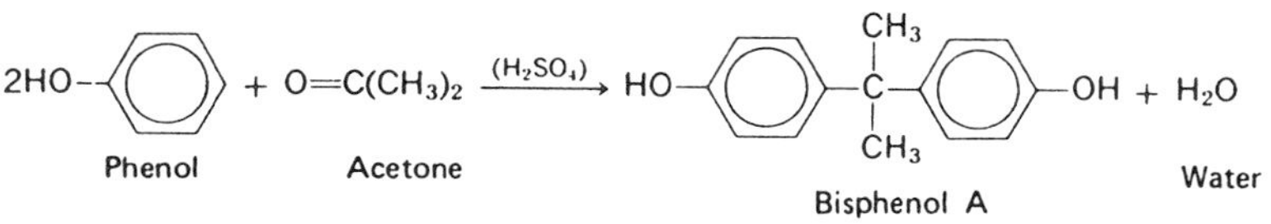

$$\text{(16.22)}$$

Epichlorohydrin (chloropropylene oxide), used for the production of epoxy resins, is obtained by the dehydrochlorination of glycerol α,β-dichlorohydrin (2,3-dichloro-1-propanol). The hydrin is produced by chlorohydrination of allyl chloride.

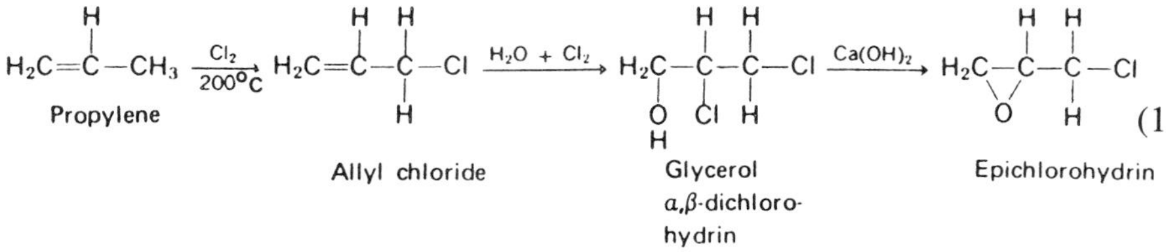

$$\text{(16.23)}$$

Methyltrichlorosilane is readily produced by the Grignard reaction of silicon tetrachloride and methylmagnesium chloride [Eq. (16.28)]. Dimethyldichlorosilane is obtained by the reaction of methylmagnesium chloride and methyltrichlorosilane [Eq. (16.25)]. Since the silicon atom is less electronegative than carbon, chlorine is much more electronegative than silicon and is a good leaving group in nucleophilic substitution reactions.

$$H_3CMgCl \quad + \quad SiCl_4 \quad \rightarrow \quad H_2CSiCl_3 \quad + \quad MgCl_2$$

Methylmagnesium	Silicon	Methyltri-	Magnesium
chloride	tetrachloride	chlorosilane	chloride

(16.24)

$$H_3CMgCl \quad + \quad H_3CSiCl_3 \quad \rightarrow \quad (H_3C)_2SiCl_2 \quad + \quad MgCl_2$$

Methylmagnesium	Methyltri-	Dimethyldi-	Magnesium
chloride	chlorosilane	chlorosilane	chloride

(16.25)

16.2 SYNTHESIS OF VINYL MONOMERS

Styrene, which is one of the most important aromatic compounds, is produced at an annual rate of about 11 billion lbs. by the catalytic vapor-phase dehydrogenation of ethylbenzene. As shown in Eq. (16.26), the latter is obtained by the Friedel-Crafts condensation of ethylene and benzene. Styrene can also be produced by the palladium acetate–catalyzed condensation of ethylene and benzene and by the dehydration of methyl phenyl carbinol obtained by the propylation of ethylbenzene hydroperoxide. Because of its toxicity, the concentration of styrene in the atmosphere must be limited to a few parts per million.

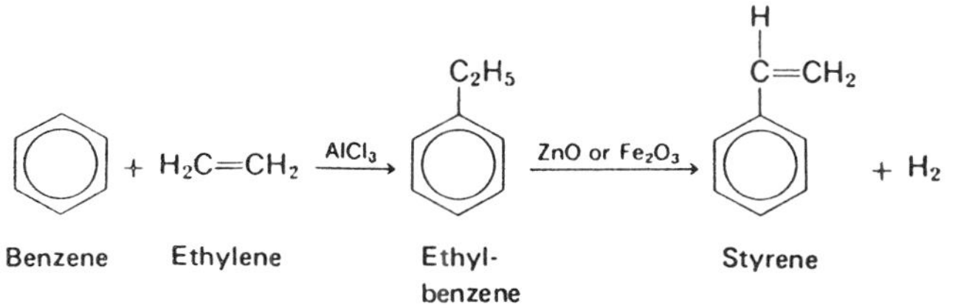

| Benzene | Ethylene | Ethyl-benzene | Styrene |

(16.26)

A process to produce styrene monomer and propylene oxide simultaneously was introduced in 1969.

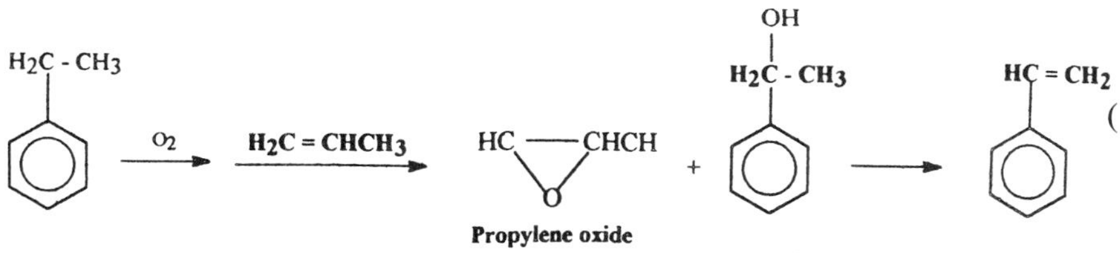

Propylene oxide

(16.27)

Vinyl chloride (chloroethene, VCM), which was formerly obtained from acetylene, is now produced at an annual rate of about 15 billion lbs. by the trans catalytic process in which chlorination of ethylene, oxychlorination of byproduct hydrogen chloride, and dehydrochlorination of ethylene dichloride take place in a single reactor.

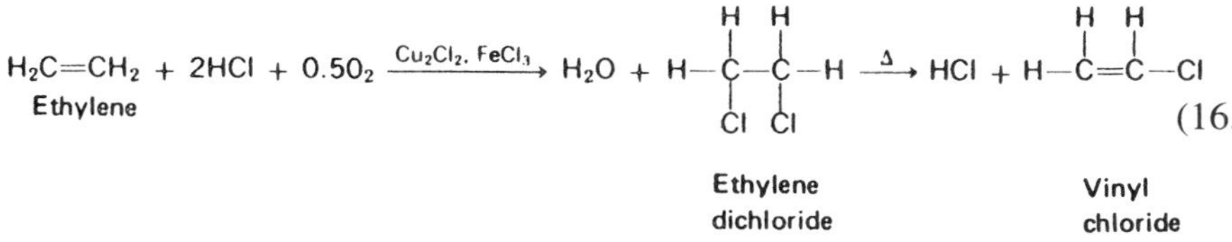

(16.28)

| Ethylene dichloride | Vinyl chloride |

VCM is also produced by the direct chlorination of ethylene and the reaction of acetylene and HCl [Eq. (16.29)]. The HCl generated in the chlorination of ethylene can be employed in the reaction with acetylene allowing a useful coupling of these two reactions [Eq. (16.30)].

$$CH_2{=}CH_2 + Cl_2 \longrightarrow ClCH_2CH_2Cl \xrightarrow[-HCl]{\Delta} CH_2{=}CHCl \tag{16.29}$$

$$HC{\equiv}CH \xrightarrow[HgCl_2]{HCl} CH_2{=}CHCl \tag{16.30}$$

Vinylidene chloride is produced by the pyrolysis of 1,1,2-trichloroethane at 400°C in the presence of lime or caustic. Since both vinyl chloride and vinylidene chloride are carcinogenic, their concentrations in air must be kept to a few parts per million or less.

$$\underset{\substack{\text{1, 1, 2-Trichloroethane}}}{\overset{\substack{H \quad H \\ | \quad | \\ }}{Cl-\underset{\substack{| \\ H}}{C}-\underset{\substack{| \\ Cl}}{C}-Cl}} \xrightarrow{400°C} \underset{\substack{\text{Vinylidene}\\\text{chloride}}}{\overset{\substack{H \quad Cl \\ | \quad | \\ }}{H-C{=}C-Cl}} + HCl \tag{16.31}$$

Vinyl acetate, which is produced annually at a rate of 2.9 billion lbs., was formerly obtained by the catalytic acetylation of acetylene. However, this monomer is now produced by the catalytic oxidative condensation of acetic acid and ethylene [Eq. (16.32)]. Other vinyl esters may be obtained by the transesterification of vinyl acetate with higher boiling carboxylic acids.

$$\underset{\text{Ethylene}}{H_2C{=}CH_2} + \underset{\text{Acetic acid}}{\overset{\substack{ }}{H_3C-\underset{\substack{\| \\ O}}{C}-OH}} \xrightarrow[\Delta, \text{ cat.}]{O_2} \underset{\text{Vinyl acetate}}{\overset{\substack{H \quad H \quad\quad O \\ | \quad | \quad\quad \| \\ }}{HC{=}C-O-C-CH_3}} \tag{16.32}$$

Acrylonitrile (vinyl cyanide) is produced at an annual rate of 700,000 tons by the ammoxidation of propylene. Since this monomer is carcinogenic, considerable care must be taken to minimize exposure to acrylonitrile.

$$\underset{\text{Propylene}}{\overset{\substack{H \quad H \quad H \\ | \quad | \quad | \\ }}{H-\underset{\substack{| \\ H}}{C}-C{=}C-H}} + \underset{\text{Ammonia}}{NH_3} + 1.5O_2 \xrightarrow[BiP(Mo_3O_{10})_4]{300\text{-}540°C} \underset{\text{Acrylonitrile}}{\overset{\substack{H \\ | \\ }}{H_2C{=}C-CN}} + 3H_2O \tag{16.33}$$

Ethylene, propylene, and butylene, are produced at annual rates of 47 billion, 25 billion, and 0.4 billion lbs., respectively, by the vapor-phase cracking of light oil fractions of petroleum feedstocks or petroleum gases. The acetylenes, produced as byproducts, are selectively absorbed, and the monomers are obtained by fractional distillation.

Tetrafluoroethylene is obtained by the thermal dehydrochlorination of chlorodifluoromethane [Eq. (16.35)]. The latter is produced by the reaction of chloroform and hydrogen fluoride [Eq. (16.34)].

$$HF + HCCl_3 \longrightarrow HCl + HCClF_2 \qquad (16.34)$$

Hydrogen   Chloroform   Hydrogen   Chlorodi-
fluoride   chloride   fluoromethane

$$2CClF_2H \xrightarrow[600^\circ]{\Delta} 2HCl + F_2C{=}CF_2 \qquad (16.35)$$

Chlorodi-   Hydrogen   Tetrafluoro-
fluoromethane   chloride   ethylene

Trifluoromonochloroethylene is obtained by the zinc metal dechlorination of trichlorotrifluoroethane. The latter is produced by the fluorination of hexachloroethane.

$$Cl_3C{-}CCl_3 \xrightarrow{SbF_3{-}SbCl_3} F_2\overset{Cl}{\underset{}{C}}{-}\overset{F}{\underset{}{CCl_2}} \xrightarrow[-Cl_2]{Zn} F_2C{=}\overset{F}{\underset{}{CCl}} + ZnCl_2 \qquad (16.36)$$

Hexachloro-   Trichlorotri-   Trifluoro-   Zinc
ethane   fluoroethane   monochloro-   chloride
                                                       ethylene

Vinylidene fluoride is produced by the thermal dehydrochlorination of 1,1,1-monochlorodifluoroethane.

$$\overset{H}{\underset{H}{HC}}{-}\overset{F}{\underset{F}{CCl}} \xrightarrow{\Delta} H_2C{=}CF_2 + HCl \qquad (16.37)$$

1,1,1-Monochlorodi-   Vinylidene   Hydrogen
fluoroethane   fluoride   chloride

Vinyl fluoride may be obtained by the catalytic hydrofluorination of acetylene.

$$HC{\equiv}CH + HF \xrightarrow{HgCl_2{-}BaCl_2} H_2C{=}\overset{H}{\underset{}{CF}} \qquad (16.38)$$

Acetylene   Hydrogen   Vinyl
               fluoride   fluoride

Vinyl ethyl ether is obtained by the ethanolysis of acetylene in the presence of potassium ethoxide.

$$HC{\equiv}CH + H{-}\overset{H}{\underset{H}{C}}{-}\overset{H}{\underset{H}{C}}{-}OH \xrightarrow[130\text{-}180^\circ C]{KOC_2H_5} H_2C{=}\overset{H}{\underset{}{C}}{-}OC_2H_5 \qquad (16.39)$$

Acetylene   Ethanol   Vinyl ethyl ether

1,3-Butadiene, which is used for the production of elastomers, is produced at an annual rate of 3.7 billion lbs. by the catalytic thermal cracking of butane and as a byproduct of other cracking reactions.

$$\underset{\text{Butane}}{\text{H}-\overset{\overset{\text{H}}{|}}{\underset{\underset{\text{H}}{|}}{\text{C}}}-\overset{\overset{\text{H}}{|}}{\underset{\underset{\text{H}}{|}}{\text{C}}}-\overset{\overset{\text{H}}{|}}{\underset{\underset{\text{H}}{|}}{\text{C}}}-\overset{\overset{\text{H}}{|}}{\underset{\underset{\text{H}}{|}}{\text{C}}}-\text{H}} \xrightarrow[\text{Cr}_2\text{O}_3-\text{Al}_2\text{O}_3]{600-700^\circ\text{C}} \underset{\text{1, 3-Butadiene}}{\text{H}_2\text{C}=\overset{\overset{\text{H}}{|}}{\text{C}}-\overset{\overset{\text{H}}{|}}{\text{C}}=\text{CH}_2} + \text{other products} \tag{16.40}$$

The isoprene monomer is not readily available from direct cracking processes. Several routes are employed for its synthesis. One route begins with the extraction of isoamylene fractions from catalytically cracked gasoline streams. Isoprene is produced by subsequent catalytic dehydrogenation.

$$\underset{}{\text{CH}_3-\overset{\overset{\text{CH}_3}{|}}{\text{C}}=\text{CH}-\text{CH}_3} + \underset{}{\text{CH}_2=\overset{\overset{\text{CH}_3}{|}}{\text{C}}-\text{CH}_2\text{CH}_3} \xrightarrow{-\text{H}_2} \underset{\text{Isoprene}}{\diagup\!\!=\!\!\diagdown} \tag{16.41}$$

Dimerization of propylene is also used to produce isoprene. Several steps are involved [Eq. (16.42)]. First, dimerization of propylene to 2-methyl-1-pentene occurs. Then isomerization to 2-methyl-2-pentene is effected. Finally, the 2-methyl-2-pentene is pyrolized to isoprene and methane. Another multistep synthesis starts with acetylene and acetone. Perhaps the most attractive route involves formaldehyde and isobutylene.

$$\underset{\text{Isobutylene}}{\diagdown\!\!=} + \underset{\text{Formaldehyde}}{2\,\text{CH}_2\text{O}} \longrightarrow \underset{}{\diagup\!\!\diagdown_{\text{O}}^{\text{O}}} \longrightarrow \underset{\text{Isoprene}}{\diagup\!\!=\!\!\diagdown} + \text{CH}_2\text{O} + \text{H}_2\text{O} \tag{16.42}$$

Chloroprene, used for the production of neoprene rubber, is obtained by the dehydrochlorination of dichlorobutene. The latter is produced by the chlorination of 1,3-butadiene, which in turn is synthesized from acetylene.

$$\underset{\text{Acetylene}}{2\text{HC}\!\equiv\!\text{CH}} \xrightarrow{\text{CuCl}} \underset{\text{1,3-Butadiene}}{\text{H}_2\text{C}=\overset{\overset{\text{H}}{|}}{\text{C}}-\overset{\overset{\text{H}}{|}}{\text{C}}=\text{CH}_2} \xrightarrow{\text{Cl}_2} \underset{\underset{\text{Dichlorobutene}}{}}{\text{H}_2\text{C}-\overset{\overset{\text{Cl}}{|}}{\underset{\underset{\text{H}}{|}}{\text{C}}}-\overset{\overset{\text{Cl}}{|}}{\text{C}}=\text{CH}_2} \xrightarrow{\Delta} \underset{\text{Chloroprene}}{\text{H}_2\text{C}=\overset{\overset{\text{Cl}}{|}}{\text{C}}-\overset{\overset{\text{H}}{|}}{\text{C}}=\text{CH}_2} \tag{16.43}$$

Acrylic acid may be prepared by the catalytic oxidative carbonylation of ethylene or by heating formaldehyde and acetic acid in the presence of potassium hydroxide.

$$\underset{\text{Ethylene}}{\text{H}_2\text{C}=\text{CH}_2} + \underset{\substack{\text{Carbon}\\\text{monoxide}}}{\text{CO}} + \underset{\text{Oxygen}}{\tfrac{1}{2}\text{O}_2} \xrightarrow[\text{FeCl}_3]{\text{ThCl}_2} \underset{\text{Acrylic acid}}{\text{H}_2\text{C}=\overset{\overset{\text{H}}{|}}{\text{C}}-\text{COOH}} \tag{16.44}$$

Methyl acrylate may be obtained by the addition of methanol to the reactants in the previous synthesis for acrylic acid or by the methanolysis of acrylonitrile.

$$H_2C{=}\overset{\displaystyle H}{\overset{|}{C}}{-}CN \ + \ H_3COH \ \xrightarrow{\ H_2O\ } \ H_2C{=}\overset{\displaystyle H}{\overset{|}{C}}{-}\underset{\displaystyle O}{\overset{\displaystyle \|}{C}}{-}OCH_3 \ + \ NH_4^+$$

(16.45)

Acrylonitrile   Methanol   Methyl acrylate   Ammonium ion

Methyl methacrylate may be prepared by the catalytic oxidative carbonylation of propylene in the presence of methanol.

$$H_2C{=}\overset{\displaystyle CH_3}{\overset{|}{CH}} \ + \ CO \ + \ \tfrac{1}{2}O_2$$

Propylene   Carbon monoxide   Oxygen

(16.46)

$$+ \ CH_3OH \ \xrightarrow[\text{FeCl}_3]{\text{ThCl}_2} \ H_2C{=}\overset{\displaystyle CH_3}{\overset{|}{C}}{-}\overset{\displaystyle O}{\overset{\|}{C}}{-}O{-}CH_3$$

Methanol   Methyl methacrylate

Other esters of acrylic and methacrylic acid may be prepared by transesterification with higher boiling alcohols.

16.3 SYNTHESIS OF FREE-RADICAL INITIATORS

Free-radical initiators are compounds containing covalent bonds that readily undergo homolytic cleavage to produce free radicals. The most widely used organic free-radical initiators are peroxides and azo compounds.

Benzoyl peroxide is produced when benzoyl chloride and sodium peroxide are stirred in water.

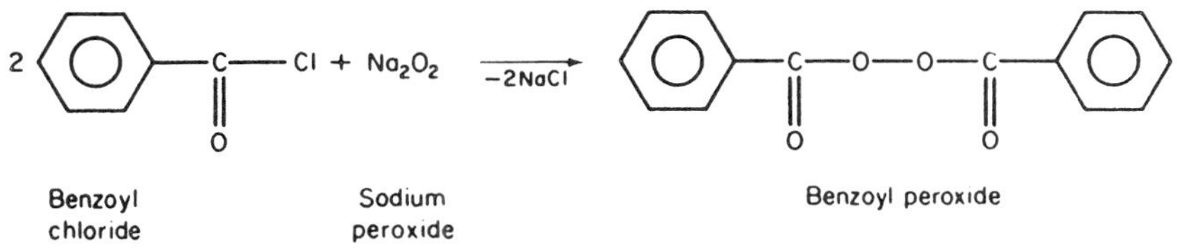

(16.47)

Benzoyl chloride   Sodium peroxide   Benzoyl peroxide

tert-Butyl hydroperoxide is produced by the acid-catalyzed addition of hydrogen peroxide to isobutylene [Eq. (16.48)]. tert-Butyl peroxide is produced when tert-butyl hydroperoxide is added to isobutylene [Eq. (16.49)].

$$(CH_3)_2C{=}CH_2 \ + \ H_2O_2 \ \xrightarrow{\ H_2SO_4\ } \ (CH_3)_3COOH$$

(16.48)

Isobutylene   Hydrogen peroxide   tert-Butyl hydroperoxide

$$(CH_3)_2C=CH_2 \quad + \quad (CH_3)_2COOH \xrightarrow{H_2SO_4} (CH_3)_3COOC(CH_3)_3 \tag{16.49}$$

Isobutylene   tert-Butyl   tert-Butyl
            hydroperoxide   peroxide

Dicumyl peroxide is produced by the air oxidation of cumene.

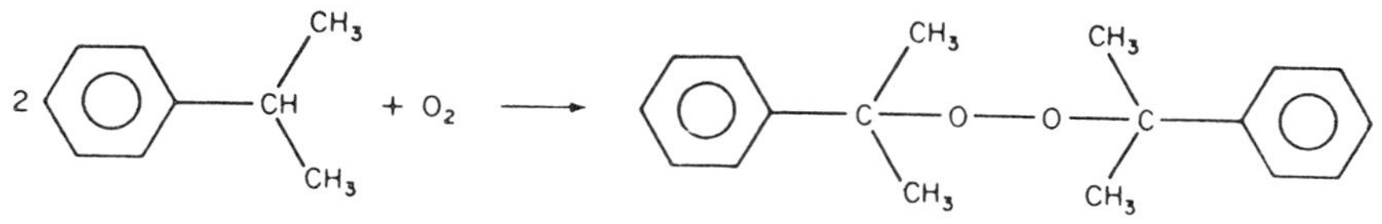

$$\tag{16.50}$$

Cumene   Oxygen   Dicumyl peroxide

All initiators are potentially explosive compounds and must be stored and handled with care. 2,2′-Azobisisobutyronitrile (AIBN, 2,2-dicyano2,2-azopropane) is obtained from the reaction of acetone with potassium cyanide and hydrozine hydrochloride. As shown in Eq. (16.51), the reaction of potassium cyanide with hydrazine hydrochloride produces hydrogen cyanide and hydrazine. The latter reacts with acetone to form acetone dihydrazone, which reacts with the former (HCN) to produce a substituted hydrazone which is then oxidized to AIBN by sodium hypochlorite. When methyl ethyl ketone is used in place of acetone, 2,2′-azobis-2-methylbutyronitrile is produced.

$$2KCN \quad + \quad Cl^-, {}^+H_3N-NH_3^+, Cl^- \xrightarrow[-KCl]{} 2HCN \quad + \quad H_2N-NH_2$$

Potassium   Hydrazine   Hydrogen   Hydrazine
cyanide   hydrochloride   cyanide

$$(CH_3)_2C=O \quad + \quad H_2N-NH_2 \longrightarrow (CH_3)_2C=N-N=C(CH_3)_2$$

Acetone   Hydrazine   Acetone dihydrazone

$$(CH_3)_2 C=N-N=C(CH_3)_2 \quad + \quad 2HCN \longrightarrow \tag{16.51}$$

Acetone dihydrazone   Hydrogen
                       cyanide

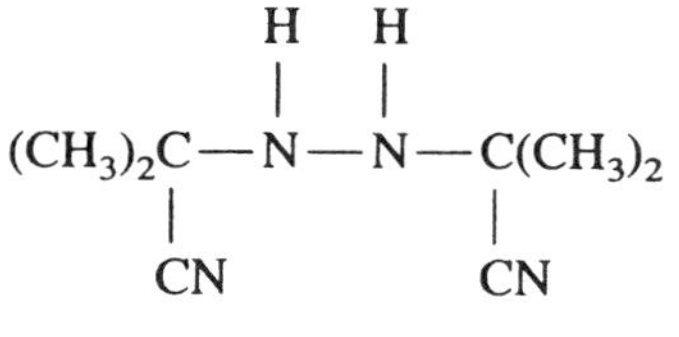

$$\begin{array}{ccc} & H & H \\ & | & | \\ (CH_3)_2C & -N-N- & C(CH_3)_2 \\ | & & | \\ CN & & CN \end{array}$$

2,2′-Hydrazobisisobutryonitrile

$$(CH_3)_2C \overset{\overset{\displaystyle H}{|}}{-N} \overset{\overset{\displaystyle H}{|}}{-N} -C(CH_3)_2 \; + \; NaOCl \; \xrightarrow[-NaCl]{-H_2O}$$

with CN groups below each $C(CH_3)_2$

2,2′-Hydrazobisisobutryonitrile   Sodium hypochlorite

$$(CH_3)_2C-N=N-C(CH_3)_2$$

with CN groups below

AIBN

The physical constants of polymer reactants and monomers are listed in Tables 16.1 and 16.2. Data on initiators are listed in Chap. 8.

16.4 MONOMER SYNTHESIS FROM BASIC FEEDSTOCKS

Most of the monomers widely employed for both vinyl and condensation polymers are derived indirectly from simple feedstock molecules. The synthesis of monomers is a lesson in inventiveness. The application of the saying that necessity is the mother of invention has led to the sequence of chemical reactions where little is wasted and by-products from one reaction are employed in another reaction as an integral starting material. Following is a brief look at some of these pathways traced from basic feedstock materials. It must be remembered that often many years of effort were involved in discovering the conditions of pressure, temperature, catalysts, etc., that must be present as one goes from the starting materials on the left-

Table 16.1 Physical Properties of Polymer Reactants

Compound	Melting point, °C	Boiling point, °C	Specific gravity	Index of refraction
Adipic acid	152	265^{10}	$1.360\frac{25}{4}$	
Epichlorohydrin	−57	117^{756}	$1.183\frac{15}{15}$	1.4397^{16}
Ethylene glycol	−15.6	198	$1.113\frac{15}{15}$	1.4318^{20}
Ethylene oxide	−111.7	10.7	$0.887\frac{7}{7}$	$1.3599^{8.4}$
Formaldehyde	−92	−21	0.815^{20}	
Furfural	−39	162	$1.159\frac{20}{4}$	1.5261^{20}
Furfuryl alcohol	. . .	169.5^{752}	$1.129\frac{25}{4}$	1.4852^{20}
Glycerol	18.2	290	$1.261\frac{20}{4}$	1.4729^{20}
Hexamethylenediamine	42	204.5		
Maleic anhydride	60	202	1.5	
Melamine	354	. . .	1.5737^{25}	
Pentaerythritol	262	276^{30}		
Phenol	43	181.8^8	$1.071\frac{25}{25}$	1.54^{45}
Phthalic acid (o)	231	. . .	$1.593\frac{20}{4}$	
Isophthalic acid (m)	347			
Terephthalic acid (p)	300 (s)			
Phthalic anhydride	132	284.5	1.527^4	
Pyromellitic dianhydride	287	397–400	1.68	
Sebacic acid	135.5	294.5^{100}	$1.207\frac{25}{4}$	$1.42^{1.33}$
Urea	132.7	. . .	$1.335\frac{20}{4}$	

Table 16.2 Physical Properties of Monomers

Compound	Melting point, $^\circ$C	Boiling point, $^\circ$C	Density	Index of refraction
Acrylamide	85	125[25]	1.122[30]	
Acrylic acid	12	141	1.0511[20]	1.4224[20]
Methyl acrylate	−75	80	0.953[20]	1.3984[20]
Acrylonitrile	−83.6	79	0.8060[20]	1.393[20]
1,3-Butadiene	−109	−4.4	0.6211[20]	1.4292[−25]
Chloroprene	. . .	59.4	0.9583[20]	1.4583[20]
Chlorotrifluoroethylene	−158	−28		
1-Butene	−185	−6.3	0.5951[20]	1.3962[20]
Isobutylene (2-methylpropene)	−141	−6.6	0.6266[−6.6]	1.3814[−25]
Ethylene	−169	−104	0.566[−102]	1.363[−100]
Tetrafluoroethylene	−142.5	−76.3	1.519[20]	
Isoprene	−146	34	0.6806[20]	1.4194[20]
Methacrylic acid	15.5	161	1.0153[20]	1.43143[20]
Methyl methacrylate	−48	101	0.936[20]	1.413[20]
Methacrylonitrile	−36	90.3	0.7998[20]	1.4007[20]
Methyl isopropenyl ketone	−54	98	0.8550[20]	1.4220[20]
Propylene	−185	−48	0.5139[−20]	
Styrene	−30.6	145.2	0.9090[20]	1.54682[20]
α-Methylstyrene	−23.2	163.4	0.9165[10]	1.5386[20]
Vinyl acetate	. . .	72.5	0.9338[20]	1.3953[20]
Vinyl chloride	−154	−14	0.99176[,15]	1.398[15]
Vinyl ethyl ether	−115	35	0.7589[20]	1.3767[20]
Vinyl fluoride	−161	−72		
Vinylidene chloride	−122	31.7	1.2129[20]	1.4249[20]
2-Vinylpyridine	. . .	80[29]	0.9985[0]	1.5494[20]

hand side to the right-hand side of the arrow. The specific conditions and alternative routes are given in Secs. 16.1 and 16.2 and are indicated in Figs. 16.1 through 16.5.

Fossil fuels refer to materials formed from the decomposition of once living matter. Because these "once" living materials contain sulfur and heavy metals such as iron and cobalt, they must be removed either prior to or subsequent to use.

The major fossil fuels are coal and petroleum. Marine organisms were typically deposited in muds and underwater, where anaerobic decay occurred. The major decomposition products are hydrocarbons, carbon dioxide, water, and ammonium. These deposits form much of the basis for petroleum resources. Many of these deposits are situated so that the evaporation of the more volatile products such as ammonia and water occurred, giving petroleum resources with little nitrogen- or oxygen-containing products. By comparison, coal is formed from plant material that has decayed to graphite carbon and methane.

Only about 5% of the fossil fuels consumed today are used as feedstocks for the production of today's synthetic carbon-based products. This includes the products produced by the chemical and drug industry with a major portion acting as the feedstocks for plastics, elastomers, coatings, fibers, etc.

The major petroleum resources contain linear, saturated hydrocarbons (alkanes), cyclic alkanes, and aromatics. For the most part, this material is considered to have a low free energy content.

Raw or crude petroleum materials are separated into groups of compounds with similar boiling points by a process called fractionation. Table 16.3 contains a brief listing of typical fraction-separated materials. Accompanying or subsequent to this fractionation occurs a process called "cracking" where the hydrocarbon

Table 16.3 Typical Straight-Chain Hydrocarbon Fractions Obtained from Distillation of Petroleum Resources

Boiling range (°C)	Average number of carbon atoms	Name	Possible uses
<30	1–4	Gas	Heating
30–180	5–10	Gasoline	Automobile fuel
180–230	11–12	Kerosene	Jet fuel, heating
230–300	13–17	Light gas oil	Diesel fuel, heating
305–400	18–25	Heavy gas oil	Heating

molecules are heated over catalysts that allow the hydrocarbon molecules to break up and then reform to structures that contain more "branching" and consequently structures that are more energetic (e.g., upon combustion they will burn with a larger amount of energy given off per weight or volume of hydrocarbon fuel). Most of our gasolines are produced by this cracking process. Petroleum processing also provides the basic feedstocks for materials that eventually end up as today's synthetic polymers.

Here we will divide the synthesis of monomers according to the number of carbons involved in the initial steps from the original fractionation and cracking process. Often important products can be obtained from several chemical sequence scenarios. The following will emphasize some of these alternate routes. As noted above, the precise reaction conditions needed for each step are the consequence of a great deal of hard work and may vary from company to company. The following focuses on tracing the chemical "stream" from selected petroleum feedstock molecules to the synthesis of important monomers.

Methane forms the basis for the single carbon fraction. The major products derived from methane with respect to polymer formation are given in Fig. 16.1. The interconnectiveness and available alternative routes for the synthesis of the materials are illustrated in Fig. 16.1. Thus, both hydrogen cyanide and methanol are produced from the methane stream, and both are involved in the synthesis of methyl methacrylate (MMA). As will be seen later, hydrogen cyanide is also involved in the synthesis of acrylic acid and methyl acrylate (Fig. 16.2) and acrylonitrile (Fig. 16.3). Methanol is also used in the synthesis of methyl acrylate and acrylic acid (Figure 16.2) and MMA from propylene (Figure 16.4). Methanol in turn serves as a basis for the production of ethanol, which in turn is utilized in the synthesis of formaldehyde and vinyl ethyl ether (Figure 16.3).

Formaldehyde, produced in the methane stream, serves as the basis of the formaldehyde-intensive resins, namely, phenyl-formaldehyde ("Bakelite"-like) resins, urea-formaldehyde resins, and melamine-formaldehyde resins. Formaldehyde is also involved in the synthesis of ethylene glycol, one of two comonomers in the production of poly(ethylene terephthalate). Formaldehyde also serves as the basic feedstock for the synthesis of polyacetals.

Another important use for methane is its conversion to synthesis gas (or syngas), a mixture of hydrogen gas and carbon monoxide as shown in Fig. 16.1. Synthesis gas can also be derived from coal. When this occurs, it is called water

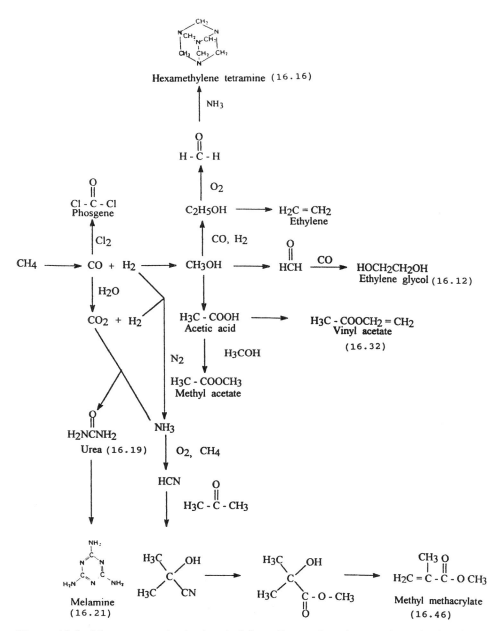

Figure 16.1 Monomer synthesis chemical flow diagram based on methane feedstock.

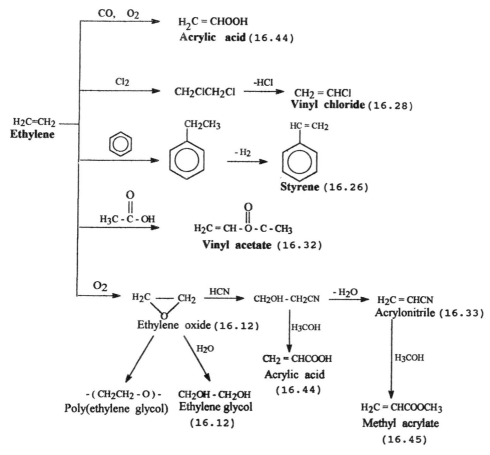

Figure 16.2 Monomer synthesis chemical flow diagram based on ethylene feedstock.

gas. Interestingly, the reaction of methane to give carbon monoxide and hydrogen can be reversed so that methane can be produced from coal through this route.

Ethylene represents the two carbon precursor (Fig. 16.2). Here, depending upon the reaction conditions, a wide variety of intermediates and products can be formed. Of course, polyethylene itself is part of the ethylene stream. Its copolymerization yields a large number of industrially important copolymers, terpolymers, graft polymers, and block polymers.

Propylene is the basic three-carbon building block considered here (Fig. 16.4). Again, its polymerization gives polypropylene and its copolymerization gives a wide variety of co- and terpolymers. Here the ingeniousness of some of the synthetic routes is illustrated in the conversion of benzene, through reaction with propylene, to cumene and the consequent oxidation forming phenol and acetone that subsequently act to form bisphenol A, a basic building block for a number of esters, as well as furnishing the native compounds themselves, both of which are involved

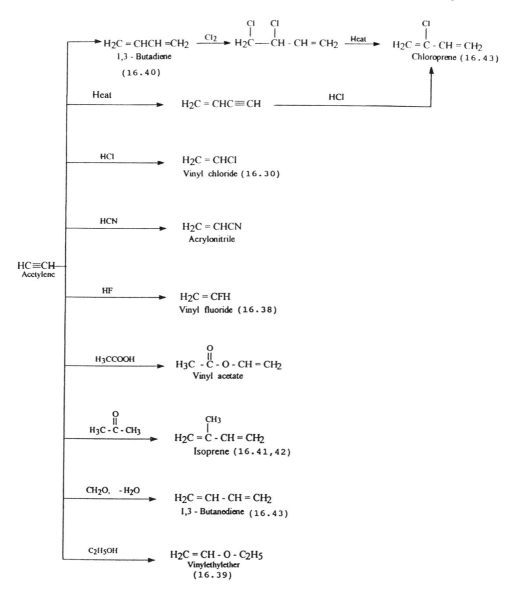

Figure 16.3 Monomer synthesis chemical flow diagram based on acetylene feedstock.

in numerous other important reactions. Phenol is involved in the synthesis of phenol-formaldehyde resins, adipic acid, and 1,6-hexanediamine (hexamethylene diamine; Fig. 16.5). Acetone, in turn, is also involved in numerous important synthetic steps and is employed as a commercial solvent. Acetone is involved in the synthesis of methyl methacrylate (Fig. 16.4) and isoprene (Fig. 16.3).

The synthesis of various acrylic esters can occur from the subsequent reaction of alcohols with acrylic acid (Fig. 16.4).

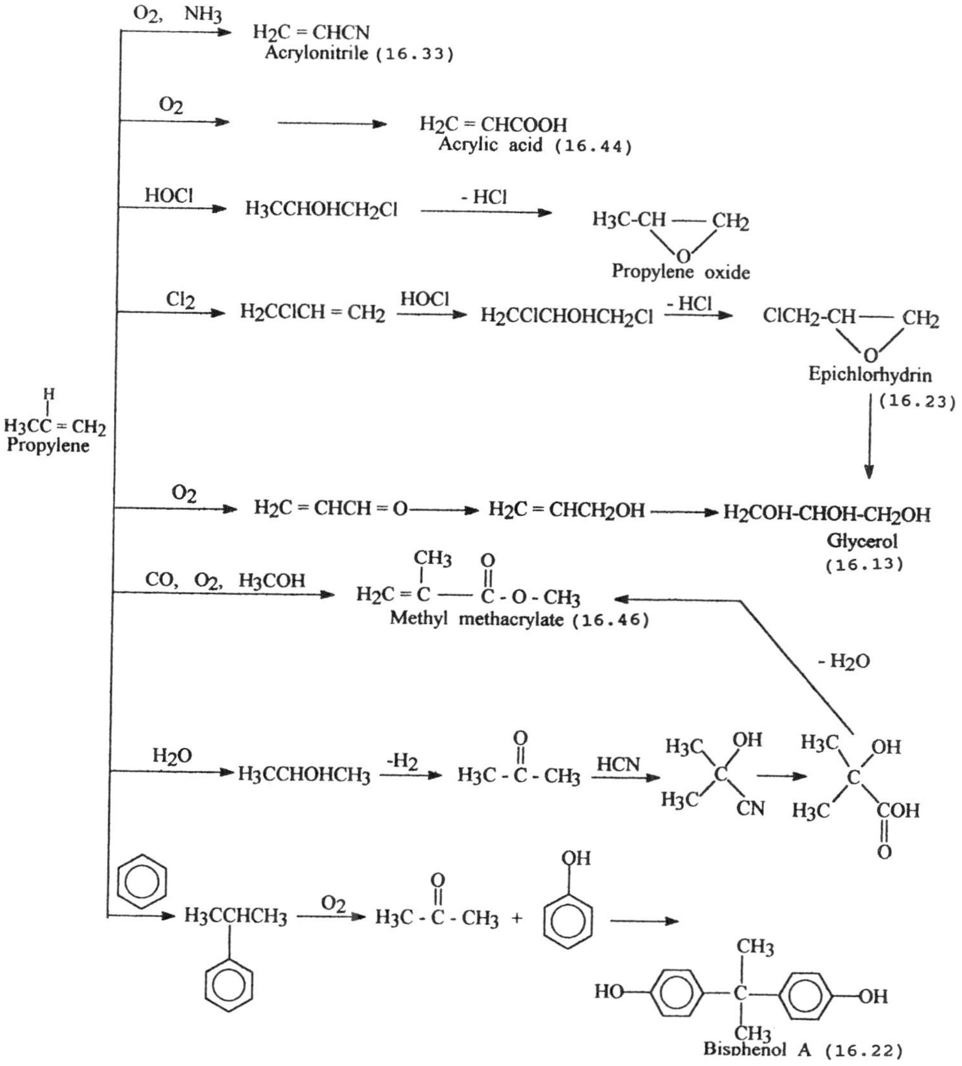

Figure 16.4 Monomer synthesis chemical flow diagram based on propylene feedstock.

The major members of the four-carbon feedstock molecules are 1,3-butadiene and isobutylene, both involved in the synthesis of a number of polymers, copolymers, and terpolymers. Butadiene is copolymerized with styrene to form SBR rubber, with acrylonitrile to form nitrile rubbers, and with both styrene and acrylonitrile to form ABS rubbers. Isobutylene is copolymerized with isoprene to form butyl rubbers.

Benzene forms the basis for a number of monomers (Fig. 16.5), including those that retain their aromatic character and those that become fully or partially saturated, such as adipic acid, which forms the basis for both many aliphatic polyesters and nylon-66.

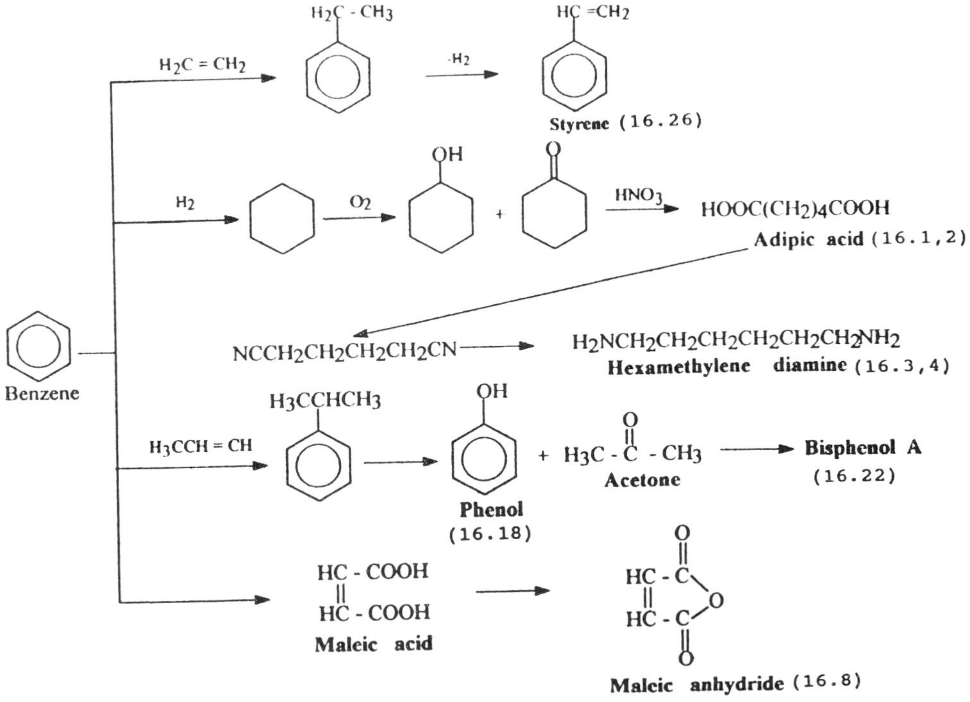

Figure 16.5   Monomer synthesis chemical flow diagram based on benzene feedstock.

Styrene is produced from benzene and ethylene, as shown in Fig. 16.5, and from the reaction of ethylbenzene through an alternative route, also producing both the styrene monomer and propylene oxide as shown in Eq. (16.27).

Other aromatics are also part of the aromatic stream. Toluene is used to produce benzene through dehydroalkylation and is used in the production of isocyanides for the production of polyurethanes. The three xylene isomers can be oxidized, forming three important reactants. Oxidation of meta-xylene gives isophthalic acid, and oxidation of para-xylene gives terephthalic acid used in the production of poly(butylene terephthalate), as well as PET. Oxidation of ortho-xylene can give phthalic anhydride used in the production of unsaturated polyesters.

Diisocyanates serve as the basis for the polyurethane industry (along with dial-cohols). These can be produced from benzene or toluene. The formation of meth-ylenediphenylisocyanate (MDI) is briefly described as follows.

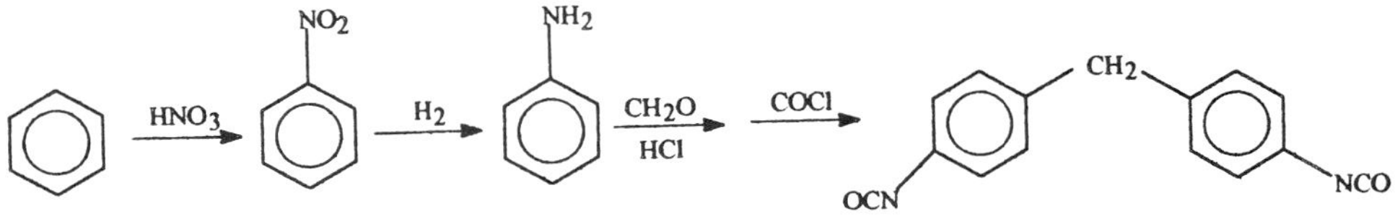

                                                               MDI   (16.52)

Cyclohexanone (Fig. 16.5) also acts as a precursor to the formation of capro-lactam, as shown in Eq. (16.11). It is employed in the synthesis of nylon-6.

Another two-carbon feedstock is acetylene. Acetylene is typically obtained from coal by converting coke into calcium carbide and then treating the calcium carbide with water. As shown in Fig. 16.3, a number of important monomers can be made from acetylene. Even so, because of the abundance of other starting materials from petroleum reserves, only some of the routes shown in Figure 16.3 are widely used.

Isoprene (2-methyl-1,3-butadiene), shown as derived from acetylene feedstock (Fig. 16.3) is commercially formed from a number of additional routes. It is also produced from isoamylene recovered from one of the gasoline streams with subsequent dehydrogenation [Eq. (16.41)]. Dimerization of propylene to 2-methyl-1-pentene, isomerization to 2-methyl-2-pentene, and finally pyrolysis to isoprene and methane is also a route to isoprene. One industrially important route starts with acetylene and acetone, while another begins with isobutylene and formaldehyde [Eq. (16.42)].

In summary, monomer synthesis from basic, readily available inexpensive feedstocks based on fossil fuels is both an art and a science developed over the past half century or so. It represents a delicate balance and interrelationship of feedstocks and so-called by-products from one reaction that are critical reactants in another reaction. Monomer and polymer synthesis continues to undergo change and improvement.

SUMMARY

1. While all polymer reactants can be produced by classic reactions, new reactions have been developed for the industrial production of many of the reactants used in large quantities.

2. The natural products furfural (from oat hulls) and ricinoleic acid (from castor oil) are used to produce adipic and sebacic acid, respectively.

3. The isomers of phthalic acid are obtained by the catalytic oxidation of xylenes.

4. ϵ-Caprolactam is produced by the classic Beckman rearrangement of cyclohexanome oxine.

5. Ethylene and propylene oxides are produced by the catalytic oxidation of ethylene and propylene, respectively.

6. 2,4-Toluene diisocyanate is produced by the phosgenation of 2,4-toluene-diamine.

7. Phenol is produced by the acidic decomposition of cumene hydroperoxide.

8. Urea is produced by the condensation of liquid ammonia and liquid carbon dioxide. Melamine is obtained by heating urea.

9. The alkylchlorosilanes are produced by the reaction of a Grignard reagent and silicon tetrachloride.

10. Styrene is produced by the catalytic dehydrogenation of ethylbenzene.

11. Vinyl chloride is produced by the oxychlorination process from ethylene and hydrogen chloride.

12. Vinyl acetate is produced by the oxidative condensation of acetic acid and ethylene.

13. Acrylonitrile is produced by the ammoxidation of propylene.

14. Ethylene, propylene, butylene, and butadiene are produced by cracking hydrocarbon feedstocks.

15. Fluorocarbon monomers are produced by the dehydrochlorination of chlorofluoro compounds.

16. Acrylic and methacrylic esters are produced by the oxidative carbonylation of ethylene and propylene, respectively.

17. Benzoyl peroxide is produced by the condensation of sodium peroxide and benzoyl chloride. tert-Butyl peroxides are produced by the addition of hydrogen peroxide to isobutylene, and dicumyl peroxide is produced by the air oxidation of cumene.

18. 2,2′-Azobisisobutyronitrile is produced by the reaction of acetone with potassium cyanide and hydrazine hydrochloride.

19. Since many reactants and monomers are toxic and initiators are unstable compounds, extreme care must be exercised in their use.

GLOSSARY

AIBN: 2,2′-Azobisisobutyronitrile.

Cannizzaro reaction: An internal oxidation-reduction reaction of aldehydes.

carbamide: Urea.

carcinogenic: Cancer producing.

chemurgic compound: Compound made from a plant source.

chloroprene: 2-Chloro-1,3-butadiene.

Friedel-Crafts condensation: Condensation that takes place in the presence of a Lewis acid, such as aluminum chloride.

Grignard reagent: RMgX.

isophthalic acid: The meta isomer.

Raney nickel: A porous nickel catalyst produced from a nickel-aluminum alloy.

TDI: 2,4-Toluene diisocyanate.

terephthalic acid: The para isomer of the dicarboxylic acid of benzene.

THF: Tetrahydrofuran.

VCM: Vinyl chloride.

EXERCISES

1. Why are there so many methods for the preparation of adipic acid?

2. Write the equations for the industrial synthesis of the following:
   a. Adipic acid
   b. Hexamethylenediamine
   c. Sebacic acid
   d. Terephthalic acid
   e. Maleic anhydride
   f. ϵ-Caprolactam

g. Ethylene oxide
h. Glycerol
i. Pentaerythritol
j. TDI
k. Hexamethylenetetramine
l. Phenol
m. Urea
n. Melamine
o. Bisphenol A
p. Epichlorohydrin
q. Methyltrichlorosilane
r. Styrene
s. Vinyl chloride
t. Vinyl acetate
u. Acrylonitrile
v. Vinyl ethyl ether
w. Methyl methacrylate
x. Benzoyl peroxide
y. tert-Butyl peroxide
z. AIBN

3. Name a reactant or monomer produced by the following:
   a. Grignard reaction
   b. Friedel-Crafts reaction
   c. Beckman rearrangement
   d. A chemurgic process
   e. A crossed Cannizzaro reaction

BIBLIOGRAPHY

Boundy, R. H., Boyer, R. F., Stroesser, S. M. (1965): *Styrene, Its Polymer, Copolymers and Derivatives*, Hafner, New York.

Braun, D., Cherdron, H., Kern, W. (1972): *Techniques of Polymer Synthesis and Characterization*, Wiley-Interscience, New York.

Deanin, R. D. (1974): *New Industrial Polymers*, ACS Symposium Series 4, Washington, D.C.

Leonard, E. C. (1970): *Vinyl and Diene Monomers*, Wiley-Interscience, New York.

Martin, L. F. (1974): *Organic Peroxide Technology*, Noyes Data Corp., Park Ridge, New Jersey.

Seymour, R. B. (1976): New sources of monomers and polymers, Polymer Eng. Sci., *16*(12): 817.

Starks, C. M. (1974): *Free Radical Telomerization*, Academic, New York.

Stille, J. K. (1968): *Industrial Organic Chemistry*, Prentice-Hall, Englewood Cliffs, New Jersey.

Stille, J. K., Campbell, T. W. (1972): *Condensation Monomers*, Wiley-Interscience, New York.

Williams, A. (1974): *Furans, Synthesis, and Applications*, Noyes Data Corp., Park Ridge, New Jersey.

Yokum, R. H., Nyquist, E. B. (1974): *Functional Monomers, Their Preparation, Polymerization and Application*, Marcel Dekker, New York.

17

Polymer Technology

Today, nearly 10,000 American companies are active in the general area of synthetic polymers. Following is a brief description of these companies divided according to their function.

Manufacturers: There are over 200 major manufacturers of general purpose polymers and numerous other manufacturers of specialty polymers.

Processors: Some companies manufacture their own polymeric materials for subsequent processing, but the majority purchase the necessary polymeric materials from other companies. Processors may specialize in the use of selected polymers, such as nylons and polycarbonates, or focus on particular techniques of processing, such as coatings, films, sheets, laminates, and bulk molded and reinforced plastics.

Fabricators and finishers: The majority of companies are involved in the fabrication and finishing of polymers, i.e., production of the end products for industrial and general public consumption. Fabrication can be divided into three broad areas: machining, forming, and fashioning. Machining includes grinding, sawing, drilling, turning on a lathe, cutting, tapping, reaming, and milling. Forming includes molding and other methods of shaping and joining by welding, gluing, screwing, or other techniques. Fashioning includes cutting, sewing, sheeting, and sealing. Fabrication sequences vary with the polymeric material and desired end product.

While much classic polymer technology was developed without the benefit of science, modern polymer technology and polymer science are closely associated. The technology of fibers, elastomers, coatings, and plastics is discussed in this chapter.

17.1 FIBERS

Since there are many natural fibers, much fiber technology was developed before the twentieth century. Long, threadlike cells of animal and vegetable origin have been used for centuries for textiles, paper, brushes, and cordage. The animal protein fibers, namely, wool and silk, are no longer competitive with other fibers unless their production is subsidized. The vegetable cellulosic fibers—cotton, kapok, abaca, agave, flax, hemp, jute, kenaf, and ramie—are still in use, but cotton is no longer the "king of fibers."

Regenerated proteins from casein (lanital), peanuts (ardil), soybeans (aralac), and zein (vicara) are used as specialty fibers, but cellulose acetate and regenerated cellulose (rayon) are used in relatively large quantities. Cellulose triacetate fibers (Tricel) are produced by the acetylation of α-cellulose, and cellulose diacetate fiber (acetate rayon) is produced by the partial deacetylation of the triacetate. Further deacetylation (saponification) yields regenerated cellulose. Table 17.1 shows the physical properties of typical fibers.

Most regenerated cellulose (rayon) is produced by the viscose process in which an aqueous solution of the sodium salt of cellulose xanthate is precipitated in an acid solution. The relatively weak fibers produced by this wet spinning process are stretched in order to produce strong (high-tenacity) rayon. The annual production of acetate rayon and rayon in the United States is 208 million lbs. and 290 million lbs., respectively.

While no truly synthetic fibers were produced before 1936, about 10 billion lbs. of these important products are now produced annually. The leading fiber is poly(ethylene terephthalate) (polyester, Dacron, Terylene, Kodel, Vycron), which is produced at an annual rate of 3.8 billion lbs. by forcing the molten polymer through small holes in a spinneret in a process called melt spinning.

Nylon-66 and nylon-6 fibers are also produced by melt spinning molten polymers at an annual rate of over 2.7 billion lbs. Acrylic fibers (Acrilan, Orlon) are produced at an annual rate of over 430 million lbs. by forcing a solution of acrylonitrile polymers in dimethylformamide (DMF) through spinnerets and evaporating the solvent from the filament. This process, which is also used for the production of cellulose acetate fiber, is called dry spinning.

Polyurethane (Perlon, Spandex), polypropylene, and polyethylene fibers are produced by the melt spinning of molten polymers. The total annual polyolefin fiber production in the United States is about 2.4 billion lbs. Polyurethane fibers are produced in large quantities in Germany, but these are used as specialty fibers in the United States. Over 600 million lbs. of these fibers are produced by the melt spinning process.

Filaments of thermoplastics such as polypropylene may also be produced by a fibrilation process in which strips of film are twisted, and these fibrils are heated and stretched. In addition to the traditional spinning and weaving processes, textiles may be produced in the form of nonwoven textiles by the fiber-bonding process. Bonding results from the addition of thermoplastics or by admixing thermoplastic fibers with cotton or rayon fibers before bonding by heat.

Table 17.1 Physical Properties of Typical Fibers[a]

Polymer	Tenacity (g/denier)	Tensile strength (kg/cm^2)	Elongation (%)
Cellulose			
Cotton	2.1–6.3	3–9000	3–10
Rayon	1.5–2.4	2–3000	15–30
High–tenacity rayon	3.0–5.0	5–6000	9–20
Cellulose diacetate	1.1–1.4	1–1500	25–45
Cellulose triacetate	1.2–1.4	1–1500	25–40
Proteins			
Silk	2.8–5.2	3–6000	13–31
Wool	1.0–1.7	1–2000	20–50
Vicara	1.1–1.2	1–1000	30–35
Nylon–66	4.5–6.0	4–6000	26
Polyester	4.4–5.0	5–6000	19–23
Polyacrylonitrile (acrylic)	2.3–2.6	2–3000	20–28
Saran	1.1–2.9	1.5–4000	20–35
Polyurethane (Spandex)	0.7	630	575
Polypropylene	7.0	5600	25
Asbestos	1.3	2100	25
Glass	7.7	2100	3.0

[a]The fineness of a fiber can be expressed by a unit called the denier, which is the mass in g of 9000 m of fiber.

17.2 ELASTOMERS

Prior to World War II, hevea rubber accounted for over 99% of all elastomers used, but synthetic elastomers account for more than 70% of all rubber used today. Natural rubber and many synthetic elastomers are available in latex form. The latex may be used, as such, for adhering carpet fibers or for dipped articles, such as gloves, but most of the latex is coagulated and the dried coagulant is used for the production of tires and mechanical goods.

Over 5.5 billion lbs. of synthetic rubber are produced annually in the United States. The principal elastomer is the copolymer of butadiene (75%) and styrene (25%) (SBR), produced at an annual rate of over 1 million tons by the emulsion polymerization of butadiene and styrene. The copolymer of butadiene and acrylonitrile (Buna-H, NBR, Hycar) is also produced by the emulsion process at an annual rate of about 200 million lbs.

Likewise, neoprene is produced by the emulsion polymerization of chloroprene at an annual rate of 126,000 tons. Butyl rubber is produced by the low-temperature cationic copolymerization of isobutylene (90%) and isoprene (10%) at an annual rate of 146,000 tons.

Polybutadiene, polyisoprene, and ethylene-propylene copolymer rubber (EPDM) are produced by anionic polymerization processes at annual rates of 550,000, 100,000, and 330,000 tons, respectively. One of the original synthetic elastomers (Thiokol), as well as polyfluorocarbon elastomers (Viton), silicone (Silastic), polyurethane (Adiprene), and phosphazenes, are specialty elastomers. The physical properties of typical elastomers are shown in Table 17.2.

Polymer-Polymer Immiscible Systems

The most frequent result of mixing two polymers is phase separation. This is a result of a combination of "like liking like" (like molecules gathering, excluding unlike polymer chains) and a less favorable, lower entropy change for polymers compared with smaller molecules. In other words, the major factor for solution of large and small molecules is an increased randomness, which works to offset the (energy-related) tendency for "like liking like." This entropy factor is generally relatively small for polymer-polymer combinations.

The tendency of polymer combinations to phase separate has been greatly exploited recently. New major groupings of polymers have been developed, including polymer blends, graft and block copolymers, and interpenetrating polymer networks (IPNs).

The emphasis on finding polymers that are complementary has led to a number of new polymer blends. The complementary nature can result from attraction due to polar-polar (dipole-dipole) interactions or hydrogen bonding between the two polymers.

The behavior of graft copolymers and IPNs is illustrated by the polymer mixture called high-impact polystyrene (HIPS). The addition of SBR [poly(butadiene-co-styrene], which is elastic, to polystyrene allows the more brittle polystyrene to add toughness, with the SBR domains acting to absorb sudden shocks. This gives an overall material with some flexibility while the polystyrene (PS) regions contribute to the material's overall strength. The varying percentage of PS and SBR and the size of PS-SBR domains can lead to materials with a wide variety of physical properties.

17.3 FILMS AND SHEETS

Films such as regenerated cellulose (cellophane) are produced by precipitating a polymeric solution after it has passed through a slit die. Other films, such as cellulose acetate, are cast from a solution of the polymer, but most films are produced by the extrusion process. Some relatively thick films and coextruded films are extruded through a flat slit die, but most thermoplastic films, such as polyethylene film, are produced by the air blowing of a warm extruded tube as it emerges from the circular die as shown in Fig. 17.1.

Films and sheeting are also produced employing calendering (Fig. 17.2). The technique is also employed to apply coatings to textiles or other supporting material. In calendering, the polymeric material is passed between a series of counter-

Table 17.2 Physical Properties of Typical Elastomers

	Pure gum vulcanizates		Carbon–black reinforced vulcanizates	
	Tensile strength (kg cm^{-2})	Elongation (%)	Tensile strength (kg cm^{-2})	Elongation (%)
Natural rubber (NR)	210	700	315	600
Styrene–butadiene rubber (SBR)	28	800	265	550
Acrylonitrile–butadiene rubber (NBR)	42	600	210	550
Polyacrylates (ABR)			175	400
Thiokol (ET)	21	300	85	400
Neoprene (CR)	245	800	245	700
Butyl rubber (IIR)	210	1,000	210	400
Polyisoprene (IR)	210	700	315	600
Ethylene–propylene rubber (EPM)		300		
Polyepichlorohydrin (CO)				
Polyfluorinated hydrocarbons (FPM)	50	600		
Silicone elastomers (SI)	70	600		
Polyurethane elastomers (AU)	350	600	420	500

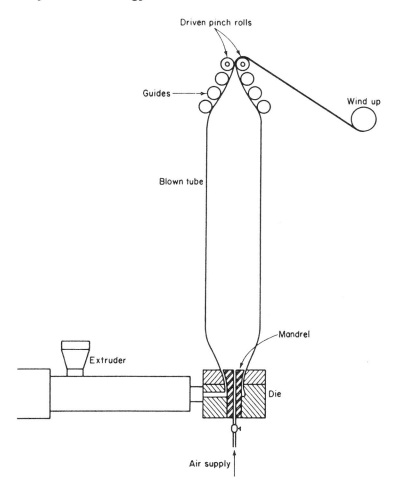

Driven pinch rolls

Guides

Wind up

Blown tube

Mandrel

Extruder

Die

Air supply

Figure 17.1 Film formation through extrusion. (From N. Bikalis, *USP Extrusion and Other Plastics Operations*, Wiley-Interscience, 1971; with permission of J. Wiley & Sons, Publishers.)

rotating rolls. The surface may be smooth or textured depending on the roller surfacing. In applying a coating to a second material, the coating compound is passed through one set of rollers while the material to be coated is passed through a second set of rollers along with the coating compound.

The most widely used films are LDPE, cellophane, PET, PVC, cellulose acetate, polyfluorocarbons, nylons, polypropylene, polystyrene, and Saran. Films of UHMWPE, polyamides, and polytetrafluoroethylene are produced by skiving molded billets. The strength of many films, such as PET, is improved by biaxial orientation.

Most of the thermoplastics used as films may also be extruded as relatively thick sheets. These sheets may also be produced by pressing a stack of film at elevated temperature (laminating) or by the calendering process.

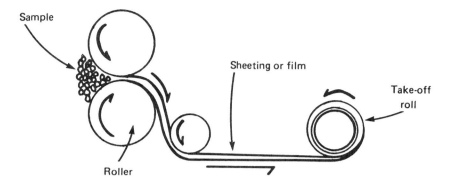

Figure 17.2 Apparatus for the production of film or sheeting employing the calendering technique.

Wire is coated by passing the wire through a plastic extruder, but most substances are coated by polymers from solutions, emulsions, or hot powders. The classic brushing process has been replaced to a large extent by roll coating, spraying, and hot powder coating. The application of polymers from water dispersions to large objects, such as automobile frames, has been improved by electrodeposition of the polymer on the metal surface.

Printing inks are highly filled solutions of resins. The classic printing inks were drying oil-based systems, but the trend in this $800 million business is toward solvent-free inks.

The permeability to gases and the tearing strength of films are shown in Table 17.3.

17.4 POLYMERIC FOAMS

Prior to 1920, the only flexible foam available was the natural sponge, but chemically foamed rubber and mechanically foamed rubber latex were introduced prior to World War II. These foams may consist of discrete unit cells (unicellular, closed cell), or they may be made up of interconnecting cells (multicellular, open cells), depending on the viscosity of the system at the time the blowing agent is introduced.

Unicellular foams are used for insulation, buoyancy, and flotation applications, while multicellular foams are used for upholstery, carpet backing, and laminated textiles. Expanded polystyrene (Styrofoam), which is produced by the extrusion of polystyrene beads containing a volatile liquid, is used to produce low-density moldings such as foamed drinking cups and insulation board. The K value for these products is on the order of 0.24 BTU.

Foamed products are also produced from PVC, LDPE, urea resins, ABS, and polyurethane (PU). The last are versatile products which range from hard (rigid) to soft (flexible). These are produced by the reaction of a polyol and a diisocyanate. Polyurethane planks are available in a wide variety of specific gravities.

The residual gases in many polymeric foams and their flammability are causes for concern. However, over 1.5 million tons of foamed plastic are produced annually, and this volume is increasing at an annual rate of about 10%.

Table 17.3 Permeability to Gases (cm/day/100 in.^{2}/mil at 25°C)

Polymer	MVT	CO$_2$	H$_2$	N$_2$	O$_2$	Initial tearing strength (g/mil)
Cellophane	0.4-134	0.4-0.6	1.2-2.2	0.5-1.6	0.2-5.0	2-20
Polyethylene (I)	1.0-1.5	2700	—	180	500	100-500
Polyethylene (III)	0.3	580	—	42	185	16-300
Rubber hydrochloride	—	288-13,500	—	—	38-2250	60-1600
Cellulose acetate	30-40	860-1000	835	30-40	117-150	1-2
Cellulose acetate butyrate	30-40	6000	—	250	950	5-10
Ethylcellulose	4.8-14.2	5000	—	600	2000	215-395
Plasticized poly(vinyl chloride)	4	100-3000	—	—	30-2000	60-1400
Saran	—	12	—	—	2.4	10-100
Poly(vinyl alcohol)	—	200	—	—	120	785-890
Poly(ethylene terephthalate)	1.7-1.8	15-25	100	0.017-1.0	6.0-8.0	12-27
Polystyrene (oriented)	7.0-10.0	900	—	—	350	5.0
Polycarbonate	11.0	1075	1600	50	300	20-25
Polyurethane	45-75	465-1650	—	41-119	75-327	220-710
Nylon-66	3-6	9.1	—	0.35	5.0	—
Nylon-6	5.4-20	10-12	110	0.9	2.6	50-90
Poly(vinyl fluoride)	3.24	11.1	58	0.25	3	12-100

17.5   REINFORCED PLASTICS AND LAMINATES

Many naturally occurring products such as wood are reinforced composites consisting of a resinous continuous phase and a discontinuous fibrous reinforcing stage. Laminates, consisting of alternate layers of phenolic resins and cloth, paper, or wood, were available prior to World War II, but reinforced plastics were not available until the 1940s.

Most modern reinforced plastics consist of unsaturated polyester resins reinforced by fibrous glass. Continuous untwisted glass strands, or rovings, may be used in the filament winding or pultrusion processes. Chopped fibrous glass stands and preformed glass mats are used for SMC, BMC, and hand layup composites.

Boron filaments, produced by the deposition of boron on tungsten wire, single crystals or whiskers of sapphire, or aluminum oxide and graphite, are used as more sophisticated reinforcements, and epoxy resins as well as thermoplastics are also used in place of unsaturated polyester resins.

While fibrous glass–reinforced polyester resins continue to be used at an annual rate in excess of 900,000 tons, the generally accepted engineering composite of the future will probably be based on graphite-reinforced epoxy resins. These composites have specific tensile strength (based on unit weight) and specific modulus 400% greater than that of wood, aluminum, steel, or fibrous glass–epoxy resin composite. These sophisticated composites have coefficients of thermal expansion less than those of many other construction materials.

Laminating is a simple binding together of different layers of materials. The binding materials are often thermosetting plastics and resins. The materials to be bound together can be paper, cloth, wood, or fibrous glass. These materials are often referred to as the reinforcing materials. Typically, sheets, impregnated by a binding material, are stacked between highly polished metal plates, subjected to high pressure and heat in a hydraulic press producing a bonded product which may be subsequently treated, depending on its final use (Fig. 17.3). The end product may be flat, rod shaped, tubular, rounded, or some other formed shape.

Reinforced plastics differ from high-pressure laminates in that little or no pressure is employed. For instance, in making formed shapes, impregnated reinforcing material is cut into a desired shape, the various layers are added to a mold, and the molding is completed by heating the mold. This process is favored over the high-pressure process because of the use of a simpler, lower-cost mold and production of strain-free products. A simple assembly for the production of reinforced materials is shown in Fig. 17.4.

17.6   MOLDED PLASTICS

Most plastics are converted to finished products by molding or extrusion. Until recently, most thermosets were molded by compression in which the molding powder was heated under pressure in the die cavity of a mold. This labor-intensive process has been replaced to some extent by transfer and injection molding. In the former process, a preformed briquette is warmed by induction heating and forced under pressure through an orifice into a heated multicavity mold. The cured moldings are ejected when the two-piece mold is reopened.

While an injection-molding press, such as the one shown in Fig. 17.5, may be used for molding thermosets, its principal use is for the rapid molding of over 3

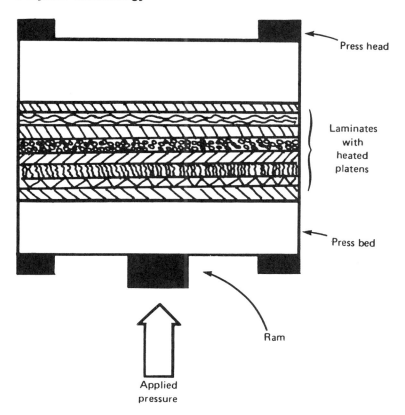

Figure 17.3 Assembly employed for the fabrication of laminates.

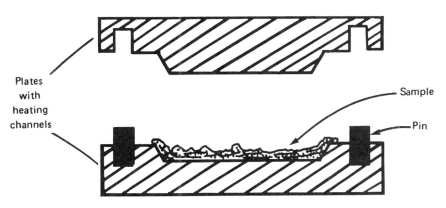

Figure 17.4 Assembly employed for production of reinforced plastics.

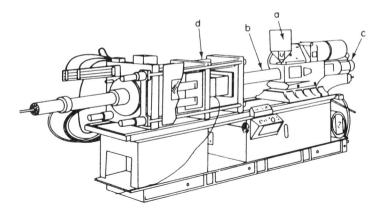

Figure 17.5 Cross-section of an injection-molding press. (From *Modern Plastics Technology* by R. Seymour, 1975, Reston Publishing Co., Reston, Virginia. Used with permission of Reston Publishing Company.)

million tons of thermoplastics annually. This press consists of a hopper (a) which feeds the molding powder to a heated cylinder (b) where the polymer is melted and forced forward by a reciprocating plunger (c) (or screw). The molten material advances toward a spreader or torpedo into a cool, closed two-piece mold (d). The cooled plastic part is ejected when the mold opens, and then the cycle is repeated. As shown in Fig. 17.6, the molten plastic passes from the nozzle through a tapered sprue, a channel or runner, and a small gate into the cooled mold cavity. The plastic

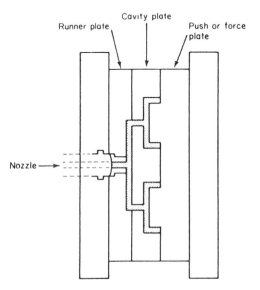

Figure 17.6 Injection mold in closed position (part d in Fig. 17.5). (Adapted from *Modern Plastics Technology* by R. Seymour, 1975, Reston Publishing Company, Reston, Virginia.)

in the narrow gate section is readily broken off and the thermoplastic materials in the sprue, runner, and gate are ground and remolded.

Labor costs of molding thermosets can be reduced by use of injection-molding techniques. BMC and plastic foams may also be injection molded. Special techniques are often used with the latter to produce a solid, dense surface. These structural foams, which are stronger on a weight basis than solid moldings, are used for furniture and similar complex objects.

As shown in Fig. 17.7, hollow plastic articles such as bottles are produced by a blow-molding process in which a heat-softened hollow plastic tube, or parison, is forced against the walls of the mold by air pressure, at a rapid rate, in an automated process. It is customary to cool the mold to increase the number of cycles per minute. Approximately 1 million tons of thermoplastics are used annually in the blow-molding process.

Warm thermoplastic sheets may be forced into a mold by a plug in order to produce articles varying in size from briefcases to boats as shown in Fig. 17.8. This thermoforming process may be as simple as laying a warm plastic sheet over a convex mold, or it may be more sophisticated, such as the plug-assisted vacuum-forming technique shown with a biaxially oriented sheet. The principal object is to obtain a part with uniform thickness.

Compression molding is the most common technique for the production of thermosetting products. This labor-intensive process is generally not used for thermoplastics. Compression molding is directly analogous to thermoforming except that molding is accomplished by simply squeezing a molding powder into a desired shape by application of pressure and heat for a specific time. Thus prepolymer, polymer flakes or pellets, mixed with appropriate additives as fillers or reinforcing materials, are added to the mold cavity. The mold is closed with application of pressure causing the mixture to conform to the mold dimensions. Heating aids in

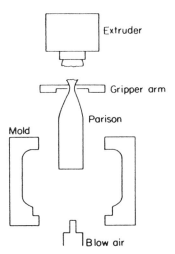

Figure 17.7 Sketch of extrusion blow-molding scheme. (From *Modern Plastics Technology* by R. Seymour, 1975, Reston Publishing Co., Reston, Virginia. Used with permission of Reston Publishing Co.)

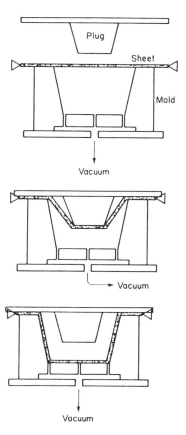

Plug

Sheet

Mold

Vacuum

Vacuum

Vacuum

Figure 17.8 Steps in plug-assisted vacuum thermoforming. (From *Modern Plastics Encyclopedia*, McGraw-Hill, New York, 1976–77; with permission of McGraw-Hill Publishers.)

this step and also forces the mixture into the desired shape before curing. A simple assembly for compression molding is shown in Fig. 17.9.

Transfer molding is similar to compression molding in that the specimen is cured into a desired mold shape under heat and pressure. It differs in that the sample is heated to a viscous mass before it reaches the mold. The heated mass is then forced into a closed mold employing a hydraulically powdered plunger. Transfer molding allows the molding of more intricate products compared to compression molding since the material is introduced into the mold in a more "flowable" form.

Rotational molding is employed to produce hollow parts by the addition of polymer to a warm mold that is rotated in an oven. Centrifugal force distributes the polymer evenly throughout the mold, and the heat melts, fuses, and sets the polymer into the desired shape. Advantages of rotational molding are the production of strain-free parts, uniform sample thickness, and relatively low mold cost.

Solvent molding occurs when a mold is immersed in a solution and withdrawn or when a mold is filled with a polymer and evaporation or cooling occurs producing an article, such as a bathing cap. Solvent molding and casting are closely related.

Applied pressure

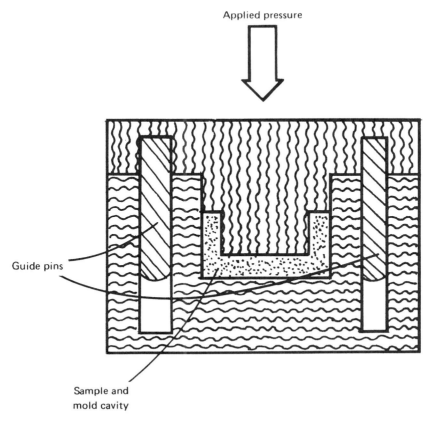

Guide pins

Sample and
mold cavity

Figure 17.9 Representation of a compression molding assembly.

17.7 CASTING

Casting is employed in making special shapes, sheets, films, tubes, and rods from
both thermoplastic and thermosetting materials. The essential difference between
most molding processes and casting is that no added pressure is employed in casting. In this technology, the polymer or prepolymer is heated to a fluid, poured into
a mold, cured at a specified temperature, and removed from the molds. Casting of
films and sheets can be done on a wheel or belt or by precipitation. In the case of
a wheel or belt, the polymer is spread to the desired thickness onto a moving belt
as the temperature is increased. The film is dried and then stripped off. "Drying"
may occur through solvent evaporation, polymerization, and cross-linking.

17.8 EXTRUSION

An extruder is an unusually versatile machine which, as shown in Fig. 17.10,
accepts granulated thermoplastic in a hopper (c), plasticates the polymer, and forces
it from the feed throat (d) through a die (f). The die may be circular for the
production of rod or pipe, it may be flat for the production of sheet, or it may have
any desired profile for the continuous production of almost any uniformly shaped

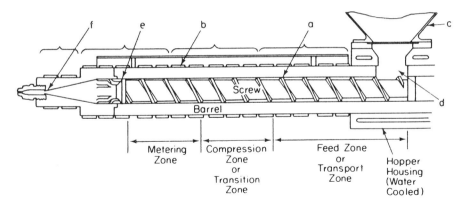

Figure 17.10 Sketch details of screw and extruder zones. (From *Modern Plastics Technology* by R. Seymour, 1975, Reston Publishing Co., Reston, Virginia. Used with permission of Reston Publishing Co.)

product. The screw (a) advances the polymer through a heated cylinder (barrel) (b) to a breaker plate and protective screen pack (e) before it enters the die (f). As shown in Fig. 17.10, the extrusion process may be divided into a feed or transport zone, a compression or transition zone, and a metering zone. Over 1 million tons of extruded pipe are produced annually in the United States. The properties of molded or extruded plastic are summarized in Table 17.4.

17.9 COATINGS

The fundamental purpose for painting is decorating, while the purpose for coating is for protection. While technically these two terms may have different meanings, the terms can be considered interchangeable, and we will follow this practice here.

Governmental edicts concerning air, water, solid particulates, and worker conditions are having real effects on the coatings industry with the generation of new coatings techniques. Paint solvents in particular are being looked at in view of increased environmental standards. The volatile organic compound (VOC) regulations under Titles I and VI of the Clean Air Act specify the phasing out of ozone-depleting chemicals—namely, chlorinated solvents—by the year 2000. Baseline solvent emissions are to be decreased. These, and related regulations, affect the emission of *all* organic volatiles, whether in coatings or other volatile-containing materials.

Paint manufacturers in the United States sell about 900 million gallons of coating material annually, or about 4 gallons per every man, woman, and child. This paint is used commercially to coat display cases, aluminum cans, washing machines, and automobiles; we use it to protect both the interior and exterior of our homes, paint baby beds, chairs, and picture frames.

Paint is typically a mixture of a liquid and one or more colorants (pigments). The liquid is called a vehicle or binder (adhesive) and may include a solvent or thinner along with the coating agent. The colored powders are called pigments.

Table 17.4 Properties of Molded Plastics

Polymer	Tensile strength (kg cm^{-2})	Flexural strength (kg cm^{-2})	Heat deflection point (@18.6 kg cm^{-2})	Dielectric constant (at 60 cycles)	Power factor (at 60 cycles)
Phenol formaldehyde resin	280	945	57		
Wood-flour filled	535	700	121	6.0	0.08
Glass-fiber filled	840	2100	232	7.1	0.05
Urea-formaldehyde, α-cellulose filled	630	980	135	8.0	0.04
Melamine-formaldehyde, α-cellulose filled	700	910		8.7	0.05
Alkyd resin, glass-filled	490	1050	130	5.7	0.010
Allyl resin, glass-filled	595	1050	232	4.4	0.03
Epoxy resin, glass-filled	1400	2450	224	4.2	0.025
Polyethylene (type I) (LDPE)	112		38	2.3	0.0005
Polyethylene (type II) (HDPE)	315	70	49	2.3	0.0005
Polypropylene	343	490	57	2.4	0.0005
Ionomers	315		38	2.4	0.002
			(@ 4.6 kg cm^{-2})		
Polystyrene	595	770	85	2.5	0.0002
Poly(methyl methacrylate)	770	1680	99	3.0	0.007
ABS copolymer	490	770	93	3.5	0.005
Polytetrafluoroethylene	245		60	2.1	0.0002
Cellulose acetate	385	630	66	5.5	0.05
Acetal resins	700	980	124	3.7	
Phenoxy resin	595	875	82	4.1	0.0012
Polycarbonate	630	945	135	3.0	0.0007
Polysulfone	700	1050	174	3.1	0.008
Poly(phenylene oxide)	770	1050	190	2.6	0.0003
Nylon-66, mineral filled	980	1000	166		
Poly(butylene terephthalate)	560	1000	66	3.29	
Polyimide	1190	1100	132	3.43	
Polymethylpentene	2800			2.12	0.00007
Poly(phenylene sulfide)	7000	14000	137		
Polyurethane	2800				0.03

Pigments may be prime or inert. Prime pigments give paint its color. These may be inorganic, such as titanium dioxide (TiO_2) for white (also contained in many paints of other colors) or oxides (Fe_2O_3, Fe_3O_4, FeO) for browns, yellows, and reds, or organic, such as phthalocyanine for greens and blues. Inert pigments such as clay, talc, calcium carbonate ($CaCO_3$), and magnesium silicate make the paint last longer (acting as fillers and extenders) and may contribute to the protective coating as do mica chips in some latex paints. The paint may also contain special agents or additives that perform specific roles.

Vehicles include liquids such as oils (both natural, modified natural) and resins and water. A latex vehicle is made by suspending synthetic resins, such as poly(methyl methacrylate), in water. This suspension is called an emulsion, and paints using such vehicles are called latex, waterborne, or emulsion paints. When the vehicle comes in contact with air, it dries or evaporates, leaving behind a solid coating. For latexes, the water evaporates, leaving behind a film of the resin.

Paints are specially formulated for specific purposes and locations. The following list includes brief descriptions of the more popular paint types.

Oil paints: Oil paints consist of a suspension of pigment in a drying oil, such as linseed oil. The film is formed by a reaction involving atmospheric oxygen which polymerizes and cross-links the drying oil. Catalysts may be added to promote the cross-linking reaction. Oil paints, once dried (cured, cross-linked), are no longer soluble, although they can be removed through polymer degradation using the appropriate paint stripper.

Oil varnishes: Varnish coatings consist of a polymer, either natural or synthetic, dissolved in a drying oil together with appropriate additives as catalysts to promote the cross-linking with oxygen. When dried, they produce a clear, tough film. The drying oil is generally incorporated, along with the dissolved polymer, into the coating.

Enamels: Enamel is an oil varnish with a pigment added. The added polymer is typically selected to provide a harder, glossier coating than the oil varnish mixture.

Lacquers: Lacquers consist of polymer solutions to which pigments have been added. The film is formed through simple evaporation of the solvent leaving the polymer film as the coating. These coatings are formed without subsequent cross-linking, thus, the surface exhibits poor resistance to some organic solvents.

Latex paints: Latex paints today account for more than one-half of the commercial paint sold. They are characterized by quick drying (generally several minutes to several hours), little odor, and easy cleanup (with water). Latex paints are polymer latexes to which pigments have been added. The film is formed by coalescence of the polymer particles on evaporation of the water. The polymer itself is not water soluble, though these paints are called waterborne coatings.

Because of the need to reduce environmental pollution, the trend is toward high-solids coatings with a solvent-volume concentration of less than 20%. Plastisols, which are solvent free, consist of polymers, such as PVC, dispersed in liquid plasticizers, such as DOP. These plastisols may be coated on substrates and fused at 160°C.

17.10 ADHESIVES

In contrast to coatings, which must adhere to one surface only, adhesives are used to join two surfaces together. Resinous adhesives were used by the Egyptians at least 6000 years ago for bonding ceramic vessels. Other adhesives, such as casein from milk, starch, sugar, and glues from animals and fish, were first used about 3500 years ago.

Adhesion occurs generally through one or more of the following mechanisms. Mechanical adhesion with interlocking occurs when the adhesive mixture flows about and into two rough substrate faces. This can be likened to a hook and eye, where the stiff plasticlike hooks get caught in the fuzzlike maze of more flexible fibers. Chemical adhesion is the bonding of primary chemical groups. Specific or secondary adhesion occurs when hydrogen bonding or polar (dipolar) bonding occurs. Viscosity adhesion occurs when movement is restricted simply due to the viscous nature of the adhesive material.

Adhesives can be divided according to the type of delivery of the adhesive or by type of polymer employed in the adhesive. Following are short summaries of adhesives divided according to these two factors.

Solvent-based adhesives: Adhesion occurs through action of the adhesive on the substrate. Solidification occurs on evaporation of the solvent. Bonding is assisted if the solvent partially interacts or, in the case of model airplane glues, actually dissolves some of the plastic (the adherent). Thus, model airplane glues often contain volatile solvents, such as toluene, which can dissolve the plastic, forming what is called a solvent weld.

Latex adhesives: These adhesives are based on polymer latexes and require that the polymers be near the T_g so that they can flow and provide good surface contact when the water evaporates. It is not surprising that the same polymers that are useful as latex paints are also useful as latex adhesives. Latex adhesives are widely employed for bonding pile to carpet backings.

Pressure-sensitive adhesives: These are actually viscous polymer melts at room temperature. The polymers must be applied at temperatures above their T_g to permit rapid flow. The adhesive is caused to flow by application of pressure. When the pressure is removed, the viscosity of the polymer is high enough to hold and adhere to the surface. Many tapes are of this type where the back is smooth and coated with a nonpolar coating so as not to bond with the "sticky" surface. The two adhering surfaces can be separated, but only with difficulty.

Hot-melt adhesives: Thermoplastics often form good adhesives simply by melting, followed by subsequent cooling after the plastic has filled surface voids. Nylons are frequently employed in this manner. Electric glue guns typically operate on this principle.

Reactive adhesives: These additives are either low molecular weight polymers or monomers that solidify by polymerization and/or cross-linking reactions after application. Cyanoacrylates, phenolics, silicone rubbers, and epoxies are examples of this type of adhesive. Plywood is formed from impregnation of thin sheets of wood with resin, with the impregnation occurring after the resin is placed between the wooden sheets.

Thermosets: Phenolic resins produced by the reaction of phenol and formaldehyde were used as adhesives by Leo Baekeland in the early 1900s. This inexpensive resin is still used for binding thin sheets to wood to produce plywood. Urea resins produced by the reactions of urea and formaldehyde have been used since 1930 as binders for wood chips in particle board.

Unsaturated polyester resins have replaced lead for auto body repair, and polyurethanes are being used to bond polyester cord to rubber in tires, to bond vinyl film to particle board, and to function as industrial sealants. Epoxy resins are used in automotive and aircraft construction and as a component of plastic cement.

Elastomers: Solutions of natural rubber (NR) have been used for laminating textiles for over a century. The Macintosh raincoat, invented in 1825, consisted of two sheets of cotton adhered by an inner layer of natural rubber.

Pressure-sensitive tape, such as Scotch Tape, consisting of a coating of a solution of a blend of natural rubber and an ester of glycerol and abietic acid (rosin) on cellophane, was developed over a half century ago. More recently, natural rubber latex and synthetic rubber (SR) have been used in place of the natural rubber solution. The requirement for pressure-sensitive adhesives is that the elastomer have a glass-transition temperature below room temperature.

Styrene-butadiene rubber (SBR) is now used as an adhesive in carpet backing and packaging. Neoprene (polychloroprene) may be blended with a terpene or phenolic resin and used as a contact adhesive for shoes and furniture.

Contact adhesives are usually applied to both surfaces, which are then pressed together. Liquid copolymers of butadiene and acrylonitrile with carboxyl end groups are used as contact adhesives in the automotive industry.

Thermoplastics, which melt without decomposition when heated, may also be used as adhesives. Polyamides and copolymers of ethylene and vinyl acetate (EVA) are used as melt adhesives. Over 200 million pounds of EVA are used annually as hot-melt adhesives. Copolymers of methyl methacrylate and other monomers are used as adhesives in the textile industry.

Anaerobic adhesives consist of mixtures of dimethacrylates and hydroperoxides (initiators) which polymerize in the absence of oxygen. They are used for anchoring bolts.

One of the most interesting and strongly bonded adhesives are cyanoacrylates (Super Glue, Krazy Glue). These monomers, such as butyl-α-cyanoacrylate, polymerize spontaneously in the presence of moist air, producing excellent adhesives. These adhesives, which have both cyano and ester polar groups in the repeating units of the polymer chain, are used in surgery and for mechanical assemblies.

$$
\begin{array}{ccc}
H & & CN \\
\diagdown & & | \\
& C = C & \\
\diagup & & \diagdown \\
H & & C-O-C_4H_9 \\
& & \| \\
& & O
\end{array}
\qquad (17.1)
$$

Butyl-α-cyanoacrylate

While naturally occurring resins formerly dominated the adhesive resin field, they now account for less than 10% of the market. Elastomers, thermosets, and thermoplastics account for 35, 30, and 20% of the adhesive resins used, respectively. SBR, phenolic, and poly(vinyl acetate) are the principal resins used in this $3 billion market.

17.11 GEOTEXTILES*

Geotextiles play a major role in geotechnical engineering. Yet 30 years ago the area of geotextiles did not exist, though forms of geotextiles have existed for thousands of years. Reinforced soil was employed by the Babylonians 3000 years ago in the construction of pyramidlike towers, or ziggurats. One famous ziggurat, the Tower of Babel, collapsed.

The Chinese for thousands of years used wood, straw, and bamboo for soil reinforcement, including construction of the Great Wall. In fact, the Chinese symbol for civil engineering can be translated "earth and wood." The Dutch have made

---

*This section is based on an article appearing in *Polymer News*, Vol. 13, No. 4, and used with permission.

extensive use of natural fibrous materials in their age-old battle with the sea. The Romans used wood and reed for foundation reinforcement.

By the 1920s, cotton fabrics were tested as a mode of strengthening road pavements in the United States. These field trials were not followed by application. The British army during World War II used armored vehicles designed to lay rolls of canvas on the ground.

The modern materials for geotextiles have been produced by the polymer (textile) industry since the early 1900s. These include poly(vinyl chloride), (PVC), polyester, nylon, and polypropylene fibers. In the 1950s, the original technology for manufacturing plastic nets was developed by the packaging industry. In the 1960s, manufacturing processes for nonwoven fabrics made from continuous synthetic filaments were developed. The stage was set for the birth of geotextiles. In 1957, nylon-woven fabric sandbags were employed at the closing of the Pluimpot, Netherlands. (As a part of this overall project, called the Delta Project, more than 10 million square meters of geotextiles have been used.) In 1958, a woven fabric of poly(vinylidene chloride) monofilaments was produced by Carthage Mills and utilized for coastal erosion control in Florida. In 1958 and 1959, synthetic sandbags were used in West Germany and Japan in the control of soil erosion. By 1966, a nonwoven fabric composed of staple fibers was initially employed as a geotextile in an asphalt overlay. Synthetic nets were initially employed in 1967 for reinforcement of soft soil in Japan. Since the early 1970s, many products have been developed, including mats (Enka, Netherlands, 1972), grids (Nethon, UK, 1981), and composites for drainage.

A geotextile can perform several functions. These include:

1. Fluid transmission—collecting a liquid or gas and conveying it within the geotextile's own plane towards an outlet.
2. Separation—acting to divide geological materials for containment and prevention of mixing.
3. Protection—alleviation or distribution of stresses and strains. For instance, a geotextile may be placed on a surface to prevent its damage by light, rain, traffic, etc., or may be placed between two materials, thus acting to dissipate the load applied to one with respect to the second.
4. Filtration—allowing liquid or small particulant material to pass through while retaining larger materials.
5. Tension member—providing tensile modulus and strength to a soil.
6. Tensioned membrane—allows for a balance in pressure differences between materials.

While geotextiles are typically employed as two-dimensional sheets, three-dimensional action is easily achieved through folding, sandwiching, and engulfing procedures. The field of geotextiles will gather in importance as we attempt to build structures and grow crops on less friendly terrain. It will also be a major factor in the containment of bulk and specialty chemicals and in the protection of our water sources.

In 1985, over 300 million square meters of geotextiles were used in more than 100,000 projects. About one-half of the sales are in the United States and Canada, two-fifths in Europe, and the remainder in Japan and other countries. In North

America, about two-thirds of the geotextiles are nonwovens and one-third wovens. In Europe about four-fifths are nonwovens and one-fifth wovens, and in Japan about three-fifths are wovens, one-fifth nets and grids, and one-fifth nonwovens.

17.12 SOLID WASTE

Polymers are giant molecules and macromolecules. They include natural organic (proteins, nucleic acids, polysaccharides such as starch and cellulose, wool, hair, skin, teeth, lignin), inorganic (sand, granite, quartz), and synthetic [plastics, elastomers (rubbers), fibers, concrete (cement), glass, paper] polymers. They compose the large bulk by weight and by volume of what is termed solid waste.

Fortunately the synthetic materials offer few immediate health problems but, in the more distant future, they offer both a resource problem (i.e., they are made utilizing nonrapidly renewable resources and their synthesis is typically energy demanding) and a disposal problem. In today's marketplace, the United States consumes about 53 billion pounds (53,000,000,000) of synthetic polymers (plastics and synthetic fibers and rubbers) annually, or 200 pounds per person. Paper products are produced at a rate almost double this yearly.

The following remarks will concentrate on the general area of solid waste, emphasizing synthetic polymeric materials—i.e., elastomers, fibers, plastics, and glass.

Biodegradable and Recyclable Products

Many states propose to deal with the problem of solid waste though legislation dealing with biodegradation and recycling. Recycling means many things. For instance, PVC can burn to give hydrogen chloride, ethylene, etc., which in turn can, after several energy intense steps, be returned to vinyl chloride and finally again to PVC. While this has been accomplished on a small scale, it has not been accomplished on anything larger than a small pilot plant scale. Further, it is both economically and energetically unfavorable. Next, PVC is not just PVC, rather, it is composed, as are most *all* commercially processed polymers, of coloring agents, antioxidants, stabilizers, plasticizers, etc.—in fact, between 20 to 40% of the PVC is not PVC but, rather, additives. Finally, collecting PVC or any homogeneous grouping of solid material is a difficult, untested task at best.

Glass is composed of the same material as is most sand, i.e., silicon dioxide. Material capable of producing commercial glass is not abundant. Fortunately, glass is readily recognizable to the public, and commercial and public collection of glass products is underway. Waste glass (also called recycled or cullet) has found many uses, ranging from filler in asphalt road material to composing 5 to 40% of "glass-making recipes."

Numerous uses for recycled rubber tires have shown to be industrially feasible. These include use of tire hides as automotive bumper guards, capped tires, and filler. Rubber by its very nature of being a three-dimensional highly cross-linked material makes true chemical recycling difficult. Commercially, the major option is to grind the rubber into small "chunks" and to apply "adhesives" to "recombine" the rubber pieces. The resulting rubber is acceptable for nondemanding applications such as rubber toys and place mats but is *not* acceptable for demanding applications such as rubber gaskets, tire hides, hoses, and belts.

"Paper" is not "paper" is not "paper." Recycled paper probably cannot be recycled again. Most "paper" today contains a wide variety of additives. The amount and type of additive determines which paper can be recycled.

The previous examples are intended to describe some of the real problems associated with the term "recyclable."

"Biodegradable" may also mean many things. For instance, two types of common drinking cups were placed on the shores of Lake Ontario. The polystyrene cup had photoactive agents placed within it. After 3 months the cup appeared largely degraded, and by 6 months it was declared degraded. If "out of sight, out of mind" applies, then the styrofoam cup was degraded. However, the photoactive agents simply allowed the cup to disappear into microscopic particles that were blown away—not truly becoming part of the "natural seascape." The typically acclaimed biodegradable paper cup was largely unchanged after 2 years on the beach. The polyethylene outer cover was erased after one-half year—yet the paper cup persisted.

It is clear that the enemy has been sighted—and the enemy is us (after Pogo).

Biodegradable Polymers

Traditional applications of synthetic polymers are based largely on their biological inertness compared with natural polymers. Now the "worm has turned," and there is interest in producing polymers that are biodegradable. As is the case with the whole solid waste question, the literature is extensive, sometimes confusing and contradictory and often of little scientific value.

Biodegradation of general nonmedical materials can be divided into microbial and chemical degradation. Chemical degradation includes wind and rain erosion, oxidation, photodegradation, acid/base water, and thermal degradation. Microbial testing includes real-site testing and laboratory testing. Fungi are used more frequently than bacteria. The degree of degradation of a polymer is typically determined by observing the evidence of colony growth, oxygen consumption, increase in cell count mass or count, or production of carbon dioxide. Most of these can give results that are open to interpretation, since polymers contain additives that may well mask what is really occurring. More reliable, yet much more time consuming, is to actually measure changes in the polymer. Changes in molecular weight distribution, solution viscosity, tensile strength, and morphological changes have all been employed. Procedures have been worked out for this and may be adaptable for some cases. Purified buffered enzymes are also employed for analysis, but it is unknown if there exist today bacteria or fungi that contain or can be made to contain such enzymes.

General Structure-Property Relationships

Polymers can be divided according to general susceptibility to either chemical or microbial degradation. A typical division would include polymers with hydrolyzable backbones. Most natural polymers are of this variety including polysaccharides, proteins, and nucleic acids, but not terpene-based materials such as the natural rubbers.

Synthetic polymers with hydrolyzable backbones include the polyurethanes, polyethers, polyesters, and nylons (polyamides). These materials are typically susceptible to acid- and/or base-induced (either chemical or microbial produced) degradation. Polymers with wholly carbon backbones or with large amounts of hydrocarbon cross-links are much less susceptible to acid- and/or base-induced degradation. Many of these polymers resist microbial and "natural" chemical-induced degradation. Poly(vinyl alcohol) is among the most readily biodegradable of these materials, being readily degraded in wastewater-activated sludges. Conversely, poly(methyl methacrylate) and polycyanoacrylates generally resist biodegradation.

Chemical Degradations

Nonagriculture commercial methods of solid waste disposal have largely involved burning, landfill, composting, or microbial disintegration and scraping or recycling. Landfill and burning has accounted for the large bulk of commercial solid-waste disposal. Landfills are no longer an alternative, and in much of the United States it has really never been a choice. Burning was largely done with little control of oxygen; thus, the adage that effluents of burning are carbon dioxide and water is approximate at best. Airborne particulants often result and, even with scrubbers, etc., undesired elements can reach our air, water, and food supply. Finally, within compacted areas free from readily available sources of oxygen, degradation is often quite slow. Thus, newspapers, raw organic matter, yard waste, etc., have been found in landfills little changed from when they were first introduced many years before.

Common backbone bond-scission reactions in polymers include the following:

1. *Homolytic bond cleavage*, to generate a pair of free radicals. This results in chain degradation or free-radical formation on a backbone that may be followed by a proton shift and subsequent chain scission.

$$-C-C- \longrightarrow -C\bullet \bullet C-$$

$$-C-C-C- \longrightarrow -C-C-C- \longrightarrow C=C + \bullet C- \quad (17.2)$$

These two reactions are initiated by high-energy radiation and heat. Exposure to UV radiation can also lead to bind scission in carbonyl-containing polymers (i.e., nylons, polyurethanes, polyesters, and polyketones).

$$-N-C-N- \longrightarrow -N-C\bullet \quad \bullet N- \quad (17.3)$$

2. *Bond rearrangement*. Certain polyketones and polyesters when exposed to UV radiation undergo bond rearrangement leading to backbone breakage.

$$R-\overset{\overset{\displaystyle O}{\|}}{C}-O-CH_2-CH_2-R' \longrightarrow R\ C\ OH \ + \ \underset{\underset{\displaystyle H}{/}}{\overset{\overset{\displaystyle H}{\backslash}}{C}}=\underset{\underset{\displaystyle R'}{\backslash}}{\overset{\overset{\displaystyle H}{/}}{C}} \qquad (17.4)$$

$$R-\overset{\overset{\displaystyle O}{\|}}{C}-CH_2-CH_2-CH_2-R' \longrightarrow R-\overset{\overset{\displaystyle O}{\|}}{C}-CH_3 \ + \ \underset{\underset{\displaystyle H}{/}}{\overset{\overset{\displaystyle H}{\backslash}}{C}}=\underset{\underset{\displaystyle R'}{\backslash}}{\overset{\overset{\displaystyle H}{/}}{C}} \quad (17.5)$$

3. *Ionic bond scission.* Most polymers containing noncarbon atoms within their backbone undergo acid- and/or base-catalyzed bond scission. Where B^- represents a base and A^+ represents an acid, we have the following illustration of ionic bond scission.

$$R-\overset{\overset{\displaystyle O}{\|}}{C}-O-R \ \overset{B^-}{\longrightarrow}\ \underset{\underset{\displaystyle B}{|}}{R-\overset{\overset{\displaystyle O^-}{|}}{C}-O-R} \ \longrightarrow \ R-\overset{\overset{\displaystyle O}{\|}}{C}-O^- \ + \ H-O-R \quad (17.6)$$

$$R-\overset{\overset{\displaystyle O}{\|}}{C}-O-R \ \overset{A^+}{\longrightarrow}\ \underset{\oplus}{R-\overset{\overset{\displaystyle AO}{|}}{C}-O-R} \ \longrightarrow \ R-\overset{\overset{\displaystyle O}{\|}}{C}-O-H \ + \ H-O-R \quad (17.7)$$

Polymers susceptible to ionic degradation include polycarbonates, siloxanes, polyesters, nylons, polyurethanes, polysaccharides, and polyethers.

4. *Oxidation.* Almost all organic polymers are subject to either extremely long-term oxidation or thermally induced oxidation. The typical final products for compounds containing carbon, nitrogen, sulfur, oxygen, and hydrogen with an excess of oxygen are carbon dioxide, nitrogen oxides (of various forms), water, and sulfur oxides. Halide-containing polymers typically generate the corresponding hydrogen halide gases but may also generate quite toxic intermediates such as ClCOCl. The sulfur oxides, nitrogen oxides, and hydrogen halides are toxic and must be "scrubbed" before gases containing them are sent into the atmosphere.

5. *Metal-catalyzed degradations.* Many commercial polymers contain metallic impurities derived from additives, metal catalysis, or introduction during workup or processing. Metal-containing pigments can be added to protect against or to sensitize materials to photodegradation. Trace amounts of transition metals may accelerate thermal degradation of polymers.

Fortunately, industrial, academic, and governmental groups are focusing attention on the problem with short-range solutions often available. Longer-term solutions await further breakthroughs, but the technology and science exists that should allow these breakthroughs to occur. Of greater concern than the scientific and technological aspects is the ability to educate ourselves sufficiently to take part in recycling efforts and the ability to perform decent solid waste conservation on a meaningful scale.

17.13 RECYCLING CODES

Most of us have seen the triangle symbol with three arrows and realize it has something to do with recycling. This symbol was initially developed by the Society of Plastics Industry as a resin/plastic identification code. It was devised largely for use on containers, but today the "chasing-arrows" triangle is being utilized as a universal symbol for recycling.

The SPI Resin Identification Code utilizes the numbers 1 through 7 and bold, capital letters as shown below.

PETE

Poly(ethylene terephthalate)—**PET**

PET is the plastic used to package the majority of soft drinks. It is also used for some liquor bottles, peanut butter jars, and edible-oil bottles. About one quarter of plastic bottles are **PET**. **PET** bottles can be clear; they are tough and hold carbon dioxide well.

HDPE

High-density polyethylene—**HDPE**

Polyethylene is available in a wide variety of molecular weights with varying branching. **HDPE** accounts for over 50% of the plastic bottle market. It is used to contain milk, juices, margarine, and some grocery snacks. **HDPE** is easily formed through application of heat and pressure; it is relatively rigid and low cost.

V

Poly(vinyl chloride)—**PVC** or **V**

Poly(vinyl chloride) can be used "pure" or as a blend to make a wide variety of products including **PVC** pipes, food packaging film, and containers for window cleaners, edible oils, and solid detergents. **PVC** accounts for only about 5% of the container market.

LDPE

Low-density polyethylene—**LDPE**

Low-density polyethylene is similar in gross structure to **HDPE** except that it contains a variety of branching, resulting in a material that has a lower softening temperature and it is amorphous, making products derived from it more porous and less tough. **LDPE** is a major material for films from which trash bags and bread bags are made. While it is more porous than **HDPE**, it does offer a good inert barrier to moisture.

PP

Polypropylene—**PP**

Polypropylene has good chemical resistance to chemicals and fatigue. Films and fibers are made from it. **PP** is the least used, of the major plastics, for household items. Some screw-on caps, lids, yogurt tubs, margarine cups, straws, and syrup bottles are made of **PP**.

PS

Polystyrene—**PS**

Polystyrene is used in a wide variety of containers, including those known as "styrofoam" plates, dishes, cups, etc. Cups, yogurt containers, egg cartons, meat trays, and plates are made from **PS**.

OTHER

Other plastics

A wide variety of other plastics are coming to the marketplace, including a number of copolymers, alloys, blends, and multilayered combinations.

17.14 SMART MATERIALS

Smart materials have been with us for some time, though the term is relatively new. Some of the first smart materials were piezoelectric materials, including poly(vinylene fluoride), that emit an electric current when pressure is applied and change volume when a current is passed through them. Most smart materials are polymeric in nature or have polymers forming integral parts of the "smart material" system.

Today's research concentrates not only on the synthesis of new smart materials, but also on the application of already existing smart materials. Much of the applications involve assemblage of smart material portions and envisioning uses for these smart materials. Thus, it is possible that, since application of pressure to a piezoelectric material causes a discharge of current, a portion of a wing could be constructed such that appropriate "warping" of a wing would result from an "electronic feedback" mechanism employing a computer coupled with a complex system of electronic sensing devices. Almost instantaneous, self-correcting changes in the overall wing shape would act to allow safer and more fuel-efficient air flights. Similar applications involving adjustable automotive wind foils are possible. Such pressure-sensing devices could also be used to measure application of "loads" through reaction of the piezoelectric sensors to stress/strain. Again, application of these computer-linked sensors would allow redistribution of "loads" and/or reconfiguration of the load-bearing structure(s) to compensate for stress/strain imbalances.

Smart materials are materials that react to an externally applied force—electrical, stress/strain (including pressure), thermal, light, and magnetic. A smart material is not smart simply because it responds to external stimuli, but it becomes smart when the interaction is used to achieve a defined engineering or scientific goal. Thus, most materials, including ceramics, alloys, and polymers, undergo volume changes as they undergo phase changes. While the best known phase changes are associated with the total changes in matter state such as melting/freezing, many more subtle phase changes occur. For polymers the best known secondary transitional phase change is the glass transition, Tg, where local segmental mobility occurs. Volume changes associated with Tg are well known and used as a measure of the crystallinity of a polymer. Along with Tg, polymers can undergo a number of other volume-changing transitions. These polymers then become smart only when the volume-changing ability is applied, such as in sensing devices. Multiple switching devices can be constructed to detect and redirect electrical signals as a function of temperature from such materials.

The use of smart materials as sensing devices and shape-changing devices has been enhanced due to the increased emphasis on composite materials that allow the introduction of smart materials as components. A typical smart material assembly might contain:

Sensor components containing smart materials that monitor changes in pressure, temperature, light, current, and/or magnetic field
Communications network that relays changes detected by the sensor components via fiber optics or conductive "wire" (including conductive polymer channels)
Actuator part that reacts to the action command.

The actuator part may also be a smart material such as a piezoelectric bar placed in a wind foil that changes orientation according to a current "ordered" by the computer center to allow better handling and more fuel-efficient operation of a machine such as an automobile.

Thus, the "smart material" may act as a sensor component, as an actuator part or by itself, as described earlier. The smart material may act on a molecular level, as in sensors, or as bulk material, such as in wind foils.

We are aware that muscles contract and expand in response to electrical, thermal, and chemical stimuli. Certain polymeric materials, including synthetic polypeptides, are known to change shape on application of electrical current, temperature, and chemical environment. For instance, selected bioelastic smart materials expand in salt solutions and may be used in desalination efforts. They are also being investigated in tissue reconstruction, as adhesive barriers to prevent adhesion growth between surgically operated body parts, and in controlled drug release where the material is designed to behave in a predetermined manner according to a specified chemical environment. The molecular design can be tailored, giving materials with varying biodegradabilities.

Most current efforts include three general types of smart materials: piezoelectric, magnetostrictive (materials that change their dimension when exposed to a magnetic field), and shape memory alloys (materials that change shape and/or volume as they undergo phases changes). Conductive polymers and liquid crystalline polymers can also be used as smart materials, since many of them undergo relatively large dimensional changes when exposed to the appropriate stimulus such as an electric field.

New technology is being combined with smart materials called micromachines, machines smaller than the width of a human hair, of the order of microns. Pressure and flow meter sensors are being investigated and commercially made.

17.15 CONSTRUCTION AND BUILDING

The use of polymeric materials as basic structural materials is widespread and of ancient origin. Today, the use of synthetic polymers in building and construction is increasing at a rapid rate. The use of plastics alone is approaching 10% of the total cost of building and construction in the United States (Table 17.5). Uses of plastics are given in Table 17.6. By weight, plastics accounts for only about 2%,

Table 17.5 Values of Plastics in Construction in the USA

Plastic	$ (billion)
Poly(vinyl chloride)	2.1
Phenolics	1.2
Polyurethanes	1.1
Polyethylene	0.5
Polystyrene	0.4
Urea and melamine	0.3
Acrylates	0.3
Other	0.7

but the total use of polymeric materials by weight is about 80% (concrete, lumber, ceramic, wood panels, plastics; Table 17.7).

As lighter and stronger polymeric materials become available, their impact on the building and construction industry and on other industries will increase. Further, as materials that perform specific tasks become available, they too will become integrated into the building and construction industry. This includes devices for gathering and storage of solar energy and smart materials including smart windows.

17.16 HIGH-PERFORMANCE THERMOPLASTICS

High-performance thermoplastics are often also referred to as advanced thermoplastics. Thermoplastics can be reformed through application of heat (sometimes also requiring application of pressure). They generally contain no or only light cross-linking (except for ionomers). Most common polymers such as polyethylene, polystyrene, poly(ethylene terephthalate), nylon-66, and polypropylene are thermoplastics.

Table 17.6 Major Plastic Applications in Building and Construction

Plastic	Use(s)
ASA	Window frames
Acrylics	Lighting fixtures and glazing
PVC and chlorinated PVC	Hot and cold water pipe; moldings, siding, window frames
Melamine and urea formaldehyde	Laminating for countertops, adhesives for wood, plywood and particle board
Phenol formaldehyde	Electrical devices and plywood adhesive
PET	Countertops and sinks
Polycarbonates	Window and skylight glazing
PE	Pipe, wire, and cable covering
Poly(ethylene oxide)	Roofing panels
PS	Insulation and sheathing
Polyurethane	Insulation and roofing systems

Table 17.7 Use of Materials in Building and Construction

Material	Weight (billion lb)	%
Concrete	250	50
Lumber	60	12
Ceramic	50	10
Wood panels	20	4
Iron and steel	15	3
Plastics	10	2
Other	95	19

Today there are many applications that require materials to be stable and melt above 200°C. These materials are referred to as high-temperature thermoplastics. As the advantages of polymeric materials become evident in new areas, the property requirements, including thermal stability, will increase, causing the polymer chemist to seek new materials or old materials produced in new ways to meet these demands.

Table 17.8 contains some of the new advanced thermoplastics that are currently available. Many of these are being utilized as lightweight replacements for metal because of their strength, high-dimensional stability, and resistance to chemicals and weathering. Poly(phenylene sulfide) has a number of automotive uses, including as an injection-molded fuel line connector and as part of the fuel filter system.

Poly(alkylene carbonate) leaves no residue when decomposed. Thus, it is used as a binder for holding ceramic and metal powders together long enough for them to be made into the desired products. These polymers also give good moisture and oxygen barriers and they are abrasion resistant, offer good clarity, and give tough films. They are being considered in food and medical packaging applications.

SUMMARY

1. Synthetic fibers are produced by forcing a solution of polymer or a molten polymer through small holes in a spinneret. The extrudate from the solution may be precipitated in the wet spinning process, or the solvent may be evaporated in the dry spinning process. The molten extrudate is cooled in the melt spinning process. Graphite filaments are produced by the pyrolysis of polyacrylonitrile filaments, pitch, or other suitable material followed by forcing the molten mass through small holes.

2. The most widely used elastomer, SBR, is obtained by coagulation of an emulsion of a copolymer of butadiene and styrene. In addition to being produced by the free-radical emulsion polymerization technique, elastomers are also produced by cationic copolymerization (butyl rubber) and by anionic polymerization (polybutadiene).

3. Films may be cast from melt or solution or extruded from a slit die, but most film is produced by air blowing a warm tubular extrudate and slitting the expanded tube.

Table 17.8 Advanced High-Temperature Thermoplastics and Their Applications

Material	Heat-deflection temperature (°C)	Properties
Poly(alkylene carbonate)	—	Leaves no degradation residue
Polyamide-imide	280	Good wear, friction and solvent resistance
Polyaniline	70	Conducts electricity
Polyarylate (aromatic polyesters)	175	Good toughness, UV stability, flame retarder
Polybenzimidazoles	440	Good hydrolytic, dimensional, and compressive stability
Polyetherimide	220	Good chemical and dimensional stability and good creep resistance
Polyethersulfone	200	Good chemical resistance and stability to hydrolysis
Polyimide	360	Good toughness
Polyketones	330	Good chemical resistance, strength and stiffness
Poly(phenylene ether)	170	Often alloyed with PS
Poly(phenylene sulfide)	260	Good dimensional stability and chemical resistance
Polyphenylsulfone	260	Good chemical resistance
Polyphthalamide	290	Good mechanical properties
Polysulfone	175	Good rigidity

4. Wire and paper may be coated by use of an extruder, but most coatings consist of solutions or aqueous dispersions of polymers. The trend in the coatings industry is toward lower concentrations of volatile organic solvents.

5. Polymeric foams may be produced by the mechanical frothing of a latex or by the use of gaseous propellants. The latter may be produced in situ in PU or added during processing.

6. Composites consisting of a reinforcing fiber and a resin have excellent strength properties. Graphite–epoxy resin composites are superior to many classic construction materials.

7. Thermosets may be compression molded by heating a prepolymer under pressure in a mold cavity. Thermoplastics are usually injection molded in a fast, automated process in which a granulated polymer is heat softened or plasticated in a barrel and forced into a closed cooled mold by a reciprocating ram or screw. The molded part is ejected when the mold opens and the process is repeated.

8. Hollow articles such as bottles are produced by air blowing a heat-softened tube or parison into a two-component mold.

9. Articles like containers or boats may be produced by the thermoforming of plastic sheets.

10. Pipe and various profiles may be produced continuously by forcing heat-softened polymer through a die and cooling the extrudate.

11. Chemical companies can perform several functions such as manufacturing, processing, fabrication, and finishing of polymer-containing materials.

12. Laminating conceptually is simply the binding together of different layers of materials. Examples include safety glass, formica, and plywood.

13. While the term "painting" emphasizes a decorative aspect and the term "coating" a protective aspect, the terms are generally used interchangeably. Paint is typically a mixture of a liquid (called a binder or vehicle) and one or more colorants (pigments) and may contain a solvent or thinner. Oil paints consist of a suspension of pigment in a drying oil. Oil varnish contains a polymer dissolved in a drying oil with appropriate additives. Enamels are oil varnishes that contain a pigment. Latexes are polymer emulsions to which pigments have been added.

14. Adhesion can occur through mechanisms that include mechanical, specific or secondary bonding, viscosity, and chemical routes. There are many types of adhesives—solvent-based, latex, pressure-sensitive, hot-melt, and reactive; thermosets, elastomers, and thermoplastic.

GLOSSARY

abaca: A hemplike fiber from the Philippines.

acetate rayon: Cellulose diacetate fibers.

Acrilan: Polyacrylonitrile-based fibers.

acrylic fibers: Fibers based on polyacrylonitrile.

adhesive: Material that binds, holding together two surfaces.

Adiprene: PU elastomer.

agave: Fibers from the leaves of the desert century plant.

Aralac: Soybean fiber.

Ardil: Peanut fiber.

biaxial orientation: The process is which a film is stretched in two directions at right angles to each other.

BMC: Bulk-molding compound.

BTU: British thermal unit.

buna-N: Acrylonitrile-butadiene copolymer.

calender: A machine for making polymeric sheet, containing counter-rotating rolls.

casting: Production of film by evaporation of a polymeric solution.

cellophane: Regenerated cellulosic film.

charge: The amount of polymer used in each molding cycle.

coextruded film: One produced by the simultaneous extrusion of two or more polymers.

Dacron: Poly(ethylene terephthalate) fiber.

DMF: Dimethylformamide.

draw: Depth of mold cavity.

drying oils: Liquids employed in coatings that will be cured, cross-linked.

dry spinning: Process for obtaining fiber by forcing a solution of a polymer through holes in a spinneret and evaporating the solvent from the extrudate.

elastomer: Rubber.

electrodeposition: The use of an electric charge to deposit polymer film or aqueous dispersion onto a metal substrate.

enamel: Oil varnish that contains a pigment.

EPDM: Curable ethylene-propylene copolymer elastomer.

extrusion: A fabrication process in which a heat-softened polymer is forced continually by a screw through a die.

fibrillation: Process for producing fiber by heating and pulling twisted film strips.

filament: A continuous thread.

filament winding: Process in which filaments are dipped in a prepolymer, wound on a mandrel, and cured.

flax: The threadlike fiber from the flax plant.

gate: Thin sections of runner at the entrance of a mold cavity.

geotextiles: Polymeric mats, sheets, and textiles employed in the control of soil, water, etc., in geological applications.

hemp: Fiber from plants of the nettle family.

hevea rubber: *Hevea brasiliensis*, natural rubber.

high-density polyethylene: Polyethylene that contains a relatively high degree of crystallinity with a low degree of branching typically made using solid-state catalysts.

Hycar: Buna-N elastomer.

jute: Plant fiber used for making burlap.

kapok: Seed fibers from the tropical silk tree.

kenaf: Cellulose fiber from the kenaf plant.

Kodel: Poly(ethylene terephthalate) fiber.

K value: A measure of thermal conductivity in BTU.

lacquers: Polymer solutions to which pigments have been added.

lamination: The plying up of sheets.

Lanital: Casein fiber.

latex: Stable dispersion of a polymer in a water.

linear low-density polyethylene: A copolymer of ethylene and about 10% of an alpha-olefin; it has densities and properties between LDPE and HDPE.

liquid crystals: Materials that undergo physical reorganization where at least one of the arrangement structures involves molecular alignment along a preferred direction causing the material to exhibit nonisotropic behavior.

mechanical goods: Industrial rubber products such as belts.

melt spinning: Process of obtaining fibers by forcing molten polymer through holes in a spinneret and cooling the filaments produced.

molding powder or compound: A premix of resin and other additives used as a molding resin.

multicellular: Open cells.

Neoprene: Polychloroprene.

nonwoven textiles: Sheet produced by binding fibers with a heated thermoplastic.

oil paints: Suspension of pigments in a drying oil.

oil varnish: A polymer dissolved in a drying oil.

parison: A short plastic tube which is heated and expanded by air in the blow-molding process.

Perlon: PU fibers.

PET: Poly(ethylene terephthalate).

phosphazene: Polyphosphonitrile elastomer.

pigment: Coloring material (colorant).

printing ink: Highly pigmented coatings used in the printing of sheets.

PU: Polyurethane.

pultrusion: Process in which filaments are dipped in a prepolymer, passed through a die, and cured.

rayon: Regenerated cellulosic fiber.

recycling codes: Designations that allow easy, quick identification of a number of plastics used in the container manufacture.

rotational molding: Polymer added to a warm, rotating mold; centrifugal force distributes the polymer evenly.

rovings: Multiple untwisted strands of filaments.

runner: Channel between the sprue and the mold cavity.

saponification: Alkaline hydrolysis of an ester.

SBR: Styrene-butadiene copolymer elastomer.

screen pack: A metal screen which prevents foreign material from reaching the die in an extruder.

silk: Natural protein fiber.

smart materials: Materials that react to an externally applied action such as the application of light, magnet, heat, electricity, and pressure in a "smart" manner.

SMC: Sheet-molding compound.

solvent molding: Immersion of a mold into a polymer solution resulting in the mold being coated with a polymeric film.

Spandex: Elastic PU fiber.

specific strength: Strength based on mass rather than area.

spinneret: A metal plate with many uniformly sized, minute holes.

sprue: Tapered orifice between nozzle and runner. This term is also applied to plastic material in the sprue.

structural foams: Polymeric foamed articles with a dense surface.

Styrofoam: Foamed polystyrene.

technology: Applied science.

tenacity: Fiber strength.

Terylene: Poly(ethylene terephthalate).

thermoforming: The shaping of a hot thermoplastic sheet.

Thiokol: Polyolefin sulfide elastomer.

transfer molding: A process in which a preheated briquette or preform is forced through an orifice into a heated mold cavity.

Tricel: Cellulose triacetate.

UHMWPE: Ultrahigh molecular weight polyethylene polymer.

ultrahigh molecular weight polyethylene: High-density PE with over 100,000 ethylene units; the size allows the chains to intertangle, increasing its strength.

ultralinear polyethylene: Polyethylene produced with soluble stereoregulating catalysts; has a narrow molecular weight distribution, high crystallinity, low branching.

unicellular: Closed coatings.

vehicle: Liquid in a coating.

Vicara: Zein fiber.

viscose process: Regeneration of cellulose fibers by precipitation of the sodium salt of cellulose xanthate in acid.

Viton: Polyfluorocarbon elastomer.

Vycron: Poly(ethylene terephthalate).

wet spinning: Obtaining fibers by precipitation of polymeric solutions.

wool: Nature protein fiber.

zein: Protein from maize (corn).

EXERCISES

1. Which is more important: (a) polymer science, or (b) polymer technology?

2. Name three important natural fibers.

3. Name an important regenerated fiber.

4. Why is secondary cellulose acetate more widely used than the tertiary cellulose acetate?

5. What is the difference between rayon and cellophane?

6. Name three important synthetic fibers.

7. Name an elastomer produced by (a) cationic, (b) anionic, (c) free-radical, and (d) step-reaction polymerization techniques.

8. How is LDPE film produced?

9. Why is there a trend toward the use of less solvent in polymeric coatings?

10. What is meant by trade sales?

11. Why is a plastisol stable at room temperature?

12. How would you produce a unicellular foam?

13. Which foam is preferable for upholstery: (a) unicellular, or (b) multicellular?

14. Why do non–flame-retardant foams burn readily?

15. What reinforced plastic has been used as an automobile body?

16. Why is graphite-reinforced eposy resin a good candidate for parts in future automobiles?

17. Why are molded thermoplastics used more than molded thermosets?

18. How could you increase the output of an injection molding press?

19. Why are structural foams used for complex furniture?

20. What are some of the advantages of a blow-molded PET bottle?

21. Why would an article be thermoformed instead of molded?

22. What is the limit to the length of an extrudate such as PVC pipe?

23. Name three popular laminates.

24. Why are the terms painting and coating often used interchangeably?

25. Why have latex waterborne coatings been popular with the general public?

26. Differentiate between oil paints, oil varnishes, latexes, enamels, and lacquers.

27. Briefly discuss the popular mechanisms related to adhesion and the general types of adhesives.

BIBLIOGRAPHY

Aldissi, M. (1989): *Inherently Conducting Polymers*, Noyes, Park Ridge, New Jersey.
Aldissi, M. (1992): *Intrinsically Conducting Polymers: An Emerging Technology*, Kluwer Academic, Dordrecht.
Bachmann, K. J. (1995): *Materials Science of Microelectronics*, VCH, New York.
Bakshi, A. (1994): *Electronic Structure of Biopolymers and Highly Conducting Polymers*, Halsted Press, New York.
Biesenberger, J. A., Sebastian, D. (1983): *Principles of Polymerization Engineering*, Wiley, New York.
Boundy, R. H., Boyer, R. F. (1952): *Styrene, Its Polymers, Copolymers, and Derivatives*, Reinhold, New York.

Brewis, D. M. (1981): *Surface Analysis and Pretreatment of Plastic and Metals*, Applied Science, Essex, England.

Briston, J. H. (1989): *Plastic Films*, 2nd ed., Longman, Essex, England.

Brown, R. P. (1987): *Physical Testing of Rubber*, 2nd ed., Elsevier, New York.

Cahn, R., Haasen, P., Kramer, E. (1996): *Processing of Polymers*, VCH, New York.

Cheremisinoff, N. P. (1993): *Guidebook to Commercial Polymers*: *Properties and Applications*, Prentice Hall, Englewood Cliffs, New Jersey.

Ching, C. (1993): *Biodegradable Polymers and Packaging*, Technomic, Lancaster, Pennsylvania.

Crawford, R. J. (1987): *Plastics Engineering*, 2nd ed., Pergamon, New York.

Dörer, K., Freitag, W., Stoye, D. (1994): *Waterborne Coatings*, Hanser-Gardner, Cincinnati.

Evans, C. W. (1981): *Practical Rubber Compounding and Processing*, Applied Science, Essex, England.

Finch, C. A. (1983): *Chemistry and Technology of Water Soluble Polymers*, Plenum, New York.

Gebelein, C. G., Williams, D., Dennan, R. (1982): *Polymers in Solar Energy Utilization*, ACS, Washington, D.C.

Golding, B. (1959): *Polymers and Resins*, Van Nostrand, Princeton, New Jersey.

Griffin, G. J. (1994): *Chemistry and Technology of Biodegradable Polymers*, Routledge Chapman & Hall, New York.

Griskey, R. (1992): *Polymer Process Engineering*, Chapman & Hall.

Grulke, E. (1993): *Introduction to Polymer Process Engineering*, Prentice Hall, Englewood Cliffs, New Jersey.

Hamid, A. (1992): *Handbook of Polymer Degradation*, Marcel Dekker, New York.

Hope, P., Folkes, M. (1993): *Polymer Blends and Alloys*, Blackie, London.

Hoyle, C., Kinstle, J. F. (1989): *Radiation Curing of Polymeric Materials*, ACS, Washington, D.C.

Kesting, R., Fritzsche, A. (1993): *Polymeric Gas Separation Membranes*, Wiley, New York.

Kresta, J. E. (1982): *Reaction Injection Molding and Fast Polymerization Reactions*, Plenum, New York.

Kroschwitz, V. (1985): *Encyclopedia of Polymer Science and Engineering*, Wiley, New York.

Lewin, M., Atlas, S., Pearce, E. E. (1982): *Flame-Retardant Polymeric Materials*, Plenum, New York.

Martino, R. (1986): *Modern Plastics Encyclopedia*, McGraw-Hill, New York.

McCrum, N. G. (1988): *Principles of Polymer Engineering*, Oxford University Press, Oxford, England.

Meyer, R. W. (1987): *Handbook of Polyester Molding Compounds and Molding Technology*, Chapman and Hall, Boston.

Miles, I., Rostami, S. (1992): *Multicomponent Polymer Systems*, Wiley, New York.

Mittal, K. L. (1983): *Adhesion Aspects of Polymeric Coatings*, Plenum, New York.

———. (1989): *Metallized Plastics*, Plenum, New York.

Mobley, D. P. (1994): *Plastics from Microbes*, Hanser-Gardner, Cincinnati.

Moore, G., Kline, D. (1983): *Properties and Processing of Polymers for Engineers*, Prentice-Hall, Englewood Cliffs, New Jersey.

Morton-Jones, D. H. (1989): *Polymer Processing*, Routledge Chapman & Hall, Boston.

Myasnikova, L., Marikhim, V. (1994): *Orientational Phenomena in Polymers*, Springer-Verlag, New York.

Pearson, J., Richardson, S. M. (1983): *Computational Analysis of Polymer Processing*, Applied Science, Essex, England.

Pearson, J. (1985): *Mechanics of Polymer Processing*, Elsevier, New York.

Pethrick, R. (1985): *Polymer Handbook*, Harwood, Chur, Switzerland.

Pollock, D. D. (1981): *Physical Properties of Materials for Engineers*, CRC, Boca Raton, Florida.

Pye, R. G. W. (1983): *Injection Mould Design*, Longman, Essex, England.

Rauwendaal, C. (1981): *Polymer Extrusion*, Oxford University Press, Oxford, England.

Rauwendaal, C. (1995): *Polymer Extrusion*, Hanser-Gardner, Cincinnati.

Rosato, D. V. (1989): *Plastics Processing Data Handbook*, Van Nostrand-Reinhold, New York.

Salamone, J. C., Riffle, J. (1992): *Contemporary Topics in Polymer Science*, Vol. 1, *Advances in New Materials*, Plenum, New York.

Scrosati, B. (1993): *Applications of Electroactive Polymers*, Chapman & Hall, New York.

Seymour, R. B. (1987): *Polymers for Engineering Applications*, American Society for Metals, Metals Park, Ohio.

————. (1987): *Polymers for Engineering Applications*, American Society for Metals, Metals Park, Ohio.

Seymour, R. B., Kirshenbaum, G. (1986): *High Performance Polymers: Their Origin and Development*, Elsevier, New York.

Shibaev, V., Lam L. (1993): *Liquid Crystalline and Mesomorphic Polymers*, Springer-Verlag, New York.

Shutov, F. A. (1986): *Integral-Structural Polymer Foams*, Springer-Verlag, New York.

Stastnajd, J., Dekee, D. (1995): *Transport Properties in Polymers* Technomic, Lancaster, Pennsylvania.

Turner, S. (1986): *Mechanical Testing of Plastics*, 2nd ed., Wiley, New York.

Vollrath, L., Haldenwanger, H. (1994): *Plastics in Automotive Engineering*, Hanser-Gardner, Cincinnati.

Whelan, A. (1989): *Developments in Plastics Technology*, Elsevier, New York.

Whelan, A., Goff, J. P. (1985): *Developments in Injection Molding*, Elsevier, New York.

Wright, R. E. (1995): *Injection Transfer Molding of Thermosets*, Hanser-Gardner, Cincinnati.

Solutions

In the case of certain questions several answers may be appropriate. Also, the depth and breadth of an answer or answers to a question may vary depending on the preference of the student/teacher. Usually, a single answer is given in brief detail, indicating at least one direction of thought. Further, the answers to some questions are not specifically contained within the text. Answers to questions also may involve taking pieces of information acquired from several chapters. Thus, the questions are meant to encourage thinking, rationalization, and integration, and possibly an extra glance through a particular chapter. Ultimately, the questions are meant to act as a teaching/learning tool.

CHAPTER 1

1. This list could include paper, wood, nylon, polyester fibers, melamine dishware, polyethylene fibers, Teflon-coated cooking utensils, starchy foods, meat, wool, hair, paint, etc.

2. (a) 1, (b) 1, (c) 2.

3. (a) Polyethylene, (b) Bakelite, (c) any protein, e.g., casein.

4. c.

5. b, c, d, e, f.

6. e, f.

7. b, c, d.

8. Possibly, wait and see or live and learn.

9. Because polymer science is a relatively new science.

10. Approximately 100%.

11. 28002.

12. Both are regenerated cellulose, which differ in physical form. Rayon is a filament and cellophane is a sheet.

13. Since the novolac resin is produced by the condensation of phenol with a small amount of formaldehyde on the acid side, it has residual unreacted phenol functional groups and is more heat stable than the resole resin, which is made using an excess of formaldehyde.

CHAPTER 2

1.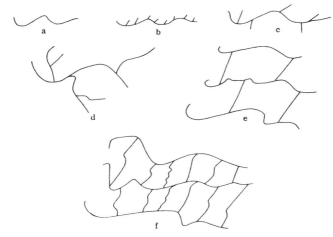

2. (a) LDPE, (b) LDPE.

3. (a) 109.5°, (b) 109.5°; zigzag chains characteristic of alkanes.

4. (a) 503.5 nm, (b) 503.5 nm.

5. d, g, i.

6. 999.

7. (a)
$$\begin{array}{c} \text{H} \quad \text{CH}_3 \\ | \quad | \\ -\text{C}-\text{C}- \\ | \quad | \\ \text{H} \quad \text{H} \end{array}$$
(b)
$$\begin{array}{c} \text{H} \quad \text{Cl} \\ | \quad | \\ -\text{H}-\text{C}- \\ | \quad | \\ \text{H} \quad \text{H} \end{array}$$
(c)
$$\begin{array}{c} \text{H} \quad \text{CH}_3 \text{ H} \quad \text{H} \\ | \quad | \quad | \quad | \\ -\text{C}-\text{C}=\text{C}-\text{C}- \\ | \qquad\qquad | \\ \text{H} \qquad\qquad \text{H} \end{array}$$

8. c.

9. d, e, f, g.

10. a.

11. a.

12. a.

13. a.

14. 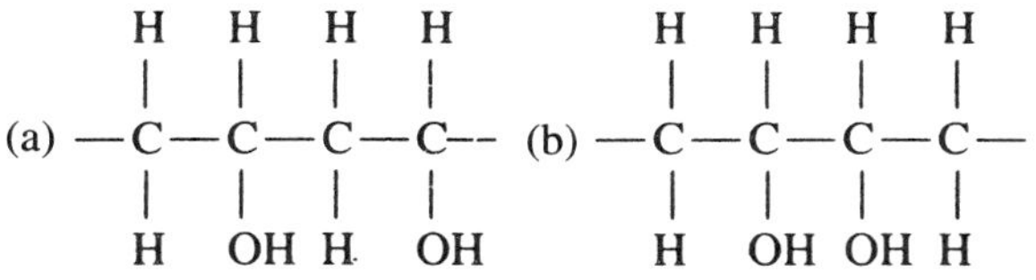 (a)

$$-\overset{\displaystyle H}{\underset{\displaystyle H}{C}}-\overset{\displaystyle H}{\underset{\displaystyle OH}{C}}-\overset{\displaystyle H}{\underset{\displaystyle H}{C}}-\overset{\displaystyle H}{\underset{\displaystyle OH}{C}}--$$

(b)

$$-\overset{\displaystyle H}{\underset{\displaystyle H}{C}}-\overset{\displaystyle H}{\underset{\displaystyle OH}{C}}-\overset{\displaystyle H}{\underset{\displaystyle OH}{C}}-\overset{\displaystyle H}{\underset{\displaystyle H}{C}}-$$

15. 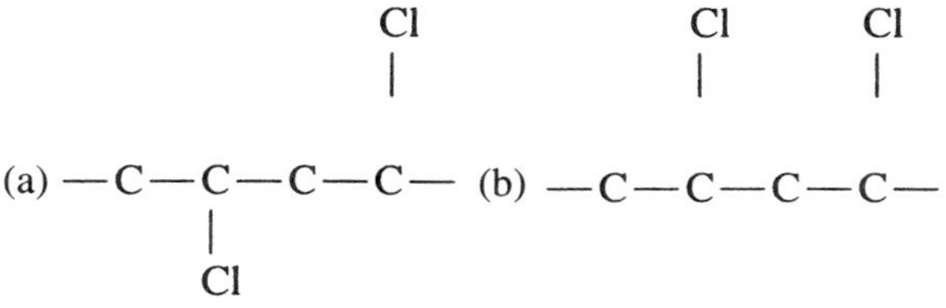 (a)

$$-C-\overset{\displaystyle }{\underset{\displaystyle Cl}{C}}-C-C-$$

(b)

$$-C-\overset{\displaystyle Cl}{C}-\overset{\displaystyle Cl}{C}-C-$$

16.

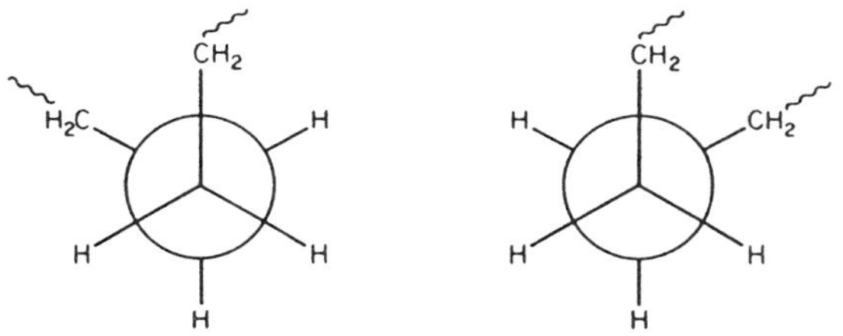

17. (a) HDPE, LDPE, hevea rubber, etc., (b) PVC, etc., (c) nylon-66, cellulose, silk, etc.

18. b.

19. a.

20. a.

21. b.

22. a.

23. 378 nm.

24. a.

25. About 50%.

26. b.

27. Intramolecular hydrogen bonds.

28. a.

29. a.

30. a.

31. Add a nucleating agent to the melt and cool rapidly in a thin layer.

32. a.

33. b.

CHAPTER 3

1. Morphology is the study of shape, while rheology is the study of flow and deformation.

2. b, d.

3. G is the shear modulus.

4. c.

5. (a) increase, (b) decrease.

6. The slope or modulus would be greater for polystyrene.

7. a has a low modulus and high viscosity, therefore η/G will be large.

9. b, since its Poisson's ratio is less than 0.5.

10. Both would have the same volume percent occupied by holes.

11. a, because of die swell.

12. At a temperature of 35°C above the T_g of polystyrene, i.e., at about 140°C.

13. a_1 and a_2 are related to the free volume.

14. The coefficient of viscosity (η) is equal to the ratio of the applied stress (s) to the applied velocity gradient $d\gamma/dt$.

15. Stress relaxation.

16. b.

17. One in which London forces are the predominant intermolecular attractions.

18. b.

19. Cohesive energy density is equal to the strength of the intermolecular forces between molecules, which is equal to the molar energy of vaporization per unit volume $\Delta E/V$.

20. b.

21. Higher.

22. $\Delta S = (\Delta H - \Delta G)/T$ at constant T.

23. A good solvent.

24. 0.

25. θ temperature.

26. b.

27. 547.6 cal cm^{-3}.

28. 0.

29. 1,000,000 g.

30. $\delta = 9.05$ H $= [1.05(133 + 28 + 735)]/104$.

31. $M = 0.33$, $(300 M)^{1/2} = 10$ H.

32. The contribution of the polar group becomes less significant as the alkyl portion increases.

33. b. $\delta = 9.2$ H.

34. a. Benzene is a better solvent, therefore the value of a is greater.

35. 0.

36. When $\alpha = 0$.

37. $\sqrt{\bar{r}^2}$ = the root mean square end-to-end distance of a polymer chain.

38. As shown in the Arrhenius equation, $\log [\eta] = \log A + E/2.3RT$, $[\eta]$ is inversely related to T.

39. b. Less is extruded in the same time under identical conditions.

CHAPTER 4

1. b, c, d, e.

2. 50,000.

3. $\overline{M}_n = 1.57 \times 10^6$, $\overline{M}_w = 1.62 \times 10^6$.

4. (a) 1, (b) 2.0.

5. $\overline{M}_n$, $\overline{M}_v$, $\overline{M}_w$, $\overline{M}_z$.

6. b, c, d.

7. $[\eta] = K\overline{M}^a$.

8. GPC, ultracentrifugation.

9. d, e, f.

10. One is the reciprocal of the other.

11. 0.5.

12. 2 (end groups).

13. 2.0.

14. 2.25×10^8.

15. c.

16. A type of double extrapolation for the determination of $\overline{M}_w$ for high molecular weight polymers in which both the concentration and the angle of the incident beam are extrapolated to zero.

17. $\overline{M}_n$.

18. Because of higher vapor pressure, the solvent evaporates faster from a pure solvent than from a solution; an application of Raoult's law.

19. The melt viscosity is proportional to the molecular weight to the 3.4 power.

20. The extremely high molecular weight polymer is much tougher.

21. When the exponent a in the Mark-Houwink equation equals 1.

22. A nonsolvent whose solubility parameter differed from that of the polymer by at least 1.8 H.

23. (a) VPO, (b) membrane osmometry.

24. Small polymer molecules may pass through a semipermeable membrane.

25. b.

26. a.

27. Answer may vary. For a *homodisperse sample*, a is better if polymer drainage is important. For a long-chained *heterodisperse* sample, b is a more accurate answer.

28. B is a constant related to the interaction of the solvent and the polymer.

CHAPTER 5

1. ASTM.

2. $TS = \dfrac{L}{A} = \dfrac{282 \text{ kg}}{1.25 \text{ cm} \times 0.32 \text{ cm}} = 705 \text{ kg cm}^{-2}.$

3. $TS = \dfrac{L}{A} = \dfrac{282 \text{ kg}}{1.27 \text{ cm} \times 0.32 \text{ cm}} = 694 \text{ kg cm}^{-1}.$

4. $CS = \dfrac{L}{A} = \dfrac{3500 \text{ kg}}{1.27 \text{ cm} \times 1.27 \text{ cm}} = 2170 \text{ kg cm}^{-2}.$

5. $\%EI = \dfrac{\Delta l}{l} = \left(\dfrac{12 - 5}{5}\right)(100) = 140\%.$

6. $E = \dfrac{TS}{El} = \dfrac{705 \text{ kg cm}^{-2}}{0.026 \text{ cm/cm}} = 27,100 \text{ kg cm}^{-2}.$

7. Creep is slow-dimensional change in polymers resulting from irreversible chain slippage or segmental motion.

8. (a) 0.5, (b) 0.3.

9. Primarily reversible bond stretching and distortion of bond angles in the polymer chain.

10. Irreversible uncoiling and slippage of chains.

11. By relative areas under the curves.

12. Increase.

13. Decrease.

14. No. The former is tougher, but this is not a linear relationship.

15. Because polymers, often being poor conductors, are so widely used in electrical applications.

16. a.

17. Because they are poor conductors of heat.

18. 18.60 kg cm^{-2}, 4.65 kg cm^{-2}.

19. Almost every organic compound will burn if the conditions are ideal for combustion. Tests such as the OI test are useful as comparative tests only.

20. a.

21. They change abruptly.

22. This is an area still in motion. Thus several answers can be properly presented.

23. 21 per propylene unit (3n-trans.-rot.).

24. about 190 to 800 μm.

25. a.

26. Preferably X-ray diffraction; nmr may also be used, etc.

27. DTA, DSC, TML, TBA, etc.

CHAPTER 6

1. Enzymes are present in the human digestive system for the hydrolysis of α linkages but not for β linkages in polysaccharides.

2. Cellobiose is a β glucoside, while maltose is an α glucoside.

3. Because of the presence of intermolecular hydrogen bonds.

4. It is the same since there is an equilibrium between the α and β conformations of D-glucose.

5. a (it is less soluble than b).

6. Three.

7. Secondary cellulose acetate.

8. It forms an amine salt.

9. a.

10. It is branched, therefore discouraging the formation of H-bonding.

11. It is a polyamide.

12. Glycine, because it has no chiral carbon atom.

13. To the positive pole.

14. Collagen has strong intermolecular hydrogen bonds.

15. It should be a linear crystalline polymer, with strong intermolecular hydrogen bonds, and high tensile strength and modulus associated with a regular repeating structure.

16. b.

17. DNA is usually greater.

18. 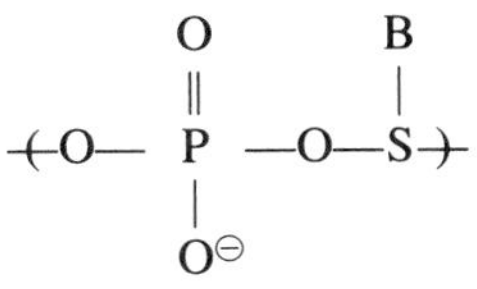 $$\left(O - \overset{\overset{\displaystyle O}{\|}}{\underset{\underset{\displaystyle O^{\ominus}}{|}}{P}} - O - \overset{\overset{\displaystyle B}{|}}{S} \right) \text{ where S = deoxyribose, S = sugar, B = base.}$$

19. b.

20. Thymine.

21. The presence of bulky pendant groups and hydrogen bonding favors the helical arrangement.

22. TAATGCAGTA.

23. The maximum of 16 dinucleotides is not enough to direct the initiation, termination, and insertion of 20 amino acids in a protein chain; 64 trinucleotides is more than ample.

24. a is a trans and b is a cis isomer of 1,4-polyisoprene.

25. 0.5.

26. Increase.

27. Contracts.

28. An elastomer is an amorphous polymer with a low T_g and low intermolecular forces which increases in entropy, order when stretched.

29. Guayule grows on arid soil in temperature areas and has a high content of natural rubber.

30. b. Isoprene is the monomer in the synthetic process, and isopentenyl-pyrosphosphate is the precursor in natural rubber.

31. It must be

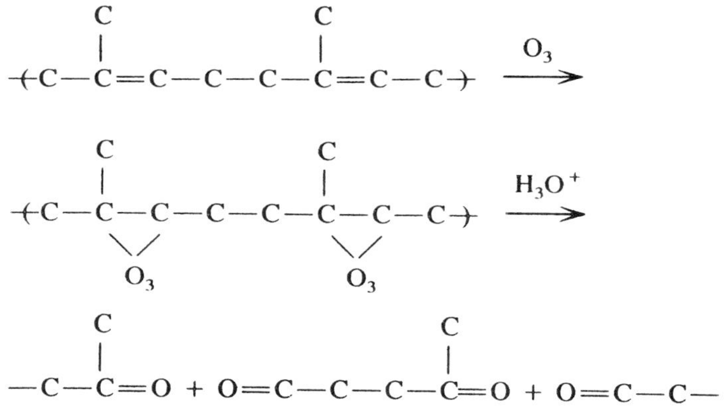

32. It is stable below T_g and contracts when heated above T_g.

33. Because of the formation of crystals.

34. Change in entropy, $-T(dS/dl)$.

35. NR, S, accelerator, antioxidant, carbon black, ZnO, and processing aids.

36. It couples with the macroradical to produce a dead polymer.

37. They often cost more, are contaminated, and are inferior to synthetic products.

38. A polar solvent, since it is a polyol.

39. A hydrocarbon, since asphalt contains aliphatic cyclic hydrocarbons.

40. b.

41. At least 60 million tons.

42. Yes, casein is no longer used as a molding resin and, hence, the article may have value as a collectors item.

CHAPTER 7

1. b, c, e.

2. No. The polymer chains would probably be too flexible.

3. a. It is less flexible.

4. 100,000.

5. log k = $(-E_a/2.3RT)$ + log A. Therefore, the value of log k increases as the first term in the equation becomes less because of an increase in the value of T.

6. The dimer.

7. A trimer or a tetramer, since two dimer molecules may react or a dimer may react with a molecule of the reactant.

8. Stretch the filament to permit the formation of more hydrogen bonds.

9. Poly(tetramethylene adipamide), or nylon-46.

10. a. A stable six-membered ring could be formed.

11. a.

12. 1 hr.

13. b.

14. 99.

15. It would be a stiff linear high molecular weight polymer with strong hydrogen bonds.

16. About 118.

17. 2.2.

18. About 0.8.

19. 2.8.

20. a. It would have better geometry.

21. It would be weak at the angle perpendicular to the stretch.

22. b.

23. a. There would be more surface available for attack.

24. Produce it from a mixture of isomers or use a mixture of esters of p-hydroxybenzoic acid and p-hydroxyphenylacetic acid.

25. Increase the number of methylene groups, e.g., nylon-1212.

26. Reduce the number of methylene groups so that it would absorb more moisture.

27. a. The amide is a stiffening group.

28. Because of the presence of bulky pendant groups.

29. b is a ladder polymer.

30. Not particularly, since the polyamine reacts to produce a polyurea. However, other propellants are used commercially in foam production.

31. Use an excess of the glycol.

32. Increase the number of methylene groups in the reactant.

33. Because of the presence of phenylene stiffening groups.

34. Because of the presence of strong polar groups.

35. The precursor, furfural, is produced from waste corn cobs.

36. An average of one functional group on each phenol is ureacted in the A-stage resin. Hence, one must add formaldehyde to produce a cross-linked polymer.

37. No; resorcinol is trifunctional and very reactive.

38. Phenolic resins were developed before there was much information on polymer science. Those involved in an art have a tendency to create new terms to describe what they do not understand.

39. Most people prefer to use light-colored dinnerware.

40. All three.

41. a. b would be a branched chain.

CHAPTER 8

1. (a) A small amount of reactants, dimer, trimer, and oligomers, plus low molecular weight polymers, (b) high molecular weight polyisobutylene plus monomer.

2. A Lewis base-cocatalyst complex. Actually, the proton is the initiator.

3. A carbonium ion.

4. A macrocarbonium ion.

5. A gegenion.

6. a.

7. b.

8. The macroions have similar charges.

9. a.

10. b.

11. (a) HDPE or PP, (b) IR, EP, or butyl rubber (IIR), (c) PP.

12. a.

13. b.

14. a, b.

15. b, d, e.

16. b.

17. $v = \overline{DP}$.

18. Polyisobutylene.

19. $R_i = kC[M]$, where C equals [catalyst-cocatalyst complex].

20. Increases the rate.

21. No observable change.

22. a.

23. c, d.

24. A cation.

25. They are equal.

26. $\overline{DP} = R_p/R_t$.

27. b. Many times higher.

28. 0%.

29. $\overline{DP}$ is independent of [C].

30. No: they are thermally unstable.

31. Cap the ends by esterification or copolymerize with dioxolane.

32. Because formaldehyde is produced in the decomposition, which is the reverse of propagation.

33. No, this is above the ceiling temperature.

34. Polymerize in the presence of oxirane.

35. Add a water-soluble linear polymer such as polyethylene oxide.

36. Because of its regular structure.

37. It is a carcinogen.

38. CO_2.

39. Convert them to solid polymers by cationic chain polymerization using sulfuric acid as the initiator.

40. A carbanion.

41. Macrocarbanions.

42. Because they are stable macrocarbanions capable of further polymerization.

43. c.

44. Nylon-6.

45.

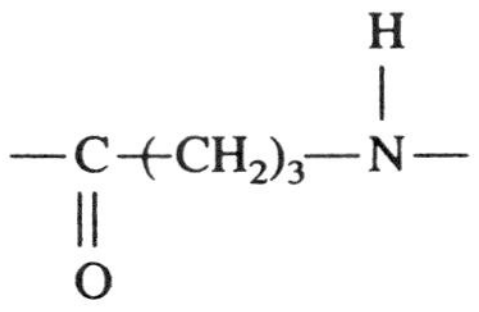

$$-\overset{\displaystyle}{\underset{\displaystyle \underset{O}{\|}}{C}}-(CH_2)_3-\overset{\displaystyle \overset{H}{|}}{N}-$$

46. $TiCl_3$ and $(C_2H_5)_2AlCl$.

47. It yields atactic polymers.

48. The propagating species in anionic chain polymerization is a butyl-terminated macroanion. Propagation takes place by the addition of monomer units to the chain end of this anion. The propagating species in complex coordination catalyst polymerization is an active center with an alkyl group from the cocatalyst as the terminal group. Propagation takes place by the insertion of the monomer between the titanium atom and the carbon atom by way of a π complex.

49. Its very low specific gravity (0.83).

50.

Cis   Trans

51.
(a)

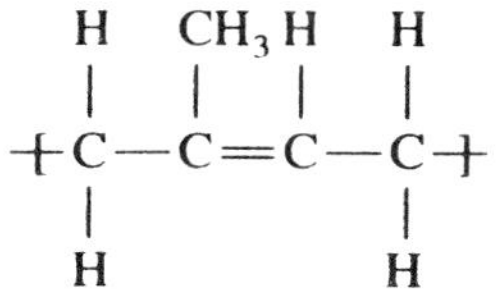

H   CH₃ H   H
| | | |
┼C — C = C — C┼
| |
H   H

(b)

H   CH₃
| |
┼C — C┼
| |
H   HC
        ‖
        CH₂

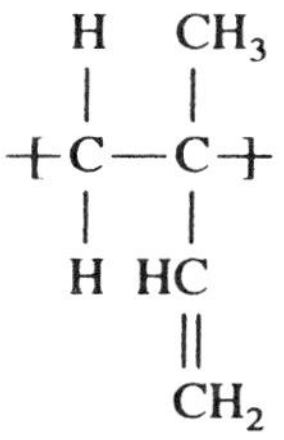

52. A chromia catalyst supported on silica (Phillips catalyst).

53. cis-Polyisoprene.

54. Ethylene-propylene copolymer.

CHAPTER 9

1. (a) H^+:BF_3OH^-, (b) Na^+:NH_2^-, (c) $(H_3C)_2$—C— N/N —C—$(CH_3)_2$.
                                                                 | |
                                                                CN   CN

2. Free-radical chain polymerization.

3. 12.5%.

4. Coupling.

5. Neither, since both will reduce $\overline{DP}$.

6. LDPE, PVC, PS, etc.

7. None, provided the system is a homogeneous solution.

8.
(a)   (b)   (c)

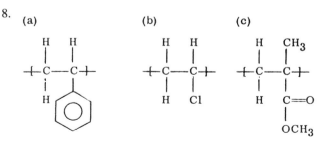

9. Since the rate of termination has decreased, more monomer may be added gradually to produce very high molecular weight polymers.

10. Initiation, which is governed by the rate of production of free radicals.

11. a.

12. Monomeric styrene.

13. $\ln 2 = 0.693$.

14. Measure the volume of N_2 produced.

15. 40 ± 10 kcal/mol.

16. One may polymerize at low temperatures and also regulate the rate of production of free radicals.

17. Neither one is a catalyst; both are good initiators.

18. a, b, c = 1×10^{-11} mol/liter, assuming monomer is still present.

19. 5 ± 3 kcal mol^{-1}.

20. $R_p \propto [I]^{1/2}$.

21. $\overline{DP}$ decreases.

22. Dioctyl phthalate or other nonvolatile dialkyl phthalates.

23. Vinyl acetate. PVA is obtained by hydrolysis of PVAc.

24. The initiator produces 2R• by homolytic cleavage.

25. Neither; it remains unchanged.

26. Ethylene has no electron-donating groups and is less polar than isobutylene.

27. Disproportionation.

28. The termination mechanism at higher temperatures is disproportionation.

29. 0 to 6 kcal/mol

30. $T_c = 61°C$. Therefore, poly(α-methylstyrene) would decompose when heated above 61°C.

31. The original 6-carbon chain segment is called the branch, and it is relatively short compared to the new chain extension.

32. b. The bond strength for C—Cl is less.

33. The $H(CH_2)_{12}S\bullet$ produced by chain transfer is an active free radical which initiates new chains.

34. No: the heat from the exothermic reaction might cause an explosion.

35. They are the same.

36. No, but one should avoid contact with the skin and ascertain that there is a minimal amount of monomer residue in the polymer.

37. It forms a ladder polymeric filament.

38. The droplets are relatively few in number, and the chances of an oligoradical entering them is extremely small.

39. The primary free radicals are water soluble.

40. The initial polymerization rate would be retarded.

41. The rate would be slower, and probably the polymerization would be by the suspension mechanism.

42. The structure of PTFE is regular. The large pendant group in FEP destroys this regularity.

43. Both $\overline{DP}$ and R_p are proportional to N/2 where N/2 is the number of active micelles.

44. b. The additional chlorine on the repeating unit increases the specific gravity.

CHAPTER 10

1. (a)

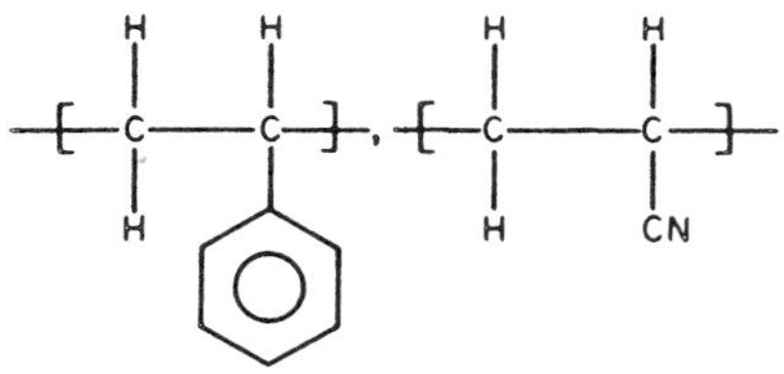

(b)

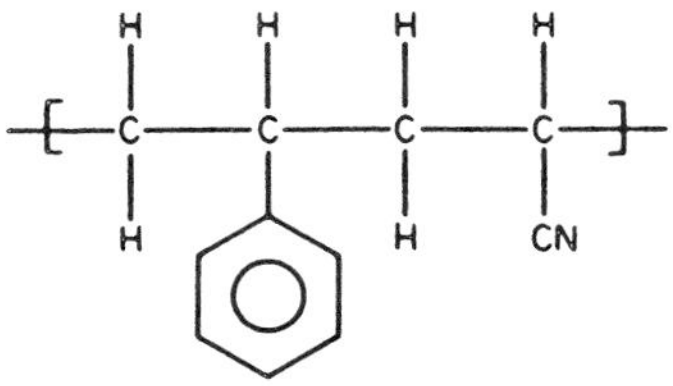

(c)

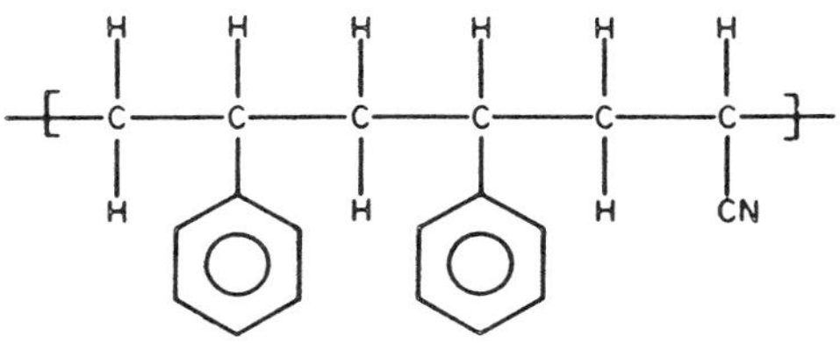

(d)

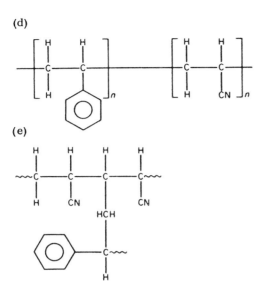

(e)

2. Equimolar.

3. $r_1 = k_{11}/k_{12}$, $r_2 = k_{22}/k_{21}$.

4. No: they are constant.

5. They vary in accordance with the Arrhenius equation.

6. Vinylidene chloride, about 76%; vinyl chloride, about 24%.

7. Styrene, etc.

8. (a) Cationic, (b) anionic or free radical, (c) Ziegler-Natta.

9. Alternating.

10. $(M_1)_n(M_2)_m(M_1)_n$.

11. $r_1r_2 \simeq 1$.

12. b.

13. A charge transfer complex is present at the lower teperature.

14. (a) 50% of each, (b) 91% S and 9% MMA, (c) 94% MMA and 6% S.

15. Isobutylene, 71.4%; isoprene, 29.6%.

16. Isobutylene, 96%.

17. Add a mixture of monomers to maintain the original ratio of reactants.

18. a. Q value is higher.

19. Vinyl chloride, 68%.

20. The presence of vinyl acetate mers in the chain breaks up the regularity characteristic of PVC so that the copolymer has a lower T_g, is more soluble in organic solvents, more flexible, and more readily processed.

21. The methacrylic acid mers present decrease the crystallinity and improve the toughness and adhesive properties. The salts act like cross-linked polymers at ordinary temperatures.

22. They are all essentially the same copolymers of butadiene and styrene.

23. They are all hydrophilic and aid dyeability.

24. B.

25. It produces a stable, molded, rubberlike product without vulcanization.

26. It is more weather resistant because of the absence of ethylenic groups.

27. An alternating copolymer is produced rapidly until the maleic anhydride is consumed. The styrene then forms a block copolymer on the SMA in a slower process.

28. Since styrene is flammable and toxic, its concentration in the atmosphere must be held to very low values, such as 1 PPM.

29. Yes: the styrene macroradicals on the grafts terminate by coupling.

30. Graft it with a lyophilic monomer such as styrene.

31. A carboxylic acid group.

CHAPTER 11

1. The formation of cyclic compounds instead of linear polymers.

2. By using a mixture of monochlore-and dichloroaldylsilanes.

3. Like the alkyl pendant groups, 6 to 7 H.

4. The silanol groups in silicic acid are polar, but the alkyl groups in silicones are nonpolar.

5. Acidify it. The silicic acid will polymerize spontaneously.

6. Failure of polymers at elevated temperatures is usually due to the breaking of covalent bonds in the main polymer chain. The O—Si—O bonds are very strong.

7.
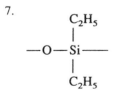

8. Phopshorus pentachloride and ammonium chloride.

9. $PCl_5 + 4H_2O \rightarrow H_3PO_4 + 5HCl$, therefore,

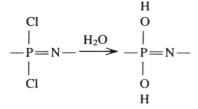

10. b.

11.

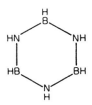

12. They decompose slowly in water.

13. Often undefined with combinations of structures.

14. Evidently below 145°C.

CHAPTER 12

1. Held together by directional covalent bonds; acts physically like many organic polymers that are above their glass transition temperature—very viscous, acts as a solid on rapid impact but as a liquid on a much elongated time scale.

2. Readily available on a large scale; inexpensive; relatively nontoxic; stands up well to most natural elements such as rain, cold, heat, mild acids and bases, and light; strong.

3. Fiberglass, colored glass, concrete, lead glasses, optical fibers, prestressed concrete, safety glass, vitreous enamels, window glass, synthetic diamonds, synthetic graphite, polymeric sulfur, silicon carbide, carbon black, etc.

4. Polymers where the structure cannot be adequately described on the basis of a single flat plane (see Table 12.1). Insoluble; many exhibit long-range disorder; difficult to structurally characterize.

5. Wide variety of applications and conditions of applications.

6. Flows like a liquid, but the flow rate is very low.

7. See Sec. 12.4.

8. Wide variety of applications that require materials that may possess glasslike properties—generally good resistance to natural elements, easily shaped, polished, and cut, many transmit light and can be colored.

9. Fiberglass and asbestos; fibers generally impart greater strength and flexibility.

10. Quartz—highly crystalline.

11. See Sec. 12.5.

12. Most of what we do and use have risks involved: taking a bath and slipping; walking across the street and being hit by a car; taking a sun bath and getting skin cancer. Some risks may be necessary, but others are not. We have many additives in our foods that prevent spoilage, but if certain additives are found to be cancer-causing we should omit the use of those additives and find suitable substitutions. We have found that the widespread use of asbestos as insulation is unacceptable and are currently using fiberglass as a proper alternative.

13. See Sec. 12.8.

14. See Figs. 12.4 and 12.5.

CHAPTER 13

1. For example, PMMA sheet, rubber bands, cellophane, clear polyethylene sheet, nylon filament, rayon.

2. Lignin.

3. Wood flour.

4. Formica table tops, plywood, fibrous glass–reinforced plastic boats, graphite–reinforced plastic golf shafts, etc.

5. Treat the surface with a silane or melt it and convert it into a fiber.

6. According to the Einstein equation, the viscosities will be equal.

7. a.

8. The intermolecular attractions impede the motion.

9. T_g increases.

10. b.

11. Advantages: more fibrous, less discoloration. Disadvantages: higher cost, more difficult to process because of bulk factor.

12. Wood flour below the surface and α-cellulose and paper near and at the surface.

13. (b) Continuous, (a) discontinuous.

14. No, it is compatible with the resin.

15. Coat the hollow spheres with metal.

16. b.

17. Light weight and flexibility.

18. Fill it with $BaSO_4$.

19. Fill it with alumina, corrundum, or silicon carbide.

20. Water is evolved when it is heated: $Al_2O_3 \cdot 3H_2O \xrightarrow{\Delta} Al_2O_3 + 3H_2O\uparrow$

21. Probably the stearate groups replace the surface carbonate groups: $CaCO_3 + 2H(CH_2)_{16}COOH \rightarrow Ca(OOC(CH_2)_{16}H)_2 + CO_2\uparrow + H_2O$

22. It serves a purpose not fulfilled by less expensive silica.

23. Silicone rubber.

24. So that the applied stress on the softer resin can be transferred to the stronger filler.

25. According to the Kelly Tyson equation, the two composites should have equal strength.

26. Use the filament winding or pultrusion process with epoxy resin–coated filaments.

27. Convenience, uniformity, speed, reproducibility, and less exposure of the workers to volatile monomers.

28. Add a surface-active agent such as TTS to the resin-filler mixture.

CHAPTER 14

1. A plasticizer such as camphor or TCP.

2. PVC decomposes at processing temperatures; therefore, it was not useful until it was plasticized by TCP and later by DOP.

3. DOP volatilized from the PVC upholstery.

4. The stiffness of polymers such as PVC is due to intermolecular attractions. A small amount of plasticizer permits chain orientation, which increases this attraction.

5. Because of the presence of residual toxic monomer and plasticizer.

6. As the size of bulky groups increases, T_g decreases. However, long pendant groups are attracted to each other and increase T_g.

7. Appropriate antioxidants are added.

8. Aging is believed to involve at least some free-radical processes. Antioxidants terminate this process, and hence the test animals live longer.

9. The water in lead-stabilized pipe might be contaminated with lead ions.

10. The diene group is a chromophoric group, and the color will change as the number of these groups increase.

11. It is essentially nontoxic and also serves as a plasticizer.

12. Because most high-energy ultraviolet radiation is filtered out indoors.

13. b.

14. No.

15. No.

16. PTFE will not burn in air but will burn in the oxygen atmosphere used in the capsule at that time.

17. Unless flame retardants are added, polyurethane foam will burn at elevated temperatures. If the neoprene protective film is punctured, the polyurethane can be ignited.

18. Laboratory tests with mice showed "tris" to be a carcinogen.

19. Probably not because of its mild toxicity.

20. Yes.

21. This is a good question that is not answered from a technical viewpoint.

22. No, prior to 1910, it was patented by Goodyear and used as a colorant. A compounder made a mistake and used a 100-fold excess. This is called serendipity.

23. Different amounts of water of hydration.

24.

25. Add carbon disulfide to piperidine at room temperature.

26. Preferably in a solution; otherwise, in individual nonmetal containers at moderate temperatures.

27. It is an initiator, but don't waste much time trying to change this or other erroneous terms.

28. They collect dust and may start fires and can shock you severely.

29. It must migrate to the surface.

30. they are less expensive and more effective on a comparative weight basis.

31. Add a foaming agent that decomposes at the temperature used to convert the plastisol to a solid (150°C).

32. The particles are coated with a surfactant. An alternative answer involves the slight crystallinity of PVC.

33. a.

34. None.

35. Bloom may be the result of the migration of lubricants to the surface.

CHAPTER 15

1. The reaction with functional groups.

2. Cross-linking through double bonds.

3. 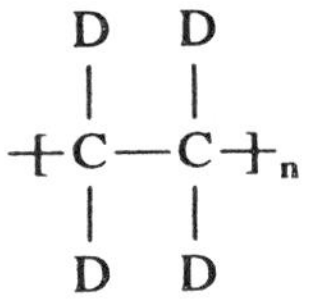
$$+\!\!\begin{array}{c} D \quad D \\ | \quad | \\ C-C \\ | \quad | \\ D \quad D \end{array}\!\!+_n$$
where $D = {}^2H$.

4. The attack is on the folds of lamellar crystals of essentially uniform thickness.

5.
$$C_6H_5-\begin{array}{c} H \\ | \\ N \end{array}-\begin{array}{c} H \\ | \\ C \\ \| \\ NH \end{array}-NC_6H_5.$$

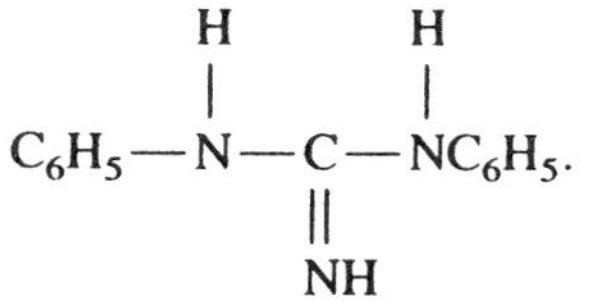

6. Expose it to high-energy radiation.

7. Hydrogenate the block copolymer of styrene and isoprene.

8. They are both alternating copolymers of propylene and ethylene.

9. It is more costly than LDPE.

10. b is a saturated polymer.

11. As an HCl scavenger in PVC.

12. Succinic aldehyde:

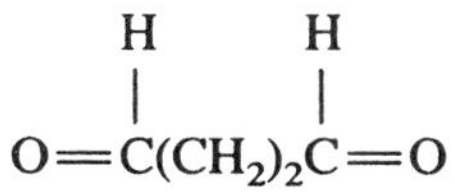

$$O{=}C(CH_2)_2C{=}O$$

with H above each terminal C.

13.
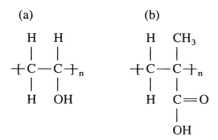

(a)

H   H
| |
$+C{-}C+_n$
| |
H   OH

(b)

H CH_3
| |
$+C{-}C+_n$
| |
H   C$=$O
        |
        OH

14. The introduction of a few methoxyl groups reduces the hydrogen-bonding forces so that the polymer with the remaining hydroxyl groups will dissolve in water.

15. It forms a protective film around dirt particles and prevents their redeposition.

16. These water-soluble polymers coat the microscopic crystals of calcium carbonate as they form and prevent their growth.

17. The maximum, i.e., about 2.8.

18. All the hydroxyl groups are not accessible in the heterogeneous acetylation reaction.

19. b has more free hydroxyl groups.

20. There are residual acetyl groups in the reactant PVA, and the acetal formation requires two 1,3-hydroxyl groups. Also, on a statistical basis some of the —OH groups become isolated.

21. It produces extremely large molecules when it forms salts with polyvalent cations such as Al^{3+}.

22. These so-called graphite fibers are used to reinforce resins such as epoxy resins.

23. None. Any vinyl alcohol would rearrange to acetaldehyde.

24. b.

25. a.

26. The phenolic hydroxyl groups react with the protein and produce a cross-linked polymer.

27. The polymerization of styrene in the molding of unsaturated polyesters such as BMC or SMC.

28. Heat the polymer and condense the monomer produced by thermal depolymerization.

29. b.

CHAPTER 16

1. Because chemical engineers are always seeking a better and more economical source of this compound, which is used on a large scale.

2. See the following equations: (a) (16.1) or (16.2), (b) (16.3) or (16.4), (c) (16.5), (d) (16.6), (e) (16.8), (f) (16.11), (g) (16.12), (h) (16.13), (i) (16.14), (j) (16.15), (k) (16.16), (l) (16.17) or (16.18), (m) (16.19), (n) (16.20) or (16.21), (o) (16.22), (p) (16.23), (q) (16.24) or (16.25), (r) (16.26), (s) (16.28), (t) (16.32), (u) (16.33), (v) (16.39), (w) (16.46), (x) (16.47), (y) (15.48) or (16.49), (z) (16.51).

3. (a) Silanes, (b) styrene (ethylbenzene), (c) ε-caprolactam, (d) adipic acid or sebacic acid, (e) pentaerythritol.

CHAPTER 17

1. They both are important and dependent on each other.

2. Cotton, wool, silk, hemp, jute, and so on.

3. Rayon, etc.

4. It is soluble in less expensive solvents, such as acetone.

5. They are both regenerated cellulose.

6. Polyester, nylon, acrylic fiber, polyurethane, polyolefins.

7. (a) Butyl rubber, (b) polybutadiene, (c) SBR, neoprene, (d) silicone, polyurethane elastomer, Thiokol.

8. By air blowing an extruded tube.

9. For prevention of environmental pollution.

10. Over-the-counter sales.

11. The suspended polymer particles are coated by a surfactant.

12. Delay the blowing step until a viscous, strong, high molecular weight polymer is present.

13. b.

14. Because of the availability of large surface areas.

15. Fibrous glass–reinforced polyester resin in the Corvette.

16. It has a high specific modulus and low coefficient of expansion.

17. Most molded thermosets are molded by high-cost compression molding.

18. Lower the temperature of the mold cavity.

19. They have high specific strength, and once an intricate design has been incorporated in the die, the intricate design, such as that resembling hard wood carving, is reproduced at low cost.

20. Low cost, light weight, recyclable, nontoxic, and less hazardous than glass.

21. The molds are relatively inexpensive, and large articles can be produced readily by the thermoforming process.

22. The length is limited by the ability to store and transport the extrudate. Actually, pipe can be made in continuous lengths by extruding on the job site. Flexible pipe such as LDPE can be coiled.

23. Formica, safety glass, plywood, Macintosh raincoat.

24. Most paints coat and most coatings contain a pigment that allows it to "look nice."

25. Relatively nontoxic, easy clean-up.

26. See Sec. 17.9.

27. See Sec. 17.10.

Appendix A: Symbols

A_0	Original concentration
A	Concentration at time T
Å	Angstrom unit (10^{08} cm)
A	Arbitrary constant
A	Area
A	Arrhenius constant
A	Lewis acid
A	Antioxidant free radical
Ac	Acetyl group
Ar	Aryl group
AA	Reactant (step reactions)
ABS	Copolymer from acrylonitrile, butadiene, and styrene
AIBN	2,2'-Azo-bis-isobutyronitrile
ANSI	American National Standards Institute (formerly American Standards Association)
AR	Acrylate
ASTM	American Society for Testing Materials
ATR	Attenuated total reflectance spectroscopy
AU	Polyurethane
AXF	Polydiphenylethane
a	Constant in the WLF equation
a	Exponent
a_T	Shift factor
B	Virial constant
BPO	Benzoyl peroxide

564

BSI	British Standards Institute
BTU	British thermal unit
b	Arbitrary constant
bp	Boiling point
C	Arbitrary constant
C	Catalyst-cocatalyst complex
C	Celsius (centigrade)
C	Concentration
C	Degree of crystallinity
CA	Cellulose acetate
CAB	Cellulose acetate butyrate
CED	Cohesive energy density
CMC	Carboxymethyl cellulose
CN	Cellulose nitrate
CO	Polyepichlorohydrin
CPVC	Critical pigment volume concentration
CR	Neoprene
C_p	Specific heat
C_s	Chain transfer constant
c	Velocity of light (3×10 cm/sec)
cal	Calorie
cm	Centimeter
cm^{-1}	Reciprocal centimeter
cm^3	Cubic centimeter
co	Copolymer
cp	Chemically pure
D	Debye units (dipole)
D	Density
D	Dextro
D	Diameter
D	Diffusion constant
DNA	Deoxyribonucleic acid
$\overline{DP}$	Average degree of polymerization
DRS	Dynamic reflectance spectroscopy
DS	Degree of substitution
DSC	Differential scanning calorimetry
DTA	Differential thermal analysis
DWV	Drain, waste, and vent pipe
d	Density
d	Dextro
d	Diameter
d	Total derivative (infinitesimal change)
E_a	Energy of activation
E	Energy content
E	Energy of vaporization

E	Young's modulus of elasticity
ECO	Epichlorohydrin elastomer
EEK	Accelerated test with sunlight at constant right angle
EGG	Einstein-Guth-Gold equation
EMMA	Equatorial mounting with mirrors (accelerated sunlight test)
EP	Epoxy resin
EPM	Poly(ethylene-co-propylene)
EPDM	Poly(ethylene-co-propylene) cross-linked
EPR	Electron paramagnetic resonance spectroscopy
ESR	Electron spin resonance spectroscopy
ET	Thiokol
ETA	Electrothermal analysis
e	Base of natural logarithms (2.718)
e	Exponential
e	Polarity factor (Alfrey-Price equation)
F	Fahrenheit
F	Mole fraction of monomers in copolymer
F	Stress (filled elastomers)
F_g	Fractional free volume (plasticizers)
FEP	Copolymer of tetrafluoroethylene and hexafluoropropylene
FPM	Polyfluorinated hydrocarbon
f	Aspect ratio
f	Efficiency factor (chain reactions)
f	Force
f	Functionality factor (step reactions)
f	Segmental friction factor
ft	Foot
G	Gauche conformation
G	Gibbs free energy
G	Modulus
G	Molar attraction constant (small)
GC	Gas chromatography
GPC	Gel permeation chromatography
GRS	Poly(butadiene-co-styrene)
g	Gauche conformation
g	Gram
g	Gravity
gr	Graft (copolymer)
H	Arbitrary constant
H	Enthalpy (heat content)
H	Hydrogen atom
H	Latent heat of transition
H	Magnetic field strength
H	Proportionality constant in light scattering
HIPS	High impact polystyrene

h	Height
h	Planck's constant (6.625×10^{-27} erg-sec)
hp	Horsepower
hr	Hour
I	Initiator (chain reactions)
I	Intensity
I	Spin of nucleus
IIR	Butyl rubber
IR	Infrared
ISO	International Standards Organization
IUPAC	International Union of Pure and Applied Chemistry
i	Incident ray
in.	Inch
it	Isotactic
K	Arbitrary constant
K	Constant in Mark-Houwink equation
K	Kelvin
K	Kinetic constant in Avrami equation
K	Rate constant
K	Thermal-conductivity factor
k	Specific rate constant
kcal	Kilocalorie
kg	Kilogram
L	Length
L	Levo
LIM	Liquid injection molding
l	Length
l	Levo
l_c	Critical fiber length
LLDPE	Linear, low-density polyethylene
ln	Natural logarithm
log	Logarithm (base 10)
M	Chain stiffener constant
M	Molecular weight
M	Monomer (chain reaction)
[M]	Monomer concentration
M	Quantum number
M·	Free-radical chain (macroradical)
Me	Methyl radical
MF	Melamine-formaldehyde resin
MR	Molar refraction
MVT	Moisture vapor transmission
MWD	Molecular weight distribution
M	Modulus

M_F	Modulus of a fiber
M_L	Longitudinal fiber
M_M	Modulus of resin matrix
M_O	Modulus of an unfilled resin
$\underline{M}_T$	Transverse modulus
$\overline{M}$	Average molecular weight
$\overline{M}_n$	Number-average molecular weight
$\overline{M}_v$	Viscosity average molecular weight
$\overline{M}_w$	Weight-average molecular weight
m	Consistency factor (power law)
m	Meta isomer
m	Meter
m	Mole fraction of reactants (copolymers)
m	Number of mers in polymer chain
ml	Milliliter
mp	Melting point
N	Nitrogen atom
N	Number of units or items
NBR	Poly(butadiene-co-acrylonitrile)
NMR	Nuclear magnetic resonance spectroscopy
NR	Natural rubber
$\overline{N}_n$	Number-average molecular weight
n	Index of flow (power law)
n	Index of refraction
n	Mole (step reactions) fraction
n	Normal (continuous chain, linear)
n	Number of mers in polymer chain
n	Numbers of theoretical plates
n	Ratio of mers in copolymer
nm	Nanometers
nmr	Nuclear magnetic resonance spectroscopy
O	Oxygen atom
OI	Oxygen index
OMP	Organometallic polymer
o	Ortho isomer
oz	Ounce
P	Phosphorus atom
P	Polymer chain
P	Polymer radical
P	Pressure
P	Resonance-stability (Alfrey-Price equation)
PA	Polyamide (nylon)
PBI	Polybenzimidazole
PBT	Poly(butylene terephthalate)
PC	Polycarbonate

PCB	Polychlorinated biphenyls
PE	Polyethylene
PEEK	Poly(ether ether ketone)
PET	Poly(ethylene terephthalate)
PF	Phenol-formaldehyde resin
PGC	Pyrolysis gas chromatography
PMMA	Poly(methyl methacrylate)
PMR	Protonmagnetic resonance spectroscopy
PNF	Polyphosphonitrilic fluorides
POM	Polyoxymethylene, polyformaldehyde, acetals
PP	Polypropylene
PPO	Poly(phenylene oxide)
PPS	Poly(phenylene sulfide)
PS	Polystyrene
PTFE	Polytetrafluoroethylene
PU	Polyurethane
PVA	Poly(vinyl alcohol)
PVAc	Poly(vinyl acetate)
PVB	Poly(vinyl butyral)
PVC	Poly(vinyl chloride)
PVDC	Poly(vinylidene chloride) (Saran)
p	Para isomer
p	Pressure
p	Probability, fractional yield (Carothers)
p	Propagation
psi	Pounds per square inch
Q	Quantity of heat, rate of heat flow
Q	Resonance-stability factor (Alfrey-Price equation)
q	Electronic charge
R	Alkyl radical
R	Gas constant (1.986 cal/mole °K)
R	Rate (chain reactions)
R	Run number (copolymers)
R·	Free radical
RF	Radio frequency
RIM	Reaction injection molding
RNA	Ribonucleic acid
r	Distance between centers of charge of dipoles
r	Radius
r	Ratio of reactants (step reaction)
r	Reactivity ratio (copolymers)
r	Refracted ray
$\bar{r}$	Average end-to-end distance
S	Entropy
S	Radius of gyration

S	Sedimentation constant
S	Solvent
S	Sulfur atom
SAM	Poly(styrene-co-acrylonitrile)
SBR	Poly(butadiene-co-styrene) elastomer
SEM	Scanning electron microscopy
SI	Silicone
SMA	Copolymer of styrene and maleic anhydride
SMC	Sheet molding compound
SN	Sulfur nitride
SPE	Society of Plastics Engineers
SPI	The Society of the Plastics Industry
SR	Synthetic rubber
S_N	Nucleophilic substitution
s	Stress
st	Syndiotactic
T	Absolute temperature (°A or °K)
T	Tentative (ASTM)
T	Trans
TAPPI	Technical Association of the Pulp and Paper Industry
TDI	Toluenediisocyanate
TGA	Thermal gravimetric analysis
TMMV	Threshold molecular weight value
TPE	Thermoplastic elastomer
TPX	Poly-4-methylpentene
T_c	Ceiling temperature
T_c	Cloud-point temperature
T_g	Glass transition temperature
T_m	Melting point
t	Termination (chain reaction)
t	Trans isomer
tit	Threodiisotactic
tst	Threosyndiotactic
tr	Transfer
UF	Urea-formaldehyde resin
UHMWPE	Ultrahigh molecular weight polyethylene polymer
UV (uv)	Ultraviolet
V	Volume
V_e	Elution volume
V_F	Fractional volume
WLF	Williams-Landel-Ferry equation
$\underline{WS}$	Polyurethane
$\overline{W}_n$	Weight average molecular weight

w	Width
w	Work
X	Ratio of reactants (copolymer)
X	Substutent (vinyl monomer)
yd	Yard
z	Critical chain length
α	Carbon atom adjacent to a functional group
α	One configuration of an isomer
α	Branching coefficient
α	First in a series
α_F	Expansion coefficient
α_c	Critical value for incipient gelation
β	One configuration of an isomer
β	Second carbon atom away from a functional group
β	Second in a series
γ	Magnetogyric ratio (nmr)
γ	Hydrogen-bonding index (Lieberman)
γ	Strain
γ	Third carbon atom away from a functional group
Δ	Change
Δ	Heat
δ	Chemical shift (nmr)
δ	Expansion factor (solution process)
δ	Fourth carbon atom away from a functional group
δ	Solubility parameter (Hildebrand)
ϵ	Fifth carbon atom away from a functional group
η	Molar absorption coefficient
η	Viscosity
η_r	Reduced viscosity
η_{rel}	Relative viscosity
η_{sp}	Specific viscosity
$[\eta]$	Intrinsic viscosity (viscosity number)
Θ	Flory critical miscibility temperature, at which polymer-solvent interaction is zero
θ	Angle of scattering
λ	Wavelength
λ	Distance from origin to vertices of tetrahedra in filler
μ	Dipole moment
μ	Measure of polymer-solvent interaction (Flory-Huggins)
μ	Micron (10^{-4} cm, 10^4 Å)
μ	Nuclear magnetic moment
ν	Average kinetic chain length of chain-reaction polymers
ν	Frequency (vibrations/sec)

ν	Kinetic chain length
π	Bond formed by the side-to-side overlap of two p_z orbitals, which accounts for the high activity of vinyl monomers
π	Osmotic pressure
ρ	Density
Σ	Summation
σ	Sigma bonds
τ	Orientation relaxation or retardation time
τ	Relaxation time
τ	Turbidity or scattered flux
τ	Universal viscosity constant (Flory)
ϕ	Fractional volume
ϕ	Jump frequency in hole filling (solution)
ϕ	Related to viscosity in Herschel-Bulkley equation
ω	Last in a series (e.g., carbon farthest away from functional group)
χ	Thickness
[ ]	Concentration

Appendix B: Trade Names

Trade or brand name	Product	Manufacturer
Abafil	Reinforced ABS	Rexall Chemical Co.
Abalyn	Abietic acid derivative	Hercules, Inc.
Abcite	Plastic sheet	E. I. du Pont de Nemours & Co., Inc.
Absafil	ABS polymers	Fiberfil
Absinol	ABS	Dart Inds.
Abson	ABS polymers	B. F. Goodrich Chemical Co.
Acelon	Cellulose acetate	May & Baker Plastics Ltd.
Acetophane	Cellulose acetate film	UCB-Sidac
Aclar	Polyfluorocarbon film	Allied Chemical Corp.
Acralen	Styrene-butadiene latex	Farbenfabriken Bayer AG
Acronal	Polyalkyl vinyl ether	General Aniline Film Corp.
Acrylacon	Fibrous glass–reinforced polymers	Rexall Chemical Co.
Acrylafil	Reinforced polymers	Rexall Chemical Co.
Acrylaglas	Fiberglass-reinforced styrene-acrylonitrile	Dart Inds.

Trade or brand name	Product	Manufacturer
Acrylicomb	Acrylic honeycomb	Dimensional Plastics Corp.
Acrylux	Acrylic	Westlake Plastics
Acrilan	Polyacrylonitrile	Chemstrand Co.
Acrylan-Rubber	Butyl acrylate-acrylonitrile copolymer	Monomer Corp.
Acrylite	Poly(methyl methacrylate)	American Cyanamid Co.
Acryloid	Resin solutions	Rohm & Haas Co.
Acrysol	Thickeners	Rohm & Haas Co.
Actol	Polyethers	Allied Chemical Corp.
Adipol	Plasticizer	FMC Corp.
Adiprene	Urethane elastomers and prepolymer	E. I. du Pont de Nemours & Co., Inc.
Admex	Plasticizers	Ashland Chemical Co.
Advastab	Antistatic agents	Cincinnati Milacron Chemicals, Inc.
Aerodux	Resorcinol-formaldehyde resin	Ciba (A. R. L.) Ltd.
Aeroflex	Polyethylene extrusions	Anchor Plastics
Aeron	Plastic-coated nylon	Flexfilm Products
Afcolene	Polystyrene and SAV copolymers	Pechiney-Saint-Gobain
Afcoryl	ABS polymers	Pechiney-Saint-Gobain
Agerite series	Antioxidants	R. T. Vanderbilt Co., Inc.
Agro	Rayon fibers	Beaunit Mills Corp.
Akulon	Nylon-6	AKU/Neth.
Alathon	Polyethylene	E. I. du Pont de Nemours & Co., Inc.
Albacar	Calcium carbonate filler	Pfizer Corp.
Albertols	Phenolic resins	Chemische Werke, Albert
Aldocryl	Acetal resin	Shell Chemical Co.
Alfane	Epoxy resin cement	The Atlas Mineral Products Co.
Algil	Styrene copolymer monofilament	Shawinigan Chemicals, Ltd.; also Polymer Corp.
Algoflon	Polytetrafluoroethylene	Montecatini
Alkathene	Polyethylene resins	Imperial Chemical Industries Ltd.

Trade or brand name	Product	Manufacturer
Alkon	Acetal copolymer	Imperial Chemical Industries Ltd.; Celanese Corp. of America
Alkor	Furan resin cement	Atlas Minerals Products Co.
Alloprene	Chlorinated natural rubber	Imperial Chemical Industries Ltd.
Alphalux	Poly(phenylene oxide)	Marbon Chemical Co.
Alsibronz	Wet ground muscovite mica	Franklin Mineral Products Co.
Alsilate	Clays	Freeport Kaolin Co.
Alsynite	Reinforced plastic panels	Reichhold Chemicals, Inc.
Amberlac	Modified alkyd resins	Rohm & Haas Co.
Amberol	Phenolic resins	Rohm & Haas Co.
Amberlite	Ion-exchange resins	Rohm & Haas Co.
Ameripol	Polyisoprene	Firestone Tire and Rubber Co.
Amer-Plate	PVC sheets	Ameron Corrosion Control
Amerith	Cellulose nitrate	Celanese Corp. of America
Amilan	Nylon	Tojo Rayon Co.
Ampcoflex	Rigid poly(vinyl chloride)	Atlas Mineral Products Co.
Ancorex	ABS extrusions	Anchor Plastics
Antarox	Low foaming wetting agent	GAF Corp.
Antiblaze	Organic phosphorus flame retardants	Mobil Chemical Co.
Antron	Nylon fiber	E. I. du Pont de Nemours & Co., Inc.
Anvyl	Vinyl extrusions	Anchor Plastics
Apogen	Epoxy resins	Apogee Chemical
Araclor	Polychlorinated polyphenyls	Monsanto
Aralac	Protein fiber	Imperial Chemical Industries Ltd.
Araldite	Epoxy resins	Ciba (A. R. L.) Ltd.
Ardil	Protein fiber	Imperial Chemical Industries Ltd.
Armite	Vulcanized fiber	Spaulding Fibre Co.
Armorite	Vinyl coating	John L. Armitage & Co.

Trade or brand name	Product	Manufacturer
Arnel	Cellulose triacetate	Celanese Corp. of America
Arnite	Poly(ethylene terephthalate)	Algemene Kuntstzijde Unie N.V.
Arochem	Modified phenolic resins	Ashland Chemical
Aroclor	Chlorinated polyphenyls	Monsanto Chemical Co.
Arodure	Urea resins	Ashland Chemical Co.
Arofene	Phenolic resins	Ashland Chemical Co.
Aroplaz	Alkyd resins	Ashland Chemical Co.
Aropol	Polyester resins	Ashland Chemical Co.
Aroset	Acrylic resins	Ashland Chemical Co.
Arothane	Polyester resins	Ashland Chemical Co.
Artfoam	Rigid urethane foam	Strux Corp.
Arylon	Polyaryl ethers	Uniroyal
Arylon T	Polyaryl ethers	Uniroyal
Astrel	Polyarylsulfone	3M Co.
Ascot	Polyolefin sheet–coated spunbounded	Appleton Coated Paper Co.
Astralit	Vinyl copolymer sheets	Dynamit Nobel/ America
Astroturf	Synthetic turf–nylon and polyethylene	Monsanto
Astyr	Butyl rubber	Montecatini
Atlac	Polyester cast resin	ICI America Inc.
Atomite	Calcium carbonate	Thompson, Weinman & Co.
Avisco	PVC film	FMC Corp.
Avistar	Polyester film	FMC Corp.
Avisun	Polypropylene	Avisum Corp.
Avron	Rayon fiber	American Viscose Corp.
Azdel	Fibrous glass–reinforced ABS copolymer sheet	Generic name
Azocel	Azodicarbonamide blowing agent	Fairmont Chemical Co.
Aztran	Poromeric sheet	B. F. Goodrich Chemical Co.
Bakelite	Phenol-formaldehyde	Union Carbide Chemicals Co.
Barden	Kaolin clay	Huber, J. M., Corp.
Barex	Barrier resin	Vistron Corp.
Baygal	Polyester for casting resins	Farbenfabriken Bayer AG

Trade or brand name	Product	Manufacturer
Baylon	Polycarbonate	Farbenfabriken Bayer AG
Baypren	Polychloroprene	Farbenfabriken Bayer AG
Beckacite	Modified phenolic	Reichhold Chemicals, Inc.; Beck, Koller & Co., Ltd.
Beckamine	Urea-formaldehyde	Reichhold Chemicals, Inc.; Beck, Koller & Co., Ltd.
Beckopox	Epoxy resins	Reichhold Chemicals, Inc.
Beckosol	Alkyd resins	Reichhold Chemicals, Inc.; Beck, Koller & Co., Ltd.
Beetle	Urea-formaldehyde resins	American Cyanamid Co.
Bemberg	Rayon fiber	Beaunit Mills Corp.
Bentone	Gelling agent	Kronos Titan GmbH
Bexone F	Poly(vinyl formal)	British Xylonite
Benvic	Poly(vinyl chloride)	Solvay & Cie S.A.
Betalux	TEF-filled acetal	Westlake Plastics
Bexphane	Polypropylene	Bakelite Xylonite Ltd.
Blanex	Polyethylene, cross-linked	Richhold Chemicals
Blapol	Polyethylene compounds	Richhold Chemicals
Blendex	ABS resin	Borg-Warner Corp.
Bolta Flex	Vinyl sheeting and film	General Tire & Rubber Co.
Bolta Thene	Rigid olefin sheets	General Tire & Rubber Co.
Boltaron	Plastic sheets	General Tire & Rubber Co.
Bonadur	Organic pigments	American Cyanamid Co.
Bondstrand	Filament wound fiberglass reinforced plastics	Ameron Corrosion Control Div.
Borofil	Boron filaments	Texaco Corp.
Boronol	Polyolefins with boron	Allied Resinous Products, Inc.
Bostik	Epoxy and polyurethane adhesives	Bostik-Finch

Trade or brand name	Product	Manufacturer
Bronco	Supported vinyl or pyroxylin	General Tire & Rubber Co.
Budene	cis-1,4-Polybutadiene	Goodyear
Busan	Flame-retardant microbiocide	Buckman Laboratories
Butacite	Poly(vinyl acetal) resins	E. I. du Pont de Nemours Co., Inc.
Bukaton	Butadiene copolymers	Imperial Chemical Industries, Ltd.
Butaprene	Styrene-butadiene elastomers	Firestone Tire & Rubber Co.
Butarez CTL	Telechelic butadiene polymer	Phillips Petroleum Co.
Buton	Butadiene-styrene resin	Enjay Chemical Co.
Bu-Tuf	Polybutene	Petrotex Chemical Corp.
Butvar	Poly(vinyl butyral) resin	Shawinigan Resins Corp.
BXL	Polysulfone	Union Carbide
Cab-O-Sil	Colloidal silica	Cabot Corp.
Cadco	Plastic rod, etc.	Cadillac Plastics
Cadon	Nylon filament	Chemstrand Corp.
Cadox	Organic peroxides	Cadet Chemical Corp.
Calwhite	Calcium carbonate	Georgia Marble Co.
Camel-Carb	Calcium carbonate	Harry T. Campbell Sons' Corp.
Capran	Nylon-6	Allied Chemical Corp.
Captax	Accelerator (2-mercapto-benzothiazole)	Goodyear Tire & Rubber Co.
Carbaglas	Fiberglass-reinforced polycarbonate	Dart Inds.
Carbitol	Solvents	Union Carbide Corp.
Carboloy	Cemented carbides	General Electric Co.
Carbomastic	Epoxy coal tar coating	Carboline Co.
Carbopol	Water-soluble resins	B. F. Goodrich Chemical Co.
Carboset	Acrylic resins	B. F. Goodrich Chemical Co.
Carbospheres	Hollow carbon spheres	Versar, Inc.
Carbowax	Poly(ethylene glycols)	Union Carbide Chemical Co.
Cariflex I	cis-1,4-Polyisoprene	Shell Chemical Co. Ltd.
Carina	Poly(vinyl chloride)	Shell Chemical Co. Ltd.
Carinex	Polystyrene	Shell Chemical Co. Ltd.

Trade or brand name	Product	Manufacturer
Carolux	Flexible filled urethane foam	North Carolina Foam
Carstab	Urethane foam catalysts	Cincinnati Milacron Chemicals, Inc.
Castcar	Cast polyolefin films	Mobil Chemical
Castethane	Castable polyurethanes	Upjohn Co.
Castomer	Urethane elastomer and coatings	Baxenden Chemical & Witco Chemical Corp.
Catalac	Phenol-formaldehyde resin	Catalin Ltd.
Celanar	Polyester film and sheeting	Celanese Plastics Co.
Celanex	Thermoplastic polyester	Celanese Plastics Co.
Celatron	Polystyrene	Celanese Plastics Co.
Celcon	Acetal copolymers	Celanese Plastics Co.
Celgard	Microporous polypropylene film	Celanese Plastics Co.
Celite	Diatomite filler	Johns-Manville Corp.
Cellasto	Microcellular urethane elastomer	North American Urethanes
Cellofoam	Polystyrene foam board	United States Mineral Products Co.
Cellonex	Cellulose acetate	Dynamit Nobel/ America
Cellon	Cellulose acetate	Dynamit Nobel
Cellosize	Hydroxyethyl cellulose	Union Carbide Corp.
Celluliner	Expanded polystyrene foam	Gilman Brothers
Cellulite	Expanded polystyrene foam	Gilman Brothers
Celluloid	Plasticized cellulose nitrate	Celanese Plastics Co.
Celogan	Blowing agents	Uniroyal, Inc.
Celpak	Rigid polyurethane foam	Dacar Chemical
Celramic	Glass nodules	Pittsburgh Corning Corp.
Cerex	Styrene copolymer	Monsanto Chemical Co.
Chemigum	Urethane elastomer	Goodyear Tire & Rubber Co.
Chem-o-sol	PVC plastisol	Chemical Products Co.
Chempro	Ion-exchange resin	Freeman Chem. Corp.

Trade or brand name	Product	Manufacturer
Celthane	Rigid polyurethane foam	Dacar Chemical
Chemfluor	Fluorocarbon plastics	Chemplast, Inc.
Chemglaze	Polyurethane-based coating	Lord Corp.
Chemgrip	Epoxy adhesives for TFE	Chemplast, Inc.
Chlorowax	Chlorinated paraffins	Diamond Alkali Co.
Cibanite	Aniline-formaldehyde resin	Ciba Products Co.
Cimglas	Fiberglass-reinforced polyester	Cincinnati Milacron
Cis-4	cis-1,4-Polybutadiene	Phillips Petroleum Co.
Clarite	PVC stabilizers	National Lead Co.
Clocel	Rigid urethane foam	Baxenden Chemical
Clopane	PVC tubing and film	Clopay Corp.
Cloudfoam	Polyurethane foam	International Foam
Co-Rexyn	Polyester resins, coatings, pastes	Interplastic Corp.
Cobex	Poly(vinyl chloride)	Bakelite Xylonite Ltd.
Cobocell	Cellulose acetate butyrate tubing	Cobon Plastics
Coboflon	Teflon tubing	Cobon Plastics
Cobothane	Ethylene–vinyl acetate tubing	Cobon Plastics
Collodion	Solution of cellulose nitrate	Generic name
Colovin	Calendered vinyl sheeting	Columbus Coated Fabrics
Conathane	Polyuethane compounds	Conap, Inc.
Conolite	Polyester laminate	Woodall Industries
Coral rubber	cis-Polyisoprene	Firestone Tire & Rubber Co.
Cordo	PVC foam and films	Ferro Corp.
Cordoflex	Polyvinylidene fluoride solutions	Ferro Corp.
Cordura	Regenerated cellulose	E. I. du Pont de Nemours & Co., Inc.
Corfam	Poromeric film	E. I. du Pont de Nemours & Co., Inc.
Corlite	Reinforced foam	Snark Products
Coro-Foam	Urethane foam	Cook Paint & Varnish
Corval	Rayon fiber	Courtaulds
Corvel	Plastic coating powders	The Polymer Corp.

Trade or brand name	Product	Manufacturer
Corvic	Vinyl polymers	Imperial Chemical Industries Otd.
Courlene	Polyethylene (fiber)	Courtaulds
Coverlight HTV	Vinyl-coated nylon fabric	Reeves Brothers
Covol	Poly(vinyl alcohol)	Corn Products Co.
Crastin	Poly(ethylene terephthalate)	Ciba Geigy
Creslan	Acrylonitrile–acrylic ester copolymers	American Cyanamid Co.
Cronar	Polyethylene	E. I. du Pont de Nemours & Co., Inc.
Cryowrap	Thermoplastic sheets and films	W. R. Grace & Co.
Cryovac	Polypropylene film	W. R. Grace & Co.
Crystalex	Acrylic resin	Rohm & Haas Co.
Crystalon	Rayon fiber	American Enka Corp.
Crystic	Polyester resins	Scott Bader Co.
Cumar	Coumarone-indene resins	Allied Chemical Corp.
Curithane	Polyaniline polyamines	Upjohn
Curon	Polyurethane foam	Reeves Brothers
Cyanaprene	Polyurethane	American Cyanamid Co.
Cyanolit	Cyanoacrylate adhesive	Leader, Denis, Ltd.
Cyasorb	Ultraviolet absorbers	American Cyanamid Co.
Cycloset	Cellulose acetate fiber	E. I. du Pont de Nemours & Co., Inc.
Cycolac	Acrylonitrile-butadiene-styrene copolymer	Borg-Warner Corp.
Cycolon	Resinous compositions	Borg-Warner Corp.
Cycoloy	Polymers with ABS alloy resins	Borg-Warner Corp.
Cycopoac	ABS and nitrile barrier	Borg-Warner Corp.
Cymac	Thermoplastic molding materials	American Cyanamid Co.
Cymel	Melamine molding compound	American Cyanamid Co.
Cyovin	Self-extinguishing ABS graft-polymer blends	Borg-Warner Corp.
Cyglas	Glass-filled polyester	American Cyanamid
Dabco	Triethylenediamine	Air Products Co.

Trade or brand name	Product	Manufacturer
Dacovin	Rigid poly(vinyl chloride)	Diamond Alkali Co.
Dacron	Polyester fiber	E. I. du Pont de Nemours & Co., Inc.
DAP	Diallyl phthalate monomer	FMC Corp.
Dapon	Diallyl phthalate prepolymer	FMC Corp.
Daponite	Dapon-fabric laminates	FMC Corp.
Daran	Poly(vinylidene chloride) emulsion coatings	W. R. Grace & Co.
Daratak	PVA emulsions	W. C. Grace
Darex	Styrene copolymer resin	W. R. Grace & Co.
Darvan	Poly(vinylidene cyanide)	Celanese Corp. of America
Darvic	Poly(vinyl chloride)	Imperial Chemical Industries, Ltd.
Davon	TFE	Davies Nitrate
Daxad	Dispersing agents	W. R. Grace & Co.
Decanox	Organic peroxides	Wallace & Tiernan, Inc.
Deenax	Antioxidants	Enjay Chemical Co.
Degalan	Poly(methyl methacrylate)	Degussa
Delrin	Acetal polymer	E. I. du Pont de Nemours & Co., Inc.
Densite	Urethane foam	Tenneco Chemical
Derakane	Polyester resin	Dow Chemical Co.
Derolite	Ion-exchange resin	Diamond Alkali Co.
Desmodur	Isocyanates for polyurethane foam	Farbenfabriken Bayer AG
Desmopan	Polyurethanes	Farbenfabriken Bayer AG
Desmophen	Polyesters and polyethers for polyurethanes	Farbenfabriken Bayer AG
Devran	Epoxy resins	Devoe & Reynolds Co.
Dexel	Cellulose acetate	British Celanese Ltd.
Dexon	Polypropylene-acrylic	Exxon Chemical
Dexsil	Polycarboranesiloxane	Olin Mathieson Corp.
Diakon	Poly(methyl methacrylate)	Imperial Chemical Industries Ltd.
Diall	Diallyl phthalate	Allied Chemical Corp.
Diaron	Melamine resins	Reichhold Chemicals

Trade or brand name	Product	Manufacturer
Dicalite	Diatomaceous earth	Dicalite/Grefco, Inc.
Dielux	Acetals	Westlake Plastics
Diene	Polybutadiene	Firestone Tire & Rubber Co.
Dimetcote	Protective coating	Americoat Corp.
Diolen	Poly(ethylene terephthalate)	ENKA-Glazstoff
Dion	Polyester resin	Diamond Alkali Co.
Dolphon	Epoxy and polyester resins	John C. Dolph
Dorvon	Polystyrene foam	Dow Chemical
Doryl	Poly(diphenyl oxide)	Westinghouse Electric Corp.
Dow Corning	Silicons	Dow Corning
Dowex	Ion-exchange resins	Dow Chemical Co.
Dralon	Polyacrylonitrile fiber	Farbenfabriken Bayer AG
Dri-Lite	Expanded polystyrene	Poly Foam
Duco	Cellulose nitrate lacquers	E. I. du Pont de Nemours & Co., Inc.
Dulac	Lacquers	Sun Chemical Corp.
Dulux	Polymeric enamels	E. I. du Pont de Nemours & Co., Inc.
Duolite	Ion-exchange resin	Diamond Alkali Co.
Duracel	Lacquers	Maas & Waldstein
Duracon	Acetal copolymers	Polyplastics
Dural	Acrylic modified PVC	Alpha Chemical & Plastics
Duralon	Furan molding resins	U.S. Stoneware Co.
Durane	Polyurethanes	Raffi & Swanson
Duramac	Alkyd resins	Commercial Solvents Corp.
Duraplex	Alkyd resins	Rohm & Haas Co.
Duraspan	Spandex fibers	Carr-Fulflex Corp.
Durelene	PVC tubing	Plastic Warehousing
Durethan	Nylon-6	Farbenfabriken Bayer AG
Durethan U	Polyurethanes	Farbenfabriken Bayer AG
Durethene	Polyethylene film	Sinclair-Koppers Co., Inc.
Durez	Phenolic resins	Hooker Chemical & Plastics
Durathon	Polybutylene resins	Witco Chemical Corp.

Trade or brand name	Product	Manufacturer
Durite	Phenolic resins	The Borden Co.
Duron	Phenolics	Firestone Foam
Dutral	Ethylene-propylene copolymer	Montecatini
Dyal	Alkyd resins	Sherwin-Williams Co.
Dyalon	Urethane elastomers	Thombert
Dyfoam	Expanded polystyrene	W. C. Grace
Dylan	Polyethylene resins	Sinclair-Koppers Co., Inc.
Dynel	Vinyl chloride–acrylonitrile copolymers	Union Carbide Corp.
Dylel	ABS copolymer	Sinclair-Koppers Co., Inc.
Dylene	Polystyrene resins	ARCO Polymer, Inc.
Dylite	Expandable polystyrene	Sinclair-Koppers Co., Inc.
Dynafilm	Polypropylene film	U.S. Industrial Chemicals Co., Div., National Distillers & Chemical Corp.
Dynel	Modacrylic fiber	Union Carbide Corp.
Dyphene	Phenol-formaldehyde resins	Sherwin-Williams Co.
Dyphos	Stabilizer for poly(vinyl chloride)	National Lead Co.
E-Foam	Epoxies	Allied Products
Easypoxy	Epoxy adhesive kits	Conap, Inc.
Ebolan	TFE materials	Chicago Gasket Co.
Ecavyl	Poly(vinyl chloride)	Kuhlmann
Eccosil	Silicon resins	Emerson & Cuming
Eccospheres	Hollow glass spheres	Emerson & Cummings, Inc.
Elastolit	Urethane engineering thermoplastics	North American Urethanes
Elastollyx	Urethane Engineering thermoplastics	North American Urethanes
Elastolur	Urethane coatings	BASF
Elastonate	Urethane isocyanate prepolymers	BASF
Elastonol	Urethane polyester polyols	North American Urethanes
Elastopel	Urethane engineering thermoplastics	North American Urethanes

Trade or brand name	Product	Manufacturer
Elastothane	Polyurethane elastomer	Thiokol Corp.
Electroglas	Cast acrylic	Glasflex Corp.
Elf	Carbon black	Cabot Corporation
El Rexene	Polyolefin resins	Rexall Chemical Co.
El Rey	Low-density polyethylene	Rexall Chemical Co.
Elvace	Acetate-ethylene copolymers	E. I. du Pont de Nemours & Co., Inc.
Elvacet	Poly(vinyl acetate) emulsion	E. I. du Pont de Nemours & Co., Inc.
Elvacite	Acrylic resins	E. I. du Pont de Nemours & Co., Inc.
Elvamide	Nylon resins	E. I. du Pont de Nemours & Co., Inc.
Elvanol	Poly(vinyl alcohol) resins	E. I. du Pont de Nemours & Co., Inc.
Elvax	Poly(ethylene-co-vinyl acetate)	E. I. du Pont de Nemours & Co., Inc.
Elvic	Poly(vinyl chloride)	Solvay
Enkalure	Nylon fiber	American Enka Corp.
Enrad	Preirradiated polyethylene	Enflo Corp.
Enrup	Thermosetting resin	United States Rubber Co.
Ensocote	PVC lacquer coatings	Uniroyal
Ensolex	Cellular plastic sheets	Uniroyal
Ensolite	Cellular plastic sheets	Uniroyal
Epibond	Epoxy adhesive resin	Furane Plastics, Inc.
Epicure	Curing agents for epoxy resins	Celanese Corp.
Epikote	Epoxy resins	Shell Chemical Co.
Epilox	Epoxy resins	Leuna
Epi-Rez	Epoxy cast resin	Celanese Corp.
Epi-Tex	Epoxy-ester resins	Hoechst-Celanese
Epikote	Epoxy resins	Shell Chemical
Epocap	Two-part epoxy compounds	Hardman
Epocast	Epoxy resins	Furane Plastics, Inc.
Epocrete	Two-part epoxy compounds	Hardman
Epocryl	Epoxy acrylate resin	Shell Chemical Co.
Epodite	Epoxy resins	Showa Highpolymer Co.

Trade or brand name	Product	Manufacturer
Epolast	Two-part epoxy compounds	Hardman
Epolene	Low-melt polyethylene	Eastman Chemical Products, Inc.
Epolite	Epoxy compounds	Hexcel Corp.
Epomarine	Two-part epoxy compounds	Hardman
Epon	Epoxy resins	Shell Chemical Co.
Eponol	Linear polyether resins	Shell Chemical Co.
Eposet	Two-part epoxy compounds	Hardman
Epotuf	Epoxy resins	Reichhold Chemical Co.
Epoxylite	Epoxy resins	Epoxylite Corp.
Escon	Polypropylene	Enjay Chemical Co.
Estane	Polyurethane resins	B. F. Goodrich Chemical Co.
Estron	Cellulose acetate filament	Eastman Chemical Products, Inc.
Ethafoam	Polyethylene foam	Dow Chemical Co.
Ethocel	Ethyl cellulose	Dow Chemical Co.
Ethofil	Fiberglass-reinforced polyethylene	Dart Industries
Ethoglas	Fiberglass-reinforced polyethylene	Dart Industries
Ethosar	Fiberglass-reinforced polyethylene	Dart Industries
Ethron	Polyethylene	Dow Chemical Co.
Ethylux	Polyethylene	Westlake Plastics
Evenglo	Polystyrene	Sinclair-Koppers Co., Inc.
Everflex	PVA copolymer emulsion	W. C. Grace
Everlon	Polyurethane foam	Stauffer Chemical
Excelite	Polyethylene tubing	Thermoplastic Processes
Exon	Poly(vinyl chloride)	Firestone Plastics
Extane	Polyurethane tubing	Pipeline Service Co.
Extrel	Plastic films	Exxon Chemical Co. U.S.A.
Extren	Fiberglass-reinforced polyester	Morrison Molded Fiber Glass Co.
Fabrifil	Chopped-rag fillers	Microfibers, Inc.
Fabrikoid	Pyroxylin-coated fabrics	E. I. du Pont de Nemours & Co., Inc.
Facilon	Reinforced PVC fabric	Sun Chemical

Trade or brand name	Product	Manufacturer
Fassgard	Vinyl-coated nylon	M. S. Fassler & Co.
Fasslon	Vinyl coating	M. S. Fassler & Co.
Felor	Nylon filaments	E. I. du Pont de Nemours & Co., Inc.
Fertene	Polyethylene	Montecatini
Fibercast	Reinforced plastic pipe	Fibercast Co.
Fiberglas	Fibrous glass	Owens-Corning Fiberglas Corp.
Fiberite	Phenolic molding compounds	Fiberite Corp.
Fibro	Rayon	Courtaulds NA, Inc.
Filabond	Unsaturated polyester	Reichhold
Filfrac	Cut cotton fiber	Rayon Processing Co. of Rhode Island
Firemaster	Fire retardants	Michigan Chemical Corp.
Firmex	Carbon black	Columbian Carbon Co.
Flakeglas	Glass flakes for reinforcements	Owens-Corning Fiberglas Corp.
Flectol	Amine-type antioxidants	Monsanto Co.
Flexane	Polyurethanes	Devcon Corp.
Flexocel	Polyurethane foam	Baxenden Chemical
Flexol	Plasticizers	Union Carbide Chemical Co.
Floranier	Cellulose	Rayonier, Inc.
Flovic	Poly(vinyl acetate)	Imperial Chemical Industries, Ltd.
Fluokem	Teflon spray	Bel-Art Products
Fluon	Polytetrafluoroethylene	Imperial Chemical Industries, Ltd.
Fluon	PTFE powders and dispersions	Imperial Chemical Industries, Ltd.
Fluorel	Poly(vinylidene fluoride)	Minnesota Mining and Mfg. Co.
Fluorglas	PTEF-impregnated materials	Dodge Industries
Fluorobestos	Asbestos-Teflon composite	Raybestos Manhattan, Inc.
Fluorocord	Fluorocarbon material	Raybestos Manhattan, Inc.
Fluorofilm	Cast Teflon films	Dilectrix Corp.
Fluoron	Polychlorotri-fluoroethylene	Stokes Molded Products
Fluoroplast	Polytetrafluoroethylene	U.S. Gasket Co.

Trade or brand name	Product	Manufacturer
Fluororay	Filled fluorocarbon	Raybestos Manhattan, Inc.
Fluorored	TFE compounds	John L. Dore Co.
Fluorosint	TFE—fluorocarbon composition	Polymer Corp.
Foamex	Poly(vinyl formal)	General Electric Co.
Foamthane	Polyurethane foam	Pittsburgh Corning Corp.
Formadall	Polyester premix	Woodall Industries
Formaldafil	Fiberglass-reinforced acetals	Dart Industries
Formaldaglass	Fiberglass-reinforced acetals	Dart Industries
Formaldasar	Fiberglass-reinforced acetals	Dart Industries
Formex	Poly(vinyl acetal)	General Electric Co.
Formica	Thermosetting laminates	Formica Corp.
Formrez el	Liquid resins for urethane elastomers	Witco Chemical Co.
Formvar	Poly(vinyl formal)	Shawinigan Resins Corp.
Forticel	Cellulose propionate	Celanese Corp. of America
Fortiflex	Polyethylene	Celanese Plastics Co.
Fortisan	Saponified cellulose acetate	Celanese Corp. of America
Fortrel	Polyester fiber	Fiber Industries, Inc.
Fostacryl	Poly(styrene-co-acrylonitrile)	Foster Grant Co.
Fostalene	Plastic	Foster Grant Co.
Fosta-Net	Polystyrene foam mesh	Foster Grant Co.
Fosta Tuf-Flex	High-impact polystyrene	Foster Grant Co.
Fostafoam	Expandable polystyrene beads	Foster Grant Co.
Fostalite	Light-stabilized polystyrene	Foster Grant Co.
Fostarene	Polystyrene	Foster Grant Co.
FPC	PVC resins compound	Firestone Tire & Rubber Co.
Freon	Blowing agents	E. I. du Pont de Nemours & Co., Inc.
Furname	Epoxy and furan resins	Atlas Mineral Products Co.
Futron	Polyester	Fusion Rubbermaid
Fyberoid	Fishpaper	Wilmington Fibre Specialty Co.

Trade or brand name	Product	Manufacturer
Fyrol	Flame retardants	Stauffer Chemical Co.
Galalith	Casein plastics	Generic name
Gama-Sperse	Calcium carbonate	Georgia Marble Co.
Gantrez	Poly(vinyl ether-comaleic anhydride)	Dyestuff Chemical Div., General Aniline & Film Corp.
Garan	Fibrous-glass roving	Johns-Manville Corp.
Garan Finish	Sizing for glass fibers	Johns-Manville Corp.
Garox	Organic peroxides	Ram Chemicals, Inc.
Gedamine	Unsaturated polyester	Charbonnages
Gelva	Poly(vinyl acetate)	Shawinigan Resins Corp.
Gelvatex	Poly(vinyl acetate) emulsions	Shawinigan Resins Corp.
Gelvatol	Poly(vinyl alcohol)	Shawinigan Resins Corp.
Gemon	Malefamide	General Electric Co.
Genaire	Poromeric film	General Tire & Rubber Co.
Genal	Thermosets	General Electric Co.
Genthane	Polyurethane elastomer	General Tire & Rubber Co.
Genetron	Fluorinated hydrocarbon monomers and polymers	Allied Chemical Co.
Gentro	Butadiene copolymer	General Tire & Rubber Co.
Geon	Poly(vinyl chloride)	B. F. Goodrich Chemical Co.
Gil-Fold	Polyethylene sheets	Gilman Brothers
Glaskyd	Glass-reinforced alkyd resin	American Cyanamid Co.
Glufil	Shell flour	Agrashell Inc.
Glyptal	Alkyd coating	General Electric Co.
Gracon	PVC	W. C. Grace
GravoFLEX	ABS sheets	Hermes Plastics
GravoPLY	Acrylic sheets	Hermes Plastics
Grex	Polyethylene	W. R. Grace & Co.
Halar	Polyfluorocarbons	Allied Chemical Co.
Halex	Polyfluorocarbon	Allied Chemical Co.
Halon	Fluorochlorocarbon	Allied Chemical Co.
Halowax	Chlorinated naphthalene	Union Carbide Corp.
Harflex	Plasticizers	Wallace & Tiernan, Inc.
Haylar	CTFE	Allied Chemical

Trade or brand name	Product	Manufacturer
Haysite	Polyester laminates	Synthane-Taylor
HB-40	Hydrogenated terphenyl	Monsanto Co.
Hercocel	Cellulose acetate	Hercules Powder Co.
Hercoflex	Phthalate plasticizers	Hercules Powder Co.
Hercolyn	Hydrogenated methyl abietate	Hercules Powder Co.
Hercose	Cellulose acetate-propionate	Hercules Powder Co.
Herculoid	Cellulose nitrate	Hercules Powder Co.
Herculon	Polypropylene	Hercules Powder Co.
Herox	Nylon	E. I. du Pont de Nemours & Co., Inc.
Het anhydride	Chlorendic anhydride	Hooker Chemical Corp.
Hetrofoam	Fire-retardant urethane foam	Hooker Chemical Corp.
Hetron	Fire-retardant polyester resins	Hooker Chemical Corp.
Heaveaplus	Copolymer of methyl methacrylate and rubber	Generic name
Hexcel	Structural honeycomb	Hexcel Products, Inc.
Hex-One	HDPE	Gulf Oil
H-film	Polyamide film from pyromellitic anhydride and 4,4-diaminodiphenyl ether	E. I. du Pont de Nemours & Co., Inc.
Hi-Blen	ABS polymers	Japanese Geon Co.
Hi-fax	High-density polyethylene	FMC Corp.; Hercules Powder Co.
Hipack	Polyethylene	Showa Highpolymer Co.
Hi-Sil	Amorphous silica	PPG Corp.
Hi-Styrolux	High-impact polystyrene	Westlake Plastics
Hitalex	Polyethylene	Hitachi Chemical Co.
Hitanol	Phenol-formaldehyde resins	Hitachi Chemical Co.
Horse Head	Zinc oxide pigments	New Jersey Zinc Co.
Hostadur	Poly(ethylene terephthalate)	Farbwerke Hoechst AG
Hostaflon C2	Polychlorotri-fluoroethylene	Farbwerke Hoechst AG
Hostaflon TF	Polytetrafluoroethylene	Farbwerke Hoechst AG
Hostalen	Polyethylene	Farbwerke Hoechst AG

Trade or brand name	Product	Manufacturer
Hostyren	Polystyrene	Hoechst
Hyamine	Cationic surfactants	Rohm & Haas Co.
Hycar	Butadiene acrylonitrile copolymer	B. F. Goodrich Chemical Co.
Hydraflex	Printing ink	Sun Chemical Corp.
Hydrepoxy	Water-based epoxies	Allied Products
Hydrin	Epichlorohydrin rubber	Goodrich-Hercules
Hydrocal	Gypsum	U.S. Gypsum Co.
Hydro Foam	Expanded phenol-formaldehyde	Smithers Co.
Hydropol	Hydrogenated polybutadiene	Phillips Petroleum Co.
Hylar	Poly(ethylene terephthalate)	E. I. du Pont de Nemours & Co., Inc.
Hylene	Organic isocyanates	E. I. du Pont de Nemours & Co., Inc.
Hypalon	Chlorosulfonated polyethylene	E. I. du Pont de Nemours & Co., Inc.
Igepal	Wetting agents	General Aniline & Film Corp.
Igepon	Surfactants, wetting agents	General Aniline & Film Corp.
Implex	Acrylic resins	Rohm & Haas Co.
Insurok	Phenol-formaldehyde molding compounds	The Richardson Co.
Intamix	Rigid PVC	Diamond Shamrock Corp.
Ionac	Ion-exchange resins	Permutit Co.
Ionol	Antioxidant	Shell Chemical Co.
Irganox	Antioxidants	Geigy Chemical Corp.
Irrathene	Irradiated polyethylene	General Electric Co.
Irvinil	PVC resins	Great American Chemical
Isoderm	Urethane integral-skinning foam	Upjohn
Isofoam	Polyurethane foam resins	Isocyanate Products, Inc.
Isomid	Polyester-polyamide film magnet wire	Schenectady Chemicals, Inc.
Isonate	Diisocyanates	Upjohn Co.
Isonol	Propoxylated amines	Upjohn Co.
Isoteraglas	Isocyianate elastomer–coated Dacron glass fabric	Natvar Corp.

Trade or brand name	Product	Manufacturer
Isothane	Polyurethane foam	Bernel Foam Products
Iupilon	Polycarbonate	Mitsubishi Edogawa Chemical Co.
Jay-Flex	Plasticizers	Enjay Chemical Co.
Jetfoam	Polyurethane foam	International foam
Jet-Kote	Furane resin coatings	Furane Plastics, Inc.
Kadox	Zinc oxide	New Jersey Zinc Co.
Kalex	Urethane resin	Di-Acro Kaufman
Kalite	Precipitated calcium carbonate	Diamond Alkali Co.
Kalmac	Calcium carbonate	Georgia Marble Co.
Kalspray	Rigid urethane foam	Baxenden Chemical
Kaofil	Coating and filler clay	Thiele Kaolin Co.
Kapsol	Plasticizers	Ohio-Apex Div., FMC Corp.
Kapton	Polyimide	E. I. du Pont de Nemours & Co., Inc.
Kardel	Polystyrene film	Union Carbide Corp.
Kaurit	Urea-formaldehyde resins	Badische Anilin & Coda-Fabrik AG
Kel-F	Trifluorochloroethylene resins	Minnesota Mining & Mfg. Co.
Keltrol	Copolymers	Textron, Inc.
Kematal	Acetal copolymers	Imperial Chemical Industries, Ltd.
Kenflex	Hydrocarbon resins	Kenrich Petrochemicals, Inc.
Ken-U-Thane	Polyurethane	Kenrich Petrochemicals, Inc.
Kessco	Plasticizers	Kessler Chemical Co., Inc.
Ketac	Ketone-aldehyde resin	American Cyanamid Co.
Kinel	Maleimide	Rhone-Poulenc
Kodacel	Cellulose acetate film	Eastman Chemical Products, Inc.
Kodaflex	Plasticizers	Eastman Chemical Products, Inc.
Kodar	Copolyesters	Eastman Chemical
Kodel	Polyester fibers	Eastman Kodak Co.
Kohinor	Vinyl resins	Pantasote Co.
Kollidon	Poly(vinyl pyrrolidone)	General Aniline & Film Corp.
Kolorbon	Rayon fiber	American Enka Corp.
Kopox	Epoxy resin	Koppers Co.

Trade or brand name	Product	Manufacturer
Korad	Acrylic film	Rohm & Haas Co.
Korez	Phenolic resin cement	Atlas Mineral Products Company
Koroseal	Poly(vinyl chloride)	B. F. Goodrich Chemical Co.
Kosmos	Carbon black	United Carbon Co.
Kotol	Resin solutions	Uniroyal, Inc.
Kralac	ABS resins	Uniroyal, Inc.
Kralastic	ABS	Uniroyal, Inc.
Kralon	High-impact styrene and ABS resins	Uniroyal, Inc.
Kraton	Butadiene block copolymers	Shell Chemical Co.
Krene	Plasticized vinyl film	Union Carbide Corp.
K-Resin	Butadiene-styrene copolymer	Phillips Petroleum Co.
Kriston	Allyl ester casting resins	B. F. Goodrich Chemical Co.
Kroniflex	Phosphate ester plasticizer	FMC Corp.
Kronisol	Dibutoxyethyl phthalate	FMC Corp.
Kronitex	Tricresyl phosphate	FMC Corp.
Kronox	Titanium dioxide	Kronis Titan
Kronox	Plasticizer	FMC Corp.
Krystal	PVC sheet	Allied Chemical
Krystaltite	PVC shrink film	Allied Chemical
Kurlon	Poly(vinyl alcohol) fibers	
Kydene	Acrylic-PVC powder	Rohm & Haas Co.
Kydex	Acrylic-poly(vinyl chloride) sheet	Rohm & Haas Co.
Kylan	Chitin	
Kynar	Poly(vinylidene fluoride)	Pennwalt Chemicals Corp.
Lamabond	Reinforced polyethylene	Columbian Carbon Co.
Lamar	Mylar vinyl laminates	Morgan Adhesives
Laminac	Polyester resins	American Cyanamid Co.
Lanital	Fiber from milk protein	Shia Viscosa
Laurox	Polymerization catalysts	Akzo Chemie Nederland BV
Laguval	Polyester resins	Farbenfabriken Bayer AG
Last-A-Foam	Plastic foam	General Plastics Mfg.

Trade or brand name	Product	Manufacturer
Lekutherm	Epoxy resins	Farbenfabriken Bayer AG
Lemac	Poly(vinyl acetate)	Borden Chemical Co.
Lemol	Poly(vinyl alcohol)	Borden Chemical Co.
Levapren	Ethylene-vinylacetate copolymers	Farbenfabriken Bayer AG
Lexan	Polycarbonate resin	General Electric Co.
Lindol	Phosphate plasticizers	Stauffer Chemical Co.
Lock Foam	Polyurethane foam	Nopco Chemical Co.
Lucidol	Benzoyl peroxide	Wallace and Tiernan, Inc.
Lucite	Poly(methyl methacrylate) and copolymers	E. I. du Pont de Nemours & Co., Inc.
Ludox	Colloidal silica	E. I. du Pont de Nemours & Co., Inc.
Lumarith	Cellulose acetate	Celanese Corp. of America
Lumasite	Acrylic sheet	American Acrylic Corp.
Lumite	Saran filaments	Chicopee Manufacturing Co.
Luperco	Organic peroxides	Pennwalt Corp.
Luperox	Organic peroxides	Pennwalt Corp.
Lustran	Molding and extrusion resins	Monsanto Chemical Co.
Lustrex	Polystyrene	Monsanto Chemical Co.
Lutonal	Poly(vinyl ethers)	Badische Anilin & Soda-Fabrik AG
Lutrex	Poly(vinyl acetate)	Foster Grant Co.
Luvican	Poly(vinyl carbazole)	Badische Anilin & Soda-Fabrik AG
Lycra	Spandex fibers	E. I. du Pont de Nemours & Co., Inc.
Macal	Cast vinyl film	Morgan Adhesives Co.
Madurik	Melamine-formaldehyde resins	Casella Farbwerke Mainkur AG
Makrofol	Polycarbonate film	Naftone, Inc.
Makrolon	Polycarbonate	Farbenfabriken Bayer AG
Marafoam	Polyurethane foam	Marblette Co.
Maraglas	Epoxy resin	Marblette Co.
Maranyl	Nylons	Imperial Chemical Industries Ltd.
Maraset	Epoxy resin	The Marblette Corp.

Trade or brand name	Product	Manufacturer
Marathane	Urethane materials	Allied Products
Maraweld	Epoxy resin	Marblette Co.
Marbon	Polystyrene and copolymers	Borg-Warner Corp.
Marlex	Polyolefin resins	Phillips Chemical Co.
Marvibond	Metal-plastics laminates	Uniroyal, Inc.
Marvinol	Poly(vinyl chloride)	Uniroyal, Inc.
Melan	Melamine resins	Hitachi Chemical Co., Ltd.
Meldin	Polyimides	Dixon Corp.
Melinex	Poly(ethylene terephthalate)	Imperial Chemical Industries Ltd.
Melit	Melamine-formaldehyde resins	Societa Italiana Pesine
Melmac	Melamine molding materials	American Cyanamid Co.
Melolam	Melamine resin	Ciba Geigy
Melurac	Melamine-urea resins	American Cyanamid Co.
Meraklon	Polypropylene	Montecatini
Merlon	Polycarbonate	Mobay Chemical Co.
Metallex	Cast acrylic sheets	Hermes Plastics
Methocel	Methylcellulose	Dow Chemical Co.
Meticone	Silicon rubber	Hermes Plastics
Metre-Set	Epoxy adhesives	Metachem Resins Corp.
Micarta	Thermosetting laminates	Westinghouse Electric Corp.
Micro-Matte	Extruded acrylic sheet with matte finish	Extrudaline, Inc.
Micronex	Carbon black	Columbian Carbon Co.
Micropel	Powdered nylon	Nypel, Inc.
Microsol	Vinyl plastisol	Michigan Chrome & Chemical
Microthene	Powdered polyethylene	U.S. Industrial Chemicals Co.
Milmar	Polyester	Morgan Adhesives
Minex	Aluminum silicate filler	American Nepheline Corp.
Mini-Vaps	Expanded PE	Malge Co.
Minit Grip	Epoxy adhesives	High Strength Plastics Corp.
Minit Man	Epoxy adhesives	Kristal Draft, Inc.
Minlon	Reinforced nylon	E. I. du Pont de Nemours & Co., Inc.

Trade or brand name	Product	Manufacturer
Mipolam	Poly(vinyl chloride)	Dynamit Nobel
Mipoplast	PVC sheets	Dynamit Nobel/ America
Mirasol	Alkyd resins	C. S. Osborn Chemicals
Mirbane	Amino resin	Showa Highpolymer Co.
Mirrex	Calendered PVC	Tenneco Chemicals
Mista Foam	Urethane foam	M. R. Plastics & Coatings
Modulene	Polyethylene resin	Muehlstein & Co.
Mogul	Carbon black	Cabot Corp.
Molplen	Polypropylene	Novamont Corp.
Moltopren	Polyurethane foam	Farbenfabriken Bayer AG
Molycor	Fiberglass-reinforced epoxy tubing	A. O. Smith, Inland, Inc.
Mondur	Organic isocyanates	Mobay Chemical Co.
Monocast	Nylon	Polymer Corp.
Montrek	Polyethyleneimine	Dow Chemical Co.
Moplen	Polypropylene	Montecatini
Mowlith	Poly(vinyl acetate)	Farbwerke Hoechst AG
Mowiol	Poly(vinyl alcohol)	Farbwerke Hoechst AG
Mowital	Poly(vinyl butyral)	Farbwerke Hoechst AG
Multrathane	Urethane elastomer	Mobay Chemical
Multron	Polyesters	Mobay Chemical Co.
Mycalex	Inorganic molded plastic	Mycalex Corp. of America
Mylar	Polyester film	E. I. du Pont de Nemours & Co., Inc.
Nacconate	Organic diisocyanate	Allied Chemical Corp.
Nadic	Maleic anhydride	Allied Chemical Corp.
Nalgon	Plasticized poly(vinyl chloride)	Nalge Co.
Napryl	Polypropylene	Pechiney-Saint-Gobain
Natene	Polyethylene	Pechiney-Saint-Gobain
Natsyn	cis-1,4-Polyisoprene	Goodyear Tire & Rubber Co.
Naugahyde	Vinyl-coated fabric	U.S. Rubber Co.
Neboy	Petroleum hydrocarbon resin	Neville Chemical Co.
NeoCryl	Acrylic resins and emulsions	Polyvinyl Chemicals
Neoprene	Polychloroprene	E. I. du Pont de Nemours & Co., Inc.

Trade or brand name	Product	Manufacturer
NeoRez	Styrene emulsions and urethane solutions	Polyvinyl Chemicals
NeoVac	PVA emulsions	Polyvinyl Chemicals
Neozone	Antioxidants	E. I. du Pont de Nemours & Co., Inc.
Nepoxide	Epoxy resin coating	Atlas Minerals & Chemicals Div., ESB
Nestrite	Phenolic and urea-formaldehyde	James Ferguson & Sons
Nevidene	Coumarone-indene resin	Neville Chemical Co.
Nevillac	Modified coumarone-indene resin	Neville Chemical Co.
Niax	Polyol polyesters	Union Carbide Corp.
Nimbus	Polyurethane foam	General Tire & Rubber Co.
Nipeon	Poly(vinyl chloride)	Japanese Geon Co.
Nipoflex	Ethylene–vinyl acetate copolymer	Toyo Soda Mfg. Co.
Nipolon	Polyethylene	Toyo Soda Mfg. Co.
Nitrocol	Nitrocellulose-based pigment dispersions	J. C. Osburn Chemicals
Noan	Styrene–methyl methacrylate copolymer	Richardson Corp.
Nob-Lock	PVC sheets	Ameron Corrosion Control
Nomex	Nylon	E. I. du Pont de Nemours & Co., Inc.
Nopcofoam	Polyurethane foams	Nopco Chemical Co.
Norchem	LDPE resin	Northern Petrochemical Co.
Nordel	Ethylene-propylene	E. I. du Pont de Nemours & Co., Inc.
Noryl	Poly(phenylene oxide)	General Electric Co.
Novacite	Altered novaculite	Malvern Minerals Co.
Novodur	ABS polymers	Farbenfabriken Bayer AG
Nuba	Modified coumarone	Neville Chemical Co.
Nuclon	Polycarbonate	Pittsburgh Plate Glass Co.
Nukem	Acid-resistant resin cements	Amercoat Corp.
Numa	Spandex fibers	American Cyanamid Corp.

Trade or brand name	Product	Manufacturer
Nupol	Thermosetting acrylic resin	Freeman Chemical
Nyglathane	Glass-filled polyurethane	Nypel, Inc.
Nylafil	Reinforced nylon	Rexall Chemical Co.
Nylaglas	Fiberglass-reinforced nylon	Dart Industries
Nylasar	Fiberglass-reinforced nylon	Dart Industries
Nylasint	Sintered nylon parts	The Polymer Corp.
Nylatron	Filled nylons	The Polymer Corp.
Nylon	Polyamides	E. I. du Pont de Nemours & Co., Inc.
Nylo-Seal	Nylon-11 tubing	Imperial-Eastman Corp.
Nylux	Nylons	Westlake Plastics
Nyplube	TFE-filled nylons	Nypel, Inc.
Nyreg	Glass-reinforced nylon	Nypel, Inc.
Oasis	Expanded phenol-formaldehyde	Smithers Co.
Olefane	Polypropylene film	Avisun Corp.
Olefil	Filled PP resin	Amoco Chemicals
Oleflo	PP resin	Amoco Chemicals
Olemer	Propylene copolymer	Avisun Corp.
Oletac	Amorphous PP	Amoco Chemicals
Opalon	Poly(vinyl chloride)	Monsanto Chemical Co.
Oppanol B	Polyisobutylene	Badische Anilin & Soda-Fabrik AG
Oppanol C	Poly(vinyl isobutylether)	Badische Anilin & Soda-Fabrik AG
Orgalacqe	Epoxy and PVC powders	Aquitaine-Organico
Orgamide R	Nylon-6	Aquitaine-Organico
Orlon	Acrylic fiber	E. I. du Pont de Nemours & Co., Inc.
Ortix	Poromeric film	Celanese Corp.
Oxiron	Epoxidized polybutadiene	
Panarez	Hydrocarbon resins	Amoco Chemical Corp.
Panda	Vinyl and urethane-coated fabric	Pandel-Bradford
Panelyte	Laminates	Thiokol Chemical Co.
Papi	Polymethylene, polyphenyl isocyanate	Upjohn Co.

Trade or brand name	Product	Manufacturer
Paracon	Polyester rubber	Bell Telephone Laboratories
Paracryl	Butadiene-acrylonitrile copolymer	U.S. Rubber Co.
Paradene	Coumarone-indene resins	Neville Chemical Co.
Paralac	Polyester resin	ICI
Paraplex	Plasticizers	Rohm & Haas Co.
Parfe	Rayon fiber	Beaunit Mills Corp.
Parlon	Chlorinated rubber	Hercules Corp.
Parylen C	Polymonochloro-p-xylene	Union Carbide Corp.
Parylen N	Polyxylene	Union Carbide Corp.
Pearlon	Polyethylene film	Visking Corp.
Pee Vee Cee	Rigid poly(vinyl chloride)	ESB Corp.
Pelaspan	Expandable polystyrene	Dow Chemical Co.
Pellethene	Thermoplastic urethane	Upjohn
Pellon Aire	Nonwoven textile	Pellon Corp.
Pentalyn	Abietic acid derivative	Hercules Co., Inc.
Penton	Chlorinated polyether resins	Hercules Co., Inc.
Perbunan N	Butadiene-acrylonitrile copolymers	Farbenfabriken Bayer AG
Percadox	Organic peroxides	Cadet Chemical Corp.
Peregal	Antistatic agents	General Aniline & Film Corp.
Perlon	Polyurethane filament	Farbenfabriken Bayer AG
PermaRex	Cast epoxy	Permali, Inc.
Permelite	Melamine resin	Melamine Plastics
Permutit	Ion-exchange resin	Permutit Co.
Perspex	Acrylic resins	Imperial Chemical Industries Ltd.
Petra	Polyester sheets	Allied Chemical
Petrothene	Polyethylene	National Distillers & Chemical Corp.
Pevalon	Poly(vinyl alcohol)	May and Baker Ltd.
Phenoxy	Polyhydroxy ether of bisphenol A	Union Carbide Corp.
Phenoweld	Phenolic adhesive	Hardman, Inc.
Philjo	Polyolefin film	Phillips-Joana Co.
Philprene	Styrene-butadiene rubber	Phillips Petroleum Co.

Trade or brand name	Product	Manufacturer
Phosgard	Phosphorus compounds	Monsanto Co.
Picco	Hydrocarbon resins	Hercules, Inc.
Piccocumaron	Hydrocarbon resins	Hercules, Inc.
Piccoflex	Acrylonitrile-styrene resins	Pennsylvania Industrial Chemical
Piccolastic	Polystyrene resin	Pennsylvania Industrial Chemical
Piccolyte	Terpene polymer resins	Hercules, Inc.
Piccotex	Vinyl-toluene copolymers	Pennsylvania Industrial Chemical
Piccoumaron	Coumarone-indene resins	Pennsylvania Industrial Chemical
Piccovar	Alkyl-aromatic resins	Pennsylvania Industrial Chemical
Pienco	Polyester resins	American Petrochemical
Pip Pip	Rubber accelerator	Generic name
Plaskon	Amino resins	Allied Chemical Corp.
Plastacel	Cellulose acetate flake	E. I. du Pont de Nemours & Co., Inc.
Plastanox	Antioxidant	American Cyanamid Corp.
Plastigel	Liquid thickeners	Plasticolors, Inc.
Plastylene	Polyethylene	Pechiney-Saint-Gobain
Plenco	Phenolic resins	Plastics Engineering Co.
Pleogen	Polyester resins and gels	Whittaker Corp.
Plexiglas	Acrylic sheets	Rohm & Haas Co.
Plexigum	Acrylate and methacrylate resins	Rohm & Haas Co.
Plicose	Polyethylene	Diamond Shamrock Corp.
Pliobond	Adhesive	Goodyear Tire & Rubber Co.
Pliofilm	Rubber hydrochloride	Goodyear Tire & Rubber Co.
Plioflex	Poly(vinyl chloride)	Goodyear Tire & Rubber Co.
Pliolite	Cyclized rubber	Goodyear Tire & Rubber Co.
Pliothene	Polyethylene rubber blends	Ametek/Westchester
Pliovic	Poly(vinyl chloride)	Goodyear Tire & Rubber Co.

Trade or brand name	Product	Manufacturer
Pluracol	Polyethers	Wyandotte Chemicals Corp.
Pluragard	Urethane foams	BASF Wyandotte Corp.
Pluronic	Polyethers	BASF Wyandotte Corp.
Plyfoam	PVC foam	
Plyocite	Phenol-impregnated materials	Reichhold Chemicals, Inc.
Plyophen	Phenolic resins	Reichhold Chemicals, Inc.
Pluronics	Block polyether diols	Wyandotte Corp.
Polex	Oriented acrylic	Southwestern Plastics
Pollopas	Urethane-formaldehyde materials	Dynamit Nobel/America
Polvonite	Cellular plastic material	Voplex Corp.
Polyallomer	Ethylene block copolymers	Eastman Chemical Products
Poly-Dap	Diallyl phthalate resins	U.S. Polymeric
Polycarbafil	Fiberglass-reinforced polycarbonates	Dart Industries
Polycure	Cross-linked polyethylene	Crooke Color & Chemical
Poly-eth	Polyethylene	Gulf Oil Corp.
Poly-eze	Ethylene copolymers	Gulf Oil Corp.
Polyfoam	Polyurethane foam	General Tire & Rubber
Polygard	Stabilizer	Goodyear Tire & Rubber Co.
Poly-Gard	Solventless epoxies	Richhold Chemicals
Polyimidal	Polyimide thermoplastics	Raychem Corp.
Polylite	Polyester resins	Reichhold Chemicals, Inc.
Polylumy	Polypropylene	Kohjin Co.
Polymet	Plastic-filled sintered metal	Polymer Corp.
Polymin	Polyethyleneimine	Badische Anilin & Soda-Fabrik AG
Polymul	Polyethylene emulsions	Diamond Shamrock Chemical
Poly-pro	Polypropylene	Gulf Oil Corp.
Polyox	Water-soluble resins	Union Carbide Corp.
Polysizer	Poly(vinyl alcohol)	Showa Highpolymer Co.
Polyteraglas	Polyester-coated Dacron glass fabric	Natvar corp.

Trade or brand name	Product	Manufacturer
Polyviol	Poly(vinyl alcohol)	Wacker Chemie GmbH
Powminco	Asbestos fibers	Powhatan Mining Co.
PPO	Poly(phenylene oxide)	Hercules, Inc.
Pro-fax	Polypropylene resins	Hercules Powder Co.
Profil	Fiberglass-reinforced polypropylene	Dart Industries
Proglas	Fiberglass-reinforced polypropylene	Dart Industries
Prohi	HDPE	Protective Lining Corp.
Propathene	Polypropylene	Imperial Chemical Industries Ltd.
Propiofan	Poly(vinyl propionate)	BASF
Propylsar	Fiberglass-reinforced polypropylene	Dart Industries
Propylus	Polypropylene	Westlake Plastics
Protectolite	Polyethylene film	Protective Lining Corp.
Protron	Ultrahigh-strength PE	Protective Lining Corp.
Purilon	Rayon	FMC Corp.
PYR-ML	Polyimide	E. I. du Pont de Nemours & Co., Inc.
Q-Cel	Inorganic hollow microspheres	Philadelphia Quartz Co.
Quadrol	Poly(hydroxy amine)	Wyandotte Chemicals, Inc.
Quelflam	Polyurethanes	Baxenden Chemical
Ravinil	Poly(vinyl chloride)	ANIC, S.P.A.
Raybrite	Alpha-cellulose filler	Rayonier, Inc.
Rayflex	Rayon	FMC Corp.
Regalite	Press-polished PVC	Tenneco Advanced Materials
REN-Shape	Epoxy	Ren Plastics
Ren-Thane	Urethane elastomers	Ren Plastics
Resiglas	Polyester resins	Kristal Draft, Inc.
Resimene	Urea and melamine resins	Monsanto Co.
Resinol	Polyolefins	Allied Resinous Products, Inc.
Resinox	Phenolic resins	Monsanto Co.
Resistoflex	Poly(vinyl alcohol)	Resistoflex Corp.
Resloom	Melamine resins	Monsanto Co.
Resolite	Urea-formaldehyde resins	Ciba Geigy
Resorasabond	Resorcinol and phenol-resorcinol	Pacific Resins & Chemicals

Trade or brand name	Product	Manufacturer
Restfoam	Urethane foam	Stauffer Chemical
Restirolo	Polystyrene	Societa Italiana Resine
Rexolene	Cross-linked polyolefin	Brand-Rex Co.
Rexolite	Polystyrene	Brand-Rex Co.
Reynolon	Plastic films	Reynolds Metals Co.
Reynosol	Urethane, PVC	Hoover Ball & Bearing Co.
Rezimac	Alkyds	Commercial Solvents Corp.
Rezyl	Alkyd varnish	Sinclair-Koppers Co., Inc.
Rhodiod	Cellulose acetate	M & B Plastics, Ltd.
Rhonite	Resins for textile finishes	Rohm & Haas Co.
Rhoplex	Acrylic emulsions	Rohm & Haas Co.
Riblene	Polyethylene	ANIC, S.P.A.
Richfoam	Polyurethane foam	E. R. Carpenter Co.
Rigidex	Polyethylene	BP Chemicals (U.K.) Ltd.
Rigidite	Acrylic and polyester resins	American Cyanamid
Rigidsol	Rigid plastisol	Watson-Standard Co.
Rigolac	Polyester resins	Showa Highpolymer Co.
Rilsan	Nylon-11	Aquitaine-Organico
Rolox	Two-part epoxies	Hardman, Inc.
Roskydal	Urea-formaldehyde resins	Farbenfabriken Bayer AG
Royalex	Cellular thermoplastic sheets	Uniroyal
Royalite	Thermoplastic sheet material	Uniroyal, Inc.
Roylar	Polyurethanes	Uniroyal, Inc.
Rucoam	Vinyl materials	Hooker Chemical
Rucon	Poly(vinyl chloride)	Hooker Chemical Corp.
Rucothane	Polyurethanes	Hooker Chemical Corp.
Rulan	Flame-retardant plastic	E. I. du Pont de Nemours & Co., Inc.
Rulon	Flame retardant	E. I. du Pont de Nemours & Co., Inc.
Ryton	Poly(phenylene sulfide)	Phillips Petroleum Co.
Saflex	Poly(vinyl butyral)	Monsanto Co.
Safom	Polyurethane foam	Monsanto Co.
Santicizer	Plasticizers	Monsanto Co.

Trade or brand name	Product	Manufacturer
Santocel	Silica aerogel fillers	Monsanto Co.
Santocure	Accelerator	Monsanto Co.
Santoflex	Antioxidants	Monsanto Co.
Santolite	Sulfonamide resin	Monsanto Co.
Santonox	Antioxidant	Monsanto Co.
Saran	Poly(vinylidene chloride)	Dow Chemical Co.
Satin Foam	Extruded polystyrene foam	Dow Chemical
Scotch	Adhesives	Minnesota Mining & Mfg. Co.
Scotchcast	Epoxy resins	Minnesota Mining & Mfg. Co.
Scotchpak	Polyester film	Minnesota Mining & Mfg. Co.
Scotchpar	Polyester film	3M Co.
Scotchweld	Adhesives	Minnesota Mining & Mfg. Co.
Seilon	Thermoplastic sheets	Seiberling Rubber Co.
Selectron	Polyester resins	PPG Corp.
Selecttrofoam	Polyurethane foam	PPG Industries
Shareen	Nylon	Courtaulds North America
Shuvin	Vinyl molding material	Reichhold Chemicals
Silastic	Silicone materials	Dow Corning Corp.
Sipon	Alkyl and aryl resin	Alcolac, Inc.
Silastomer	Silicones	Midland Silicones Ltd.
Silbon	Rayon paper	Kohjin Co.
Silene	Calcium silicate	PPG Corp.
Silocet	Silicon rubber	ICI
Silvacon	Lignin extenders	Weyerhauser Co.
Siponate	Alkyl and aryl sulfonates	Alcolac, Inc.
Sirfen	Phenol-formaldehyde resins	Societa Italiana Resine
Sir-pel	Poromeric film	Georgia Bonded Fibers
Sirtene	Polyethylene	Societa Italiana Resine
Skinwich	Polyurethane integral-skinning foam	Upjohn
Softlite	Ionomer foam	Gilman Brothers
Solarflex	Chlorinated polyethylene	Pantasote Co.
Solithane	Urethane prepolymers	Thiokol Corp.
Sonite	Epoxy resin	Smooth-On, Inc.

Trade or brand name	Product	Manufacturer
Solvar	Poly(vinyl acetate)	Shawinigan Resins Corp.
Solvic	Poly(vinyl chloride)	Solvay & Cie
Soreflon	Polytetrafluoroethylene	Rhone-Poulenc
Spandal	Polyurethane laminates	Baxenden Chemical
Spandex	Polyurethane filaments	E. I. du Pont de Nemours & Co., Inc.
Spandofoam	Polyurethane foam	Baxenden Chemical
Spandoplast	Expanded polystyrene	Baxenden Chemical
Spectran	Polyester	Monsanto Textiles
Spenkel	Polyurethane resin	Textron Inc.
S-polymers	Butadiene-styrene copolymer	Esso Labs
Spraythane	Urethane resin	Thiokol Chemical Corp.
Staflex	Vinyl plasticizers	Reichhold Chemical, Inc.
Standlite	Phenol-formaldehyde resins	Hitachi Chemical Co.
Starex	Poly(vinyl acetate)	International Latex & Chemical Corp.
Statex	Carbon black	Columbian Carbon Co.
Structo-Foam	Foamed polystyrene slab	Stauffer Chemical Co.
Strux	Cellular cellulose	Aircraft Specialties
Sty-Grade	Degradable additive for polymers	Bio-Degradable Plastics, Inc.
Stylafoam	Coated polystyrene sheets	Gilman Brothers
Stymer	Styrene copolymer	Monsanto Co.
Stypol	Urea-formaldehyde resins	Freeman
Styrafil	Fiberglass-reinforced polystyrene	Dart Industries, Inc.
Styraglas	Fiberglass-reinforced polystyrene	Dart Industries, Inc.
Styrex	Resin	Dow Chemical Co.
Styrocel	Polystyrene (expandable)	Styrene Products Ltd.
Styroflex	Biaxially oriented polystyrene film	Natvar Corp.
Styrofoam	Extruded expanded polystyrene; foam	Dow Chemical Co.
Styrolus	Polystyrene	Westlake Plastics
Styron	Polystyrene	Dow Chemical Co.

Trade or brand name	Product	Manufacturer
Styronol	Polystyrene	Allied Resinous Prods.
Sulfasar	Fiberglass-reinforced polysulfone	Dart Industries
Sulfil	Fiberglass-reinforced polysulfone	Dart Industries
Sullvac	Acrylonitrile-butadiene-styrene copolymer	O'Sullivan Rubber Corp.
Sunlon	Nylon resins	Sun Chemical Corp.
Super Aeroflex	Linear PE	Anchor Plastic Co.
Super Coilife	Epoxy potting resin	Westinghouse Electric
Super Dylan	High-density polyethylene	Arco Polymer Co.
Superflex	Grafted high-impact polystyrene	Gordon Chemical
Superflow	Polystyrene	Gordon Chemical
Surflex	Ionomer film	Flex-O-Glass, Inc.
Surlyn	Ionomer resins	E. I. du Pont de Nemours & Co., Inc.
Syn-U-Tex	Urethane-formaldehyde and melamine-formaldehyde	Celanese Coatings Co.
Swedcast	Acrylic sheet	Swedlow, Inc.
Sylgard	Silicone casting resins	Dow Corning Corp.
Sylplast	Urea-formaldehyde resins	Sylvan Plastics, Inc.
Synpro	Metallic stearates	Dart Industries, Inc.
Syntex	Alkyd resins	Celanese Corp.
Synthane	Laminated plastic products	Synthane Corp.
Syretex	Styrenated alkyd resins	Celanese Coatings Co.
TanClad	Spray or dip plastisol	Tamite Inds.
TDI	Tolylene diisocyanate	E. I. du Pont de Nemours & Co., Inc.
Tedlar	Polyvinyl fluorocarbon resins	E. I. du Pont de Nemours & Co., Inc.
Teflon	Fluorocarbon resins	E. I. du Pont de Nemours & Co., Inc.
Teflon FEP	TFE copolymer	E. I. du Pont de Nemours & Co., Inc.
Teflon TFE	Polytetrafluoroethylene	E. I. du Pont de Nemours & Co., Inc.
Teglac	Alkyd coatings	American Cyanamid Co.
Tego	Phenolic resins	Rohm & Haas Co.

Trade or brand name	Product	Manufacturer
Tempra	Rayon fiber	American Enka Corp.
Tempreg	Low-pressure laminate	U.S. Plywood Corp.
Tenamene	Antioxidants	Eastman Kodak Co.
Tenite	Cellulose derivatives	Eastman Kodak Co.
Tenn Foam	Polyurethane foam	Morristown Foam Corp.
Tenox	Antioxidant	Eastman Chemical Products, Inc.
Teracol	Poly(oxytetramethylene glycol)	E. I. du Pont de Nemours & Co., Inc.
Tere-Cast	Polyester casting resins	Reichhold Chemicals
Terluran	ABS polymers	Badische Anilin & Soda-Fabrik AG
Terucello	Carboxymethyl cellulose	Showa Highpolymer Co.
Terylem	Poly(ethylene terephthalate)	ICI
Terylene	Polyester fiber	ICI
Tetran	Tetrafluoroethylene	Pennsalt Chemical Corp.
Tetra-Phen	Phenolic resins	Georgia-Pacific
Tetra-Ria	Amino resins	Georgia-Pacific
Tetraloy	Filled TFE molding resins	Whitford Chemical
Tetronic	Polyethers	Wyandotte Chemical Corp.
Texicote	Poly(vinyl acetate)	Scott Bader Co.
Texileather	Pyroxylin-leather cloth	General Tire & Rubber Co.
Texin	Urethane elastomer	Mobay Chemical Co.
Textolite	Laminated plastic	General Electric Co.
Thermaflow	Reinforced polyesters	Atlas Powder Co.
Thermalux	Polysulfones	Westlake Plastics
Thermasol	Vinyl plastisols and organosols	Lakeside Plastics
Thermax	Carbon black	Commercial Solvents Corp.
Thermco	Expanded polystyrene	Holland Plastics
Thiokol	Poly(ethylene sulfide)	Thiokol Corp.
Thornel	Graphite filaments	Union Carbide Corp.
Thurane	Polyurethane foam	Dow Chemical Co.
Tinuvin	Ultraviolet stabilizers	Geigy Industrial Chemicals, Div., Geigy Chemical Corp.

Trade or brand name	Product	Manufacturer
Ti-Pure	Titanium dioxide pigments	E. I. du Pont de Nemours & Co., Inc.
Titanox	Titanium dioxide pigments	Titanium Pigment Corp.
T-Lock	PVC sheets	Amercoat Corp.
Topel	Rayon fiber	Courtaulds
TPX	Poly-4-methylpentene-1	Imperial Chemical Industries, Ltd.
Trans-4	trans-1,4-Polybutadiene	Phillips Petroleum Co.
Tran-Stay	Polyester film	Transiwrap Co.
Trem	Viscosity depressant	Nopco Chemical Div., Diamond Shamrock Chemical Co.
Trevarno	Resin-impregnated cloth	Coast Mfg. & Supply Corp.
Tri-Foil	TFE-coated aluminum foil	Tri-Point Inds.
Trilon	TFE	Dynamit Nobel/America
Triocel	Rayon acetate	Celanese Fibers
Trithene	Trifluorochloroethylene	Union Carbide Corp.
Trolen	Polyethylene	Dynamit Nobel AG
Trolitan	Phenol-formaldehyde resins	Dynamite Nobel/America
Trosifol	Poly(vinyl butyral) film	Dynamite Nobel/America
Trosiplast	Poly(vinyl chloride)	Dynamit Nobel AG
Trulon	Poly(vinyl chloride) resin	Olin Corp.
Tuads	Accelerator	R. T. Vanderbilt Co.
Tuffak	Polycarbonates	Rohm & Haas
Tuftane	Polyurethane	B. F. Goodrich Chemical
Tusson	Rayon fiber	Beaunit Mills Corp.
Tybrene	ABS polymers	Dow Chemical Co.
Tygon	Vinyl copolymer	U.S. Stoneware Co.
Tylose	Cellulose ethers	Farbwerke Hoechst AG
Tynex	Nylon bristles and filaments	E. I. du Pont de Nemours & Co., Inc.
Tyril	Styrene-acrylonitrile copolymer	Dow Chemical Co.
Tyrilfoam	Styrene/Acrylonitrile foam	Dow Chemical Co.

Trade or brand name	Product	Manufacturer
Tyrin	Chlorinated polyethylene	Dow Chemical Co.
Ucon	Lubricants	Union Carbide Corp.
Udel	Plastic film	Union Carbide Corp.
Uformite	Urea resins	Rohm & Haas Co.
Ultramid	Nylons	Badische Anilin & Soda-Fabrik AG
Ultrathene	Finely divided polyolefins	National Distillers & Chemical Corp.
Ultrapas	Melamine-formaldehyde resins	Dynamit Nobel AG
Ultron	Vinyl film	Monsanto Co.
Ultryl	Poly(vinyl chloride)	Phillips Petroleum Co.
Unifoam	Polyurethane foam	William T. Burnett & Co.
Unipoxy	Epoxy resins and adhesives	Kristal Kraft Co.
Unitane	Titanium dioxide	American Cyanamid Co.
Unox	Epoxides	Union Carbide Corp.
Updown	Polychloroprene foam	
Urac	Urea-formaldehyde resins	American Cyanamid Co.
Urafil	Fiberglass-reinforced polyurethane	Dart Industries
Uraglas	Fiberglass-reinforced polyurethane	Dart Industries
Uralite	Polyurethanes	Hexcel Corp.
Urapol	Polyurethane elastomeric coatings	Gordon Chemicals
Urapac	Rigid polyurethane	North American Urethanes
Urapol	Polyurethane elastomeric coatings	Poly Resins
Urecoll	Urea-formaldehyde resins	BASF Corp.
Uscolite	ABS copolymer	U.S. Rubber Co.
U-Thane	Rigid insulation polyurethane	Upjohn
Uvex	Cellulose acetate butyrate	Eastman Kodak Co.
Uvinul series	Ultraviolet light absorbers	General Aniline & Film Corp.

Trade or brand name	Product	Manufacturer
Valox	Poly(butylene terephthalate)	General Electric Co.
Valsof	PE emulsions	United Merchants & Manfs.
Vanstay	Stabilizers	R. T. Vanderbilt Co.
Varcum	Phenolic resins	Reichhold Chemicals, Inc.
Varex	Polyester resins	McClosky Varnish Co.
Varkyd	Alkyd and modified alkyd resins	McCloskey Varnish Co.
Varsil	Silicon-coated fiberglass	New Jersey Wood Finishing Co.
Vazo	Azobisisobutyronitrile	E. I. du Pont de Nemours & Co., Inc.
V del	Polysulfone resins	Union Carbide Corp.
Vectra	PP fibers	Exxon Chemical
Velene	Polystyrene-foam laminates	Scott Paper Co.
Velon	Poly(vinyl chloride)	Firestone Tire & Rubber Co.
Verel	Modacrylic staple fibers	Eastman Chemical Products, Inc.
Versamid	Polyamide resins	General Mills, Inc.
Versel	Polyester thermoplastic	Allied Chemical Corp.
Versi-Ply	Co-extruded film	Pearson Inds.
Vespel	Polymellitimide	E. I. du Pont de Nemours & Co., Inc.
Vestamid	Nylon-12	Chemische Werke Huls AG
Vestolit	Poly(vinyl chloride)	Chemische Werke Huls AG
Vestyron	Polystyrene	Chemische Werke Huls AG
VGB	Acetaldehyde-aniline accelerator	Uniroyal Corp.
Vibrathane	Polyurethane intermediates	Uniroyal Corp.
Vibrin	Polyester resins	Uniroyal Corp.
Vibrin-Mat	Polyester-glass molding material	W. R. Grace
Vibro-Flo	Epoxy and polyester coating powders	Armstrong Products
Vicara	Protein fiber	Virginia-Caroline Chem. Corp.

Trade or brand name	Product	Manufacturer
Viclan	Poly(vinylidene chloride)	Imperial Chemical Industries, Ltd.
Vicron	Fine calcium carbonate	Pfizer Minerals, Pigments & Metals
Videne	Polyester film	Goodyear Tire & Rubber Co.
Vinac	Poly(vinyl acetate) emulsions	Air Reduction Co.
Vinapas	Poly(vinyl acetate)	Wacker Chemie GmbH
Vinidur	Poly(vinyl chloride)	BASF Corp.
Vinoflex	Poly(vinyl chloride)	BASF Corp.
Vinol	Poly(vinyl alcohol)	Air Reduction Co.
Vinsil	Rosin derivative	Hercules, Inc.
Vinylite	Poly(vinyl chloride-co-vinyl acetate)	Union Carbide Corp.
Vinyon	Poly(vinyl chloride-co-acrylonitrile)	Union Carbide Corp.
Vipla	Poly(vinyl chloride)	Montecatini Edison S.p.A.
Viscalon	Rayon fiber	American Enka Corp.
Viskon	Nonwoven fabrics	Union Carbide Corp.
Vistanex	Polyisobutylene	Enjay Chemical Co.
Vitel	Polyester resins	Goodyear Tire & Rubber Co.
Vithane	Polyurethanes	Goodyear Tire & Rubber Co.
Viton	Copolymer of vinylidene fluoride and hexafluoropropylene	E. I. du Pont de Nemours & Co., Inc.
Vituf	Polyester resins	Goodyear Tire & Rubber
Volara	Closed-cell LDPE foam	Voltek, Inc.
Volaron	Closed-cell LDPE foam	Voltek, Inc.
Volasta	Closed-cell medium density PE foam	Voltek, Inc.
Voranol	Polyurethane resins	Dow Chemical
Vulcaprene	Polyurethane	Imperial Chemical Industries, Ltd.
Vulkollan	Urethane elastomer	Mobay Chemical Co.
Vult-Acet	PVA latexes	General Latex & Chemical
Vultafoam	Polyurethane foam	General Latex & Chemical

Trade or brand name	Product	Manufacturer
Vultathane	Polyurethane coatings	General Latex & Chemical
Vybak	Poly(vinyl chloride)	Bakelite Xylonite Ltd.
Vycron	Polyester fiber	Beaunit Mills Corp.
Vydax	Release agent	E. I. du Pont de Nemours & Co., Inc.
Vydyne	Nylon resins	Monsanto Co.
Vygen	Poly(vinyl chloride)	General Tire & Rubber Co.
Vynaclor	PVC emulsions	B. F. Goodrich Chemical
Vynaloy	Vinyl sheets	B. F. Goodrich Chemical
Vynex	Rigid vinyl sheeting	Nixon-Baldwin Chemicals, Inc.
Vyram	Rigid poly(vinyl chloride)	Monsanto Co.
Vyrene	Spandex fiber	U.S. Rubber Co.
Webril	Nonwoven fabric	The Kendall Co.
Weldfast	Epoxy and polyester adhesives	Fibercast Co.
Wellamid	Nylon-66 and -6 molding resins	Wellman, Inc.
Welvic	Poly(vinyl chloride)	Imperial Chemical Industries, Ltd.
Whirlclad	Plastic coatings	The Polymer Corp.
Whirlsint	Powdered polymers	The Polymer Group
Whitcon	Fluoroplastic lubricants	Whitford Chemical Corp.
Wicaloid	Styrene/Butadiene emulsions	Ott Chemical Co.
Wicaset	PVC emulsions	Ott Chemical Co.
Wilfex	Vinyl plastisols	Flexible Products Co.
Wing-stay	Alkylated phenol antioxidants	Goodyear Tire & Rubber Co.
Wintrol	Retarders	Stepan Chemical Co.
XT Polymer	Acrylics	American Cynamid Co.
Xylon	Nylon-66 and Nylon-6	Dart Industries
Xylonite	Cellulose nitrate	B. X. Plastics, Inc.
Zantrel	Rayon fiber	American Enka Corp.
Zee	Polyethylene wrap	Crown Zellerbach Corp.
Zefran	Acrylic fiber	Dow Chemical Co.
Zelan	Water repellent	E. I. du Pont de Nemours & Co., Inc.

Trade or brand name	Product	Manufacturer
Zelec	Lubricant and release agent	E. I. du Pont de Nemours & Co., Inc.
Zelux	PE films	Union Carbide Corp.
Zendel	Polyethylene	Union Carbide Corp.
Zerlon	Acrylic/Styrene copolymer	Dow Chemical
Zerok	Protective coatings	Atlas Minerals & Chemicals Div.
Zetafax	Poly(ethylene-co-acrylic acid)	Dow Chemical Co.
Zetafin	Poly(ethylene-co-ethyl acrylate)	Dow Chemical Co.
Zytel series	Nylons	E. I. du Pont de Nemours & Co., Inc.

Appendix C: Sources of Laboratory Exercises

The Education Committees of the Divisions of Polymer Chemistry and Polymeric Materials: Science and Engineering strongly advise that laboratory experiences illustrating principles presented in the lecture material be included in introductory courses of polymer chemistry. The extent and type of these laboratory experiences will vary from teacher to teacher and from course to course and may include lecture demonstrations, group experiments, and individual laboratory exercises.

There is no single, dominating, polymer laboratory textbook. A listing of recent books written specifically as laboratory manuals follows. This is followed by a list, divided into several categories, of *Journal of Chemical Education* articles related to polymers. To minimize problems, all exercises should emphasize safety-related aspects and should be performed by the instructor before asking the students to perform them.

Teachers just beginning to master polymer lecture and laboratory are encouraged initially to emphasize the lecture portion by utilizing simple exercises demonstrating solution of polymers, increase in viscosity of dilute polymer solutions, and synthesis of a condensation and a vinyl polymer. Help should be available from surrounding chemical industry personnel involved with polymers or from a nearby school employing a person experienced with polymers. Such associations can become mutually beneficial.

Again, the *safety* and *toxicological* aspects associated with each exercise must be stressed. Some of the monomers may be quite toxic, but it should be emphasized to students that the resulting polymers are *not* typically toxic.

LABORATORY MANUALS

1. D. Braun, H. Cherdron, and W. Kern, *Techniques of Polymer Synthesis and Characterization,* Wiley-Interscience, New York, 1972.
2. E. A. Collins, J. Bares, and F. W. Billmeyer, *Experiments in Polymer Science,* Wiley-Interscience, New York, 1973.
3. E. M. McCaffery, *Laboratory Preparation for Macromolecular Chemistry,* McGraw-Hill, New York, 1970.
4. W. R. Sorenson and T. W. Campbell, *Preparative Methods of Polymer Chemistry,* 2nd ed., Wiley-Interscience, New York, 1968.

JOURNAL OF CHEMICAL EDUCATION ARTICLES

General

1. J. Benson, Viscometric determination of the isoelectronic point of a protein, *40*:468 (1963).
2. F. Rodriguez, Simple models for polymer stereochemistry, *45*:507 (1968).
3. H. Kaye, Disposable models for the demonstration of configuration and conformation of vinyl polymers, *48*:201 (1971).
4. I. Nicholson, Disposable macromolecular model kits, *46*:671 (1969).
5. P. H. Mazzocchi, Demonstration-ordered polymers, *50*:505 (1973).
6. C. E. Carraher, Polymer models, *47*:581 (1970).
7. F. Rodriguez, Demonstrating rubber elasticity, *50*:764 (1973).
8. D. Napper, Conformation of macromolecules, *67*:305 (1969).
9. D. Smith and J. Raymonda, Polymer molecular weight distribution, *49*:577 (1972).
10. F. Billmeyer, P. Geil, and K. van der Weg, Growth and observation of spherulites in polyethylene, *37*:206 (1960).
11. P. Morgan, Models for linear polymers, *37*:206 (1960).
12. H. Hayman, Models illustrating the helix-coil transition in polypeptides, *41*:561 (1964).
13. G. Gorin, Models of the polypeptide α-helix and of protein molecules, *41*:44 (1964).
14. H. Pollard, Polyethylene and pipecleaner models of biological polymers, *43*:327 (1966).
15. W. Van Doorne, J. Kuipers, and W. Hoekstra, A computer program for the distribution of end-to-end distances in polymer molecules, *53*:353 (1976).
16. R. Seymour and G. A. Stahl, Plastics, separation of waste. An experiment in solvent fractionation, *53*:653 (1976).
17. T. L. Daines and K. W. Morse, The chemistry involved in the preparation of a paint pigment. An experiment for the freshman laboratory, *53*:117 (1976).
18. S. S. Taylor and J.E. Dixon, Affinity chromatography of lactate dehydrogenase. A biochemistry experiment, *55*:675 (1978).
19. S. Krause, Macromolecular solutions as an integral part of beginning physical chemistry, *55*:174 (1978).
20. G. A. Krulik, Electroless plating of plastics, *55*:361 (1978).
21. M. Gorodetsky, Electroplating of polyethylene, *55*:66 (1978).

22. W. C. Penker, Recycling disposable plastics for laboratory use, *54*:245 (1977).
23. S. D. Daubert and S. F. Sontum, Computer simulation of the determination of amino acid sequences in polypeptides, *54*:35 (1977).
24. G. E. Dirreen and B. Z. Shakhashiri, The preparation of polyurethane foam: A lecture demonstration, *54*:431 (1977).
25. Optical rotation and the DNA helix-to-coil transition. An undergraduate project, *51*:591 (1974).
26. A. Factor, The chemistry of polymer burning and flame retardance, *51*:453 (1974).
27. M. B. Hocking and G. W. Rayner Canham, Polyurethane foam demonstrations: The unappreciated toxicity of toluene-2,4-diisocyanate, *51*:A580 (1974).
28. R. Har-zri and J. T. Wittes, Calculation of the number of cis-trans isomers in a "symmetric" polyene, *52*:545 (1975).
29. G. A. Hiegel, A simple model of an α-helix, *52*:231 (1975).
30. B. Morelli and L. Lampugnani, Ion-exchange resins—A Simple apparatus, *52*:572 (1975).
31. M. E. Mrvosh and K. E. Daugherty, The low cost construction of inorganic polymer models using polyurethane, *52*:239 (1975).
32. E. J. Barrett, Biopolymer models of nucleic acids, *52*:168 (1979).
33. W. D. Wilson and M. W. Davidson, Isolation and characterization of bacterial DNA: A project-oriented laboratory in physical biochemistry, *56*:204 (1979).
34. D. E. Powers, W. C. Harris, and V. F. Kalasinsky, Laboratory automation in the undergraduate curriculum: Determination of polyethylene chain branching by computerized IR methods, *56*:128 (1979).
35. R. E. Baudreau, A. Heaney, and D. L. Weller, A sedimentation experiment using a preparative ultracentrifuge, *52*:128 (1975).
36. M. W. Davidson and W. D. Wilson, Stand polarity: Antiparalleled molecular interactions in nucleic acids, *52*:323 (1975).
37. P. Ander, An introduction to polyelectrolytes via the physical chemistry laboratory, *56*:481 (1979).
38. C. E. Carraher, Resistivity measurements, *54*:576 (1977).
39. C. Arends, stress-strain behavior of rubber, *37*:41 (1960).
40. C. Carraher, Reaction vessel with stirring and atmosphere controls, *46*:314 (1969).
41. K. Chapman and J. Fleming, Field test evaluation report on introduction to polymer chemistry, *58*:904 (1981).
42. D. Jewett and J. Lawless, A method for applying a permanent teflon coating to ground glass joints, *58*:903 (1981).
43. S. Butler and S. Malott, Textile chemistry for the artist, *58*:295 (1981).
44. A. Werner, Synthetic materials in art conservation, *58*:321 (1981).
45. H. Wittcoff, Benzene—The polymer former, *58*:270 (1981).
46. K. Kramer, An aid to molecular sequence studies: Use of ceramic magnets to visualize sequences of peptides, proteins, and nucleic acids, *58*:72 (1981).
47. J. Silverman, Radiation processing: The industrial applications of radiation chemistry, *58*:168 (1981).

48. C. M. Melliar-Smith, Optical fibers and solid state chemistry, *57*:574 (1980).
49. J. Webb and F. Van Bockxmeer, Transferrins—Illustrative metal-binding proteins, *57*:689 (1980).
50. L. Petrakis and D. W. Grandy, Coal analysis, characterization and petrography, *57*:689 (1980).
51. H. Morita, Peat and its organic chemistry, *57*:695 (1980).
52. H. Wittcoff, Propylene—A basis for creative chemistry, *57*:707 (1980).
53. C. Carraher and R. Deanin, Core curriculum in introductory courses of polymer chemistry, *57*:436 (1980).
54. P. Sattsangi, S. Sattsangi, and H. Grosman, Isolation of soybean agglutinin (SBA) from soy meal, *59*:977 (1982).
55. L. H. Bowman, Classroom participation—Introduction to polymer science, *58*:548 (1981).
56. N. Senozan and R. Hunt, Hemoglobin: Its occurrence, structure and adaptation, *59*:173 (1982).
57. R. Tiede, Glass fibers—Are they the solution? *59*:198 (1982).
58. R. B. Seymour, Recommended ACS syllabus for introductory courses in polymer chemistry, *59*:652 (1982).
59. M. Coleman and W. Varnell, A computational method designed to aid in the teaching of copolymer composition and microstructure as a function of conversion, *59*:847 (1982).
60. H. Mark, Polymer chemistry in Europe and America—How it all began, *58*:527 (1981).
61. C. S. Marvel, The development of polymer chemistry in America—The early years, *58*:535 (1981).
62. R. Quirk, Stereochemistry and macromolecules: Principles and applications, *58*:540 (1981).
63. F. Harris, Introduction to polymer chemistry, *58*:837 (1981).
64. J. McGrath, Chain reaction polymerization, *58*:844 (1981).
65. J. K. Stille, Step-growth polymerization, *58*:862 (1981).
66. T. Ward, Molecular weight and molecular weight distributions in synthetic polymers, *58*:867 (1981).
67. P. H. Geil, The morphology of crystalline polymers, *58*:879 (1981).
68. P. Geil and S. Carr, EMMSE: Education modules for materials science and engineering, *58*:908 (1981).
69. W. Mattice, Macromolecules in undergraduate physical chemistry, *58*:911 (1981).
70. J. McGrath, Block and graft copolymers, *58*:914 (1981).
71. C. Carraher, Organometallic polymers, *58*:921 (1981).
72. J. Preston, High-strength/High-modulus fibers from aromatic polymers, *58*:935 (1981).
73. C. Bazuin and A. Eisenberg, Ion-containing polymers: Ionomers, *58*:938 (1981).
74. O. Olabisi, Interpretations of polymer-polymer miscibility, *58*:944 (1981).
75. P. E. Cassidy, Polymers for extreme service conditions, *58*:951 (1981).
76. J. Gittith, The chemistry of coatings, *58*:956 (1981).

77. P. Dubin, A polymer-solution "thermometer": A demonstration of the thermodynamic consequences of specific polymer-solfent interactions, *58*:866 (1981).

78. F. D. Williams, POLYMERLAB: A computer-generated problem, *60*:45 (1983).

79. J. J. Eisch and Karl Ziegler, Master advocate for the unity of pure and applied research, *60*:1009 (1983).

80. L. Mathias, The laboratory for introductory polymer courses, *60*:990 (1983).

81. C. Carraher, R. Seymour, E. Pearce, G. Donaruma, N. Miller, C. Gebelein, L. Sperling, F. Rodriguez, G. Kirshenbaum, R. Ottenbrite, R. Hester, and B. Bulkin, Polymer core course committees, *60*:971 (1983).

82. C. Carraher, J. Campbell, M. Hanson, C. Schnildknecht, S. Israel, N. Miller, and E. Hellmuth, Polymer chemistry for introductory general chemistry courses, *60*:973 (1983).

83. J. Jeffreys, Drawing a proton alpha-helix, *60*:549 (1983).

84. G. Patel and N.-L. Yang, Polydiacetylenes: An ideal color system for teaching polymer science, *60*:181 (1983).

85. A. Frimer, The semipermeability of biological membranes: An undergraduate laboratory experiment, *62*:89 (1985).

86. S. Batra, Polyester—Making bees and other innovative insect chemists, *62*: 121 (1985).

87. E. Levy and T. Wampler, Identification and differentiation of synthetic polymers by pyrolysis capillary gas chromatography, *63*:A64 (1986).

88. L. Sperling, C. Carraher, L. Mandelkern, H. Coker, R. Blumenstein, F. Fowkes, E. Hellmuth, D. Karl, J. E. Mark, W. Mattice, F. Rodriguez, C. Rogers, and R. Stein, Polymer principles in the undergraduate physical chemistry course, Part I, *62*:780 (1985); Part II, *62*:1030 (1985).

89. E. Caldwell and M. Eftinck, Preparation and characterization of myosin proteins, *62*:900 (1983).

90. I. Bertini, C. Luchinat, and R. Monnanni, Zinc enzymes, *62*:924 (1985).

91. J. Groves, Key elements of the chemistry of cytochrome P-450, *62*:928 (1985).

92. I. Lulav and D. Samuel, The chemistry of polymers, proteins, and nucleic acids: A short course on macromolecules for secondary schools, *62*:1075 (1985).

93. F. Rodriguez, L. H. Sperling, R. Cohen, D. R. Paul, N. Peppas, S. Rosen, M. T. Shaw, and M. Tirrell, Polymer principles for the chemical engineering curriculum, *62*:1079 (1985).

94. J. Dunach, Polymerization evaluation by spectrophotometric measurements, *62*:450 (1985).

95. D. Balasubramanian and B. Chandani, Poly(ethylene glycol): A poor chemist's crown, *60*:77 (1983).

96. A. Horta, A course on macromolecules, *62*:286 (1985).

97. A. Blumberg, Membranes and films from polymers, *63*:414 (1986).

98. C. Carraher and R. Seymour, Physical aspects of polymer structure: A dictionary of terms, *63*:418 (1986).

99. J. A. Campbell, Paper: A modified natural polymer, *63*:420 (1986).

100. J. Cleary, Diapers and polymers, *63*:422 (1986).

101. L. Mathias and R. Storey, Polymer science in a governor's school: Teaching an advanced topic to gifted high school students, *63*:424 (1986).
102. J. Robyt, Graphic representation of oligosaccharide and polysaccharide structures containing hexopyranose units, *63*:560 (1986).
103. W. Scovell, Supercoiled DNA, *63*:562 (1986).
104. F. Mayo, The evolution of free radical chemistry at Chicago, *63*:97 (1986).
105. C. Walling, The development of free radical chemistry, *63*:99 (1986).
106. B. Goodall, The history and current state of the art of propylene polymerization catalysts, *63*:191 (1986).
107. F. H. Juergens, A. R. Ellis, G. H. Dieckmann, and R. I. Perkins, Levitating a magnet using a superconductive material. *64*:851 (1987).
108. R. Baker and J. C. Thompson, Superconductive powder, A simple demonstration of high T_c, *64*:853 (1987).
109. D. C. Harris, M. E. Hills, and T. A. Hewston, Superconductivity in $TBa_2Cu_3O_{8-x}$, preparation, iodometric analysis and classroom demonstration, *64*:847 (1987).
110. A. B. Ellis, Superconductors: Better levitation through chemistry, *64*:836 (1987).
111. F. A. Matsen, Three theories of superconductivity, *64*:842 (1987).
112. W. C. Hoyt, Analogy for polymers, freight trains, *65*:718 (1988).
113. B. J. Wuensch, Materials scientists and engineers, The teaching of crystallography to (SYMP), *65*:494 (1988).
114. S. R. Turner and R. C. Daly, Polymers in microlithograhy, *64*:322 (1988).
115. S. F. Russo, Storage of polyacrylamide gels, *65*:370 (1988).
116. H. Mark, Polymer science and engineering—facts and trends, *65*:334 (1988).
117. R. B. Seymour, Polymers are everywhere, *65*:327 (1988).
118. W. S. Richardson, III and L. Burns, HPLC of the polypeptides in a hydrolysate of egg-white lysozyme: An experiment for the undergraduate biochemistry laboratory, *65*:162 (1988).
119. G. B. Kauffman, Products of chemistry. Wallace Hume Carothers and nylon, the first completely synthetic fiber, *65*:803 (1988).
120. J. Estelrich and R. Pouplana, The purification of a blood group A glycoprotein: An affinity chromatography experiment, *65*:556 (1988).
121. E. M. Ross, Receptor-G protein-effector: The design of a biochemical switchboard, *65*:937 (1988).
122. K. A. Parson, Recombinant DNA technology: A topics course for undergraduates, *65*:325 (1988).
123. C. L. Evans and F. J. Torre, SDS-polyacrylamide gel electropheresis of snake venoms, *65*:1011 (1988).
124. C. E. Carraher, Jr. and R. B. Seymour, Structure-organic aspects (definitions) polymer definitions, *65*:314 (1988).
125. A. T. Jacob and A. B. Ellis, A double-decker levitation experiment using a sandwich of superconductors, *65*:1094 (1988).
126. W. H. Chan, K. S. Lam, and W. K. Yu, Antioxidants in plastic: An instrumental analysis project, *66*:172 (1989).
127. C. M. Lovett, Jr., T. N. Fitzgibbon, and R. Chang, Effect of UV irradiation on DNA as studied by its thermal denaturation, *66*:526 (1989).
128. F. J. Waller, Fluoropolymers: Part I, *66*:487 (1989).

129. J. W. Diehl, Ionomers, *66*:901 (1989).
130. D. M. Snyder, An overview of oriental lacquer: Art and chemistry of the original high-tech coating, *66*:977 (1989).
131. G. Schuerman and R. Bruzan, Chemistry of (PROD) paint, *66*:327 (1989).
132. H. A. Carter, The acidity of paper, Chemistry in the comics: Part 3, *66*:883 (1989).
133. T. Abraham and F. Rodriquez, Polymer principles (LTE), *66*:790 (1989).
134. D. Fain, Polystyrene film for IR spectroscopy, zero-cost, *66*:171 (1989).
135. R. B. Seymour, Alkanes: Abundant, pervasive, important, and essential, *66*:59 (1989).
136. R. B. Seymour, Alkenes and their derivatives: The alchemists dream come true, *66*:670 (1989).
137. M. J. Minch, The UOP experience, Are high school students ready for recombinant DNA?, *66*:64 (1989).
138. F. Rodriguez, Demonstrations of polymer principles: IV. Mechanical properties, classroom, *67*:784 (1990).
139. G. B. Kauffman, S. W. Mason, and R. B. Seymour, Polynorbornene, happy and unhappy balls: Neoprene (PROD), *67*:198 (1990).
140. T. M. Letcher and N. S. Jutseke, A closer look at cotton, rayon and polyester fibers, *67*:361 (1990).
141. M. P. Tarazona and E. Saiz, Happy polymer party, *67*:238 (1990).
142. R. B. Seymour and G. B. Kauffman, Piezoelectric polymers: Direct converters of work to electricity, *67*:763 (1990).
143. D. Sahu, A. Langner, and T. F. George, A guide to computational methods in superconductivity theory, BCS primer, *67*:738 (1990).
144. B. Riedl and P. D. Kamdem, Some aspects of chemistry teaching and research in wood science, *67*:543 (1990).
145. R. B. Seymour, Chemicals in everyday life, *64*:63 (1987).
146. R. R. Sinden, DNA: Biological significance, supercoiled (CB), *64*:294 (1987).
147. A. C. Wilbraham, DNA model for every student, *64*:806 (1987).
148. A. Feigenbaum and D. Scholler, Expanded polystyrene: An experiment for undergraduate students, *64*:810 (1987).
149. D. R. Burfield, Polymer glass transition temperatures, *64*:875 (1987).
150. S. F. Wang and S. J. Grossman, Plastic materials for insulating applications, *64*:39 (1987).
151. W. H. Chan, Plasticizers in PVC: A combined IR and GC approach, *64*:897 (1987).
152. R. D. Deanin, Chemistry of plastics, *64*:45 (1987).
153. C. E. Carraher, Jr., G. Hess, and L. H. Sperling, Polymer nomenclature—or what's in a name?, *64:*36 (1987).
154. M. Sherman, Polymers, polymers, everywhere!: A workshop for pre-high school teachers and students, *64*:868 (1987).
155. C. E. Carraher, Jr. and R. B. Seymour, Properties and testing: Definitions, polymer definitions, *64*:866 (1987).
156. J. Allan, Fiber identification: A colorful experiment for all ages (CK), *67*:256 (1990).

157. T. M. Letcher and N. S. Lutseke, Fibers, A closer look at cotton, rayon, and polyester, *67*:361 (1990).

158. H. J. M. Bowen, Identification of polymers in university class experiments, *67*:75 (1990).

159. S. F. Russo and J. U. Dahlberg, Immunoassay for human transferrin, an enzyme, *67*:175 (1990).

160. N. K. Mathur and C. K. Narang, Chitin and chitosan, versatile polysaccharides from marine animals, *67*:938 (1990).

161. G. B. Kauffman and R. B. Seymour, Elastomers: I, natural rubber, *67*:422 (1990).

162. D. Weller and P. Gariepy, Field inversion of agarose gel electrophoresis of DNA, *68*:81 (1991).

163. S. Farrell and D. Choo, A versatile and inexpensive enzyme purification experiment for undergraduate biochemistry labs, *68*:692 (1991).

164. S. Farrell, L. Farrell and L. Dircks, You too can be a molecular biologist: Basic cloning experiment for the undergraduate biochemistry lab, *68*:707 (1991).

165. M. Vestling, Insulin: HPLC mapping of protease digestion products: A biochemistry laboratory experiment, *68*:958 (1991).

166. H. Sommerfeld and R. Blume, Biodegradable films: Based on partially hydrolyzed corn starch or potato starch, *69*:A151 (1992).

167. J. P. Droske, Incorporating polymeric materials topics into the undergraduate chemistry core curriculum, *69*:1014 (1992).

168. S. C. Kohlwein, Biological membranes: Function and assembly, *69*:3 (1992).

169. J. McHale, R. Schaeffer, and R. Salomon, Demonstrating high temperature superconductivity in the chemistry lab through the Meissner effect, *69*:1031 (1992).

170. R. Seymour and G. Kauffman, Elastomers: III. Thermoplastic elastomers, *69*:967 (1992).

171. R. Seymour and G. Kauffman, Polymer blends: Superior products from inferior materials, *69*:646 (1992).

172. R. Seymour and G. Kauffman, Formaldehyde: A simple compound with many uses, *69*:457 (1992).

173. R. Seymour and G. Kauffman, The rise and fall of celluloid, *69*:311 (1992).

174. R. Lewis, M. Choquette, E. Darden, M. Gilbert, D. Martinez, C. Myhaver, K. Schlichter, R. Woudenberg, and K. Zawistowski, Interfacial polymerizations: microscale polymer laboratory experiments for undergraduate students, *69*:A215 (1992).

175. F. Rodriquez, Classroom demonstrations of polymer principles, Part V. Polymer fabrication, *69*:915 (1992).

176. M. Rodriquez and W. Flurkey, A biochemistry project to study mushroom tyrosinase: Enzyme localization, isoenzymes, and detergent activation, *69*:767 (1992).

177. M. Sherman, Producing fibers of poly(vinyl alcohol): An alternative method, *69*:883 (1992).

178. R. Seymour and G. Kauffman, Products of chemistry: Polyurethanes: A close of modern, versatile materials, *69*:909 (1992).

179. U. Pindur, M. Haber, and K. Sattler, Antitumor active drugs as intercalators of deoxyribonucleic acid: Molecular models of intercalation complexes, *70*: 263 (1993).
180. M. P. Stevens, Polymer additives. I. Mechanical property modifiers, *70*:444 (1993); II. Chemical and aesthetic property modifiers, *70*:535 (1993); III. Surface property and processing modifiers, *70*:713 (1993).
181. J. Fortman, Pictorial analogies. V. Polymers, *70*:403 (1993).
182. L. Woodward and M. A. Bernard, Evening polymer programs to pique the interests of youngsters and adults, *70*:1006 (1993).
183. C. Vijayan and M. Ravikumar, Computer series—Fractal nature of polymer conformation, *70*:830 (1993).
184. A. Levinson, Identifying plastics by density, *70*:174 (1993).
185. K. Kolb and D. Kolb, Identifying plastics by density, *70*:174 (1993).
186. K. Lau, K. Wong, S. Yeung, and F. Chau, Fiber optic sensors: Their principles and fabrication, *70*:336 (1993).
187. G. Kauffman, The first semi-synthetic fiber product, Rayon, *70*:887 (1993).
188. G. Stroebel, J. Whitesell, and R. Kriegel, Fibers: An improved method, slime and poly(vinyl alcohol), *70*:893 (1993).
189. R. Seymour and G. Kauffman, The ubiquity and longevity of fibers, *70*:449 (1993).
190. R. Bruzan and D. Baker, Plastic density determination by titration, *70*:397 (1993).
191. O. Teleman, B. Jonsson, and S. Engstrom, LUSim—A Macintosh program for simulation of molecular and polymer systems, *70*:641 (1993).
192. A. Blumberg, Identifying polymers through combustion and density, *70*:339 (1993).
193. R. Utecht, A kinetic study of yeast alcohol dehydrogenase, *71*:346 (1994).
194. P. Weber and D. Buck, A fast and simple method for the determination of the amino acid composition of proteins, capillary electrophoresis, *71*:609 (1994).
195. K. R. Williams, Analysis of ethylene-vinyl acetate copolymers: A combined TGA/FTIR experiment, *71*:A195 (1994).
196. D. Katz, A bag of slime, a novel lab procedure, *71*:891 (1994).
197. R. Seymour and G. Kauffman, Chemical magic: Polymers from a nonexistent monomer, *71*:582 (1994).
198. P. Rasmussen, Combustion toxicity and flammability of polymeric materials, *71*:809 (1994).
199. K. Strothkamp and R. Strothkamp, Fluorescence measurements of ethidium binding to DNA, *71*:77 (1994).
200. J. P. Byrne, Rubber elasticity: A simple method for measurement of thermodynamic properties, *71*:531 (1994).
201. J. S. Anderson, Polymers and material science: A course for nonscience majors, *71*:1044 (1994).
202. F. Marino, Use of tangle links to show globular protein structure, *71*:741 (1994).
203. N. Glickstein, The wonder in spider thread chemistry, *71*:948 (1994).
204. W. Peng and B. Riedl, Thermosetting resins, *72*:587 (1995).
205. J. Meister, Polymer literature and samples for classroom use, *72*:593 (1995).

Synthesis

1. A. Silkha, M. Albeck, and M. Frankel, Anionic polymerization of vinyl monomers, *35*:345 (1958).
2. P. Ander, Dependence of molecular weight of polystyrene on initiator concentration, *47*:233 (1970).
3. D. Armitage, M. Hughes, and A. Sindern, Preparation of "bouncing putty," *50*:434 (1973).
4. W. Rose, Preparation of terephthaloyl chloride. Prelude to ersatz nylon, *44*: 283 (1967).
5. S. Wilen, C. Kemer, and I. Waltcher, Polystyrene—A multistep synthesis, *38*:304 (1961).
6. E. Senogles and L. Woolf, Polymerization kinetics. Dead-end radical polymerization, *44*:157 (1967).
7. J. Bradbury, Polymerization kinetics and viscometric characterization of polystyrene, *40*:465 (1963).
8. W. R. Sorenson, Polymer synthesis in the undergraduate organic laboratory, *42*:8 (1965).
9. C. E. Carraher, Synthesis of poly(β-alanine), *55*:668 (1978).
10. G. Ceska, Emulsion polymerization and film formation of dispersed polymeric particles, *50*:767 (1973).
11. E. McCaffery, Kinetics of condensation polymerization, *46*:59 (1969).
12. C. Carraher, Synthesis of furfuryl alcohol and furoic acid, *55*:269 (1978).
13. P. Morgan and S. Kwolek, The nylon rope trick, *36*:182 (1959).
14. C. E. Carraher, Synthesis of caprolactam and nylon 6, *55*:51 (1978).
15. G. R. Pettit and G. R. Pettit, III, Preparation of a polysulfide rubber, *55*:472 (1978).
16. A. S. Wilson and V. R. Petersen, Bakelite demonstration: A safer procedure, *55*:652 (1978).
17. M. Morcellet, J. Morcellet, M. Delporte, and J. Esterez, Synthesis and characterization of vinyl pyridine styrene copolymers, *55*:22 (1978).
18. J. Siberman and R. Silberman, Synthesis of stabilized semi-crystalline polymer foam, *55*:797 (1978).
19. R. J. Mazza, Free radical polymerization of styrene. A radiotracer experiment, *52*:476 (1975).
20. D. E. Kranbuehl, T. V. Harris, A. K. Howe, and D. W. Thompson, Organometallic catalyzed synthesis and characterization of polyethylene. An advanced laboratory experiment, *52*:261 (1975).
21. M. N. Iskander and P. A. Jones, Peptide synthesizer, synthesis of an enkephalinase inhibitor, Gly-Gly-Phe-Leu, by continuous-flow, *67*:170 (1990).
22. R. J. Palma, Sr. and E. Sargent, A revision of a general chemistry experiment on an inorganic polymerization-condensation reaction, *67*:614 (1990).
23. Y. Bedard and B. Riedl, Synthesis of a phenol-formaldehyde thermosetting polymer, *67*:977 (1990).
24. A. J. Buglass and J. W. Waterhouse, Polymer-supported oxidation reactions: Two contrasting experiments for the undergraduate laboratory, *64*:371 (1987).

25. D. Villemin, Polymerization of norbornene and phenylacetylene, an undergraduate experiment in homogeneous catalysis: Synthesis of phenylethoxycarbene tungsten pentacarbonyl, *64*:183 (1987).
26. L. J. Mathias and T. Viswanathan, Polymerizations: An organic and polymer laboratory experiment. Ring-opening and ring-forming, *64*:639 (1987).
27. A. Feigenbaum and D. Scholler, Polystyrene: An experiment for undergraduate students, expanded, *64*:810 (1987).
28. L. J. Mathias and T. Viswanathan, Ring-opening and ring-forming polymerizations: An organic and polymer laboratory experiment, *64*:639 (1987).
29. R. F. Jordan, Olefin polymerization catalysts, cationic metal-alkyl (SYMP), *65*:285 (1988).
30. A. E. Holboke and R. P. Pinnell, Sulfonation of polystyrene: Preparation and characterization of an ion exchange resin in the organic laboratory, *66*:613 (1989).
31. E. F. Silversmith, Free-radical polymerization of acrylamide, *69*:763 (1992).
32. D. M. Snyder, Illustration of M_n and M_w in chain-growth polymerization using a simplified model: An undergraduate polymer chemistry laboratory exercise, *69*:422 (1992).
33. W. Timmer and J. Villalobos, The polymerase chain reaction, *70*:273 (1993).
34. J. Vebrel, Y. Grohens, A. Kadmiri, and E. Gowling, The study of the cross-linking of styrene and an unsaturated polyester: The chemistry of canoe manufacture, differential scanning calorimetry, *70*:501 (1993).
35. J. Marentette and G. R. Brown, Polymer spherulites: II. Crystallization kinetics, *70*:539 (1993); I. Birefringence and morphology, *70*:435 (1993).
36. T. Viswanathan and J. Jethmalani, The aqueous ring-opening metathesis polymerization of furan-maleic anhydride adduct: Increased catalytic activity using a recyclable transition metal catalyst, *70*:165 (1993).
37. P. Djurovich and R. Watts, A simple and reliable chemical preparation of $YBa_2Cu_3O_{7-x}$ superconductors: An experiment in high temperature superconductivity for an advanced undergraduate laboratory, *70*:497 (1993).
38. B. Dewprashad and E. Eisenbraun, Fundamentals of epoxy formulation, *71*:290 (1994).
39. H. Andrews-Henry, Polystyrene kinetics by infrared: An experiment for physical and organic chemistry laboratories, *71*:357 (1994).
40. B. C. Sherman, W. Euler, and R. K. Force, Polyaniline—a conducting polymer: Electrochemical synthesis and electrochromic properties, *71*:A94 (1994).
41. M. R. Mueller and J. D. Britt, Formation of titanium gels from titanium alkoxides: A demonstration of kinetically controlled polymerization, *71*:890 (1994).

Reactions of and on Polymers

1. W. F. Berkowitz, Acid hydrolysis of nylon 66, *47*:536 (1970).
2. C. E. Carraher, Generation of poly(vinyl alcohol) and arrangement of structural units, *55*:473 (1978).
3. J. Vinson, Hydrolysis of latex paint in dimentyl sulfoxide, *46*:877 (1969).

4. D. N. Buchanan and R. W. Kleinman, Peptide hydrolysis and amino acid analysis. A first year organic or biochemistry experiment, *53*:255 (1976).
5. D. Blackman, Acid-catalyzed hydrolysis of starch, *55*:722 (1978).
6. J. H. Ross, Polymer crosslinking and gel formation without heating, *54*:110 (1977).
7. K. Rungruangsak and B. Panijpan, The mechanism of action of salivary amylase, *56*:423 (1979).
8. N. S. Allen and J. F. McKellar, Polymer photooxidation: An experiment to demonstrate the effect of additives, *56*:273 (1979).
9. M. Pickering and R. Crabtree, Protein denaturation: A physical chemistry laboratory project, *58*:513 (1981).
10. A. Marty, M. Boiret, and M. Deumie, How to illustrate ligand-protein binding in a class experiment, *63*:365 (1986).
11. J. Barton, Protein denaturation and tertiary structure, *63*:367 (1986).
12. H. Conlon and D. Walt, Immobilization of enzymes in polymer supports, *63*:368 (1986).
13. J. W. Hill, A model for denaturing a protein by heat, *63*:370 (1986).
14. S. Burgmayer and E. Stiefel, Molybdenum enzymes, cofactors and model systems, *62*:942 (1983).
15. T. Pinkerton, C. Desilets, D. Hoch, M. Mikelsons, and G. Wilson, Bioinorganic activity of technetium radiopharmaceuticals, *62*:965 (1985).
16. K. Suslick and T. Reinert, The synthetic analogs of O_2-binding heme proteins, *62*:974 (1985).
17. K. Karlin and Y. Gultneh, Bioinorganic chemical modeling of dioxygen-activating copper proteins, *62*:983 (1985).
18. J. Valentine and D. de Freitas, Copper-zinc superoxide dismutase, *62*:990 (1985).
19. D. McMillin, The taming of the blue, *62*:997 (1985).
20. A. Splittgerber, The catalytic function of enzymes, *62*:1008 (1985).
21. K. Ekpenyoung, Monomer reactivity ratios: Acrylic acid-methymethacrylate copolymerization in dimethylsulfoxide, *62*:173 (1985).
22. A. Bard, Chemical modification of electrodes, *60*:302 (1983).
23. R. Scott, A. G. Mauk, and H. B. Gray, Experimental approaches to studying biological electron transfer, *62*:932 (1985).
24. M. D. Jones and J. T. Fayerman, Industrial applications of recombinant DNA technology, *64*:337 (1987).
25. M. G. Roig, Chemical modification of catalytically essential amino acid residues in enzymes, *64*:77 (1987).
26. J. F. Sebastian, Reversible activators of (CB) enzymes, *64*:1031 (1987).
27. K. A. Clingman and J. Hajdu, Reactions of lipolytic enzymes: An undergraduate biochemistry experiment, *64*:358 (1987).
28. G. M. McCorkle and S. Altman, RNA's as catalysts: A new class of enzymes (CB), *64*:221 (1987).
29. M. H. Caruthers, Chemical synthesis of (CB) DNA, *66*:577 (1989).
30. F. Blatter and E. Schumacher, The preparation of pure zeolite NaY and its conversion to high-silica faujasite: An experiment for laboratory courses in inorganic chemistry, *67*:519 (1990).

31. P. Phinyocheep and I. M. Tang, Determination of the hole concentration (copper valency) in superconductors: A laboratory using the potentiometric titration method, *71*:A115 (1994).
32. H. Burrows, H. Ellis, and C. Odilora, The dehydrochlorination of PVC: An introductory experiment in gravimetric analysis, *72*:448 (1995).

Physical Properties

1. D. P. Miller, Infinitely variable individualized unknowns for molecular weight by viscosity, *57*:200 (1980).
2. E. M. Pearce, C. Wright, and B. Bordoloi, Competency based modular experiments in polymer science and technology, *57*:375 (1980).
3. M. Lauffer, Motion in viscous liquids, *58*:250 (1981).
4. D. Hill and J. O'Donnell, Some properties of poly(methyl methacrylate) studied by radiation degradation: An interdisciplinary student experiment, *58*:74 (1981).
5. G. L. Wilkes, An overview of the basic rheological behavior of polymer fluids with an emphasis on polymer melts, *58*:880 (1981).
6. J. Aklonis, Mechanical properties of polymers, *58*:892 (1981).
7. J. E. Mark, Rubber elasticity, *58*:898 (1981).
8. M. Bader, Rubber elasticity—A physical chemistry experiment, *58*:285 (1981).
9. L. H. Sperling, Molecular motion in polymers: Mechanical behavior of polymers near the glass-rubber transition temperature, *59*:942 (1982).
10. J. Perrin and G. Martin, The viscosity of polymeric fluids, *60*:516 (1983).
11. L. Mathias, Evaluation of a viscosity—Molecular weight relationship: An undergraduate-graduate polymer experiment, *60*:422 (1983).
12. J. Ross, Demonstration of solvent differences by visible polymer swelling, *60*:169 (1983).
13. L. Natarajan, M. Robinson, and R. Blankenship, A physical chemistry laboratory experiment, linear dichroism of cyanine dyesin stretched polyvinyl alcohol films, *60*:241 (1983).
14. G. Henderson, D. Campbell, V. Kuzmicz, and L. H. Sperling, Gelation as a physically crosslinked elastomer, *62*:269 (1985).
15. L. Mathias and D. R. Moore, A simple, inexpensive molecular weight measurement for water-soluble polymers using microemulsions, *62*:545 (1985).
16. D. Armstrong, J. Marx, D. Kyle, and A. Alak, Synthesis and a simple molecular weight determination of polystyrene, *62*:705 (1985).
17. H. Cloutier and R. E. Prud'homme, Rapid identification of thermoplastic polymers, *62*:815 (1985).
18. J. Kovac, Molecular size and Raoult's Law, *62*:1090 (1985).
19. E. Casassa, A. Sarquis, and C. Van Dyke, The gelation of polyvinyl alcohol with borax: A novel class participation experiment involving the preparation and properties of a "slime," *63*:57 (1986).
20. A. M. Sarquis, Dramatization of polymeric bonding using slime, *63*:60 (1986).
21. B. Shakhashiri, G. Dirreen, and L. Williams, The preparation and properties of polybutadiene (jumping rubber), *57*:738 (1980).

22. L. Wade and L. Stell, A laboratory introduction to polymeric reagents, *57*: 438 (1980).

23. J. Malmin, Modifications in the synthesis of caprolactam and nylon 6, *57*: 742 (1980).

24. S. Moss, Polymerization using the rotating-sector method, free-radical: A computer-based study, *59*:1021 (1982).

25. K. Ekpenyong and R. Okonkwo, Determination of acrylonitrile/methyl-methacrylate copolymer composition by infrared spectroscopy, *60*:431 (1983).

26. G. East and S. Hassell, An alternative procedure for the nylon rope trick, *60*:69 (1983).

27. R. Wing and B. Shasha, Encapsulation of organic chemicals within a starch matrix: An undergraduate laboratory experiment, *60*:247 (1983).

28. J. Wilson and J. K. Hamilton, Wood cellulose as a chemical feedstock for the cellulose esters industry, *63*:49 (1986).

29. U. S. Aithal and T. J. Aminabhavi, Measurement of diffusivity of organic liquids through polymer membranes: A simple and inexpensive laboratory experiment, *67*:82 (1990).

30. G. L. Hardgrove, D. A. Tarr, and G. L. Miessler, Polymers in the physical chemistry laboratory. An integrated experimental program, *67*:979 (1990).

31. J. P. Queslel and J. E. Mark, Advances in rubber elasticity and characterization of elastomeric networks, *64*:491 (1987).

32. C. E. Carraher, Jr., Thermodynamics and the bounce, *64*:43 (1987).

33. M. D. Schuh, Enhanced unfolding of myoglobin: A biophysical chemistry experiment, microcomputer-analyzed initial rate kinetics of the benzene, *65*: 740 (1988).

34. L. Oliver Smith, Polymerization distribution experiment simulation (CS), *65*: 795 (1988).

35. W. H. Chan, K. S. Lam, and W. K. Yu, Antioxidants in Plastic: An instrumental analysis project, *66*:172 (1989).

36. W. Andrade, Polethylene film used as windows in infrared analysis of solid, liquid, or gaseous samples, *66*:865 (1989).

37. W. Guo, B. M. Fung, and R. E. Frech, Polymeric fast ion conductor: A physical chemistry experiment, *66*:783 (1989).

38. R. C. Plumb, Antique windowpanes and the flow of supercooled liquids, *66*: 994 (1989).

39. S. B. Clough, Stretched elastomers: A case of decreasing length upon heating, *64*:42 (1987).

40. L. J. Mathias, Converting sunlight to mechanical energy: A polymer example of entropy, *64*:889 (1987).

41. J. Khurma, A new viscometer design for polymer solutions, *68*:63 (1991).

42. C. Hamann, D. Myers, K. Rittle, E. Wirth, and O. Moe, Quantitative determination of the amino acid composition of a protein using gas chromatography–mass spectrometry, *68*:438 (1991).

43. S. Turchi, C. David, and J. Edwards, A study in enzyme kinetics using an ion-specific electrode, *68*:687 (1991).

44. T. Viswanathan, F. Watson, and D. Yang, Undergraduate organic and poly-
     mer lab experiments that exemplify structure determination by NMR, *68*:
     685 (1991).

45. M. Tarazona and E. Salz, An experimental introduction to molecular weight
     averages of polymers: A simple experiment that uses paper clips, *69*:765
     (1992).

46. D. McLaughlin, Size-exclusion chromatography: Separating large molecules
     from small, *69*:280 (1992).

47. F. L. Pilar, Weight-average molecular weights: How to pick a football team,
     69:280 (1992).

48. J. Richards, Viscosity and the shapes of macromolecules: A physical chem-
     istry experiment using molecular-level models in the interpretation of mac-
     roscopic data obtained from simple measurements, *70*:685 (1993).

49. G. Paradossi, Size and shape of macromolecules: Calculation of the scatter-
     ing function for simple geometries, *70*:440 (1993).

50. K. R. Williams and U. Bernier, The determination of number-average mo-
     lecular weight: A polymer experiment for lower-division chemistry students,
     71:265 (1994).

51. M. Mougan, A. Coello, A. Jover, F. Meijide, and J. Vazquez Tato, Spectro-
     fluorimeters as light-scattering apparatus: Application to polymers molecular
     weight determination, *72*:284 (1995).

Appendix D: Syllabus

Three interrelated questions can be addressed when considering the construction of course syllabi. These questions are (a) topics to be covered, (b) order in which these topics should be covered, and (c) proportion of time spent on each topic. Just as with other areas of chemistry, such as general and organic chemistry, there exists a healthy variety of topics, extent of coverage of each topic, and a certain order to follow.

The problem with respect to polymer science is compounded in that there exists a wide variety of introductory polymer courses with respect to duration. Typically the duration of such courses fall within the range of 30 to 90 hours of lecture (1 quarter to 2 semesters—3 hours credit), with 45 hours being the most common strength.

One major assumption agreed upon by most academic and industrial polymer scientists and associated education committees is that there should be both a core of material common to introductory courses and a portion of optional material reflecting individual interests of teachers, student bodies, and local preferences and circumstances. (For instance, if a student population within an introductory course is largely premedicine, then topics related to natural polymers might be emphasized, while student populations high in material sciences and engineering might benefit from greater emphasis on polymer rheology and dynamics.) This assumption was utilized in the generation of the first ACS Standardized Examination in Polymer Chemistry (1978) and is continuing to be utilized in deliberations made by the Joint Committee on Polymer Education.

The Polymer Syllabus Committee developed a proposal listing broad topic areas and suggested times to be devoted to each topic area. This listing is given below, along with associated chapters in the present text for ready identification.

Basically, the Syllabus Committee has proposed that all lecture courses include portions of the first seven topics with the level and extent of coverage guided by such factors as available class time, additional topics covered, interest of instructor, student interests, class composition, etc. It must be emphasized that the "optional topics" listed are not to be considered limiting and that additional topics can be introduced, again dependent on factors such as available time.

Topics	Amount of course time (%)	Chapter
Major Topics		
Introduction	5	1
Morphology	10	2
Stereochemistry		
Molecular interactions		
Crystallinity		
Molecular Weights	10	3, 4
Average molecular weight		
Fractionation of polydisperse systems		
Characterization techniques		
Testing and Characterization of Polymers	10	5
Structure-property relationships		
Physical tests		
Instrumental characterizations		
Step-Reaction Polymerizations	10	7
Kinetics		
Polymers produced by step polymerizations		
Chain-Reaction Polymerization	10	8, 9
Kinetics of ionic chain-reaction polymerization		
Kinetics of free-radical chain-reaction polymerization		
Polymers produced by chain-reaction polymerization		
Copolymerization	10	10
Kinetics		
Types of copolymers		
Polymer blends		
Principal copolymers		
Optional Topics	35	
Rheology (Flow Properties—Viscoelasticity)		3
Solubility		3
Natural and Biomedical Polymers		6
Additives		7, 14
Fillers		
Plasticizers		
Stabilizers		
Flame retardants		
Colorants		
Reactions of Polymers		15
Synthesis of Polymer Reactants		10
Polymer Technology		15

Topics	Amount of course time (%)	Chapter
Plastics		
Elastomers		
Fibers		
Coatings and adhesives		

The Syllabus Committee, and other committees derived from the Joint Polymer Education Committee, have considered the question of order of topics, concluding that almost any order of topics is possible and practical in given situations—i.e., the decision was made not to agree on a given order of topic introduction. The present text is written to permit the various chapters to be considered in almost any order without significant loss of teaching and learning effectiveness. Thus there exists both a freedom in choice of topics, depth of coverage, and order for presentation of topics, as well as some guidelines regarding the important subjects every polymer science teacher should consider.

ELECTRONIC EDUCATION

The amount of polymer-related information that appears on the World Wide Web, WWW, is rapidly increasing. In general, polymer-related material can be obtained from "surfing" the "POLY.COM". Two "Web-sites" developed in cooperation of PolyEd for particular polymer education use are

WWW.MATH.UMASS.EDU:80/POLYMERSCIENCE   and
WWW.PSRC.USM.EDU/POLYCLASS

Additional polymer-related education material is available from the National Information Center for Polymer Education, Dr. John Droske, Director, University of Wisconsin-Stevens Point, Stevens Point, WI 54481 (TEL: 715-346-3703).

Appendix E: Polymer Core Course Committees

The ACS Committee on Professional Training in its 1979—80 publication "Undergraduate Professional Education in Chemistry: Criteria and Evaluation Procedures" notes: "In view of the current importance of inorganic chemistry, biochemistry, and polymer chemistry, advanced courses in these areas are especially recommended and students should be strongly encouraged to take one or more of them. Furthermore, the basic aspects of these three important areas should be included at some place in the core material." It is with respect to the directive given in the final sentence quoted above that a series of committees was developed under the Joint Polymer Education Committee. Topics related to polymer chemistry (principles, illustrations, and application) should be included in all of the core courses within the training of an undergraduate chemistry major at an appropriate level.

Six committees, hereafter called the Core Course Committees, were formed in 1980. An overall philosophy was developed of trying, wherever possible, to use polymers to enhance topics essential to the core courses consistent with the general CPT philosophy of a constant volume (time) curricula for chemistry majors. Realistic expectations were sought. Polymer topics, principles, and illustrations were to be introduced to help, complement, and complete a student's formal training, but *not* at the exclusion of other essential core course concepts.

Eight objectives were generated as guidelines for the Core Course Reports. These are:

1. Determine general topics to be included in each core course.
2. Determine depth of coverage of each topic.
3. List short, specific polymer illustrations which can be used in teaching key concepts to students.

4. Find timely illustrations of polymer science and include enough specific information so that the examples can be used directly by teachers.
5. Determine a general estimate as to the amount of time that should be spent on polymer-related topics in each course.
6. Evaluate available texts.
7. Assess overall state of affairs in each course.
8. Develop examples of questions and answers—multiple guess, short answers, essays.

Over one-half of the Core Course Reports have been published and are available either in the *Journal of Chemical Education* or from the office of the Division of Polymer Chemistry, Polytechnic Institution of New York, Brooklyn, New York 11201. Those published to date are: Introduction—J. Chem. Ed. *60*(11):971 (1983); General—*60*(11):973 (1983); Samples—*61*:161 (1984); Inorganic—*61*:230 (1984); Physical—Part I—*62*:780 (1985); and Part II—2:1030 (1985).

Appendix F: Polymer Models

Unfortunately we are not able to readily observe molecules, even macromolecules, on a molecular level. Thus, we resort to models to convey molecular shapes and sizes. Following is a brief discussion of simple models employed to illustrate giant molecules.

Many of the most often employed models are quite simple and are used to illustrate the one-dimensionality of linear polymers. Items such as rope, wire, string, thread, and yarn are typically used. Extensions include calculations based on the diameter of a string, etc., and the corresponding lengths of string involved.

Physical cross-linking and general amorphous random chain shapes are readily illustrated by folding together several strings, bowl of long spaghetti, etc. Removal of one piece of string involves movement and possible entanglement (physical cross-linking) with accompanying portions of string. Chemical cross-links can be demonstrated by physically attaching several chains together using paper clips, strips of tape, wire, additional small pieces of string, etc. The concepts of number of repeat units (degree of polymerization) for both vinyl (using similarly colored beads) and condensation (using alternately beads of two different colors), and types of copolymers (alternating, branched, graft, random, or block) can be readily illustrated through use of pop beads. Thus random copolymers are represented through use of pop beads of two colors, popped together in a random sequence, alternating copolymers through arrangement of the beads by alternating color and block copolymers through attachment of sequences of same-colored beads. Representation of graft copolymers and chemical cross-linking requires a second hole to be appropriately drilled into some of the beads. The concept of the preferential head-to-tail orientation can also be presented through use of pop beads with the hole end representing the head or tail portion, etc. Ball-and-stick models can be employed to illustrate most of the previously mentioned concepts but are often reserved for difficult-to-demonstrate concepts such as conformation, configuration, and tacticity (atactic, isotactic, and syndiotactic).

Concepts of polymer drag in viscous fluids, physical cross-linking, etc., can be illustrated through use of numbers of wires bent at predetermined lengths at an almost tetrahedral angle. Styrofoam balls, marbles, etc., can be added to assist in illustrating free volume and viscous drag. The wire represents the atoms present in the polymer backbone. A randomization of the bonding direction will give wire-polymers that can also be used to describe average end-to-end distances and other random-flight aspects.

Most synthetic polymers, when allowed, will form shapes dependent primarily on size and electronic nature such as to minimize steric and electronic constraints and to maximize electronic associations. The major space-saving or sterically min-imizing shape for linear macromolecules is that of a helix. A simple rope or string drawn in two and twisted readily illustrates the double helix associated with double-helical DNAs, whereas a simple twisting of the rope illustrates the helical nature of many synthetic isotactic vinyl polymers. The second major shape employed by nature in space-limiting situations is that of a puckered or pleated sheet. (A ready illustration of this is a pleated skirt.) This geometry allows ready, rapid low exten-sions, but inhibits more radical mobility (stretching). The helical, space-saving shape can be further illustrated by additions onto the chains by way of toothpicks, pieces of paper, etc., to represent side chains of vinylic substitutions.

Paper models have been effectively employed to illustrate tacticity, chain fold-ing, and a number of other geometric concepts. These often require special folding skills.

Appendix G: Structures of Common Polymers

SYNTHETIC

Acrylonitrile-butadiene-styrene terpolymer (ABS)

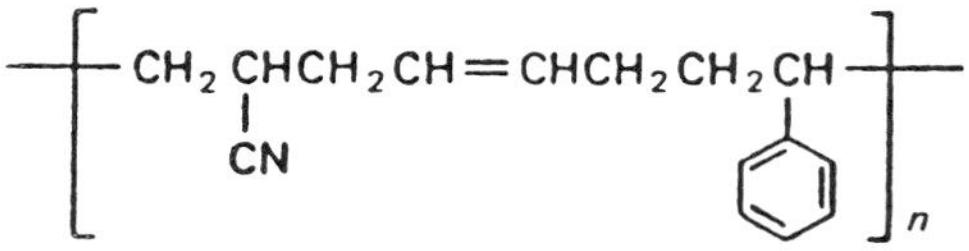

$$\left[CH_2\,CHCH_2\,CH=CHCH_2\,CH_2\,CH \right]_n$$

(with CN substituent on first CH and phenyl ring on last CH)

Butyl rubber

$$\left[CH_2-\underset{\underset{CH_3}{|}}{\overset{\overset{CH_3}{|}}{C}}-CH_2\,CH=CCH_2 \right]_n$$

(with CH_3 substituent on the $=C$)

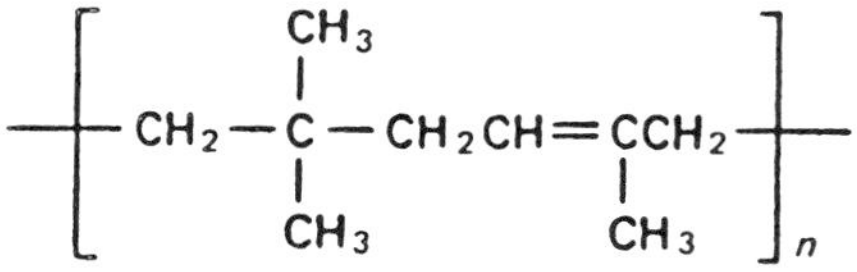

Ethylene-methacrylic acid copolymers (Ionomers)

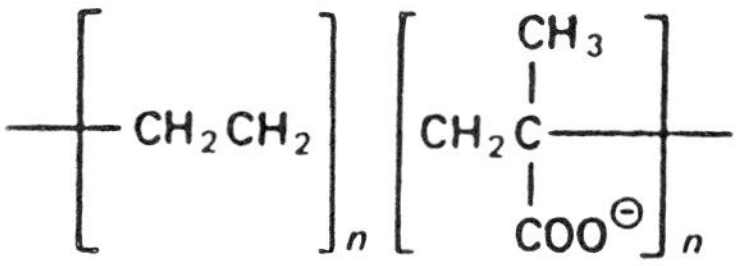

$$\left[CH_2\,CH_2 \right]_n \left[CH_2\,\underset{\underset{COO^{\ominus}}{|}}{\overset{\overset{CH_3}{|}}{C}} \right]_n$$

The diagrams represent idealized structures in the form of regularly repeating units. This is especially true for the natural polymers and copolymers.

Ethylene-propylene elastomers

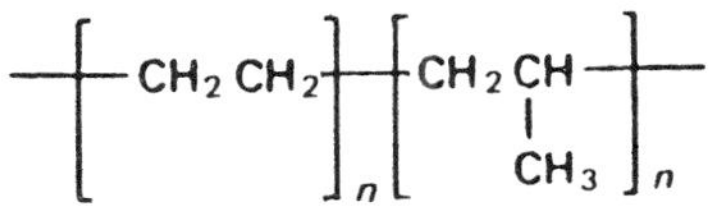

Melamine-formaldehyde resins (MF)

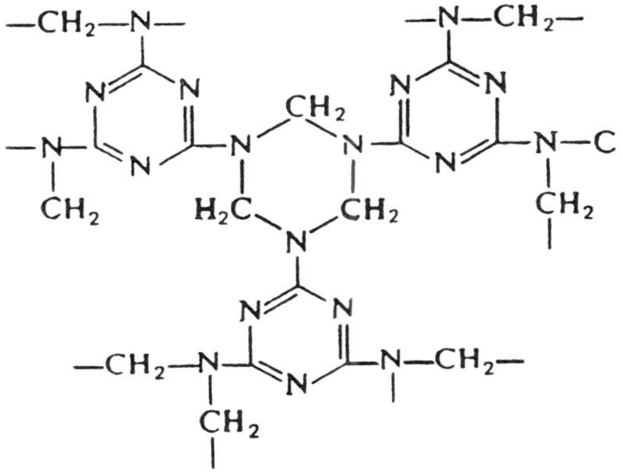

Nitrile Rubber (NBR)

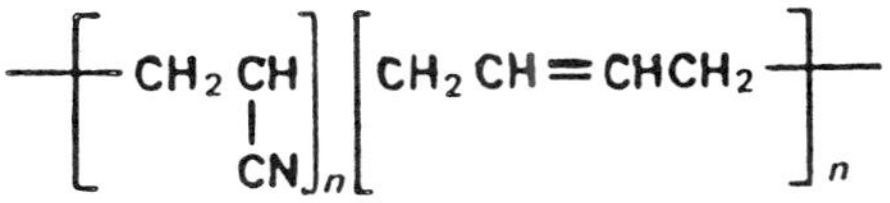

Phenol-formaldehyde resins (PF)

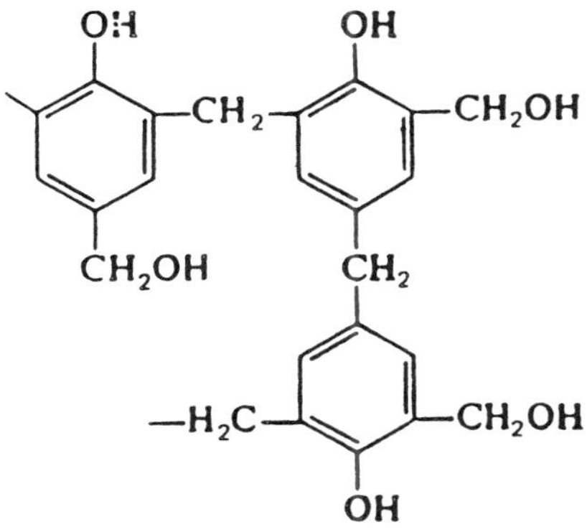

Polyacetaldehyde

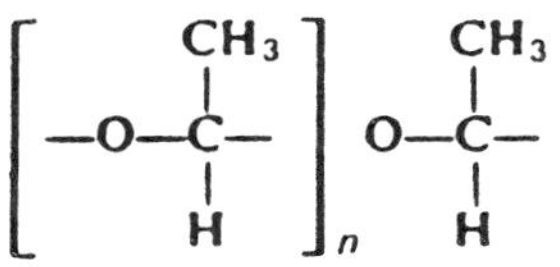

$$\left[-O-\underset{\underset{H}{|}}{\overset{\overset{CH_3}{|}}{C}}- \right]_n \quad O-\underset{\underset{H}{|}}{\overset{\overset{CH_3}{|}}{C}}-$$

Polyacrolein

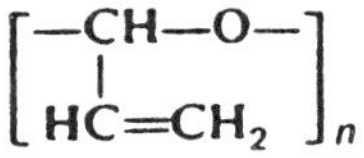

$$\left[\begin{array}{c} -CH-O- \\ | \\ HC=CH_2 \end{array} \right]_n$$

Polyacrylamide

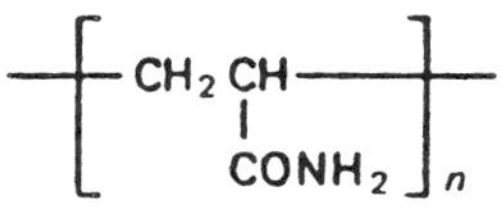

$$-\left[\begin{array}{c} CH_2\ CH- \\ | \\ CONH_2 \end{array} \right]_n-$$

Poly(acrylamide oxime)

$$[-CH_2-CH-]_n$$
$$\underset{NH_2 \qquad NOH}{\overset{|}{C}}$$

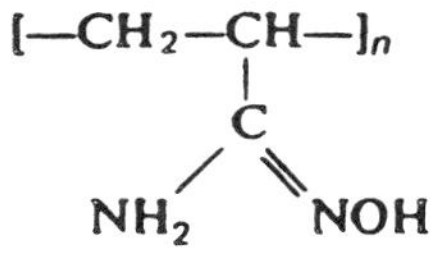

Poly(acrylic anhydride)

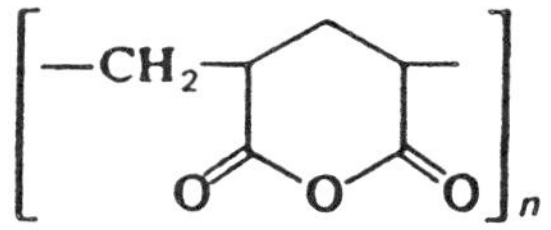

$$\left[-CH_2 \underset{O \qquad O \qquad O}{\bigcirc} \right]_n$$

Polyacrylonitrile

$$\left[CH_2-\underset{\underset{CN}{|}}{CH} \right]_n$$

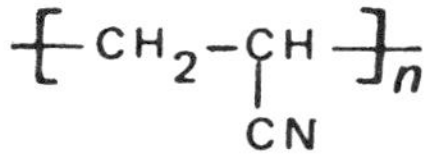

Poly(β-alanine) (3-Nylon)

$$\left[CH_2-CH_2-\overset{\overset{O}{||}}{C}-NH \right]_n$$

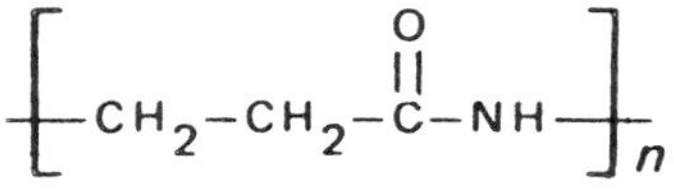

Polyallene

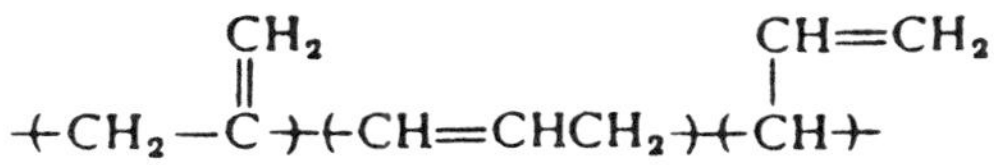

$$\underset{\displaystyle +CH_2-\overset{\displaystyle\overset{CH_2}{\|}}{C}\,+\,(CH\!=\!CHCH_2\,)\,+\,\overset{\displaystyle\overset{CH=CH_2}{|}}{CH}+}{}$$

Polybenzobenzimidazole from pyromellitic dianhydride
and tetraminobiphenyl

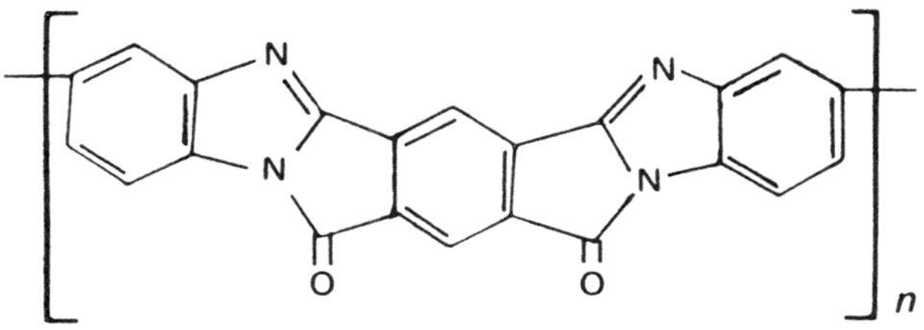

Poly(γ-benzyl-L-glutamate)

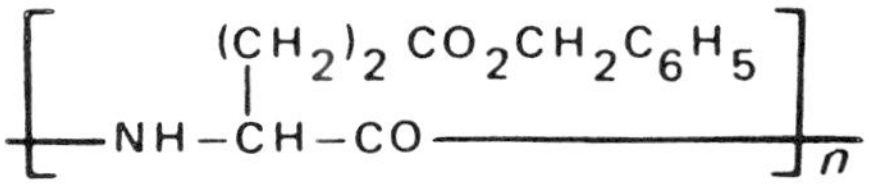

$$\left[\,-NH-\underset{\displaystyle\underset{}{\overset{\displaystyle\overset{(CH_2)_2\,CO_2CH_2C_6H_5}{|}}{CH}}}{}-CO\,\right]_n$$

Poly(1,3-bis(p-carboxyphenoxy)propane anhydride)

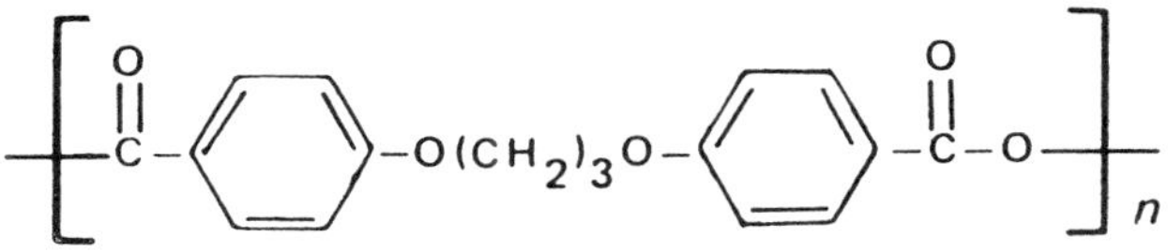

Polybutadiene (Butadiene rubber, BR)

$$-\!\left[CH_2CH\!=\!CHCH_2\right]_n\!-$$

1,2-Polybutadiene

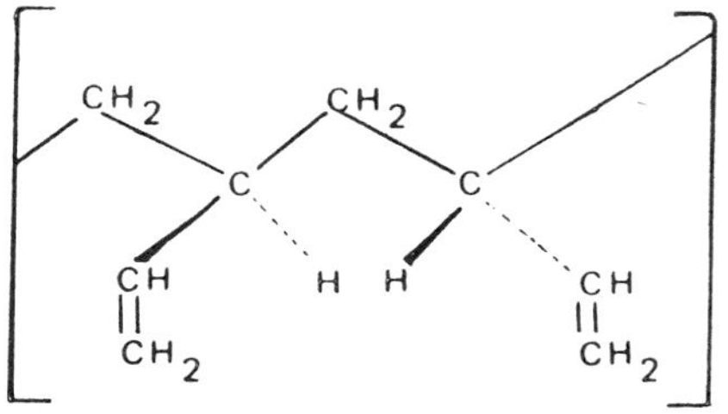

trans-1,4-Polybutadiene

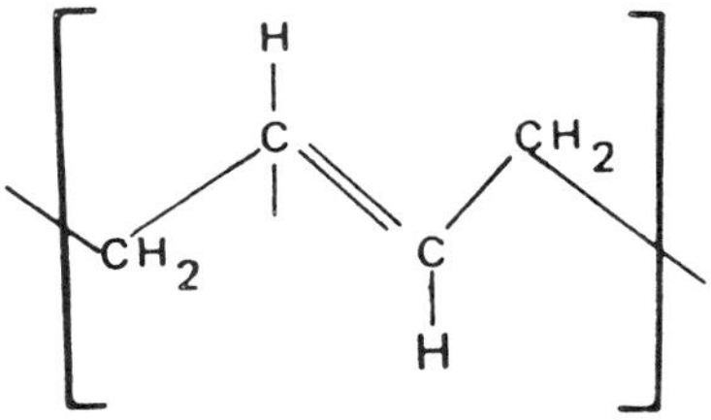

Poly(butylene terephthalate) (PBT)

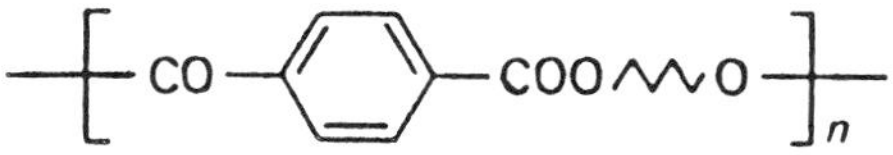

Poly-n-butyraldehyde

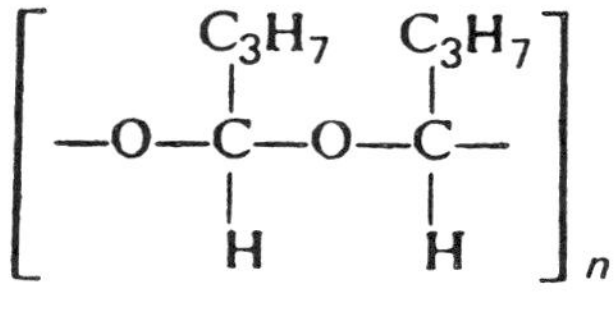

Polychloroprene

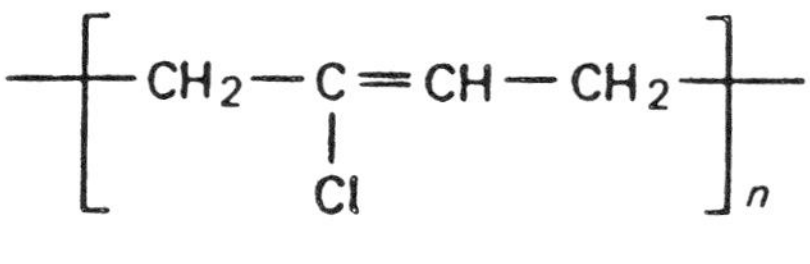

Poly(1,4-cyclohexanedicarbinyl terephthalate)
or Poly(oxyterephthaloyloxymethylene-1,4-cyclohexylenemethylene)

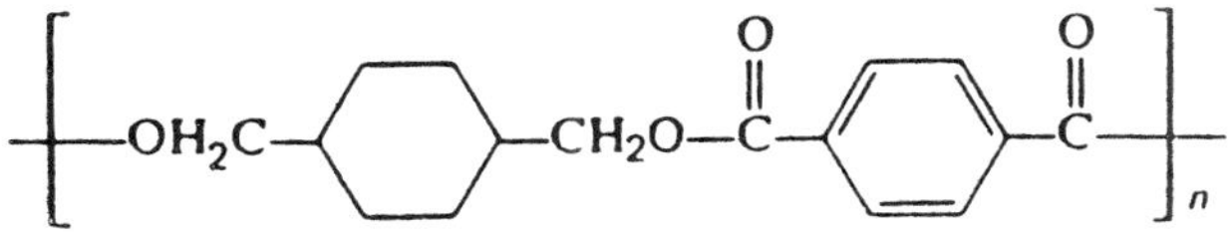

Poly(decamethyleneoxamide)

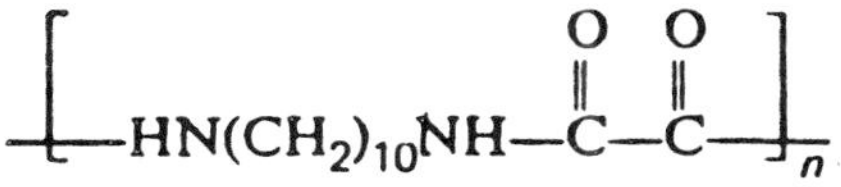

Polydecamethylenesebacamide (10-10 Polyamide)

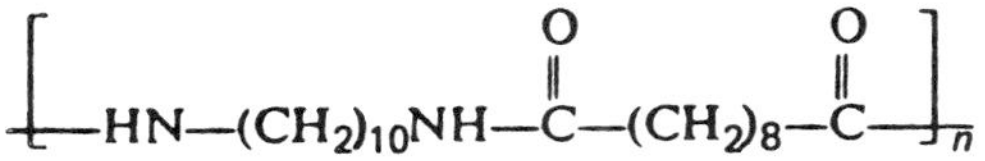

Polydecamethyleneurea

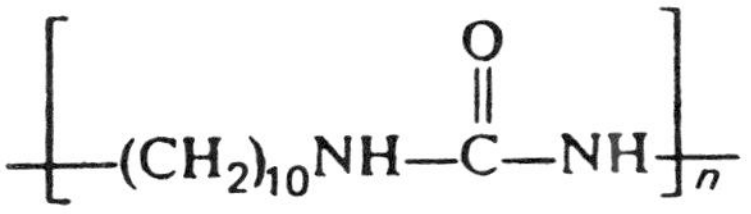

Polydichloroacetaldehyde

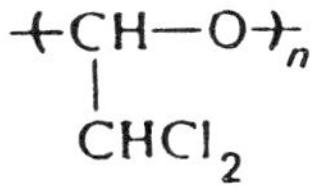

Polydiethynylbenzene
or Poly(1,3-phenylene-1,3-butadiyne-1,4-diyl)

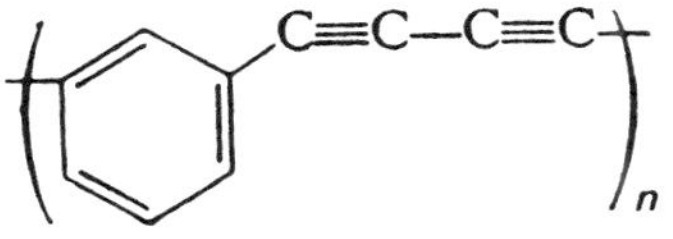

Poly(3, 3'-dimethoxy-4-4'-biphenylene carbodiimide)

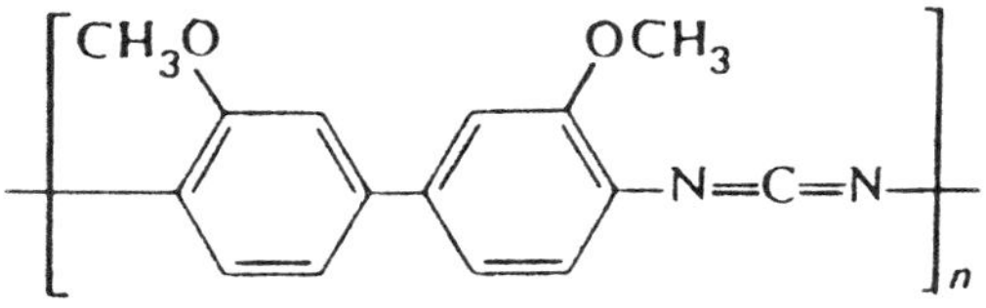

$$\left[\begin{array}{c} \text{CH}_3\text{O} \quad\quad \text{OCH}_3 \\ \\ -\!\!\!\!-\!\!\!\!\bigcirc\!\!-\!\!\bigcirc\!\!-\!\!\text{N}\!\!=\!\!\text{C}\!\!=\!\!\text{N}\!\!- \end{array} \right]_n$$

Poly(2, 3-dimethylbutadiene)

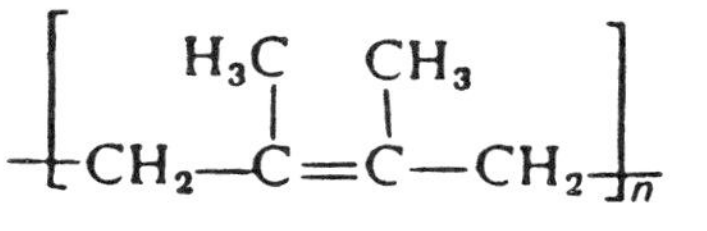

$$\left[\begin{array}{c} \text{H}_3\text{C} \quad \text{CH}_3 \\ | \quad\quad | \\ \text{CH}_2\!\!-\!\!\text{C}\!\!=\!\!\text{C}\!\!-\!\!\text{CH}_2 \end{array} \right]_n$$

Poly(3, 5-dimethyl-1, 4-phenylene sulfonate)

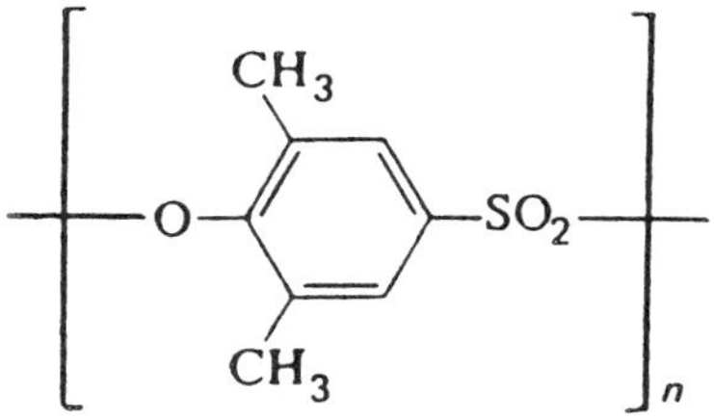

$$\left[\begin{array}{c} \text{CH}_3 \\ | \\ -\text{O}\!\!-\!\!\bigcirc\!\!-\!\!\text{SO}_2\!\!- \\ | \\ \text{CH}_3 \end{array} \right]_n$$

Poly(2, 5-dimethylpiperazine terephthalamide)

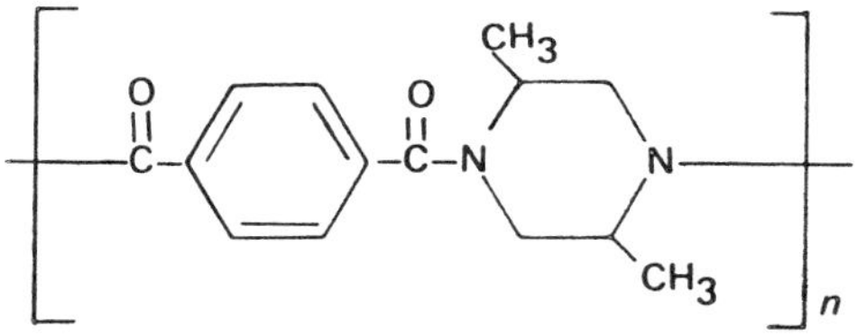

$$\left[\begin{array}{c} \text{O} \quad\quad\quad \text{O} \quad \text{CH}_3 \\ \| \quad\quad\quad \| \quad | \\ -\text{C}\!\!-\!\!\bigcirc\!\!-\!\!\text{C}\!\!-\!\!\text{N}\!\!\diamond\!\!\text{N}\!\!- \\ \quad\quad\quad\quad\quad\quad | \\ \quad\quad\quad\quad\quad \text{CH}_3 \end{array} \right]_n$$

Polydimethylsilmethylene

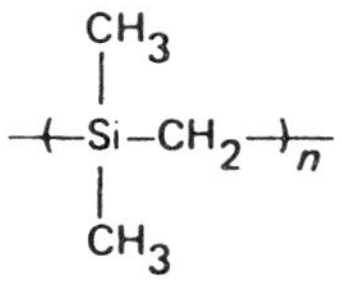

$$\begin{array}{c} \text{CH}_3 \\ | \\ -\!\!\left(\!\!\text{Si}\!\!-\!\!\text{CH}_2\!\!\right)_n \\ | \\ \text{CH}_3 \end{array}$$

Polyethylene (PE)

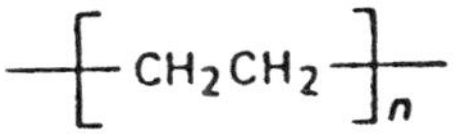

$$-\!\left[\!-CH_2CH_2-\!\right]_n\!-$$

Poly(ethylene-co-propylene adipate)

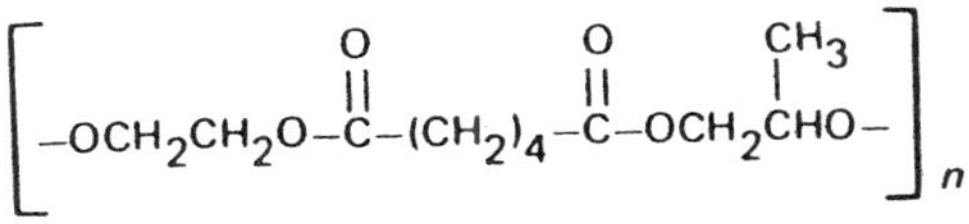

$$\left[-OCH_2CH_2O-\overset{\displaystyle O}{\overset{\|}{C}}-(CH_2)_4-\overset{\displaystyle O}{\overset{\|}{C}}-OCH_2\overset{\displaystyle CH_3}{\overset{|}{C}}HO-\right]_n$$

Poly(ethylene glycol) (PEG)

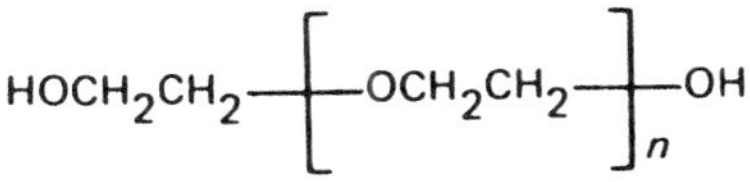

$$HOCH_2CH_2-\!\left[\!-OCH_2CH_2-\!\right]_n\!-OH$$

Poly(ethylene methylene bis(4-phenylcarbamate))

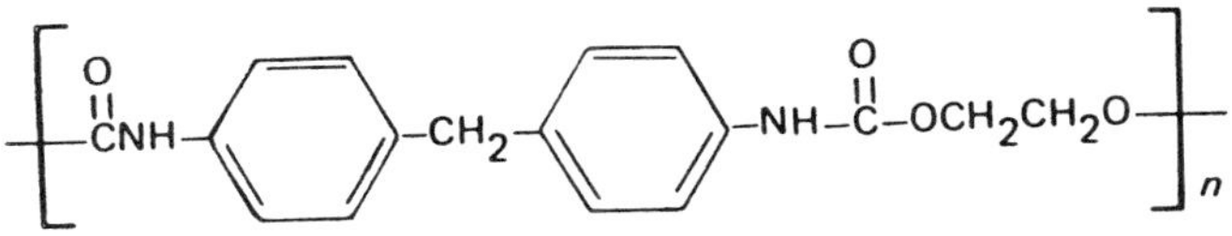

$$\left[\!-\overset{\displaystyle O}{\overset{\|}{C}}NH-\!\!\bigcirc\!\!-CH_2-\!\!\bigcirc\!\!-NH-\overset{\displaystyle O}{\overset{\|}{C}}-OCH_2CH_2O-\!\right]_n$$

Poly(ethylene N,N'-piperazinedicarboxylate)

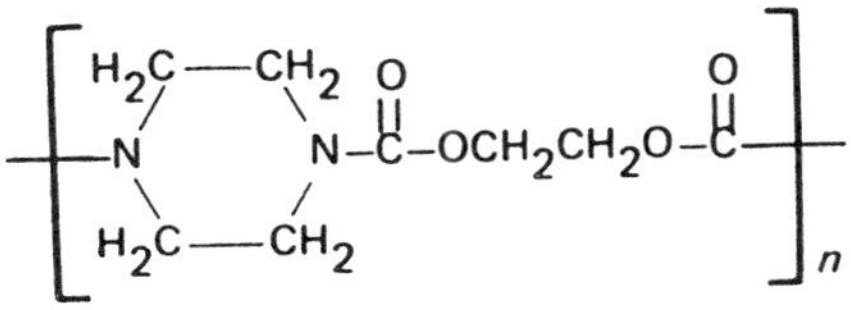

$$\left[\begin{array}{c} H_2C-\!\!-CH_2 \\ -N \qquad\quad N-\overset{\displaystyle O}{\overset{\|}{C}}-OCH_2CH_2O-\overset{\displaystyle O}{\overset{\|}{C}}- \\ H_2C-\!\!-CH_2 \end{array}\right]_n$$

Poly(ethylene terephthalamide)

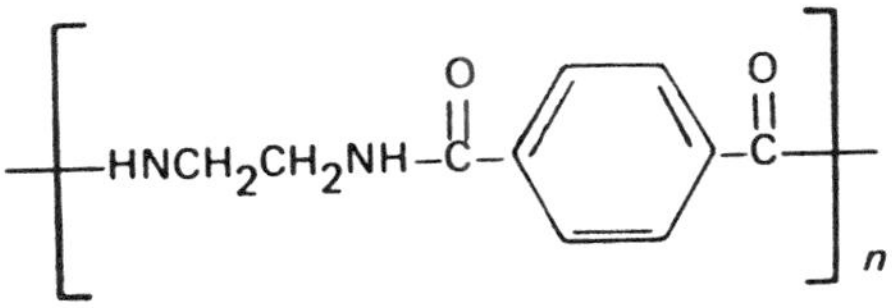

$$\left[\!-HNCH_2CH_2NH-\overset{\displaystyle O}{\overset{\|}{C}}-\!\!\bigcirc\!\!-\overset{\displaystyle O}{\overset{\|}{C}}-\!\right]_n$$

Poly(ethylene terephthalate) (PET)

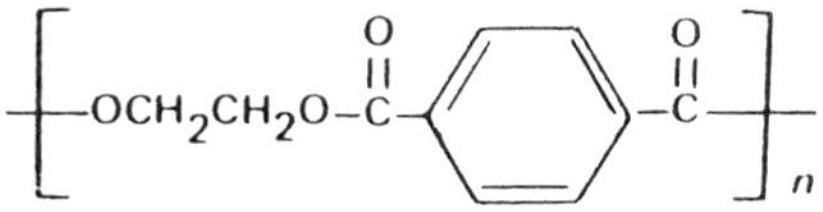

$$\left[-OCH_2CH_2O-\overset{\overset{O}{\|}}{C}-\underset{}{\bigcirc}-\overset{\overset{O}{\|}}{C}-\right]_n$$

Poly(ethylene tetrasulfide)

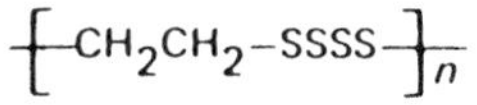

$$\left[-CH_2CH_2-SSSS-\right]_n$$

Poly-α-L-glutamic acid

$$\left[-NH\ \underset{\underset{(CH_2)_2CO_2H}{|}}{CH}\ CO-\right]_n$$

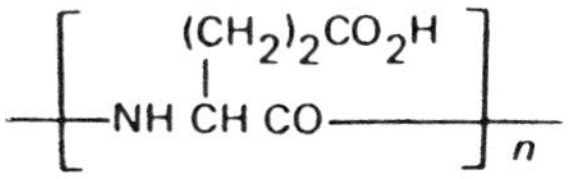

Poly(glyceryl phthalate)

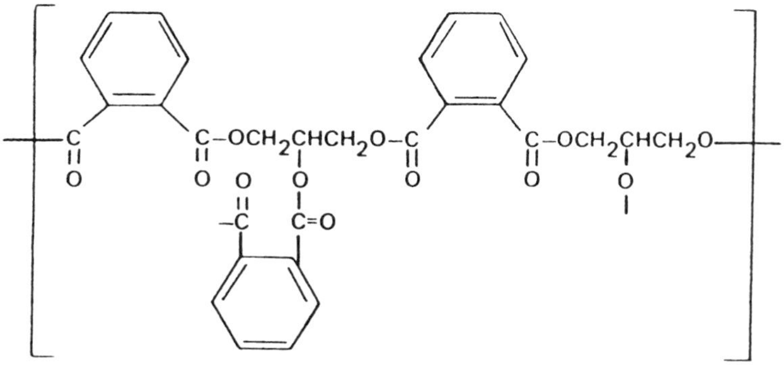

Poly(glycolic ester)

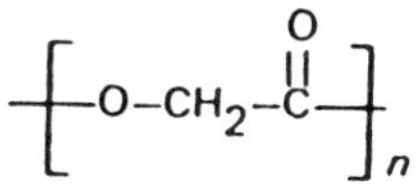

$$\left[-O-CH_2-\overset{\overset{O}{\|}}{C}-\right]_n$$

Poly(hexamethyleneadipamide) (6-6 Nylon)

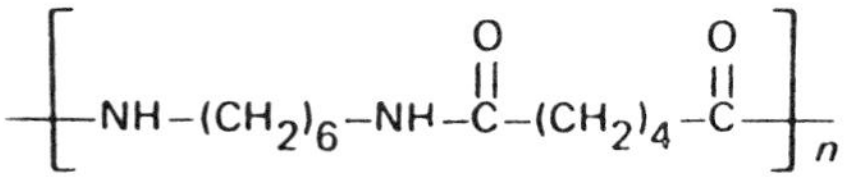

$$\left[-NH-(CH_2)_6-NH-\overset{\overset{O}{\|}}{C}-(CH_2)_4-\overset{\overset{O}{\|}}{C}-\right]_n$$

Poly(hexamethylene m-benzenedisulfonamide)

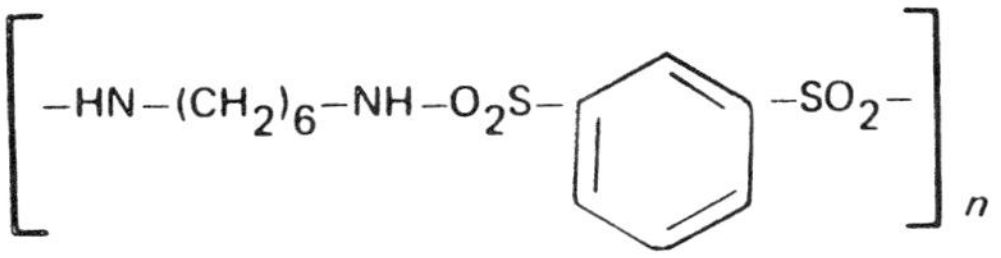

$$\left[-HN-(CH_2)_6-NH-O_2S- \underset{\bigcirc}{} -SO_2- \right]_n$$

Poly(hexamethylenedi-n-butylmalonamide)

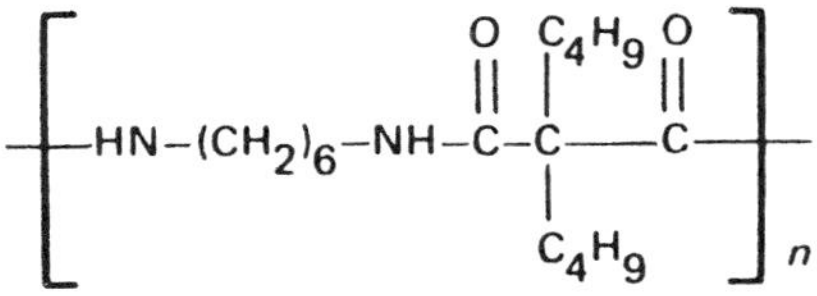

$$\left[-HN-(CH_2)_6-NH-\overset{O}{\underset{\|}{C}}-\overset{C_4H_9}{\underset{\underset{C_4H_9}{|}}{C}}-\overset{O}{\underset{\|}{C}} \right]_n$$

Poly(N, N'-hexamethylene-2, 5-diketopiperazine)

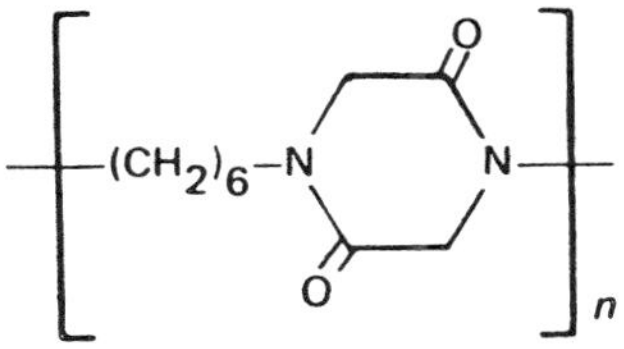

$$\left[-(CH_2)_6-N \overset{O}{\diagdown \diagup} N- \right]_n$$

Polyhexamethylenesebacamide (6-10 Nylon)

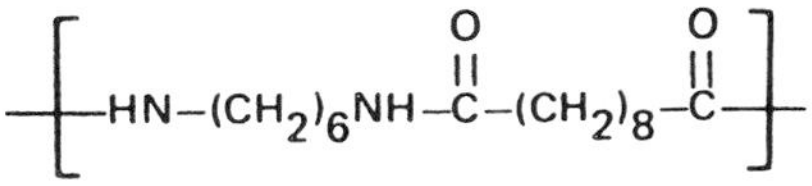

$$\left[-HN-(CH_2)_6NH-\overset{O}{\underset{\|}{C}}-(CH_2)_8-\overset{O}{\underset{\|}{C}} \right]$$

Poly(hexamethylene thioether)

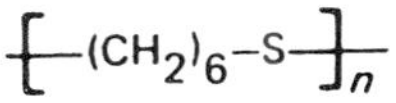

$$\left[-(CH_2)_6-S- \right]_n$$

Polyhexamethyleneurea

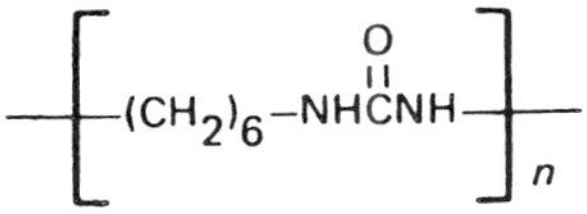

$$\left[-(CH_2)_6-NH\overset{O}{\underset{\|}{C}}NH- \right]_n$$

Polyisobutylene (PIB)

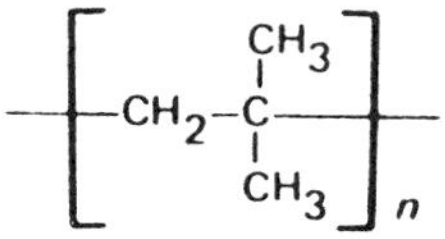

$$\left[-CH_2-\underset{\underset{CH_3}{|}}{\overset{\overset{CH_3}{|}}{C}}- \right]_n$$

Polyisoprene

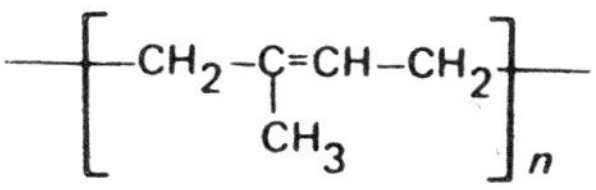

$$\left[-CH_2-\underset{\underset{CH_3}{|}}{C}=CH-CH_2- \right]_n$$

3,4-Polyisoprene

$$+CH_2-\underset{\underset{\underset{CH_2}{\parallel}}{C}}{CH}+$$

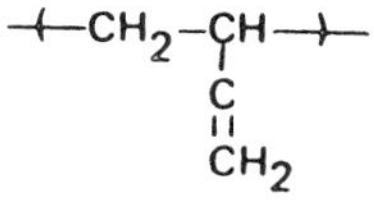

trans-1,4-Polyisoprene

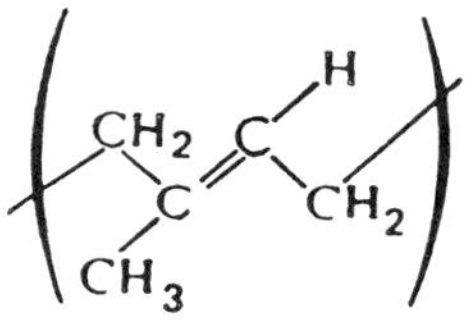

Poly(methyl acrylate)

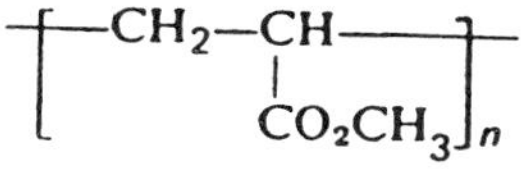

$$\left[-CH_2-\underset{\underset{CO_2CH_3}{|}}{CH}- \right]_n$$

Polymethyleneadipamide (1-6 Nylon)

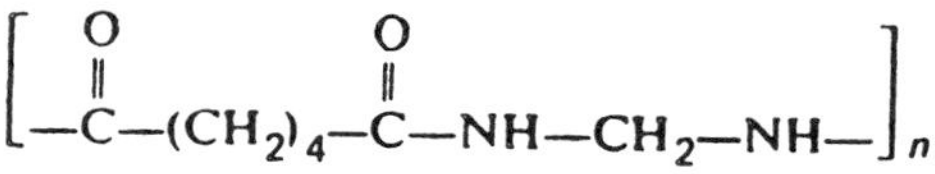

$$\left[\overset{\overset{O}{\parallel}}{-C}-(CH_2)_4-\overset{\overset{O}{\parallel}}{C}-NH-CH_2-NH- \right]_n$$

Poly[methylene bis(4-phenylurea)]

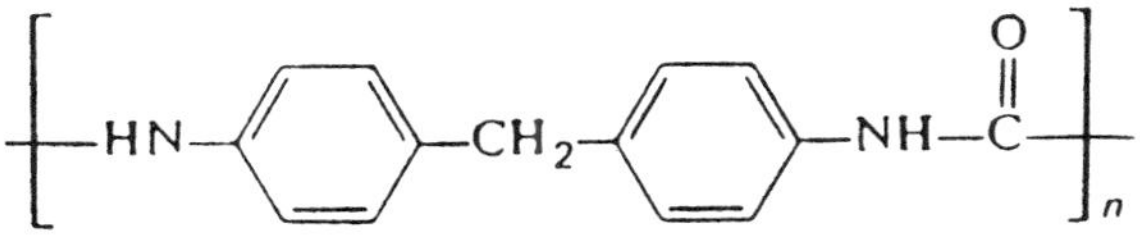

$$\left[-HN-\!\!\!\bigcirc\!\!\!-CH_2-\!\!\!\bigcirc\!\!\!-NH-\overset{\overset{\displaystyle O}{\|}}{C} \right]_n$$

Poly[methylene bis(4-phenylurethylene)]

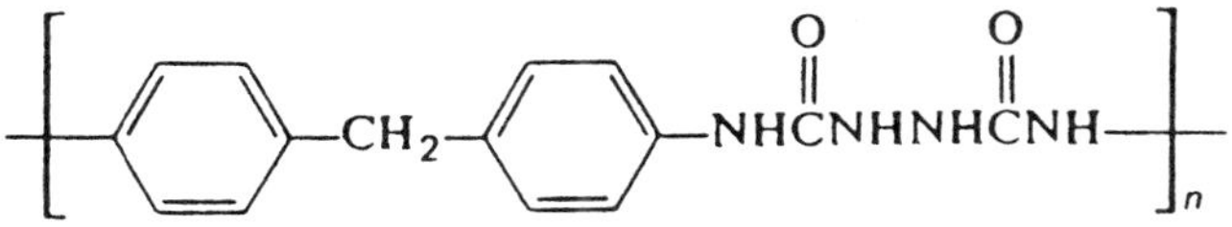

$$\left[-\!\!\!\bigcirc\!\!\!-CH_2-\!\!\!\bigcirc\!\!\!-NH\overset{\overset{\displaystyle O}{\|}}{C}NHNH\overset{\overset{\displaystyle O}{\|}}{C}NH- \right]_n$$

Poly(4-methyl-1-hexene)

$$CH_3CH_2 \quad CH_3$$
$$CH$$
$$CH_2$$
$$-\!\!\left[CH_2-CH \right]_n$$

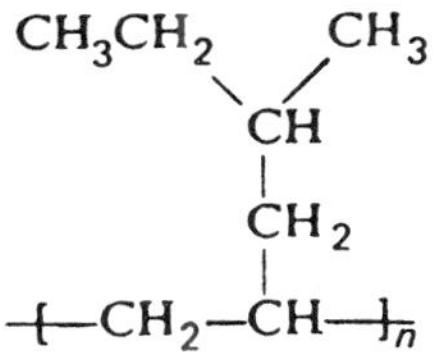

Poly(methyl methacrylates) (PMMA)

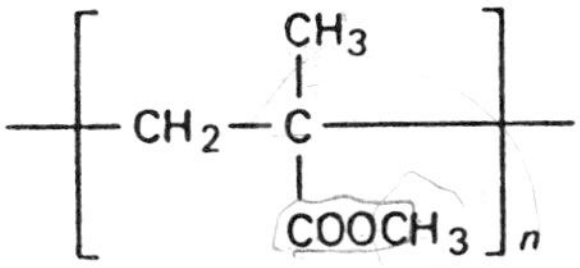

$$\left[-CH_2-\overset{\overset{\displaystyle CH_3}{|}}{\underset{\underset{\displaystyle COOCH_3}{|}}{C}}- \right]_n$$

Poly(4-methyl-1-pentene)

$$CH_2-CH(CH_3)_2$$
$$-\!\!\left[CH_2-CH \right]_n$$

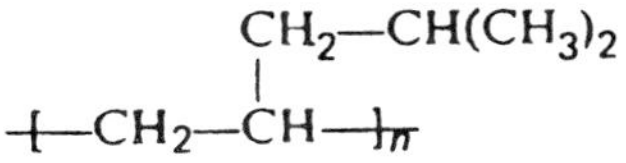

Poly(methyl vinyl ketone)

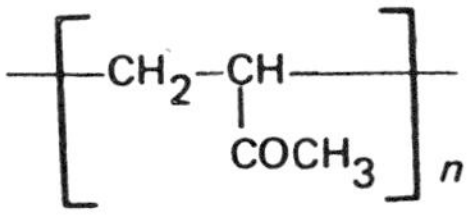

$$\left[-CH_2-\overset{}{\underset{\underset{\displaystyle COCH_3}{|}}{CH}}- \right]_n$$

Polynonamethylenepyromellitimide

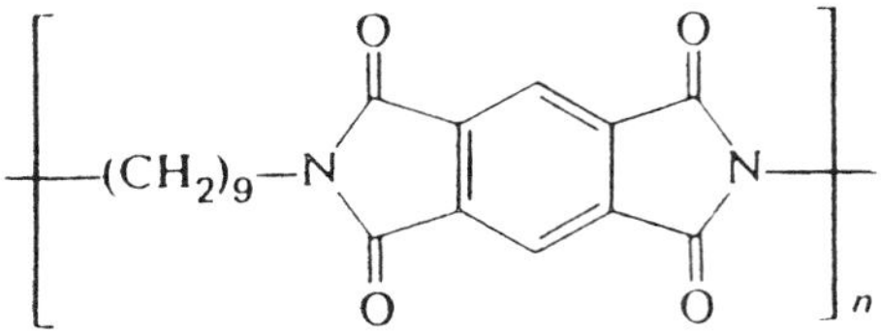

Poly(3,5-octamethylene-4-amino-1,2,4-triazole)

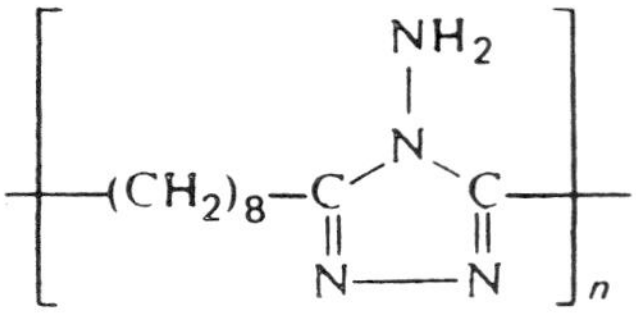

Poly(2,2'-octamethylene-5,5'-dibenzimidazole)

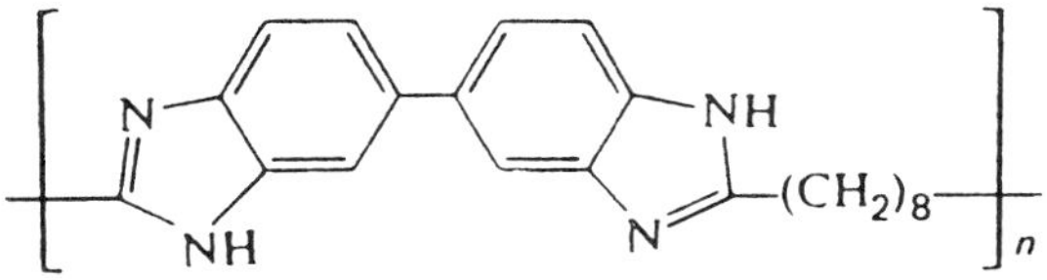

Poly(4-oxaheptamethyleneurea)

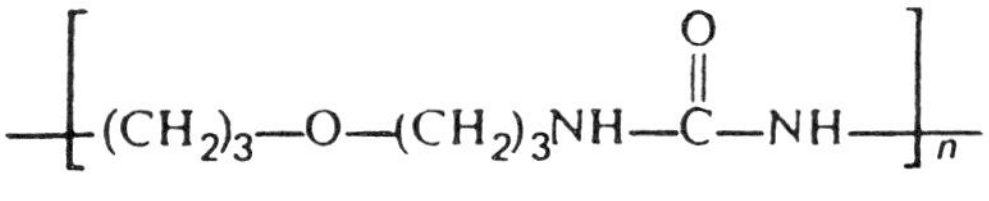

Poly[2,2'-(4,4'-oxydiphenylene)-6,6'-oxydiquinoxaline]

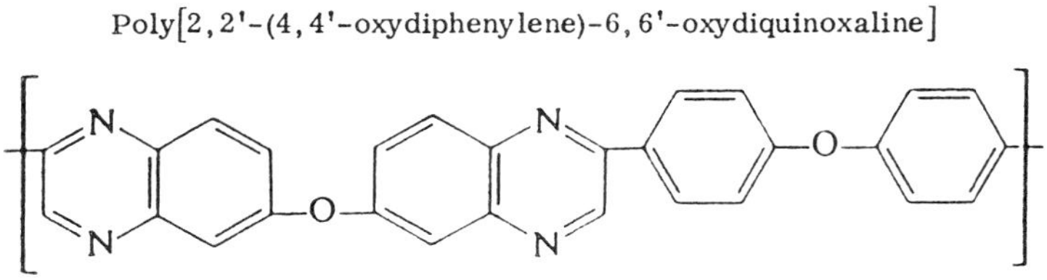

Polyoxymethylene

Polyacetal

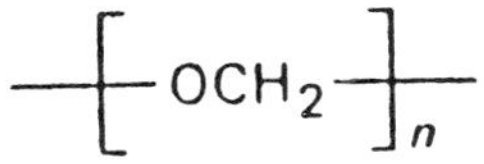

$$\left[\!\!\left[-OCH_2-\right]\!\!\right]_n$$

Poly(m-phenyl carboxylate)

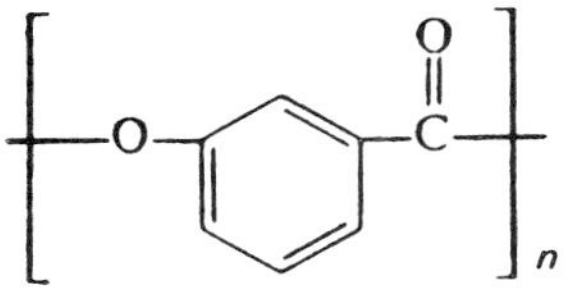

Poly(1,4-phenylene adipate)

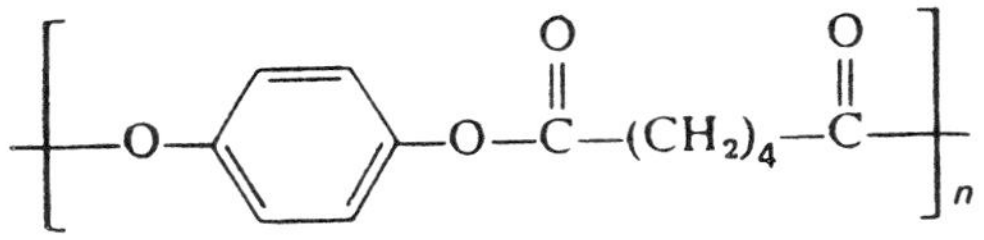

Poly[2,2'(m-phenylene)-5,5'(6,6'-benzimidazole)]

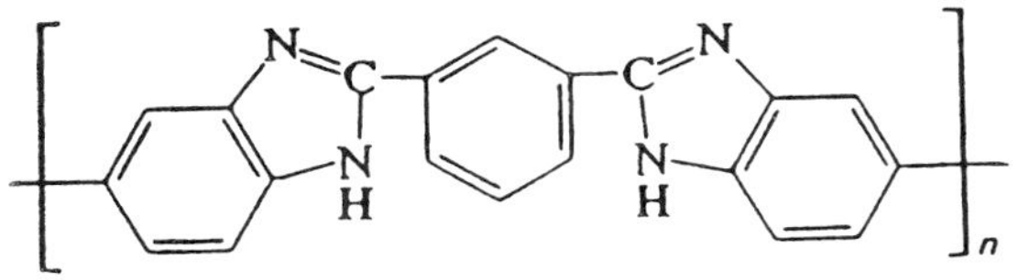

Poly[p-phenylene bis(dimethylsiloxane)]

[Poly(tetramethyl-p-silylphenylenesiloxane]

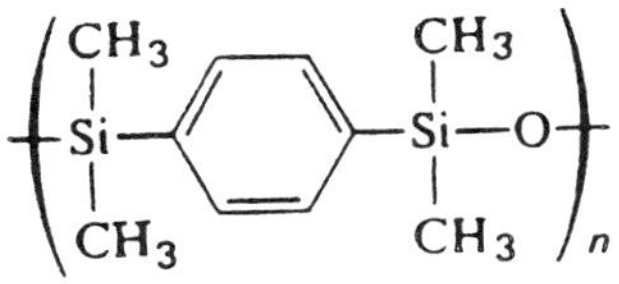

Poly(phenylene oxide) (PPO)

Poly(2,6-dimethyl-p-phenylene ether)

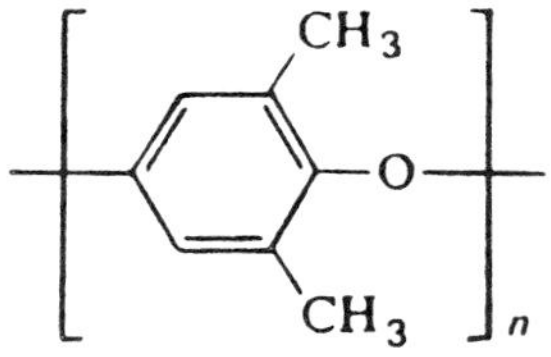

Poly(1,4-phenylene sebacate)

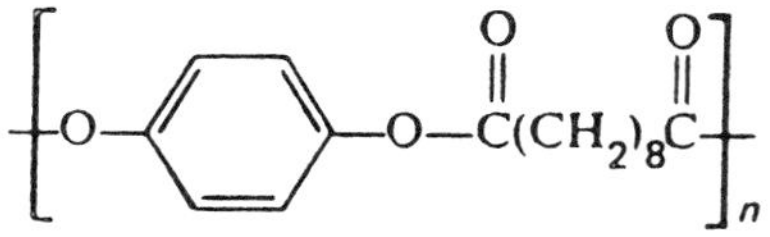

Poly(phenylene sulfide) (PPS)

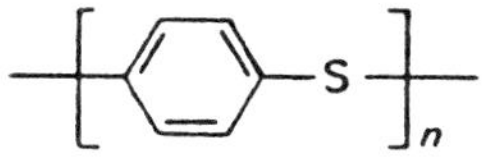

Poly(phenyl vinylene carbonate)

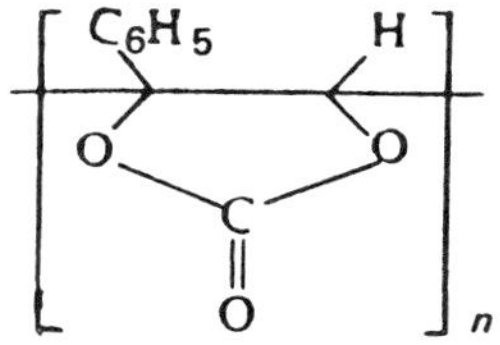

Poly(N,N'-phthaloyl-trans-2,5-dimethylpiperazine)

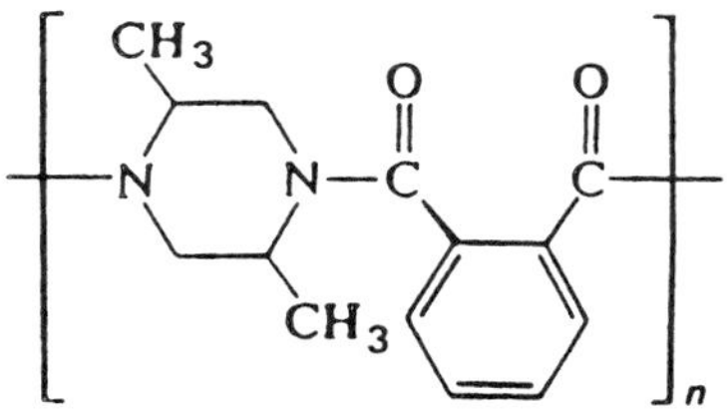

Poly[2,2-propanebis(4-phenyl carbonate)]

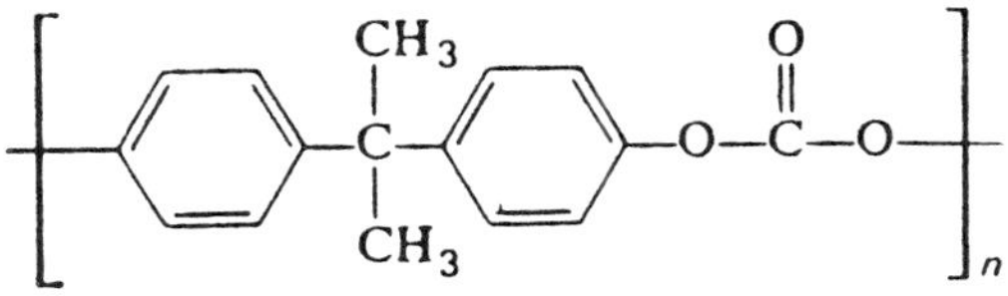

Poly[2,2-propanebis(4-phenyl phenylphosphonate)]

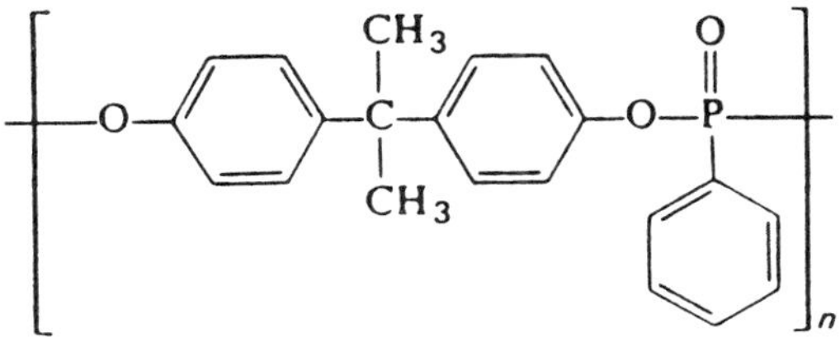

Polypropanesultam

$$+(CH_2)_3SO_2NH +_n$$

Polypropylene (PP)

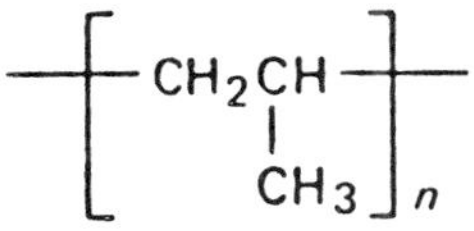

Poly(propylene glycol) (PPG)

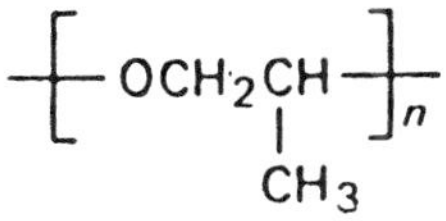

Poly(sebacic anhydride)

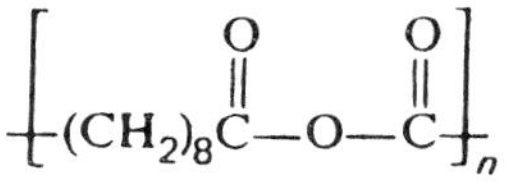

$$\left[(CH_2)_8 \overset{\overset{\displaystyle O}{\|}}{C}-O-\overset{\overset{\displaystyle O}{\|}}{C} \right]_n$$

Polystyrene (PS)

$$\left[CH_2-\underset{\underset{\displaystyle \bigcirc}{}}{CH} \right]_n$$

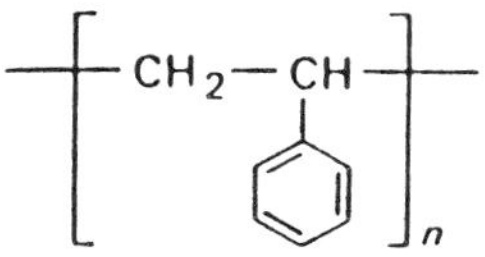

Polytetrafluoroethylene (PTFE)

$$\left[CF_2CF_2 \right]_n$$

Poly(tetrafluoro-1,2,3,4-tetrathiane)

$$\left[CF_2CF_2SSSS \right]_n$$

Poly(tetrafluoro-1,2,3-trithiolane)

$$\left[CF_2CF_2SSS \right]_n$$

Poly(tetramethylene hexamethylenedicarbamate)

$$\left[(CH_2)_4-O-\overset{\overset{\displaystyle O}{\|}}{C}-NH(CH_2)_6NH-\overset{\overset{\displaystyle O}{\|}}{C}-O \right]_n$$

Poly(tetramethylene isophthalate)

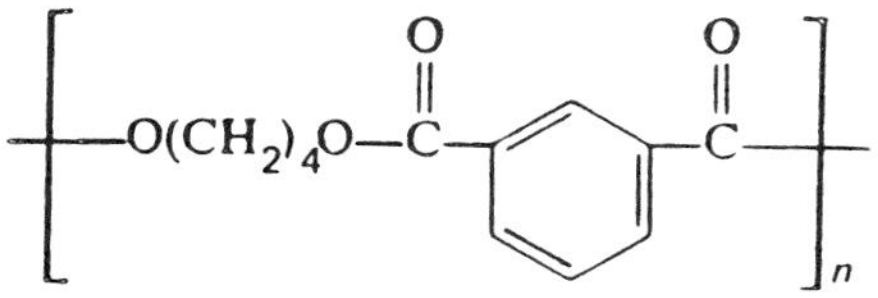

$$\left[O(CH_2)_4O-\overset{\overset{\displaystyle O}{\|}}{C}-\bigcirc-\overset{\overset{\displaystyle O}{\|}}{C} \right]_n$$

Poly(tetramethylene sebacate)

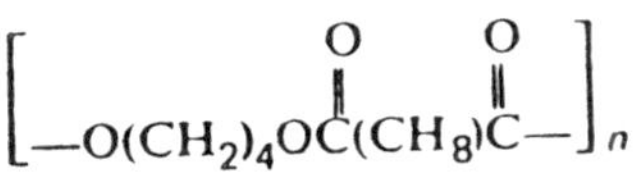

$$\left[-O(CH_2)_4O\overset{\overset{\displaystyle O}{\|}}{C}(CH_8)\overset{\overset{\displaystyle O}{\|}}{C}-\right]_n$$

Polythiazyl

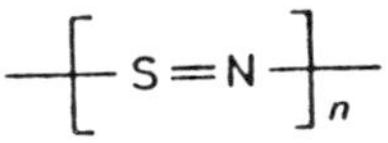

$$-\left[-S=N-\right]_n$$

Poly(11-undecanoamide)

$$\left[-(CH_2)_{10}-\overset{\overset{\displaystyle O}{\|}}{C}--NH-\right]_n$$

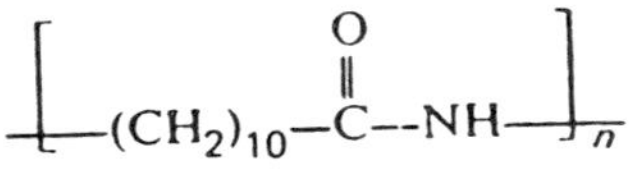

Poly(vinyl acetate) (PVAc)

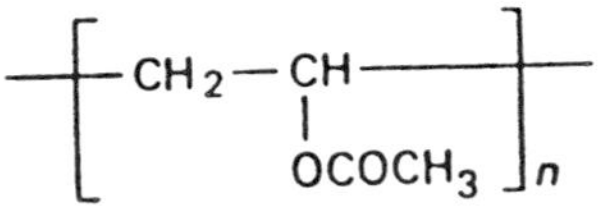

$$\left[-CH_2-\underset{\underset{\displaystyle OCOCH_3}{|}}{CH}-\right]_n$$

Poly(vinyl alcohol) (PVA)

$$\left[-CH_2\underset{\underset{\displaystyle OH}{|}}{CH}-\right]_n$$

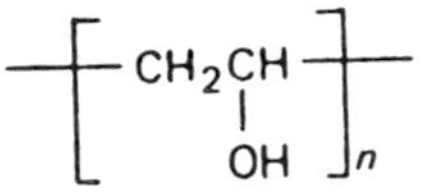

Poly(vinyl t-butyl ether)

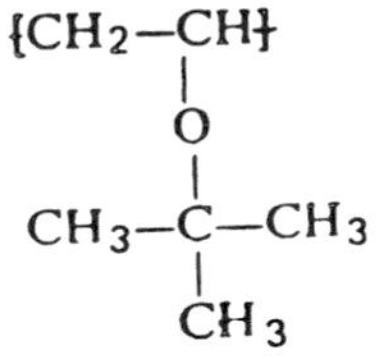

$$\{CH_2-CH\}$$
$$|$$
$$O$$
$$|$$
$$CH_3-\underset{\underset{\displaystyle CH_3}{|}}{\overset{\overset{\displaystyle}{}}{C}}-CH_3$$

Poly(vinyl butyral) (PVB)

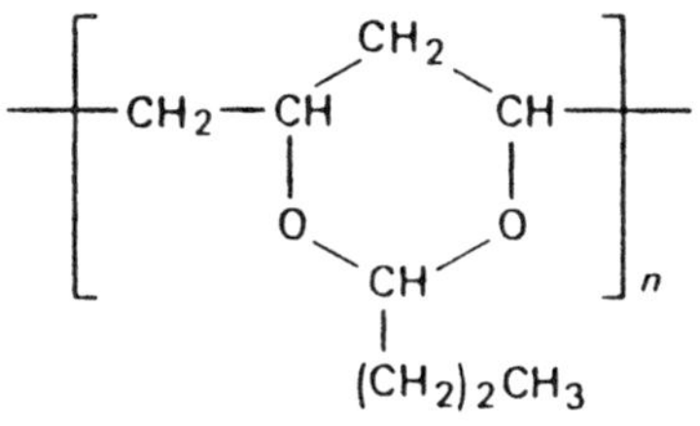

$$\left[CH_2-CH \overset{\overset{\displaystyle CH_2}{\diagup\quad\diagdown}}{\underset{\underset{\displaystyle CH}{O\diagdown\quad\diagup O}}{}} CH \right]_n$$

$$(CH_2)_2CH_3$$

Poly(vinyl butyrate)

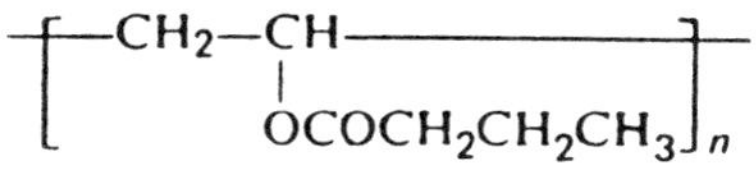

$$\left[CH_2-CH \atop OCOCH_2CH_2CH_3 \right]_n$$

Poly(vinyl carbazole)

$$\left[CH_2-CH \atop N \right]_n$$

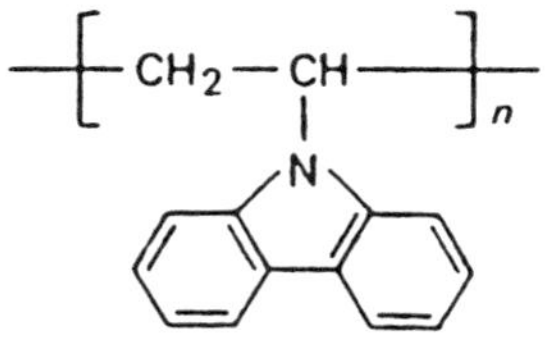

Poly(vinyl chloride) (PVC)

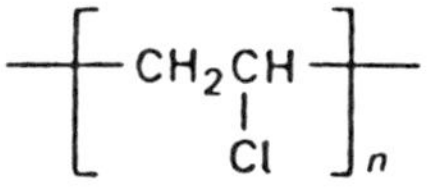

$$\left[CH_2CH \atop Cl \right]_n$$

Poly(vinyl chloride-co-vinyl acetate)

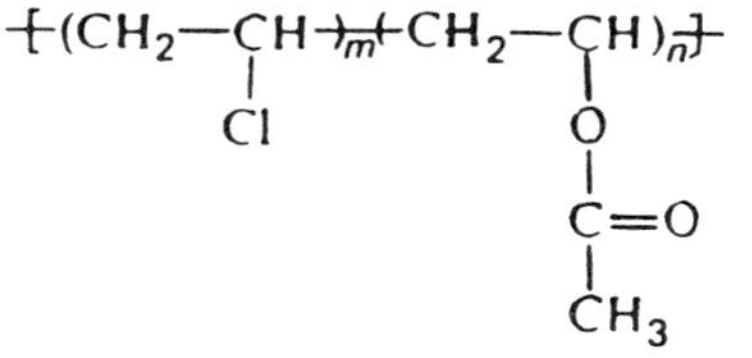

$$\left[(CH_2-CH)_m (CH_2-CH)_n \right]$$
$$Cl \qquad O$$
$$C=O$$
$$CH_3$$

Poly(vinyl formal) (PVF)

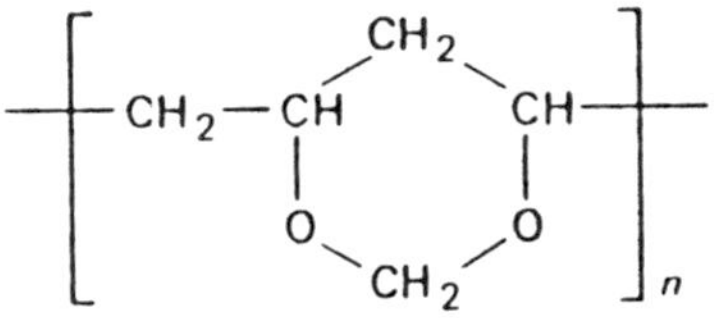

Poly(vinylidene chloride)

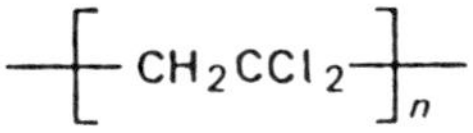

$$-\left[-CH_2CCl_2-\right]_n-$$

Poly(vinyl isobutyl ether)

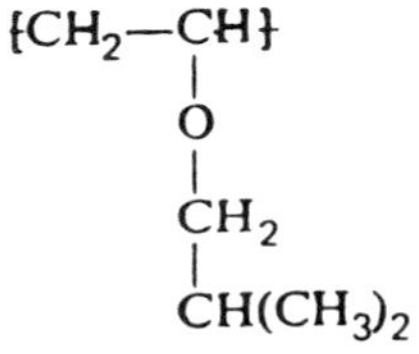

{CH₂—CH}
|
O
|
CH₂
|
CH(CH₃)₂

Poly(vinyl pyridine)

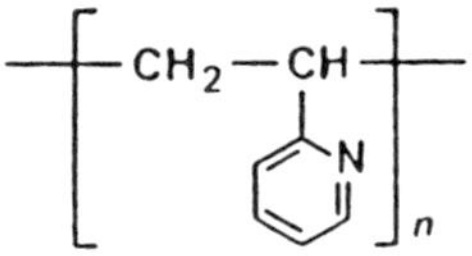

Poly(vinyl pyrrolidone)

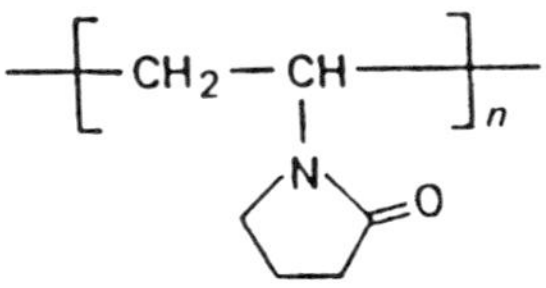

Poly(p-xylylene thiodipropionate)

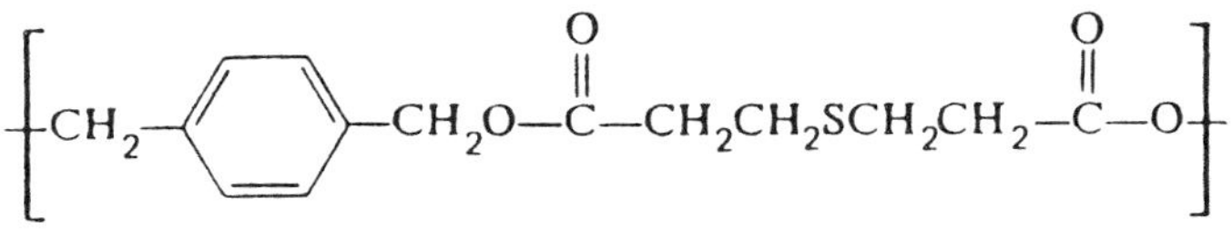

$$\left[CH_2 - \bigcirc - CH_2O - \overset{\overset{\displaystyle O}{\|}}{C} - CH_2CH_2SCH_2CH_2 - \overset{\overset{\displaystyle O}{\|}}{C} - O \right]$$

Poly(1,4-xylylenyl)-2-methylpiperazine

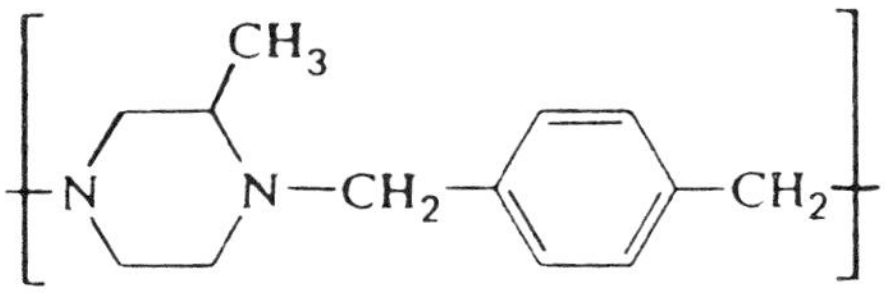

$$\left[\begin{array}{c} CH_3 \\ N \bigcirc N - CH_2 - \bigcirc - CH_2 \end{array} \right]$$

Potassium polymetaphosphate

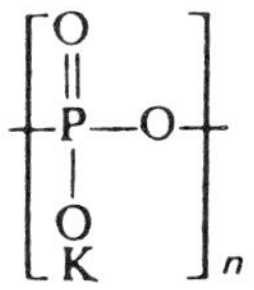

$$\left[\begin{array}{c} O \\ \| \\ P - O \\ | \\ O \\ K \end{array} \right]_n$$

Styrene–acrylonitrile copolymer (SAN)

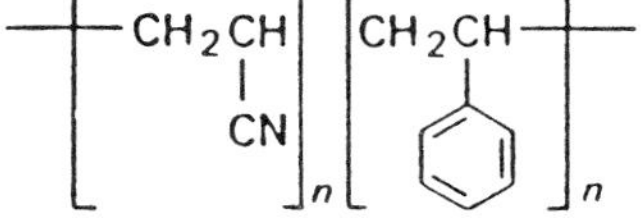

$$\left[\begin{array}{c} CH_2CH \\ | \\ CN \end{array} \right]_n \left[\begin{array}{c} CH_2CH \\ | \\ \bigcirc \end{array} \right]_n$$

Styrene-butadiene rubber (SBR)

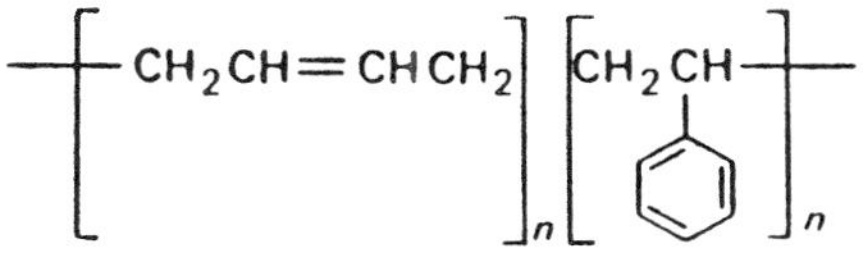

Urea-formaldehyde resins (UF)

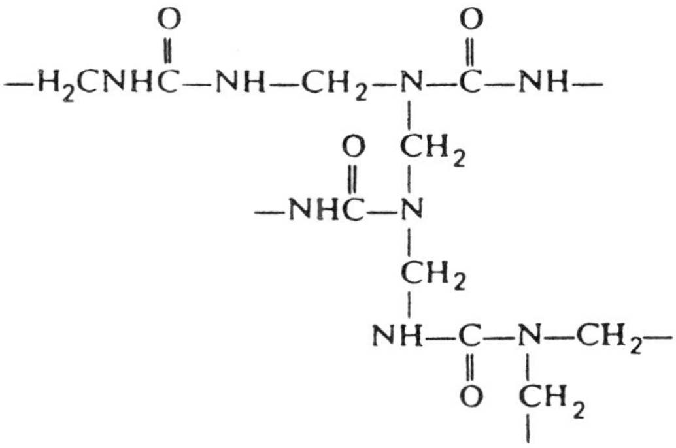

POLYSACCHARIDES: PARTIAL STRUCTURES

Amylopectin—details of branch-point

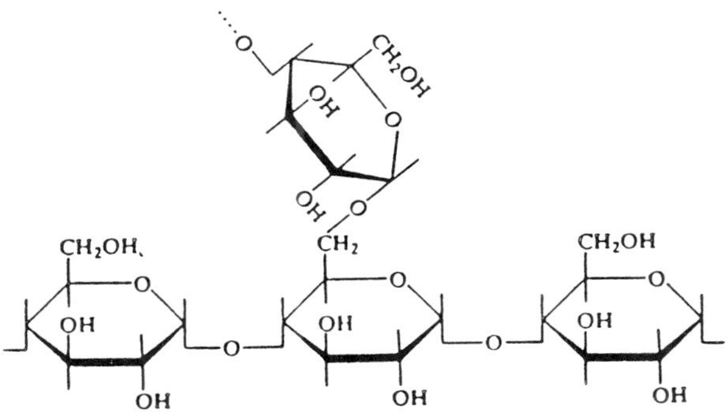

Linear amylose

Cellulose

Inulin

Levan

Mannan

Xylan

Chitin

Appendix H: Mathematical Values and Units

Prefixes for Multiples and Submultiples

Multiple or submultiple		Prefix	SI symbol
10^{12}	1000 000 000 000	tera	T
10^{9}	1000 000 000	giga	G
10^{6}	1000 000	mega	M
10^{3}	1000	kilo	k
10^{2}	100	hecto	h
10^{1}	10	deka	da
10^{0}	1		
10^{-1}	0.1	deci	d
10^{-2}	0.01	centi	c
10^{-3}	0.001	milli	m
10^{-6}	0.000 001	micro	μ
10^{-9}	0.000 000 001	nano	n
10^{-12}	0.000 000 000 001		
10^{-15}		femto	f
10^{-18}		atto	a

Conversion Factors

Unit	Value
1 Ångstrom	$= 10^{-10}$ m $= 10^{-8}$ cm $= 10^{-7}$ mm $= 10^{-4}$ μm $= 10^{-1}$ nm
1 atmosphere	$= 1.013 \times 10^{5}$ Pa $= 1.01325$ bar
1 bar	$= 10^{5}$ Pa $= 0.9869$ atm $= 14.5038$ p.s.i.
1 calorie (N.B.S.)	$= 4.184$j $= 3.9657 \times 10^{-3}$ B.T.U.
1.24 electron volts (eV)	$= 28.59$ kcal mol^{-1} $(= 1000$ mm$)$
1 gigapascal (GPa)	$= 10^{9}$ Pa $= 10^{4}$ bar $= 10$ kbar
1 g/cm^{3}	$= 10^{3}$ kg m^{-3} $= 1$ Mg/m^{3}
1 kilobar (kbar)	$= 10^{8}$ Pa $= 986.9$ atm
1 kilopascal (kPa)	$= 1000$ Pa $= 10^{-2}$ bar
1 kilocalorie (N.B.S.)	$= 4.184 \times 10^{10}$ erg $= 4.84$j $= 3.966$ B.T.U.
1 liter	$= 1.0567$ quart (U.S.)
1 megagram m^{-3} (Mg/m^{3})	$= 10^{3}$ kg/m^{3} $= 1$ g/cm^{3}
1 meter	$= 3.28084$ feet
1 millimeter	$= 0.0394$ inch
1 micrometer (μm)	$= 10^{4}$ Å $= 10^{-3}$ mm $= 10^{-6}$ m $= 1$ micron (μ)
1 nanometer (nm)	$= 10$ Å $= 10^{-9}$ m
1 ounce (avoir)	$= 28.35$ g
1 point (gem)	$= 10^{-2}$ carat $= 0.04$ pearl grain
1 poise	$= 10^{-1}$ N s m^{-2}
1 pound (avoir)	$= 453.59$ g
1 stoke	$= 10^{-4}$ m^{2} s^{-1}

Units of Measurement

Quantity	Unit	SI symbol	Formula
Acceleration			m/s^2
Amount of substance	mole	mol	
Area			m^2
Bulk modulus			N/m^2
Chemical potential	joule	J	Nm
Compressibility			$1/Pa$
Density			kg/m^3
Electric charge	coulomb	C	A s
Electrical capacitance	farad	F	A s/V
Electrical conductivity	siemens	S	A/V
Electrical current	ampere	A	
Electrical field strength			V/m
Electrical inductance	henry	H	V s/A
Electrical resistance	ohm	Ω	V/A
Electromotive force	volt	V	W/A
Energy	joule	J	N m
Enthalpy	joule	J	N m
Entropy			J/K
Force	newton	N	$kg\ m/s^2$
Frequency	hertz	Hz	$(cycle)/s^2$
Gibbs free energy	joule	J	N M
Heat capacity			J/K
Heat flow			$J/s\ m^2$
Helmholtz free energy	joule	J	N m
Illuminance	lux	lx	lm/m^2
Length	meter	m	
Luminance			cd/m^2
Luminous flux	lumen	lm	cd sr
Luminous intensity	candela	cd	
Magnetic field strength			A/m
Magnetic flux	weber	Wb	V s
Magnetic flux density	tesla	T	Wb/m^2
Magnetic permeability			H/m
Magnetic permittivity			F/m
Mass	kilogram	kg	
Power	watt	W	J/s
Pressure	pascal	Pa	N/m^2
Resistivity			$\Omega\ m$
Shear modulus			N/m^2
Surface tension			N/m

Units of Measurement Continued

Quantity	Unit	SI symbol	Formula
Temperature	kelvin	K	
Thermal conductivity			W/mk
Thermal expansion			1/K
Time	second	s	
Velocity			m/s
Viscosity (dynamic)			Ns/m^2
Viscosity (kinematic)			m^2/s
Voltage	volt	V	W/A
Volume			m^3
Wavelength			m
Wave number			1/m
Work	joule	J	N/m
Young's modulus			N/m^2

Physical Constants

Quantity		Numerical value (SI)	Numerical value (cgs)
Acceleration due to gravity at earth's surface			
Equator	g	9.7805 m s^{-2}	$9.7805 \times 10^2 \text{ cm s}^{-2}$
Avogadro's constant	N_A	$6.02252 \times 10^{23} \text{ mol}^{-1}$	$6.02252 \times 10^{23} \text{ mol}^{-1}$
Boltzmann's constant	k	$1.380622 \times 10^{-23} \text{ J K}^{-1}$	$1.380622 \times 10^{-16} \text{ erg K}^{-1}$
Electron charge		$1.6021917 \times 10^{-19} \text{ C}$	$1.6021917 \times 10^{-20} \text{ emu}$
Faraday's constant	F	$9.64870 \times 10^4 \text{ mol}^{-1}$	$9.64870 \times 10^3 \text{ emu mol}^{-1}$
Gas constant	R	$8.31432 \text{ J mol}^{-1} \text{ K}^{-1}$	$1.9872 \text{ cal mol}^{-1} \text{ K}^{-1}$
Gravitational constant	G	$6.6732 \times 10^{-11} \text{ N m}^2 \text{ kg}^{-2}$	$6.6732 \times 10^{-8} \text{ dyne cm}^2 \text{ g}^{-2}$
Permittivity of a vacuum	ϵ_0	$8.8419413 \times 10^{-12} \text{ F m}^{-1}$	$1.0 \text{ dyne cm}^2 \text{ statcoul}^{-2}$
Permeability of a vacuum	μ_0	$1.2566371 \times 10^{-6} \text{ H m}^{-1}$	
Planck's constant	h	$6.626196 \times 10^{-34} \text{ J s}$	$6.626196 \times 10^{-27} \text{ erg s}$
Velocity of light in a vacuum	c	$2.99792458 \times 10^8 \text{ m s}^{-1}$	$2.99792458 \times 10^{10} \text{ cm s}^{-1}$

Appendix I: Comments on Health

Synthetic polymers, like most biological polymers, are nontoxic under normal and intended use. Most of the additives employed in today's polymer industry are also relatively nontoxic. Even so, care should be exercised when dealing with many of the monomers of synthetic polymers and when dealing with polymeric materials under extreme conditions such as the high temperatures that may be employed in the processing of some polymeric materials or when polymers are involved in commercial or domestic fires.

FIRE

Fire hazards involve not only burning—most deaths occur from the ingestion of volatiles produced by the fire. Carbon monoxide, the major cause of death, causes unconsciousness in less than 3 minutes due to its preferential attack on hemoglobin (see Sec. 15.10).

Interestingly, one important observation concerning burning in general is whether colored smoke is produced. Some materials burn producing lots of darkly colored smoke. Some of this colored smoke may be due to the production of aromatic systems including fused ring systems which may or may not contain harmful chemical species, including respiratory toxins and cancer-causing agents. For instance, PVC often emits hydrogen chloride as a major degradation product. As the hydrogen chloride is eliminated, "conjugation" of the backbone occurs accompanied by the formation of fused ring compounds. Emission of hydrogen chloride can occur at temperatures employed for the processing of some polymers. Because of the elimination of hydrogen chloride by PVC, care is exercised in the processing of PVC products by manufacturers who do not want to put their workers at risk. In contrast, polyacrylonitrile does not typically emit hydrogen cyanide upon burning. Instead, internal cyclization occurs, resulting in the formation of major amounts of char. In fact, the pyrolysis of polyacrylonitrile forms the basis for the production of carbon fibers.

MEASURES OF TOXICITY

Toxicity involves the affect of various materials on living objects such as bacteria, fungi, yeast, plants, mice, fish, and humans. Tests to determine the toxicity of various substances are typically done either in situ or in vitro. In situ tests are carried out at the location where the toxicity is in question. For example, in situ tests of the ability of cancer drugs to inhibit human ovarian cancer are carried out within the living beings. In vitro tests are examinations made away from the location where the toxicity is in question. Thus, testing various agents on human cancer cell lines in a petri dish constitutes an in vitro examination.

While we are mainly concerned with the affects of various agents on humans, we often employ test animals, such as specially genetically developed mice and rats, as measures of an agent's toxicity. While it is assumed that there is a relationship between the type and level of toxicity to these mice and rats and the toxicity to humans, this relationship is not always apparent. Even so, such animal tests are generally foundational in the assignment of allowed exposure levels to specific agents. It is known that the toxicity varies between species, amount administered, type of administration, and duration of exposure. This difference is a consequence of a number of factors, including the tendencies of certain agents to affect certain parts of the body including varying tendencies for the agents to become concentrated within specific organs and tissues and their mechanism(s) of action. Table I.1 lists toxicity definitions utilized in toxicology today.

As noted before, while commercially available synthetic polymers are relatively nontoxic, the monomers themselves vary greatly in toxicity. (For instance, ethyleneimine, an acknowledged cancer-causing agent, has a LD_{50} (oral/rat) of 15 mg/kg while polyethyleneimine has a LD_{50} (oral/rat) of 3300 mg/kg. This points out

Table I.1 Descriptions of Toxic Measures

TDLo/Toxic Dose Low—The lowest dose introduced by any route other than inhalation over any period of time that produces any toxic effect in humans or to produce carcinogenic, teratogenic, mutagenic, or neoplastic effects in humans or animals.

TCLo/Toxic Concentration Low—Any concentration in air that causes any toxic effect in humans or produces a carcinogenic, teratogenic, mutagenic, or neoplastigenic toxic effect in humans or animals.

LDLo/Lethal Dose Low—The lowest dose introduced by any route other than by inhalation over any time to have caused death in humans or the lowest single dose to have caused death in animals.

LD_{50}/Lethal Dose Fifty—A calculated dose expected to cause the death of 50% of a tested population from exposure by any route other than inhalation.

LCLo/Lethal Concentration Low—The lowest concentration in air to have caused death in a human or death in animals when exposed for 24 hours or less.

LC_{50}/Lethal Concentration Fifty—A calculated concentration of a substance in air that would cause death in 50% of a test population from exposure for 24 hours or less.

---

the need for monomers and other potentially toxic agents to be removed from polymers.) Table I.2 contains the time-weighted average (TWA) for some monomers as cited by the United States Occupational Standards. For comparison, entries for some well-known toxic compounds have been added.

CUMULATIVE AFFECTS

While exposure of the general public to toxins is dangerous and to be avoided, exposure of people that deal with commercial chemicals on a daily basis is even more important. Such people must take special care to avoid exposure to dangerous chemicals, since many of these accumulate in our bodies over a long time, slowly building to levels that may be unhealthy.

Most of the toxic and undesirable chemicals of a decade ago have been eliminated from the common workplace. This includes halogenated hydrocarbons such as carbon tetrachloride and aromatic hydrocarbons such as benzene. Further, chemicals that are known to be potentially toxic, such as some monomers, are being eliminated from polymeric materials to within the limits of detection.

We are also becoming more aware of the effects of nonchemical hazards within the workplace. These hazards are being dealt with as they become known. Some of these possible hazards include exposure to certain radiations employed in materials processing, machine and other noise, machine operation, and exposure to other potential safety situations.

Industrial recognition of customer and employee safety is a major factor behind the international programs known as ISO 9000 and ISO 14000 (see Sec. 1.3). Further, a number of national agencies and associations deal with aspects of the environmental and personal safety issues, including the Environmental Protection Agency, National Fire Protection Association, Occupational Safety and Health Administration, Department of Transportation, Food and Drug Administration, and the National Institute for Occupational Safety and Health.

Table I.2 TWA Values for Selected Monomers and Additional Recognized Toxins

Chemical	TWA (ppm)
Acetic anhydride	5
Acrylonitrile	20
Benzene	10
1,3-Butadiene	1000
Carbon monoxide	50
Chloroprene	25
1,2-Ethylenediamine	10
Ethylene oxide	50
Formaldehyde	3
Hydrazine	1
Hydrogen cyanide	10
Phenol	5
Styrene	100
Vinyl chloride	500

ENVIRONMENT

Today, industry and business recognize that part of doing business is taking care of the environment. Along with various government agencies and associated laws, business and industry are finding that good environment practices are good business. Advances continue with respect to lowering potentially harmful emissions as solid waste or into the water and air shared by all of us. Chemical industries are taking the lead in this clean-up process (see also ISO 9000 & 14000; Appendix J). A combination of watchful vigilance and trust is needed to continue this effort.

Appendix J: Comments on ISO 9000 and 14000

The International Organization for Standardization (ISO) has members in about 100 countries working to develop common global standards. The ISO 9000 series encompasses the product development sequence from strategic planning to customer service. Currently, it is a series of five quality system standards, with two of the standards focusing on guidance and three contractual standards.

ISO 9000 certification is often obtained to promote a company's perceived quality level, for supplier control, and to promote certain management practices—often Total Quality Management (TQM) management practices. It is used as a global standardizing "tool" with respect to business and industry in its broadest sense, including banking, volunteer organizations, and most aspects of the chemical (including the polymer) industry.

ISO 9000 requires what is called a third-party assessment but involves developing first and second-party strategies. First party refers to the supplier company that requests ISO 9000 certification. Second party refers to the customer whose needs have been met by the first party through the use of quality management procedures achieved through ISO 9000 compliance. Third party refers to an outside reviewer that certifies that the first party has satisfied ISO 9000 procedures.

While ISO 9000 is a management tool, it affects the way industry does business and deals with quality control issues, such as how machinery and parts manufactured by a company are monitored for quality. It focuses on satisfying the customer—the purchaser of the raw materials, manufactured parts, or assembled items—and includes the eventual end-customer—the general public. It is an attempt to assure quality goods.

ISO 14000 is a series of standards intended to assist in managing the impact of manufactured materials including finished products and original feedstocks. It addresses the need to have one internationally accepted environmental management system that involves cradle-to-grave responsibility for manufactured materials, emphasizing the impact of products, operations, and services on the environment.

Index